Monographs in Theoretical Computer Science
An EATCS Series

For further volumes:
http://www.springer.com/series/776

Donald Sannella • Andrzej Tarlecki

Foundations of Algebraic Specification and Formal Software Development

Springer

Prof. Donald Sannella
The University of Edinburgh
School of Informatics
Informatics Forum
10 Crichton St.
Edinburgh, EH8 9AB
United Kingdom

Prof. Andrzej Tarlecki
Institute of Informatics
Faculty of Mathematics,
Informatics and Mechanics
University of Warsaw
ul. Banacha 2
02-097 Warsaw, Poland
and
Institute of Computer Science
Polish Academy of Sciences
ul. Ordona 21
01-237 Warsaw, Poland

Series Editors
Prof. Dr. Juraj Hromkovič
ETH Zentrum
Department of Computer Science
Swiss Federal Institute of Technology
8092 Zürich, Switzerland

Prof. Dr. Grzegorz Rozenberg
Leiden Institute of Advanced
Computer Science
University of Leiden
Niels Bohrweg 1
2333 CA Leiden, The Netherlands

Prof. Dr. Arto Salomaa
Turku Centre of Computer Science
Lemminkäisenkatu 14 A
20520 Turku, Finland

ISSN 1431-2654
ISBN 978-3-642-44065-6 ISBN 978-3-642-17336-3 (eBook)
DOI 10.1007/978-3-642-17336-3
Springer Heidelberg Dordrecht London New York

ACM Codes: F.3, D.2

Printed on acid-free paper

Springer is part of Springer Science+Business Media (www.springer.com)

To Monika and Teresa

Preface

As its title promises, this book provides foundations for software specification and formal software development from the perspective of work on algebraic specification. It concentrates on developing basic concepts and studying their fundamental properties rather than on demonstrating how these concepts may be used in the practice of software construction, which is a separate topic.

The foundations are built on a solid mathematical basis, using elements of universal algebra, category theory and logic. This mathematical toolbox provides a convenient language for precisely formulating the concepts involved in software specification and development. Once formally defined, these notions become subject to mathematical investigation in their own right. The interplay between mathematics and software engineering yields results that are mathematically interesting, conceptually revealing, and practically useful, as we try to show.

Some of the key questions that we address are: What is a specification? What does a specification mean? When does a software system satisfy a specification? When does a specification guarantee a property that it does not state explicitly? How does one prove this? How are specifications structured? How does the structure of specifications relate to the modular structure of software systems? When does one specification correctly refine another specification? How does one prove correctness of refinement steps? When can refinement steps be composed? What is the role of information hiding? We offer answers that are simple, elegant and general while at the same time reflecting software engineering principles.

The theory we present has its origins in work on algebraic specifications starting in the early 1970s. We depart from and go far beyond this starting point in order to overcome its limitations, retaining two prominent characteristics.

The first is the use of many-sorted algebras consisting of a collection of sets of data values together with functions over those sets, or similar structures, as models of software systems. This level of abstraction fits with the view that the correctness of the input/output behaviour of a software system takes precedence over all its other properties. Certain fundamental software engineering concepts, such as information hiding, have direct counterparts on the level of such models.

The second is the use of logical axioms, usually in a logical system in which equality has a prominent role, to describe the properties that the functions are required to satisfy. This property-oriented approach allows the use of formal systems of rules to reason about specifications and the relationship between specifications and software systems. Still, the theory we present is semantics-oriented, regarding models as representations of reality. The level of syntax and its manipulation, including axioms and formal proof rules, merely provide convenient means of dealing with properties of (classes of) such models.

Our primary source of software engineering intuition is the relatively simple world of first-order functional programming, and in particular functional programming with modules as in Standard ML. It is simpler than most other programming paradigms, it offers the most straightforward fit with the kinds of models we use, and it provides syntax ("functors") that directly supports a methodology of software development by stepwise refinement. Even though some aspects of more elaborate programming paradigms are not directly reflected, the fundamental concepts we study are universal and are relevant in such contexts as well.

This book contains five kinds of material.

The requisite mathematical underpinnings:
: Chapters 1 and 3 are devoted to the basic concepts of universal algebra and category theory, respectively. This material finds application in many different areas of theoretical computer science and these chapters may be independently used for teaching these subjects. Our aim is to provide a generally accessible summary rather than an expository introduction. We omit many standard concepts and results that are not needed for our purposes and include refinements to classical universal algebra that are required for its use in modelling software. Most of the proofs are left to the reader as exercises.

Traditional algebraic specifications:
: Chapter 2 presents the standard material that forms the basis of work on algebraic specifications. From the point of view of an algebraist, much of this would be viewed as part of universal algebra. Additionally, Section 2.7 explores some of the ways in which these basics may be modified to cope with different aspects of software systems. Again, this chapter is a summary rather than an expository introduction, and many proofs are omitted.

Elements of the theory of institutions:
: In Chapter 4 we introduce the notion of an *institution*, developed as a formalisation of the concept of a logical system. This provides a suitable basis for a general theory of formal software specification and development. Chapter 10 contains some more advanced developments in the theory of institutions.

Formal specification and development:
: Chapters 5–8 constitute the core of this book. Chapter 5 develops a theory of specification in an arbitrary institution. Special attention is paid to the issue of structure in specifications. Chapter 6 is devoted to the topic of parameterisation, both of algebras and of specifications themselves. Chapter 7 presents a theory of formal software development by stepwise refinement of specifications. Chapter 8

introduces the concept of *behavioural equivalence* and studies its role in software specification and development.

Proof methods:

Chapter 9 complements the model-theoretic picture from the previous chapters by giving the corresponding proof methods, including calculi for proving consequences of specifications and correctness of refinement steps.

The dependency between chapters and sections is more or less linear, except that Chapter 10 does not depend on Chapter 9. This dependency is not at all strict. This is particularly relevant to Chapter 3 on category theory: anyone who is familiar with the concepts of category, functor and pushout may omit this chapter, returning to it if necessary to follow some of the details of later chapters. On first reading one may safely omit the following sections, which are peripheral to the main topic of the book or contain particularly advanced or speculative material: 2.6, 2.7, 3.5 except for 3.5.1, 4.1.2, 4.4.2, 4.5, 6.3, 6.4, 6.5, 8.2.3, 8.5.3, 9.5, 9.6 and Chapter 10.

This book is self-contained, although mathematical maturity and some acquaintance with the problems of software engineering would be an advantage. In the mathematical material, we assume a very basic knowledge of set theory (set, membership, Cartesian product, function, etc. — see for instance [Hal70]), but we recall all of the set-theoretic notation we use in Section 1.1. Likewise, we assume a basic knowledge of the notation and concepts of first-order logic and proof calculi; see for instance [End72]. In the examples that directly relate to programming, we assume some acquaintance with simple concepts of functional programming. No advanced features are used and so these examples should be self-explanatory to anyone with experience using a programming language with types and recursion.

In an attempt to give a complete treatment of the topics covered without going on at much greater length, quite a few important results are relegated to exercises with the details left for the reader to fill in. Fairly detailed hints are provided in many cases, and in the subsequent text there is no dependence on details of the solutions that are not explicitly given in these hints.

This book is primarily a monograph, with researchers and advanced students as its target audience. Even though it is not intended as a textbook, we have successfully used some parts of it for teaching, as follows:

Universal algebra and category theory:

A one-semester course based on Chapters 1 and 3.

Basic algebraic specifications:

A one-semester course for undergraduates based on Chapters 1 and 2.

Advanced algebraic specifications:

An advanced course that follows on from the one above based on Chapters 4–7.

Institutions:

A graduate course with follow-up seminar on abstract model theory based on most of Chapter 4 and parts of Chapter 10.

The material in this book has roots in the work of the entire algebraic specification community. The basis for the core chapters is our own research papers, which are here expanded, unified and taken further. We attempt to indicate the origins of the most important concepts and results, and to provide appropriate bibliographical references and pointers to further reading, in the final section of each chapter. The literature on algebraic specification and related topics is vast, and we make no claim of completeness. We apologize in advance for possible omissions and misattributions.

Acknowledgements

All of this material has been used in some form in courses at the University of Edinburgh, the University of Warsaw, and elsewhere, including summer schools and industrially oriented training courses. We are grateful to all of our students in these courses for their attention and feedback.

This book was written while we were employed by the University of Edinburgh, the University of Warsaw, and the Institute of Computer Science of the Polish Academy of Sciences. We are grateful to our colleagues there for numerous discussions and for the atmosphere and facilities which supported our work. The underlying research and travel was partly supported by grants from the British Council, the Committee for Scientific Research (Poland), the Engineering and Physical Sciences Research Council (UK), the European Commission, the Ministry of Science and Higher Education (Poland), the Scottish Informatics and Computer Science Alliance and the Wolfson Foundation.

We are grateful to the entire algebraic specification community, which has provided much intellectual stimulation and feedback over the years. We will not attempt to list the numerous members of that community who have been particularly influential on our thinking, but we give special credit to our closest collaborators on these topics, and in particular to Michel Bidoit, Till Mossakowski and Martin Wirsing. Our Ph.D. students contributed to the development of our ideas on some of the topics here, and we particularly acknowledge the contributions of David Aspinall, Tomasz Borzyszkowski, Jordi Farrés-Casals and Wiesław Pawłowski.

We are grateful for discussion and helpful comments on the material in this book and the research on which it is based. In addition to the people mentioned above, we would like to acknowledge Jiři Adámek, Jorge Adriano Branco Aires, Thorsten Altenkirch, Egidio Astesiano, Hubert Baumeister, Jan Bergstra, Pascal Bernard, Gilles Bernot, Didier Bert, Julian Bradfield, Victoria Cengarle, Maura Cerioli, Rocco De Nicola, Răzvan Diaconescu, Luis Dominguez, Hans-Dieter Ehrich, Hartmut Ehrig, John Fitzgerald, Michael Fourman, Harald Ganzinger, Marie-Claude Gaudel, Leslie Ann Goldberg, Joseph Goguen, Jo Erskine Hannay, Robert Harley, Bob Harper, Rolf Hennicker, Claudio Hermida, Piotr Hoffman, Martin Hofmann, Furio Honsell, Cliff Jones, Jan Jürjens, Shin-ya Katsumata, Ed Kazmierczak, Yoshiki Kinoshita, Spyros Komninos, Bernd Krieg-Brückner, Sławomir Lasota, Jacek Leszczyłowski, John Longley, David MacQueen, Tom Maibaum, Lambert Meertens, José Meseguer, Robin Milner, Eugenio Moggi, Bernhard Möller, Brian Monahan, Peter Mosses, Tobias Nipkow, Fernando Orejas, Marius Petria, Gordon Plotkin, Axel Poigné,

John Power, Horst Reichel, Grigore Roşu, David Rydeheard, Oliver Schoett, Lutz Schröder, Douglas Smith, Stefan Sokołowski, Thomas Streicher, Eric Wagner, Lincoln Wallen and Marek Zawadowski. We apologize for any omissions. We are grateful to Stefan Kahrs, Bartek Klin and Till Mossakowski for their thoughtful and detailed comments on a nearly final version which led to many improvements, to Mihai Codescu for helpfully checking examples for errors, to Ronan Nugent of Springer and to Springer's copyeditor.

Finally, we would like to express our very special appreciation to Rod Burstall and Andrzej Blikle, our teachers, supervisors, and friends, who introduced us to this exciting area, brought us to scientific maturity, and generously supported us in our early careers.

Edinburgh and Warsaw,
September 2011

Don Sannella
Andrzej Tarlecki

Contents

Introduction

Software is everywhere and affects nearly all aspects of daily life. Software systems range in size from tiny (for instance, embedded software in simple devices, or the solution to a student's first programming exercise) to enormous (for instance, the World Wide Web, regarded as a single distributed system). The quality of software systems is highly variable, and everybody has suffered to some extent as a consequence of imperfect software.

This book is about one approach, called *algebraic specification*, to understanding and improving certain aspects of software quality. Algebraic specification is one of a collection of so-called *formal methods* which use ideas from logic and mathematics to model, analyse, design, construct and improve software. It provides means for precisely defining the problem to be solved and for ensuring the correctness of a constructed solution. The purpose of this book is to provide mathematically well-developed foundations for various aspects of this activity.

The material presented here is sufficient to support the entirely formal development of modular software systems from specifications of their required behaviour, with proofs of correctness of individual steps in the development ensuring correctness of the composed system. Although such a strict formal approach is infeasible in practice for real software systems, it serves as a useful reference point for the evaluation of less formal means for improving quality of software.

The following sections discuss some of the basic motivations which underlie this approach to formal software specification and development.

0.1 Modelling software systems as algebras

In order to be useful for the intended purpose, a software system should satisfy a wide range of requirements. For instance, it should be:

Efficient: The system should be tolerably efficient with respect to its usage of time, memory, bandwidth and other resources.

Robust: Small changes to the system should not dramatically affect its quality.

Reliable: The system should not break down under any circumstances. Incorrect user input should be recognized and explicitly rejected, and faults in the system's environment should be dealt with in a reasonable fashion.

Secure: The system should be protected against unauthorized use. It should be possible to restore any data that is lost or corrupted by an attack, and confidential data should be protected from disclosure.

User-friendly: The system should be easy to use, even without extensive prior knowledge or experience with it.

Well documented: The system's functionality, design and implementation should all be appropriately documented.

All of these properties are very important, although in practice some of them may be sacrificed, or even unachievable in absolute terms. But above all, the system must be:

Correct: The system must exhibit the required externally visible input/output behaviour.

Of course, there are various degrees of correctness, and in practice large systems contain bugs. In spite of this grim reality, it is clear that correctness — or at least, a close approximation to correctness — is the primary goal, and this is the property on which work on formal specification and development concentrates.

Software systems are complex objects. When we are interested in input/output behaviour only, it is useful to abstract away from the concrete details of code and algorithms and model software using mathematical functions, focussing solely on the relationship between inputs and outputs.

For example, consider the following four function definitions:

```
fun f1(n) =
  if n<=1 then 1 else f1(n-1)+f1(n-2)

fun f2(n) =
  if n<=1 then 1
  else let fun g(n) =
             if n=1 then (1,1)
             else let val (u,v) = g(n-1) in (u+v,u) end
       in let val (u,v) = g(n-1) in u+v end
       end

fun f3(n) =
  if n<=1 then 1
  else let val muv = ref (1,1,1)
       in (while let val (m,u,v) = !muv in m < n end do
                muv := let val (m,u,v) = !muv in (m+1,u+v,u) end;
           let val (m,u,v) = !muv in u end)
       end

public static nat f4(nat n) {
  nat u = 1, v = 1;
  for (nat m = 1; m < n; m++) {
```

```
        u = u + v;
        v = u - v;
    }
    return u;
}
```

Each of these definitions of the Fibonacci function over the set of natural numbers $\mathbb{N} = \{0, 1, 2, \ldots\}$ is different and has different properties. First of all, `f1`, `f2`, and `f3` are in Standard ML while `f4` is in Java; we ignore the fact that neither Standard ML nor Java has a type of natural numbers. Next, `f1` and `f2` use recursion and are purely functional while `f3` and `f4` are iterative and use assignment. The functions `f3` and `f4` actually encode the same algorithm in two different notations, an iterative version of the recursive algorithm that `f2` encodes using a local auxiliary function `g`. Also, `f1` runs in time that is exponential in n while `f2`, `f3` and `f4` require only linear time. However, the most important feature of these definitions is that they all encode the Fibonacci function $\mathit{fib}\colon \mathbb{N} \to \mathbb{N}$ defined in the usual way:

$$\begin{aligned} \mathit{fib}(0) &= 1 \\ \mathit{fib}(1) &= 1 \\ \mathit{fib}(n+2) &= \mathit{fib}(n+1) + \mathit{fib}(n) \end{aligned}$$

Before defining *fib*, it was natural to indicate that it takes elements of $\mathbb{N}$ as input and delivers elements of $\mathbb{N}$ as output. We do not really want to consider *fib* in isolation from the set of natural numbers; we view the four function definitions above as defining the function *fib* over $\mathbb{N}$, bundling data and function together:

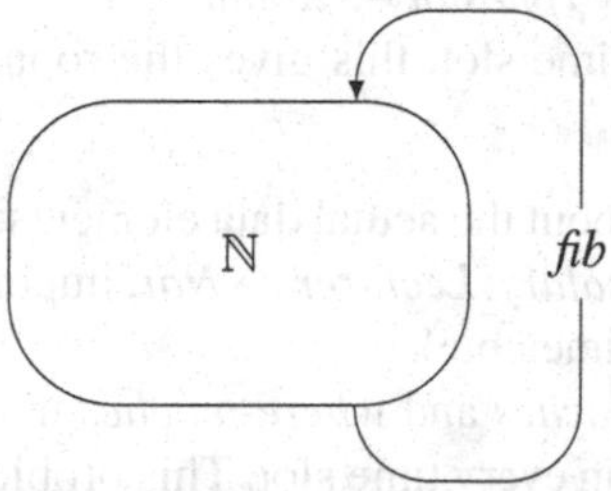

This simple example illustrates the way in which we will model every software system as an *algebra*, that is, a set of data together with a number of functions over this set.[1] In order to deal with systems that manipulate several kinds or *sorts* of data it is necessary to use so-called *many-sorted* or *heterogeneous* algebras that contain a number of different sets of data (rather than just a single set) with functions between these sets. Functions and data types that are defined and used in a software

[1] By *software system* we mean a collection of type definitions and function definitions in a language like C or Standard ML. A software system in the sense of a traditional imperative language is a software system in this sense together with a sequence of statements (the `main` function, in C) making reference to the defined types and functions; these are not themselves made available to the user. In object-oriented languages, a software system is a collection of objects; again, this may be viewed as a software system in our sense since it essentially defines a family of types and functions, the latter capturing the objects' methods by taking the (global) state of the objects as an additional argument and returning the updated global state as an additional result.

system have names, such as `f2`, `+` and `nat` above. These names are used to refer to components of the algebra in order to compute with them, to reason about them and to build larger systems over them. The set of names associated with an algebra is called its *signature*. The formal definitions of these concepts appear in Chapter 1.

Example 0.1.1 (Timetable). A teaching timetable for a university records an assignment of lecturers to courses and courses to rooms and time slots. Such a timetable may be viewed as an algebra with the following sorts:

Course: Data elements of this sort represent courses offered by the university, e.g. Medieval History.

Lecturer: Data elements of this sort represent lecturers working in the university, e.g. Kowalski.

Timeslot: Data elements of this sort represent hours during the week, e.g. Thursday 9–10 am.

Room: Data elements of this sort represent lecture rooms in the university, e.g. JCMB 3315.

The algebra includes functions for operating on this data, for example:

$\mathit{who\text{-}teaches}: \mathit{Course} \to \mathit{Lecturer}$

For any course, this gives the lecturer who teaches it.

$\mathit{what\text{-}teaches}: \mathit{Lecturer} \times \mathit{Timeslot} \to \mathit{Course}$

For any lecturer and time slot, this gives the course taught by the lecturer at that time.

$\mathit{where\text{-}teaches}: \mathit{Lecturer} \times \mathit{Timeslot} \to \mathit{Room}$

For any lecturer and time slot, this gives the room where the lecturer is at that time.

We have been vague about the actual data elements; also there are more functions than we have listed (e.g. $\mathit{salary}: \mathit{Lecturer} \to \mathit{Nat}$, important for the university and the lecturers but not for the timetable).

The functions *what-teaches* and *where-teaches* unrealistically require that every lecturer teaches a course in every time slot. This problem may be resolved by adding an element *nothing* to the set *Course* and an element *nowhere* to the set *Room*, and adjusting some of the details below; see Sections 2.7.3–2.7.5 for much more on this and other options.

Extending this timetable algebra to give timetable and registration information for students as well would involve adding new sorts and new functions. The new sorts would be:

Student: Data elements of this sort represent students enrolled in the university.

Bool: The two boolean values *true* and *false*.

The new functions would include:

$\mathit{enrolled}: \mathit{Student} \times \mathit{Course} \to \mathit{Bool}$

For any student and course, this states whether or not the student is enrolled in that course.

what-attends: $Student \times Timeslot \rightarrow Course$
For any student and time slot, this gives the course attended by the student at that time.
where-attends: $Student \times Timeslot \rightarrow Room$
For any student and time slot, this gives the room where the student is supposed to be at that time.

All of these functions assume a static, fixed timetable with an unchanging assignment of lecturers and students to courses and time slots. Adding functions that change the timetable would require us to introduce a new sort *Timetable* having all possible timetables as its data elements, with functions like *enrol*: $Student \times Course \times Timetable \rightarrow Timetable$, and then the functions above would take the timetable as an additional argument. □

0.2 Specifications

Any attempt to build a software system must begin with some description of the task the system is supposed to perform. Such a description need not tightly constrain every single aspect of the required system; for example, a result is often only required up to a certain accuracy in numerical problems, the efficiency of a system is usually constrained only to fall within certain limits, and details of input/output format may often be left to the programmer. Another example is the description of the task to be performed by a compiler: the compiler should generate correct code, but the exact code to be generated is not prescribed. However loose, such a description characterises which actual software systems would be acceptable for the intended purpose and which would not be.

As discussed in the previous section, we concentrate on the functional behaviour of software systems, modelling them as algebras. Hence, a description of a class of systems amounts to a characterisation of a class of algebras. The term *specification* is used to refer to a formal object, normally in textual form, that defines such a class. It is natural to expect that every specification unambiguously defines both a signature and a class of algebras over that signature, since part of the purpose of the specification is to indicate the names of the types and functions to be provided. In this indirect way, a specification describes a class of software systems that are its acceptable *realisations*. These are the systems whose functional behaviour is captured by one of the algebras in the class defined by the specification.

The standard way of describing a class of algebras is by listing the properties they are to satisfy. Such properties may be expressed as sentences in some logical system such as equational logic or first-order logic. These sentences are called *axioms*. For any given algebra and axiom, the semantics of the logical system determines whether the algebra satisfies the axiom or not. A set of axioms thus describes a class of algebras, namely the class of all algebras that satisfy all the axioms in that set.[2] Work on algebraic specification, to which the material in this book belongs, is based on these two fundamental principles: first, software systems may be modelled as algebras; second, properties of algebras may be described using axioms. This style of specification, which is covered in Chapter 2, naturally separates the issue of describing *what* the system is to do (given by the axioms) from that of describing *how* those requirements are achieved (given by the algorithms and data structures in the system).

Example 0.2.1 (Timetable, continued). Consider the project of assigning courses to rooms and time slots in a university. This can be viewed as the task of constructing an algebra like the one described in Example 0.1.1. The prerequisites for this are complete lists of the courses, lecturers, time slots and rooms and information concerning which lecturers are able to teach which courses. The latter may be expressed using axioms such as:

$$\textit{who-teaches}(\textit{Calculus}) = \textit{Smith} \vee \textit{who-teaches}(\textit{Calculus}) = \textit{Kowalski}$$

The main problem is to make sure that the assignment is done in such a way that no conflicts arise. The signature given in Example 0.1.1 guarantees that lecturers are not required to be in two places at once, since *what-teaches* and *where-teaches* are *functions*, i.e. they map any tuple of inputs (lecturer and time slot, in this case) to exactly one output (a course or a room respectively). However, we must ensure that:

1. No room is simultaneously occupied by two different courses.
2. Lecturers are sent to the courses they are assigned to teach.
3. All courses have time slots allocated to them.

These requirements are formally expressed by the following axioms:

1. $\forall t{:}\textit{Timeslot}, c, c'{:}\textit{Course} \bullet$
   $$\textit{where-teaches}(\textit{who-teaches}(c), t) = \textit{where-teaches}(\textit{who-teaches}(c'), t) \Rightarrow c = c'$$
2. $\forall l{:}\textit{Lecturer}, t{:}\textit{Timeslot} \bullet\ \textit{who-teaches}(\textit{what-teaches}(l, t)) = l$
3. $\forall c{:}\textit{Course} \bullet\ \exists t{:}\textit{Timeslot} \bullet\ \textit{what-teaches}(\textit{who-teaches}(c), t) = c$

Extending the problem to include timetable and registration information for students as well involves adding new axioms expressing the consistency between lecturers' and students' schedules. That is:

4. Students are sent to the courses in which they are enrolled:
   $\forall s{:}\textit{Student}, t{:}\textit{Timeslot} \bullet\ \textit{enrolled}(s, \textit{what-attends}(s, t)) = \textit{true}$
5. Students are sent to each course they are enrolled for, each time it is taught:
   $$\forall s{:}\textit{Student}, l{:}\textit{Lecturer}, t{:}\textit{Timeslot} \bullet\ \textit{enrolled}(s, \textit{what-teaches}(l, t)) = \textit{true} \Rightarrow \textit{what-attends}(s, t) = \textit{what-teaches}(l, t)$$

---

[2] The use of the two distinct terms "set" and "class" has a mathematical justification, to be discussed in Section 3.1.1.1.

6. Students and lecturers are sent to the same place for the same course:
$$\forall l{:}\mathit{Lecturer}, s{:}\mathit{Student}, t{:}\mathit{Timeslot} \bullet$$
$$\mathit{where\text{-}teaches}(l,t) = \mathit{where\text{-}attends}(s,t) \Leftrightarrow$$
$$\mathit{what\text{-}teaches}(l,t) = \mathit{what\text{-}attends}(s,t) \qquad \square$$

The process of constructing such a specification is a subject in itself. Capturing desired properties in the form of axioms is sometimes difficult, as is deciding when a given set of axioms captures all of the desired properties. It follows that specifications can contain bugs, just as software systems can, and correctness of a system with respect to a specification is a matter of consistency between two independent definitions, either or both of which may contain errors. But once a set of axioms has been written down, theorem proving tools can be used to *validate* them by exploring their consequences, with unexpected consequences or lack of expected consequences triggering a revision of the axioms.

Up to now the discussion has concentrated on describing the requirements that a software system is to fulfill. A specification of this kind plays the role of a *contract* between a client (the customer) and the programmer (or programming team) responsible for building the system. On one hand, this contract records the features that the programmer has to ensure. On the other hand, it records the features of the system on which the client may rely. It is important that this contract be exhaustive in the sense that it record *all* the expected properties of the system; the programmer is not required to provide any features that are not explicitly stated in the contract, and the client should not rely on such features either. Any actual system will satisfy properties that are not mentioned in the contract. For example, some release of a compiler may happen to ensure that uninitialised variables are set to 0. But if this is not stated in the language definition (which is the compiler specification) it would be dangerous to rely on this feature since it may change in the next release of the compiler.

It is commonly accepted that large systems should be organized into modules that encapsulate logically coherent units. Such units may be modelled as algebras in exactly the same way as discussed above for complete systems. Specifications are required here as well to describe the interfaces between modules. These specifications constrain the programmer responsible for implementing each module to providing the required features in the same way as the specification of the overall system constrains the programming team as a whole. The clients here are other modules in the system, which may use data and/or functions that this module supplies.

As before, it is important that the interface specification record all the expected properties of the module. This means that programmers responsible for other modules are not allowed to take advantage of accidental features of modules on which they rely. Thus interface specifications serve two main purposes. First, they provide a means of communication between a module implementor and the outside world. At the same time, they serve to *prevent* undesirable communication by defining exactly those details on which others are allowed to depend, thus abstracting away from the internal details of the module implementation. A special form of such information hiding, supported by modern programming languages like Standard ML and Java, is *data abstraction*, where the exact representation of data is kept hidden.[3]

If this discipline is strictly adhered to, then programmers are free to change internal details of their module implementations without restriction, provided that the module interface specification is still satisfied. Another practical advantage of carefully specifying module interfaces is that these specifications provide the documentation necessary to support the reuse of modules in the construction of other systems.

The problems of scale which led to the introduction of modular structuring of large software systems affect large specifications as well. The specification of a large system involves thousands of properties that the system is required to satisfy. If these properties are simply listed one by one in the form of axioms, the specification would be completely unmanageable: it would be difficult to construct, and nearly impossible to understand and use. It is even a non-trivial task to understand the relatively short list of axioms in Example 0.2.1 and ensure that all the desired properties are included. The remedy to this problem is to structure such specifications into units of logically related properties which are then combined to build more complex specifications. In the example the list of axioms has been divided into two groups of related axioms to ease understanding. Mechanisms for structuring specifications are covered in Chapter 5.

The structure of a specification is not just a superficial feature of its presentation. It is important not only for understanding specifications but also for all aspects of their use. For example, in proving that certain additional properties are consequences of those explicitly stated in the specification, the structure of the specification may be exploited in guiding the search for a proof. Similarly, the structure of the specification of a software system may play a useful role in the way that the system is decomposed into modules.

0.3 Software development

As discussed above, and according to the traditional waterfall model of the software life cycle, the specification of a system is the starting point for its subsequent development. Once the specification of a software system is agreed on, the programmer is committed to building a system exhibiting a behaviour that conforms to that required by the specification. The usual way to proceed is to construct the system by whatever means are available, making informal reference to the specification in the process, and then verifying in some way that the result does indeed realise the specification. Other life cycle models take a different view, but in those that do not reject the need for specifications outright, the role of the specification as the definition of correct system behaviour remains, along with the need for verification. In

[3] In Standard ML, data abstraction is achieved using opaque signature ascription. The term “data abstraction” tends to be avoided in Java, but interfaces and access modifiers provide the required support.

the so-called V-model, the specification has an additional role, that of providing an abstract view of the final implemented system.

The most widespread verification method is testing, which checks that in certain selected cases the behaviour exhibited by the system satisfies the constraints imposed by the specification. Testing a system built to satisfy the timetable specification of Example 0.2.1 would involve checking whether the axioms hold for chosen values of variables. For instance, one might check that the axiom

$$\forall l{:}\mathit{Lecturer}, t{:}\mathit{Timeslot} \bullet \mathit{who\text{-}teaches}(\mathit{what\text{-}teaches}(l,t)) = l$$

holds for l = Kowalski and t = Thursday 9–10 am. This has the disadvantage that correctness can be ensured in this fashion only when the system operates on a fixed and finite set of data and exhaustive testing of all combinations is carried out. Model checking is one way of rapidly conducting such exhaustive testing.

An alternative to testing is to provide a formal proof that the system is correct with respect to the specification. For the timetable example this would amount to proving that the system satisfies the axioms listed above. However, after many years of work on software verification it now seems to be more or less widely accepted that full proofs of correctness will probably never be feasible for systems of realistic size. On the other hand, proofs of selected properties of critical parts of important systems *are* done by some software developers.

From a practical point of view, the main ground for pessimism is the huge gap between the high-level specification of requirements and the low-level details of the realisation, including the specific data representation and algorithms used and the coding of these in a particular programming language. The fact that transparency and readability are usually sacrificed for the sake of efficiency makes the gap even wider.

This leads to the idea that software systems should be developed from specifications in such a way that the result is guaranteed to be correct by construction. The approach we follow here is to develop a system from its specification via a series of small refinement steps, inspired by the programming discipline of stepwise refinement. Each refinement step captures a single design decision, for instance a choice between several functions that satisfy the specification, between several algorithms that implement the same function, or between several ways of efficiently representing a given data type. If each of these individual refinement steps can be proved correct then the resulting system is guaranteed to satisfy the original specification. Each of these proofs is orders of magnitude easier than a correctness proof for the resulting system since each refinement step is small. In principle it would be possible to combine all the individual correctness proofs to yield a proof of the correctness of the system with respect to the specification, but in practice this would never be necessary. Formal development of systems from specifications is covered in Chapters 7 and 8.

Even if we consider the very simple problem of developing a software system that realises the specification given in Example 0.2.1 (disregarding the extension to handle students), it is difficult and unnatural to come up with the definitions of all three functions simultaneously. One would tend to define them one after another, perhaps starting with the decision of which lecturer will teach which course, then assigning time slots to the courses, and finally arranging rooms for courses. The definition of each of these functions constrains the choices available at subsequent steps since the axioms in the specification impose certain compatibility properties.

This methodology does not prevent us from making bad design decisions. For example, for some choices of *who-teaches* and *what-teaches* there may be no way to define *where-teaches* such that the specification is satisfied because of limitations on the number of rooms. This means that backtracking may be necessary during software development.

In the course of refining a large specification it will be necessary to decompose it into appropriately chosen smaller units. The refinement of these units may proceed separately, possibly involving further decomposition. This will result in a collection of modules that can be combined to yield a correct software system. Decomposition and refinement steps may be freely interleaved during the development process.

Once a system has been built in this fashion, the development history which includes all of the intermediate specifications (and possibly even the proofs of correctness) constitutes very complete design documentation. This facilitates later maintenance of the system. Even if the original specification is changed in the course of maintenance, it is normally possible to use this documentation to trace which parts of the system this change affects, localizing the fragment of the system that must be changed.

The rosy picture painted above neglects the fact that all stages of the software development process are arduous and error-prone. Coming up with a formal specification that accurately reflects all the vague and informal requirements of the customer is difficult; ideas for refinement steps are hard to come by and their formalisation is often a struggle as well; an advantageous decomposition of the problem is often difficult to find; and proofs of correctness are laborious. This leaves a lot of scope for the skill of designers and programmers. The scale of the formal objects involved and the need for meticulous accuracy and attention to detail make these creative tasks infeasible for humans to perform with pencil and paper. Many of these problems may be resolved through the use of computer-based tools to support the software development process. The most obvious candidates for this are mechanical theorem provers and proof checkers as well as some means of keeping track of all the different bits, how they interrelate and what remains to be done.

0.4 Generality and abstraction

The motivation for focussing on the functional behaviour of software systems and abstracting away from the concrete details of code and algorithms was discussed in Section 0.1. This led to the decision to model software systems as algebras. There are, however, many important aspects of the functional behaviour of systems that are not captured by this model, for example:

Non-termination: Systems do not always terminate on all inputs. The functional behaviour of such a system does not directly determine a (total) function.

Exceptions: Some operations fail on certain inputs, yielding an exceptional result or an error message. Although such results may be viewed as data values, they must be distinguished in some way. This is not accommodated by the standard definition of an algebra.

Input/output: Systems may interact with their environment during execution and this interaction is part of the functional behaviour of the system. The fact that input and output may be interleaved means that ordinary functions do not accurately model such systems.

Such aspects of systems, and their combinations (for example in reactive systems, which are designed to run forever and to react to input stimuli), are modelled by changing the notion of algebra (e.g. using so-called coalgebras to model reactive systems). Often only relatively minor enrichments are required. For instance partial algebras, where functions may be undefined on some arguments, can be used to model non-termination. Such adjustments are also necessary if we want to handle all of the relevant concepts that are present in programming languages, such as polymorphism, higher-order functions, lazy evaluation, imperative features, concurrency and mobility. Some of these elaborations are discussed in Section 2.7.

Moreover, each of these aspects of behaviour can be specified in different ways. This amounts to a choice between alternative logical systems for writing the axioms in specifications. As a very simple example, for specifying partial algebras using equational axioms there are two standard choices: strong equality with definedness formulae, or existential equality (see Section 2.7.4). Coalgebras can be specified using different modal logics. Even for ordinary algebras, there is a choice of whether to use purely equational axioms or full first-order or even higher-order logic, with trade-offs between expressive power and ease of reasoning.

There are at least three ways to proceed with the formation of a theory of software specification and development given these complications. The first is to start by devising a notion of algebra that accommodates all of the aspects of system behaviour and all relevant concepts of programming languages we can think of, with a logic for writing axioms that is rich enough to conveniently specify all of these aspects of behaviour in all of their combinations. Then we erect an appropriate theory on top of this basis. One problem with this approach is that the whole strategy breaks down when a new aspect of behaviour emerges or a new feature of programming languages becomes popular. For example, we would have to start again from scratch if we had not taken concurrency into account and it became necessary to add this later. Another problem is that the huge variety of features that would have to be considered would make the basic concepts of the theory very complicated indeed. This would yield an unwieldy theory in which one would be unable to see the forest for the trees.

Another possibility is to consider each target programming language separately and design a notion of algebra appropriate for modelling software systems built using just the particular features of this language, with an appropriate choice of logical notation for writing axioms. This has the obvious disadvantage that we must start afresh for each programming language we consider, or even for different dialects of the same language.

Many aspects of the theory of system specification and development actually turn out to be independent of the particular details of the notion of algebra and the logical system used. This is illustrated by the fact that if we erect a complete theory for several different programming languages as described above, we will find ourselves repeating the same work time after time with only relatively minor modifications. Thus a third possibility is to develop a generic theory which is *parameterised* by the notion of algebra to be used and the definition of what it means for an algebra to satisfy an axiom. Given a particular choice of the notion of algebra, the theory can simply be *instantiated* to adapt it to that choice. Analogously, it is possible to parameterise the theory by the notion of axiom, which enables the use of different logical systems in writing specifications, and by the notion of signature to accommodate different type systems.

This third approach is the one adopted in this book. The theory presented is almost entirely independent of the particular aspects of functional behaviour of systems under consideration, of how these are described by axioms, and of the features of the underlying type system. This general view leads to reusable concepts and results and ultimately to reusable tools, which can be instantiated in particular situations as required. The resulting uniform framework exposes the essential concepts inherent in specification and development, and separates them from the sordid details of specific situations. The foundations required to support this theory are developed in Chapter 4 and then applied in subsequent chapters.

Working at this level of generality necessitates the use of mathematical tools that are appropriate for formulating general definitions and proving general facts. The language and concepts of *category theory* are convenient for dealing with the kind of generality involved. The basic concepts of category theory that are required are presented in Chapter 3.

Despite the advantages of generality, it is necessary to examine specific instantiations for the purposes of both presentation and motivation. Achieving understanding requires examination of concrete situations and examples, and these in turn demonstrate the need for developments in the general theory. Much of the time it will be sufficient to consider the simple situation in which systems are modelled as "standard" algebras, ignoring their inadequacy for the aspects of systems mentioned above, with axioms written using equations and sometimes propositional connectives and first-order quantifiers. This is the situation that is treated in Chapters 1 and 2. Examples that are meant to appeal to the reader's programming intuition are sprinkled throughout the later chapters, using an instantiation of the emerging theory to a context that is akin to a purely functional first-order subset of the Standard ML programming language, based on definitions and notations in Examples 4.1.25 and 6.1.9 and Exercise 7.3.5.

0.5 Formality

Algebraic specification as it is presented here is close to the "hard-core", uncompromising end of the spectrum of existing work on formal methods. Indeed, one of its advantages over competing approaches is that it has complete mathematical foundations. Thus, a claim of correctness of a software system or component with respect to a precise algebraic specification of the problem, when backed with all the relevant formal proofs, amounts to a complete justification without reliance on informal reasoning, guesswork or crossed fingers.

Of course, software is almost never developed in this way in practice. One reason is that 100% confidence in correctness is hardly ever necessary, and achieving it involves an enormous amount of hard work. Furthermore, experience suggests that failure of a proof of correctness is often the result of an error in the specification itself. Achieving confidence that the original specification of a problem spells out what is actually required will generally involve human interaction and other processes that are necessarily informal and error-prone.

In practice, shortcuts are normal and formal proofs are rarely attempted. Even when they are attempted, proofs are often sketched informally to a level of detail that is sufficient to check the main points of importance rather than done in full detail to completion using a proof assistant. Such a mode of use of formal methods is referred to as *rigorous methods*. Sometimes certain critical components of a system, or certain important properties, will be selected for special attention. Such a component might be one containing a complicated and important algorithm, or one that protects the system from catastrophic failure. An important property might be exception freedom or freedom from deadlock, or a security property. The degree of confidence that is justified in the outcome depends on an appropriate choice of the components and/or properties of greatest importance, and the care that is taken with informal or incomplete proofs. The above points notwithstanding, the power and sophistication of automated theorem proving tools and the computing power available to engineers have increased over time to the point where it is becoming feasible to formally verify whole systems or components of systems, and such proofs are increasingly being done in practice, especially for hardware.

From this point of view, the material in this book may be seen as providing a reference point for less formal means of improving quality of software, including rigorous methods. Another approach that puts major emphasis on the trade-off between practical benefits achieved and effort required is called *lightweight formal methods*. Here, some of the techniques of formal methods are used to improve the quality of software via early detection and removal of errors, without any expectation that they can be entirely eliminated. Lightweight formal methods rely on the use of automated analysis tools to provide cost-effective programming support.

In this book we will not explicitly point out opportunities for relaxing formality, and to a large extent that is a matter of engineering judgement in particular circumstances. Neither will we discuss to what extent the material presented provides opportunities for the provision of useful automated analyses. In general we do not provide algorithms or present decidability or complexity results for the decision problems discussed. There is a clear trade-off between expressibility of notations on the one hand and ease of automation on the other; our approach here is firmly on the side of expressibility, with compromises in favour of automation left as a separate (but practically important) issue.

0.6 Outlook

In the previous sections we have outlined the motivations that underlie the algebraic approach to specification and formal software development. We have also discussed the need for a general approach that abstracts away from specific aspects of software systems.

This book presents mathematical foundations for algebraic specification and software development that support these practically motivated ideas. It concentrates on developing basic concepts and studying their fundamental properties rather than on demonstrating how these concepts may be used in the practice of software construction, which is a separate topic. This provides the necessary foundation for further work towards practical software production, on at least the following levels:

- More user-oriented notations and theories could be developed on top of the rudiments presented here. For example, high-level user-friendly specification languages could be defined, based on the primitive operations presented in Chapter 5.
- A computationally tractable and practically useful subset of the notations and concepts presented here could be selected and used in the style of lightweight formal methods as discussed above.
- Tools, techniques, hints and heuristics could be developed to support and guide the user's specification and development activity.

All of these aspects are beyond the scope of this book, and we deliberately avoid dealing with problems arising at these levels here. In particular we do not present "how-to" guidelines for building specifications or for validating them against real-life requirements, or for coming up with design decisions in software development. Some of the more substantial examples provide some hints in this direction.

The CASL specification language ([BM04], [Mos04]) is an attempt at a user-friendly specification notation, underpinned by many of the ideas presented here, with methodological guidelines for use of the features it provides. The material in this book and languages like CASL are not the end of the story, and they do not by any means solve all of the problems encountered in engineering practical software systems. But they do provide a solid basis for coming to grips with some of the key technical problems in software development.

Chapter 1
Universal algebra

The most basic assumption in work on algebraic specification is that programs are modelled as algebras. This point of view abstracts from the concrete details of code and algorithms, and regards the input/output behaviour of functions and the representation of data as of primary importance. Representing programs in terms of sets (of data values) and ordinary mathematical functions over these sets greatly simplifies the task of reasoning about program correctness. See Section 0.1 for some illustrative examples and more introductory discussion on this point.

The branch of mathematics that deals with algebras in this general sense (as opposed to the study of specific classes of algebras, such as groups and rings) is called *universal algebra* or sometimes *general algebra*. However, work on universal algebra by mathematicians has concentrated almost exclusively on the special case of single-sorted algebras with first-order total functions. The generalisation to *many-sorted* or *heterogeneous* algebras is required to model programs that manipulate several kinds or *sorts* of data; further generalisations are necessary to handle programs that fail to terminate on some inputs, that generate exceptions during execution, and so on. This chapter summarizes the basic concepts and results of many-sorted universal algebra that will be required for the rest of this book. Some extensions useful for modelling more complex programs will be discussed later, in Section 2.7. In this chapter, all proofs are left as exercises for the reader.

1.1 Many-sorted sets

When using an algebra to model a program which manipulates several sorts of data, it is natural to partition the underlying set of values in the algebra so that there is one set of values for each sort of data. It is often convenient to manipulate such a family of sets as a unit, in such a way that operations on this unit respect the "typing" of data values.

The following definitions and notational conventions allow us to manipulate sorted families of sets (of functions, of relations, etc.) in the same way as ordinary sets (functions, relations, etc.). Ordinary sets (functions, relations, etc.) correspond to the degenerate case in which there is just one sort, so these definitions also serve to recall the notation and terminology of set theory to be used throughout this book. Let S be a set; the notation $\langle X_s \rangle_{s \in S}$ is a standard shorthand for the family of objects X_s indexed by $s \in S$, i.e. the function with domain S which maps each $s \in S$ to X_s.

Throughout this section, let S be a set (of sorts).

Definition 1.1.1 (Many-sorted set). An S*-sorted set* is an S-indexed family of sets $X = \langle X_s \rangle_{s \in S}$, which is *empty* if X_s is empty for all $s \in S$. The empty S-sorted set will be written (ambiguously) as $\varnothing$. The S-sorted set X is *finite* if X_s is finite for all $s \in S$ and there is a finite set $\widehat{S} \subseteq S$ such that $X_s = \varnothing$ for all $s \in S \setminus \widehat{S}$.

Let $X = \langle X_s \rangle_{s \in S}$ and $Y = \langle Y_s \rangle_{s \in S}$ be S-sorted sets. Union, intersection, Cartesian product, set difference, disjoint union, inclusion (subset) and equality of X and Y are defined componentwise as follows:

$X \cup Y = \langle X_s \cup Y_s \rangle_{s \in S}$
$X \cap Y = \langle X_s \cap Y_s \rangle_{s \in S}$
$X \times Y = \langle X_s \times Y_s \rangle_{s \in S}$, with pairs written $\langle x, y \rangle \in X_s \times Y_s$ when $x \in X_s$ and $y \in Y_s$
$X \setminus Y = \langle X_s \setminus Y_s \rangle_{s \in S}$
$X \uplus Y = \langle X_s \uplus Y_s \rangle_{s \in S}$ (where $X_s \uplus Y_s = (\{1\} \times X_s) \cup (\{2\} \times Y_s)$)
$X \subseteq Y$ iff (if and only if) $X_s \subseteq Y_s$ for all $s \in S$
$X = Y$ iff $X \subseteq Y$ and $Y \subseteq X$ (equivalently, iff X and Y are equal as functions).

We write $X \subset Y$ when $X \subseteq Y$ and $X \neq Y$. □

Exercise 1.1.2. Give a formal explanation of the above statement that "Ordinary sets ... correspond to the degenerate case [of many-sorted sets] in which there is just one sort". How many $\varnothing$-sorted sets are there? □

Notation. It will be very convenient to pretend that $X \subseteq X \uplus Y$ and $Y \subseteq X \uplus Y$. Although this is never actually the case, it allows us to treat disjoint union in the same way as ordinary union, the difference being that when $X \cap Y \neq \varnothing$, $X \uplus Y$ contains two "copies" of the common elements and keeps track of which copy is from X and which is from Y. To see that this does not cause problems, observe that there are injective S-sorted functions (see the next definition) $i1{:}X \to X \uplus Y$ and $i2{:}Y \to X \uplus Y$ defined by $i1_s(x) = \langle 1, x \rangle$ for all $s \in S$ and $x \in X_s$ and similarly for $i2$. A pedant would be able to correct what follows by simply inserting the functions $i1$ and/or $i2$ where appropriate in expressions involving $\uplus$. □

Exercise 1.1.3. Extend the above definitions of union, intersection, product and disjoint union to operations on I-indexed families of S-sorted sets, for an arbitrary index set I. For example, the definition for product is $(\prod \langle X_i \rangle_{i \in I})_s = \{f{:}I \to \bigcup_{i \in I}(X_i)_s \mid f(i) \in (X_i)_s \text{ for all } i \in I\}$ for each $s \in S$. If $I = 1, \ldots, n$ then we use the simpler notation $X_1 \times \cdots \times X_n$, with tuples $\langle x_1, \ldots, x_n \rangle$, as a generalisation of the case $n = 2$ above; for $n = 0$, the empty tuple is written $\langle\rangle$. □

Definition 1.1.4 (Many-sorted function). Let $X = \langle X_s \rangle_{s \in S}$ and $Y = \langle Y_s \rangle_{s \in S}$ be S-sorted sets. An S*-sorted function* $f{:}X \to Y$ is an S-indexed family of functions $f =$

$\langle f_s : X_s \to Y_s \rangle_{s \in S}$; X is called the *domain* (or *source*) of f, and Y is called its *codomain* (or *target*). Application of f to $x \in |X|_s$ is written as $f_s(x)$ or sometimes as $f(x)$; if X is a product then we write $f_s(x_1,\ldots,x_n)$ or $f(x_1,\ldots,x_n)$ rather than $f_s(\langle x_1,\ldots,x_n \rangle)$.

An S-sorted function $f : X \to Y$ is an *identity* (an *inclusion*, a *surjection*, an *injection*, a *bijection*, etc.) if for every $s \in S$, the function $f_s : X_s \to Y_s$ is an identity (an inclusion, a surjection, an injection, a bijection, etc.). The identity S-sorted function on X will be written as $id_X : X \to X$. If $X \subseteq Y$, the inclusion will often be written $X \hookrightarrow Y$.

If $f : X \to Y$ and $g : Y \to Z$ are S-sorted functions then their *composition* $f;g : X \to Z$ is the S-sorted function defined by $f;g = \langle f_s;g_s \rangle_{s \in S}$. That is, if $s \in S$ and $x \in X_s$ then $(f;g)_s(x) = g_s(f_s(x))$.[1]

Let $f : X \to Y$ be an S-sorted function and $X' \subseteq X$, $Y' \subseteq Y$ be S-sorted sets. The *image of X' under f* is the S-sorted set $f(X') = \langle f_s(X'_s) \rangle_{s \in S} \subseteq Y$, where $f_s(X'_s) = \{f_s(x) \mid x \in X'_s\} \subseteq Y_s$ for all $s \in S$. The *coimage of Y' under f* is the S-sorted set $f^{-1}(Y') = \langle f_s^{-1}(Y'_s) \rangle_{s \in S} \subseteq X$, where $f_s^{-1}(Y'_s) = \{x \in X_s \mid f_s(x) \in Y'_s\} \subseteq X_s$ for all $s \in S$. □

All functions in this book are total except where they are explicitly designated as partial. When $f : X \to Y$ is partial, then $dom(f) \subseteq X$ is the S-sorted set of those elements $x \in X_s$ for which $f_s(x)$ is defined, for all $s \in S$.

Definition 1.1.5 (Many-sorted binary relation). Let $X = \langle X_s \rangle_{s \in S}$ and $Y = \langle Y_s \rangle_{s \in S}$ be S-sorted sets. An *S-sorted binary relation between X and Y*, written $R \subseteq X \times Y$, is an S-indexed family of binary relations $R = \langle R_s \subseteq X_s \times Y_s \rangle_{s \in S}$. For $s \in S$, $x \in X_s$ and $y \in Y_s$, $x \, R_s \, y$ (sometimes written $x \, R \, y$) means $\langle x,y \rangle \in R_s$.

If $R \subseteq X \times Y$ is an S-sorted binary relation then its *inverse* is the S-sorted binary relation $R^{-1} \subseteq Y \times X$ such that for $s \in S$, $\langle y,x \rangle \in (R^{-1})_s$ iff $\langle x,y \rangle \in R_s$. If $Q \subseteq Y \times Z$ is also an S-sorted binary relation, then the *composition of R and Q* is the S-sorted binary relation $R;Q \subseteq X \times Z$ such that for $s \in S$, $\langle x,z \rangle \in (R;Q)_s$ iff there exists some $y \in Y_s$ such that $\langle x,y \rangle \in R_s$ and $\langle y,z \rangle \in Q_s$. □

The generalisation of many-sorted binary relations to n-ary relations, for $n \geq 0$, is obvious. As usual, many-sorted functions may be viewed as special many-sorted relations. Both may also be viewed as special cases of many-sorted sets, to which the set-theoretic operations and relations in Definition 1.1.1 apply.

Definition 1.1.6 (Kernel of a many-sorted function). Let $f : X \to Y$ be an S-sorted function. The *kernel of f* is the S-sorted binary relation $ker(f) = \langle ker(f_s) \rangle_{s \in S} \subseteq X \times X$ where $ker(f_s) = \{\langle x,y \rangle \mid x,y \in X_s \text{ and } f_s(x) = f_s(y)\} \subseteq X_s \times X_s$ is the kernel of f_s for all $s \in S$. □

Definition 1.1.7 (Many-sorted equivalence). Let $X = \langle X_s \rangle_{s \in S}$ be an S-sorted set. An S-sorted binary relation $R \subseteq X \times X$ is an *S-sorted equivalence (relation) on X* if it is:

[1] This "diagrammatic" order of composition and the semicolon notation will be used consistently throughout this book.

- reflexive: $x \, R_s \, x$;
- symmetric: $x \, R_s \, y$ implies $y \, R_s \, x$; and
- transitive: $x \, R_s \, y$ and $y \, R_s \, z$ implies $x \, R_s \, z$

for all $s \in S$ and $x, y, z \in X_s$. The symbol $\equiv$ is often used for (S-sorted) equivalence relations.

Let $\equiv$ be an S-sorted equivalence on X. If $s \in S$ and $x \in X_s$ then the *equivalence class of x modulo* $\equiv$ is the set $[x]_{\equiv_s} = \{y \in X_s \mid x \equiv_s y\}$. The *quotient of X modulo* $\equiv$ is the S-sorted set $X/{\equiv} = \langle X_s/{\equiv_s} \rangle_{s \in S}$ where $X_s/{\equiv_s} = \{[x]_{\equiv_s} \mid x \in X_s\}$ for all $s \in S$. □

Example 1.1.8. Let $S = \{s_1, s_2\}$, and let X and Y be two S-sorted sets defined as follows:

$$X = \langle X_s \rangle_{s \in S} \text{ where } X_{s_1} = \{\Box, \triangle\} \text{ and } X_{s_2} = \{\clubsuit, \heartsuit, \spadesuit\},$$
$$Y = \langle Y_s \rangle_{s \in S} \text{ where } Y_{s_1} = \{1,2,3\} \text{ and } Y_{s_2} = \{1,2,3\}.$$

Let $f : X \to Y$ be the S-sorted function such that

$$f_{s_1} = \{\Box \mapsto 1, \triangle \mapsto 3\},$$
$$f_{s_2} = \{\clubsuit \mapsto 1, \heartsuit \mapsto 2, \spadesuit \mapsto 2\}.$$

(i.e. $f_{s_1}(\Box) = 1$ and $f_{s_1}(\triangle) = 3$; analogously for f_{s_2}). Then the kernel of f is the S-sorted equivalence relation $ker(f) = \langle ker(f_s) \rangle_{s \in S}$ where

$$ker(f_{s_1}) = \{\langle \Box, \Box \rangle, \langle \triangle, \triangle \rangle\},$$
$$ker(f_{s_2}) = \{\langle \clubsuit, \clubsuit \rangle, \langle \heartsuit, \heartsuit \rangle, \langle \heartsuit, \spadesuit \rangle, \langle \spadesuit, \heartsuit \rangle, \langle \spadesuit, \spadesuit \rangle\}.$$

The quotient of X modulo $ker(f)$ is the S-sorted set $X/ker(f) = \langle X_s/ker(f_s) \rangle_{s \in S}$ where

$$X_{s_1}/ker(f_{s_1}) = \{\{\Box\}, \{\triangle\}\},$$
$$X_{s_2}/ker(f_{s_2}) = \{\{\clubsuit\}, \{\heartsuit, \spadesuit\}\}.$$ □

Exercise 1.1.9. Show that if $f : X \to Y$ is an S-sorted function, then $ker(f)$ is an S-sorted equivalence on X. □

Exercise 1.1.10. Show that if $\equiv$ is an S-sorted equivalence on X then for all $s \in S$ and $x, y \in X_s$, $[x]_{\equiv_s} = [y]_{\equiv_s}$ iff $x \equiv_s y$. □

Notation. Subscripts selecting components of S-sorted sets (functions, relations, etc.) are often omitted when there is no danger of confusion. Then Exercise 1.1.10 would read "... for all $s \in S$ and $x, y \in X_s$, $[x]_{\equiv} = [y]_{\equiv}$ iff $x \equiv y$" or even "... for all $x, y \in X$, $[x]_{\equiv} = [y]_{\equiv}$ iff $x \equiv y$". □

1.2 Signatures and algebras

The functions and data types defined by a program have names. These names are used to refer to and reason about the parts of the program, and to build larger programs which rely on the functionality the program provides. The connection between a program and an algebra used to model it is provided by these names, which are attached to the corresponding components of the algebra. The set of names associated with an algebra is called its signature. The signature of an algebra defines the *syntax* of the algebra by characterising the ways in which its components may legally be combined; the algebra itself supplies the *semantics* by assigning interpretations to the names in the signature.

Definition 1.2.1 (Many-sorted signature). A *(many-sorted) signature* is a pair $\Sigma = \langle S, \Omega \rangle$, where:

- S is a set (of sort names); and
- Ω is an $(S^* \times S)$-sorted set (of operation names)

where S^* is the set of finite (including empty) sequences of elements of S. We will sometimes write $sorts(\Sigma)$ for S and $ops(\Sigma)$ for Ω. Σ is a *subsignature* of a signature $\Sigma' = \langle S', \Omega' \rangle$ if $S \subseteq S'$ and $\Omega_{w,s} \subseteq \Omega'_{w,s}$ for all $w \in S^*, s \in S$. □

Many-sorted signatures will be referred to as *algebraic* signatures when it is necessary to distinguish them from other kinds of signatures to be introduced later.

Notation. Saying that $f : s_1 \times \cdots \times s_n \to s$ is in $\Sigma = \langle S, \Omega \rangle$ means that $s_1 \ldots s_n \in S^*$, $s \in S$ and $f \in \Omega_{s_1 \ldots s_n, s}$. Then f is said to have *arity* $s_1 \ldots s_n$ and *result sort* s. The abbreviation $f : s$ will be used for the constant operation $f : \varepsilon \to s$ (ε is the empty sequence). □

This definition of signature does not accommodate programs containing higher-order functions, or functions returning multiple results. A possible extension for handling higher-order functions is briefly discussed in Section 2.7.6. As for functions with multiple results, a function $f : s_1 \times \cdots \times s_n \to t_1 \times \cdots \times t_m$ may be viewed as a family of m functions:

$$f_1 : s_1 \times \cdots \times s_n \to t_1 \qquad \ldots \qquad f_m : s_1 \times \cdots \times s_n \to t_m.$$

Generalising the definition of signature to handle such functions in a more direct way is easy but makes subsequent developments somewhat messier in a non-interesting way.

The definition above *does* permit overloaded operation names, since it is possible to have both $f : s_1 \times \cdots \times s_n \to s$ and $f : t_1 \times \cdots \times t_m \to t$ in a signature Σ, where $s_1 \ldots s_n s \neq t_1 \ldots t_m t$. A more restrictive definition of signature, adequate for most purposes, would have a set Ω of operation names (and a set S of sort names) with functions $arity : \Omega \to S^*$ and $sort : \Omega \to S$. These two definitions are equivalent if each operation name in Ω is taken to be tagged with its arity and result sort.

In the rest of this section, let $\Sigma = \langle S, \Omega \rangle$ be a signature.

Definition 1.2.2 (Many-sorted algebra). A Σ-*algebra* A consists of:

- an S-sorted set $|A|$ of *carrier sets* (or *carriers*); and
- for each $f{:}\, s_1 \times \cdots \times s_n \to s$ in Σ, a function (or *operation*) $(f{:}\, s_1 \times \cdots \times s_n \to s)_A{:}\, |A|_{s_1} \times \cdots \times |A|_{s_n} \to |A|_s$. □

If A is a Σ-algebra and s is a sort name in Σ then $|A|_s$, the carrier set of sort s in A, is the universe of data values of sort s; accordingly, we often refer to the elements of carrier sets as *values*. If $f{:}\, s_1 \times \cdots \times s_n \to s$ is in Σ then the operation $(f{:}\, s_1 \times \cdots \times s_n \to s)_A$ is a function on the corresponding carrier sets of A. If $n = 0$ (i.e. $f{:}\, s$), then $|A|_{s_1} \times \cdots \times |A|_{s_n}$ is a singleton set containing the empty tuple $\langle\rangle$, and then $(f{:}\, s)_A$ may be viewed as a constant denoting the value $(f{:}\, s)_A(\langle\rangle) \in |A|_s$. Notice that $(f{:}\, s_1 \times \cdots \times s_n \to s)_A$ is a *total* function so algebras as defined here are only appropriate for modelling programs containing total functions. See Sections 2.7.3–2.7.5 for several ways of extending the definitions to cope with partial functions. Note also that there is no restriction on the cardinality of $|A|_s$; in particular, $|A|_s$ may be empty and need not be countable.

Notation. Let A be a Σ-algebra and let $f{:}\, s_1 \times \cdots \times s_n \to s$ be in Σ. We always write f_A in place of $(f{:}\, s_1 \times \cdots \times s_n \to s)_A$ when there is no danger of confusion. When $n = 0$ (i.e. $f{:}\, s$), we write $(f{:}\, s)_A$ or f_A in place of $(f{:}\, s)_A(\langle\rangle)$. □

Exercise 1.2.3. If $\Omega_{\varepsilon,s} \neq \varnothing$ for some $s \in S$, then there are no $\langle S, \Omega\rangle$-algebras having an empty carrier of sort s. Characterise signatures for which all algebras have non-empty carriers of all sorts. □

Example 1.2.4. Let $S1 = \{Shape, Suit\}$ and $\Omega 1_{\varepsilon,Shape} = \{box\}$, $\Omega 1_{\varepsilon,Suit} = \{hearts\}$, $\Omega 1_{Shape,Shape} = \{boxify\}$, $\Omega 1_{Shape\,Suit,Suit} = \{f\}$, and $\Omega 1_{w,s} = \varnothing$ for all other $w \in S1^*, s \in S1$. Then $\Sigma 1 = \langle S1, \Omega 1\rangle$ is a signature with sort names *Shape* and *Suit* and operation names $box{:}\, Shape$, $hearts{:}\, Suit$, $boxify{:}\, Shape \to Shape$ and $f{:}\, Shape \times Suit \to Suit$. We can present $\Sigma 1$ in tabular form as follows (this notation will be used later with the obvious meaning):

$$\begin{aligned}\Sigma 1 = \ &\textbf{sorts}\ Shape, Suit\\ &\textbf{ops}\ \ box{:}\, Shape\\ &\qquad\ \ hearts{:}\, Suit\\ &\qquad\ \ boxify{:}\, Shape \to Shape\\ &\qquad\ \ f{:}\, Shape \times Suit \to Suit.\end{aligned}$$

We define a $\Sigma 1$-algebra $A1$ as follows:

$$\begin{aligned}&|A1|_{Shape} = \{\Box, \triangle\},\\ &|A1|_{Suit} = \{\clubsuit, \heartsuit, \spadesuit\},\\ &box_{A1} = \Box \in |A1|_{Shape},\\ &hearts_{A1} = \heartsuit \in |A1|_{Suit},\\ &boxify_{A1}{:}\, |A1|_{Shape} \to |A1|_{Shape} = \{\Box \mapsto \Box, \triangle \mapsto \Box\},\end{aligned}$$

and $f_{A1}{:}\, |A1|_{Shape} \times |A1|_{Suit} \to |A1|_{Suit}$ is defined by the following table:

f_{A1}	♣	♡	♠
□	♣	♠	♡
△	♡	♠	♠

(NOTE: Reference will be made to $\Sigma 1$ and $A1$ in examples throughout the rest of this chapter.) □

Definition 1.2.5 (Subalgebra). Let A and B be Σ-algebras. B is a *subalgebra* of A if:

- $|B| \subseteq |A|$; and
- for $f\colon s_1 \times \cdots \times s_n \to s$ in Σ and $b_1 \in |B|_{s_1}, \ldots, b_n \in |B|_{s_n}$, $f_B(b_1, \ldots, b_n) = f_A(b_1, \ldots, b_n)$.

B is a *proper* subalgebra of A if it is a subalgebra of A and $|B| \neq |A|$. A subalgebra of A is determined by an S-sorted subset $|B|$ of $|A|$ which is closed under the operations of Σ, i.e. such that for each $f\colon s_1 \times \cdots \times s_n \to s$ in Σ and $b_1 \in |B|_{s_1}, \ldots, b_n \in |B|_{s_n}$, $f_A(b_1, \ldots, b_n) \in |B|_s$. □

If B is a (proper) subalgebra of A then B is "smaller" than A in the sense that it contains fewer *data values* than A. Both A and B are Σ-algebras though, so A and B contain interpretations for exactly the same sort and operation names.

Exercise 1.2.6. Let A be a Σ-algebra. Show that the intersection of any family of (carriers of) subalgebras of A is a (carrier of a) subalgebra of A. Use this to show that for any $X \subseteq |A|$, there is a least subalgebra of A that contains X. This is called the *subalgebra of A generated by X*. Give an explicit construction of this algebra. HINT: Consider the family of S-sorted sets $X_i \subseteq |A|$, $i \geq 0$, where $X_0 = X$ and X_{i+1} is obtained from X_i by adding the results of applying the operations of A to arguments in X_i. □

Definition 1.2.7 (Reachable algebra). Let A be a Σ-algebra. A is *reachable* if A has no proper subalgebra (equivalently, if A is generated by $\varnothing$). □

By Exercise 1.2.6, every algebra has a unique reachable subalgebra.

Example 1.2.8. Let $\Sigma 1 = \langle S1, \Omega 1\rangle$ and $A1$ be as defined in Example 1.2.4. Define a $\Sigma 1$-algebra $B1$ by

$$
\begin{aligned}
&|B1|_{Shape} = \{\Box\},\\
&|B1|_{Suit} = \{\heartsuit, \spadesuit\},\\
&box_{B1} = \Box \in |B1|_{Shape},\\
&hearts_{B1} = \heartsuit \in |B1|_{Suit},\\
&boxify_{B1}\colon |B1|_{Shape} \to |B1|_{Shape} = \{\Box \mapsto \Box\},\\
&f_{B1}\colon |B1|_{Shape} \times |B1|_{Suit} \to |B1|_{Suit} = \{\langle \Box, \heartsuit\rangle \mapsto \spadesuit, \langle \Box, \spadesuit\rangle \mapsto \heartsuit\}.
\end{aligned}
$$

$B1$ is the subalgebra of $A1$ generated by $\varnothing$. That is, $B1$ is the reachable subalgebra of $A1$. □

Definition 1.2.9 (Product algebra). Let A and B be Σ-algebras. The *product algebra* $A \times B$ is the Σ-algebra defined as follows:

- $|A \times B| = |A| \times |B|$; and
- for each $f{:}\, s_1 \times \cdots \times s_n \to s$ in Σ and $\langle a_1, b_1 \rangle \in |A \times B|_{s_1}, \ldots, \langle a_n, b_n \rangle \in |A \times B|_{s_n}$, $f_{A \times B}(\langle a_1, b_1 \rangle, \ldots, \langle a_n, b_n \rangle) = \langle f_A(a_1, \ldots, a_n), f_B(b_1, \ldots, b_n) \rangle \in |A \times B|_s$.

This generalises to the product $\prod \langle A_i \rangle_{i \in I}$ of a family of Σ-algebras, indexed by an arbitrary set I (possibly empty), as follows:

- $|\prod \langle A_i \rangle_{i \in I}| = \prod \langle |A_i| \rangle_{i \in I}$; and
- for each $f{:}\, s_1 \times \cdots \times s_n \to s$ in Σ and $f_1 \in |\prod \langle A_i \rangle_{i \in I}|_{s_1}, \ldots, f_n \in |\prod \langle A_i \rangle_{i \in I}|_{s_n}$, $f_{\prod \langle A_i \rangle_{i \in I}}(f_1, \ldots, f_n)(i) = f_{A_i}(f_1(i), \ldots, f_n(i))$ for all $i \in I$. □

Exercise 1.2.10. Definition 1.2.9 shows how two Σ-algebras can be combined to form a new Σ-algebra by taking the Cartesian product of their carriers. According to Exercise 1.2.6, the same can be done (with subalgebras of a fixed algebra) using intersection. Try to formulate definitions of *union* and *disjoint union* of algebras where $|A \cup B| = |A| \cup |B|$ and $|A \uplus B| = |A| \uplus |B|$ respectively. What happens? □

1.3 Homomorphisms and congruences

A homomorphism between algebras is the analogue of a function between sets, and a congruence relation on an algebra is the analogue of an equivalence relation on a set. An algebra has more structure than a set, so homomorphisms and congruences are required to respect the additional structure (i.e. the behaviour of the operations). Homomorphisms and congruences are important basic tools for relating algebras and constructing new algebras from old ones.

Throughout this section, let $\Sigma = \langle S, \Omega \rangle$ be a signature.

Definition 1.3.1 (Homomorphism). Let A and B be Σ-algebras. A *Σ-homomorphism $h{:}\, A \to B$* is an S-sorted function $h{:}\, |A| \to |B|$ which respects the operations of Σ, i.e. such that for all $f{:}\, s_1 \times \cdots \times s_n \to s$ in Σ and $a_1 \in |A|_{s_1}, \ldots, a_n \in |A|_{s_n}$, $h_s(f_A(a_1, \ldots, a_n)) = f_B(h_{s_1}(a_1), \ldots, h_{s_n}(a_n))$. A Σ-homomorphism $h{:}\, A \to B$ is an *identity* (an *inclusion*, *surjective*, etc.) if it is an identity (an inclusion, surjective, etc.) when viewed as an S-sorted function. □

Notation. If $h{:}\, A \to B$ is a Σ-homomorphism, then $|h|{:}\, |A| \to |B|$ denotes h viewed as an S-sorted function. The only difference between h and $|h|$ is that in the case of $|h|$ we have "forgotten" that the additional condition required of a homomorphism is satisfied; in particular, $h(a)$ and $|h|(a)$ are the same for any $a \in |A|$. □

Informally, the homomorphism condition says that the behaviour of the operations in A is reflected in that of the operations in B. This condition can be expressed in the form of a diagram as follows:

$$\begin{array}{ccc}
|A|_{s_1} \times \cdots \times |A|_{s_n} & \xrightarrow{h_{s_1} \times \cdots \times h_{s_n}} & |B|_{s_1} \times \cdots \times |B|_{s_n} \\
\Big\downarrow f_A & & \Big\downarrow f_B \\
|A|_s & \xrightarrow[h_s]{} & |B|_s
\end{array}$$

where we define $(h_{s_1} \times \cdots \times h_{s_n})(a_{s_1}, \ldots, a_{s_n}) = \langle h_{s_1}(a_{s_1}), \ldots, h_{s_n}(a_{s_n}) \rangle$ for all $a_1 \in |A|_{s_1}, \ldots, a_n \in |A|_{s_n}$. The homomorphism condition amounts to the requirement that this diagram *commute*, i.e. that composing the functions on the top and right-hand side arrows gives the same result as composing the functions on the left-hand side and bottom arrows. Such commutative diagrams will be used heavily in later chapters, particularly in Chapter 3.

Example 1.3.2. Let $\Sigma 1 = \langle S1, \Omega 1 \rangle$ and $A1$ be as defined in Example 1.2.4. Define a $\Sigma 1$-algebra $C1$ by

$|C1|_{Shape} = |C1|_{Suit} = \{1,2,3\}$,
$box_{C1} = 1 \in |C1|_{Shape}$,
$hearts_{C1} = 2 \in |C1|_{Suit}$,
$boxify_{C1} \colon |C1|_{Shape} \to |C1|_{Shape} = \{1 \mapsto 1, 2 \mapsto 3, 3 \mapsto 1\}$,

and $f_{C1} \colon |C1|_{Shape} \times |C1|_{Suit} \to |C1|_{Suit}$ is defined by the following table:

f_{C1}	1	2	3
1	1	2	3
2	2	1	2
3	2	2	1

Let $h1 \colon |A1| \to |C1|$ be the $S1$-sorted function such that

$h1_{Shape} = \{\Box \mapsto 1, \triangle \mapsto 3\}$,
$h1_{Suit} = \{\clubsuit \mapsto 1, \heartsuit \mapsto 2, \spadesuit \mapsto 2\}$.

It is easy to verify that $h1 \colon A1 \to C1$ is a $\Sigma 1$-homomorphism by checking the following:

$$\begin{aligned}
h1_{Shape}(box_{A1}) &= box_{C1} \\
h1_{Suit}(hearts_{A1}) &= hearts_{C1} \\
h1_{Shape}(boxify_{A1}(\Box)) &= boxify_{C1}(h1_{Shape}(\Box)) \\
h1_{Shape}(boxify_{A1}(\triangle)) &= boxify_{C1}(h1_{Shape}(\triangle)) \\
h1_{Suit}(f_{A1}(\Box, \clubsuit)) &= f_{C1}(h1_{Shape}(\Box), h1_{Suit}(\clubsuit)) \\
h1_{Suit}(f_{A1}(\Box, \heartsuit)) &= f_{C1}(h1_{Shape}(\Box), h1_{Suit}(\heartsuit)) \\
h1_{Suit}(f_{A1}(\Box, \spadesuit)) &= f_{C1}(h1_{Shape}(\Box), h1_{Suit}(\spadesuit))
\end{aligned}$$

$$h1_{Suit}(f_{A1}(\triangle,\clubsuit)) = f_{C1}(h1_{Shape}(\triangle),h1_{Suit}(\clubsuit))$$
$$h1_{Suit}(f_{A1}(\triangle,\heartsuit)) = f_{C1}(h1_{Shape}(\triangle),h1_{Suit}(\heartsuit))$$
$$h1_{Suit}(f_{A1}(\triangle,\spadesuit)) = f_{C1}(h1_{Shape}(\triangle),h1_{Suit}(\spadesuit)).$$ □

Exercise 1.3.3. Let A be a Σ-algebra. Show that $id_{|A|}{:}A \to A$ (the identity S-sorted function) is a Σ-homomorphism. Let $h{:}A \to B$ and $h'{:}B \to C$ be Σ-homomorphisms. Show that $|h|;|h'|{:}|A| \to |C|$ is a Σ-homomorphism $h;h'{:}A \to C$. □

Exercise 1.3.4. Let $h{:}A \to B$ be a Σ-homomorphism, and let A' be a subalgebra of A. Let the *image of A' under h* be the Σ-algebra $h(A')$ defined as follows:

- $|h(A')| = |h|(|A'|)$; and
- for each $f{:}s_1 \times \cdots \times s_n \to s$ in Σ, $f_{h(A')}(h_{s_1}(a_1),\ldots,h_{s_n}(a_n)) = h_s(f_{A'}(a_1,\ldots,a_n))$ for all $a_1 \in |A'|_{s_1},\ldots,a_n \in |A'|_{s_n}$.

Show that $h(A')$ is a well-defined Σ-algebra (in particular, that for each $f{:}s_1 \times \cdots \times s_n \to s$ in Σ, the function $f_{h(A')}{:}|h(A')|_{s_1} \times \cdots \times |h(A')|_{s_n} \to |h(A')|_s$ is well defined) and that it is a subalgebra of B. Formulate a definition of the *coimage* of a subalgebra B' of B under h, and show that it is a subalgebra of A. □

Exercise 1.3.5. Let $h{:}A \to B$ be a Σ-homomorphism, and suppose $X \subseteq |A|$. Show that the subalgebra of B generated by $|h|(X) \subseteq |B|$ is the image of the subalgebra of A generated by X. Show that it follows that if $h{:}A \to B$ is surjective and A is reachable then B is reachable. □

Exercise 1.3.6. Let B be a reachable Σ-algebra. Show that for any Σ-algebra A, there is at most one Σ-homomorphism $h{:}B \to A$, and that any Σ-homomorphism $h{:}A \to B$ is surjective. □

Definition 1.3.7 (Isomorphism). Let A and B be Σ-algebras. A Σ-homomorphism $h{:}A \to B$ is a *Σ-isomorphism* if it has an inverse, i.e. there is a Σ-homomorphism $h^{-1}{:}B \to A$ such that $h;h^{-1} = id_{|A|}$ and $h^{-1};h = id_{|B|}$. (**Exercise:** Show that if h^{-1} exists then it is unique.) Then A and B are called *isomorphic* and we write $h{:}A \cong B$ or just $A \cong B$. □

Exercise 1.3.8. Let $h{:}A \cong B$ and $h'{:}B \cong C$ be Σ-isomorphisms. Show that their composition is a Σ-isomorphism $h;h'{:}A \cong C$. Show that $\cong$ (as a binary relation on Σ-algebras) is reflexive and symmetric, and is therefore an equivalence relation. □

Two isomorphic algebras are typically regarded as indistinguishable for all practical purposes. It is easy to see why: the only way they can differ is in the particular choice of data values in the carriers. The size of the carriers and the way the operations behave on the values in the carriers is exactly the same. For this reason we are often satisfied with a definition of an algebra "up to isomorphism", i.e. a description of an isomorphism class of algebras in a context where one would expect a definition of a single algebra. An example of this is in Fact 1.4.10 below. The notion of isomorphism can be generalised to other kinds of structures, where it embodies exactly the same concept of indistinguishability. See Chapter 3 for this generalisation and for many more examples of definitions of objects "up to isomorphism".

Example 1.3.9. Let $\Sigma 1 = \langle S1, \Omega 1\rangle$ and $A1$ be as defined in Example 1.2.4. Define a $\Sigma 1$-algebra $D1$ by

$$\begin{aligned}
&|D1|_{Shape} = \{\Box, \triangle\},\\
&|D1|_{Suit} = \{1,2,3\},\\
&box_{D1} = \triangle \in |D1|_{Shape},\\
&hearts_{D1} = 2 \in |D1|_{Suit},\\
&boxify_{D1}: |D1|_{Shape} \to |D1|_{Shape} = \{\Box \mapsto \triangle, \triangle \mapsto \triangle\},
\end{aligned}$$

and $f_{D1}: |D1|_{Shape} \times |D1|_{Suit} \to |D1|_{Suit}$ is defined by the following table:

f_{D1}	1	2	3
$\Box$	2	3	3
$\triangle$	1	3	2

Let $i1: |A1| \to |D1|$ be the $S1$-sorted function such that

$$\begin{aligned}
&i1_{Shape} = \{\Box \mapsto \triangle, \triangle \mapsto \Box\},\\
&i1_{Suit} = \{\clubsuit \mapsto 1, \heartsuit \mapsto 2, \spadesuit \mapsto 3\}.
\end{aligned}$$

This defines a $\Sigma 1$-homomorphism $i1: A1 \to D1$, which is a $\Sigma 1$-isomorphism; so $A1 \cong D1$. □

Exercise 1.3.10. Show that a homomorphism is an isomorphism iff it is bijective. □

Exercise 1.3.11. Show that there is an injective homomorphism $h: A \to B$ iff A is isomorphic to a subalgebra of B. □

Example 1.3.12. Let $\Sigma = \langle S, \Omega\rangle$ be the signature

sorts s
ops $a: s$
$f: s \to s$

and define Σ-algebras A and B by

$$\begin{aligned}
&|A|_s = \mathbb{N} \text{ (the natural numbers)},\\
&a_A = 0 \in |A|_s,\\
&f_A: |A|_s \to |A|_s = \{n \mapsto n+1 \mid n \in \mathbb{N}\},
\end{aligned}$$

$$\begin{aligned}
&|B|_s = \{n \in \mathbb{N} \mid \text{the Turing machine with Gödel number } n \text{ halts on all inputs}\},\\
&a_B = \text{the smallest } n \in |B|_s,\\
&f_B: |B|_s \to |B|_s = \{n \in |B|_s \mapsto \text{the smallest } m \in |B|_s \text{ such that } m > n\}.
\end{aligned}$$

Let $i: |A| \to |B|$ be the S-sorted function such that

$$i_s(n) = \text{the } (n+1)\text{th smallest element of } |B|_s$$

for all $n \in |A|_s$. The function i_s is well defined since $|B|_s$ is infinite. This defines a Σ-homomorphism $i{:}A \to B$ which is an isomorphism.

Although $A \cong B$, the Σ-algebras A and B are not "the same" from the point of view of computability: everything in A is computable, in contrast to B ($|B|_s$ is not recursively enumerable and f_B is not computable). Isomorphisms capture *structural* similarity, ignoring what the values in the carriers are and what the operations actually compute. This example shows that, for some purposes, properties stronger than structural similarity are important. □

Definition 1.3.13 (Congruence). Let A be a Σ-algebra. A *Σ-congruence on A* is an (S-sorted) equivalence $\equiv$ on $|A|$ which respects the operations of Σ: for all $f{:}s_1 \times \cdots \times s_n \to s$ in Σ and $a_1, a_1' \in |A|_{s_1}, \ldots, a_n, a_n' \in |A|_{s_n}$, if $a_1 \equiv_{s_1} a_1'$ and ... and $a_n \equiv_{s_n} a_n'$ then $f_A(a_1,\ldots,a_n) \equiv_s f_A(a_1',\ldots,a_n')$. □

Exercise 1.3.14. Show that the intersection of any family of Σ-congruences on A is a Σ-congruence on A. Use this to show that for any S-sorted binary relation R on $|A|$ there is a least (with respect to $\subseteq$) Σ-congruence on A that includes R.

Show that the kernel of any Σ-homomorphism $h{:}A \to B$ is a Σ-congruence on A.

Show that a surjective Σ-homomorphism is an isomorphism iff its kernel is the identity. □

Definition 1.3.15 (Quotient algebra). Let A be a Σ-algebra, and let $\equiv$ be a Σ-congruence on A. The *quotient algebra of A modulo $\equiv$* is the Σ-algebra $A/{\equiv}$ defined by:

- $|A/{\equiv}| = |A|/{\equiv}$; and
- for each $f{:}s_1 \times \cdots \times s_n \to s$, $f_{A/\equiv}([a_1]_{\equiv_{s_1}},\ldots,[a_n]_{\equiv_{s_n}}) = [f_A(a_1,\ldots,a_n)]_{\equiv_s}$ for all $a_1 \in |A|_{s_1},\ldots,a_n \in |A|_{s_n}$. □

Exercise 1.3.16. Show that $A/{\equiv}$ in Definition 1.3.15 is a well-defined Σ-algebra. □

Example 1.3.17. Let $\Sigma 1 = \langle S1, \Omega 1\rangle$ and $A1$ be as defined in Example 1.2.4, and let $\equiv = \langle \equiv_s \rangle_{s \in S1}$ be the $S1$-sorted binary relation on $|A1|$ defined by

$$\begin{aligned}
&\equiv_{Shape} = \{\langle \square,\square\rangle, \langle \triangle,\triangle\rangle\},\\
&\equiv_{Suit} = \{\langle \clubsuit,\clubsuit\rangle, \langle \heartsuit,\heartsuit\rangle, \langle \heartsuit,\spadesuit\rangle, \langle \spadesuit,\heartsuit\rangle, \langle \spadesuit,\spadesuit\rangle\}.
\end{aligned}$$

This defines a congruence on $A1$. $A1/{\equiv}$ is the $\Sigma 1$-algebra defined by

$$\begin{aligned}
&|A1/{\equiv}|_{Shape} = \{\{\square\},\{\triangle\}\},\\
&|A1/{\equiv}|_{Suit} = \{\{\clubsuit\},\{\heartsuit,\spadesuit\}\},\\
&box_{A1/\equiv} = \{\square\} \in |A1/{\equiv}|_{Shape},\\
&hearts_{A1/\equiv} = \{\heartsuit,\spadesuit\} \in |A1/{\equiv}|_{Suit},\\
&boxify_{A1/\equiv}{:}\, |A1/{\equiv}|_{Shape} \to |A1/{\equiv}|_{Shape} = \{\{\square\} \mapsto \{\square\}, \{\triangle\} \mapsto \{\square\}\},
\end{aligned}$$

and $f_{A1/\equiv}{:}\, |A1/{\equiv}|_{Shape} \times |A1/{\equiv}|_{Suit} \to |A1/{\equiv}|_{Suit}$ is defined by the following table:

$f_{A1/\equiv}$	$\{\clubsuit\}$	$\{\heartsuit,\spadesuit\}$
$\{\Box\}$	$\{\clubsuit\}$	$\{\heartsuit,\spadesuit\}$
$\{\triangle\}$	$\{\heartsuit,\spadesuit\}$	$\{\heartsuit,\spadesuit\}$

□

Exercise 1.3.18. Let $\equiv$ be a Σ-congruence on A, and let $h_s(a) = [a]_{\equiv_s}$ for $s \in S$, $a \in |A|_s$. Show that $\langle h_s : |A|_s \to (|A|/\equiv)_s \rangle_{s \in S}$ is a Σ-homomorphism $h: A \to A/\equiv$ with $ker(h) = \equiv$. □

Exercise 1.3.19. Let $h: A \to B$ be a Σ-homomorphism. Show that $A/ker(h)$ is isomorphic to $h(A)$. HINT: The isomorphism is given by $[a]_{ker(h_s)} \mapsto h_s(a)$ for $s \in S$, $a \in |A|_s$. □

Exercise 1.3.20. Let $\equiv$ be a Σ-congruence on A. Show that for any Σ-homomorphism $h: A \to B$ such that $\equiv \subseteq ker(h)$, there exists a unique Σ-homomorphism $g: A/\equiv \to B$ such that $h_s(a) = g_s([a]_{\equiv_s})$ for all $s \in S$, $a \in |A|_s$. □

Exercise 1.3.21. Show that there is a surjective homomorphism $h: A \to B$ iff there is a congruence $\equiv$ on A such that B is isomorphic to $A/\equiv$. □

Exercise 1.3.22. Let A be a Σ-algebra, let $\equiv$ be a congruence on A and let B be a subalgebra of $A/\equiv$. Show that there is a subalgebra C of A and congruence $\equiv'$ on C such that $B = C/\equiv'$. □

Exercise 1.3.23. Let $h: A \to B$ be a Σ-homomorphism. Show that there is a unique Σ-congruence $\equiv$ on A and a unique injective Σ-homomorphism $g: A/\equiv \to B$ such that $h_s(a) = g_s([a]_{\equiv_s})$ for all $s \in S$, $a \in |A|_s$. □

1.4 Term algebras

For any signature Σ there is a special Σ-algebra whose values are just well-formed terms (i.e. expressions) built from the operation names in Σ. A Σ-algebra of terms with variables is similarly determined by a signature $\Sigma = \langle S, \Omega \rangle$ and an S-sorted set of variables. These algebras are rather boring insofar as modelling programs is concerned — the term algebra models a program which does no real computation. But the homomorphisms from these algebras to *other* algebras turn out to be very useful technical tools, as shown by the definitions below.

Throughout this section, let $\Sigma = \langle S, \Omega \rangle$ be a signature and let X be an S-sorted set (of variables), where $x \in X_s$ for $s \in S$ means that the variable x is of sort s (written $x{:}s$). Note that "overloading" of variable names is permitted here, since there is no requirement that X_s and $X_{s'}$ be disjoint for $s \neq s' \in S$.

Definition 1.4.1 (Term algebra). The *Σ-algebra $T_\Sigma(X)$ of terms with variables X* is the Σ-algebra defined as follows:

- $|T_\Sigma(X)|$ is the least (with respect to $\subseteq$) S-sorted set of words (sequences) over the alphabet

$$S \cup \bigcup_{\substack{w \in S^* \\ s \in S}} \Omega_{w,s} \cup \bigcup_{s \in S} X_s \cup \{:, (, ,,)\}$$

such that:

- the word "$x{:}s$" $\in |T_\Sigma(X)|_s$ for all $s \in S$ and $x \in X_s$; and
- for all $f{:}s_1 \times \cdots \times s_n \to s$ in Σ and all words $t_1 \in |T_\Sigma(X)|_{s_1}, \ldots, t_n \in |T_\Sigma(X)|_{s_n}$, the word "$f(t_1,\ldots,t_n){:}s$" $\in |T_\Sigma(X)|_s$.

- for all $f{:}s_1 \times \cdots \times s_n \to s$ in Σ and all words $t_1 \in |T_\Sigma(X)|_{s_1}, \ldots, t_n \in |T_\Sigma(X)|_{s_n}$, $f_{T_\Sigma(X)}(t_1,\ldots,t_n) =$ (the word) "$f(t_1,\ldots,t_n){:}s$" $\in |T_\Sigma(X)|_s$.

(Quotation marks are used here solely to emphasize that terms are words, and are not part of the words they delimit.) If $s \in S$ and $t \in |T_\Sigma(X)|_s$ then t is a *Σ-term of sort s with variables X*; the *free variables of t*, $FV(t) \subseteq X$, is the S-sorted set of variables that actually occur in t: for $s \in S$ and $x \in X_s$, $x \in FV(t)_s$ if t contains the subword "$x{:}s$".

The *Σ-algebra of ground terms* is the Σ-algebra $T_\Sigma = T_\Sigma(\varnothing)$ of terms without variables. If $s \in S$ and $t \in |T_\Sigma|_s$ then t is a *ground Σ-term.* □

The values of $T_\Sigma(X)$ are "fully typed" terms formed using the variables in X and the operation names in Σ, and the operations of $T_\Sigma(X)$ just build complicated terms from simpler terms. Note that a term $t \in |T_\Sigma(X)|$ need not contain all the variables in X, and that some variables may occur more than once in t. T_Σ is also called the *Σ-word algebra*, and its carriers $|T_\Sigma|$ are sometimes called the *Herbrand universe for Σ*.

Example 1.4.2. Let $\Sigma 1 = \langle S1, \Omega 1 \rangle$ be as defined in Example 1.2.4. Then $T_{\Sigma 1}$ is the $\Sigma 1$-algebra defined by

$$\begin{aligned}
|T_{\Sigma 1}|_{Shape} = \{ & \text{“}box(){:}Shape\text{”}, \\
& \text{“}boxify(box(){:}Shape){:}Shape\text{”}, \\
& \text{“}boxify(boxify(box(){:}Shape){:}Shape){:}Shape\text{”}, \\
& \ldots \}, \\
|T_{\Sigma 1}|_{Suit} = \{ & \text{“}hearts(){:}Suit\text{”}, \\
& \text{“}f(box(){:}Shape, hearts(){:}Suit){:}Suit\text{”}, \\
& \text{“}f(boxify(box(){:}Shape){:}Shape, hearts(){:}Suit){:}Suit\text{”}, \\
& \text{“}f(box(){:}Shape, f(box(){:}Shape, hearts(){:}Suit){:}suit){:}suit\text{”}, \\
& \ldots \}
\end{aligned}$$

where the operations of $T_{\Sigma 1}$ are the term formation operations

$$\begin{aligned}
& box_{T_{\Sigma 1}} = \text{“}box(){:}Shape\text{”} \in |T_{\Sigma 1}|_{Shape}, \\
& hearts_{T_{\Sigma 1}} = \text{“}hearts(){:}Suit\text{”} \in |T_{\Sigma 1}|_{Suit}, \\
& boxify_{T_{\Sigma 1}}{:}|T_{\Sigma 1}|_{Shape} \to |T_{\Sigma 1}|_{Shape} \\
& \quad = \{ \text{“}box(){:}Shape\text{”} \mapsto \text{“}boxify(box(){:}Shape){:}Shape\text{”}, \\
& \qquad \text{“}boxify(box(){:}Shape){:}Shape\text{”} \mapsto \\
& \qquad\qquad \text{“}boxify(boxify(box(){:}Shape){:}Shape){:}Shape\text{”}, \\
& \qquad \ldots \},
\end{aligned}$$

and similarly for $f{:}\, Shape \times Suit \to Suit$. □

Notation. Sort decorations (e.g. "$:Shape$" in "$box():Shape$") are often unambiguously determined, and they will usually be omitted when this is the case. When $\Omega_{\varepsilon,s} \cap X_s = \varnothing$ for some $s \in S$, then variables of sort s cannot be confused with constants (nullary operations) of sort s and so we will usually drop the parentheses "()" in the latter. We will omit quotation marks whenever it is clear from the context that we are dealing with terms. Finally, in examples we will use infix notation for binary operations when convenient. □

Example 1.4.2 (revisited). We repeat Example 1.4.2, making use of these notational conventions.

Let $\Sigma 1 = \langle S1, \Omega 1 \rangle$ be as defined in Example 1.2.4. Then $T_{\Sigma 1}$ is the $\Sigma 1$-algebra defined by

$$
\begin{aligned}
|T_{\Sigma 1}|_{Shape} &= \{box, boxify(box), boxify(boxify(box)), \ldots\},\\
|T_{\Sigma 1}|_{Suit} &= \{hearts, f(box, hearts), f(boxify(box), hearts),\\
&\qquad f(box, f(box, hearts)), \ldots\}
\end{aligned}
$$

where the operations of $T_{\Sigma 1}$ are the term formation operations

$$
\begin{aligned}
&box_{T_{\Sigma 1}} = box \in |T_{\Sigma 1}|_{Shape},\\
&hearts_{T_{\Sigma 1}} = hearts \in |T_{\Sigma 1}|_{Suit},\\
&boxify_{T_{\Sigma 1}} : |T_{\Sigma 1}|_{Shape} \to |T_{\Sigma 1}|_{Shape}\\
&\qquad = \{box \mapsto boxify(box), boxify(box) \mapsto boxify(boxify(box)), \ldots\},
\end{aligned}
$$

and similarly for $f{:}\, Shape \times Suit \to Suit$. □

Example 1.4.3. The notational conventions above will almost always be applicable. They cannot be adopted from the outset (i.e. in Definition 1.4.1) because of the relatively rare examples where confusion can arise. For example, let $\Sigma 2 = \langle S2, \Omega 2 \rangle$ be the signature with sorts s, s_1, s_2 and operations $a{:}\, s_1$, $a{:}\, s_2$, $f{:}\, s_1 \to s$ and $f{:}\, s_2 \to s$ (no mistake here, repetition of names is deliberate).

According to the definition, $|T_{\Sigma 2}|_s = \{$"$f(a():s_1):s$", "$f(a():s_2):s$"$\}$. If all sort decorations were omitted then both of the terms in this set would become "$f(a())$" and so $|T_{\Sigma 2}|_s$ would have just this single element. The "outer" decoration can be omitted but the "inner" decoration is required; thus, e.g. "$f(a():s_1)$".

Similarly, if X is an $S2$-sorted set of variables such that $a \in X_{s_1}$, then "$f(a():s_1)$" and "$f(a{:}\, s_1)$" are different terms in $|T_{\Sigma 2}(X)|_s$, so the convention of writing "$a():s_1$" as "$a{:}\, s_1$" cannot be used.

Since the definitions permit variables and operation names like $f(a():s_1)$ and even " or , or (), the custom of writing terms as sequences of symbols without explicit separators can cause confusion. Luckily, such names never arise in practice and so for the purposes of this book this problem can safely be forgotten. □

Fact 1.4.4. *For any Σ-algebra A and S-sorted function $v{:}\, X \to |A|$ there is exactly one Σ-homomorphism $v^{\#}{:}\, T_\Sigma(X) \to A$ that extends v, i.e. such that $v^{\#}_s(\iota_X(x)) = v_s(x)$ for all $s \in S$, $x \in X_s$, where $\iota_X{:}\, X \to |T_\Sigma(X)|$ is the embedding that maps each variable in X to its corresponding term.*

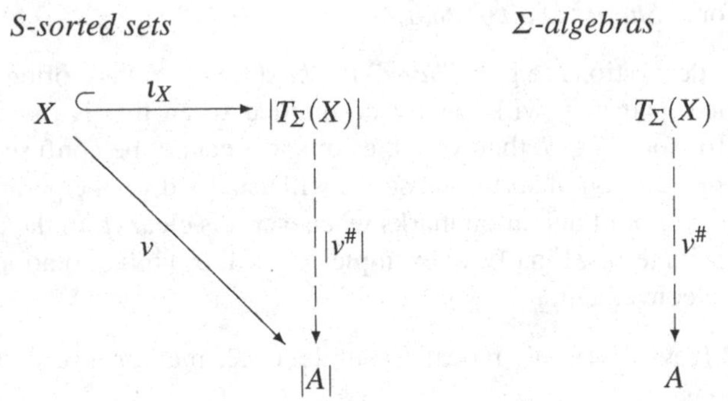

□

The existence and uniqueness of $v^\#$ follow easily from the requirements that $v^\#$ extends v (this fixes the value of $v^\#$ for any variable as a term in $|T_\Sigma(X)|$) and that $v^\#$ is a Σ-homomorphism (this determines the value of $v^\#$ for any term $f(t_1,\ldots,t_n) \in |T_\Sigma(X)|$ as a function of the values of $v^\#$ for its immediate subterms $t_1,\ldots,t_n \in |T_\Sigma(X)|$). The homomorphism which results is the function which evaluates Σ-terms based on the assignment of values in A to variables in X given by v.

Definition 1.4.5 (Term evaluation). Let A be a Σ-algebra A and let $v{:}X \to |A|$ be an S-sorted function. By Fact 1.4.4 there is a unique Σ-homomorphism $v^\#{:}\,T_\Sigma(X) \to A$ that extends v. Let $s \in S$ and let $t \in |T_\Sigma(X)|_s$ be a Σ-term of sort s; the *value of t in A under the valuation* v is $v^\#(t) \in |A|_s$. When $t \in |T_\Sigma|_s$ the value of t does not depend on v; then the *value of t in A* is $\varnothing^\#(t)$ where $\varnothing{:}\varnothing \to |A|$ is the empty function. To make the algebra explicit, we write $t_A(v)$ for $v^\#(t)$, and t_A for $t_A(\varnothing)$ when t is ground. □

Exercise 1.4.6. Let $t \in |T_\Sigma(X)|$ be a Σ-term and let A be a Σ-algebra. Show that if $v{:}X \to |A|$ and $v'{:}X \to |A|$ coincide on $FV(t)$, then $t_A(v) = t_A(v')$. This follows from another fact: for any $t \in |T_\Sigma(X)|$, $X \subseteq Y$ (so that $t \in |T_\Sigma(Y)|$) and $v{:}Y \to |A|$, we have $t_A(v) = t_A(\iota;v)$, where $\iota{:}X \hookrightarrow Y$ is the inclusion (and so $\iota;v{:}X \to |A|$). □

Exercise 1.4.7. Define evaluation of terms in an inductive fashion. Convince yourself that the result is the same as that given by Definition 1.4.5. □

Exercise 1.4.8. Let $h{:}A \to B$ be a Σ-homomorphism, let $v{:}X \to |A|$ be an S-sorted function, and let $t \in |T_\Sigma(X)|$ be a Σ-term. Using Fact 1.4.4, prove that $h(v^\#(t)) = (v;h)^\#(t)$. Compare this with a proof of the same property based on your inductive definition of term evaluation from Exercise 1.4.7. □

Exercise 1.4.9. Functions $\theta{:}X \to |T_\Sigma(Y)|$ are sometimes called *substitutions* (of terms in $T_\Sigma(Y)$ for variables in X). Using Fact 1.4.4, define the Σ-term $t[\theta] \in |T_\Sigma(Y)|$ resulting from applying the substitution θ to a Σ-term $t \in |T_\Sigma(X)|$. Show that $t[\iota_X] = t$ for any $t \in |T_\Sigma(X)|$, where ι_X maps each variable in X to its corresponding term in $|T_\Sigma(X)|$. Define the composition $\theta;\theta'$ of substitutions $\theta{:}X \to |T_\Sigma(Y)|$ and $\theta'{:}Y \to |T_\Sigma(Z)|$, and show that $(t[\theta])[\theta'] = t[\theta;\theta']$ for any Σ-term t and substitutions θ and θ'. □

Notation. Suppose $u \in |T_\Sigma(Y)|_s$ for some sort $s \in S$. Then $[x \mapsto u]$ (or just $x \mapsto u$), when used as a substitution $\{x{:}s\} \cup X \to |T_\Sigma(X \cup Y)|$, is shorthand for the function $\{x{:}s \mapsto u\} \cup \{z \mapsto z \mid z \in X, z \neq x{:}s\}$. For $t \in |T_\Sigma(\{x{:}s\} \cup X)|$, $t[x \mapsto u] \in |T_\Sigma(X \cup Y)|$ thus stands for the term obtained by substituting u for x in t. This notation generalises straightforwardly to $[x_1 \mapsto u_1, \ldots, x_n \mapsto u_n]$ and $t[x_1 \mapsto u_1, \ldots, x_n \mapsto u_n]$ provided $x_1, \ldots, x_n$ are distinct variables. □

Fact 1.4.10. *The property of $T_\Sigma(X)$ in Fact 1.4.4 defines $T_\Sigma(X)$ up to isomorphism: if B is a Σ-algebra and $\eta{:}X \to |B|$ is an S-sorted function such that for any Σ-algebra A and S-sorted function $v{:}X \to |A|$ there is a unique Σ-homomorphism $v^\$ {:} B \to A$ such that $\eta;|v^\$| = v$, then B is isomorphic to $T_\Sigma(X)$, where $\eta^\#{:}T_\Sigma(X) \to B$ is an isomorphism with inverse $\iota_X^\$ {:} B \to T_\Sigma(X)$.*

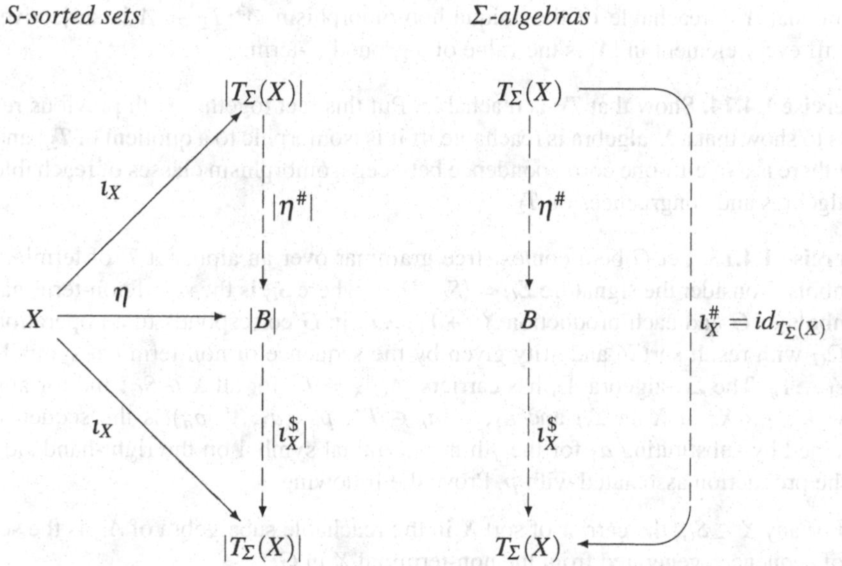

□

Fact 1.4.4 says that the definition of $T_\Sigma(X)$ fixes the definition of the term evaluation function "for free" (see Definition 1.4.5). Fact 1.4.10 says that this property is unique (up to isomorphism) to $T_\Sigma(X)$, so in fact the explicit definition of $T_\Sigma(X)$ is superfluous — it would be enough to define $T_\Sigma(X)$ as "the" (unique up to isomorphism) Σ-algebra for which Definition 1.4.5 makes sense. $T_\Sigma(X)$ is a particular example of a *free object* — see Section 3.5 for more on this topic.

Example 1.4.11. Let $\Sigma 1 = \langle S1, \Omega 1 \rangle$ be as defined in Example 1.2.4. Then $T_{\Sigma 1}$ is the $\Sigma 1$-algebra described in Example 1.4.2. Let $T1$ be the $\Sigma 1$-algebra defined by

$$|T1|_{Shape} = \{box, box\,boxify, box\,boxify\,boxify, \ldots\},$$
$$|T1|_{Suit} = \{hearts, box\,hearts\,f, box\,boxify\,hearts\,f, box\,box\,hearts\,f\,f, \ldots\}$$

where the operations of $T1$ are the postfix term formation operations

$$\begin{aligned}
box_{T1} &= box \in |T1|_{Shape},\\
hearts_{T1} &= hearts \in |T1|_{Suit},\\
boxify_{T1}&: |T1|_{Shape} \to |T1|_{Shape}\\
&= \{box \mapsto box\, boxify, box\, boxify \mapsto box\, boxify\, boxify, \ldots\},
\end{aligned}$$

and similarly for $f: Shape \times Suit \to Suit$. Then $T1$ satisfies the property of $T_{\Sigma 1}$ in Fact 1.4.4 (the fact that $X = \varnothing$ here makes this easy to check — there is only one function $v: \varnothing \to |A1|$ for any $\Sigma 1$-algebra $A1$), so by Fact 1.4.10 (where $\eta: \varnothing \to |T1|$ is the empty function) $T1$ is isomorphic to $T_{\Sigma 1}$. The isomorphism $\varnothing^{\#}: T_{\Sigma 1} \to T1$ converts a $\Sigma 1$-term to its postfix form. □

Exercise 1.4.12. Prove Facts 1.4.4 and 1.4.10. □

Exercise 1.4.13. Let A be a Σ-algebra and let $\varnothing: \varnothing \to |A|$ be the empty function. Show that A is reachable iff the unique homomorphism $\varnothing^{\#}: T_\Sigma \to A$ is surjective, i.e., iff every element in $|A|$ is the value of a ground Σ-term. □

Exercise 1.4.14. Show that T_Σ is reachable. Put this fact together with previous results to show that a Σ-algebra is reachable iff it is isomorphic to a quotient of T_Σ, and that there is a one-to-one correspondence between isomorphism classes of reachable Σ-algebras and congruences on T_Σ. □

Exercise 1.4.15. Let G be a context-free grammar over an alphabet T of terminal symbols. Consider the signature $\Sigma_G = \langle S_G, \Omega_G \rangle$, where S_G is the set of non-terminal symbols of G and each production $X \to Y_1 \ldots Y_n$ in G corresponds to an operation in Ω_G with result sort X and arity given by the sequence of non-terminal symbols in $Y_1 \ldots Y_n$. The Σ_G-algebra A_G has carriers $|A_G|_X = T^*$ for all $X \in S_G$, and for any $p: X_1 \times \cdots \times X_n \to X$ in Σ_G and $a_1, \ldots, a_n \in T^*$, $p_{A_G}(a_1, \ldots, a_n)$ is the sequence obtained by substituting a_j for the jth non-terminal symbol on the right-hand side of the production associated with p. Prove the following:

1. For any $X \in S_G$, the carrier of sort X in the reachable subalgebra of A_G is the set of sequences generated from the non-terminal X in G.
2. The algebra T_{Σ_G} is isomorphic to the algebra of parse trees of G.
3. The grammar G is unambiguous iff the reachable subalgebra of A_G is isomorphic to T_{Σ_G}. □

1.5 Changing signatures

A signature morphism defines a mapping from the sort and operation names in one signature to those in another signature in such a way that the arity and result sort of operations are respected. (This requirement is analogous to the requirement that homomorphisms respect the behaviour of the operations.) Signature morphisms will be used extensively in later chapters to mediate constructions involving multiple signatures. The crucial point that makes these constructions work is that a signature morphism from Σ to Σ' induces translations of syntax (terms — later, also logical formulae) and semantics (algebras and homomorphisms) between Σ and Σ'.

Two kinds of signature morphisms are introduced in this section. The first kind will be used throughout the rest of the book. The second kind, *derived signature morphisms*, provide one example of a way in which a basic definition could be modified. Such a modification would not affect later definitions and results, since these depend only on the induced translations of terms, algebras and homomorphisms.

1.5.1 Signature morphisms

Definition 1.5.1 (Signature morphism). Let $\Sigma = \langle S, \Omega \rangle$ and $\Sigma' = \langle S', \Omega' \rangle$ be signatures. A *signature morphism* $\sigma: \Sigma \to \Sigma'$ is a pair $\sigma = \langle \sigma_{sorts}, \sigma_{ops} \rangle$ where $\sigma_{sorts}: S \to S'$ and σ_{ops} is a family of functions respecting the arities and result sorts of operation names in Σ, that is $\sigma_{ops} = \langle \sigma_{w,s}: \Omega_{w,s} \to \Omega'_{\sigma^*_{sorts}(w), \sigma_{sorts}(s)} \rangle_{w \in S^*, s \in S}$ (where for $w = s_1 \dots s_n \in S^*$, $\sigma^*_{sorts}(w) = \sigma_{sorts}(s_1) \dots \sigma_{sorts}(s_n)$). A signature morphism $\sigma: \Sigma \to \Sigma'$ is a *signature inclusion* $\sigma: \Sigma \hookrightarrow \Sigma'$ if σ_{sorts} is an inclusion and $\sigma_{w,s}$ is an inclusion for all $w \in S^*, s \in S$. □

Signature morphisms as defined above will be referred to as *algebraic* signature morphisms when it is necessary to distinguish them from other kinds of signature morphisms to be introduced later. Note that σ_{sorts} and (the functions constituting) σ_{ops} are not required to be either surjective or injective.

Notation. When $\sigma: \Sigma \to \Sigma'$, both σ_{sorts} and σ_{ops} (and its components $\sigma_{w,s}$ for all $w \in S^*, s \in S$) will be denoted by σ. □

Example 1.5.2. Let $\Sigma = \langle S, \Omega \rangle$ be the signature

sorts *Polygon*, *Figure*, *Trump*
ops *square*: *Polygon*
boxify: *Polygon* $\to$ *Polygon*
boxify: *Polygon* $\to$ *Figure*
g: *Figure* $\times$ *Trump* $\to$ *Trump*

Let $\Sigma 1 = \langle S1, \Omega 1 \rangle$ be the signature defined in Example 1.2.4.

Define $\sigma_{sorts}: S \to S1$ and $\sigma_{ops} = \langle \sigma_{w,s}: \Omega_{w,s} \to \Omega 1_{\sigma^*_{sorts}(w), \sigma_{sorts}(s)} \rangle_{w \in S^*, s \in S}$ by

$$\sigma_{sorts} = \{Polygon \mapsto Shape, Figure \mapsto Shape, Trump \mapsto Suit\},$$
$$\sigma_{\varepsilon,Polygon} = \{square \mapsto box\},$$
$$\sigma_{Polygon,Polygon} = \{boxify \mapsto boxify\},$$
$$\sigma_{Polygon,Figure} = \{boxify \mapsto boxify\},$$
$$\sigma_{Figure\ Trump,Trump} = \{g \mapsto f\},$$

and $\sigma_{w,s} = \varnothing$ for all other $w \in S^*, s \in S$. Then $\sigma: \Sigma \to \Sigma 1$ is a signature morphism. □

Exercise 1.5.3. Given two signature morphisms $\sigma{:}\Sigma \to \Sigma'$ and $\sigma'{:}\Sigma' \to \Sigma''$, let $(\sigma;\sigma')_{sorts} = \sigma_{sorts};\sigma'_{sorts}$ and $(\sigma;\sigma')_{ops} = \sigma_{ops};\sigma'_{ops}$ (or rather, to be more precise: $(\sigma;\sigma')_{w,s} = \sigma_{w,s};\sigma'_{\sigma^*_{sorts}(w),\sigma_{sorts}(s)}$ for $w \in S^*, s \in S$). Show that this defines a signature morphism $\sigma;\sigma'{:}\Sigma \to \Sigma''$. □

In the rest of this section, let $\sigma{:}\Sigma \to \Sigma'$ be a signature morphism, where $\Sigma = \langle S, \Omega\rangle$ and $\Sigma' = \langle S', \Omega'\rangle$. As will be defined below, any such signature morphism gives rise to a translation of Σ-terms to Σ'-terms, and of Σ'-algebras and homomorphisms to Σ-algebras and homomorphisms. Note that the direction of translation of algebras and homomorphisms is "backwards" with respect to the direction of the signature morphism, as the following figure indicates.

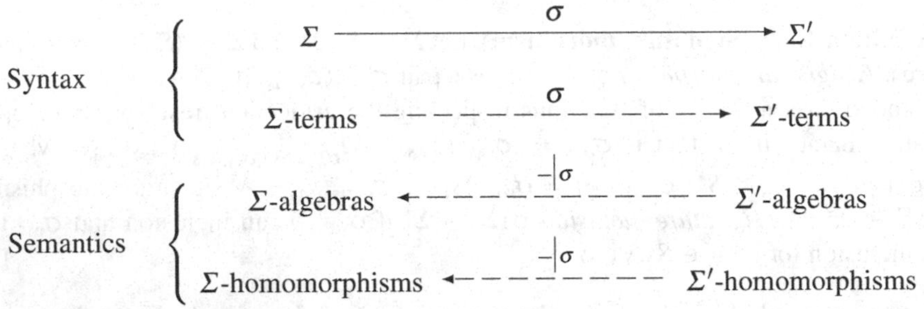

Definition 1.5.4 (Reduct algebra). Let A' be a Σ'-algebra. The *σ-reduct of A'* is the Σ-algebra $A'|_\sigma$ defined as follows:

- $|A'|_\sigma|_s = |A'|_{\sigma(s)}$ for all $s \in S$; and
- for all $f{:}s_1 \times \cdots \times s_n \to s$ in Σ,
  $$f_{A'|_\sigma}{:}|A'|_\sigma|_{s_1} \times \cdots \times |A'|_\sigma|_{s_n} \to |A'|_\sigma|_s = \sigma(f)_{A'}{:}|A'|_{\sigma(s_1)} \times \cdots \times |A'|_{\sigma(s_n)} \to |A'|_{\sigma(s)}.$$ □

If Σ is a subsignature of Σ', $\sigma{:}\Sigma \hookrightarrow \Sigma'$ is the signature inclusion, and A' is a Σ'-algebra, then $A'|_\sigma$ is just A' with some carriers and/or operations removed.

Notation. We sometimes write $A'|_\Sigma$ for $A'|_\sigma$ when $\sigma{:}\Sigma \to \Sigma'$ is obvious, such as when σ is a signature inclusion. □

Example 1.5.5. Let $\sigma{:}\Sigma \to \Sigma1$ be the signature morphism defined in Example 1.5.2 and let $A1$ be the $\Sigma1$-algebra defined in Example 1.2.4. Then $A1|_\sigma$ is the Σ-algebra such that

$$\begin{aligned}
&|A1|_\sigma|_{Polygon} = |A1|_{Shape} = \{\square, \triangle\},\\
&|A1|_\sigma|_{Figure} = |A1|_{Shape} = \{\square, \triangle\},\\
&|A1|_\sigma|_{Trump} = |A1|_{Suit} = \{\clubsuit, \heartsuit, \spadesuit\},\\
&square_{A1|_\sigma} = box_{A1} = \square,\\
&boxify_{A1|_\sigma}{:}|A1|_\sigma|_{Polygon} \to |A1|_\sigma|_{Polygon}\\
&\qquad = boxify_{A1}{:}|A1|_{Shape} \to |A1|_{Shape} = \{\square \mapsto \square, \triangle \mapsto \square\},\\
&g_{A1|_\sigma}{:}|A1|_\sigma|_{Figure} \times |A1|_\sigma|_{Trump} \to |A1|_\sigma|_{Trump}\\
&\qquad = f_{A1}{:}|A1|_{Shape} \times |A1|_{Suit} \to |A1|_{Suit} = \{\langle\square, \clubsuit\rangle \mapsto \clubsuit, \langle\square, \heartsuit\rangle \mapsto \spadesuit, \ldots\}.
\end{aligned}$$ □

Exercise 1.5.6. A Σ-algebra A can be seen as a function taking names in Σ to their interpretations, having the composition $\sigma;A$ as its σ-reduct; spell out the details. □

Exercise 1.5.7. Let $\sigma:\Sigma \to \Sigma'$ be a signature morphism that is surjective on sort names, and let A' be a Σ'-algebra. Show that if $A'|_\sigma$ is reachable then A' is reachable. Give counterexamples showing that the opposite implication does not hold, and that the implication itself does not hold if some sort names in Σ' are not in the image of Σ under σ. □

Definition 1.5.8 (Reduct homomorphism). Let $h':A' \to B'$ be a Σ'-homomorphism. The *σ-reduct of h'* is the Σ-homomorphism $h'|_\sigma:A'|_\sigma \to B'|_\sigma$ defined as the S-sorted function $h'|_\sigma:|A'|_\sigma| \to |B'|_\sigma|$ such that $(h'|_\sigma)_s = h'_{\sigma(s)}$ for all $s \in S$. (**Exercise:** Show that $h'|_\sigma:A'|_\sigma \to B'|_\sigma$ is indeed a Σ-homomorphism.) □

Exercise 1.5.9. Define the σ-reduct $\equiv'|_\sigma$ of a Σ'-congruence $\equiv'$ on a Σ'-algebra A', and prove that it is a Σ-congruence on $A'|_\sigma$. Show that σ-reduct distributes over quotient, i.e. $(A'/\equiv')|_\sigma = (A'|_\sigma)/(\equiv'|_\sigma)$. □

The following definition of the translation of terms along a signature morphism $\sigma:\Sigma \to \Sigma'$ may look somewhat daunting, but its simple upshot is to translate each term $t \in |T_\Sigma(X)|$ to the Σ'-term obtained by replacing each operation name from Σ by its image under σ. Some care must be taken in the treatment of variables: since variables for different sorts are not required to be distinct, to make sure they are not inadvertently identified by the translation, for each sort s' in Σ' we have to take a disjoint union of the sets of variables of sorts mapped to s'.

Definition 1.5.10 (Term translation). Let X be an S-sorted set of variables. Define $X' = \langle X'_{s'} \rangle_{s' \in S'}$ to be the S'-sorted set such that

$$X'_{s'} = \biguplus_{\sigma(s)=s'} X_s \quad \text{for each } s' \in S'.$$

Then $(T_{\Sigma'}(X'))|_\sigma$ is a Σ-algebra. Let $i:X \to |(T_{\Sigma'}(X'))|_\sigma|$ be the obvious embedding (if not for the disjoint union in the definition of X' and the explicit decoration of variables with sorts in terms, i would coincide with ι_X, which maps each variable to its corresponding term). Then by Fact 1.4.4 there is a unique Σ-homomorphism $\widehat{\sigma}:T_\Sigma(X) \to (T_{\Sigma'}(X'))|_\sigma$ extending i:

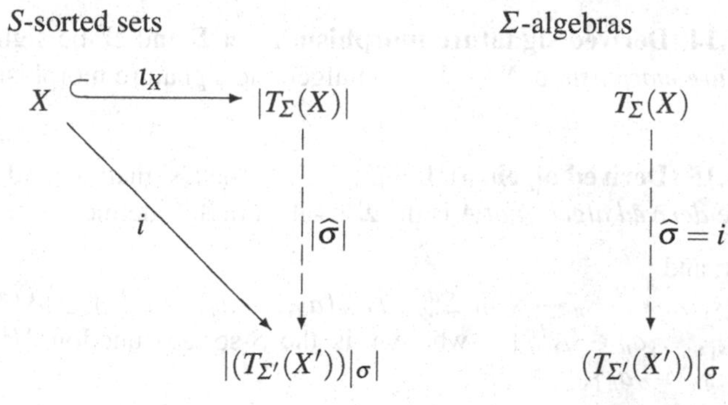

The *translation of a Σ-term $t \in |T_\Sigma(X)|$ by σ* is the Σ'-term $\widehat{\sigma}(t) \in |T_{\Sigma'}(X')|$. To keep the notation simple, we will write just $\sigma(t)$ for $\widehat{\sigma}(t)$. □

Example 1.5.11. Let $\sigma{:}\Sigma \to \Sigma 1$ be the signature morphism defined in Example 1.5.2, where $\Sigma = \langle S, \Omega \rangle$ and $\Sigma 1 = \langle S1, \Omega 1 \rangle$. Let X be the S-sorted set of variables $x{:}\,\mathit{Polygon}, x{:}\,\mathit{Figure}, y{:}\,\mathit{Figure}, z{:}\,\mathit{Trump}$. The $S1$-sorted set of variables X' in Definition 1.5.10 is then $x_1{:}\,\mathit{Shape}, x_2{:}\,\mathit{Shape}, y{:}\,\mathit{Shape}, z{:}\,\mathit{Suit}$, and

$$\sigma(g(\mathit{boxify}(x{:}\,\mathit{Polygon}), g(x{:}\,\mathit{Figure}, z))) = f(\mathit{boxify}(x_1), f(x_2, z)),$$
$$\sigma(g(x{:}\,\mathit{Figure}, g(\mathit{boxify}(\mathit{boxify}(\mathit{square})), z))) = f(x_2, f(\mathit{boxify}(\mathit{boxify}(\mathit{box})), z)),$$

and so on. □

Exercise 1.5.12. Let $t \in |T_\Sigma|$ be a ground Σ-term and let A' be a Σ'-algebra. Show that the value of t is invariant under change of signature, i.e. $\sigma(t)_{A'} = t_{A'|_\sigma}$.

Formulate and prove a more general version of this result in which t may contain variables. □

1.5.2 Derived signature morphisms

A derived signature morphism from Σ to Σ' is like an algebraic signature morphism from Σ to Σ' except that operation names in Σ are mapped to *terms* over Σ'. This allows operation names in Σ to be mapped to combinations of operations in Σ', and also handles the case where the order of arguments of the corresponding operations in Σ and Σ' are different.

Definition 1.5.13 (Derived signature). Let $\Sigma = \langle S, \Omega \rangle$ be a signature. For any sequence $s_1 \ldots s_n \in S^*$, let $I_{s_1 \ldots s_n}$ be the S-sorted set $\boxed{1}{:}\,s_1, \ldots, \boxed{n}{:}\,s_n$. The *derived signature of* Σ is the signature $\Sigma^{der} = \langle S, \Omega^{der} \rangle$ where for each $s_1 \ldots s_n \in S^*$ and $s \in S$, $\Omega^{der}_{s_1 \ldots s_n, s} = |T_\Sigma(I_{s_1 \ldots s_n})|_s$. □

In the derived signature of Σ, a Σ-term t of sort s with variables $I_{s_1 \ldots s_n}$ represents an operation $t{:}\,s_1 \times \cdots \times s_n \to s$. The variable $\boxed{i}{:}\,s_i$ in $I_{s_1 \ldots s_n}$ thus stands for the ith argument of t. Note that a "bare" variable $\boxed{i} \in |T_\Sigma(I_{s_1 \ldots s_n})|_{s_i}$ is an operation $i{:}\,s_1 \times \cdots \times s_n \to s_i$ in Σ^{der}, corresponding to a projection function.

Definition 1.5.14 (Derived signature morphism). Let Σ and Σ' be signatures. A *derived signature morphism* $\delta{:}\,\Sigma \to \Sigma'$ is an algebraic signature morphism $\delta{:}\,\Sigma \to (\Sigma')^{der}$. □

Definition 1.5.15 (Derived algebra). Let $\Sigma = \langle S, \Omega \rangle$ be a signature, and let A be a Σ-algebra. The *derived algebra of* A is the Σ^{der}-algebra A^{der} defined as follows:

- $|A^{der}| = |A|$; and
- for each $t{:}\,s_1 \times \cdots \times s_n \to s$ in Σ^{der}, $t_{A^{der}}(a_1, \ldots, a_n) = t_A(v) \in |A^{der}|_s$ for all $a_1 \in |A^{der}|_{s_1}, \ldots, a_n \in |A^{der}|_{s_n}$, where v is the S-sorted function $\{(\boxed{1}{:}\,s_1) \mapsto a_1, \ldots, (\boxed{n}{:}\,s_n) \mapsto a_n\}$. □

In the rest of this section, let $\delta\colon\Sigma \to \Sigma'$ be a derived signature morphism. The following corresponds to Definition 1.5.4 for algebraic signature morphisms; later exercises correspond to Definitions 1.5.8 and 1.5.10 and related results.

Definition 1.5.16 (Reduct algebra w.r.t. a derived signature morphism). Let A' be a Σ'-algebra. The *δ-reduct of A'* is the Σ-algebra $A'|_\delta$ defined as follows:

- $|A'|_\delta|_s = |A'|_{\delta(s)}$ for all $s \in S$; and
- for all $f\colon s_1 \times \cdots \times s_n \to s$ in Σ, $f_{A'|_\delta}\colon |A'|_\delta|_{s_1} \times \cdots \times |A'|_\delta|_{s_n} \to |A'|_\delta|_s = \delta(f)_{(A')^{der}}$.

Equivalently, $A'|_\delta$ is the Σ-algebra $(A')^{der}|_\delta$, viewing δ as the algebraic signature morphism $\delta\colon\Sigma \to (\Sigma')^{der}$. □

Exercise 1.5.17 (Reduct homomorphism w.r.t. a derived signature morphism). What is the δ-reduct $h'|_\delta$ of a Σ'-homomorphism $h'\colon A' \to B'$? Prove that $h'|_\delta\colon A'|_\delta \to B'|_\delta$ is a Σ-homomorphism. □

Exercise 1.5.18 (Term translation w.r.t. a derived signature morphism). Let $t \in |T_\Sigma(X)|$ be a Σ-term, where X is an S-sorted set of variables. Define $\delta(t)$, the translation of t by δ (the result should be a Σ'-term). □

Example 1.5.19. Let $\Sigma = \langle S, \Omega\rangle$ be the signature defined in Example 1.5.2, and let $\Sigma 1 = \langle S1, \Omega 1\rangle$ be the signature defined in Example 1.2.4. Let $\delta\colon\Sigma \to \Sigma 1$ be the derived signature morphism defined by

$$\begin{aligned}
&\delta_{sorts} = \{Polygon \mapsto Shape, Figure \mapsto Shape, Trump \mapsto Suit\},\\
&\delta_{\varepsilon,Polygon} = \{square \mapsto boxify(box)\},\\
&\delta_{Polygon,Polygon} = \{boxify \mapsto \boxed{1}{:}Shape\},\\
&\delta_{Polygon,Figure} = \{boxify \mapsto boxify(boxify(\boxed{1}{:}Shape))\},\\
&\delta_{Figure\,Trump,Trump} = \{g \mapsto f(boxify(\boxed{1}{:}Shape), f(\boxed{1}{:}Shape, \boxed{2}{:}Suit))\},
\end{aligned}$$

and $\delta_{w,s} = \varnothing$ for all other $w \in S^*, s \in S$.

Let $A1$ be the $\Sigma 1$-algebra defined in Example 1.2.4. Then $A1|_\delta$ is the Σ-algebra such that

$$\begin{aligned}
&|A1|_\delta|_{Polygon} = |A1|_\delta|_{Figure} = \{\square, \triangle\},\\
&|A1|_\delta|_{Trump} = \{\clubsuit, \heartsuit, \spadesuit\},\\
&square_{A1|_\delta} = \square,\\
&boxify_{A1|_\delta}\colon |A1|_\delta|_{Polygon} \to |A1|_\delta|_{Polygon} = \{\square \mapsto \square, \triangle \mapsto \triangle\}\\
&boxify_{A1|_\delta}\colon |A1|_\delta|_{Polygon} \to |A1|_\delta|_{Figure} = \{\square \mapsto \square, \triangle \mapsto \square\},
\end{aligned}$$

and $g_{A1|_\delta}\colon |A1|_\delta|_{Figure} \times |A1|_\delta|_{Trump} \to |A1|_\delta|_{Trump}$ is defined by the following table:

$g_{A1\|_\delta}$	$\clubsuit$	$\heartsuit$	$\spadesuit$
$\square$	$\clubsuit$	$\heartsuit$	$\spadesuit$
$\triangle$	$\spadesuit$	$\heartsuit$	$\heartsuit$

Let X be the S-sorted set of variables $x{:}\,Polygon, x{:}\,Figure, y{:}\,Figure, z{:}\,Trump$. A correct solution to Exercise 1.5.18 would translate $g(boxify(x{:}\,Polygon), g(x{:}\,Figure, z))$ (a Σ-term with variables X) to

$$f(boxify(\underbrace{boxify(boxify(x_1))}_{=\delta(boxify(x:Polygon))}), f(\underbrace{boxify(boxify(x_1))}_{=\delta(boxify(x:Polygon))}, \underbrace{f(boxify(x_2), f(x_2,z))}_{=\delta(g(x:Figure,z))})).$$

□

Exercise 1.5.20. Repeat Exercise 1.5.12 for the case of derived signature morphisms. □

Exercise 1.5.21. A more complex definition of derived signature morphism $\delta{:}\,\Sigma \to \Sigma'$ would allow a sort name s in Σ to be mapped to a *Cartesian product* $s'_1 \times \cdots \times s'_n$ of sorts $s'_1, \ldots, s'_n$ in Σ'. Give versions of the above definitions which permit this. □

Exercise 1.5.22. Another variation on the definition of derived signature morphism would permit operation names in Σ to be mapped to recursively defined functions in terms of the operation names in Σ'. Give versions of the above definitions which would allow this. HINT: Look at a book like [Sch86] before attempting this exercise. □

1.6 Bibliographical remarks

This chapter presents the basic notions of universal algebra that are required in the sequel. There is a vast literature on universal algebra as a branch of mathematics, and the concepts and results we need here are a tiny fraction of this. Applications of universal algebra in computer science are widespread, going back at least to [BL69].

For much more on universal algebra, see e.g. [Grä79] or [Coh65] but note that both of these handle only the single-sorted case. A presentation of some of this material for a computer scientist audience is [Wec92]; see also [MT92], where applications to some topics in computer science other than the ones covered in this book are indicated.

The style of presentation here is relaxed but it might still be too dense for some readers, who might prefer the gentler style, with proofs of many of the results which we omit here, in [GTW76], [EM85], [MG85] or [LEW96].

The generalisation from single-sorted to many-sorted algebras originates with [Hig63] and [BL70]. The first applications to computer science came later [Mai72], becoming prominent with [GTW76]. The generalisation is straightforward from a purely mathematical standpoint, but there are a few subtle issues that will surface in later chapters. For instance, we admit empty carrier sets in Definition 1.2.2, unlike most logic books and, for instance, [BT87] and [Mos04]. Admitting empty carrier sets requires more care in the presentation of rules for reasoning — see Exercises 2.4.9 and 2.4.10 below — but it also makes some results smoother — see Exercise 2.5.18.

There are different definitions of many-sorted signature in the literature. The one here is quite general, allowing overloading of operation names, and originates with [GTWW73] and [Gog74]. In some early papers, signatures are called "operator domains". Definitions that do not permit overloading are used in [EM85] and [Wir90], but as remarked after Definition 1.2.1, these definitions are equivalent if each operation name is taken to be tagged with its arity and result sort.

Signature morphisms emerged around 1978 in the context of early work on the semantics of parameterised specifications in the style of Definition 6.3.5 below, see [Ehr78] and [GB78]; Definition 1.5.1 is from the latter. A number of variants of and restrictions to this notion have been considered. One possible simplifying assumption is to restrict attention to injective signature morphisms as in [BHK90], or to bijective signature morphisms, which are sometimes referred to as "renamings". The notion of reduct, but only with respect to a signature inclusion, arises in universal algebra. The generalisation from signature morphisms to derived signature morphisms originates in [GTW76], and is related to the even more general notion of (theory) interpretation in logic [End72]. Since the 1970s, derived signature morphisms have made only sporadic appearances in the algebraic specification literature; see for instance [SB83] and [HLST00].

Chapter 2
Simple equational specifications

A specification is an unambiguous description of a signature Σ and a class of Σ-algebras. Because we model programs as algebras, a specification amounts to a characterisation of a class of programs. Each of these programs is regarded as a correct realisation of the specification.

Given a signature Σ (which, if finite, may be presented by simply listing its sort names and its operation names with their arities and result sorts), there are two basic techniques that may be used for describing a class of Σ-algebras. The first is to simply give a list of all the algebras in the class. Unfortunately, we are almost always interested in *infinite* classes of algebras, where this technique is useless (although sometimes this may be made to work if we can present a finite number of algebras that in some precise way represent the entire class we want). The second is to describe the functional behaviour of the algebras in the class by listing the properties (axioms) they are to satisfy. This is the fundamental specification technique used in work on algebraic specification and the one that will be studied in this chapter. The simplest and most common case is when the properties are expressed in the form of universally quantified equations; in most of this chapter, we restrict attention to this case. Section 2.7 indicates other forms of axioms that may be of use, along with some possible variations on the definitions of Chapter 1, and further possibilities will be discussed in Chapter 4. Since most of the results in this chapter are fairly standard and proofs are readily available in the literature, most proofs are left as exercises for the reader.

Chapters 5 and 8 will cover additional techniques for describing classes of algebras. All of these involve taking a class of algebras and performing a simple operation to obtain another class of algebras, often over a different signature. Using such methods, complex specifications of classes of complex algebras may be built from small and easily understood units.

2.1 Equations

Any given signature characterises the class of algebras over that signature. Although this fixes the names of sorts and operations, it is an exceedingly limited form of description since each such class contains a wide range of different algebras. Any two algebras taken from such a class may have carrier sets of different cardinalities and containing different elements; even if both algebras happen to have "matching" carrier sets, the results produced by applying operations may differ. For most applications it is necessary to focus on a subclass of algebras, obtained by imposing *axioms* which serve as constraints on the permitted behaviour of operations. One particularly simple form of axioms is equations, which constrain behaviour by asserting that two given terms have the same value. Equations have limited expressive power, but this disadvantage is to some extent balanced by the simplicity and convenience of reasoning in equational logic (see Sections 2.4 and 2.6).

Variables in equations will be taken from a fixed but arbitrary infinite set $\mathscr{X}$. We require $\mathscr{X}$ to be closed under finite disjoint union: if $\langle X_i \rangle_{i \in I}$ is finite and $X_i \subseteq \mathscr{X}$ for all $i \in I$, then $\biguplus \langle X_i \rangle_{i \in I} \subseteq \mathscr{X}$. We use variable names like x, y, z in examples, and so we assume that these are all in $\mathscr{X}$. Throughout this section, let $\Sigma = \langle S, \Omega \rangle$ be a signature.

Definition 2.1.1 (Equation). A *Σ-equation* $\forall X \bullet t = t'$ consists of:

- a finite S-sorted set X (of variables), such that $X_s \subseteq \mathscr{X}$ for all $s \in S$; and
- two Σ-terms $t, t' \in |T_\Sigma(X)|_s$ for some sort $s \in S$.

A Σ-equation of the form $\forall \varnothing \bullet t = t'$ is called a *ground (Σ-)equation*, and will sometimes be written $t = t'$. □

The explicit quantification over X in a Σ-equation $\forall X \bullet t = t'$ is essential, as will become clear in Section 2.4. It is nevertheless common practice to leave quantification implicit, writing $t = t'$ in place of $\forall FV(t) \cup FV(t') \bullet t = t'$, but we will not follow this practice except for ground equations.

Definition 2.1.2 (Satisfaction). A Σ-algebra A *satisfies* (or *is a model of*) a Σ-equation $\forall X \bullet t = t'$, written $A \models_\Sigma \forall X \bullet t = t'$, if for every ($S$-sorted) function $v: X \to |A|$, $t_A(v) = t'_A(v)$.

A satisfies (or is a model of) a set $\mathscr{E}$ of Σ-equations, written $A \models_\Sigma \mathscr{E}$, if $A \models_\Sigma e$ for every equation $e \in \mathscr{E}$. A class $\mathscr{A}$ of Σ-algebras satisfies a Σ-equation e, written $\mathscr{A} \models_\Sigma e$, if $A \models_\Sigma e$ for every $A \in \mathscr{A}$. Finally, a class $\mathscr{A}$ of Σ-algebras satisfies a set $\mathscr{E}$ of Σ-equations, written $\mathscr{A} \models_\Sigma \mathscr{E}$, if $A \models_\Sigma \mathscr{E}$ for every $A \in \mathscr{A}$ (equivalently, if $\mathscr{A} \models_\Sigma e$ for every $e \in \mathscr{E}$, i.e. $A \models_\Sigma e$ for every $A \in \mathscr{A}$ and $e \in \mathscr{E}$). □

Notation. We sometimes write $\models$ in place of $\models_\Sigma$ when Σ is obvious. □

Occasionally we will say that an equation e *holds* in an algebra A when $A \models e$, and similarly for sets of equations and classes of algebras.

Exercise 2.1.3. Recall $\Sigma 1$ and $A1$ from Example 1.2.4. Give some $\Sigma 1$-equations (both ground and non-ground) that are satisfied by $A1$. Give some $\Sigma 1$-equations (both ground and non-ground) that are *not* satisfied by $A1$. □

Exercise 2.1.4. If $\forall X \bullet t = t'$ is a Σ-equation and $X \subseteq X'$ (and $X'_s \subseteq \mathscr{X}$ for all $s \in S$), it follows from Definition 2.1.1 that $\forall X' \bullet t = t'$ is also a Σ-equation. Show that $A \models_\Sigma \forall X \bullet t = t'$ implies that $A \models_\Sigma \forall X' \bullet t = t'$. Give a counterexample showing that the converse does *not* hold. (HINT: Consider $X_s = \varnothing$ and $|A|_s = \varnothing$ for some $s \in S$.) Show that it *does* hold if Σ has only one sort. □

Exercise 2.1.5. Show that surjective Σ-homomorphisms preserve satisfaction of Σ-equations: if $h: A \to B$ is a surjective Σ-homomorphism then $A \models_\Sigma e$ implies $B \models_\Sigma e$ for any Σ-equation e. Show that injective Σ-homomorphisms reflect satisfaction of Σ-equations: if $h: A \to B$ is an injective Σ-homomorphism then $B \models_\Sigma e$ implies $A \models_\Sigma e$ for any Σ-equation e. Conclude that Σ-isomorphisms preserve and reflect satisfaction of Σ-equations. □

Exercise 2.1.6. Give an alternative definition of $A \models_\Sigma \forall X \bullet t = t'$ via the satisfaction of $t = t'$ viewed as a ground equation over an enlarged signature. HINT: Definition 2.1.2 involves quantification over valuations $v: X \to |A|$. Consider how this might be replaced by quantification over algebras having a signature obtained from Σ by adding a constant for each variable in X. □

A signature morphism $\sigma: \Sigma \to \Sigma'$ gives rise to a translation of Σ-equations to Σ'-equations. This is essentially a simple matter of applying the translation on terms induced by σ to both sides of the equation.

Definition 2.1.7 (Equation translation). Let $\forall X \bullet t = t'$ be a Σ-equation, and let $\sigma: \Sigma \to \Sigma'$ be a signature morphism. Recall from Definition 1.5.10 that we then have $\sigma(t), \sigma(t') \in |T_{\Sigma'}(X')|$ where

$$X'_{s'} = \biguplus_{\sigma(s)=s'} X_s \qquad \text{for each } s' \in S'.$$

The *translation of* $\forall X \bullet t = t'$ *by* σ is then the Σ'-equation $\sigma(\forall X \bullet t = t') = \forall X' \bullet \sigma(t) = \sigma(t')$. (The fact that $\mathscr{X}$ is closed under finite disjoint union guarantees that this is indeed a Σ'-equation.) □

An important result which brings together some of the main definitions above is the following:

Lemma 2.1.8 (Satisfaction Lemma [BG80]). *If $\sigma: \Sigma \to \Sigma'$ is a signature morphism, e is a Σ-equation and A' is a Σ'-algebra, then $A' \models_{\Sigma'} \sigma(e)$ iff $A'|_\sigma \models_\Sigma e$.* □

When e is a ground Σ-equation, it is easy to see that this follows directly from the property established in Exercise 1.5.12. When σ is injective (on both sort and operation names), it seems intuitively clear that the Satisfaction Lemma should hold, since the domain of quantification of variables is unchanged, the only difference between e and $\sigma(e)$ is the names used for sorts and operations, and the only difference between A' and $A'|_\sigma$ (apart from sort/operation names) is that A' might provide interpretations for sort and operation names which do not appear in $\sigma(e)$ and so cannot affect its satisfaction. When σ is non-injective the Satisfaction Lemma still holds, but this is less intuitively obvious (particularly when σ is non-injective on sort names).

Exercise 2.1.9. Take a signature morphism $\sigma\colon \Sigma \to \Sigma'$ which is non-injective on sort and operation names, a Σ-equation involving the sort and operation names for which σ is not injective, and a Σ'-algebra, and check that the Satisfaction Lemma holds in this case. □

Exercise 2.1.10. Prove the Satisfaction Lemma, using Exercise 1.5.12. □

Exercise 2.1.11. Define the translation of a Σ-equation by a derived signature morphism $\delta\colon \Sigma \to \Sigma'$, and convince yourself that the Satisfaction Lemma also holds for this case. □

The Satisfaction Lemma says that the translations of syntax (terms, equations) and semantics (algebras) induced by signature morphisms are coherent with the definition of satisfaction. Said another way, the manner in which satisfaction of equations by algebras varies according to the signature at hand fits exactly with these translations. Further discussion of the property embodied in the Satisfaction Lemma may be found in Section 4.1.

2.2 Flat specifications

A signature together with a set of equations over that signature constitutes a simple form of specification. We refer to these as *flat* (meaning *unstructured*) specifications in order to distinguish them from the *structured* specifications to be introduced in Chapter 5, formed from simpler specifications using specification-building operations. As we shall see later, it is possible in some (but not all) cases to "flatten" a structured specification to yield a flat specification describing the same class of algebras. Throughout this section, let Σ be a signature.

Definition 2.2.1 (Presentation). A *presentation* (also known as a *flat specification*) is a pair $\langle \Sigma, \mathscr{E} \rangle$ where $\mathscr{E}$ is a set of Σ-equations (called the *axioms* of $\langle \Sigma, \mathscr{E} \rangle$). A presentation $\langle \Sigma, \mathscr{E} \rangle$ is sometimes referred to as a *Σ-presentation*. □

The term "presentation" is chosen to emphasize the syntactic nature of the concept. The idea is that a presentation *denotes* (or *presents*) a semantic object which is inconvenient to describe directly. A reasonable objection to the definition above is that it fails to include restrictions to ensure that presentations are truly syntactic objects, namely that Σ and $\mathscr{E}$ are *finite*, or at least effectively presentable in some other sense (e.g. recursive or recursively enumerable). Although it would be possible to impose such a restriction, we refrain from doing so in order to avoid placing undue emphasis on issues of this kind.

Definition 2.2.2 (Model of a presentation). A *model* of a presentation $\langle \Sigma, \mathscr{E} \rangle$ is a Σ-algebra A such that $A \models_\Sigma \mathscr{E}$. $Mod[\langle \Sigma, \mathscr{E} \rangle]$ is the class of all models of $\langle \Sigma, \mathscr{E} \rangle$. □

Taking $\langle \Sigma, \mathscr{E} \rangle$ to denote the semantic object $Mod[\langle \Sigma, \mathscr{E} \rangle]$ is sometimes called taking its *loose semantics*. The word "loose" here refers to the fact that this is not always (in fact, hardly ever) an isomorphism class of algebras: $A, B \in Mod[\langle \Sigma, \mathscr{E} \rangle]$ does *not* imply that $A \cong B$. In Section 2.5 we will consider the so-called *initial semantics* of presentations in which a further constraint is imposed on the models of a presentation, forcing every presentation to denote an isomorphism class of algebras.

Example 2.2.3. Let $\textsc{Bool} = \langle \Sigma_{\textsc{Bool}}, \mathscr{E}_{\textsc{Bool}} \rangle$ be the presentation below.[1]

> **spec** $\textsc{Bool} =$ **sorts** *Bool*
>
> > > **ops**   *true*: *Bool*
> > > > *false*: *Bool*
> > > > $\neg\,\_\_ : Bool \to Bool$
> > > > $\_\_ \wedge \_\_ : Bool \times Bool \to Bool$
> > > > $\_\_ \Rightarrow \_\_ : Bool \times Bool \to Bool$
> > > $\forall p, q : Bool$
> > > > - $\neg true = false$
> > > > - $\neg false = true$
> > > > - $p \wedge true = p$
> > > > - $p \wedge false = false$
> > > > - $p \wedge \neg p = false$
> > > > - $p \Rightarrow q = \neg(p \wedge \neg q)$

Define $\Sigma_{\textsc{Bool}}$-algebras $A1$, $A2$ and $A3$ as follows:

$|A1|_{Bool} = \{\star\}$	$|A2|_{Bool} = \{\clubsuit, \heartsuit, \spadesuit\}$	$|A3|_{Bool} = \{tt, f\!f\}$
$true_{A1} = \star$	$true_{A2} = \clubsuit$	$true_{A3} = tt$
$false_{A1} = \star$	$false_{A2} = \heartsuit$	$false_{A3} = f\!f$

$$\neg_{A1} = \{\star \mapsto \star\} \qquad \neg_{A2} = \{\clubsuit \mapsto \heartsuit, \quad \heartsuit \mapsto \clubsuit, \quad \spadesuit \mapsto \spadesuit\} \qquad \neg_{A3} = \{tt \mapsto f\!f, \quad f\!f \mapsto tt\}$$

$\wedge_{A1}$	$\star$
$\star$	$\star$

$\wedge_{A2}$	$\clubsuit$	$\heartsuit$	$\spadesuit$
$\clubsuit$	$\clubsuit$	$\heartsuit$	$\heartsuit$
$\heartsuit$	$\heartsuit$	$\heartsuit$	$\heartsuit$
$\spadesuit$	$\spadesuit$	$\heartsuit$	$\heartsuit$

$\wedge_{A3}$	tt	$f\!f$
tt	tt	$f\!f$
$f\!f$	$f\!f$	$f\!f$

$\Rightarrow_{A1}$	$\star$
$\star$	$\star$

$\Rightarrow_{A2}$	$\clubsuit$	$\heartsuit$	$\spadesuit$
$\clubsuit$	$\clubsuit$	$\heartsuit$	$\clubsuit$
$\heartsuit$	$\clubsuit$	$\clubsuit$	$\clubsuit$
$\spadesuit$	$\clubsuit$	$\spadesuit$	$\clubsuit$

$\Rightarrow_{A3}$	tt	$f\!f$
tt	tt	$f\!f$
$f\!f$	tt	tt

---

[1] Here and in the sequel we use notation from OBJ [KKM88] and CASL [Mos04] to introduce infix, prefix and "mixfix" operations. We also follow CASL by itemizing axioms in specifications, marking them with • and introducing universal quantification over the variables only once for the entire list of axioms. Although the meaning of an axiom can be affected by adding quantification over variables that it does not contain — see Exercise 2.1.4 — this pathology does not arise in any of our examples.

Each of these algebras is a model of BOOL. (NOTE: Reference will be made to BOOL and to its models $A1$, $A2$ and $A3$ in later sections of this chapter. The name BOOL has been chosen for the same reason as `bool` is used for the type of truth values in programming languages; it is technically a misnomer since this is not a specification of Boolean algebras; see Example 2.2.4 below.)

Exercise. Show that the models defined and in fact all the models of BOOL satisfy $\forall p{:}Bool \bullet \neg(p \wedge \neg false) = \neg p$. Define a model of BOOL that does not satisfy $\forall p{:}Bool \bullet \neg\neg p = p$. □

Example 2.2.4. Let $\mathrm{BA} = \langle \Sigma_{\mathrm{BA}}, \mathscr{E}_{\mathrm{BA}} \rangle$ be the following presentation.

$$
\begin{array}{lll}
\textbf{spec}\ \mathrm{BA} = & \textbf{sorts} & Bool \\
& \textbf{ops} & \mathsf{true}{:}\,Bool \\
& & \mathsf{false}{:}\,Bool \\
& & \neg\_\_{:}\,Bool \to Bool \\
& & \_\_\vee\_\_{:}\,Bool \times Bool \to Bool \\
& & \_\_\wedge\_\_{:}\,Bool \times Bool \to Bool \\
& & \_\_\Rightarrow\_\_{:}\,Bool \times Bool \to Bool \\
& \multicolumn{2}{l}{\forall p,q,r{:}\,Bool} \\
& & \bullet\ p \vee (q \vee r) = (p \vee q) \vee r \\
& & \bullet\ p \wedge (q \wedge r) = (p \wedge q) \wedge r \\
& & \bullet\ p \vee q = q \vee p \\
& & \bullet\ p \wedge q = q \wedge p \\
& & \bullet\ p \vee (p \wedge q) = p \\
& & \bullet\ p \wedge (p \vee q) = p \\
& & \bullet\ p \vee (q \wedge r) = (p \vee q) \wedge (p \vee r) \\
& & \bullet\ p \wedge (q \vee r) = (p \wedge q) \vee (p \wedge r) \\
& & \bullet\ p \vee \neg p = \mathsf{true} \\
& & \bullet\ p \wedge \neg p = \mathsf{false} \\
& & \bullet\ p \Rightarrow q = \neg p \vee q
\end{array}
$$

Models of BA are called *Boolean algebras*. One such model is the following two-valued Boolean algebra $\mathbb{B}$:

$$
\begin{array}{l}
|\mathbb{B}|_{Bool} = \{tt, ff\}, \\
\mathsf{true}_{\mathbb{B}} = tt, \\
\mathsf{false}_{\mathbb{B}} = ff, \\
\neg_{\mathbb{B}} = \{tt \mapsto ff, ff \mapsto tt\}
\end{array}
$$

and

$\vee_{\mathbb{B}}$	*tt*	*ff*
tt	*tt*	*tt*
ff	*tt*	*ff*

$\wedge_{\mathbb{B}}$	*tt*	*ff*
tt	*tt*	*ff*
ff	*ff*	*ff*

$\Rightarrow_{\mathbb{B}}$	*tt*	*ff*
tt	*tt*	*ff*
ff	*tt*	*tt*

This is essentially the same as $A3$ in Example 2.2.3. Note that $A1$ can be turned into a (trivial) Boolean algebra in a similar way, but this is not the case with $A2$.

Exercise. Given a Boolean algebra B, define a relation $\leq_B \subseteq |B| \times |B|$ by $a \leq_B b$ iff $a \vee_B b = b$. Show that $\leq_B$ is a partial order with $true_B$ and $false_B$ as its greatest and least elements respectively, and with $a \vee_B b$ yielding the least upper bound of a, b and $a \wedge_B b$ yielding their greatest lower bound. (In fact, $\langle |B|, \leq_B \rangle$ is a distributive lattice with top and bottom elements and complement $\neg_B$.) □

Exercise 2.2.5. Show that all Boolean algebras (the models of BA as introduced in Exercise 2.2.4) satisfy the *de Morgan laws*:

$$\forall p,q{:}Bool \bullet \neg(p \vee q) = \neg p \wedge \neg q$$
$$\forall p,q{:}Bool \bullet \neg(p \wedge q) = \neg p \vee \neg q.$$
□

The following characterisation of the expressive power of flat equational specifications is one of the classical theorems of universal algebra.

Definition 2.2.6 (Equationally definable class). A class $\mathscr{A}$ of Σ-algebras is *equationally definable* if $\mathscr{A} = Mod[\langle \Sigma, \mathscr{E} \rangle]$ for some set $\mathscr{E}$ of Σ-equations. □

Definition 2.2.7 (Variety). A class $\mathscr{A}$ of Σ-algebras is *closed under subalgebras* if for any $A \in \mathscr{A}$ and subalgebra B of A, $B \in \mathscr{A}$. Similarly, $\mathscr{A}$ is *closed under homomorphic images* if for any $A \in \mathscr{A}$ and Σ-homomorphism $h{:}A \to B$, $h(A) \in \mathscr{A}$, and $\mathscr{A}$ is *closed under products* if for any family $\langle A_i \in \mathscr{A} \rangle_{i \in I}$, $\prod \langle A_i \rangle_{i \in I} \in \mathscr{A}$.

A non-empty class of Σ-algebras which is closed under subalgebras, homomorphic images, and products is called a *variety*. □

Proposition 2.2.8. *Any equationally definable class $\mathscr{A}$ of Σ-algebras is a variety.* □

Exercise 2.2.9. Prove Proposition 2.2.8: show that for any presentation $\langle \Sigma, \mathscr{E} \rangle$, $Mod[\langle \Sigma, \mathscr{E} \rangle]$ is closed under subalgebras, homomorphic images and products. For example, formalise the following argument to show closure under subalgebras: if $A \models_\Sigma e$ and B is a subalgebra of A then $B \models_\Sigma e$ since removing values from the carriers of an algebra does not affect the truth of universally quantified assertions about its behaviour. Closure under products and under homomorphic images are not much more difficult to prove. □

Theorem 2.2.10 (Birkhoff's Variety Theorem [Bir35]). *If Σ is a signature with a finite set of sort names then a class $\mathscr{A}$ of Σ-algebras is a variety iff $\mathscr{A}$ is equationally definable.* □

The "if" part of this theorem is (a special case of) Proposition 2.2.8. A complete proof of the "only if" part is beyond the scope of this book; the curious reader should consult [Wec92].

Example 2.2.11. Consider the signature

$$\begin{aligned} \Sigma = \textbf{sorts}\ & s \\ \textbf{ops}\ & 0{:}s \\ & \_\_ \times \_\_ {:} s \times s \to s \end{aligned}$$

and the class $\mathscr{A}$ of Σ-algebras satisfying the familiar cancellation law:

if $a \neq 0$ and $a \times b = a \times c$ then $b = c$.

The Σ-algebra A such that $|A|_s$ is the set of natural numbers and $\times_A$ is ordinary multiplication is in $\mathscr{A}$. The Σ-algebra B such that $|B|_s = \{0,1,2,3\}$ and $\times_A$ is multiplication modulo 4 is not in $\mathscr{A}$. (**Exercise:** Why not?) Since B is a homomorphic image of A, this shows that $\mathscr{A}$ is not a variety and hence is not equationally definable. □

Exercise 2.2.12. Formulate a definition of what it means for a class of Σ-algebras to be closed under homomorphic coimages. Are varieties closed under homomorphic coimages? □

Exercise 2.2.13. Formulate definitions of what it means for a class of Σ-algebras to be closed under quotients and under isomorphisms. Show that closure under both quotients and isomorphisms is equivalent to closure under homomorphic images. □

The assumption in Theorem 2.2.10 that the set of sort names in Σ is finite cannot easily be omitted:

Exercise 2.2.14. A family $\mathscr{B}$ of Σ-algebras is *directed* if any two algebras $B_1, B_2 \in \mathscr{B}$ are subalgebras of some $B \in \mathscr{B}$. Define the *union* $\bigcup \mathscr{B}$ of such a family to be the least Σ-algebra such that each $B \in \mathscr{B}$ is a subalgebra of $\bigcup \mathscr{B}$ (the carrier of $\bigcup \mathscr{B}$ is the union of the carriers of all algebras in $\mathscr{B}$, and the values of operations on arguments are inherited from the algebras in $\mathscr{B}$; this is well defined since $\mathscr{B}$ is directed). Prove that since we consider equations with finite sets of variables only, then for any presentation $\langle \Sigma, \mathscr{E} \rangle$, $Mod[\langle \Sigma, \mathscr{E} \rangle]$ is *closed under directed unions*, that is, given any *directed* family of algebras $\mathscr{B} \subseteq Mod[\langle \Sigma, \mathscr{E} \rangle]$, its union $\bigcup \mathscr{B}$ is also in $Mod[\langle \Sigma, \mathscr{E} \rangle]$.

A generalisation of Theorem 2.2.10 that we hint at here without a proof is that for *any* signature Σ, a class of Σ-algebras is equationally definable iff it is a variety that is closed under directed unions. □

Exercise 2.2.15. Consider a signature with an infinite set of sort names and no operations. Let $\mathscr{A}_{fin}$ be the class of all algebras over this signature that have non-empty carriers for a finite set of sorts only, and let $\mathscr{A}$ be the closure of $\mathscr{A}_{fin}$ under products and subalgebras (this adds algebras where the carrier of each sort is either a singleton or empty). Check that $\mathscr{A}$ is a variety. Prove, however, that $\mathscr{A}$ is not definable by any set of equations. HINT: Use Exercise 2.2.14. □

Exercise 2.2.16. Modify the definition of equation (Definition 2.1.1) so that infinite sets of variables are allowed; it is enough to consider sets of variables that are finite for each sort, but may be non-empty for infinitely many sorts. Extend the notion of satisfaction (Definition 2.1.2) to such generalised equations in the obvious way. Check that the class $\mathscr{A}$ defined in Exercise 2.2.15 is definable by such equations.

HINT: Consider all equations of the form $\forall X \cup \{x,y{:}s\} \bullet x = y$, for all sorts s and sets X of variables such that $X_{s'} \neq \varnothing$ for infinitely many sorts s'.

Another generalisation of Theorem 2.2.10 that we want to hint at here is that for *any* signature Σ a class of Σ-algebras is definable by such generalised equations iff it is a variety. The proof of the "if" part is as easy as for ordinary equations (Proposition 2.2.8). The proof of the "only if" part is quite similar to that of the finitary case. □

A final remark to clarify the nuances in the many-sorted versions of Theorem 2.2.10 is that the theorem holds for *any* signature (also with an infinite set of sort names) when we restrict attention to algebras with non-empty carriers of all sorts: all varieties of such algebras (with closure under subalgebras limited to subalgebras with non-empty carriers) are definable by equations with a finite set of variables.

2.3 Theories

Any equationally definable class of algebras has many different presentations; in practice the choice of presentation is determined by various factors, including the need for simplicity and understandability and the desire for elegance. On the other hand, such a class uniquely determines the largest set of equations that defines it, called its theory. Since this is an infinite set, it is not a useful way of presenting the class. However, it is a useful set to consider since it contains all axioms in all presentations of the class, together with all their consequences.

Throughout this section, let Σ be a signature.

Definition 2.3.1 (*$Mod_\Sigma(\mathscr{E})$, $Th_\Sigma(\mathscr{A})$, $Cl_\Sigma(\mathscr{E})$, $Cl_\Sigma(\mathscr{A})$*)**.** Given any set $\mathscr{E}$ of Σ-equations, $Mod_\Sigma(\mathscr{E})$ (the *models of* $\mathscr{E}$) denotes the class of all Σ-algebras satisfying all the Σ-equations in $\mathscr{E}$:

$$Mod_\Sigma(\mathscr{E}) = \{A \mid A \text{ is a } \Sigma\text{-algebra and } A \models_\Sigma \mathscr{E}\} \quad (= Mod[\langle \Sigma, \mathscr{E} \rangle]).$$

For any class $\mathscr{A}$ of Σ-algebras, $Th_\Sigma(\mathscr{A})$ (the *theory of* $\mathscr{A}$) denotes the set of all Σ-equations satisfied by each Σ-algebra in $\mathscr{A}$:

$$Th_\Sigma(\mathscr{A}) = \{e \mid e \text{ is a } \Sigma\text{-equation and } \mathscr{A} \models_\Sigma e\}.$$

A set $\mathscr{E}$ of Σ-equations is *closed* if $\mathscr{E} = Th_\Sigma(Mod_\Sigma(\mathscr{E}))$. The *closure* of a set $\mathscr{E}$ of Σ-equations is the (closed) set $Cl_\Sigma(\mathscr{E}) = Th_\Sigma(Mod_\Sigma(\mathscr{E}))$. Analogously, a class $\mathscr{A}$ of Σ-algebras is *closed* if $\mathscr{A} = Mod_\Sigma(Th_\Sigma(\mathscr{A}))$, and the *closure* of $\mathscr{A}$ is $Cl_\Sigma(\mathscr{A}) = Mod_\Sigma(Th_\Sigma(\mathscr{A}))$. □

Proposition 2.3.2. *For any sets $\mathscr{E}$ and $\mathscr{E}'$ of Σ-equations and classes $\mathscr{A}, \mathscr{B}$ of Σ-algebras:*

1. If $\mathscr{E} \subseteq \mathscr{E}'$ then $Mod_\Sigma(\mathscr{E}) \supseteq Mod_\Sigma(\mathscr{E}')$.

2. *If* $\mathscr{B} \supseteq \mathscr{A}$ *then* $Th_\Sigma(\mathscr{B}) \subseteq Th_\Sigma(\mathscr{A})$.
3. $\mathscr{E} \subseteq Th_\Sigma(Mod_\Sigma(\mathscr{E}))$ *and* $Mod_\Sigma(Th_\Sigma(\mathscr{A})) \supseteq \mathscr{A}$.
4. $Mod_\Sigma(\mathscr{E}) = Mod_\Sigma(Th_\Sigma(Mod_\Sigma(\mathscr{E})))$ *and* $Th_\Sigma(\mathscr{A}) = Th_\Sigma(Mod_\Sigma(Th_\Sigma(\mathscr{A})))$.
5. $Cl_\Sigma(\mathscr{E})$ *and* $Cl_\Sigma(\mathscr{A})$ *are closed.*

Proof. **Exercise.** HINT: Properties 4 and 5 follow from properties 1–3. □

For any signature Σ, the functions Th_Σ and Mod_Σ constitute what is known in lattice theory as a Galois connection.

Definition 2.3.3 (Galois connection). A *Galois connection* is given by two partially ordered sets A and M (in Proposition 2.3.2, A is the set of all sets of Σ-equations, and M is the "set" of all classes of Σ-algebras, both ordered by inclusion) and maps $\_^*: A \to M$ and $\_^+: M \to A$ (here Mod_Σ and Th_Σ) satisfying properties corresponding to 2.3.2(1)–2.3.2(3). An element $a \in A$ (or $m \in M$) is called *closed* if $a = (a^*)^+$ (or $m = (m^+)^*$). □

Some useful properties — including ones corresponding to 2.3.2(4) and 2.3.2(5) — hold for any Galois connection.

Exercise 2.3.4. For any Galois connection and any $a, b \in A$ and $m \in M$, show that the following properties hold:

1. $a \leq_A m^+$ iff $a^* \geq_M m$.
2. If a and b are closed then $a \leq_A b$ iff $a^* \geq_M b^*$. (Show that the "if" part fails if a or b is not closed.)

Here, $\leq_A$ and $\leq_M$ are the orders on A and M respectively. □

Exercise 2.3.5. For any Galois connection such that A and M have binary least upper bounds ($\sqcup_A$, $\sqcup_M$) and greatest lower bounds ($\sqcap_A$, $\sqcap_M$), and for any $a, b \in A$, show that the following properties hold:

1. $(a \sqcup_A b)^* = a^* \sqcap_M b^*$.
2. $(a \sqcap_A b)^* \geq_M a^* \sqcup_M b^*$.

HINT: $\sqcup_A$ satisfies the following properties for any $a, b, c \in A$:

- $a \leq_A a \sqcup_A b$ and $b \leq_A a \sqcup_A b$.
- If $a \leq_A c$ and $b \leq_A c$ then $a \sqcup_A b \leq_A c$.

And analogously for $\sqcap_A$, $\sqcup_M$ and $\sqcap_M$.

State and prove analogues to 1 and 2 for any $m, n \in M$, and instantiate all these general properties for the Galois connection between sets of Σ-equations and classes of Σ-algebras. □

Definition 2.3.6 (Semantic consequence). A Σ-equation e is a *semantic consequence* of a set $\mathscr{E}$ of Σ-equations, written $\mathscr{E} \models_\Sigma e$, if $e \in Cl_\Sigma(\mathscr{E})$ (equivalently, if $Mod_\Sigma(\mathscr{E}) \models_\Sigma e$). □

Notation. We write $\mathscr{E} \models e$ instead of $\mathscr{E} \models_\Sigma e$ when the signature Σ is obvious. □

The use of the double turnstile ($\models$) here is the same as its use in logic: $\mathscr{E} \models e$ if the equation e is satisfied in every algebra which satisfies all the equations in $\mathscr{E}$. Here, $\mathscr{E}$ is a set of *assumptions* and e is a *conclusion* which *follows from* $\mathscr{E}$. We refer to this as *semantic* (or *model-theoretic*) consequence to distinguish it from a similar relation defined by means of "syntactic" inference rules in the next section.

Example 2.3.7. Recall Example 2.2.3. The exercise there shows the following:

$$\mathscr{E}_{\mathrm{BOOL}} \models_{\Sigma_{\mathrm{BOOL}}} \forall p{:}Bool\bullet \neg(p \wedge \neg false) = \neg p$$
$$\mathscr{E}_{\mathrm{BOOL}} \not\models_{\Sigma_{\mathrm{BOOL}}} \forall p{:}Bool\bullet \neg\neg p = p$$

Then, referring to Example 2.2.4, Exercise 2.2.5 shows that the de Morgan laws are semantical consequences of the set of axioms $\mathscr{E}_{\mathrm{BA}}$. □

Exercise 2.3.8. Prove that semantic consequence is preserved by translation along signature morphisms: for any signature morphism $\sigma{:}\Sigma \to \Sigma'$, set $\mathscr{E}$ of Σ-equations, and Σ-equation e,

$$\text{if } \mathscr{E} \models_{\Sigma} e \text{ then } \sigma(\mathscr{E}) \models_{\Sigma'} \sigma(e).$$

Equivalently, $\sigma(Cl_{\Sigma}(\mathscr{E})) \subseteq Cl_{\Sigma'}(\sigma(\mathscr{E}))$. Show that neither the reverse implication nor the reverse inclusion hold in general. □

Exercise 2.3.9. Let $\sigma{:}\Sigma \to \Sigma'$ be a signature morphism and let $\mathscr{E}'$ be a closed set of Σ'-equations. Show that $\sigma^{-1}(\mathscr{E}')$ is a closed set of Σ-equations. □

See Section 4.2 for some further results on semantic consequence and translation along signature morphisms, presented in a more general context.

Definition 2.3.10 (Theory). A *theory* is a presentation $\langle \Sigma, \mathscr{E} \rangle$ such that $\mathscr{E}$ is closed. A presentation $\langle \Sigma, \mathscr{E} \rangle$ (where $\mathscr{E}$ need not be closed) *presents* the theory $\langle \Sigma, Cl_{\Sigma}(\mathscr{E}) \rangle$. A theory $\langle \Sigma, \mathscr{E} \rangle$ is sometimes referred to as a Σ-*theory*. □

A theory morphism between two theories is a signature morphism between their signatures that maps the equations in the source theory to equations belonging to the target theory.

Definition 2.3.11 (Theory morphism). For any theories $\langle \Sigma, \mathscr{E} \rangle$ and $\langle \Sigma', \mathscr{E}' \rangle$, a *theory morphism* $\sigma{:}\langle \Sigma, \mathscr{E} \rangle \to \langle \Sigma', \mathscr{E}' \rangle$ is a signature morphism $\sigma{:}\Sigma \to \Sigma'$ such that $\sigma(e) \in \mathscr{E}'$ for every $e \in \mathscr{E}$; if, moreover, σ is a signature inclusion $\sigma{:}\Sigma \hookrightarrow \Sigma'$ then $\sigma{:}\langle \Sigma, \mathscr{E} \rangle \hookrightarrow \langle \Sigma', \mathscr{E}' \rangle$ is a *theory inclusion*. □

Exercise 2.3.12. Let $\sigma{:}\langle \Sigma, \mathscr{E} \rangle \to \langle \Sigma', \mathscr{E}' \rangle$ and $\sigma'{:}\langle \Sigma', \mathscr{E}' \rangle \to \langle \Sigma'', \mathscr{E}'' \rangle$ be theory morphisms. Show that $\sigma;\sigma'{:}\Sigma \to \Sigma''$ is a theory morphism $\sigma;\sigma'{:}\langle \Sigma, \mathscr{E} \rangle \to \langle \Sigma'', \mathscr{E}'' \rangle$. □

Proposition 2.3.13. *Let $\sigma{:}\Sigma \to \Sigma'$ be a signature morphism, $\mathscr{E}$ be a set of Σ-equations and $\mathscr{E}'$ be a set of Σ'-equations. Then the following conditions are equivalent:*

1. σ is a theory morphism $\sigma{:}\langle \Sigma, Cl_{\Sigma}(\mathscr{E}) \rangle \to \langle \Sigma', Cl_{\Sigma'}(\mathscr{E}') \rangle$.

2. $\sigma(\mathscr{E}) \subseteq Cl_{\Sigma'}(\mathscr{E}')$.
3. For every $A' \in Mod_{\Sigma'}(\mathscr{E}')$, $A'|_\sigma \in Mod_\Sigma(\mathscr{E})$.

Proof. **Exercise**. HINT: Use the Satisfaction Lemma, Lemma 2.1.8. □

The fact that 2.3.13(2) implies 2.3.13(1) gives a shortcut for checking if a signature morphism is a theory morphism: one need only check, for each axiom in some *presentation* of the source theory, that the translation of that axiom is in the target theory. The equivalence between 2.3.13(1) and 2.3.13(3) is similar in spirit to the Satisfaction Lemma, demonstrating a perfect correspondence between translation of syntax (axioms) along a signature morphism and translation of semantics (models) in the opposite direction. This equivalence shows that there is a model-level alternative to the axiom-level phrasing of Definition 2.3.11; in fact, we will take this alternative in the case of structured specifications (Chapter 5), where there is no equivalent axiom-level characterisation (Exercise 5.5.4).

Example 2.3.14. Let Σ be the signature

$$\begin{aligned} \Sigma = \textbf{sorts}\ & s, BBool \\ \textbf{ops}\ & tttt{:}BBool \\ & ffff{:}BBool \\ & not{:}BBool \to BBool \\ & and{:}BBool \times BBool \to BBool \\ & \_\_\le\_\_{:}s \times s \to BBool \end{aligned}$$

and recall the presentation $\text{BOOL} = \langle \Sigma_{\text{BOOL}}, \mathscr{E}_{\text{BOOL}} \rangle$ from Example 2.2.3. Define a signature morphism $\sigma{:}\Sigma \to \Sigma_{\text{BOOL}}$ by

$$\begin{aligned} &\sigma_{sorts} = \{s \mapsto Bool, BBool \mapsto Bool\}, \\ &\sigma_{\varepsilon,BBool} = \{tttt \mapsto true, ffff \mapsto false\}, \\ &\sigma_{BBool,BBool} = \{not \mapsto \neg\}, \\ &\sigma_{BBool\,BBool,BBool} = \{and \mapsto \wedge\}, \\ &\sigma_{s\,s,BBool} = \{\le \mapsto \Rightarrow\}. \end{aligned}$$

Let $\mathscr{E}$ be the set of Σ-equations

$$\mathscr{E} = \{\forall x{:}s \bullet x \le x = tttt, \forall p{:}BBool \bullet and(p, tttt) = p\}.$$

Then $Cl_\Sigma(\mathscr{E})$ includes Σ-equations, such as $\forall p{:}BBool, x{:}s \bullet and(p, x \le x) = p$, that were not in $\mathscr{E}$. Similarly, by Example 2.3.7, $Cl_{\Sigma_{\text{BOOL}}}(\mathscr{E}_{\text{BOOL}})$ includes the Σ_{BOOL}-equation $\forall p{:}Bool \bullet \neg(p \wedge \neg false) = \neg p$, but it does *not* include $\forall p{:}Bool \bullet \neg\neg p = p$. The presentations $\langle \Sigma, Cl_\Sigma(\mathscr{E}) \rangle$ and $\langle \Sigma_{\text{BOOL}}, Cl_{\Sigma_{\text{BOOL}}}(\mathscr{E}_{\text{BOOL}}) \rangle$ are theories — the latter is the theory presented by BOOL. The signature morphism $\sigma{:}\Sigma \to \Sigma_{\text{BOOL}}$ is a theory morphism $\sigma{:}\langle \Sigma, Cl_\Sigma(\mathscr{E}) \rangle \to \langle \Sigma_{\text{BOOL}}, Cl_{\Sigma_{\text{BOOL}}}(\mathscr{E}_{\text{BOOL}}) \rangle$.

Recalling Example 2.2.4, the theory presented by BA is $\langle \Sigma_{\text{BA}}, Cl_{\Sigma_{\text{BA}}}(\mathscr{E}_{\text{BA}}) \rangle$, the theory of Boolean algebras, with $Cl_{\Sigma_{\text{BA}}}(\mathscr{E}_{\text{BA}})$ including, for instance, the de Morgan laws (Exercise 2.2.5). The obvious signature morphism $\iota{:}\Sigma_{\text{BOOL}} \to \Sigma_{\text{BA}}$ is a theory morphism $\iota{:}\langle \Sigma_{\text{BOOL}}, Cl_{\Sigma_{\text{BOOL}}}(\mathscr{E}_{\text{BOOL}}) \rangle \to \langle \Sigma_{\text{BA}}, Cl_{\Sigma_{\text{BA}}}(\mathscr{E}_{\text{BA}}) \rangle$.

These two theory morphisms can be composed, yielding the theory morphism $\sigma;\iota{:}\langle \Sigma, Cl_\Sigma(\mathscr{E}) \rangle \to \langle \Sigma_{\text{BA}}, Cl_{\Sigma_{\text{BA}}}(\mathscr{E}_{\text{BA}}) \rangle$. □

Exercise 2.3.15. Give presentations $\langle\Sigma,\mathscr{E}\rangle$ and $\langle\Sigma',\mathscr{E}'\rangle$ and a theory morphism $\sigma:\langle\Sigma, Cl_\Sigma(\mathscr{E})\rangle \to \langle\Sigma', Cl_{\Sigma'}(\mathscr{E}')\rangle$ such that $\sigma(\mathscr{E}) \not\subseteq \mathscr{E}'$. Note that this does *not* contradict the equivalence between 2.3.13(1) and 2.3.13(2). □

2.4 Equational calculus

As we have seen, each presentation $\langle\Sigma,\mathscr{E}\rangle$ determines a theory $\langle\Sigma, Cl_\Sigma(\mathscr{E})\rangle$, where $Cl_\Sigma(\mathscr{E})$ contains $\mathscr{E}$ together with all of its semantic consequences. An obvious question at this point is how to determine whether or not a given Σ-equation $\forall X \bullet t = t'$ belongs to the set $Cl_\Sigma(\mathscr{E})$, i.e. how to decide if $\mathscr{E} \models_\Sigma \forall X \bullet t = t'$. The definition of $Cl_\Sigma(\mathscr{E})$ does not provide an effective method: according to this, testing $\mathscr{E} \models_\Sigma \forall X \bullet t = t'$ involves constructing the (infinite!) class $Mod_\Sigma(\mathscr{E})$ and checking whether or not $\forall X \bullet t = t'$ is satisfied by each of the algebras in this class, that is, checking for each algebra $A \in Mod_\Sigma(\mathscr{E})$ and function $v: X \to |A|$ (there may be infinitely many such functions for a given A) that $t_A(v) = t'_A(v)$. An alternative is to proceed "syntactically" by means of *inference rules* which allow the elements of $Cl_\Sigma(\mathscr{E})$ to be *derived* from the axioms in $\mathscr{E}$ via a sequence of formal proof steps.

Throughout this section, let Σ be a signature.

Definition 2.4.1 (Equational calculus). A Σ-equation e is a *proof-theoretic consequence* of (or is *provable* from) a set $\mathscr{E}$ of Σ-equations, written $\mathscr{E} \vdash_\Sigma e$, if this can be derived by application of the following inference rules:

(axiom) $$\frac{}{\mathscr{E} \vdash_\Sigma \forall X \bullet t = t'} \quad \forall X \bullet t = t' \in \mathscr{E}$$

(reflexivity) $$\frac{}{\mathscr{E} \vdash_\Sigma \forall X \bullet t = t} \quad X_s \subseteq \mathscr{X} \text{ for all } s \in S, \text{ and } t \in |T_\Sigma(X)|$$

(symmetry) $$\frac{\mathscr{E} \vdash_\Sigma \forall X \bullet t = t'}{\mathscr{E} \vdash_\Sigma \forall X \bullet t' = t}$$

(transitivity) $$\frac{\mathscr{E} \vdash_\Sigma \forall X \bullet t = t' \qquad \mathscr{E} \vdash_\Sigma \forall X \bullet t' = t''}{\mathscr{E} \vdash_\Sigma \forall X \bullet t = t''}$$

(congruence) $$\frac{\mathscr{E} \vdash_\Sigma \forall X \bullet t_1 = t'_1 \quad \cdots \quad \mathscr{E} \vdash_\Sigma \forall X \bullet t_n = t'_n}{\mathscr{E} \vdash_\Sigma \forall X \bullet f(t_1,\ldots,t_n) = f(t'_1,\ldots,t'_n)} \quad \begin{array}{l} f: s_1 \times \cdots \times s_n \to s \text{ in } \Sigma, \\ t_i, t'_i \in |T_\Sigma(X)|_{s_i} \text{ for } i \leq n \end{array}$$

(instantiation) $$\frac{\mathscr{E} \vdash_\Sigma \forall X \bullet t = t'}{\mathscr{E} \vdash_\Sigma \forall Y \bullet t[\theta] = t'[\theta]} \quad \theta: X \to |T_\Sigma(Y)|$$

□

Exercise 2.4.2 (Admissibility of weakening and cut). Prove that if $\mathscr{E} \vdash_\Sigma \forall X \bullet t = t'$ and $\mathscr{E} \subseteq \mathscr{E}'$ then $\mathscr{E}' \vdash_\Sigma \forall X \bullet t = t'$. (HINT: Simple induction on the structure of the derivation of $\mathcal{E} \vdash_\Sigma \forall X \bullet t = t'$.) This shows that the following rule is admissible[2]:

$$\text{(weakening)} \quad \frac{\mathcal{E} \vdash_\Sigma \forall X \bullet t = t'}{\mathcal{E} \cup \mathcal{E}' \vdash_\Sigma \forall X \bullet t = t'}$$

Prove that if $\mathcal{E} \vdash_\Sigma e$ and $\{e\} \cup \mathcal{E}' \vdash_\Sigma e'$ then $\mathcal{E} \cup \mathcal{E}' \vdash_\Sigma e'$. (HINT: Use induction on the structure of the derivation of $\{e\} \cup \mathcal{E}' \vdash_\Sigma e'$; for the case of the axiom rule, use the fact that weakening is admissible.) This shows that the following rule is admissible:

$$\text{(cut)} \quad \frac{\mathcal{E} \vdash_\Sigma e \qquad \{e\} \cup \mathcal{E}' \vdash_\Sigma e'}{\mathcal{E} \cup \mathcal{E}' \vdash_\Sigma e'}$$

Check that your proof can be generalised to show that if $\mathcal{E} \vdash e'$ and $\mathcal{E}'_e \vdash e$ for each $e \in \mathcal{E}$ then $\bigcup_{e \in \mathcal{E}} \mathcal{E}'_e \vdash e'$. □

Exercise 2.4.3 (Consequence is preserved by translation). Show that for any signature morphism $\sigma : \Sigma \to \Sigma'$, set $\mathcal{E}$ of Σ-equations, and Σ-equation e, if $\mathcal{E} \vdash_\Sigma e$ then $\sigma(\mathcal{E}) \vdash_{\Sigma'} \sigma(e)$. □

Example 2.4.4. Recall the presentation $\mathrm{BOOL} = \langle \Sigma_{\mathrm{BOOL}}, \mathcal{E}_{\mathrm{BOOL}} \rangle$ given in Example 2.2.3. The following is a derivation of $\mathcal{E}_{\mathrm{BOOL}} \vdash_{\Sigma_{\mathrm{BOOL}}} \forall p{:}Bool \bullet \neg(p \wedge \neg false) = \neg p$:

$$\frac{\begin{array}{c} \nabla P \\ \mathcal{E}_{\mathrm{BOOL}} \vdash_{\Sigma_{\mathrm{BOOL}}} \forall p{:}Bool \bullet \\ \neg(p \wedge \neg false) = \neg(p \wedge true) \end{array} \qquad \dfrac{\overline{\mathcal{E}_{\mathrm{BOOL}} \vdash_{\Sigma_{\mathrm{BOOL}}} \forall p{:}Bool \bullet p \wedge true = p}}{\mathcal{E}_{\mathrm{BOOL}} \vdash_{\Sigma_{\mathrm{BOOL}}} \forall p{:}Bool \bullet \neg(p \wedge true) = \neg p}}{\mathcal{E}_{\mathrm{BOOL}} \vdash_{\Sigma_{\mathrm{BOOL}}} \forall p{:}Bool \bullet \neg(p \wedge \neg false) = \neg p}$$

where P is the derivation

$$\frac{\dfrac{\overline{\mathcal{E}_{\mathrm{BOOL}} \vdash_{\Sigma_{\mathrm{BOOL}}} \forall p{:}Bool \bullet p = p} \qquad \dfrac{\overline{\mathcal{E}_{\mathrm{BOOL}} \vdash_{\Sigma_{\mathrm{BOOL}}} \neg false = true}}{\mathcal{E}_{\mathrm{BOOL}} \vdash_{\Sigma_{\mathrm{BOOL}}} \forall p{:}Bool \bullet \neg false = true}}{\mathcal{E}_{\mathrm{BOOL}} \vdash_{\Sigma_{\mathrm{BOOL}}} \forall p{:}Bool \bullet p \wedge \neg false = p \wedge true}}{\mathcal{E}_{\mathrm{BOOL}} \vdash_{\Sigma_{\mathrm{BOOL}}} \forall p{:}Bool \bullet \neg(p \wedge \neg false) = \neg(p \wedge true)}$$

Exercise. Tag each step above with the inference rule being applied. □

Exercise 2.4.5. Give a derivation of $\mathcal{E}_{\mathrm{BOOL}} \vdash_{\Sigma_{\mathrm{BOOL}}} \forall p{:}Bool \bullet p \Rightarrow p = true$.

---

[2] A rule is *admissible* in a formal system of rules if its conclusion is derivable in the system provided that all its premises are derivable. This holds in particular if the rule is *derivable* in the system, that is, if it can be obtained by composition of the rules in the system.

A considerably more serious challenge is to give derivations for the de Morgan laws from the axioms of Boolean algebra (see Example 2.2.4 and Exercise 2.2.5). □

On its own, the equational calculus is nothing more than a game with symbols; its importance lies in the correspondence between the two relations $\models_\Sigma$ and $\vdash_\Sigma$. As we shall see, this correspondence is exact: $\vdash_\Sigma$ is both *sound* and *complete* for $\models_\Sigma$. Soundness ($\mathscr{E} \vdash_\Sigma e$ implies $\mathscr{E} \models_\Sigma e$) is a vital property for any formal system: it ensures that the inference rules cannot be used to derive an incorrect result.

Theorem 2.4.6 (Soundness of equational calculus). *Let $\mathscr{E}$ be a set of Σ-equations and let e be a Σ-equation. If $\mathscr{E} \vdash_\Sigma e$ then $\mathscr{E} \models_\Sigma e$.* □

Exercise 2.4.7. Prove Theorem 2.4.6. Use induction on the depth of the derivation of $\mathscr{E} \vdash_\Sigma e$, showing that each rule in the system preserves the indicated property. □

Example 2.4.8. By Theorem 2.4.6, the formal derivation in Example 2.4.4 justifies the claim in Example 2.3.7 that $\mathscr{E}_{\mathrm{BOOL}} \models_{\Sigma_{\mathrm{BOOL}}} \forall p{:}Bool\bullet \neg(p \wedge \neg \mathit{false}) = \neg p$. On the other hand, since $\mathscr{E}_{\mathrm{BOOL}} \not\models_{\Sigma_{\mathrm{BOOL}}} \forall p{:}Bool\bullet \neg\neg p = p$, there can be no proof in the equational calculus for $\mathscr{E}_{\mathrm{BOOL}} \vdash_{\Sigma_{\mathrm{BOOL}}} \forall p{:}Bool\bullet \neg\neg p = p$. □

It is a somewhat counterintuitive fact (see [GM85]) that simplifying the calculus by omitting explicit quantifiers in equations yields an unsound system. This is due to the fact that algebras may have empty carrier sets. Any equation that includes a quantified variable $x{:}s$ will be satisfied by any algebra having an empty carrier for s, even if x appears on neither side of the equation. The instantiation rule is the only one that can be used to change the set of quantified variables; it is designed to ensure that quantified variables are eliminated only when it is sound to do so.

Exercise 2.4.9. Formulate a version of the equational calculus without explicit quantifiers on equations and show that it is unsound. (HINT: Consider the signature Σ with sorts s, s' and operations $f{:}\, s \to s'$, $a{:}\, s'$, $b{:}\, s'$, and set $\mathscr{E} = \{f(x) = a, f(x) = b\}$ of Σ-equations. Show that $\mathscr{E} \vdash_\Sigma a = b$ in your version of the calculus. Then give a Σ-algebra $A \in Mod_\Sigma(\mathscr{E})$ such that $A \not\models_\Sigma a = b$.) Pinpoint where this proof of unsoundness breaks down for the version of the equational calculus given in Definition 2.4.1. □

Exercise 2.4.10. Show that the equational calculus without explicit quantifiers is sound when the definition of Σ-algebra is changed to require all carrier sets to be non-empty, or when either of the following constraints on Σ is imposed:

1. Σ has only one sort.
2. All sorts in Σ are *non-void*: for each sort name s in Σ, $|T_\Sigma|_s \neq \varnothing$. □

Exercise 2.4.11. Give an example of a signature Σ which satisfies neither 2.4.10(1) nor 2.4.10(2) for which the equational calculus without explicit quantifiers is sound. □

Completeness ($\mathscr{E} \models_\Sigma e$ implies $\mathscr{E} \vdash_\Sigma e$) is typically more difficult to achieve than soundness: it means that the rules in the system are powerful enough to derive all correct results. It is not as important as soundness, in the sense that a complete but unsound system is useless while (as we shall see in the sequel) a sound but incomplete system is often the best that can be obtained. The equational calculus happens to be complete for $\models_\Sigma$:

Theorem 2.4.12 (Completeness of equational calculus). *Let $\mathscr{E}$ be a set of Σ-equations and let e be a Σ-equation. If $\mathscr{E} \models_\Sigma e$ then $\mathscr{E} \vdash_\Sigma e$.*

Proof sketch. Suppose $\mathscr{E} \models_\Sigma \forall X \bullet t = t'$. Define $\equiv \subseteq |T_\Sigma(X)| \times |T_\Sigma(X)|$ by $u \equiv u' \iff \mathscr{E} \vdash_\Sigma \forall X \bullet u = u'$; then $\equiv$ is a Σ-congruence on $T_\Sigma(X)$. $T_\Sigma(X)/{\equiv} \models_\Sigma \mathscr{E}$, so $T_\Sigma(X)/{\equiv} \models_\Sigma \forall X \bullet t = t'$, and thus $t \equiv t'$, i.e. $\mathscr{E} \vdash_\Sigma \forall X \bullet t = t'$. □

Exercise 2.4.13. Fill in the gaps in the proof of Theorem 2.4.12. □

There are several different but equivalent versions of the equational calculus. The following exercise considers various alternatives to the congruence and instantiation rules.

Exercise 2.4.14. Show that the version of the equational calculus in Definition 2.4.1 is equivalent to the system obtained when the congruence and instantiation rules are replaced by the following single rule:

$$\text{(substitutivity)} \quad \frac{\mathscr{E} \vdash_\Sigma \forall X \bullet t = t' \qquad \text{for each } x \in X,\ \mathscr{E} \vdash_\Sigma \forall Y \bullet \theta(x) = \theta'(x)}{\mathscr{E} \vdash_\Sigma \forall Y \bullet t[\theta] = t'[\theta']} \quad \theta, \theta' : X \to |T_\Sigma(Y)|$$

Show that this is equivalent to the system having the following more restricted version of the substitutivity rule:

$$\text{(substitutivity}') \quad \frac{\mathscr{E} \vdash_\Sigma \forall X \cup \{x{:}s\} \bullet t = t' \qquad \mathscr{E} \vdash_\Sigma \forall Y \bullet u = u'}{\mathscr{E} \vdash_\Sigma \forall X \cup Y \bullet t[x \mapsto u] = t'[x \mapsto u']} \quad u, u' \in |T_\Sigma(Y)|_s$$

(HINT: The equivalence relies on the fact that the set of quantified variables in an equation is finite.) Finally, show that both of the following rules may be derived in any of these systems:

$$\text{(abstraction)} \quad \frac{\mathscr{E} \vdash_\Sigma \forall X \bullet t = t'}{\mathscr{E} \vdash_\Sigma \forall X \cup Y \bullet t = t'} \quad Y_s \subseteq \mathscr{X} \text{ for all } s \in S$$

$$\text{(concretion)} \quad \frac{\mathscr{E} \vdash_\Sigma \forall X \cup \{x{:}s\} \bullet t = t'}{\mathscr{E} \vdash_\Sigma \forall X \bullet t = t'} \quad t, t' \in |T_\Sigma(X)| \text{ and } |T_\Sigma(X)|_s \neq \varnothing$$

□

A consequence of the soundness and completeness theorems is that the equational calculus constitutes a *semi-decision procedure* for $\models_\Sigma$: enumerating all derivations will eventually produce a derivation for $\mathscr{E} \vdash_\Sigma e$ if $\mathscr{E} \models_\Sigma e$ holds, but if $\mathscr{E} \not\models_\Sigma e$ then this procedure will never terminate. This turns out to be the best we can achieve:

Theorem 2.4.15. *There is no decision procedure for* $\models_\Sigma$.

Proof. Follows immediately from the undecidability of the word problem for semigroups [Pos47]. □

Mechanised proof search techniques can be applied with considerable success to the discovery of derivations (and under certain conditions, discussed in Section 2.6, a decision procedure *is* possible), but Theorem 2.4.15 shows that such techniques can provide no more than a partial solution.

2.5 Initial models

The class of algebras given by the loose semantics of a Σ-presentation contains too many algebras to be very useful in practice. In particular, Birkhoff's Variety Theorem guarantees that this class will always include degenerate Σ-algebras having a single value of each sort in Σ, as well as (nearly always) Σ-algebras that are not reachable. This unsatisfactory state of affairs is a consequence of the limited power of equational axioms. A standard way out is to take the so-called *initial semantics* of presentations, which selects a certain class of "best" models from among all those satisfying the axioms. Various alternatives to this approach will be presented in the sequel.

Throughout this section, let $\langle \Sigma, \mathscr{E} \rangle$ be a presentation.

Exercise 2.5.1. Verify the above claim concerning Birkhoff's Variety Theorem, being specific about the meaning of "nearly always". □

There are two features that render certain models of presentations unfit for use in practice. The mnemonic terms "junk" and "confusion" were coined in [BG81] to characterise these:

Definition 2.5.2 (Junk and confusion). Let A be a model of $\langle \Sigma, \mathscr{E} \rangle$. We say that *A contains junk* if it is not reachable, and that *A contains confusion* if it satisfies a ground Σ-equation that is not in $Cl_\Sigma(\mathscr{E})$. □

The intuition behind these terms should be readily apparent: "junk" refers to useless values which could be discarded without being missed, and "confusion" refers to the values of two ground terms being unnecessarily identified (confused).

Example 2.5.3. Recall the presentation $\text{BOOL} = \langle \Sigma_{\text{BOOL}}, \mathscr{E}_{\text{BOOL}} \rangle$ and its models $A1$, $A2$ and $A3$ from Example 2.2.3. $A1$ contains confusion ($A1 \models_{\Sigma_{\text{BOOL}}} \mathit{true} = \mathit{false} \notin Cl_{\Sigma_{\text{BOOL}}}(\mathscr{E}_{\text{BOOL}})$) but not junk; $A2$ contains junk (there is no ground Σ_{BOOL}-term t such that $t_{A2} = \spadesuit \in |A2|_{\mathit{Bool}}$) but not confusion; $A3$ contains neither junk nor confusion. There are models of BOOL containing both junk and confusion. (**Exercise:** Find one.) □

Exercise 2.5.4. Consider the following specification of the natural numbers with addition:

$$
\begin{array}{lll}
\textbf{spec}\ \mathrm{NAT} = & \textbf{sorts} & Nat \\
& \textbf{ops} & 0{:}Nat \\
& & succ{:}Nat \to Nat \\
& & \_\_+\_\_{:}Nat \times Nat \to Nat \\
& \forall m,n{:}Nat & \\
& & \bullet\ 0+n=n \\
& & \bullet\ succ(m)+n = succ(m+n)
\end{array}
$$

List some of the models of NAT. Which of these contain junk and/or confusion? (NOTE: For reference later in this section, Σ_{NAT} refers to the signature of NAT and $\mathscr{E}_{\mathrm{NAT}}$ refers to its axioms.) □

Exercise 2.5.5. According to Exercise 1.3.5, surjective homomorphisms reflect junk. Show that injective homomorphisms preserve junk and reflect confusion, and that all homomorphisms preserve confusion. It follows that isomorphisms preserve and reflect junk and confusion. □

Examples like the ones above suggest that often the algebras of interest are those which contain neither junk nor confusion. Recall Exercise 1.4.14, which characterised reachable Σ-algebras as those which are isomorphic to a quotient of T_Σ. Accordingly, the algebras we want are all isomorphic to quotients of T_Σ; by Exercise 2.5.5 it is enough to consider just these quotient algebras themselves. Of course, not all quotients $T_\Sigma/\equiv$ will be models of $\langle \Sigma, \mathscr{E} \rangle$: this will only be the case when $\equiv$ identifies enough terms that the equations in $\mathscr{E}$ are satisfied. But if $\equiv$ identifies "too many" terms, $T_\Sigma/\equiv$ will contain confusion. There is exactly one Σ-congruence that yields a model of $\langle \Sigma, \mathscr{E} \rangle$ containing no confusion:

Definition 2.5.6 (Congruence generated by a set of equations). The relation $\equiv_{\mathscr{E}} \subseteq |T_\Sigma| \times |T_\Sigma|$ is defined by $t \equiv_{\mathscr{E}} t' \iff \mathscr{E} \models_\Sigma t = t'$, for all $t,t' \in |T_\Sigma|$. $\equiv_{\mathscr{E}}$ is called the *Σ-congruence generated by $\mathscr{E}$*. □

Exercise 2.5.7. Prove that $\equiv_{\mathscr{E}}$ is a Σ-congruence on T_Σ. □

Theorem 2.5.8 (Quotient construction). *$T_\Sigma/\equiv_{\mathscr{E}}$ is a model of $\langle \Sigma, \mathscr{E} \rangle$ containing no junk and no confusion.* □

Exercise 2.5.9. Prove Theorem 2.5.8. HINT: Note that $T_\Sigma/\equiv_{\mathscr{E}}$ contains no junk by Exercise 1.4.14. Then show that for any term $t \in T_\Sigma(X)$ and substitution $\theta{:}X \to T_\Sigma$, $t_{T_\Sigma/\equiv_{\mathscr{E}}}(\theta') = [t[\theta]]_{\equiv_{\mathscr{E}}}$, where $\theta'(x) = [\theta(x)]_{\equiv_{\mathscr{E}}}$ for $x \in X$. Use this to show that $T_\Sigma/\equiv_{\mathscr{E}}$ satisfies all the equations in $\mathscr{E}$ and contains no confusion. □

Example 2.5.10. Recall again the presentation $\mathrm{BOOL} = \langle \Sigma_{\mathrm{BOOL}}, \mathscr{E}_{\mathrm{BOOL}} \rangle$ from Example 2.2.3. The model $T_{\Sigma_{\mathrm{BOOL}}}/\equiv_{\mathscr{E}_{\mathrm{BOOL}}}$ of BOOL is defined as follows:

$$
\begin{aligned}
|T_{\Sigma_{\mathrm{BOOL}}}/{\equiv_{\mathscr{E}_{\mathrm{BOOL}}}}|_{Bool} &= \{[true]_{\equiv_{\mathscr{E}_{\mathrm{BOOL}}}}, [false]_{\equiv_{\mathscr{E}_{\mathrm{BOOL}}}}\},\\
true_{T_{\Sigma_{\mathrm{BOOL}}}/{\equiv_{\mathscr{E}_{\mathrm{BOOL}}}}} &= [true]_{\equiv_{\mathscr{E}_{\mathrm{BOOL}}}},\\
false_{T_{\Sigma_{\mathrm{BOOL}}}/{\equiv_{\mathscr{E}_{\mathrm{BOOL}}}}} &= [false]_{\equiv_{\mathscr{E}_{\mathrm{BOOL}}}},\\
\neg_{T_{\Sigma_{\mathrm{BOOL}}}/{\equiv_{\mathscr{E}_{\mathrm{BOOL}}}}} &= \{[true]_{\equiv_{\mathscr{E}_{\mathrm{BOOL}}}} \mapsto [false]_{\equiv_{\mathscr{E}_{\mathrm{BOOL}}}}, [false]_{\equiv_{\mathscr{E}_{\mathrm{BOOL}}}} \mapsto [true]_{\equiv_{\mathscr{E}_{\mathrm{BOOL}}}}\},
\end{aligned}
$$

$\wedge_{T_{\Sigma_{\mathrm{BOOL}}}/{\equiv_{\mathscr{E}_{\mathrm{BOOL}}}}}$	$[true]_{\equiv_{\mathscr{E}_{\mathrm{BOOL}}}}$	$[false]_{\equiv_{\mathscr{E}_{\mathrm{BOOL}}}}$
$[true]_{\equiv_{\mathscr{E}_{\mathrm{BOOL}}}}$	$[true]_{\equiv_{\mathscr{E}_{\mathrm{BOOL}}}}$	$[false]_{\equiv_{\mathscr{E}_{\mathrm{BOOL}}}}$
$[false]_{\equiv_{\mathscr{E}_{\mathrm{BOOL}}}}$	$[false]_{\equiv_{\mathscr{E}_{\mathrm{BOOL}}}}$	$[false]_{\equiv_{\mathscr{E}_{\mathrm{BOOL}}}}$

$\Rightarrow_{T_{\Sigma_{\mathrm{BOOL}}}/{\equiv_{\mathscr{E}_{\mathrm{BOOL}}}}}$	$[true]_{\equiv_{\mathscr{E}_{\mathrm{BOOL}}}}$	$[false]_{\equiv_{\mathscr{E}_{\mathrm{BOOL}}}}$
$[true]_{\equiv_{\mathscr{E}_{\mathrm{BOOL}}}}$	$[true]_{\equiv_{\mathscr{E}_{\mathrm{BOOL}}}}$	$[false]_{\equiv_{\mathscr{E}_{\mathrm{BOOL}}}}$
$[false]_{\equiv_{\mathscr{E}_{\mathrm{BOOL}}}}$	$[true]_{\equiv_{\mathscr{E}_{\mathrm{BOOL}}}}$	$[true]_{\equiv_{\mathscr{E}_{\mathrm{BOOL}}}}$

where

$$
\begin{aligned}
[true]_{\equiv_{\mathscr{E}_{\mathrm{BOOL}}}} &= \{true, \neg false, true \wedge true, \neg(false \wedge true),\\
&\qquad \neg(false \wedge \neg false), false \Rightarrow false, \ldots\},\\
[false]_{\equiv_{\mathscr{E}_{\mathrm{BOOL}}}} &= \{false, \neg true, true \wedge false, \neg(true \wedge true),\\
&\qquad \neg(true \wedge \neg false), true \Rightarrow false, \ldots\}.
\end{aligned}
$$

The carrier set $|T_{\Sigma_{\mathrm{BOOL}}}/{\equiv_{\mathscr{E}_{\mathrm{BOOL}}}}|_{Bool}$ has just two elements since the axioms in $\mathscr{E}_{\mathrm{BOOL}}$ can be used to reduce each ground Σ_{BOOL}-term to *true* or *false*, and $true \not\equiv_{\mathscr{E}_{\mathrm{BOOL}}} false$. Note that the "syntactic" nature of $T_{\Sigma_{\mathrm{BOOL}}}$ is preserved in $T_{\Sigma_{\mathrm{BOOL}}}/{\equiv_{\mathscr{E}_{\mathrm{BOOL}}}}$, e.g. for each $x \in [true]_{\equiv_{\mathscr{E}_{\mathrm{BOOL}}}}$, "$\neg x$" $\in [false]_{\equiv_{\mathscr{E}_{\mathrm{BOOL}}}} = \neg_{T_{\Sigma_{\mathrm{BOOL}}}/{\equiv_{\mathscr{E}_{\mathrm{BOOL}}}}}([true]_{\equiv_{\mathscr{E}_{\mathrm{BOOL}}}})$. □

Exercise 2.5.11. Recall the presentation $\mathrm{NAT} = \langle \Sigma_{\mathrm{NAT}}, \mathscr{E}_{\mathrm{NAT}} \rangle$ from Exercise 2.5.4. Construct the model $T_{\Sigma_{\mathrm{NAT}}}/{\equiv_{\mathscr{E}_{\mathrm{NAT}}}}$ of NAT. □

Exercise 2.5.12. Show that $\equiv_{\mathscr{E}}$ is the only Σ-congruence making Theorem 2.5.8 hold. □

The special properties of $T_\Sigma/{\equiv_{\mathscr{E}}}$ described by Theorem 2.5.8 can be captured very succinctly by saying that $T_\Sigma/{\equiv_{\mathscr{E}}}$ is a so-called *initial model* of $\langle \Sigma, \mathscr{E} \rangle$.

Definition 2.5.13 (Initial model of a presentation). A Σ-algebra A is *initial in* a class $\mathscr{A}$ of Σ-algebras if $A \in \mathscr{A}$ and for every $B \in \mathscr{A}$ there is a unique Σ-homomorphism $h{:}A \to B$. An *initial model of* $\langle \Sigma, \mathscr{E} \rangle$ is a Σ-algebra that is initial in $Mod[\langle \Sigma, \mathscr{E} \rangle]$. $IMod[\langle \Sigma, \mathscr{E} \rangle]$ is the class of all initial models of $\langle \Sigma, \mathscr{E} \rangle$. □

In the next chapter we will see that this definition can be generalised to a much wider context than that of algebras and homomorphisms.

Theorem 2.5.14 (Initial model theorem). *$T_\Sigma/{\equiv_{\mathscr{E}}}$ is an initial model of $\langle \Sigma, \mathscr{E} \rangle$.*

Proof sketch. $T_\Sigma/{\equiv_{\mathscr{E}}}$ is a model of $\langle\Sigma,\mathscr{E}\rangle$ by Theorem 2.5.8. For $B \in Mod[\langle\Sigma,\mathscr{E}\rangle]$, let $\varnothing^\sharp\colon T_\Sigma \to B$ be the unique homomorphism from the algebra of ground Σ-terms to B. Since $B \models_\Sigma \mathscr{E}$, we have $\equiv_{\mathscr{E}} \subseteq ker(\varnothing^\sharp)$, and by Exercise 1.3.20 there is a homomorphism $h\colon T_\Sigma/{\equiv_{\mathscr{E}}} \to B$ which is unique by Exercise 1.3.6. (**Exercise:** Fill in the gaps in this proof.) □

Example 2.5.15. Recall the presentation $\mathrm{BOOL} = \langle\Sigma_{\mathrm{BOOL}},\mathscr{E}_{\mathrm{BOOL}}\rangle$ and its models $A1$, $A2$ and $A3$ from Example 2.2.3, and its model $T_{\Sigma_{\mathrm{BOOL}}}/{\equiv_{\mathscr{E}_{\mathrm{BOOL}}}}$ from Example 2.5.10, which is an initial model by Theorem 2.5.14. Σ_{BOOL}-homomorphisms from $T_{\Sigma_{\mathrm{BOOL}}}/{\equiv_{\mathscr{E}_{\mathrm{BOOL}}}}$ to $A1$, $A2$ and $A3$ are as follows:

$$\begin{array}{ll}
h1\colon T_{\Sigma_{\mathrm{BOOL}}}/{\equiv_{\mathscr{E}_{\mathrm{BOOL}}}} \to A1 & h1_{Bool} = \{[true]_{\equiv_{\mathscr{E}_{\mathrm{BOOL}}}} \mapsto \star, [false]_{\equiv_{\mathscr{E}_{\mathrm{BOOL}}}} \mapsto \star\},\\
h2\colon T_{\Sigma_{\mathrm{BOOL}}}/{\equiv_{\mathscr{E}_{\mathrm{BOOL}}}} \to A2 & h2_{Bool} = \{[true]_{\equiv_{\mathscr{E}_{\mathrm{BOOL}}}} \mapsto \clubsuit, [false]_{\equiv_{\mathscr{E}_{\mathrm{BOOL}}}} \mapsto \heartsuit\},\\
h3\colon T_{\Sigma_{\mathrm{BOOL}}}/{\equiv_{\mathscr{E}_{\mathrm{BOOL}}}} \to A3 & h3_{Bool} = \{[true]_{\equiv_{\mathscr{E}_{\mathrm{BOOL}}}} \mapsto 1, [false]_{\equiv_{\mathscr{E}_{\mathrm{BOOL}}}} \mapsto 0\}.
\end{array}$$

(**Exercise:** Check uniqueness.)

$A1$ is not an initial model: for example, there is no homomorphism from $A1$ to $A2$, nor from $A1$ to $A3$. In general, models containing confusion cannot be initial since homomorphisms preserve confusion (Exercise 2.5.5). Similarly, $A2$ is not an initial model: for example, there is no homomorphism from $A2$ to $A3$, since there is no value in $|A3|_{Bool}$ to which such a homomorphism could map the "extra" value $\spadesuit \in |A2|_{Bool}$. On the other hand, $A3$ is initial: for example, there is a unique homomorphism $g1\colon A3 \to A1$ (where $g1_{Bool}(1) = g1_{Bool}(0) = \star$), there is a unique homomorphism $g2\colon A3 \to A2$ (where $g2_{Bool}(1) = \clubsuit$ and $g2_{Bool}(0) = \heartsuit$), and there is a unique homomorphism $g\colon A3 \to T_{\Sigma_{\mathrm{BOOL}}}/{\equiv_{\mathscr{E}_{\mathrm{BOOL}}}}$ (where $g_{Bool}(1) = [true]_{\equiv_{\mathscr{E}_{\mathrm{BOOL}}}}$ and $g_{Bool}(0) = [false]_{\equiv_{\mathscr{E}_{\mathrm{BOOL}}}}$). □

Exercise 2.5.16. Recall the model you constructed in Exercise 2.5.11 of the specification NAT of natural numbers with addition. Show that there is a unique homomorphism from this model to each of the models you considered in Exercise 2.5.4. □

Exercise 2.5.17. Using Theorem 2.5.14, show that T_Σ is an initial model of $\langle\Sigma,\varnothing\rangle$. Contemplate how this relates to Fact 1.4.4 and Definition 1.4.5. □

Exercise 2.5.18. Note that initial models of $\langle\Sigma,\mathscr{E}\rangle$ may have empty carriers for some sorts. Show that this is necessary: give an example of a presentation $\langle\Sigma,\mathscr{E}\rangle$ such that no algebra is initial in the class of its models that have non-empty carriers of all sorts. Link this with Exercise 1.2.3. □

Taking a presentation $\langle\Sigma,\mathscr{E}\rangle$ to denote the class $IMod[\langle\Sigma,\mathscr{E}\rangle]$ of its initial models is called taking its *initial semantics*. We know from Theorem 2.5.14 that $IMod[\langle\Sigma,\mathscr{E}\rangle]$ is never empty. Although the motivation for wishing to exclude models containing junk and confusion was merely to weed out certain kinds of degenerate cases, the effect of this constraint is to restrict attention to an isomorphism class of models:

Exercise 2.5.19. Show that any two initial models of a presentation are isomorphic. Conclude that the initial models of a presentation are exactly those containing no junk and no confusion. □

For some purposes, restricting attention to an isomorphism class of models is clearly inappropriate. The following exercise demonstrates what can go wrong.

Exercise 2.5.20. Consider the addition of a subtraction operation $\_\_-\_\_: Nat \times Nat \rightarrow Nat$ to the specification NAT in Exercise 2.5.4, with the axioms $\forall m{:}Nat \bullet m - 0 = m$ and $\forall m,n{:}Nat \bullet succ(m) - succ(n) = m - n$. These axioms do not fix the value of $m - n$ when $n > m$; assume that we are willing to accept any value in this case, perhaps because we are certain for some reason that it will never arise. Construct an initial model of this specification. Why is this model unsatisfactory? Can you think of a better model? What is the problem with restricting to an isomorphism class of models of this specification? □

The phenomenon illustrated here arises in cases where operations are not defined in a *sufficiently complete* way. Roughly speaking, a definition of an operation is sufficiently complete when the value produced by the operation is defined for all of the possible values of its arguments. See Definition 6.1.22 below for a proper formulation of this property in a more general context.

One may argue that Exercise 2.5.20 is unconvincing, since the lack of sufficient completeness arises there because we do not really need $m - n$ to be defined as a natural number when $n > m$, and that this can be dealt with using one of the approaches to partial functions below (Sections 2.7.3, 2.7.4, or 2.7.5). However, the same phenomenon arises in other cases as well:

Exercise 2.5.21. Give a specification of natural numbers with a function that for each natural number n chooses an arbitrary number that is greater than n. HINT: You may first extend the specification NAT of Exercise 2.5.4 with a sort *Bool* with operations and axioms as in BOOL in Example 2.2.3, and add a binary operation $\_\_<\_\_: Nat \times Nat \rightarrow Bool$ with the following axioms:

$$\forall n{:}Nat \bullet 0 < succ(n) = true$$
$$\forall m{:}Nat \bullet succ(m) < 0 = false$$
$$\forall m,n{:}Nat \bullet succ(m) < succ(n) = m < n$$

The required function $ch: Nat \rightarrow Nat$ may now be constrained by the obvious axiom $\forall n{:}Nat \bullet n < ch(n) = true$.

Clearly, the definition of *ch* cannot be sufficiently complete. Construct the initial model of the resulting specification and check that it is not satisfactory. Referring to other algebraic approaches presented in Sections 2.7.3, 2.7.4, and 2.7.5 below, check that none of them offers a satisfactory solution either. □

The above exercise indicates one of the most compelling reasons for considering alternatives to initial semantics: requiring specifications to define all operations in a sufficiently complete way is much too restrictive in many practical cases. Such a requirement is also undesirable for methodological reasons, since it forces the specifier of a problem to make decisions which are more appropriately left to the implementor.

The comments above notwithstanding, there are certain common situations in which initial semantics is appropriate and useful. In particular, the implicit "no junk" constraint conveniently captures the "that's all there is" condition which is needed in inductive definitions of syntax.

Example 2.5.22. Consider the following specification of syntax for simple arithmetic expressions:

$$\begin{array}{lll}\textbf{spec}\ \textsc{Expr} = & \textbf{sorts} & \mathit{Expr} \\ & \textbf{ops} & x{:}\,\mathit{Expr} \\ & & y{:}\,\mathit{Expr} \\ & & 0{:}\,\mathit{Expr} \\ & & \mathit{plus}{:}\,\mathit{Expr}\times\mathit{Expr}\to\mathit{Expr} \\ & & \mathit{minus}{:}\,\mathit{Expr}\times\mathit{Expr}\to\mathit{Expr} \\ & \forall e,e'{:}\,\mathit{Expr} & \\ & & \bullet\ \mathit{plus}(e,e') = \mathit{plus}(e',e)\end{array}$$

The axiom requires the *syntax* of addition to be commutative. In the initial semantics of EXPR, the "no junk" condition ensures that the only expressions (values of sort *Expr*) are those built from 0, x and y using *plus* and *minus*. The "no confusion" condition ensures that no undesired identification of expressions occurs: for example, the syntax of addition is not associative and the syntax of subtraction is not commutative. □

Exercise 2.5.23. Write a specification of (finite) sets of natural numbers. The operations should include $\varnothing{:}\mathit{NatSet}$, $\mathit{singleton}{:}\mathit{Nat}\to\mathit{NatSet}$ and $\_\_\cup\_\_{:}\mathit{NatSet}\times\mathit{NatSet}\to\mathit{NatSet}$. □

The "no junk" condition is more powerful than it might appear to be at first glance. Imposing the constraint that every value be expressible as a ground term makes it possible to use induction on the structure of terms to prove properties of all the values in an algebra. This means that for reasoning about models of specifications containing no junk, such as initial models, it is sound to add an induction rule scheme to the equational calculus presented in the previous section. Since the form of the induction rule scheme varies according to the signature of the specification at hand, this is best illustrated by means of examples.

Example 2.5.24. Recall the presentation $\textsc{Nat} = \langle\Sigma_{\textsc{Nat}},\mathscr{E}_{\textsc{Nat}}\rangle$ of natural numbers with addition given in Exercise 2.5.4. To simplify notation, let x and y stand for variable names such that $x{:}\mathit{Nat}$ and $y{:}\mathit{Nat}$ are not in $\Sigma_{\textsc{Nat}}$ and $x{:}\mathit{Nat}$ does not appear in the $\mathit{sorts}(\Sigma_{\textsc{Nat}})$-sorted set of variables X used below. The following induction rule scheme is sound for reachable models of NAT (and for reachable models of all other $\Sigma_{\textsc{Nat}}$-presentations):

$$\frac{\mathscr{E}\vdash_{\Sigma_{\textsc{Nat}}} P(0) \qquad \mathscr{E}\cup\{P(x)\}\vdash_{\Sigma_{\textsc{Nat}}\cup\{x:\mathit{Nat}\}} P(\mathit{succ}(x)) \\ \mathscr{E}\cup\{P(x),P(y)\}\vdash_{\Sigma_{\textsc{Nat}}\cup\{x,y:\mathit{Nat}\}} P(x+y)}{\mathscr{E}\vdash_{\Sigma_{\textsc{Nat}}} \forall x{:}\mathit{Nat}\bullet P(x)}$$

Here, $P(x)$ stands for a $(\Sigma_{\textsc{Nat}}\cup\{x{:}\mathit{Nat}\})$-equation, $\forall X\bullet t = t'$; think of this as a $\Sigma_{\textsc{Nat}}$-equation with free variable $x{:}\mathit{Nat}$. Then $P(0)$ stands for the $\Sigma_{\textsc{Nat}}$-equation

$\forall X \bullet t[x \mapsto 0] = t'[x \mapsto 0]$, $P(succ(x))$ stands for the $(\Sigma_{\text{NAT}} \cup \{x{:}Nat\})$-equation $\forall X \bullet t[x \mapsto succ(x)] = t'[x \mapsto succ(x)]$, and analogously for $P(y)$ and $P(x+y)$, and $\forall x{:}Nat \bullet P(x)$ stands for the Σ_{NAT}-equation $\forall X \cup \{x{:}Nat\} \bullet t = t'$. The following additional inference rule is needed to infer equations over $\Sigma_{\text{NAT}} \cup \{x{:}Nat\}$ and $\Sigma_{\text{NAT}} \cup \{x, y{:}Nat\}$ from Σ_{NAT}-equations:

$$\frac{\mathscr{E} \vdash_{\Sigma} \forall X \bullet t = t'}{\mathscr{E} \vdash_{\Sigma \cup \Sigma'} \forall X \bullet t = t'}$$

Exercise. Show that adding the two inference rules above to the equational calculus gives a system that is sound for reachable models of Σ_{NAT}-presentations.

The inference rule scheme above can be used for proving theorems such as associativity and commutativity of $+$. But note that the axioms for $+$ fully define it in terms of 0 and *succ*: it is possible to prove by induction on the structure of terms that for every ground Σ_{NAT}-term t there is a ground Σ_{NAT}-term t' such that t' does not contain the $+$ operation and $\mathscr{E}_{\text{NAT}} \vdash_{\Sigma_{\text{NAT}}} t = t'$. (**Exercise:** Prove it. Note that this is a proof at the meta-level *about* $\vdash$, not a derivation at the object level *using* $\vdash$.) This shows that the third premise of the above induction rule scheme is redundant. Eliminating it gives the following scheme, which is more obviously related to the usual form of induction for natural numbers:

$$\frac{\mathscr{E} \cup \mathscr{E}_{\text{NAT}} \vdash_{\Sigma_{\text{NAT}}} P(0) \qquad \mathscr{E} \cup \mathscr{E}_{\text{NAT}} \cup \{P(x)\} \vdash_{\Sigma_{\text{NAT}} \cup \{x:Nat\}} P(succ(x))}{\mathscr{E} \cup \mathscr{E}_{\text{NAT}} \vdash_{\Sigma_{\text{NAT}}} \forall x{:}Nat \bullet P(x)}$$

Taking $P(x)$ to be $\forall n, p{:}Nat \bullet x + (n + p) = (x + n) + p$, we have the following derivation, which proves that addition is associative in initial models of NAT (**Exercise:** Supply the derivations P_1 and P_2):

$$\frac{\begin{array}{c} \bigtriangledown P_1 \\ \mathscr{E}_{\text{NAT}} \vdash_{\Sigma_{\text{NAT}}} \forall n, p{:}Nat \bullet \\ 0 + (n + p) = (0 + n) + p \end{array} \qquad \begin{array}{c} \bigtriangledown P_2 \\ \mathscr{E}_{\text{NAT}} \cup \{\forall n, p{:}Nat \bullet x + (n + p) = (x + n) + p\} \\ \vdash_{\Sigma_{\text{NAT}} \cup \{x:Nat\}} \forall n, p{:}Nat \bullet \\ succ(x) + (n + p) = (succ(x) + n) + p \end{array}}{\mathscr{E}_{\text{NAT}} \vdash_{\Sigma_{\text{NAT}}} \forall x, n, p{:}Nat \bullet x + (n + p) = (x + n) + p}$$

There are models of NAT containing junk which do not satisfy $\forall x, n, p{:}Nat \bullet x + (n + p) = (x + n) + p$. Hence, this equation is not in $Cl_{\Sigma_{\text{NAT}}}(\mathscr{E}_{\text{NAT}})$ and induction is required for its derivation. □

Exercise 2.5.25. Recall again the presentation $\text{BOOL} = \langle \Sigma_{\text{BOOL}}, \mathscr{E}_{\text{BOOL}} \rangle$ from Example 2.2.3. Give an induction rule scheme that is sound for reachable models of Σ_{BOOL}-presentations. (HINT: There will be five premises, one for each operation in BOOL.) Show that three of the premises are redundant (HINT: eliminate one operation at a time), which gives the following rule scheme:

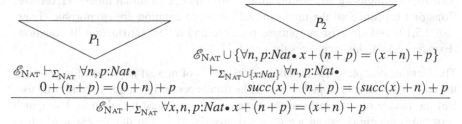

$$\frac{\mathscr{E} \cup \mathscr{E}_{\text{BOOL}} \vdash_{\Sigma_{\text{BOOL}}} P(\mathit{true}) \qquad \mathscr{E} \cup \mathscr{E}_{\text{BOOL}} \vdash_{\Sigma_{\text{BOOL}}} P(\mathit{false})}{\mathscr{E} \cup \mathscr{E}_{\text{BOOL}} \vdash_{\Sigma_{\text{BOOL}}} \forall x{:}\mathit{Bool} \bullet P(x)}$$

Use this to prove that $\forall p{:}\mathit{Bool} \bullet \neg\neg p = p$ holds in initial models of BOOL. Prove that the axiom $\forall p{:}\mathit{Bool} \bullet p \wedge \neg p = \mathit{false}$ is redundant for the initial semantics of BOOL, that is:

$$\mathscr{E}_{\text{BOOL}} \setminus \{\forall p{:}\mathit{Bool} \bullet p \wedge \neg p = \mathit{false}\} \vdash_{\Sigma_{\text{BOOL}}} \forall p{:}\mathit{Bool} \bullet p \wedge \neg p = \mathit{false}. \qquad \square$$

Adding an induction rule scheme appropriate to the signature at hand to the equational calculus gives a system that is sound for reasoning about initial models of specifications, and is more powerful than the equational calculus on its own. However, the resulting system is not always complete. In fact, it turns out that completeness is unachievable in general: there is *no* sound proof system that is complete for reasoning about initial models of arbitrary specifications. In order to prove that this is the case, it is necessary to formalise what we mean by the term "proof system". For our purposes it will suffice to assume that any proof system has a recursively enumerable set of theorems. See [Chu56] for a discussion of the philosophical considerations (e.g. finiteness of proofs, decidability of the correctness of individual proof steps) underlying this assumption.

Theorem 2.5.26 (Incompleteness for initial semantics). *There is a presentation* $\langle \Sigma, \mathscr{E} \rangle$ *such that there is no proof system which is sound and complete with respect to satisfaction of equations in the class of initial models of* $\langle \Sigma, \mathscr{E} \rangle$.

Proof. As a consequence of Matiyasevich's theorem, the set of equations which hold in the standard model of the natural numbers (with 0, *succ*, $+$, $\times$ and $-$, such that $m - n = 0$ when $n \geq m$) is not recursively enumerable [DMR76, Sect. 8]. Therefore, this cannot be the set of theorems produced by any proof system. It is easy to construct a (single-sorted) presentation having this as an initial model. (**Exercise:** Construct it.) Since all the initial models of a presentation are isomorphic (Exercise 2.5.19) and since isomorphisms preserve and reflect satisfaction of equations (Exercise 2.1.5), this completes the proof. □

The fact that completeness cannot be achieved is of no real importance in practice: the equational calculus together with induction is perfectly adequate for normal use. But the failure of completeness does mean that care must be taken to distinguish between semantic consequence ($\models$) and provability ($\vdash$) in theoretical work. It is important to recognize that semantic consequence is the relation of primary importance, since it is based directly on satisfaction, which embodies *truth*. Provability is merely an *approximation* to truth, albeit one that is of great importance for practical use since it is based on mechanical syntactic manipulation. The failure of completeness means that the approximation cannot be exact, but by being sound it errs on the side of safety.

Exercise 2.5.27. Show that the equational calculus (without added induction rule schemes) is complete with respect to satisfaction of *ground* equations in initial models of specifications. □

The additional specification techniques introduced in Chapter 5 will lead to a widening of the gap between satisfaction and provability. In particular, even completeness with respect to satisfaction of ground equations will be impossible to retain.

A generalisation of the concept of initial model is needed to give a fully satisfactory specification of classes of models that are naturally parametric with respect to some basic data. An example is the definition of terms in Section 1.4, which is parametric in an S-sorted set of variables. Another is the specification of sets (see Exercise 2.5.23): it should be possible to specify sets without building in a specification of the kind of values in the sets (in this case, natural numbers).

Exercise 2.5.28. Suppose that all information about the natural numbers is removed from the specification of sets you gave in Exercise 2.5.23, by deleting operations on natural numbers like *succ* and changing the sort name *Nat* to *Elem*. Construct an initial model of the resulting specification. Why is this model unsatisfactory? □

The required concept is that of a *free* model extending a given algebra, which captures the idea of initiality *relative to* a fixed part of the model. See Section 3.5 for the details, Section 4.3 for the use of this concept in the context of specifications, and Chapter 6 for much more on the general topic of parameterisation.

2.6 Term rewriting

Although there is no decision procedure for $\models_\Sigma$ (Theorem 2.4.15), there is a class of specifications for which consequence can be decided. The idea is similar to the one behind the strategy used in mathematics for proving that an equation follows from a set of equational axioms: one applies the axioms in an attempt to reduce both sides of the equation to a common result, and if this is successful then the equation follows from the axioms. An essential ingredient of this strategy is the use of equations as directed *simplification* or *rewrite rules*.

Throughout this section, let $\Sigma = \langle S, \Omega \rangle$ be a signature, and let X be an S-sorted set of variables such that $X_s \subseteq \mathscr{X}$ for all $s \in S$.

Assumption. For simplicity of presentation, we assume throughout this section that either Σ has only one sort, or all sorts in Σ are non-void (see Exercise 2.4.10). Under this assumption, the version of the equational calculus without explicit quantifiers is sound, and all references to the calculus below are to this version. See Exercises 2.6.11 and 2.6.26 for hints on how to do away with this assumption. □

Definition 2.6.1 (Context). A *Σ-context for sort $s \in S$* is a term $C \in |T_\Sigma(X \uplus \{\Box{:}s\})|$ containing one occurrence of the distinguished variable $\Box$. We write $C[\Box]$ to suggest that C should be viewed as a term with a hole in it. Substitution of a term $t \in |T_\Sigma(X)|_s$ in $C[\Box]$ gives the term $C[\Box{:}s \mapsto t] \in |T_\Sigma(X)|$, written $C[t]$. □

Definition 2.6.2 (Rewrite rule). A *Σ-rewrite rule r of sort $s \in S$* consists of two Σ-terms $t, t' \in |T_\Sigma(X)|_s$, written $t \to t'$. A Σ-rewrite rule $r = t \to t'$ of sort s determines a set of *reduction steps* $C[t[\theta]] \to_r C[t'[\theta]]$ for all Σ-contexts $C[\square]$ for sort s and substitutions $\theta: X \to |T_\Sigma(X)|$; this defines the relation $\to_r \subseteq |T_\Sigma(X)| \times |T_\Sigma(X)|$, the *one-step reduction relation generated by r*. The inverse of one-step reduction $\to_r$ is *one-step expansion*, written ${}_r\!\leftarrow$. $\square$

A reduction step $u \to_r u'$ according to a rewrite rule $r = t \to t'$ is an application of an *instance* $t[\theta] \to t'[\theta]$ of r to replace the *subterm* $t[\theta]$ of u (corresponding to the "hole" in $C[\square]$) by $t'[\theta]$. The subterm $t[\theta]$ of u is called a *redex* (short for "reducible expression").

Definition 2.6.3 (Term rewriting system). A *Σ-term rewriting system R* is a set of Σ-rewrite rules. The *set of Σ-equations determined by R* is $Eq(R) = \{t = t' \mid t \to t' \in R\}$; by the assumption, we can dispense with explicit quantification of variables in equations. The *one-step reduction relation generated by R* is the relation

$$\to_R \; = \bigcup_{r \in R} \to_r \qquad (\subseteq |T_\Sigma(X)| \times |T_\Sigma(X)|).$$

The inverse of one-step reduction $\to_R$ is *one-step expansion*, written ${}_R\!\leftarrow$. $\square$

Given a set $\mathscr{E}$ of Σ-equations, a Σ-term rewriting system R will be of greatest relevance to $\mathscr{E}$ when $Cl_\Sigma(\mathscr{E}) = Cl_\Sigma(Eq(R))$. One way to obtain such an R is to use the equations themselves as rewrite rules by selecting an *orientation* for each equation $t = t'$: either $t \to t'$ or $t' \to t$. For reasons that will become clear below, the most useful orientation is the one in which the right-hand side of the rule is "simpler" than the left-hand side. It is not always obvious how to measure simplicity of terms — in fact, this is a major issue in the theory of term rewriting — and sometimes there is no satisfactory orientation, as in the case of an equation such as $n + m = m + n$.

In the rest of this section, let R be a Σ-term rewriting system.

Definition 2.6.4 (Reduction $\to_R^*$ and convertibility $\sim_R$). The *reduction relation* $\to_R^* \subseteq |T_\Sigma(X)| \times |T_\Sigma(X)|$ *generated by R* is the transitive reflexive closure of $\to_R$. In other words, $t \to_R^* t'$ if $t = t'$ or there exist terms $t_1, \ldots, t_n \in |T_\Sigma(X)|$, $n \geq 0$, such that $t \to_R t_1 \to_R \cdots \to_R t_n \to_R t'$; then we say that *t reduces to t'*. The inverse of reduction $\to_R^*$ is *expansion*, written ${}_R^*\!\leftarrow$. The *convertibility relation* $\sim_R \subseteq |T_\Sigma(X)| \times |T_\Sigma(X)|$ *generated by R* is the symmetric transitive reflexive closure of $\to_R$. In other words, $t \sim_R t'$ if $t = t'$ or there exist terms $t_1, \ldots, t_n \in |T_\Sigma(X)|$, $n \geq 0$, such that $t \to_R t_1$ or $t\, {}_R\!\leftarrow t_1$, and $t_1 \to_R t_2$ or $t_1\, {}_R\!\leftarrow t_2$, and ..., and $t_n \to_R t'$ or $t_n\, {}_R\!\leftarrow t'$; then we say that *t converts to t'*. $\square$

Exercise 2.6.5. Check that $\sim_R$ is a Σ-congruence on $T_\Sigma(X)$. $\square$

Example 2.6.6. Recall again the presentation $\mathrm{BOOL} = \langle \Sigma_{\mathrm{BOOL}}, \mathscr{E}_{\mathrm{BOOL}} \rangle$ from Example 2.2.3. The following Σ_{BOOL}-term rewriting system RBOOL obviously satisfies $Cl_{\Sigma_{\mathrm{BOOL}}}(\mathscr{E}_{\mathrm{BOOL}}) = Cl_{\Sigma_{\mathrm{BOOL}}}(Eq(\mathrm{RBOOL}))$:

$$\begin{aligned}\mathrm{RBOOL} = \{&\neg true \to false,\ \neg false \to true,\ p \wedge true \to p,\ p \wedge false \to false,\\ &p \wedge \neg p \to false,\ p \Rightarrow q \to \neg(p \wedge \neg q)\}.\end{aligned}$$

(Observe that in the rule $p \Rightarrow q \to \neg(p \wedge \neg q)$, the right-hand side is not obviously simpler than the left-hand side.) For example, we have (the redex reduced by each step is underlined)

$$
\begin{aligned}
\neg(p \wedge (\underline{q \Rightarrow \neg \mathit{false}})) &\to_{\mathrm{RBool}} \neg(p \wedge \neg(q \wedge \neg\underline{\neg \mathit{false}})) \\
&\to_{\mathrm{RBool}} \neg(p \wedge \neg(q \wedge \underline{\neg \mathit{true}})) \\
&\to_{\mathrm{RBool}} \neg(p \wedge \neg(\underline{q \wedge \mathit{false}})) \\
&\to_{\mathrm{RBool}} \neg(p \wedge \underline{\neg \mathit{false}}) \\
&\to_{\mathrm{RBool}} \neg(\underline{p \wedge \mathit{true}}) \\
&\to_{\mathrm{RBool}} \neg p
\end{aligned}
$$

so $\neg(p \wedge (q \Rightarrow \neg \mathit{false})) \to^*_{\mathrm{RBool}} \neg p$, and

$$
\begin{aligned}
\neg(p \wedge (q \Rightarrow \mathit{false})) \;{}_{\mathrm{RBool}}{\leftarrow}\; & \neg(p \wedge (q \Rightarrow \neg \mathit{true})) \\
\to_{\mathrm{RBool}}\; & \neg(p \wedge \neg(q \wedge \neg\neg \mathit{true})) \\
\to_{\mathrm{RBool}}\; & \neg(p \wedge \neg(q \wedge \neg \mathit{false})) \\
{}_{\mathrm{RBool}}{\leftarrow}\; & \neg(p \wedge \neg((q \wedge \mathit{true}) \wedge \neg \mathit{false})) \\
\to_{\mathrm{RBool}}\; & \neg(p \wedge \neg((q \wedge \mathit{true}) \wedge \mathit{true})) \\
\to_{\mathrm{RBool}}\; & \neg(p \wedge \neg(q \wedge \mathit{true}))
\end{aligned}
$$

so $\neg(p \wedge (q \Rightarrow \mathit{false})) \sim_{\mathrm{RBool}} \neg(p \wedge \neg(q \wedge \mathit{true}))$. □

Exercise 2.6.7. Recall the presentation $\mathrm{Nat} = \langle \Sigma_{\mathrm{Nat}}, \mathscr{E}_{\mathrm{Nat}} \rangle$ from Exercise 2.5.4. Looking at the equations in $\mathscr{E}_{\mathrm{Nat}}$, give a Σ_{Nat}-term rewriting system RNat such that $Cl_{\Sigma_{\mathrm{Nat}}}(\mathscr{E}_{\mathrm{Nat}}) = Cl_{\Sigma_{\mathrm{Nat}}}(Eq(\mathrm{RNat}))$, and practice reducing and converting some Σ_{Nat}-terms using RNat. □

The convertibility relation generated by R coincides with equality provable from $Eq(R)$. This fact is captured by the following two theorems.

Theorem 2.6.8 (Soundness of convertibility). *If $t \sim_R t'$ then $Eq(R) \vdash_\Sigma t = t'$.*

Proof sketch. Consider a reduction step $C[t[\theta]] \to_R C[t'[\theta]]$. This corresponds to a derivation involving an application of the axiom rule, to derive $Eq(R) \vdash t = t'$; an application of instantiation, to derive $Eq(R) \vdash t[\theta] = t'[\theta]$; and repeated applications of reflexivity and congruence, to derive $Eq(R) \vdash C[t[\theta]] = C[t'[\theta]]$. The definition of $\sim_R$ as the symmetric transitive reflexive closure of $\to_R$ corresponds directly to applications of the symmetry, transitivity and reflexivity rules. (**Exercise:** Fill in the gaps in this proof.) □

Lemma 2.6.9. *Suppose $t, t' \in |T_\Sigma(X)|_s$ for $s \in S$. If $t \sim_R t'$ then*

1. *$C[t] \sim_R C[t']$ for any Σ-context $C[\square]$ for sort s.*
2. *$t[\theta] \sim_R t'[\theta]$ for any substitution $\theta : X \to |T_\Sigma(X)|$.*

Proof. **Exercise:** Do it. □

Theorem 2.6.10 (Completeness of convertibility). *If $Eq(R) \vdash_\Sigma t = t'$ then $t \sim_R t'$.*

Proof sketch. By induction on the depth of the derivation of $Eq(R) \vdash_\Sigma t = t'$. The most interesting case is when the last step is an application of the congruence rule:

$$\frac{Eq(R) \vdash_\Sigma t_1 = t'_1 \qquad \cdots \qquad Eq(R) \vdash_\Sigma t_n = t'_n}{Eq(R) \vdash_\Sigma f(t_1,\ldots,t_n) = f(t'_1,\ldots,t'_n)}$$

where $f{:}s_1 \times \cdots \times s_n \to s$. By the inductive assumption, $t_1 \sim_R t'_1$ and ... and $t_n \sim_R t'_n$. Then, by repeated application of Lemma 2.6.9(1), we have $f(t_1,t_2\ldots,t_n) \sim_R f(t'_1,t_2\ldots,t_n) \sim_R \cdots \sim_R f(t'_1,t'_2\ldots,t'_n)$ (using first the context $f(\Box{:}s_1,t_2\ldots,t_n)$, then $f(t'_1,\Box{:}s_2,\ldots,t_n)$, then ..., then $f(t'_1,t'_2,\ldots,\Box{:}s_n)$). When the last step of the derivation of $Eq(R) \vdash_\Sigma t = t'$ is an application of the instantiation rule, the result follows directly by Lemma 2.6.9(2). (**Exercise:** Complete the proof.) □

Exercise 2.6.11. Try to get rid of the need for the assumption on Σ made at the beginning of this section in all the definitions and results above. This will involve rewriting terms of the form $(X)t$ using rewrite rules of the form $\forall X \bullet t \to t'$, in both cases with explicit variable declarations. □

Given the exact correspondence between convertibility and provable equality, a decision procedure for $t \sim_R t'$ amounts to a decision procedure for $\mathscr{E} \vdash_\Sigma t = t'$, provided $Cl_\Sigma(\mathscr{E}) = Cl_\Sigma(Eq(R))$. The problem with testing $t \sim_R t'$ by simply applying the definition is that the "path" from t to t' may include both reduction steps and expansion steps, and may be of arbitrary length. But when R satisfies certain conditions, it is sufficient to test just a *single* path having the special form $t \to_R^* t'' \,{}_R^*\!\!\leftarrow t'$, which yields a simple and efficient decision procedure for convertibility.

Definition 2.6.12 (Normal form). A Σ-term $t \in T_\Sigma(X)$ is a *normal form (for R)* if there is no term t' such that $t \to_R t'$. □

Definition 2.6.13 (Termination). A Σ-term rewriting system R is *terminating* (or *strongly normalising*) if there is no infinite reduction sequence $t_1 \to_R t_2 \to_R \cdots$; that is, whenever $t_1 \to_R t_2 \to_R \cdots$, there is some (finite) $n \geq 1$ such that t_n is a normal form. □

The usual way to show that a term rewriting system R is terminating is to demonstrate that each rule in R reduces the complexity of terms according to some carefully chosen measure.

Definition 2.6.14 (Confluence). A Σ-term rewriting system R is *confluent* (or is *Church-Rosser*) if whenever $t \to_R^* t_1$ and $t \to_R^* t_2$, there is a term t_3 such that $t_1 \to_R^* t_3$ and $t_2 \to_R^* t_3$. □

Definition 2.6.15 (Completeness). A Σ-term rewriting system R is *complete* if it is both terminating and confluent. □

Completeness of a term rewriting system should not be confused with completeness of a proof system, as in for example Theorem 2.6.10 above.

Exercise 2.6.16. Suppose that R is a complete Σ-term rewriting system, and let $t \in |T_\Sigma(X)|$ be a Σ-term. Show that there is a unique normal form $NF_R(t) \in |T_\Sigma(X)|$ such that $t \to_R^* NF_R(t)$.

HINT: An *abstract reduction system* consists of a set A together with a binary relation $\to \subseteq A \times A$. A Σ-term rewriting system R is a particular example, where $A = |T_\Sigma(X)|$ and $\to$ is $\to_R$. Concepts such as normal form and confluence make sense in the context of any abstract reduction system, and the required property holds in this more abstract setting. □

Example 2.6.17. The term rewriting system RBOOL from Example 2.6.6 is both terminating and confluent, and is therefore complete. As the reduction sequence in Example 2.6.6 shows, $NF_{\text{RBOOL}}(\neg(p \wedge (q \Rightarrow \neg \mathit{false}))) = \neg p$.

The term rewriting system $\text{RBOOL}' = \text{RBOOL} \cup \{p \wedge q \to q \wedge p\}$ is not terminating: $p \wedge q \to_{\text{RBOOL}'} q \wedge p \to_{\text{RBOOL}'} p \wedge q \to_{\text{RBOOL}'} q \wedge p \to_{\text{RBOOL}'} \cdots$.

The term rewriting system $\text{RBOOL}'' = \text{RBOOL} \cup \{(p \wedge q) \wedge r \to p \wedge (q \wedge r)\}$ is not confluent: $(p \wedge \neg p) \wedge q \to_{\text{RBOOL}''} \mathit{false} \wedge q$ and $(p \wedge \neg p) \wedge q \to_{\text{RBOOL}''} p \wedge (\neg p \wedge q)$, and both $\mathit{false} \wedge q$ and $p \wedge (\neg p \wedge q)$ are normal forms. □

Exercise 2.6.18. Is your term rewriting system RNAT from Exercise 2.6.7 complete? If not, find an alternative term rewriting system for NAT that is complete. □

Exercise 2.6.19. A Σ-term rewriting system R is *weakly confluent* if whenever $t \to_R t_1$ and $t \to_R t_2$, there is a term t_3 such that $t_1 \to_R^* t_3$ and $t_2 \to_R^* t_3$. Find a term rewriting system that is weakly confluent but not confluent. (HINT: Weak confluence plus termination implies confluence, so don't bother looking at terminating term rewriting systems.) Weak confluence is a much easier condition to check than confluence, so the usual way to prove that a term rewriting system is confluent is to show that it is weakly confluent and terminating. □

In view of the obvious analogy between reduction and computation, $NF_R(t)$ can be thought of as the *value* of t; since $NF_R(t)$ need not be a ground term, this is a more general notion of computation than the usual one.

Exercise 2.6.20. Convince yourself that $NF_R\colon |T_\Sigma(X)| \to |T_\Sigma(X)|$ is computable for any finite complete term rewriting system R — perhaps try to implement it in your favourite programming language. □

Theorem 2.6.21 (Decision procedure for convertibility). *If R is complete, then $t \sim_R t'$ iff $NF_R(t) = NF_R(t')$.* □

Exercise 2.6.22. Prove Theorem 2.6.21. HINT: The proof does not depend on the definition of $\to_R$, but only on the assumption that R is complete. □

Since $t \sim_R t'$ iff $Eq(R) \vdash_\Sigma t = t'$ (by soundness and completeness of convertibility) iff $Eq(R) \models_\Sigma t = t'$ (by soundness and completeness of the equational calculus), Theorem 2.6.21 constitutes a decision procedure for consequence:

Corollary 2.6.23 (Decision procedure for $Eq(R) \models_\Sigma t = t'$). *If R is complete, then $Eq(R) \models_\Sigma t = t'$ iff $NF_R(t) = NF_R(t')$.* □

Example 2.6.24. Since the term rewriting system RBOOL from Example 2.6.6 is complete (see Example 2.6.17), Corollary 2.6.23 can be used to prove that $Eq(\text{RBOOL}) \models_{\Sigma_{\text{BOOL}}} \neg(p \wedge (q \Rightarrow \neg \mathit{false})) = p \Rightarrow (p \wedge \neg p)$: this follows since we have $NF_{\text{RBOOL}}(\neg(p \wedge (q \Rightarrow \neg \mathit{false}))) = \neg p = NF_{\text{RBOOL}}(p \Rightarrow (p \wedge \neg p))$. Since $Cl_{\Sigma_{\text{BOOL}}}(\mathscr{E}_{\text{BOOL}}) = Cl_{\Sigma_{\text{BOOL}}}(Eq(\text{RBOOL}))$, this proves that $\mathscr{E}_{\text{BOOL}} \models_{\Sigma_{\text{BOOL}}} \neg(p \wedge (q \Rightarrow \neg \mathit{false})) = p \Rightarrow (p \wedge \neg p)$.

Exercise. Give a derivation of $\mathscr{E}_{\text{BOOL}} \vdash_{\Sigma_{\text{BOOL}}} \neg(p \wedge (q \Rightarrow \neg \mathit{false})) = p \Rightarrow (p \wedge \neg p)$ in the equational calculus. Compare this with the above proof. □

Exercise 2.6.25. Recall your complete term rewriting system for NAT from Exercise 2.6.18. Relying on Corollary 2.6.23, use this to prove that $\mathscr{E}_{\text{NAT}} \models_{\Sigma_{\text{NAT}}} succ(succ(0)) + succ(n) = succ(succ(succ(n)))$, and that $\mathscr{E}_{\text{NAT}} \not\models_{\Sigma_{\text{NAT}}} succ(m) + succ(n) = succ(succ(m+n))$. □

Exercise 2.6.26. Let $t \to t'$ be a Σ-rewrite rule of sort s. The following restrictions are often imposed:

- $t \notin X_s$; and
- $FV(t') \subseteq FV(t)$.

Show that, if these restrictions are imposed on rewrite rules, then Corollary 2.6.23 holds even without the assumption on Σ made at the beginning of this section. (These restrictions seem harmless since almost no complete term rewriting system contains rules that violate them.) □

Exercise 2.6.27. Equality of terms in the equational theory of a rewriting systems is also decidable under somewhat weaker requirements than those in Corollary 2.6.23. A term rewriting system R is *weakly normalising* if for each term t there is a finite reduction sequence in R leading from t to a normal form. R is *semi-complete* if it is weakly normalising and confluent.

Generalising Exercise 2.6.16, show that if R is a semi-complete Σ-term rewriting system, then for any Σ-term $t \in |T_\Sigma(X)|$ there is a unique normal form $NF_R(t) \in |T_\Sigma(X)|$ such that $t \to_R^* NF_R(t)$. Moreover, convince yourself that the function $NF_R \colon |T_\Sigma(X)| \to |T_\Sigma(X)|$ is then computable. Finally, show that the property captured by Corollary 2.6.23 holds for all semi-complete term rewriting systems R. □

By Corollary 2.6.23, the problem of deciding consequence $\mathscr{E} \models_\Sigma e$ is reduced to the problem of finding a finite complete term rewriting system R such that $Cl_\Sigma(\mathscr{E}) = Cl_\Sigma(Eq(R))$. Clearly, by Theorem 2.4.15, this is not always possible. But the *Knuth-Bendix completion algorithm* can sometimes be used to produce such an R given $\mathscr{E}$ together with an order relation on terms. The algorithm works by pinpointing causes of failure of (weak) confluence and adding rules to correct them, where the supplied term ordering is used to orient these new rules. The algorithm is iterative and may fail to terminate; it may also fail because the ordering supplied is inadequate.

The Knuth-Bendix completion algorithm can also be used to reason about initial models of specifications, using a method known as *inductionless induction* or *proof by consistency*. This method is based on the observation that an equation $t = t'$

holds in the initial models of $\langle\Sigma,\mathscr{E}\rangle$ iff there is no ground equation $u=u'$ such that $\mathscr{E}\not\models u=u'$ and $\mathscr{E}\cup\{t=t'\}\models u=u'$. (**Exercise:** Prove this fact.) Given a complete term rewriting system R such that $Cl_\Sigma(\mathscr{E})=Cl_\Sigma(Eq(R))$ (perhaps produced using the Knuth-Bendix algorithm), the Knuth-Bendix algorithm is used to produce a complete term rewriting system R' for $\mathscr{E}\cup\{t=t'\}$ by extending R. It is then possible to test if R and R' have the same normal forms for ground Σ-terms; if so, then $t=t'$ holds in the initial models of $\langle\Sigma,\mathscr{E}\rangle$.

2.7 Fiddling with the definitions

In principle, the specification framework presented in the preceding sections is powerful enough for any conceivable computational application. This is made precise by a theorem in [BT87] (cf. [Vra88]) which states that for every reachable *semi-computable* Σ-algebra A there is a presentation $\langle\Sigma',\mathscr{E}'\rangle$ with finite $\mathscr{E}'$ such that $A=A'|_\Sigma$ for some initial model $A'\in IMod[\langle\Sigma',\mathscr{E}'\rangle]$. (See [BT87] for the definition of semi-computable algebra.) In spite of this fact, there are several reasons why this framework is inconvenient for use in practice.

One deficiency becomes apparent as soon as one attempts to write specifications that are somewhat larger than the examples we have seen so far. In order to be understandable and usable, large specifications must be built up incrementally from smaller specifications. Specification mechanisms designed to cope with such problems of scale are presented in Chapter 5. These methods also solve the problem illustrated by Exercise 2.5.20; see Exercise 5.1.11.

Another difficulty arises from the relatively low level of equational logic as a language for describing constraints to be satisfied by the operations of an algebra. When using equational axioms, it is often necessary to write a dozen equations to express a property that can be formulated much more clearly using a single axiom in some more powerful logic. Some properties that are easy to express in more powerful systems are not expressible at all using equations. Similar awkwardness is caused by the limitations of the type system used here, in comparison with the polymorphic type systems of modern programming languages such as Standard ML [Pau96]. Finally, the present framework is only able to cope conveniently with algebras comprised of *total* and *deterministic* functions operating on data values built by *finitary* compositions of such functions, a limitation which rules out its use for very many programs of interest.

All these difficulties can be addressed by making appropriate modifications to the standard framework presented in the preceding sections. An example was already given in Section 1.5.2 where it was shown how signature morphisms could be replaced by derived signature morphisms. This section is devoted to a sketch of some other possible modifications. The presentation is very brief and makes no attempt to be truly comprehensive; the interested reader will find further details (and further citations) in the cited references.

2.7.1 Conditional equations

The most obvious kind of modification is to replace the use of equational axioms by formulae in a more expressive language. Some care is required since a number of the results presented above depend on the use of equational axioms. A relatively unproblematic choice is to use equations that apply only when certain pre-conditions (expressed as equations) are satisfied.

Let $\Sigma = \langle S, \Omega \rangle$ be a signature.

Definition 2.7.1 (Conditional equation). A *conditional Σ-equation* $\forall X \bullet t_1 = t'_1 \wedge \ldots \wedge t_n = t'_n \Rightarrow t_0 = t'_0$ consists of

- a finite S-sorted set X (of variables), such that $X_s \subseteq \mathscr{X}$ for all $s \in S$; and
- for each $0 \leq j \leq n$ (where $n \geq 0$), two Σ-terms $t_j, t'_j \in |T_\Sigma(X)|_{s_j}$ for some sort $s_j \in S$.

A Σ-algebra A *satisfies* a conditional Σ-equation $\forall X \bullet t_1 = t'_1 \wedge \ldots \wedge t_n = t'_n \Rightarrow t_0 = t'_0$ if for every (S-sorted) function $v: X \to |A|$, if $(t_1)_A(v) = (t'_1)_A(v)$ and $\ldots$ and $(t_n)_A(v) = (t'_n)_A(v)$, then $(t_0)_A(v) = (t'_0)_A(v)$. □

Note that variables in the conditions ($t_1 = t'_1 \wedge \ldots \wedge t_n = t'_n$) that do not appear in the consequent ($t_0 = t'_0$) can be seen as existentially quantified: for example, the conditional equation $\forall a, b{:}t \bullet a \times b = 1 \Rightarrow a \times a^{-1} = 1$ is equivalent to the formula $\forall a{:}t \bullet (\exists b{:}t \bullet a \times b = 1) \Rightarrow a \times a^{-1} = 1$ in ordinary first-order logic.

Exercise 2.7.2. Define the translation of conditional Σ-equations by a signature morphism $\sigma: \Sigma \to \Sigma'$. □

The remaining definitions of Sections 2.1–2.5 require only superficial changes, and most results go through with appropriate modifications.

Let $\langle \Sigma, \mathscr{E} \rangle$ be a presentation, where $\mathscr{E}$ is a set of conditional Σ-equations. $Mod[\langle \Sigma, \mathscr{E} \rangle]$ is not always a variety, as is (almost) shown by Example 2.2.11; in this sense, the power of conditional equations is strictly greater than that of ordinary equations.

Exercise 2.7.3. The cancellation law given in Example 2.2.11 is not a conditional equation. Give a version of this example that uses only conditional equations. HINT: Equality can be axiomatised as an operation $eq: s \times s \to Bool$. □

In spite of this increase in expressive power, there is a proof system that is sound and complete with respect to conditional equational consequence [Sel72], and the quotient construction can be used to construct an initial model of $\langle \Sigma, \mathscr{E} \rangle$ [MT92] (cf. Lemma 3.3.12 below). Term rewriting with conditional rewrite rules is possible, but there are some complications; see [Klo92] and [Mid93].

Exercise 2.7.4. [Sel72] gives a proof system that is sound and complete for conditional equational consequence in the single-sorted case. Extend this to the many-sorted case, where explicit quantifiers are required for the same reason as in the equational calculus. □

Exercise 2.7.5. Recall Exercise 2.5.21 concerning the specification of a function $ch\colon Nat \to Nat$ that for each natural number n chooses an arbitrary number that is greater than n. Modify this, using a conditional equation to make ch choose an arbitrary number that is *less* than n when $0 < n$. □

Example 2.7.6. Let $\mathrm{HA} = \langle \Sigma_{\mathrm{HA}}, \mathscr{E}_{\mathrm{HA}} \rangle$ be the following presentation.[3]

$$
\begin{array}{lll}
\textbf{spec}\ \mathrm{HA} = \textbf{sorts}\ Bool & & \\
\quad \textbf{ops} & \mathsf{true}\colon Bool & \\
 & \mathsf{false}\colon Bool & \\
 & \neg\_\_\colon Bool \to Bool & \\
 & \_\_\vee\_\_\colon Bool \times Bool \to Bool & \\
 & \_\_\wedge\_\_\colon Bool \times Bool \to Bool & \\
 & \_\_\Rightarrow\_\_\colon Bool \times Bool \to Bool & \\
\quad \forall p,q,r\colon Bool & & \\
 & \bullet\ p \vee (q \vee r) = (p \vee q) \vee r & \\
 & \bullet\ p \wedge (q \wedge r) = (p \wedge q) \wedge r & \\
 & \bullet\ p \vee q = q \vee p & \\
 & \bullet\ p \wedge q = q \wedge p & \\
 & \bullet\ p \vee (p \wedge q) = p & \\
 & \bullet\ p \wedge (p \vee q) = p & \\
 & \bullet\ p \vee \mathsf{true} = \mathsf{true} & \\
 & \bullet\ p \vee \mathsf{false} = p & \\
 & \bullet\ (p \vee (r \wedge q) = p) \ \Rightarrow\ ((q \Rightarrow p) \vee r = (q \Rightarrow p)) & \\
 & \bullet\ ((q \Rightarrow p) \vee r = (q \Rightarrow p)) \ \Rightarrow\ (p \vee (r \wedge q) = p) & \\
 & \bullet\ \neg p = (p \Rightarrow \mathsf{false}) &
\end{array}
$$

Models of HA are called *Heyting algebras*.

Exercise. Recall the presentation BA of Boolean algebras in Example 2.2.4. Show that every Boolean algebra is a Heyting algebra. Then repeat the exercise in Example 2.2.4, building for every Heyting algebra H a lattice $\langle |H|, \leq_H \rangle$ with top and bottom elements. Check that the conditional axioms concerning the implication $\Rightarrow$ can now be captured by requiring that $r \wedge q \leq_H p$ is equivalent to $r \leq_H q \Rightarrow p$. Show that the lattice is distributive.

Give an example of a Heyting algebra that is not Boolean. Check which of the axioms of the presentation BA do not follow from HA.

Prove that an *equational* presentation with the same models as HA can be given. HINT: Use Theorem 2.2.10. Or consider the following properties of the implication: $p \Rightarrow p = \mathsf{true}$, $q \wedge (q \Rightarrow p) = q \wedge p$, $p \vee (q \Rightarrow p) = q \Rightarrow p$, and $q \Rightarrow (p \wedge r) = (q \Rightarrow p) \wedge (q \Rightarrow r)$. □

---

[3] We use the same symbol $\Rightarrow$ for implication in conditional equations and for an operation in the presentation below — the usual symbols are used for other propositional connectives as well, as in Example 2.2.4. We use extra space around the implication symbol in the conditional equations below in order to make them easier to read.

2.7.2 Reachable semantics

In Section 2.5, the motivation given for taking a presentation $\langle\Sigma,\mathcal{E}\rangle$ to denote the class $IMod[\langle\Sigma,\mathcal{E}\rangle]$ of its initial models was the desire to exclude models containing junk and confusion. The need to exclude models containing confusion stems mainly from the use of equational axioms, which make it impossible to rule out degenerate models having a single value of each sort in Σ. If a more expressive language is used for axioms, or if degenerate models are ruled out by some other means, then models containing confusion need not be excluded.

Example 2.7.7. Consider the following specification of sets of natural numbers (a variant of the one in Exercise 2.5.23):

$$
\begin{array}{lll}
\textbf{spec } \mathrm{SETNAT} = & \textbf{sorts} & Bool, Nat, NatSet \\
& \textbf{ops} & true: Bool \\
& & false: Bool \\
& & \_\_\vee\_\_: Bool \times Bool \to Bool \\
& & 0: Nat \\
& & succ: Nat \to Nat \\
& & eq: Nat \times Nat \to Bool \\
& & \varnothing: NatSet \\
& & add: Nat \times NatSet \to NatSet \\
& & \_\_\in\_\_: Nat \times NatSet \to Bool \\
& \multicolumn{2}{l}{\forall p: Bool, m, n: Nat, S: NatSet} \\
& & \bullet\ p \vee true = true \\
& & \bullet\ p \vee false = p \\
& & \bullet\ eq(n,n) = true \\
& & \bullet\ eq(0, succ(n)) = false \\
& & \bullet\ eq(succ(n), 0) = false \\
& & \bullet\ eq(succ(m), succ(n)) = eq(m,n) \\
& & \bullet\ n \in \varnothing = false \\
& & \bullet\ m \in add(n,S) = eq(m,n) \vee m \in S
\end{array}
$$

There are many different models of SETNAT, including algebras having a single value of each sort. Suppose we restrict attention to algebras that do not satisfy the equation $true = false$; this excludes such degenerate models (see the exercise below). Consider the following two equations:

Commutativity of *add*:

$$\forall m, n{:}Nat, S{:}NatSet \bullet add(m, add(n,S)) = add(n, add(m,S))$$

Idempotency of *add*:

$$\forall n{:}Nat, S{:}NatSet \bullet add(n, add(n,S)) = add(n,S)$$

The models of SETNAT that do not satisfy $true = false$ may be classified according to which of these two equations they satisfy:

"List-like" algebras: *add* is neither commutative nor idempotent.
"Set-like" algebras: *add* is both commutative and idempotent.

"Multiset-like" algebras: *add* is commutative but not idempotent.
"List-like" algebras without repeated adjacent entries: *add* is idempotent but not commutative.

There are also "hybrid" models of SETNAT, e.g. those in which *add* is commutative but is only idempotent for $n \neq 0$. The initial models of SETNAT are "list-like" algebras. Adding the commutativity and idempotency requirements to SETNAT as additional axioms would eliminate all but the "set-like" algebras.

Exercise. Show that restricting attention to models of SETNAT that do not satisfy $true = false$ eliminates all but "sensible" realisations of sets of natural numbers, by forcing $eq(succ^m(0), succ^n(0)) = true$ iff $m = n$ iff $succ^m(0) = succ^n(0)$, and $a \in add(a_1, add(a_2, \ldots, add(a_p, \varnothing)\ldots)) = true$ iff $eq(a, a_1) = true$ or ... or $eq(a, a_p) = true$, for $m, n, p \geq 0$. Note that m, n and p are ordinary integers here, *not* values of the sort *Nat*, and $succ^m(0)$ means $\underbrace{succ(\ldots succ}_{m \text{ times}}(0)\ldots)$. □

Consideration of examples like the one above suggests various alternatives to taking the initial semantics of specifications. One choice is to require signatures to include the sort *Bool* and the constants *true* and *false*, and to exclude models satisfying $true = false$. This might be called taking the *non-degenerate loose semantics* of specifications. Another choice is to additionally exclude models containing junk:

Definition 2.7.8 (Reachable semantics). Let $\Sigma = \langle S, \Omega \rangle$ be a signature such that $Bool \in S$ and $true{:}Bool$ and $false{:}Bool$ are in Ω. A *reachable non-degenerate model* of a presentation $\langle \Sigma, \mathscr{E} \rangle$ is a reachable Σ-algebra A such that $A \models_\Sigma \mathscr{E}$ and $A \not\models_\Sigma true = false$. $RMod[\langle \Sigma, \mathscr{E} \rangle]$ is the class of all reachable non-degenerate models of $\langle \Sigma, \mathscr{E} \rangle$. Taking $\langle \Sigma, \mathscr{E} \rangle$ to denote $RMod[\langle \Sigma, \mathscr{E} \rangle]$ is called taking its *reachable semantics*. □

The motivation for excluding models containing junk is the same as in the case of initial semantics. $RMod[\langle \Sigma, \mathscr{E} \rangle]$ is not always an isomorphism class of models, as Example 2.7.7 demonstrates (the classification given there was for *all* models that do not satisfy $true = false$, but it also applies to the reachable models in this class). There is still a problem when operations are not defined in a sufficiently complete way, although the problem is less severe than in the case of initial semantics.

Exercise 2.7.9. Reconsider the problem posed in Exercise 2.5.20, by writing a reachable model specification of natural numbers including a subtraction operation $\_\_ - \_\_ {:} Nat \times Nat \to Nat$ together with the axioms $\forall m{:}Nat \bullet m - 0 = m$ and $\forall m, n{:}Nat \bullet succ(m) - succ(n) = m - n$. Recall from Exercise 2.5.20 the assumption that we are willing to accept any value for $m - n$ when $n > m$, which is why the axioms do not constrain the value of $m - n$ in this case. List some of the reachable non-degenerate models of this specification, and decide whether the models you considered in Exercise 2.5.20 are reachable non-degenerate models (ignoring the difference in signatures). From an intuitive point of view, is this an adequate class of models for this specification? □

Exercise 2.7.10. Definition 2.7.8 permits algebras $A \in RMod[\langle\Sigma,\mathscr{E}\rangle]$ with values of sort *Bool* other than $true_A$ and $false_A$. This is ruled out if all operations delivering results in sort *Bool* are defined in a sufficiently complete way to yield either *true* or *false* on each argument that is definable by a ground term. Check that the specification SETNAT in Example 2.7.7 ensures this property, and so all of its reachable non-degenerate models have a two-element carrier of sort *Bool*. Give an example of a specification for which this is not the case. □

The equational calculus is sound for reasoning about the reachable semantics of presentations, since $RMod[\langle\Sigma,\mathscr{E}\rangle] \subseteq Mod[\langle\Sigma,\mathscr{E}\rangle]$ for any presentation $\langle\Sigma,\mathscr{E}\rangle$. It is sound to add induction rule schemes such as those given in Section 2.5; these are sound for any class of reachable models. Completeness is unachievable, for exactly the same reason as in the case of initial semantics; the proof of Theorem 2.5.26 can be repeated in this context almost without change. Finally, the techniques of term rewriting presented in Section 2.6 remain sound.

Initial semantics cannot be used for specifications with axioms that are more expressive than (infinitary) conditional equations [Tar86b], in the sense that initial models of such specifications are not guaranteed to exist. To illustrate the problem, the following example shows what can go wrong when the language of axioms is extended to permit disjunctions of equations.

Example 2.7.11. Consider the following specification:

spec STATUS = **sorts** *Status*
ops *single*: *Status*
married: *Status*
widowed: *Status*
• $widowed = single \vee widowed = married$

where disjunction of equations has the obvious interpretation. There are three kinds of algebras in *Mod*[STATUS]:

1. Those satisfying $single = widowed = married$.
2. Those satisfying $single = widowed \neq married$.
3. Those satisfying $single \neq widowed = married$.

None of these is an initial model of STATUS: there are no homomorphisms from algebras in the first class to algebras in either of the other two classes, and no homomorphisms in either direction between algebras in the second and third classes. □

In contrast, reachable semantics can be used for specifications with axioms of any form (once a definition of satisfaction of such axioms by algebras has been given, of course).

Another alternative to initial semantics deserves brief mention.

Definition 2.7.12 (Final semantics). Let $\Sigma = \langle S,\Omega\rangle$ be a signature such that $Bool \in S$ and $true{:}\,Bool$ and $false{:}\,Bool$ are in Ω. A Σ-algebra $A \in RMod[\langle\Sigma,\mathscr{E}\rangle]$ is a *final* (or *terminal*) *model of* $\langle\Sigma,\mathscr{E}\rangle$ if for every $B \in RMod[\langle\Sigma,\mathscr{E}\rangle]$ there is a unique Σ-homomorphism $h{:}\,B \to A$. Taking $\langle\Sigma,\mathscr{E}\rangle$ to denote the class of its final models is called taking its *final semantics*. □

As in the case of initial semantics, the final models of a presentation form an isomorphism class. Recall that a model of a presentation is initial iff it contains no junk and no confusion (Exercise 2.5.19). We can give a similar characterisation of final models as the models containing no junk and *maximal confusion*: a final model A satisfies as many ground equations as possible, subject to the restriction that $A \not\models true = false$ (imposed on all reachable non-degenerate models).

Example 2.7.13. Recall the specification SETNAT from Example 2.7.7, and the classification of models of SETNAT according to the commutativity and idempotence of *add*. The final models of SETNAT are in the class of "set-like" algebras, in which *add* is both commutative and idempotent. (**Exercise:** Why?) □

Not all presentations with equational axioms have final models, but it is possible to impose conditions on the form of presentations that guarantee the existence of final models [BDP+79].

Exercise 2.7.14. Find a variation on the specification STATUS in Example 2.7.11 that has no final models. □

When reachable or final semantics of presentations is used with equational or conditional equational axioms, sometimes more operations are required in specifications than in the case of initial semantics. These additional operations are needed to provide ways of "observing" values of sorts other than *Bool*, in order to avoid models that are degenerate on these other sorts. For example, the presence of the operation *eq* in Example 2.7.7 ensures that $succ^m(0) = succ^n(0)$ only if $m = n$ in all models that do not satisfy $true = false$; it would not be needed if we were interested only in the initial models of SETNAT. Such operations are not required if inequations are allowed as axioms.

Exercise 2.7.15. Recall the presentation NAT given in Exercise 2.5.4. Augment this with the sort *Bool* and constants $true, false\colon Bool$ (to make reachable and final semantics applicable), and show that final models of the resulting specification have a single value of sort *Nat*. Add an operation $even\colon Nat \to Bool$, with the following axioms:

$\forall n{:}Nat$
- $even(succ(succ(n))) = even(n)$
- $even(succ(0)) = false$
- $even(0) = true$

Show that final models of the resulting specification have exactly two values of sort *Nat*. Replace *even* with $\_\_\leq\_\_\colon Nat \times Nat \to Bool$, with appropriate axioms, and show that final models of the resulting specification satisfy $succ^m(0) = succ^n(0)$ iff $m = n$. (We have already seen that this is the case if $eq\colon Nat \times Nat \to Bool$ is added in place of $\leq$.) □

Although the inclusion of additional operations tends to make specifications longer, it is not an artificial device. In practice, one would expect each sort to come with an assortment of operations for creating, manipulating and observing values of that sort, so specifications such as NAT are less natural than NAT augmented with operations like $\leq$ and/or *eq*.

2.7.3 Dealing with partial functions: error algebras

An obvious inadequacy of the framework(s) presented above stems from the use of *total* functions in algebras to interpret the operation names in a signature. Since partial functions are not at all uncommon in computer science applications — a very simple example being the predecessor function $pred{:}\,Nat \to Nat$, which is undefined on 0 — a great deal of work has gone into ways of lifting this restriction. Three main approaches are discussed below:

Error algebras (this subsection): Predecessor is regarded as a total function, with $pred(0)$ specified to yield an *error* value.
Partial algebras (Section 2.7.4): Predecessor is regarded as a partial function.
Order-sorted algebras (Section 2.7.5): Predecessor is regarded as a total function on a sub-domain that excludes the value 0.

A fourth approach is to use ordinary (total) algebras, leaving the value of $pred(0)$ unspecified. This is more an attempt to avoid the issue than a solution, and it is workable only in frameworks that deal adequately with definitions that are not sufficiently complete; see Exercises 2.5.20, 2.7.9, and 5.1.11.

The most obvious way of adding error values to algebras does not work, as the following example demonstrates.

Example 2.7.16. Consider the following specification of the natural numbers, where $pred(0)$ is specified to yield an error:

$$
\begin{array}{lll}
\textbf{spec}\ \textsc{NatPred} = & \textbf{sorts} & Nat \\
& \textbf{ops} & 0{:}\,Nat \\
& & succ{:}\,Nat \to Nat \\
& & pred{:}\,Nat \to Nat \\
& & error{:}\,Nat \\
& & \_\_+\_\_{:}\,Nat \times Nat \to Nat \\
& & \_\_\times\_\_{:}\,Nat \times Nat \to Nat \\
& \forall m,n{:}\,Nat & \\
& & \bullet\ pred(succ(n)) = n \\
& & \bullet\ pred(0) = error \\
& & \bullet\ 0+n = n \\
& & \bullet\ succ(m)+n = succ(m+n) \\
& & \bullet\ 0 \times n = 0 \\
& & \bullet\ succ(m) \times n = (m \times n)+n
\end{array}
$$

Initial models of NATPRED will have many "non-standard" values of sort *Nat*, in addition to the intended one (*error*). For example, the axioms of NATPRED do not force the ground terms $pred(error)$ and $pred(error)+0$ to be equal to any "normal" value, or to *error*. (**Exercise:** Give an initial model of NATPRED.) A possible solution to this is to add axioms that collapse these non-standard values to a single point:

$$
\begin{array}{ll}
\textbf{spec } \mathrm{NATPRED} = \textbf{sorts} & Nat \\
\textbf{ops} & \ldots \\
& \forall m,n{:}Nat \\
& \bullet \ldots \\
& \bullet\ succ(error) = error \\
& \bullet\ pred(error) = error \\
& \bullet\ error + n = error \\
& \bullet\ n + error = error \\
& \bullet\ error \times n = error \\
& \bullet\ n \times error = error
\end{array}
$$

Unfortunately, NATPRED now has only trivial models: $error = 0 \times error = 0$, and so $error = succ(error) = succ(0)$, $error = succ(error) = succ(succ(0))$, and so on. □

The above example suggests that a more delicate treatment is required. A number of approaches have been proposed; here we follow [GDLE84], which is fairly powerful without sacrificing simplicity and elegance. The main ideas of this approach are:

- Error values are distinguished from non-error ("OK") values.
- In an *error signature*, operations that may produce errors when given OK arguments (*unsafe* operations) are distinguished from those that always preserve OK-ness (*safe* operations).
- In an *error algebra*, each carrier is partitioned into an error part and an OK part. Safe operations are required to produce OK results for OK arguments, and homomorphisms are required to preserve OK-ness.
- In equations, variables that can take OK values only (*safe* variables) are distinguished from variables that can take any value (*unsafe* variables). Assignments of values to variables are required to map safe variables to OK values.

Definition 2.7.17 (Error signature). An *error signature* is a triple $\Sigma = \langle S, \Omega, safe \rangle$ where

- $\langle S, \Omega \rangle$ is an ordinary signature; and
- *safe* is an $S^* \times S$-sorted set of functions $\langle safe_{w,s}{:}\, \Omega_{w,s} \to \{tt, ff\} \rangle_{w \in S^*, s \in S}$.

An operation $f{:}\, s_1 \times \cdots \times s_n \to s$ in Σ is *safe* if $safe_{s_1 \ldots s_n, s}(f) = tt$; otherwise it is *unsafe*. □

Example 2.7.16 (revisited). An appropriate error signature for NATPRED would be the following:

$$
\begin{array}{ll}
\Sigma\mathrm{NATPRED} = \textbf{sorts} & Nat \\
\textbf{ops} & 0{:}Nat \\
& succ{:}Nat \to Nat \\
& pred{:}Nat \to Nat,\ unsafe \\
& error{:}Nat,\ unsafe \\
& \_\_+\_\_{:}Nat \times Nat \to Nat \\
& \_\_\times\_\_{:}Nat \times Nat \to Nat
\end{array}
$$

Obviously, *error* is unsafe, and *pred* is unsafe since it produces an error when applied to 0; all the remaining operations are safe. (By convention, the safe operations are those that are not explicitly marked as unsafe.) □

In the rest of this section, let $\Sigma = \langle S, \Omega, \mathit{safe} \rangle$ be an error signature.

Definition 2.7.18 (Error algebra). An *error Σ-algebra A* consists of

- an ordinary Σ-algebra A; and
- an S-sorted set of functions $OK = \langle OK_s : |A|_s \to \{\mathit{tt}, \mathit{ff}\} \rangle_{s \in S}$

such that safe operations preserve OK-ness: for every $f : s_1 \times \cdots \times s_n \to s$ in Σ such that $\mathit{safe}_{s_1 \ldots s_n, s}(f) = \mathit{tt}$ and $a_1 \in |A|_{s_1}, \ldots, a_n \in |A|_{s_n}$ such that $OK_{s_1}(a_1) = \cdots = OK_{s_n}(a_n) = \mathit{tt}$, $OK_s(f_A(a_1, \ldots, a_n)) = \mathit{tt}$. A value $a \in |A|_s$ for $s \in S$ is an *OK value* if $OK_s(a) = \mathit{tt}$; otherwise it is an *error value*. □

Definition 2.7.19 (Error homomorphism). Let A and B be error Σ-algebras. An *error Σ-homomorphism $h : A \to B$* is an S-sorted function $h : |A| \to |B|$ with the usual homomorphism property (for all $f : s_1 \times \cdots \times s_n \to s$ in Σ and $a_1 \in |A|_{s_1}, \ldots, a_n \in |A|_{s_n}$, $h_s(f_A(a_1, \ldots, a_n)) = f_B(h_{s_1}(a_1), \ldots, h_{s_n}(a_n))$) such that h preserves OK-ness: for every $s \in S$ and $a \in |A|_s$ such that $OK_s(a) = \mathit{tt}$ (in A), $OK_s(h_s(a)) = \mathit{tt}$ (in B). □

Definition 2.7.20 (Error variable set). An *error S-sorted variable set X* consists of an S-sorted set X such that $X_s \subseteq \mathscr{X}$ for all $s \in S$, and an S-sorted set of functions $\mathit{safe} = \langle \mathit{safe}_s : X_s \to \{\mathit{tt}, \mathit{ff}\} \rangle_{s \in S}$. A variable $x{:}s$ in X is *safe* if $\mathit{safe}_s(x) = \mathit{tt}$; otherwise it is *unsafe*. An *assignment* of values in an error Σ-algebra A to an error S-sorted variable set X is an S-sorted function $v : X \to |A|$ assigning OK values to safe variables: for every $x{:}s$ in X such that $\mathit{safe}_s(x) = \mathit{tt}$, $OK_s(v_s(x)) = \mathit{tt}$. □

Definition 2.7.21 (Error algebra of terms). Let X be an error S-sorted variable set. The *error Σ-algebra $ET_\Sigma(X)$ of terms with variables X* is defined in an analogous way to the ordinary term algebra $T_\Sigma(X)$, with the following partition of the S-sorted set of terms into OK and error values:

> For all sorts $s \in S$ and Σ-terms $t \in |ET_\Sigma(X)|_s$, if t contains an unsafe variable or operation then $OK_s(t) = \mathit{ff}$; otherwise $OK_s(t) = \mathit{tt}$.

We adopt the same notational conventions for terms as before, dropping sort decorations, etc., when there is no danger of confusion. Let ET_Σ denote $ET_\Sigma(\varnothing)$. □

The definitions of term evaluation, error equation, satisfaction of an error equation by an error algebra, error presentation, model of an error presentation, semantic consequence, and initial model are analogous to the definitions given earlier in the standard many-sorted algebraic framework (Definitions 1.4.5, 2.1.1, 2.1.2, 2.2.1, 2.2.2, 2.3.6 and 2.5.13 respectively). Because assignments are required to map safe variables to OK values, an error equation may be satisfied by an error algebra even if it is not satisfied when error values are substituted for safe variables.

Exercise 2.7.22. Spell out the details of these definitions. □

As before, every error presentation has an isomorphism class of initial models, and an analogous quotient construction gives an initial model.

Definition 2.7.23 (Congruence generated by a set of equations). Let $\mathscr{E}$ be a set of error Σ-equations. The Σ-congruence $\equiv_{\mathscr{E}}$ on ET_Σ is defined by $t \equiv_{\mathscr{E}} t' \iff \mathscr{E} \models_\Sigma t = t'$ for all $t,t' \in |ET_\Sigma|$. $\equiv_{\mathscr{E}}$ is called the *Σ-congruence generated by $\mathscr{E}$*. (NOTE: A Σ-congruence on an error Σ-algebra A is just an ordinary Σ-congruence on the ordinary Σ-algebra underlying A.) □

Definition 2.7.24 (Quotient error algebra). Let A be an error Σ-algebra, and let $\equiv$ be a Σ-congruence on A. The definition of $A/\equiv$, the *quotient error algebra of A modulo* $\equiv$, is analogous to that of the ordinary quotient algebra $A/\equiv$, with the following partition of congruence classes into OK and error values:

> For all sorts $s \in S$ and congruence classes $[a]_{\equiv_s} \in |A/\equiv|_s$, if there is some $b \in [a]_{\equiv_s}$ such that $OK_s(b) = tt$ (in A), then $OK_s([a]_{\equiv_s}) = tt$ (in $A/\equiv$); otherwise $OK_s([a]_{\equiv_s}) = ff$. □

Note that if there are both OK and error values in a congruence class, the class is regarded as an OK value in the quotient.

Theorem 2.7.25 (Initial model theorem). *The error Σ-algebra $ET_\Sigma/\equiv_{\mathscr{E}}$ is an initial model of the error presentation $\langle\Sigma,\mathscr{E}\rangle$.* □

Exercise 2.7.26. Sketch a proof of Theorem 2.7.25. HINT: Take inspiration from the proof of Theorem 2.5.14. □

Exercise 2.7.27. Try to find conditions analogous to "no junk" and "no confusion" that characterise the initial models of an error presentation. □

Example 2.7.16 (revisited). Using the approach outlined above, here is an improved version of the specification NATPRED:

spec NATPRED = **sorts** *Nat*
ops $0{:}Nat$
$succ{:}Nat \to Nat$
$pred{:}Nat \to Nat$, *unsafe*
$error{:}Nat$, *unsafe*
$\_\_+\_\_{:}Nat \times Nat \to Nat$
$\_\_\times\_\_{:}Nat \times Nat \to Nat$
$\forall m,n{:}Nat$
• $pred(succ(n)) = n$
• $pred(0) = error$
• $0+n = n$
• $succ(m)+n = succ(m+n)$
• $0 \times n = 0$
• $succ(m) \times n = (m \times n)+n$

(By convention, variables in equations are safe unless otherwise indicated.) In initial models of NATPRED, the error values of sort *Nat* correspond exactly to "error messages", i.e. ground terms containing at least one occurrence of *error*. These terms can be regarded as recording the sequence of events that took place since the error occurred. The record is accurate since the initial models of NATPRED do *not* satisfy equations like $0 \times error = 0$, in contrast to the initial models of the earlier version. To collapse the error values to a single point without affecting the OK values, axioms can be added as follows:

spec NATPRED = **sorts** *Nat*
ops ...
$\forall m, n{:}Nat, k{:}Nat{:}unsafe$
- ...
- $pred(error) = error$
- $succ(error) = error$
- $error + k = error$
- $k + error = error$
- $error \times k = error$
- $k \times error = error$

It is also possible to specify *error recovery* using this approach:

spec NATPRED = **sorts** *Nat*
ops ...
$recover{:}Nat \to Nat$
$\forall m, n{:}Nat, k{:}Nat{:}unsafe$
- ...
- $recover(error) = 0$
- $recover(n) = n$

In initial models of this version of NATPRED, *recover* is the identity on *Nat* except that $recover(error)$ gives the OK value 0. □

Although only initial semantics of error presentations has been mentioned above, the alternatives of reachable and final semantics apply as in the standard case. The key points of the standard framework not mentioned here (e.g. analogues to the soundness, completeness and incompleteness theorems) carry over to the present framework as well.

Exercise 2.7.28. Find a definition of error signature morphism which makes the Satisfaction Lemma hold, taking the natural definition of the σ-reduct $A'|_\sigma$ of an error Σ'-algebra A' induced by an error signature morphism $\sigma{:}\Sigma \to \Sigma'$. □

Although the approach to error specification presented above is quite attractive, there are examples that cannot be treated in this framework.

Exercise 2.7.29. Consider the following specification of *bounded natural numbers*:

spec BOUNDEDNAT = **sorts** *Nat*
ops $0{:}Nat$
$succ{:}Nat \to Nat$, *unsafe*
$overflow{:}Nat$, *unsafe*
• $succ(succ(succ(succ(succ(succ(0)))))) = overflow$

The intention is to specify a (very) restricted subset of the natural numbers, where an attempt to compute a number larger than 5 results in overflow. Show that an initial model of BOUNDEDNAT will have only one OK value. Change BOUNDEDNAT so that its initial models have six OK values (corresponding to $0, succ(0), \ldots, succ^5(0)$). What if the bound is 2^{32} rather than 5? □

2.7.4 Dealing with partial functions: partial algebras

An obvious way to deal with partial functions is to simply change the definition of algebra to allow operation names to be interpreted as partial functions. But for many of the basic notions in the framework that depend on the definition of algebra, beginning with the concepts of subalgebra and homomorphism, there are several ways to extend the usual definition to the partial case. Choosing a coherent combination of these definitions is a delicate matter. Here we follow the approach of [BW82b].

Throughout this section, let $\Sigma = \langle S, \Omega \rangle$ be a signature.

Definition 2.7.30 (Partial algebra). A *partial* Σ*-algebra* A is like an ordinary Σ-algebra, except that each $f{:}s_1 \times \cdots \times s_n \to s$ in Σ is interpreted as a *partial* function $(f{:}s_1 \times \cdots \times s_n \to s)_A{:}|A|_{s_1} \times \cdots \times |A|_{s_n} \to |A|_s$. The *(total)* Σ*-algebra underlying* A is the Σ-algebra $A_\perp$ defined as follows:

- $|A_\perp|_s = |A|_s \uplus \{\perp_s\}$ for every $s \in S$; and
- $(f{:}s_1 \times \cdots \times s_n \to s)_{A_\perp}(a_1, \ldots, a_n) =$
$$\begin{cases} \perp_s & \text{if } a_j = \perp_{s_j} \text{ for some } 1 \le j \le n \\ (f{:}s_1 \times \cdots \times s_n \to s)_A(a_1, \ldots, a_n) & \text{if this is defined} \\ \perp_s & \text{otherwise} \end{cases}$$
for every $f{:}s_1 \times \cdots \times s_n \to s$ and $a_1 \in |A_\perp|_{s_1}, \ldots, a_n \in |A_\perp|_{s_n}$. □

We employ the same notational conventions as before. Note that according to this definition, the value of a constant need not be defined: a constant $c{:}s$ is associated in an algebra A with a partial function $c_A{:}\{\langle\rangle\} \to |A|_s$, where $\{\langle\rangle\}$ is the nullary Cartesian product.

Definition 2.7.31 (Homomorphism). Let A and B be partial Σ-algebras. A *weak* Σ*-homomorphism* $h{:}A \to B$ is an S-sorted (total) function $h{:}\,|A| \to |B|$ such that for all $f{:}s_1 \times \cdots \times s_n \to s$ in Σ and $a_1 \in |A|_{s_1}, \ldots, a_n \in |A|_{s_n}$,

if $f_A(a_1, \ldots, a_n)$ is defined then $f_B(h_{s_1}(a_1), \ldots, h_{s_n}(a_n))$ is defined, and
$h_s(f_A(a_1, \ldots, a_n)) = f_B(h_{s_1}(a_1), \ldots, h_{s_n}(a_n))$.

If moreover h satisfies the condition

if $f_B(h_{s_1}(a_1),\dots,h_{s_n}(a_n))$ is defined then $f_A(a_1,\dots,a_n)$ is defined

then h is called a *strong Σ-homomorphism.* □

Other possibilities would be generated by allowing homomorphisms to be partial functions.

Exercise 2.7.32. Consider a partial Σ-algebra A and its underlying total Σ-algebra $A_\perp$. Given any Σ-congruence $\equiv$ on $A_\perp$, removing all pairs involving $\perp$ yields a *strong Σ-congruence on A*. Check that such strong congruences are exactly kernels of strong Σ-homomorphisms; cf. Exercises 1.3.14 and 1.3.18. Check that strong congruences are equivalence relations that preserve and reflect definedness of operations and are closed under defined operations. Kernels of weak Σ-homomorphisms are *weak Σ-congruences*: equivalence relations that are closed under defined operations. Spell out these definitions in detail. For any partial Σ-algebra A and weak Σ-congruence $\equiv$ on A, generalise Definition 1.3.15 to define the *quotient of A by $\equiv$*, written $A/{\equiv}$. Note that an operation is defined in $A/{\equiv}$ on a tuple of equivalence classes provided that in A it is defined on at least one tuple of their respective elements. Check which of Exercises 1.3.18–1.3.23 carry over. □

Definition 2.7.33 (Term evaluation). Let X be an S-sorted set of variables, let A be a partial Σ-algebra, and let $v{:}X \to |A|$ be a (total) S-sorted function assigning values in A to variables in X. Since $|A| \subseteq |A_\perp|$, this is an S-sorted function $v_\perp{:}X \to |A_\perp|$, and by Fact 1.4.4 there is a unique (ordinary) Σ-homomorphism $v^\#_\perp : T_\Sigma(X) \to A_\perp$ which extends $v_\perp$. Let $s \in S$ and let $t \in |T_\Sigma(X)|_s$ be a Σ-term of sort s; the *value of t in A under the valuation v* is $v^\#_\perp(t)$ if $v^\#_\perp(t) \neq \perp_s$, and is undefined otherwise. □

Satisfaction of an equation $\forall X \bullet t = t'$, where the values of t and/or t' may be undefined, can be defined in several different ways. Following [BW82b], we use *strong* equality (also known as *Kleene* equality), whereby the equality holds if (for any assignment of values to variables) the values of t and t' either are both defined and equal, or are both undefined. The usual interpretation of definitional equations in recursive function definitions (see for instance Example 4.1.25 and Exercise 4.1.30 below) makes them hold as strong equations. An alternative is *existential equality* (where $=$ is usually written $\stackrel{e}{=}$), whereby the equality holds only when the values of t and t' are defined and equal. When strong equality is used, there is a need for an additional form of axiom called a *definedness formula*: $\forall X \bullet \mathit{def}(t)$ holds if for any assignment of values to variables, the value of t is defined. These are superfluous with existential equality since $\forall X \bullet \mathit{def}(t)$ holds iff $\forall X \bullet t \stackrel{e}{=} t$ holds. Definedness formulae with $X = \varnothing$ are called *ground* and are often written without quantification as $\mathit{def}(t)$.

Exercise 2.7.34. Formalise the definitions of satisfaction of equations (using strong equality) and of definedness formulae. □

Using both equations and definedness formulae as axioms, the definitions of presentation, model of a presentation, semantic consequence, isomorphism, and initial model (with respect to *weak* homomorphisms) are analogous to those given earlier.

Exercise 2.7.35. Spell out the details of these definitions. Note though that not all of the properties of these notions carry over from the standard algebraic framework; for instance, a (weak) bijective homomorphism need not be an isomorphism of partial algebras. □

Theorem 2.7.36 (Initial model theorem). *Any presentation $\langle \Sigma, \mathscr{E} \rangle$ has an initial model I, characterised by the following properties:*

- *I contains no junk;*
- *I is* minimally defined, *i.e. for all $t \in |T_\Sigma|$, t_I is defined only if $\mathscr{E} \models_\Sigma def(t)$; and*
- *I contains no confusion, i.e. for all $t, t' \in |T_\Sigma|_s, s \in S$, t_I and t'_I are defined and equal only if $\mathscr{E} \models_\Sigma t = t'$.*

Proof sketch. Let $\Sigma_\perp$ be the signature obtained by adding a constant $\perp_s : s$ to Σ for each sort $s \in S$. Define a congruence $\sim \subseteq |T_{\Sigma_\perp}| \times |T_{\Sigma_\perp}|$ as follows: for $t_1, t_2 \in |T_{\Sigma_\perp}|_s$ for some $s \in S$, $t_1 \sim t_2$ iff any of the following conditions holds:

1. t_1 contains $\perp_{s'}$ and t_2 contains $\perp_{s''}$ for some $s', s'' \in S$;
2. t_1 contains $\perp_{s'}$ for some $s' \in S$, $t_2 \in |T_\Sigma|_s$ (so t_2 does not contain $\perp_{s''}$ for any $s'' \in S$) and $\mathscr{E} \not\models def(t_2)$, or vice versa;
3. $t_1, t_2 \in |T_\Sigma|_s$, and either $\mathscr{E} \not\models def(t_1)$ and $\mathscr{E} \not\models def(t_2)$ or $\mathscr{E} \models t_1 = t_2$.

I is constructed by taking the quotient of $T_{\Sigma_\perp}$ by $\sim$, and then regarding congruence classes containing the constants $\perp_s$ as undefined values. □

Exercise 2.7.37. Complete the above proof by showing that

- $\sim$ is a congruence on $T_{\Sigma_\perp}$;
- $I \models \mathscr{E}$;
- I is an initial model of $\langle \Sigma, \mathscr{E} \rangle$; and
- I has the properties promised in Theorem 2.7.36.

Show that any model of $\langle \Sigma, \mathscr{E} \rangle$ satisfying the properties in Theorem 2.7.36 is isomorphic to I and is therefore an initial model of $\langle \Sigma, \mathscr{E} \rangle$. □

Exercise 2.7.38. Suppose that we modify Theorem 2.7.36 by replacing the phrase "t_I and t'_I are defined and equal" with "$I \models_\Sigma t = t'$". Give a counterexample showing that this version of the theorem is false. □

Exercise 2.7.39. A partial Σ-algebra $A \in Mod[\langle \Sigma, \mathscr{E} \rangle]$ is a *strongly initial model of* $\langle \Sigma, \mathscr{E} \rangle$ if for every minimally defined $B \in Mod[\langle \Sigma, \mathscr{E} \rangle]$ containing no junk, there is a unique strong Σ-homomorphism $h: A \to B$. Show that I is an initial model of $\langle \Sigma, \mathscr{E} \rangle$ iff I is a strongly initial model of $\langle \Sigma, \mathscr{E} \rangle$. □

Again, reachable and final semantics are applicable for partial algebras as well as initial semantics, and the key points of the standard framework carry over with appropriate changes (for instance, the equational calculus must be modified to deal with definedness formulae as well as equations).

Example 2.7.16 (revisited). Here is a version of the specification NATPRED in which *pred* is specified to be a partial function:

$$\begin{array}{lll}
\textbf{spec}\ \mathrm{NATPRED} = & \textbf{sorts}\ \mathit{Nat} & \\
& \textbf{ops} & 0{:}\,\mathit{Nat} \\
& & \mathit{succ}{:}\,\mathit{Nat} \rightarrow \mathit{Nat} \\
& & \mathit{pred}{:}\,\mathit{Nat} \rightarrow \mathit{Nat} \\
& & __+__{:}\,\mathit{Nat} \times \mathit{Nat} \rightarrow \mathit{Nat} \\
& & __\times__{:}\,\mathit{Nat} \times \mathit{Nat} \rightarrow \mathit{Nat} \\
& \forall m,n{:}\,\mathit{Nat} & \\
& & \bullet\ \mathit{def}(0) \\
& & \bullet\ \mathit{def}(\mathit{succ}(n)) \\
& & \bullet\ \mathit{pred}(\mathit{succ}(n)) = n \\
& & \bullet\ 0+n = n \\
& & \bullet\ \mathit{succ}(m)+n = \mathit{succ}(m+n) \\
& & \bullet\ 0 \times n = 0 \\
& & \bullet\ \mathit{succ}(m) \times n = (m \times n)+n
\end{array}$$

In initial models of NATPRED, all operations behave as expected, and all are total except for *pred*, which is undefined only on 0.

Exercise. Show that $\forall m,n{:}\mathit{Nat} \bullet \mathit{def}(m+n)$ and $\forall m,n{:}\mathit{Nat} \bullet \mathit{def}(m \times n)$ are consequences of the definedness axioms for 0 and *succ* and the equations defining $+$ and $\times$ in reachable models of NATPRED. You will need to use induction, so first formulate an appropriate induction rule scheme and convince yourself that it is sound.

Exercise. Suppose that the axiom $\mathit{def}(0)$ were removed from NATPRED. Describe the initial models of the resulting presentation. □

2.7.5 Partial functions: order-sorted algebras

Any partial function amounts to a total function on a restricted domain. The idea of *order-sorted algebra* is to avoid partial functions by enabling the domain of each function to be specified exactly. This is done by introducing *subsorts*, which correspond to subsets at the level of values, and requiring operations to behave in an appropriate fashion when applied to a value of a subsort or when expected to deliver a value of a supersort. A number of different approaches to order-sorted algebra have been proposed, and their relative merits are a matter for debate. Here we follow the approach of [GM92].

Definition 2.7.40 (Order-sorted signature). An *order-sorted signature* is a triple $\Sigma = \langle S, \leq, \Omega \rangle$ where $\langle S, \Omega \rangle$ is an ordinary signature and $\leq$ is a partial order on the set S of sort names, such that whenever $f{:}\, s_1 \times \cdots \times s_n \rightarrow s$ and $f{:}\, s'_1 \times \cdots \times s'_n \rightarrow s'$ are operations (having the same name and same number of arguments) in Ω and $s_i \leq s'_i$ for all $1 \leq i \leq n$, then $s \leq s'$. When $s \leq s'$ for $s, s' \in S$, we say that s is a *subsort* of s' (or equivalently, s' is a *supersort* of s). The subsort ordering is extended to sequences of sorts of equal length in the usual way: $s_1 \dots s_n \leq s'_1 \dots s'_n$ if $s_i \leq s'_i$ for all $1 \leq i \leq n$. □

The restriction on Ω ([GM92] calls this condition *monotonicity*) is a fairly natural one, keeping in mind that the subsort ordering corresponds to subset on the value level: restricting a function to a subset of its domain may diminish, but not enlarge, its codomain. Note that an effect of this restriction is to rule out overloaded constants.

Throughout the rest of this section, let $\Sigma = \langle S, \leq, \Omega \rangle$ be an order-sorted signature, and let $\widehat{\Sigma} = \langle S, \Omega \rangle$ be the (ordinary) signature underlying Σ.

Definition 2.7.41 (Order-sorted algebra). An *order-sorted Σ-algebra A* is an ordinary $\widehat{\Sigma}$-algebra, such that:

- for all $s \leq s'$ in Σ, $|A|_s \subseteq |A|_{s'}$; and
- whenever $f{:}s_1 \times \cdots \times s_n \to s$ and $f{:}s'_1 \times \cdots \times s'_n \to s'$ are operations (having the same name and same number of arguments) in Ω and $s_1 \dots s_n \leq s'_1 \dots s'_n$, the function $(f{:}s_1 \times \cdots \times s_n \to s)_A{:}\,|A|_{s_1} \times \cdots \times |A|_{s_n} \to |A|_s$ is the set-theoretic restriction of the function $(f{:}s'_1 \times \cdots \times s'_n \to s')_A{:}\,|A|_{s'_1} \times \cdots \times |A|_{s'_n} \to |A|_{s'}$. □

An effect of the second restriction ([GM92] calls this condition *monotonicity* as well) is to prevent ambiguity in the evaluation of terms; see below.

Definition 2.7.42 (Order-sorted homomorphism). Let A and B be order-sorted Σ-algebras. An *order-sorted Σ-homomorphism* $h{:}A \to B$ is an ordinary $\widehat{\Sigma}$-homomorphism such that $h_s(a) = h_{s'}(a)$ for all $a \in |A|_s$ whenever $s \leq s'$. When h has an inverse, it is an *order-sorted Σ-isomorphism* and we write $A \cong B$. □

Let X be an S-sorted set (of variables) such that X_s and $X_{s'}$ are disjoint for $s \neq s'$.

Definition 2.7.43 (Order-sorted term algebra). The *order-sorted Σ-algebra $T_\Sigma(X)$ of terms with variables X* is just like $T_{\widehat{\Sigma}}(X)$, except that for any term $t \in |T_\Sigma(X)|_s$ such that $s \leq s'$, we also have $t \in |T_\Sigma(X)|_{s'}$. Let $T_\Sigma = T_\Sigma(\varnothing)$. □

Exercise 2.7.44. Check that $T_\Sigma(X)$ is an order-sorted Σ-algebra. □

Example 2.7.45. One way of reformulating NATPRED as an order-sorted specification (see below) will involve introducing a sort *NzNat* (non-zero natural numbers) such that $NzNat \leq Nat$, with operations $0{:}\,Nat$ and $succ{:}\,Nat \to NzNat$. According to the definition of order-sorted term algebra, the term $succ(0)$ has sort *Nat* as well as *NzNat*, which means that $succ(succ(0))$ is well formed (and has sort *Nat* as well as *NzNat*). □

As the above example demonstrates, a given term may appear in more than one carrier of $T_\Sigma(X)$. The following condition on Σ ensures that this does not lead to ambiguity.

Definition 2.7.46 (Regular order-sorted signature). Σ is *regular* if for any $f{:}s_1 \times \cdots \times s_n \to s$ in Σ and $s'_1 \dots s'_n \leq s_1 \dots s_n$, there is a least $s^*_1 \dots s^*_n s^*$ such that $s'_1 \dots s'_n \leq s^*_1 \dots s^*_n$ and $f{:}s^*_1 \times \cdots \times s^*_n \to s^*$ is in Σ. □

Theorem 2.7.47 (Terms have least sorts). *If Σ is regular, then for every term $t \in |T_\Sigma(X)|$ there is a least sort $s \in S$, written $sort(t)$, such that $t \in |T_\Sigma(X)|_s$.* □

Exercise 2.7.48. Prove Theorem 2.7.47. What happens when X is an *arbitrary* S-sorted set, i.e. if we remove the restriction that X_s and $X_{s'}$ are disjoint for $s \neq s'$? □

Now the definition of term evaluation is analogous to the usual one.

Fact 2.7.49. *Suppose that Σ is regular. Then, for any order-sorted Σ-algebra A and S-sorted function $v{:}X \to |A|$, there is exactly one order-sorted Σ-homomorphism $v^\#{:}T_\Sigma(X) \to A$ which extends v, i.e. such that $v^\#_s(x) = v_s(x)$ for all $s \in S$, $x \in X_s$.* □

Exercise 2.7.50. Define term evaluation. □

Definition 2.7.51 (Order-sorted equation; satisfaction). Suppose that Σ is regular, and let the equivalence relation $\equiv$ be the symmetric transitive closure of $\leq$. *Order-sorted Σ-equations* $\forall X \bullet t = t'$ are as usual, except that we require $sort(t) \equiv sort(t')$ (in other words, $sort(t)$ and $sort(t')$ are in the same *connected component* of $\langle S, \leq \rangle$) instead of $sort(t) = sort(t')$. An order-sorted Σ-algebra A *satisfies* an order-sorted Σ-equation $\forall X \bullet t = t'$, written $A \models_\Sigma \forall X \bullet t = t'$, if the value of t in $|A|_{sort(t)}$ and the value of t' in $|A|_{sort(t')}$ coincide for every S-sorted function $v{:}X \to |A|$. □

A problem with this definition is that satisfaction of order-sorted Σ-equations is not preserved by order-sorted Σ-isomorphisms (compare Exercise 2.1.5). The following condition on Σ ensures that this anomaly does not arise.

Definition 2.7.52 (Coherent order-sorted signature). $\langle S, \leq \rangle$ is *filtered* if for any $s, s' \in S$ there is some $s'' \in S$ such that $s \leq s''$ and $s' \leq s''$. $\langle S, \leq \rangle$ is *locally filtered* if each of its connected components is filtered. Σ is *coherent* if $\langle S, \leq \rangle$ is locally filtered and Σ is regular. □

Exercise 2.7.53. Find Σ, A, B and e such that Σ is regular, $A \models_\Sigma e$ and $A \cong B$ but $B \not\models_\Sigma e$. Show that if Σ is coherent then this is impossible. □

The definitions of order-sorted presentation, model of an order-sorted presentation, semantic consequence, and initial model are analogous to those given earlier. For every order-sorted presentation $\langle \Sigma, \mathscr{E} \rangle$ such that Σ is coherent, an initial model may be constructed as a quotient of T_Σ [GM92]. There is a version of the equational calculus that is sound and complete for coherent signatures [GM92], and the use of term rewriting for proof as discussed in Section 2.6 is sound, provided that each rewrite rule $t \to t'$ is *sort decreasing*, i.e. $sort(t') \leq sort(t)$ [KKM88].

Example 2.7.16 (revisited). Here is a version of the specification NATPRED in which *pred* is specified to be a total function on the non-zero natural numbers:

$$
\begin{aligned}
\textbf{spec } \textrm{NATPRED} = \textbf{sorts } & \mathit{NzNat} \leq \mathit{Nat} \\
\textbf{ops } & 0{:}\,\mathit{Nat} \\
& \mathit{succ}{:}\,\mathit{Nat} \to \mathit{NzNat} \\
& \mathit{pred}{:}\,\mathit{NzNat} \to \mathit{Nat} \\
& \_\_+\_\_{:}\,\mathit{Nat} \times \mathit{Nat} \to \mathit{Nat} \\
& \_\_\times\_\_{:}\,\mathit{Nat} \times \mathit{Nat} \to \mathit{Nat} \\
\forall m,n{:}\,\mathit{Nat} & \\
& \bullet\ \mathit{pred}(\mathit{succ}(n)) = n \\
& \bullet\ 0+n = n \\
& \bullet\ \mathit{succ}(m)+n = \mathit{succ}(m+n) \\
& \bullet\ 0 \times n = 0 \\
& \bullet\ \mathit{succ}(m) \times n = (m \times n)+n
\end{aligned}
$$

In this version of NATPRED, there are terms that are not well formed in spite of the fact that each operator application seems to be to a value in its domain. For example, consider the following "term":

$$\mathit{pred}(\mathit{succ}(0)+\mathit{succ}(0)).$$

According to the signature of NATPRED, $\mathit{succ}(0)+\mathit{succ}(0)$ is a term of sort *Nat*; it is not a term of sort *NzNat* in spite of the fact that its value is non-zero. In the term algebra, *pred* applies only to terms of sort *NzNat*; thus the application of *pred* to $\mathit{succ}(0)+\mathit{succ}(0)$ is not defined. One way of getting around this problem might be to add additional operators to the signature of NATPRED:

$$
\begin{aligned}
\textbf{spec } \textrm{NATPRED} = \textbf{sorts } & \mathit{NzNat} \leq \mathit{Nat} \\
\textbf{ops } & \ldots \\
& \_\_+\_\_{:}\,\mathit{NzNat} \times \mathit{Nat} \to \mathit{NzNat} \\
& \_\_+\_\_{:}\,\mathit{Nat} \times \mathit{NzNat} \to \mathit{NzNat} \\
& \_\_\times\_\_{:}\,\mathit{NzNat} \times \mathit{NzNat} \to \mathit{NzNat} \\
\ldots &
\end{aligned}
$$

Then $\mathit{succ}(0)+\mathit{succ}(0)$ is a term of sort *NzNat*, as desired. Unfortunately, this signature is not regular. (**Exercise:** Why not? What can be done to make it regular?)

An alternative is to use a so-called *retract*, an additional operation for converting from a sort to one of its subsorts:

$$
\begin{aligned}
\textbf{spec } \textrm{NATPRED} = \textbf{sorts } & \mathit{NzNat} \leq \mathit{Nat} \\
\textbf{ops } & \ldots \\
& r{:}\,\mathit{Nat} \to \mathit{NzNat} \\
\forall m,n{:}\,\mathit{Nat}, k{:}\,\mathit{NzNat} & \\
& \bullet\ \ldots \\
& \bullet\ r(k) = k
\end{aligned}
$$

Now, the term $\mathit{pred}(r(\mathit{succ}(0)+\mathit{succ}(0)))$ is well formed, and is equal to $\mathit{succ}(0)$ in all models of NATPRED. In the words of [GM92], inserting the retract r into $\mathit{pred}(r(\mathit{succ}(0)+\mathit{succ}(0)))$ gives it "the benefit of the doubt", and the term is "vindicated" by the fact that it is equal to a term that does not contain r. The term *pred*$(r(0))$ is also well formed, but in the initial model of NATPRED this term is equal only to other terms containing the retract *r*, and can thus be regarded as an error message. The use of retracts (which can be inserted automatically) is well behaved under certain conditions on order-sorted presentations [GM92].

Another version of NATPRED is obtained by using an *error supersort* for the codomain of *pred* rather than a subsort for its domain:

$$
\begin{array}{lll}
\textbf{spec } \text{NATPRED} = & \textbf{sorts} & Nat \le Nat? \\
& \textbf{ops} & 0{:}\,Nat \\
& & succ{:}\,Nat \to Nat \\
& & pred{:}\,Nat \to Nat? \\
& & \_\_+\_\_{:}\,Nat \times Nat \to Nat \\
& & \_\_\times\_\_{:}\,Nat \times Nat \to Nat \\
& \forall m,n{:}\,Nat & \\
& & \bullet\ pred(succ(n)) = n \\
& & \bullet\ 0+n = n \\
& & \bullet\ succ(m)+n = succ(m+n) \\
& & \bullet\ 0\times n = 0 \\
& & \bullet\ succ(m)\times n = (m\times n)+n
\end{array}
$$

The sort *Nat*? may be thought of as *Nat* extended by the addition of an error value corresponding to *pred*(0).

Here we have the same problem with ill-formed terms as before; an example is the term *succ*(*pred*(*succ*(0))). Again, retracts solve the problem. In this case, the required retract is the operation $r{:}\,Nat? \to Nat$, defined by the axiom $\forall n{:}Nat \bullet r(n) = n$. □

Exercise 2.7.54. Try to view the error algebra approach presented in Section 2.7.3 as a special case of order-sorted algebra. □

2.7.6 Other options

The previous sections have mentioned only a few of the ways in which the standard framework can be improved to make it more suitable for particular kinds of applications. A great many other variations are possible; a few of these are sketched below.

Example 2.7.55 (First-order predicate logic). Signatures may be modified to enable them to include (typed) *predicate names* in addition to operation names, e.g. $\_\_\le\_\_{:}\,Nat \times Nat$. Atomic formulae are then formed by applying predicates to terms; in *first-order predicate logic with equality*, the predicate $\_\_=\_\_{:}\,s\times s$ is implicitly available for any sort *s*. Formulae are built from atomic formulae using logical connectives and quantifiers. Algebras are modified to include relations on their carriers to interpret predicate names; the interpretation of the built-in equality predicate (if available) may be forced to be the underlying equality on values, or it may merely be required to be a congruence relation. Homomorphisms are required to respect predicates as well as operations. The satisfaction of a *sentence* (a formula without free variables) by an algebra is as in first-order logic. See Example 4.1.12 for details of the version of first-order predicate logic with equality we will use. Presentations involving predicates and first-order axioms are appropriate for the specification of programs in *logic programming languages* such as Prolog, where the Horn clause fragment of first-order logic is used for writing the programs themselves. Note that such presentations may have no models at all, but even if they have some models, they may have no initial models (see Example 2.7.11) or no final models (see Exercise 2.7.14), or even no reachable models. (**Exercise:** Give a specification with first-order axioms having some models but no reachable model.) □

Example 2.7.56 (Higher-order functions). Higher-order functions (taking functions as parameters and/or returning functions as results) can be accommodated by interpreting certain sort names as (subsets of) function spaces. Given a set S of (base) sorts, let $S^{\rightarrow}$ be the closure of S under formation of function types: $S^{\rightarrow}$ is the smallest set such that $S \subseteq S^{\rightarrow}$ and for all $s_1,\ldots,s_n,s \in S^{\rightarrow}$, $s_1 \times \cdots \times s_n \rightarrow s \in S^{\rightarrow}$. Then a *higher-order signature* Σ is a pair $\langle S,\Omega\rangle$ where Ω is an $S^{\rightarrow}$-indexed set of operation names. This determines an ordinary signature $\Sigma^{\rightarrow}$ comprised of the sort names $S^{\rightarrow}$ and the operation names in Ω together with operation names $apply\colon (s_1 \times \cdots \times s_n \rightarrow s) \times s_1 \times \cdots \times s_n \rightarrow s$ for every $s_1,\ldots,s_n,s \in S^{\rightarrow}$. Note that, except for the various instances of *apply*, all the operations in $\Sigma^{\rightarrow}$ are constants, albeit possibly of "functional" sort. A *higher-order Σ-algebra* is just an ordinary (total) $\Sigma^{\rightarrow}$-algebra, and analogously for the definitions of higher-order Σ-homomorphism, reachable higher-order Σ-algebra, higher-order Σ-term, higher-order Σ-equation, satisfaction of a higher-order Σ-equation by a higher-order Σ-algebra, and higher-order presentation. A higher-order Σ-algebra A is *extensional* if for all sorts $s_1 \times \cdots \times s_n \rightarrow s \in S^{\rightarrow}$ and values $f,g \in |A|_{s_1\times\cdots\times s_n\rightarrow s}$, $f = g$ whenever $apply_A(f,a_1,\ldots,a_n) = apply_A(g,a_1,\ldots,a_n)$ for all $a_1 \in |A|_{s_1},\ldots,a_n \in |A|_{s_n}$. Any extensional higher-order algebra is isomorphic to an (extensional) algebra A, where every carrier $|A|_{s_1\times\cdots\times s_n\rightarrow s}$ is a subset of the function space $|A|_{s_1} \times \cdots \times |A|_{s_n} \rightarrow |A|_s$ and all the operations $apply_A$ are the usual function application. A higher-order Σ-algebra A is a *model* of a presentation $\langle\Sigma,\mathscr{E}\rangle$ if $A \models_\Sigma \mathscr{E}$, A is extensional, and A is reachable. The reachability requirement (no junk) means that $|A|_{s_1\times\cdots\times s_n\rightarrow s}$ will almost never be the full function space $|A|_{s_1} \times \cdots \times |A|_{s_n} \rightarrow |A|_s$: only the functions that are denotable by ground terms will be present in $|A|_{s_1\times\cdots\times s_n\rightarrow s}$. Higher-order (equational) presentations always have initial models [MTW88]. □

Example 2.7.57 (Polymorphic types). Standard ML [Pau96] and some other programming languages define *polymorphic types* such as α `list` (instances of which include `bool list` and `(bool list) list`) and *polymorphic values* of those types, such as `head`: $\forall\alpha\bullet\alpha$ `list` $\rightarrow \alpha$ (which is then applicable to values of types such as `bool list` and `(bool list) list`, yielding results of types `bool` and `bool list`, respectively). To specify such types and functions, signatures are modified to contain *type constructors* in place of sort names; for example, `list` is a unary type constructor and `bool` is a nullary type constructor. Terms built using these type constructors and *type variables* (such as α above) are the *polymorphic types* of the signature. The set Ω of operation names is then indexed by non-empty sequences of polymorphic types, where $f \in \Omega_{t_1 \dots t_n,t}$ means $f{:}\forall FV(t_1) \cup \dots \cup FV(t_n) \cup FV(t) \bullet t_1 \times \cdots \times t_n \to t$. There are various choices for algebras over such signatures. Perhaps the most straightforward choice is to require each algebra A to incorporate a (single-sorted) *algebra of carriers* $Carr(A)$, having sets interpreting types as values and with an operation to interpret each type constructor. Then, for each operation $f \in \Omega_{t_1 \dots t_n,t}$ and for each instantiation of type variables $i{:}V \to |Carr(A)|$, A has to provide a function $f_{A,i}{:}i^{\#}(t_1) \times \cdots \times i^{\#}(t_n) \to i^{\#}(t)$. Various conditions may be imposed to ensure that the interpretation of polymorphic operations is *parametric* in the sense of [Str67], by requiring $f_{A,i}$ and $f_{A,i'}$ to be appropriately related for different type variable instantiations i, i'; see Exercise 3.4.40 for a hint in this direction. Another choice would be to interpret each type as the set of equivalence classes of a *partial equivalence relation* on a model of the untyped λ-calculus [BC88]. Axioms contain (universal) quantifiers for type variables in addition to quantifiers for ordinary variables, as in System F [Gir89]; alternatively, type variable quantification may be left implicit, as in Extended ML [KST97]. □

Example 2.7.58 (Non-deterministic functions). Non-deterministic functions may be handled by interpreting operation names in algebras as relations, or equivalently as set-valued functions. Homomorphisms are required to preserve possible values of functions: for any homomorphism $h{:}A \to B$ and operation $f{:}s_1 \times \cdots \times s_n \to s$, if a is a possible value of $f_A(a_1,\dots,a_n)$ then $h_s(a)$ is a possible value of $f_B(h_{s_1}(a_1),\dots,h_{s_n}(a_n))$. Universally quantified inclusions between sets of possible values may be used as axioms: $t \subseteq t'$ means that every possible value of t is a possible value of t'. □

Example 2.7.59 (Recursive definitions). Following [Sco76], partial functions may be specified as least solutions of recursive equations, where "least" is with respect to an ordering on the space of functions of a given type. To accommodate this, we can use *continuous algebras*, i.e. ordinary (total) Σ-algebras with carriers that are complete partially ordered sets (so-called *cpos*) and with operation names interpreted as *continuous functions* on these sets. See Example 3.3.14. The "bottom" element $\perp$ of the carrier for a sort, if it exists, represents the completely undefined value of that sort. The order on carriers induces an order on (continuous) functions in the usual fashion. A homomorphism between continuous algebras is required to be continuous as a function between cpos. It is possible to define a language of axioms that allows direct reference to least upper bounds of chains (see Example 4.1.22), and/or to the order relation itself. Such techniques may also be used to specify infinite data types such as *streams*. □

2.8 Bibliographical remarks

Much of the material presented here is well known, at least in its single-sorted version, in universal algebra as a branch of mathematics. Standard references are [Grä79] and [Coh65]. We approach this material from the direction of computer science — see [Wec92] and [MT92] — and present the fundamentals of equational specifications as developed in the 1970s ([Zil74], [Gut75], [GTW76]); see also [EM85] for an extended monograph-style presentation.

The simplest and most limited form of a specification is a "bare" signature, and this is what is used to characterise classes of algebras (program modules) in modularisation systems for programming languages — see, e.g., Standard ML [MTHM97], [Pau96], where such characterisations are in fact called signatures, *type classes* in Haskell [Pey03] and *concepts* in C++ [C++09]. Presentations correspond to Extended ML signatures [ST85] and to C++ concepts containing axioms.

The first appearance of the Satisfaction Lemma (Lemma 2.1.8) in the algebraic specification literature was in [BG80], echoing the semantic consequences of the definition of (theory) interpretations in logic [End72]. This fundamental link between syntax and semantics will become one of the cornerstones of later development starting in Chapter 4.

One topic that is only touched upon here (see, e.g., Theorem 2.2.10) is the expressive power of specifications. See [BT87] for a comprehensive survey of what is known about the expressive power of the framework presented in this chapter. The main theorem is the one mentioned at the beginning of Section 2.7.

We make a distinction between presentations and theories that is not present in some other work. This distinction surfaces in the definition of theory morphisms (Definition 2.3.11). For two presentations (not necessarily theories) $\langle\Sigma,\mathscr{E}\rangle$ and $\langle\Sigma',\mathscr{E}'\rangle$, [Gan83] takes a signature morphism $\sigma\colon\Sigma\to\Sigma'$ to be a specification morphism $\sigma\colon\langle\Sigma,\mathscr{E}\rangle\to\langle\Sigma',\mathscr{E}'\rangle$ if $\sigma(\mathscr{E})\subseteq\mathscr{E}'$. Such a σ is referred to as an "axiom-preserving theory morphism" in [Mes89]. Exercise 2.3.15 shows that this is not equivalent to our definition of theory morphism between the theories presented by those presentations. Another possibility is to require σ to map only the *ground* equations in $\mathscr{E}$ to equations in $Cl_{\Sigma'}(\mathscr{E}')$, as in [Ehr82]. These alternative definitions seem unsatisfactory since they make little or no sense on the level of models, in contrast to the relationship between theory and model levels for theory morphisms given by Proposition 2.3.13. We will later (Definition 5.5.1) define *specification morphisms*, as a generalisation of morphisms between presentations, relying on this relationship.

The many-sorted equational calculus is presented in [GM85] together with a proof that it is sound and complete. This builds on the standard equational calculus [Bir35], but the modifications needed to deal with empty carriers in the many-sorted context came as a surprise at the time. Our choice of rules in Section 2.4 is different from this standard version but the two systems are equivalent (Exercise 2.4.14) and the proofs of soundness and completeness are analogous.

The initial algebra approach to specification (Section 2.5) is the classical one. It originated with the seminal paper [GTW76], and was further developed by Hartmut Ehrig and his group; see [EM85] for a comprehensive account.

Example 2.5.24 and Exercise 2.5.25 point at useful ways to make inductive proofs easier by providing derived induction rule schemes, as possible, for instance, in the logics of Larch [GH93] and CASL [Mos04] and their proof support systems (LP [GG89] and HETS [MML07], respectively); see also Chapter 6 of [Far92].

The proof of the incompleteness theorem for initial semantics (Theorem 2.5.26) from [MS85] follows [Nou81] where it was used to show that the equational calculus with a specific induction rule scheme is not complete. An alternative to adding induction rules to the equational calculus is to restrict attention to so-called ω-complete presentations; these are presentations $\langle \Sigma, \mathscr{E} \rangle$ for which the equational calculus itself yields all of the Σ-equations that hold in initial models of $\langle \Sigma, \mathscr{E} \rangle$ [Hee86]. Then the problem becomes one of finding an ω-complete presentation corresponding to a given presentation. By the incompleteness theorem, this is not always possible.

There is a substantial body of theory on term rewriting systems; Section 2.6 is only the tip of the iceberg. For much more on the topic, and for the details of the Knuth-Bendix completion algorithm [KB70] that have been omitted in Section 2.6, see [DJ90], [Klo92], [BN98], [Kir99] and [Ter03]. See [KM87] or [DJ90] for a discussion of proof by consistency, which originated with [Mus80]. Like most work in this area, all these restrict attention to the single-sorted case. See [EM85] for a treatment of the many-sorted case, up to the soundness and completeness theorems for conversion, without our simplifying assumption (cf. Exercise 2.6.11).

In the case of reachable and final semantics, it is usual to look at reachable or final *extensions* of algebras (alternative terminology: hierarchical specifications), rather than at the reachable or final interpretation of a completed specification. See [BDP$^+$79] or [WB82] for reachable semantics, and [GGM76] or [Wan79] for final semantics. Under appropriate conditions, the reachable models of a presentation form a complete lattice, with the initial model at one extreme and the final model at the other; see [GGM76] and [BWP84]. For such hierarchical specifications, an incompleteness theorem that is even stronger than Theorem 2.5.26 may be proved: no sound proof system can derive all *ground* equational consequences of such specifications; see [MS85].

The first attempt to specify errors by distinguishing error values from OK values was [Gog78]. More details of the approach outlined in Section 2.7.3 can be found in [GDLE84]. The final semantics of error presentations is discussed in [Gog85]. See [BBC86] for an alternative approach which is able to deal with examples like the one discussed in Exercise 2.7.29.

More details of the approach to partial algebras outlined in Section 2.7.4 can be found in [BW82b]. Weak Σ-homomorphisms are called total Σ-homomorphisms there. Alternative approaches to the specification of partial algebras are presented in [Rei87] and [Kre87], and more recently in [Mos04]. See [Bur86] for a comprehensive analysis of the various alternative definitions of the basic notions.

See [GM92], further refined in [Mes09], for more on the approach to order-sorted algebra in Section 2.7.5. Alternative approaches include [Gog84], [Poi90] and [Smo86], which is sometimes referred to as "universal" order-sorted algebra to distinguish it from "overloaded" order-sorted algebra as presented here. A universal order-sorted algebra contains a single universe of values, where a sort corresponds to a subset of the universe and each operation name identifies a (single) function on the universe. A compromise is in rewriting logic [Mes92] as implemented in Maude [CDE$^+$02]. See [Mos93] and [GD94a] for surveys comparing the different approaches. [GD94a] discusses how some of the definitions and results in Section 2.7.5 can be generalised by dropping or weakening the monotonicity requirements on order-sorted signatures and order-sorted algebras. Yet a different approach to subsorting is taken in CASL [Mos04] where subsort coercions may be arbitrary injective functions rather than merely inclusions.

First-order predicate logic has been used as a framework for algebraic specification in various approaches; see for instance CIP-L [BBB$^+$85] and CASL [Mos04]. See [Poi86], [MTW88], [Mei92] and [Qia93] for different approaches to the algebraic specification of higher-order functions. Frameworks that cater for the specification of polymorphic types and functions are described in [Mos89], [MSS90] and [KST97]. See [Nip86] for more on algebras with non-deterministic operations; for a different approach using relation algebra, see [BS93]. See [WM97] for a comprehensive overview. Soundness and completeness of term rewriting for non-deterministic specifications is studied in [Hus92]. Continuous algebras and the use of Scott-style domain-theoretic techniques in algebraic specification were first discussed in [GTWW77]. See [Sch86] or [GS90] for much more on domain theory itself. Although these and other extensions to the standard framework have been explored separately, the few attempts that have been made to combine such extensions (see, e.g., [AC89] and [Mos04]) have tended to reveal new problems.

Chapter 3
Category theory

One of the main purposes of this book is to present a general, abstract theory of specifications that is independent of the exact details of the semantic structures (algebras) used to model particular aspects of program behaviour. Appropriate mathematical tools are required to support the development of such a theory. The basics of category theory provide us with just what we need: a simple, yet powerful language that allows definitions and results to be formulated at a sufficiently general, abstract level.

The most fundamental "categorical dogma" is that for many purposes it does not really matter exactly what the objects we study are; more important are their mutual relationships. Hence, objects should never be considered on their own; they should always come equipped with an appropriate notion of a *morphism* between them. In many typical examples, the objects are sets with some additional structure imposed on them, and their morphisms are maps that preserve this structure. "Categorical dogma" states that the interesting properties of objects may be formulated purely in terms of morphisms, without referring to the internal structure of objects at all. As a very simple example, consider the following two definitions.

Definition. Given two sets A and B, the *Cartesian product* of A and B is the set $A \times B$ that consists of all the pairs of elements from A and B, respectively: $A \times B = \{\langle a,b \rangle \mid a \in A, b \in B\}$. □

Definition. Given two sets A and B, a *product* of A and B is a set P together with two functions $\pi_1 : P \to A$ and $\pi_2 : P \to B$ such that for any set C with functions $f : C \to A$ and $g : C \to B$ there exists a unique function $h : C \to P$ such that $h;\pi_1 = f$ and $h;\pi_2 = g$.

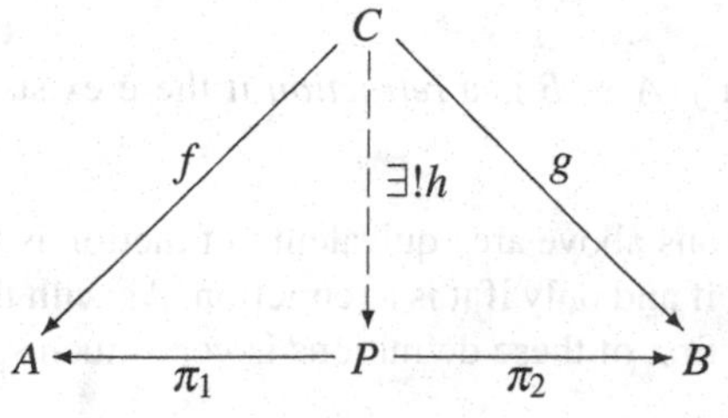

□

It is easy to see that the Cartesian product of any two sets is a product in the sense of the latter definition, where the functions π_1 and π_2 are the projections on the first and second components respectively (HINT: Define $h{:}C \to A \times B$ by $h(c) = \langle f(c), g(c) \rangle$ for all $c \in C$). Moreover, although a product P of two sets A and B does not have to be their Cartesian product $A \times B$ since the elements of P do not have to be pairs of objects from A and B, P is always isomorphic to $A \times B$: there is a one-to-one correspondence between elements of P and of $A \times B$. Thus, the two definitions may be viewed as equivalent for many purposes.

The reader may feel that the former definition (of the Cartesian product) is far simpler than the latter (of a product). Indeed, to most of us, brought up to consider set-theoretic concepts as the basis of all mathematics, this is in fact the case. However, the former definition suffers from a serious deficiency: it is formulated in terms of elements and the membership relation for sets, which constitute the specific internal structure of sets. Consequently, it is very specifically oriented towards defining the Cartesian product of sets and of sets only. If we now wanted to define the Cartesian product of, say, algebras (cf. Definition 1.2.9), we would have to reformulate this definition substantially (in this case, by adding definitions of operations for product algebras). To define the Cartesian product of structures of yet another kind, yet another different version of this definition would have to be explicitly stated. It is desirable to avoid such repetition of the same story for different specific kinds of objects whenever possible.

The latter definition (of a product) is quite different from this point of view. It does not refer to the internal structure of sets at all; it defines a product of two sets entirely in terms of its relationships with these sets and with other sets. To obtain a definition of a product of two algebras, it is enough to replace "set" by "algebra" and "function" by "homomorphism". The same would apply to other kinds of structures, as long as there is an appropriate notion of a morphism between them.

The conclusion we draw from this example is that, first of all, objects of any kind should be considered together with an appropriate notion of a morphism between them, and then that the structure imposed on the collection of objects by these morphisms should be exploited to formulate definitions at an appropriate level of generality and abstraction.

Let us have a look at another example:

Definition. A function $f{:}A \to B$ is *surjective* if for every $b \in B$ there exists $a \in A$ such that $b = f(a)$. □

Definition. A function $f{:}A \to B$ is an *epimorphism* if for any functions $g, g'{:}B \to C$, $f;g = f;g'$ implies $g = g'$. □

Definition. A function $f{:}A \to B$ is a *retraction* if there exists a function $g{:}B \to A$ such that $g;f = id_B$. □

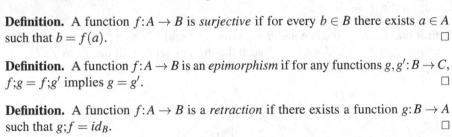

All the three definitions above are equivalent: a function is surjective if and only if it is an epimorphism, if and only if it is a retraction. As with the previous example, one may argue that the first of these definitions is very much specific to sets, and so not abstract and not general enough. The two other definitions lack this deficiency: they do not refer to the internal structure of sets, but use functions (set morphisms) to define the concept. However, the two definitions when applied to other kinds of objects (and their morphisms) may well turn out not to be equivalent. We cannot say that one of them is "right" and the other is "wrong"; they simply incorporate different aspects of what for sets is the property of "being surjective". The lesson to draw from this is that one has to be cautious when generalising a certain property to a more abstract setting. An attempt to formulate a definition at a more general level should provide us with a better understanding of the essence of the property being defined; it may well turn out, however, that there is more than one essence in it, giving several non-equivalent ways to reformulate the definition in a more abstract way.

Finding an adequate generalisation is not always easy. Sometimes even very simple notions we are accustomed to viewing as fundamental are difficult to formulate in categorical terms, as they depend in an essential way on the internal structure of the objects under consideration, which is exactly what we want to abstract from. The usual set-theoretic union operation is an example of such a notion.

Once we succeed in providing a more general version of a certain notion, it may be instantiated in many different ways. Often, an adequate generalisation of an important specific concept leads to interesting instantiations in the contexts of objects (and morphisms between them) different from the ones we started with. Indeed, interesting instantiations in other contexts may be regarded as a test of the adequacy of the generalisation.

A more wide-ranging polemic on the advantages of category theory presented at a rather intuitive level may be found in [Gog91a].

With these remarks in mind, this chapter introduces the basic concepts and results of category theory. It is not our intention to provide a full-blown introductory text on category theory; although a few concepts are introduced which will not be used elsewhere in this book, we consciously refrain from discussing many important but more involved concepts and results. Our aim in this chapter is to provide a brief but comprehensive overview of the basics of category theory, both in order to make this book self-contained and to provide a handy reference.

3.1 Introducing categories

3.1.1 Categories

Definition 3.1.1 (Category). A *category* $\mathbf{K}$ consists of:

- a collection $|\mathbf{K}|$ of $\mathbf{K}$-*objects*;
- for each $A, B \in |\mathbf{K}|$, a collection $\mathbf{K}(A,B)$ of $\mathbf{K}$-*morphisms from A to B*; and

- for each $A,B,C \in |\mathbf{K}|$, a *composition operation*[1] $\_;\_ : \mathbf{K}(A,B) \times \mathbf{K}(B,C) \to \mathbf{K}(A,C)$

such that:

1. for all $A,B,A',B' \in |\mathbf{K}|$, if $\langle A,B\rangle \neq \langle A',B'\rangle$ then $\mathbf{K}(A,B) \cap \mathbf{K}(A',B') = \varnothing$;
2. (*existence of identities*) for each $A \in |\mathbf{K}|$, there is a morphism $id_A \in \mathbf{K}(A,A)$ such that $id_A;g = g$ for all morphisms $g \in \mathbf{K}(A,B)$ and $f;id_A = f$ for all morphisms $f \in \mathbf{K}(B,A)$; and
3. (*associativity of composition*) for any $f \in \mathbf{K}(A,B)$, $g \in \mathbf{K}(B,C)$ and $h \in \mathbf{K}(C,D)$, $f;(g;h) = (f;g);h$. □

Notation. We will refer to *objects* and *morphisms* instead of **K**-objects and **K**-morphisms when **K** is clear from the context. We write $f : A \to B$ (in **K**) for $A,B \in |\mathbf{K}|$, $f \in \mathbf{K}(A,B)$. For any $f : A \to B$, we will refer to A as the *source* or *domain*, and to B as the *target* or *codomain* of f. The collection of all morphisms of **K** will be (ambiguously) denoted by **K** as well, i.e. $\mathbf{K} = \bigcup_{A,B \in |\mathbf{K}|} \mathbf{K}(A,B)$. □

The above is just one of several possible equivalent definitions of a category. For example, the identities, the existence of which is required in (2), are sometimes considered as part of the structure of a category.

Exercise 3.1.2. Prove that in any category, identities are unique. □

The notion of a category is very general. Accepting the categorical dogma that objects of any kind come equipped with a notion of morphism between them, it is difficult to think of a collection of objects and accompanying morphisms that do not form a category. Almost always there is a natural operation of morphism composition, which obeys two of the basic requirements above: it has identities and is associative. Perhaps the first requirement, which allows us to unambiguously identify the source and target of any morphism, is the most technical and hence least intuitively appealing. But even in cases where the same entity may be viewed as a morphism between different objects, this entity can always be equipped with an explicit indication of the source and target of the morphism (cf. Example 3.1.6), thus satisfying this requirement.

In the rest of this subsection we give a number of examples of categories. We start with some rather trivial examples, mainly of formal interest, and only then define some more typically considered categories. Further examples, which are often more complex, may be found in the following sections of this chapter (and in later chapters; see, e.g., Section 10.3).

Example 3.1.3 (Preorder category). A binary relation $\leq \subseteq X \times X$ is a *preorder on* X if:

- $x \leq x$ for all $x \in X$; and

[1] We will use the semicolon (;) to denote composition of morphisms in any category, just as we used it for composition of functions and homomorphisms in the preceding chapters. Composition will always be written in diagrammatic order: $f;g$ is to be read as "f followed by g".

- $x \leq y \wedge y \leq z \Rightarrow x \leq z$ for all $x, y, z \in X$.

A *preorder* category is a category that has at most one morphism with any given source and target.

Every preorder $\leq \subseteq X \times X$ gives rise to a preorder category $\mathbf{K}_{\leq}$ where $|\mathbf{K}_{\leq}| = X$ and $\mathbf{K}_{\leq}(x,y)$ has exactly one element if $x \leq y$ and is empty otherwise.

This definition does not identify the category $\mathbf{K}_{\leq}$ unambiguously, since different elements may be used as morphisms in $\mathbf{K}_{\leq}(x,y)$ for $x \leq y$. However, we will not worry here about the exact nature of morphisms (or objects) in a category, and we will treat this and similar definitions below as sufficient. More formally, all categories satisfying the above requirements are isomorphic in the technical sense to be discussed in Section 3.4 (cf. Definition 3.4.69).

Here are some trivial examples of preorder categories:

0: (the empty category)

1:

2:

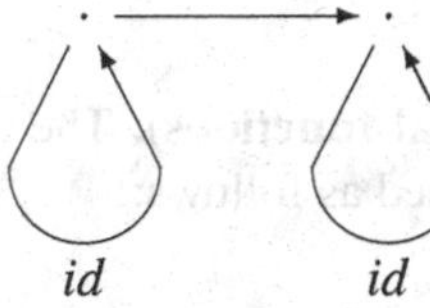

3: (+ identities)

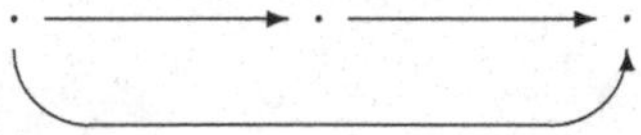

4: (+ identities)

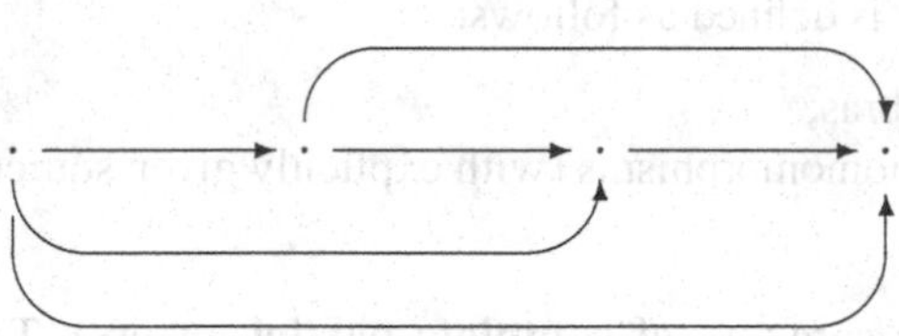

⋮ ⋮

Exercise. How many morphisms does **n** have? □

Example 3.1.4 (Discrete category). A category $\mathbf{K}$ is *discrete* whenever for all $A, B \in |\mathbf{K}|$, $\mathbf{K}(A,B)$ is empty if $A \neq B$ and contains exactly one element (the identity) otherwise.

Any collection of objects X gives rise to a discrete category $\mathbf{K}_X$ where $|\mathbf{K}_X| = X$. □

Example 3.1.5 (Monoid category). A category $\mathbf{K}$ is a *monoid* if $\mathbf{K}$ has exactly one object.

A set X together with a function $\_;\_ : X \times X \to X$ and a distinguished element $id \in X$ is a *monoid* $\langle X, ;, id \rangle$ if $(x;y);z = x;(y;z)$ and $id;x = x;id = x$ for all $x, y, z \in X$. Every monoid $\langle X, ;, id \rangle$ gives rise to a monoid (category) having morphisms X and composition ;. □

Example 3.1.6 (Set, the category of sets). The category **Set** of sets with functions as morphisms is defined as follows:

Objects of **Set**: sets;

Morphisms of **Set**: functions; however, to ensure that the requirements stated in Definition 3.1.1 are satisfied (disregarding the particular mathematical representation of the concept of a function one uses), we will always consider functions with explicitly given domain and codomain. Thus, a morphism in the category **Set** with source A and target B is a triple $\langle A, f, B \rangle$, where $f : A \to B$ is a function. □

Example 3.1.7 ($\mathbf{Set}^S$, the category of S-sorted sets). For any set S, the category $\mathbf{Set}^S$ of S-sorted sets is defined as follows:

Objects of $\mathbf{Set}^S$: S-sorted sets;

Morphisms of $\mathbf{Set}^S$: S-sorted functions (with explicitly given source and target sets). □

Example 3.1.8 (Pfn, the category of sets with partial functions). The category **Pfn** of sets with partial functions as morphisms is defined as follows:

Objects of **Pfn**: sets;

Morphisms of **Pfn**: partial functions (with explicitly given source and target sets). □

Example 3.1.9 ($\mathbf{Alg}(\Sigma)$, the category of Σ-algebras). For any signature Σ, the category $\mathbf{Alg}(\Sigma)$ of Σ-algebras is defined as follows:

Objects of $\mathbf{Alg}(\Sigma)$: Σ-algebras;

Morphisms of $\mathbf{Alg}(\Sigma)$: Σ-homomorphisms (with explicitly given source and target algebras). □

Example 3.1.10 (CPO, the category of complete partial orders). The category **CPO** of complete partial orders[2] and continuous functions between them is defined as follows:

Objects of **CPO**: complete partial orders, i.e. partially ordered sets $\langle X, \leq \rangle$ such that any countable chain $x_0 \leq x_1 \leq \ldots$ in $\langle X, \leq \rangle$ has a least upper bound $\bigsqcup_{i \geq 0} x_i$;

Morphisms of **CPO**: continuous functions, i.e. functions that preserve least upper bounds of countable chains. □

---

[2] Cpos and continuous functions as defined here are often referred to as ω-cpos and ω-continuous functions, respectively. Note though that we do not require a cpo to have a least element.

Exercise 3.1.11. Complete the above examples by formalising composition in the obvious way. Indicate identities and prove associativity of composition. □

Example 3.1.12 (AlgSig, the category of algebraic signatures). The category **AlgSig** of (algebraic) signatures is defined as follows:

Objects of **AlgSig**: signatures;
Morphisms of **AlgSig**: signature morphisms;
Composition in **AlgSig**: for any $\sigma:\Sigma \to \Sigma'$ and $\sigma':\Sigma' \to \Sigma''$, their composition $\sigma;\sigma':\Sigma \to \Sigma''$ is given by $(\sigma;\sigma')_{sorts} = \sigma_{sorts};\sigma'_{sorts}$ and $(\sigma;\sigma')_{ops} = \sigma_{ops};\sigma'_{ops}$, cf. Exercise 1.5.3. □

Exercise 3.1.13 (AlgSig^{der}, the category of signatures with derived morphisms). Recall the concept of a derived signature morphism from Definition 1.5.14. Define the category **AlgSig**^{der} of algebraic signatures with derived signature morphisms. Use Exercise 1.5.18 to define composition of derived signature morphisms. □

Example 3.1.14 (T$_\Sigma$, the category of substitutions over a signature Σ). Recall (cf. Section 1.4) that for any signature $\Sigma = \langle S, \Omega \rangle$ and S-sorted set of variables X, $T_\Sigma(X)$ is the algebra of terms over Σ with variables X. $T_\Sigma(X)$ is characterised up to isomorphism by the property that for any Σ-algebra A, any S-sorted map $v:X \to |A|$ uniquely extends to a Σ-homomorphism $v^{\#}:T_\Sigma(X) \to A$ (Facts 1.4.4 and 1.4.10).

For any algebraic signature Σ, the category $\mathbf{T}_\Sigma$ of substitutions over Σ is defined as follows (cf. Exercise 1.4.9):

Objects of $\mathbf{T}_\Sigma$: S-sorted sets (of variables);
Morphisms of $\mathbf{T}_\Sigma$: for any sets X and Y, a morphism θ from X to Y is a substitution of terms with variables Y for variables X, i.e. an S-sorted function $\theta:X \to |T_\Sigma(Y)|$;
Composition in $\mathbf{T}_\Sigma$: given any sets X, Y and Z, and morphisms $\theta:X \to Y$ and $\theta':Y \to Z$ in $\mathbf{T}_\Sigma$, i.e. functions $\theta:X \to |T_\Sigma(Y)|$ and $\theta':Y \to |T_\Sigma(Z)|$, their composition $\theta;\theta':X \to Z$ is the function $\theta;\theta':X \to |T_\Sigma(Z)|$ defined by $(\theta;\theta')_s(x) = (\theta')^{\#}_s(\theta_s(x))$ for all $s \in S, x \in X_s$. □

Exercise 3.1.15 (T$_\Sigma/\mathscr{E}$, the category of substitutions over Σ modulo equations $\mathscr{E}$). Generalise the above definition of the category of substitutions by considering terms up to an equivalence generated by a set of equations. That is, for any algebraic signature $\Sigma = \langle S, \Omega \rangle$ and set $\mathscr{E}$ of Σ-equations, for any S-sorted set of variables X define two terms $t_1, t_2 \in |T_\Sigma(X)|_s$ (for any sort $s \in S$) to be equivalent, written $t_1 \equiv t_2$, if $\mathscr{E} \vdash_\Sigma \forall X \bullet t_1 = t_2$ (cf. Section 2.4). Now, by analogy with the category of substitutions, define the category $\mathbf{T}_\Sigma/\mathscr{E}$ to have S-sorted sets as objects and substitutions modulo $\mathscr{E}$ as morphisms. A substitution of terms modulo $\mathscr{E}$ with variables Y for variables X is an S-sorted function $\theta:X \to (|T_\Sigma(Y)|/{\equiv})$. Composition in $\mathbf{T}_\Sigma/\mathscr{E}$ is defined analogously as in $\mathbf{T}_\Sigma$, by choosing a representative of each of the equivalence classes assigned to variables: given $\theta:X \to (|T_\Sigma(Y)|/{\equiv})$ and $\theta':Y \to (|T_\Sigma(Z)|/{\equiv})$, $\theta;\theta':X \to (|T_\Sigma(Z)|/{\equiv})$ maps any $x \in X$ to $(\theta')^{\#}(t)$, where $\theta(x) = [t]_{\equiv}$ (show that the result does not depend on the choice of the representative $t \in \theta(x)$). □

Exercise 3.1.16 ($\mathbf{T}_{\Sigma,\mathscr{E}}$, the algebraic $\langle\Sigma,\mathscr{E}\rangle$-theory). Building on the definition of the category of substitutions modulo a set of equations sketched above, abstract away from the actual names of variables used in the objects of $\mathbf{T}_\Sigma/\mathscr{E}$ by listing them in some particular order, as in derived signatures (cf. Definition 1.5.13). That is, for any algebraic signature $\Sigma = \langle S,\Omega\rangle$ and set $\mathscr{E}$ of Σ-equations, define the category $\mathbf{T}_{\Sigma,\mathscr{E}}$ with sequences $s_1 \dots s_n \in S^*$ of sort names as objects. A morphism in $\mathbf{T}_{\Sigma,\mathscr{E}}$ from $s_1 \dots s_n \in S^*$ to $s'_1 \dots s'_m \in S^*$ is an n-tuple $\langle [t_1]_\equiv, \dots, [t_n]_\equiv\rangle$ of terms modulo $\mathscr{E}$, where the equivalence $\equiv$ is sketched in Exercise 3.1.15 above, and for $i = 1,\dots,n$, $t_i \in |T_\Sigma(I_{s'_1\dots s'_m})|_{s_i}$, with $I_{s'_1\dots s'_m} = \{\boxed{1}{:}s'_1,\dots,\boxed{m}{:}s'_m\}$. The composition in $\mathbf{T}_{\Sigma,\mathscr{E}}$ is given by substitution on representatives of equivalence classes (the position of a term in a tuple identifies the variable it is to be substituted for). $\mathbf{T}_{\Sigma,\mathscr{E}}$ is usually referred to as the *algebraic theory* over Σ generated by $\mathscr{E}$.[3] □

3.1.1.1 Foundations

In Chapters 1 and 2 we have followed normal mathematical practice and used the term "*class*" (as in Bernays-Gödel set theory) for collections that are possibly too "large" to be sets. In the above, and in the definition of a category in particular, we have instead very cautiously used the non-technical term *collection*, and talked of *collections* of objects and morphisms. This allowed us to gloss over the issue of the choice of appropriate set-theoretical foundations for category theory. Even a brief look at the examples above indicates that we could not have been talking here just of *sets* (in the sense of Zermelo-Fraenkel set theory): we want to consider categories like **Set**, where the collection of objects consists of all sets, and so cannot be a set itself. Using classes might seem more promising, since if we replace the term "collection" by "class" in Definition 3.1.1 then at least examples of categories like **Set** would be covered. However, this is not enough either, since even in this simple presentation of the basics of category theory we will encounter some categories (like **Cat**, the category of "all" categories, and functor categories defined later in this chapter) where objects themselves are proper classes and the collection of objects forms a "conglomerate" (a collection of classes that is too "large" to be a class; cf. [HS73]). See [Bén85] for a careful analysis of the basic requirements imposed on a set theory underlying category theory.

Perhaps the most traditional solution to the problem of set-theoretic foundations for category theory is sketched in [Mac71]. The idea is to work within a hierarchy of *set universes* $\langle \mathbf{U}_n\rangle_{n\geq 0}$, where each universe $\mathbf{U}_n$, $n \geq 0$, is closed under the standard set-theoretic operations, and is an element of the next universe in the hierarchy, $\mathbf{U}_n \in \mathbf{U}_{n+1}$. Then there is a notion of category corresponding to each level of the hierarchy, and one is required to indicate at which level of the hierarchy one is working at any given moment.

---

[3] In the literature, the algebraic theory over Σ generated by $\mathscr{E}$ is often defined with substitutions considered as morphisms in the opposite direction, i.e. as the category $\mathbf{T}^{op}_{\Sigma,\mathscr{E}}$ opposite to $\mathbf{T}_{\Sigma,\mathscr{E}}$ (cf. Definition 3.1.22 below).

However, in our view such pedantry would hide the intuitive appeal of "naive" category theory. We will therefore ignore the issue of set-theoretic foundations for category theory in the sequel, with just one exception: we define what it means for a category to be (locally) small and use this to occasionally warn the reader about potential foundational hazards.

Definition 3.1.17 (Small category). A category $\mathbf{K}$ is *locally small* if for any $A, B \in |\mathbf{K}|$, $\mathbf{K}(A,B)$ is a set (an element of the lowest-level universe $\mathbf{U}_0$); $\mathbf{K}$ is *small* if in addition $|\mathbf{K}|$ is a set as well. □

3.1.2 Constructing categories

In the examples of the previous subsection, each category was constructed "from scratch" by explicitly defining its objects and morphisms and their composition. Category theory also provides numerous ways of modifying a given category to yield a different one, and of putting together two or more categories to obtain a more complicated one. Some of the simplest examples are given in this subsection.

3.1.2.1 Subcategories

Definition 3.1.18 (Subcategory). A category $\mathbf{K1}$ is a *subcategory* of a category $\mathbf{K2}$ if $|\mathbf{K1}| \subseteq |\mathbf{K2}|$ and $\mathbf{K1}(A,B) \subseteq \mathbf{K2}(A,B)$ for all objects $A, B \in |\mathbf{K1}|$, with composition and identities in $\mathbf{K1}$ the same as in $\mathbf{K2}$. $\mathbf{K1}$ is a *full* subcategory of $\mathbf{K2}$ if additionally $\mathbf{K1}(A,B) = \mathbf{K2}(A,B)$ for all $A, B \in |\mathbf{K1}|$. $\mathbf{K1}$ is a *wide* subcategory of $\mathbf{K2}$ if $|\mathbf{K1}| = |\mathbf{K2}|$. □

For any category $\mathbf{K}$, any collection $X \subseteq |\mathbf{K}|$ of objects of $\mathbf{K}$ determines a full subcategory $\mathbf{K}|_X$ of $\mathbf{K}$, defined by $|\mathbf{K}|_X| = X$. Whenever convenient, if $\mathbf{K}$ is evident from the context, we will identify collections $X \subseteq |\mathbf{K}|$ with $\mathbf{K}|_X$.

Example 3.1.19 (FinSet, the category of finite sets). The category **FinSet** of finite sets is defined as follows:

Objects of **FinSet**: finite sets;
Morphisms and composition in **FinSet**: as in **Set**.

FinSet is a full subcategory of **Set**. □

Example 3.1.20. The category of single-sorted signatures is a full subcategory of the category **AlgSig** of (many-sorted) signatures.

The discrete category of sets is a subcategory of the category of sets with inclusions as morphisms, which is a subcategory of the category of sets with injective functions as morphisms, which is a subcategory of **Set**, which is a wide subcategory of **Pfn** (the category of sets with partial functions).

For any signature Σ and set $\mathscr{E}$ of Σ-equations, the class $Mod_\Sigma(\mathscr{E})$ of Σ-algebras that satisfy $\mathscr{E}$ determines a full subcategory of $\mathbf{Alg}(\Sigma)$, which we denote by $\mathbf{Mod}_\Sigma(\mathscr{E})$. □

Exercise 3.1.21. Give an example of two categories $\mathbf{K1}$, $\mathbf{K2}$ such that $|\mathbf{K1}| \subseteq |\mathbf{K2}|$, $\mathbf{K1}(A,B) \subseteq \mathbf{K2}(A,B)$ for all objects $A,B \in |\mathbf{K1}|$, composition in $\mathbf{K1}$ is the same as in $\mathbf{K2}$, but $\mathbf{K1}$ is *not* a subcategory of $\mathbf{K2}$. □

3.1.2.2 Opposite categories and duality

One of the fundamental theorems of lattice theory (cf. e.g. [DP90]) is the so-called *duality principle*. Any statement in the language of lattice theory has a dual, obtained by systematically replacing greatest lower bounds by least upper bounds and vice versa. The duality principle states that the dual of any theorem of lattice theory is a theorem as well. In a sense, this allows the number of proofs in lattice theory to be cut by half: proving a fact gives its dual "for free". A very similar phenomenon occurs in category theory; in fact, the duality principle of lattice theory may be viewed as a consequence of a more general duality principle of category theory. Replacing greatest lower bounds by least upper bounds and vice versa is generalised here to the process of "reversing morphisms".

Definition 3.1.22 (Opposite category). The *opposite category* of a category $\mathbf{K}$ is the category $\mathbf{K}^{op}$ where:

Objects of $\mathbf{K}^{op}$: $|\mathbf{K}^{op}| = |\mathbf{K}|$;
Morphisms of $\mathbf{K}^{op}$: $\mathbf{K}^{op}(A,B) = \mathbf{K}(B,A)$ for all $A,B \in |\mathbf{K}^{op}|$;
Composition in $\mathbf{K}^{op}$: for $f \in \mathbf{K}^{op}(A,B)$ (i.e. $f \in \mathbf{K}(B,A)$) and $g \in \mathbf{K}^{op}(B,C)$ (i.e. $g \in \mathbf{K}(C,B)$), $f;g \in \mathbf{K}^{op}(A,C)$ is $g;f \in \mathbf{K}(C,A)$.

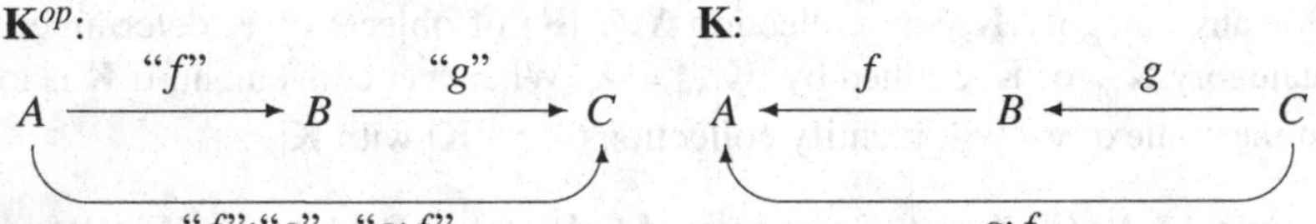

□

Exercise 3.1.23. Check that:

1. $\mathbf{K}^{op}$ is a category.
2. $(\mathbf{K}^{op})^{op} = \mathbf{K}$.
3. Identities in $\mathbf{K}^{op}$ are the same as in $\mathbf{K}$. □

If W is a categorical concept (property, statement, etc.) then its *dual*, *co-W*, is obtained by reversing all the morphisms in W. This idea may be formalised in two ways. The first is to introduce a formal language of category theory, and then define the operation of forming a dual as an operation on formal statements in this language. The other is to formally interpret *co-W* in a category $\mathbf{K}$ as W in the category $\mathbf{K}^{op}$. Since formalising the language of category theory is beyond the scope of this book (but cf. [Mac71] or [Hat82]), we take the second option here and will rely on an intuitive understanding of duality in the sequel. For example, consider the following property of objects in a category:

$$P(X): \text{ for any object } Y \text{ there is a morphism } f{:}Y \to X.$$

Then:

$$co\text{-}P(X): \text{ for any object } Y \text{ there is a morphism } f{:}X \to Y.$$

Note that indeed $co\text{-}P(X)$ in any category $\mathbf{K}$ amounts to $P(X)$ in $\mathbf{K}^{op}$.

Since any category is the opposite of a certain category (namely, of its opposite), the following fact holds:

Fact 3.1.24 (Duality principle). *If W holds for all categories then* co-W *holds for all categories as well.* □

3.1.2.3 Product categories

Definition 3.1.25 (Product category). For any two categories **K1** and **K2**, the *product category* $\mathbf{K1} \times \mathbf{K2}$ is defined by:

Objects of $\mathbf{K1} \times \mathbf{K2}$: $|\mathbf{K1} \times \mathbf{K2}| = |\mathbf{K1}| \times |\mathbf{K2}|$ (the Cartesian product);
Morphisms of $\mathbf{K1} \times \mathbf{K2}$: for all $A, A' \in |\mathbf{K1}|$ and $B, B' \in |\mathbf{K2}|$,
$\mathbf{K1} \times \mathbf{K2}(\langle A,B\rangle, \langle A',B'\rangle) = \mathbf{K1}(A,A') \times \mathbf{K2}(B,B')$;
Composition in $\mathbf{K1} \times \mathbf{K2}$: for $f{:}A \to A'$ and $f'{:}A' \to A''$ in **K1**, $g{:}B \to B'$ and $g'{:}B' \to B''$ in **K2**, $\langle f,g\rangle;\langle f',g'\rangle = \langle f;f', g;g'\rangle$. □

Exercise 3.1.26. Identify the category to which each semicolon in the above definition of composition in $\mathbf{K1} \times \mathbf{K2}$ refers. Then show that $\mathbf{K1} \times \mathbf{K2}$ is indeed a category. □

Exercise 3.1.27. Define $\mathbf{K}^n$, where $\mathbf{K}$ is a category and $n \geq 1$. What would you suggest for $n = 0$? □

3.1.2.4 Morphism categories

Definition 3.1.28 (Morphism category). For any category **K**, the *category* $\mathbf{K}^{\to}$ *of* **K**-*morphisms* is defined by:

Objects of $\mathbf{K}^{\to}$: **K**-morphisms;
Morphisms of $\mathbf{K}^{\to}$: a morphism in $\mathbf{K}^{\to}$ from $f{:}A \to A'$ (in **K**) to $g{:}B \to B'$ (in **K**) is a pair $\langle k,k'\rangle$ of **K**-morphisms where $k{:}A \to B$ and $k'{:}A' \to B'$ such that $k;g = f;k'$;
Composition in $\mathbf{K}^{\to}$: $\langle k,k'\rangle;\langle l,l'\rangle = \langle k;l, k';l'\rangle$. □

The requirement in the definition of a morphism in $\mathbf{K}^{\to}$ may be more illustratively restated as the requirement that the following diagram commutes in the category **K**:

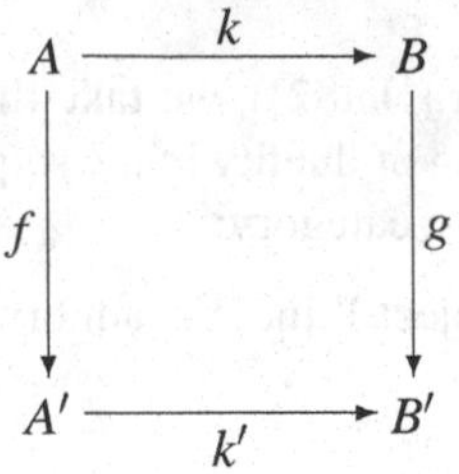

For now, we will rely on an intuitive understanding of the concept of a diagram in a category; see Section 3.2.5 for a formal definition. We say that a diagram in a category *commutes* (or, *is commutative*) if for any two paths with the same source and target nodes, the composition of morphisms along each of the two paths yields the same result.

Drawing diagrams and *"chasing"* a diagram in order to prove that it is commutative is one of the standard and intuitively most appealing techniques used in category theory. For example, to justify Definition 3.1.28 above it is essential to show that the composition of two morphisms in $\mathbf{K}^{\rightarrow}$ as defined there yields a morphism in $\mathbf{K}^{\rightarrow}$. This may be done by *pasting together* two diagrams like the one above along a common edge, obtaining the following diagram:

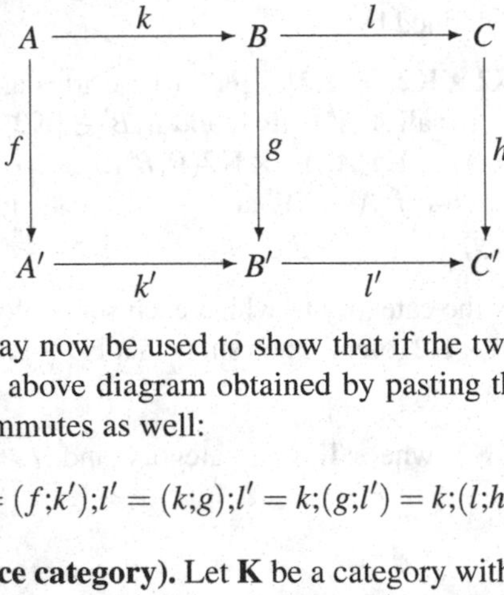

A simple argument may now be used to show that if the two simpler diagrams are commutative then the above diagram obtained by pasting them together along the edge labelled by g commutes as well:

$$f;(k';l') = (f;k');l' = (k;g);l' = k;(g;l') = k;(l;h) = (k;l);h.$$

Definition 3.1.29 (Slice category). Let $\mathbf{K}$ be a category with $A \in |\mathbf{K}|$. The *category* $\mathbf{K}{\downarrow}A$ *of* $\mathbf{K}$*-objects over* A (or, the *slice of* $\mathbf{K}$ *over* A) is defined by:

Objects of $\mathbf{K}{\downarrow}A$: pairs $\langle X, f\rangle$ where $X \in |\mathbf{K}|$ and $f \in \mathbf{K}(X,A)$;
Morphisms of $\mathbf{K}{\downarrow}A$: a morphism from $\langle X, f\rangle$ to $\langle Y, g\rangle$ is a $\mathbf{K}$-morphism $k{:}X \rightarrow Y$ such that $k;g = f$:

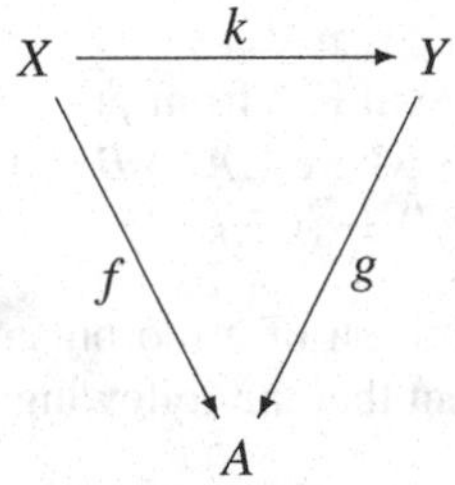

Composition in $\mathbf{K}{\downarrow}A$: as in $\mathbf{K}$. □

Exercise 3.1.30. Show that $\mathbf{K}{\downarrow}A$ may be constructed as a subcategory of $\mathbf{K}^{\rightarrow}$. Is it full? □

Exercise 3.1.31. Define $\mathbf{K}{\uparrow}A$, the category of $\mathbf{K}$-objects *under* A. Compare $(\mathbf{K}{\downarrow}A)^{op}$, $\mathbf{K}^{op}{\downarrow}A$ and $(\mathbf{K}^{op}{\downarrow}A)^{op}$ with $\mathbf{K}{\uparrow}A$. □

3.1.3 Category-theoretic definitions

In this section we will give a few simple examples of how certain special morphisms may be characterised in a style that is typical for category-theoretic definitions. As indicated in the introduction to this chapter, the idea is to abstract away from the "internal" properties of objects and morphisms, characterising them entirely in categorical language by referring only to arbitrary objects and morphisms of the category under consideration. Such definitions may be formulated for an arbitrary category, and then instantiated to a particular one when necessary. We will also indicate a few basic properties of the concepts we introduce that hold in any category.

Throughout this section, let $\mathbf{K}$ be an arbitrary but fixed category. Morphisms and objects we refer to below are those of $\mathbf{K}$, unless explicitly qualified otherwise.

3.1.3.1 Epimorphisms and monomorphisms

Definition 3.1.32 (Epimorphism). A morphism $f{:}A \to B$ is an *epimorphism* (or is *epi*) if for all $g{:}B \to C$ and $h{:}B \to C$, $f;g = f;h$ implies $g = h$.

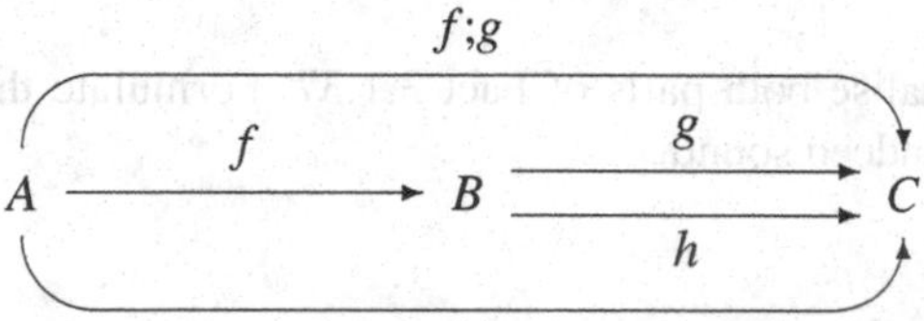

□

Example 3.1.33. In **Set**, f is epi iff f is surjective. □

There are "natural" categories in which epimorphisms need not be surjective. For example:

Exercise 3.1.34. Recall the category **CPO** of complete partial orders and continuous functions introduced in Example 3.1.10. Give an example of a continuous function that is an epimorphism in **CPO** even though it is not surjective. Try to characterise epimorphisms in this category. □

Definition 3.1.35 (Monomorphism). A morphism $f{:}B \to A$ is a *monomorphism* (or is *mono*) if for all $g{:}C \to B$ and $h{:}C \to B$, $g;f = h;f$ implies $g = h$.

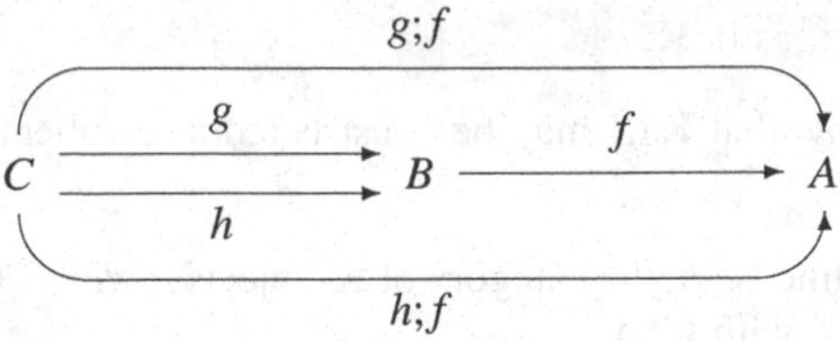

□

Example 3.1.36. In **Set**, f is mono iff f is injective. □

Note that mono means the same as co-epi, i.e. f is mono in $\mathbf{K}$ iff f is epi in $\mathbf{K}^{op}$.

Fact 3.1.37.

1. *If $f{:}A \to B$ and $g{:}B \to C$ are mono then $f;g{:}A \to C$ is mono.*
2. *For any $f{:}A \to B$ and $g{:}B \to C$, if $f;g{:}A \to C$ is mono then f is mono.*

Proof. The proof is rather straightforward, and significantly more complex proofs will be omitted in the rest of this chapter. We present it here explicitly only as a simple example of the style of argument, very common in category-theoretic proofs, exploiting the most basic properties of composition in an arbitrary category.

1. According to Definition 3.1.35, we have to show that for any $h,h'{:}D \to A$ if $h;(f;g) = h';(f;g)$ then $h = h'$. So, suppose $h;(f;g) = h';(f;g)$. Then, since composition is associative, $(h;f);g = (h';f);g$. Consequently, since g is mono, by Definition 3.1.35, $h;f = h';f$. Thus, using the fact that f is mono, we can indeed conclude that $h = h'$.
2. Similarly as in the previous case: suppose that for some $h,h'{:}D \to A$, $h;f = h';f$. Then also $(h;f);g = (h';f);g$, and so $h;(f;g) = h';(f;g)$. Now, since $f;g$ is mono, it follows directly from the definition that indeed $h = h'$.

□

Exercise 3.1.38. Dualise both parts of Fact 3.1.37. Formulate the dual proofs and check that they are indeed sound. □

3.1.3.2 Isomorphic objects

Definition 3.1.39 (Isomorphism). A morphism $f{:}A \to B$ is an *isomorphism* (or is *iso*) if there is a morphism $f^{-1}{:}B \to A$ such that $f;f^{-1} = id_A$ and $f^{-1};f = id_B$. The morphism $f^{-1}{:}B \to A$ is called the *inverse* of f, and the objects A and B are called *isomorphic*. We write $f{:}A \cong B$ or just $A \cong B$.

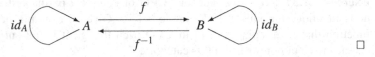

□

Exercise 3.1.40. Show that the inverse of a morphism, if it exists, is unique. □

Note that iso means the same as co-iso, that is, isomorphism is a *self-dual* concept.

Exercise 3.1.41. Check that if $f{:}A \to B$ and $g{:}B \to C$ are iso then $f;g{:}A \to C$ is iso as well. □

In **Set**, a morphism is iso iff it is both epi and mono. However, this property does not carry over to an arbitrary category:

Exercise 3.1.42. Show that if f is iso then f is both epi and mono. The converse is not true in general; give a counterexample. □

Exercise 3.1.43. We say that a morphism $f{:}A \to B$ is a *retraction* if there is a morphism $g{:}B \to A$ such that $g;f = id_B$. Dually, a morphism $f{:}A \to B$ is a *coretraction* if there is a morphism $g{:}B \to A$ such that $f;g = id_A$. Show that:

1. A morphism is iso iff it is both a retraction and a coretraction.
2. Every retraction is epi.
3. A morphism is iso iff it is an epi coretraction.

Dualise the above facts. □

It is easy to see that any two isomorphic objects have the same "categorical properties". Intuitively, such objects have abstractly the same structure and so are indistinguishable within the given category (which does not mean that isomorphic objects cannot have different "non-categorical" properties; cf. Example 1.3.12). Indeed, an isomorphism and its inverse determine one-to-one mappings between morphisms going into and coming out of isomorphic objects. Hence, categorical definitions of objects define them only "up to isomorphism". The following section provides typical examples of this phenomenon.

3.2 Limits and colimits

In this section we show how certain special objects in an arbitrary category together with their "characteristic" morphisms may be defined in purely categorical terms by so-called *universal properties*; we hope that the reader will recognize the pattern in the example definitions below. Sections 3.2.1–3.2.4 present some typical instances of this, introducing the most commonly used cases of the general *limit construction* and its dual, which are then presented in their full generality in Section 3.2.5. In most of the cases in this section we will explicitly spell out the duals of the concepts introduced, since many of them have interesting instances in some common categories (and are traditionally given independent names).

Throughout this section, let **K** be an arbitrary but fixed category. Morphisms and objects we refer to are those of **K**, unless explicitly qualified otherwise.

3.2.1 Initial and terminal objects

Definition 3.2.1 (Initial object). An object $I \in |\mathbf{K}|$ is *initial in* $\mathbf{K}$ if for each $A \in |\mathbf{K}|$ there is exactly one morphism from I to A. □

Example 3.2.2. The empty set $\varnothing$ is initial in **Set**, as well as in **Pfn**. The algebra T_Σ of ground Σ-terms is initial in $\mathbf{Alg}(\Sigma)$, for any signature $\Sigma \in |\mathbf{AlgSig}|$.

Recall the definition of an initial model of an equational specification (Definition 2.5.13). For any signature Σ and set $\mathscr{E}$ of Σ-equations, the initial model of $\langle \Sigma, \mathscr{E} \rangle$ (which exists by Theorem 2.5.14) is an initial object in the category $\mathbf{Mod}_\Sigma(\mathscr{E})$ (as defined in Example 3.1.20). □

Exercise 3.2.3. What is an initial object in **AlgSig**? Look for initial objects in other categories. □

Fact 3.2.4.

1. *Any two initial objects in* $\mathbf{K}$ *are isomorphic.*
2. *If I is initial in* $\mathbf{K}$ *and I' is isomorphic to I then I' is initial in* $\mathbf{K}$ *as well.*

Proof. The proof is rather straightforward. We present it here explicitly only as a simple example of the style of argument, very common in category-theoretic proofs, which exploits universality (a special case of which is the property used in the definition of an initial object). The requirement that there *exist* a morphism satisfying a certain property is used to construct some diagrams, and then the *uniqueness* of this morphism is used to show that the diagrams constructed commute.

1. Suppose that $I, I' \in |\mathbf{K}|$ are two initial objects in $\mathbf{K}$. Then, by the initiality of I, there exists a morphism $f : I \to I'$. Similarly, by the initiality of I', there exists a morphism $g : I' \to I$. Thus, we have constructed the following diagram:

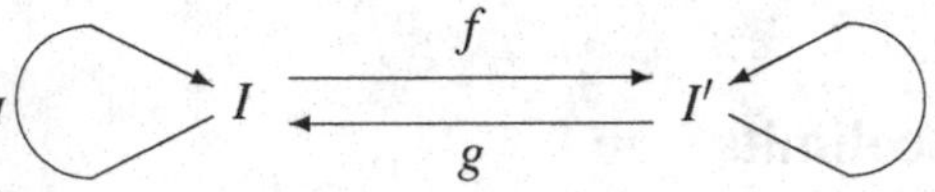

Now, by the initiality of I, there is a *unique* morphism from I to I, and so $id_I = f;g$. Similarly, $id_{I'} = g;f$. Thus f is an isomorphism (with inverse g) and I and I' are indeed isomorphic.

2. Suppose that $I \in |\mathbf{K}|$ is initial in $\mathbf{K}$, and let $i : I \to I'$ be an isomorphism with inverse $i^{-1} : I' \to I$. Consider an arbitrary object $A \in |\mathbf{K}|$. By the "existence part" of the initiality property of I, we know that there exists a morphism $f : I \to A$. Hence, there exists a morphism from I' to A as well, namely $i^{-1};f : I' \to A$. Then, let $f' : I' \to A$ be an arbitrary morphism from I' to A. By the "uniqueness part" of the initiality property of I, $f = i;f'$, and so $i^{-1};f = i^{-1};(i;f') = (i^{-1};i);f' = id_{I'};f' = f'$. This shows that $i^{-1};f$ is the only morphism from I' to A, and so I' is indeed initial in $\mathbf{K}$. □

The last fact indicates that the initiality property identifies an object up to isomorphism. As argued in Section 3.1.3.2, in category theory this is the most exact characterisation of an object we may expect. In the following we will speak of "the" initial object, meaning an initial object identified up to isomorphism. We adopt the same convention in the many similar cases introduced in the sequel.

3.2.1.1 Dually:

Definition 3.2.5 (Terminal object). An object $1 \in |\mathbf{K}|$ is *terminal in* $\mathbf{K}$ if for each $A \in |\mathbf{K}|$ there is exactly one morphism from A to 1. □

Note that terminal means the same as co-initial.

Exercise 3.2.6. Are there any terminal objects in **Set**, $\mathbf{Alg}(\Sigma)$ or **AlgSig**? What about terminal objects in $\mathbf{AlgSig}^{der}$?

Check that $\varnothing$ is terminal in **Pfn**; hence the initial and terminal objects in **Pfn** coincide.

Recall the definition of a terminal (final) model of an equational specification (Definition 2.7.12). Restate it using the notion of a terminal object as defined above. □

Exercise 3.2.7. Dualise Fact 3.2.4. □

3.2.2 *Products and coproducts*

Definition 3.2.8 (Product). A *product* of two objects $A, B \in |\mathbf{K}|$ is an object $A \times B \in |\mathbf{K}|$ together with a pair of morphisms $\pi_A : A \times B \to A$ and $\pi_B : A \times B \to B$ such that for any object $C \in |\mathbf{K}|$ and pair of morphisms $f : C \to A$ and $g : C \to B$ there is exactly one morphism $\langle f, g \rangle : C \to A \times B$ such that the following diagram commutes:

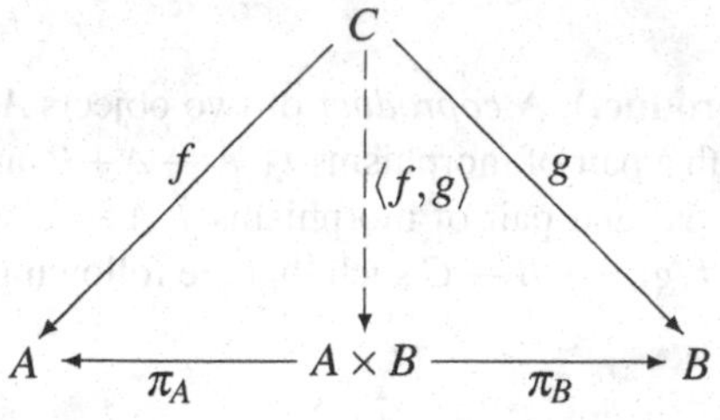

□

Example 3.2.9. In **Set**, the Cartesian product of A and B is a product $A \times B$, where π_A, π_B are the projection functions. For any signature Σ, products in $\mathbf{Alg}(\Sigma)$ are defined analogously (cf. Definition 1.2.9). □

Exercise 3.2.10. Define a product of two sets in **Pfn**. HINT: In general it is *not* their Cartesian product; you need to add "pairs" that are missing one element. □

Exercise 3.2.11. What is the product of two objects in a preorder category? □

Exercise 3.2.12. Show that any two products of $A, B \in |\mathbf{K}|$ are isomorphic. □

Exercise 3.2.13. Suppose that $A, B \in |\mathbf{K}|$ have a product. Given $f: C \to A$ and $g: C \to B$, and hence $\langle f, g \rangle: C \to A \times B$, show that for any $h: D \to C$, $h;\langle f, g \rangle = \langle h;f, h;g \rangle$, and that for any $k: C \to A \times B$, $k = \langle k;\pi_A, k;\pi_B \rangle$. □

Exercise 3.2.14. Prove that:

1. $A \times B \cong B \times A$ for any $A, B \in |\mathbf{K}|$.
2. $(A \times B) \times C \cong A \times (B \times C)$ for any $A, B, C \in |\mathbf{K}|$. HINT: The following diagram might be helpful:

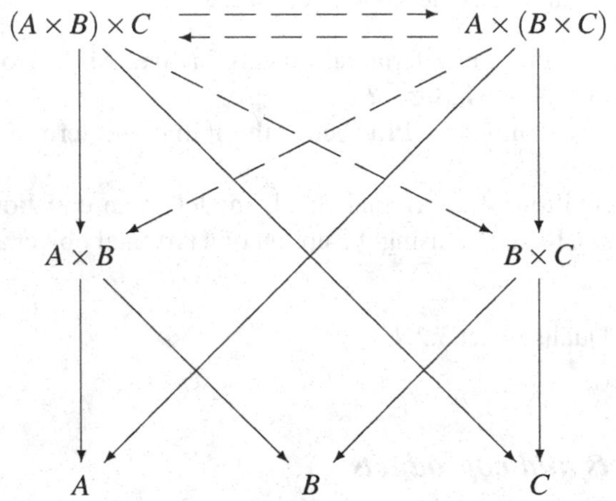

□

Exercise 3.2.15. Define the product of an arbitrary family of **K**-objects. What is the product of the empty family? □

3.2.2.1 Dually:

Definition 3.2.16 (Coproduct). A *coproduct* of two objects $A, B \in |\mathbf{K}|$ is an object $A + B \in |\mathbf{K}|$ together with a pair of morphisms $\iota_A: A \to A + B$ and $\iota_B: B \to A + B$ such that for any object $C \in |\mathbf{K}|$ and pair of morphisms $f: A \to C$ and $g: B \to C$ there is exactly one morphism $[f, g]: A + B \to C$ such that the following diagram commutes:

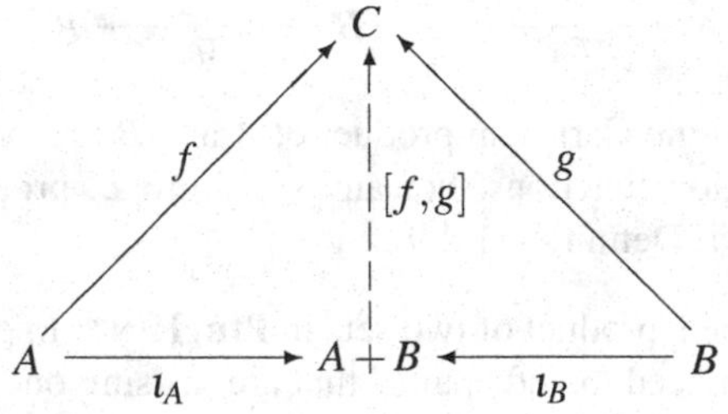

□

Example 3.2.17. In **Set**, the disjoint union of sets A and B is their coproduct $A+B$, where ι_A, ι_B are the injections. Similarly, in **AlgSig**, the (componentwise) disjoint union of algebraic signatures Σ and Σ' is their coproduct $\Sigma+\Sigma'$, where ι_A, ι_B are the obvious injections. □

Exercise 3.2.18. Dualise the exercises for products. □

Exercise 3.2.19. For any algebraic signature $\Sigma = \langle S, \Omega \rangle$ and two S-sorted sets X and Y, show that their disjoint union $X \uplus Y$ is the coproduct of X and Y in the category $\mathbf{T}_\Sigma$ of substitutions over Σ (recall Example 3.1.14), where the coproduct injections are the identity substitutions (of the corresponding variables from $X \uplus Y$ for variables in X and in Y, respectively). Generalise this to the category $\mathbf{T}_\Sigma/\mathscr{E}$ of substitutions over Σ modulo a set $\mathscr{E}$ of Σ-equations (cf. Exercise 3.1.15). Finally, characterise coproducts in the category $\mathbf{T}_{\Sigma,\mathscr{E}}$, the algebraic theory over Σ generated by $\mathscr{E}$ (Exercise 3.1.16). □

3.2.3 *Equalisers and coequalisers*

We have defined above products and coproducts for arbitrary pairs of objects in a category. In this section we deal with constructions for pairs of morphisms constrained to being *parallel*, i.e. pairs of morphisms that have the same source and the same target.

Definition 3.2.20 (Equaliser). An *equaliser* of two parallel morphisms $f{:}A \to B$ and $g{:}A \to B$ is an object $E \in |\mathbf{K}|$ together with a morphism $h{:}E \to A$ such that $h;f = h;g$, and such that for any object $E' \in |\mathbf{K}|$ and morphism $h'{:}E' \to A$ satisfying $h';f = h';g$ there is exactly one morphism $k{:}E' \to E$ such that $k;h = h'$:

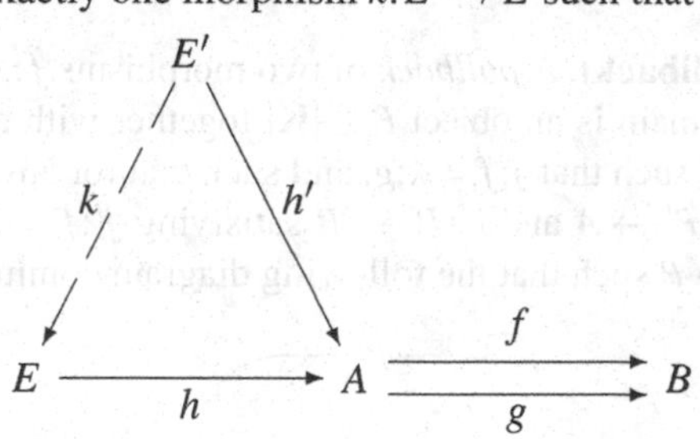

□

Exercise 3.2.21. Show that an equaliser of $f{:}A \to B$ and $g{:}A \to B$ is unique up to isomorphism. □

Exercise 3.2.22. Show that every equaliser (to be more precise, its morphism part) is mono, and every epi equaliser is iso. □

Exercise 3.2.23. Construct equalisers of pairs of parallel morphisms in **Set**. Then, for any signature Σ, construct equalisers of pairs of parallel morphisms in $\mathbf{Alg}(\Sigma)$. HINT: For any two functions $f,g{:}A \to B$ consider the set $\{a \in A \mid f(a) = g(a)\} \subseteq A$.

Define equalisers in **Pfn**, adapting the hint above if necessary. □

3.2.3.1 Dually:

Definition 3.2.24 (Coequaliser). The dual notion to equaliser is *coequaliser*. The diagram now looks as follows:

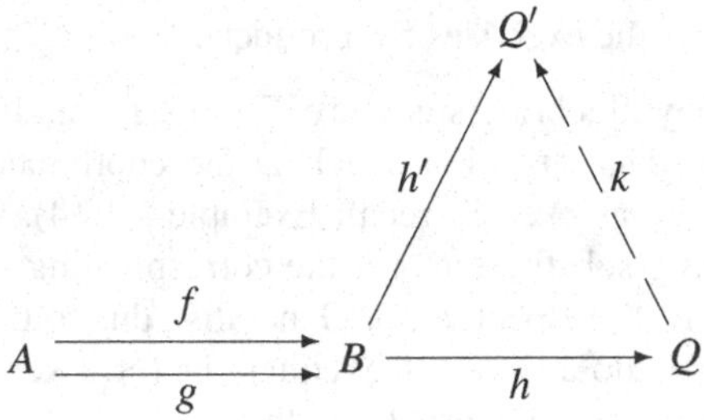

Exercise. Formulate explicitly the definition of a coequaliser. Then dualise the exercises for equalisers. □

Exercise 3.2.25. What is the coequaliser of two morphisms in **Set**? What is the coequaliser of two morphisms in **AlgSig**? What is the coequaliser of two morphisms in $\mathbf{Alg}(\Sigma)$? HINT: Given two functions $f,g{:}A \to B$ consider the quotient of B by the least equivalence relation $\equiv$ on B such that for all $a \in A$, $f(a) \equiv g(a)$. □

Exercise 3.2.26. What is the coequaliser of two morphisms in the category of substitutions $\mathbf{T}_\Sigma$? □

3.2.4 Pullbacks and pushouts

Definition 3.2.27 (Pullback). A *pullback* of two morphisms $f{:}A \to C$ and $g{:}B \to C$ having the same codomain is an object $P \in |\mathbf{K}|$ together with a pair of morphisms $j{:}P \to A$ and $k{:}P \to B$ such that $j;f = k;g$, and such that for any object $P' \in |\mathbf{K}|$ and pair of morphisms $j'{:}P' \to A$ and $k'{:}P' \to B$ satisfying $j';f = k';g$ there is exactly one morphism $h{:}P' \to P$ such that the following diagram commutes:

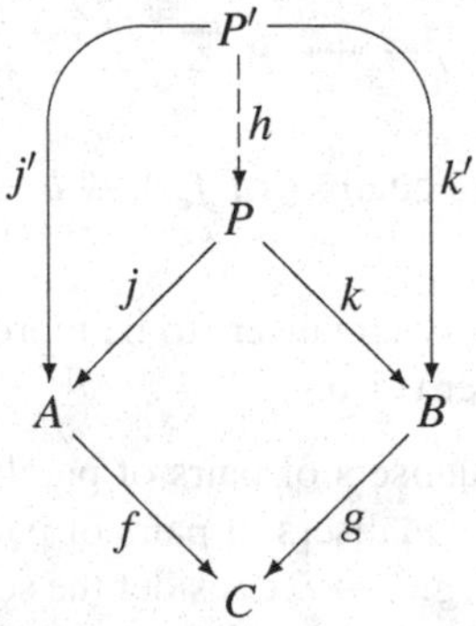

□

Exercise 3.2.28. Show that a pullback of $f{:}A \to C$ and $g{:}B \to C$ is unique up to isomorphism. □

Exercise 3.2.29. Show that if **K** has products (of all pairs of objects) and equalisers (of all pairs of parallel morphisms) then it has pullbacks as well (i.e. all pairs of morphisms with a common target have pullbacks in **K**).

HINT: To construct a pullback of $f{:}A \to C$ and $g{:}B \to C$, first construct the product $A \times B$ with projections $\pi_A{:}A \times B \to A$ and $\pi_B{:}A \times B \to B$ and then the equaliser $h{:}P \to A \times B$ of $\pi_A;f{:}A \times B \to C$ and $\pi_B;g{:}A \times B \to C$. □

Exercise 3.2.30. Construct the pullback of two morphisms in **Set**, and then in **Alg**(Σ), **AlgSig**, and **Pfn**. □

Exercise 3.2.31. Prove that if **K** has a terminal object and all pullbacks (i.e. any pair of **K**-morphisms with a common target has a pullback in **K**) then:

1. **K** has all (binary) products.
2. **K** has all equalisers. HINT: Get the equaliser of $f,g{:}A \to B$ from the pullback of $\langle id_A, f\rangle, \langle id_A, g\rangle{:}A \to A \times B$ (see Definition 3.2.8 for notation). □

Exercise 3.2.32. Show that pullbacks translate monomorphisms to monomorphisms: if

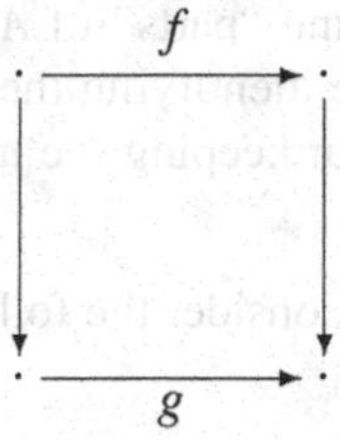

is a pullback square and g is mono, then f is mono as well. □

Exercise 3.2.33. Consider the following diagram:

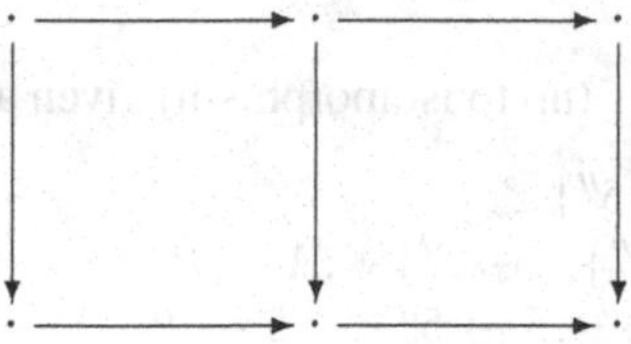

Prove that:

1. If the two squares are pullbacks then the outer rectangle is a pullback.
2. If the diagram commutes and the outer rectangle and right-hand square are both pullbacks then so is the left-hand square. □

3.2.4.1 Dually:

Definition 3.2.34 (Pushout). The dual notion to pullback is *pushout*. The diagram now looks as follows:

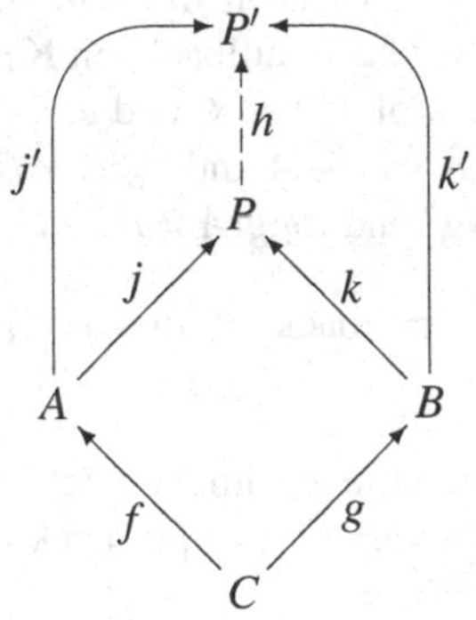

Exercise. Spell out the definition of a pushout explicitly. Then dualise the exercises for pullbacks. □

Pushouts provide a basic tool for putting together structures of various kinds. Given two objects A and B, a pair of morphisms $f{:}C \to A$ and $g{:}C \to B$ indicates a common source from which some "parts" of A and B come. The pushout of f and g puts together A and B while identifying the parts coming from the common source as indicated by f and g, but keeping the new parts disjoint (cf. the dual of Exercise 3.2.29).

Example 3.2.35. Working in **Set**, consider the following:

$$
\begin{aligned}
&A = \{1,2,3\}\\
&B = \{3,4,5\}\\
&C = \{\clubsuit\}\\
&f = \{\clubsuit \mapsto 2\} \quad :C \to A\\
&g = \{\clubsuit \mapsto 4\} \quad :C \to B
\end{aligned}
$$

Then the pushout object P is (up to isomorphism) given as follows:

$$
\begin{aligned}
&P = \{1', \{2'{=}4''\}, 3', 3'', 5''\}\\
&j = \{1 \mapsto 1', 2 \mapsto \{2'{=}4''\}, 3 \mapsto 3'\} \quad :A \to P\\
&k = \{3 \mapsto 3'', 4 \mapsto \{2'{=}4''\}, 5 \mapsto 5''\} \quad :B \to P
\end{aligned}
$$
□

Example 3.2.36. The general comments above about the use of pushouts for putting together objects in categories apply in particular when one wants to combine algebraic signatures, as we will frequently do throughout the rest of the book. As a very simple example of a pushout in the category **AlgSig** of algebraic signatures, consider the signature Σ_{NAT} of natural numbers defined in Exercise 2.5.4. Then, let $\Sigma_{\mathrm{NAT}fib}$ be its extension by a new operation name *fib*:*Nat* $\to$ *Nat* and $\Sigma_{\mathrm{NAT}mult}$ its extension by another operation name *mult*:*Nat* $\times$ *Nat* $\to$ *Nat*. We then have two signature inclusions:

$$\Sigma_{\mathrm{NAT}_{fib}} \longleftarrow \Sigma_{\mathrm{NAT}} \hookrightarrow \Sigma_{\mathrm{NAT}_{mult}}$$

Their pushout in **AlgSig** yields a signature $\Sigma_{\mathrm{NAT}_{fib,mult}}$ which (up to isomorphism) consists of the shared signature Σ_{NAT} (once, no repetitions!) together with each of the operations added by the two extensions.

This is deceptively simple though, involving only single-sorted signature inclusions that introduce different operation names.

Exercise. Give examples of pushouts in **AlgSig** with signatures involving more than one sort, operation names that coincide, and signature morphisms that are not injective on sorts and/or on operation names. □

3.2.5 The general situation

The definitions introduced in the previous subsections followed a common, more general pattern. As an example, consider again the definition of a pullback (Definition 3.2.27; the notation below refers to the diagram there). Given a diagram in the category at hand (the two morphisms f and g of which we construct the pullback), we consider an object P in this category together with morphisms going from the object to the nodes of the diagram (j, k and an implicit $c{:}P \to C$) such that all the resulting paths starting from P commute ($j;f = c = k;g$ — hence c may remain implicit). Moreover, from among all such objects we choose the one that is in a sense "closest" to the diagram: for any object P' with morphisms from it to the diagram nodes (j', k' and an implicit c') satisfying the required commutativity property ($j';f = c' = k';g$), P' may be uniquely projected onto the chosen object P (via a morphism h) so that all the resulting paths starting from P' commute ($h;j = j'$ and $h;k = k'$, which also implies $h;c = c'$). This is usually referred to as the *universal property* of pullbacks and, more generally, of arbitrary *limits* as defined below. The (dual) universal property of pushouts and, more generally, of arbitrary *colimits* as defined below, may be described by looking at objects with morphisms going from the nodes of a diagram into them. We will formalise this in the rest of this section.

Definition 3.2.37 (Graph). Let Σ_G be the following signature:

sorts *Node*, *Edge*
ops $source{:}\,Edge \to Node$
$target{:}\,Edge \to Node$

A Σ_G-algebra is called a *graph*. (Note that these graphs may have multiple edges between any two nodes; such graphs are sometimes called *multigraphs*.) The category **Graph** of graphs is $\mathbf{Alg}(\Sigma_G)$. Given a graph G, we write $e{:}n \to m$ as an abbreviation for $n,m \in |G|_{Node}$, $e \in |G|_{Edge}$, $source_G(e) = n$ and $target_G(e) = m$. □

Exercise 3.2.38. Construct an initial object, coproducts, coequalisers and pushouts in **Graph**. □

Exercise 3.2.39. Define formally the category **Path**(G) of paths in a graph G, where:

Objects of **Path**(G): $|G|_{Node}$;
Morphisms of **Path**(G): paths in G, i.e. finite sequences $e_1 \ldots e_n$ of elements of $|G|_{Edge}$ such that $source_G(e_{i+1}) = target_G(e_i)$ for $i < n$. Notice that we have to allow for $n = 0$, for each node. □

A diagram in **K** is a graph having nodes labelled with **K**-objects and edges labelled with **K**-morphisms with the appropriate source and target. Formally:

Definition 3.2.40 (Diagram). A *diagram* D in **K** consists of:

- a graph $G(D)$;
- for each node $n \in |G(D)|_{Node}$, an object $D_n \in |\mathbf{K}|$; and
- for each edge $e\colon n \to m$ in $G(D)$, a morphism $D_e\colon D_n \to D_m$.

A diagram D is *connected* if its graph $G(D)$ is connected (that is, any two nodes in $G(D)$ are linked by a sequence of edges disregarding their direction, or, formally, if the total relation on the set of nodes of $G(D)$ is the only equivalence between the nodes that links all nodes having an edge between them). □

Every small category **K** gives rise to a graph $G(\mathbf{K})$ with all **K**-objects as nodes and all **K**-morphisms as edges, and a diagram $D(\mathbf{K})$ that labels the nodes and edges of $G(\mathbf{K})$ by themselves.

Definition 3.2.41 (Cone and cocone). A *cone* α *over a diagram* D *in* **K** is a **K**-object X together with a family of **K**-morphisms $\langle \alpha_n\colon X \to D_n \rangle_{n \in |G(D)|_{Node}}$ such that for every edge $e\colon n \to m$ in the graph $G(D)$ the following diagram commutes:

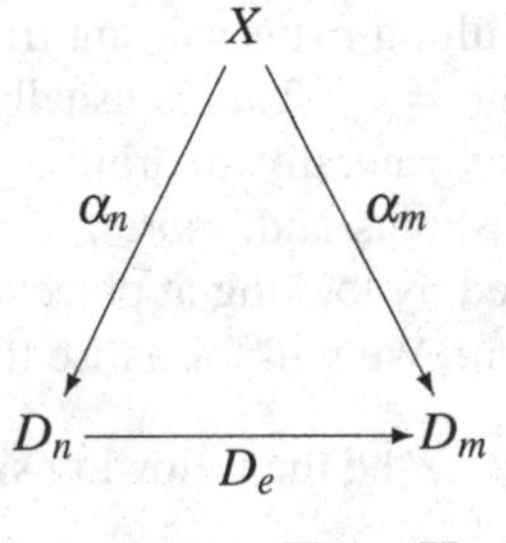

Dually, a *cocone* α *over a diagram* D *in* **K** is a **K**-object X together with a family of **K**-morphisms $\langle \alpha_n\colon D_n \to X \rangle_{n \in |G(D)|_{Node}}$ such that for every edge $e\colon n \to m$ in the graph $G(D)$ the following diagram commutes:

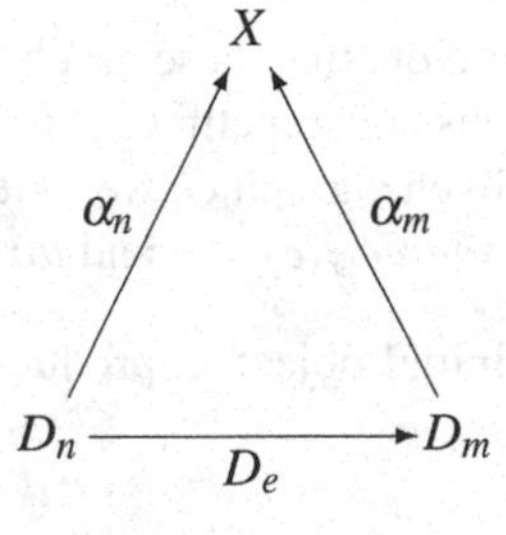

□

In the following we will write cones simply as families $\langle \alpha_n : X \to D_n \rangle_{n \in |G(D)|_{Node}}$, omitting any separate mention of the apex X, and similarly for cocones. The notation is not quite justified when the diagram (and hence the family) is empty; this will not lead to any misunderstanding.

Let D be a diagram in $\mathbf{K}$ with $|G(D)|_{Node} = N$ and $|G(D)|_{Edge} = E$.

Definition 3.2.42 (Limit and colimit). A *limit of D in* $\mathbf{K}$ is a cone $\langle \alpha_n : X \to D_n \rangle_{n \in N}$ such that for any cone $\langle \alpha'_n : X' \to D_n \rangle_{n \in N}$ there is exactly one morphism $h: X' \to X$ such that for every $n \in N$ the following diagram commutes:

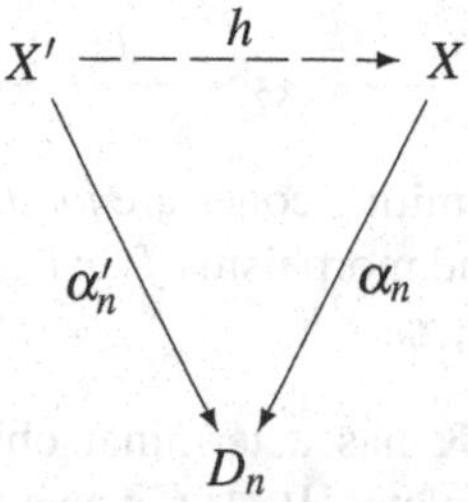

If $\langle \alpha_n : X \to D_n \rangle_{n \in N}$ is a limit of D, we will refer to X as the *limit object* of D (or sometimes just the *limit* of D), and to the morphisms α_n, $n \in N$, as the limit *projections*.

Dually, a *colimit of D in* $\mathbf{K}$ is a cocone $\langle \alpha_n : D_n \to X \rangle_{n \in N}$ such that for any cocone $\langle \alpha'_n : D_n \to X' \rangle_{n \in N}$ there is exactly one morphism $h: X \to X'$ such that for every $n \in N$ the following diagram commutes:

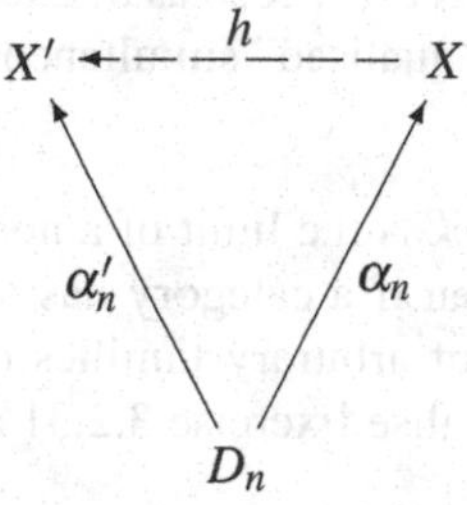

If $\langle \alpha_n : D_n \to X \rangle_{n \in N}$ is a colimit of D, we will refer to X as the *colimit object* of D (or sometimes just the *colimit* of D), and to the morphisms α_n, $n \in N$, as the colimit *injections*. □

Definition 3.2.43 (Completeness and cocompleteness). A category $\mathbf{K}$ is *(finitely) complete* if every (finite) diagram in $\mathbf{K}$ has a limit. Dually, $\mathbf{K}$ is *(finitely) cocomplete* if every (finite) diagram in $\mathbf{K}$ has a colimit. □

Exercise 3.2.44. Define formally the category $\mathbf{Cone}(D)$ of cones over a diagram D, where:

Objects of $\mathbf{Cone}(D)$: cones over D;

Morphisms of $\mathbf{Cone}(D)$: a morphism between cones $\alpha = \langle \alpha_n : X \to D_n \rangle_{n \in N}$ and $\alpha' = \langle \alpha'_n : X' \to D_n \rangle_{n \in N}$ is a $\mathbf{K}$-morphism $h: X \to X'$ such that $\alpha_n = h;\alpha'_n$ for $n \in N$.

Prove that the limit of D is a terminal object in **Cone**(D). Note that this implies that a limit of any diagram is unique up to isomorphism.

Present the category of objects over an object (cf. Definition 3.1.29) as the category of cones over a certain diagram. □

Exercise 3.2.45. Show that products, terminal objects, equalisers and pullbacks in **K** are limits of simple diagrams in **K**. □

Exercise 3.2.46. Construct in **Set** a limit of the diagram

$$A_0 \xleftarrow{f_0} A_1 \xleftarrow{f_1} A_2 \xleftarrow{f_2} A_3 \xleftarrow{f_3} \cdots$$ □

Exercise 3.2.47. Show that all limiting cones are *jointly monomorphic*: given a limit $\langle \alpha_n : X \to D_n \rangle_{n \in |G(D)|_{Node}}$ of D and morphisms $f, g : Y \to X$, we have $f = g$ whenever for all $n \in |G(D)|_{Node}$, $f;\alpha_n = g;\alpha_n$. □

Exercise 3.2.48. Show that if **K** has a terminal object, binary products and all equalisers then it is finitely complete. HINT: Given a finite diagram in **K**, first build the product of all its objects, and then gradually turn it into a limit by "equalising" the triangles formed by product projections and morphisms in the diagram.

Use Exercise 3.2.31 to conclude that if **K** has a terminal object and all pullbacks then it is finitely complete. □

Exercise 3.2.49. Show that if **K** has products of arbitrary families of objects and all equalisers then it is complete. HINT: Proceed as in Exercise 3.2.48, but notice that all the triangles involved may be "equalised" simultaneously in one step; cf. [Mac71], Theorem V.2.1. □

Exercise 3.2.50. A *wide pullback* is the limit of a non-empty family of morphisms with a common target. Show that if a category has a terminal object and all wide pullbacks then it has products of arbitrary families of objects, and then conclude that it is complete. HINT: Generalise Exercise 3.2.31 and use Exercise 3.2.49. □

Exercise 3.2.51. Recall that for any category **K** and object $A \in |\mathbf{K}|$, $\mathbf{K}{\downarrow}A$ is the slice category of objects over A (Definition 3.1.29).

Notice that $\mathbf{K}{\downarrow}A$ has a terminal object. Then show that binary products in $\mathbf{K}{\downarrow}A$ are essentially given by the pullbacks in **K** (of morphisms to A), and similarly, arbitrary non-empty products in $\mathbf{K}{\downarrow}A$ are essentially given by wide pullbacks in **K**. Check also that any (wide) pullback in $\mathbf{K}{\downarrow}A$ is given by the corresponding (wide) pullback in **K** (no morphisms to A added).

Conclude that $\mathbf{K}{\downarrow}A$ is finitely complete if **K** has all pullbacks, and $\mathbf{K}{\downarrow}A$ is complete if **K** has all wide pullbacks. □

Exercise 3.2.52. Dualise the above exercises. □

Exercise 3.2.53. Show that:

1. **Set** is complete and cocomplete.

2. **FinSet** is finitely complete and finitely cocomplete, but is neither complete nor cocomplete.
3. **Alg**(Σ) is complete for any signature Σ. (It is also cocomplete, but the proof is harder — give it a try!)
4. **AlgSig** is cocomplete. (Is it complete?)

HINT: Use Exercise 3.2.49 and its dual, and the constructions of (co)products and (co)equalisers in these categories hinted at in Examples 3.2.9, 3.2.17 and Exercises 3.2.23, 3.2.25. Check that, given a diagram D with nodes N and edges E in **Set**, its limit is (up to isomorphism) the set of families $\langle d_n \rangle_{n \in N}$ that are compatible with D in the sense that $d_n \in D_n$ for each $n \in N$ and $d_m = D_e(d_n)$ for each edge $e\colon n \to m$, with the obvious projections. Check that its colimit is (up to isomorphism) the quotient of the disjoint union $\biguplus_{n \in N} D_n$ by the least equivalence relation that is generated by all pairs $\langle d_n, D_e(d_n) \rangle$ for $e\colon n \to m$ in E and $d_n \in D_n$. □

Exercise 3.2.54. Show that $\mathbf{AlgSig}^{der}$ is not finitely cocomplete. HINT: Consider a morphism mapping a binary operation to the projection on the first argument and another morphism mapping the same operation to the projection on the second argument. Can such a pair of morphisms have a coequaliser? □

Exercise 3.2.55. When is a preorder category (finitely) complete and cocomplete? □

3.3 Factorisation systems

In this section we will interrupt our presentation of the basic concepts of category theory and try to illustrate how they can be used to formulate some well-known ideas at a level of generality and abstraction that ensures their applicability in many specific contexts.

The concept on which we concentrate here is that of *reachability* (cf. Section 1.2). Recall that the original definition of a reachable algebra used the notion of a subalgebra (cf. Definition 1.2.7). Keeping in mind that in the categorical framework we deal with objects identified up to isomorphism, we slightly generalise the standard formulation and, for any signature $\Sigma \in |\mathbf{AlgSig}|$, say that a Σ-algebra B is a subalgebra of A if there exists an *injective* Σ-homomorphism from B to A. A dual notion is that of a *quotient*: a Σ-algebra B is a quotient of a Σ-algebra A if there exists a *surjective* Σ-homomorphism from A to B. Now, a Σ-algebra A is *reachable* if it has no proper subalgebra (i.e. every subalgebra of A is isomorphic to A), or equivalently, if it is a quotient of the algebra T_Σ of ground Σ-terms (cf. Exercise 1.4.14). In this formulation, the above definitions may be used to introduce a notion of reachability in an arbitrary category. However, we need an appropriate generalisation of the concept of injective and surjective homomorphisms. A first attempt might be to use arbitrary epimorphisms and monomorphisms for this purpose, but it soon turns out that these concepts are not "fine enough" to ensure the properties we are after. An appropriate refinement of these is given if the category is equipped with a *factorisation system*.

Definition 3.3.1 (Factorisation system). Let **K** be an arbitrary category. A *factorisation system* for **K** is a pair $\langle \mathbf{E}, \mathbf{M} \rangle$, where:

- **E** is a collection of epimorphisms in **K** and **M** is a collection of monomorphisms in **K**;
- each of **E** and **M** is closed under composition and contains all isomorphisms in **K**;
- every morphism in **K** has an $\langle \mathbf{E}, \mathbf{M} \rangle$-*factorisation*: for each $f \in \mathbf{K}$, $f = e_f;m_f$ for some $e_f \in \mathbf{E}$ and $m_f \in \mathbf{K}$;

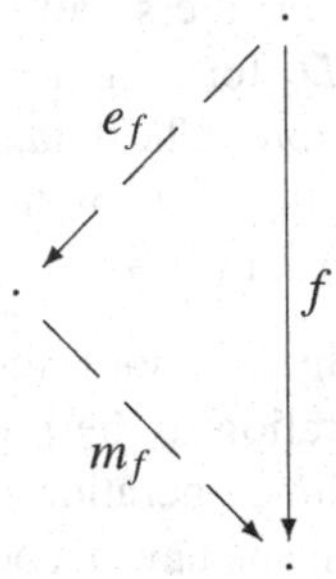

- $\langle \mathbf{E}, \mathbf{M} \rangle$-factorisations are unique up to isomorphism: for any $e, e' \in \mathbf{E}$ and $m, m' \in \mathbf{M}$, if $e;m = e';m'$ then there exists an isomorphism i such that $e;i = e'$ and $i;m' = m$.

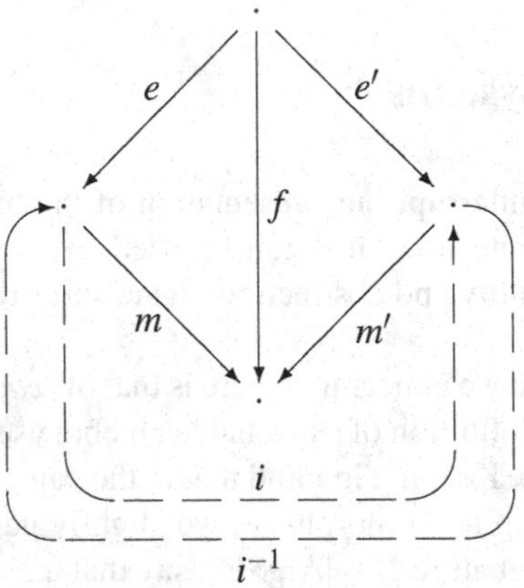

□

Example 3.3.2. Set has a factorisation system $\langle \mathbf{E}, \mathbf{M} \rangle$, where **E** is the collection of all surjective functions and **M** is the collection of all injective functions. □

Example 3.3.3. For any algebraic signature Σ, $\mathbf{Alg}(\Sigma)$ has a factorisation system[4] $\langle \mathbf{TE}_\Sigma, \mathbf{TM}_\Sigma \rangle$, where $\mathbf{TE}_\Sigma$ is the collection of all surjective Σ-homomorphisms and $\mathbf{TM}_\Sigma$ is the collection of all injective Σ-homomorphisms; see Exercise 1.3.23. □

Consider an arbitrary category **K** equipped with a factorisation system $\langle \mathbf{E}, \mathbf{M} \rangle$.

[4] "T" in $\mathbf{TE}_\Sigma$ and $\mathbf{TM}_\Sigma$ indicates that we are dealing with ordinary *total* algebras here, as opposed to partial and continuous algebras with the factorisation systems discussed below.

Lemma 3.3.4 (Diagonal fill-in lemma). *For any morphisms f_1, f_2, e, m in* **K**, *where $e \in \mathbf{E}$ and $m \in \mathbf{M}$, if $f_1;m = e;f_2$ then there exists a unique morphism g such that $e;g = f_1$ and $g;m = f_2$.*

Proof sketch. The required "diagonal" is given by $g = e_{f_2};i;m_{f1}$, as illustrated by the diagram below; its uniqueness follows easily since e is an epimorphism.

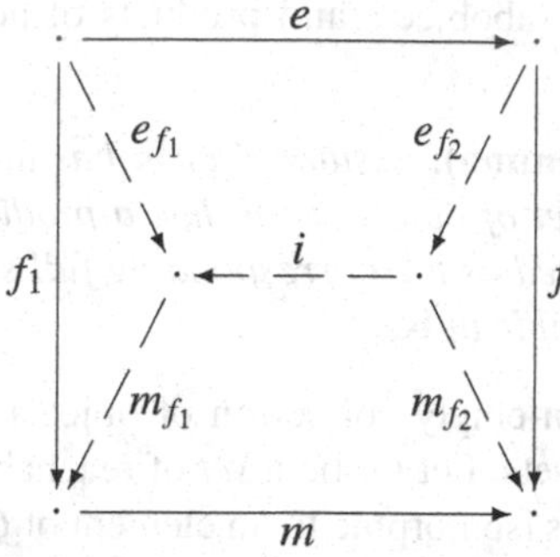

□

Exercise 3.3.5. Show that if $e \in \mathbf{E}$ and $e;f \in \mathbf{M}$ for some morphism $f \in \mathbf{K}$, then e is an isomorphism. Dually, if $m \in \mathbf{M}$ and $f;m \in \mathbf{E}$ for some morphism $f \in \mathbf{K}$, then m is an isomorphism. □

Definition 3.3.6 (Subobject and quotient). Let $A \in |\mathbf{K}|$. A *subobject* of A is an object $B \in |\mathbf{K}|$ together with a morphism $m: B \to A$ such that $m \in \mathbf{M}$. A *quotient* of A is an object $B \in |\mathbf{K}|$ together with a morphism $e: A \to B$ such that $e \in \mathbf{E}$. □

Definition 3.3.7 (Reachable object). An object $A \in |\mathbf{K}|$ is *reachable* if it has no proper subobject, i.e. if every morphism $m \in \mathbf{M}$ with target A is an isomorphism. □

The category $\mathbf{Alg}(\Sigma)$ of Σ-algebras and the notion of a reachable algebra provide an instance of the general concept of reachability introduced in the above definition. The following theorem gives more general versions of well-known facts often laboriously proved in the standard algebraic framework.

Theorem 3.3.8. *Assume that* **K** *has an initial object Λ. Then:*

1. *An object $A \in |\mathbf{K}|$ is reachable iff it is a quotient of the initial object Λ.*
2. *Every object in $|\mathbf{K}|$ has a reachable subobject which is unique up to isomorphism.*
3. *If $A \in |\mathbf{K}|$ is reachable then for every $B \in |\mathbf{K}|$ there exists at most one morphism from A to B.*
4. *If $A \in |\mathbf{K}|$ is reachable and $f \in \mathbf{K}$ is a morphism with target A then $f \in \mathbf{E}$.* □

Exercise 3.3.9. Prove the theorem and identify the familiar facts about reachable algebras generalised here. □

One of the main results of Chapter 2, Theorem 2.5.14, states that any equational specification has an initial model. This is just a special case of a more general result which we formulate and prove for an arbitrary category with "reachability structure" satisfying an additional, technical property that any object has up to isomorphism only a *set* of quotients.

Definition 3.3.10 (Co-well-powered category). $\mathbf{K}$ is $\mathbf{E}$-*co-well-powered* if for any $A \in |\mathbf{K}|$ there exists a *set* of morphisms $E \subseteq \mathbf{E}$ such that for every morphism $e \in \mathbf{E}$ with source A there exist a morphism $e' \in E$ and an isomorphism i such that $e = e';i$. □

Definition 3.3.11 (Quasi-variety). A collection of objects $Q \subseteq |\mathbf{K}|$ is a *quasi-variety* if it is closed under subobjects and products of non-empty sets of objects in Q. □

Lemma 3.3.12 (Initiality lemma). *Assume that* $\mathbf{K}$ *has an initial object and is* $\mathbf{E}$-*co-well-powered, and any set of objects in* $\mathbf{K}$ *has a product. Then any non-empty quasi-variety in* $\mathbf{K}$ *(considered as the corresponding full subcategory of* $\mathbf{K}$*) has an initial object which is reachable in* $\mathbf{K}$.

Proof. Let $Q \subseteq |\mathbf{K}|$ be a non-empty collection of objects closed under subobjects and products of non-empty sets. Let Q_r be a *set* of reachable objects in Q such that every reachable object in Q is isomorphic to an element of Q_r (such a set exists since $\mathbf{K}$ is $\mathbf{E}$-co-well-powered). The reachable subobject of the product of Q_r (which is unique up to isomorphism) is a reachable initial object in Q. □

It is now easy to check that in the context of Example 3.3.3 every class of Σ-algebras definable by a set of Σ-equations is a non-empty quasi-variety, and hence Lemma 3.3.12 directly implies Theorem 2.5.14.

We conclude this section with two examples of categories naturally equipped with a notion of reachability which is an instance of the general concept introduced above.

Example 3.3.13. Recall Definitions 2.7.30 and 2.7.31 of partial Σ-algebras and Σ-homomorphisms between them. For any signature Σ, define the category of partial Σ-algebras, $\mathbf{PAlg}(\Sigma)$, as follows:

Objects of $\mathbf{PAlg}(\Sigma)$: partial Σ-algebras;
Morphisms of $\mathbf{PAlg}(\Sigma)$: weak Σ-homomorphisms.

Define also the subcategory $\mathbf{PAlg_{str}}(\Sigma)$ of partial Σ-algebras with *strong* homomorphisms between them, as follows:

Objects of $\mathbf{PAlg_{str}}(\Sigma)$: partial Σ-algebras;
Morphisms of $\mathbf{PAlg_{str}}(\Sigma)$: strong Σ-homomorphisms.

The category $\mathbf{PAlg}(\Sigma)$ of partial Σ-algebras with weak Σ-homomorphisms has a factorisation system $\langle \mathbf{PE}_\Sigma, \mathbf{PM}_\Sigma \rangle$, where $\mathbf{PE}_\Sigma$ is the collection of all epimorphisms in $\mathbf{PAlg}(\Sigma)$ and $\mathbf{PM}_\Sigma$ is the collection of all monomorphisms in $\mathbf{PAlg}(\Sigma)$ that are strong Σ-homomorphisms.

Exercise. Characterise epimorphisms in $\mathbf{PAlg}(\Sigma)$ (they are not surjective in general) and prove that $\langle \mathbf{PE}_\Sigma, \mathbf{PM}_\Sigma \rangle$ is indeed a factorisation system for $\mathbf{PAlg}(\Sigma)$. Check then that factorisation of a strong Σ-homomorphism in $\langle \mathbf{PE}_\Sigma, \mathbf{PM}_\Sigma \rangle$ consists of strong Σ-homomorphisms. Conclude that strong homomorphisms in $\mathbf{PE}_\Sigma$ and $\mathbf{PM}_\Sigma$, respectively, form a factorisation system for $\mathbf{PAlg_{str}}(\Sigma)$. □

Example 3.3.14. For any signature Σ, define the category of continuous Σ-algebras, $\mathbf{CAlg}(\Sigma)$, as follows:

Objects of $\mathbf{CAlg}(\Sigma)$: *continuous Σ-algebras*, which are just like ordinary (total) Σ-algebras, except that their carriers are required to be complete partial orders and their operations are continuous functions (cf. Exercise 3.1.10);

Morphisms of $\mathbf{CAlg}(\Sigma)$: *continuous Σ-homomorphisms*: given any continuous Σ-algebras $A, B \in |\mathbf{CAlg}(\Sigma)|$, a continuous Σ-homomorphism from A to B is a Σ-homomorphism $h: A \to B$ which is continuous as a function between complete partial orders. We say that h is *full* if it reflects the ordering, i.e. for all $a, a' \in |A|_s$, $h(a) \leq_B h(a')$ implies $a \leq_A a'$.

The category $\mathbf{CAlg}(\Sigma)$ of continuous Σ-algebras has a factorisation system $\langle \mathbf{CE}_\Sigma, \mathbf{CM}_\Sigma \rangle$, where $\mathbf{CM}_\Sigma$ is the collection of all full monomorphisms in $\mathbf{CAlg}(\Sigma)$ and $\mathbf{CE}_\Sigma$ is the collection of all *strongly dense* epimorphisms in $\mathbf{CAlg}(\Sigma)$. A continuous Σ-homomorphism $h: A \to B$ is strongly dense if B has no proper continuous subalgebra which contains the set-theoretic image of $|A|$ under h. (Note that the expected notion of a continuous subalgebra is determined by the chosen collection of factorisation monomorphisms $\mathbf{CM}_\Sigma$.) This is equivalent to the requirement that every element of $|B|$ is the least upper bound of a countable chain of least upper bounds of countable chains of ... of elements in the set-theoretic image of $|A|$ under h. Consequently, given a strongly dense continuous homomorphism $h: A \to B$, every element of $|B|$ is the least upper bound of a subset (not necessarily a chain though) of the set-theoretic image of $|A|$ under h, which yields the key argument to show that $\mathbf{CAlg}(\Sigma)$ is $\mathbf{CE}_\Sigma$-co-well-powered.

Exercise. Prove that $\langle \mathbf{CE}_\Sigma, \mathbf{CM}_\Sigma \rangle$ is indeed a factorisation system for $\mathbf{CAlg}(\Sigma)$. Also, try to construct an example of an epimorphism in $\mathbf{CAlg}(\Sigma)$ which is not strongly dense. □

Exercise 3.3.15. Characterise reachable algebras in $\mathbf{PAlg}(\Sigma)$ and in $\mathbf{CAlg}(\Sigma)$. Instantiate the facts listed in Theorem 3.3.8 to these categories. □

3.4 Functors and natural transformations

As explained in the introduction to this chapter, for category theorists it is tantamount to heresy to consider objects in the absence of morphisms between them. Up to now we have departed from this dogma in our study of categories themselves; in the previous sections of this chapter we have worked with categories without introducing any notion of a morphism between them. We hasten here to correct this lapse: morphisms between categories are *functors*, to be introduced in this section. And to atone we will also introduce *natural transformations*, which are morphisms between functors.

3.4.1 Functors

A category consists of a collection of objects and a collection of morphisms with structure given by the choice of sources and targets of morphism, by the definition of composition and by the identities that are assumed to exist. As in other standard cases of collections with additional structure, morphisms between categories are maps between the collections of objects and morphisms, respectively, that preserve this structure.

Definition 3.4.1 (Functor). A *functor* $\mathbf{F}:\mathbf{K1} \to \mathbf{K2}$ from a category $\mathbf{K1}$ to a category $\mathbf{K2}$ consists of:

- a function $\mathbf{F}_{Obj}: |\mathbf{K1}| \to |\mathbf{K2}|$; and
- for each $A,B \in |\mathbf{K1}|$, a function $\mathbf{F}_{A,B}: \mathbf{K1}(A,B) \to \mathbf{K2}(\mathbf{F}_{Obj}(A),\mathbf{F}_{Obj}(B))$

such that:

- $\mathbf{F}$ preserves identities: $\mathbf{F}_{A,A}(id_A) = id_{\mathbf{F}_{Obj}(A)}$ for all objects $A \in |\mathbf{K}|$; and
- $\mathbf{F}$ preserves composition: for all morphisms $f:A \to B$ and $g:B \to C$ in $\mathbf{K1}$, $\mathbf{F}_{A,C}(f;g) = \mathbf{F}_{A,B}(f);\mathbf{F}_{B,C}(g)$. □

Notation. We use $\mathbf{F}$ to refer to both $\mathbf{F}_{Obj}$ and $\mathbf{F}_{A,B}$ for all $A,B \in |\mathbf{K1}|$. □

In the literature, functors as defined above are sometimes referred to as *covariant* functors. A *contravariant* functor is then defined in the same way except that it "reverses the direction of morphisms", i.e. a contravariant functor $\mathbf{F}:\mathbf{K1} \to \mathbf{K2}$ maps a $\mathbf{K1}$-morphism $f:A \to B$ to a $\mathbf{K2}$-morphism $\mathbf{F}(f):\mathbf{F}(B) \to \mathbf{F}(A)$. Even though we will use this terminology sometimes, no new formal definition is required: a contravariant functor from $\mathbf{K1}$ to $\mathbf{K2}$ is a (covariant) functor from $\mathbf{K1}^{op}$ to $\mathbf{K2}$ (cf. Examples 3.4.7 and 3.4.29 below).

Example 3.4.2 (Identity functor). A functor $\mathbf{Id_K}:\mathbf{K} \to \mathbf{K}$ is defined in the obvious way. □

Example 3.4.3 (Inclusion functor). If $\mathbf{K1}$ is a subcategory of $\mathbf{K2}$ then the inclusion $\mathbf{I}:\mathbf{K1} \hookrightarrow \mathbf{K2}$ is a functor. □

Example 3.4.4 (Constant functor). For any $A \in |\mathbf{K2}|$, $\mathbf{C}_A:\mathbf{K1} \to \mathbf{K2}$ is a functor, where $\mathbf{C}_A(B) = A$ for any $B \in |\mathbf{K1}|$ and $\mathbf{C}_A(f) = id_A$ for any $\mathbf{K1}$-morphism f. □

Example 3.4.5 (Opposite functor). For any functor $\mathbf{F}:\mathbf{K1} \to \mathbf{K2}$, there is a functor $\mathbf{F}^{op}:\mathbf{K1}^{op} \to \mathbf{K2}^{op}$ which is the "same" as $\mathbf{F}$, but is considered between the opposite categories. □

Example 3.4.6 (Power set functor). $\mathcal{P}:\mathbf{Set} \to \mathbf{Set}$ is a functor, where $\mathcal{P}(X) = \{Y \mid Y \subseteq X\}$ for any set X, and for any function $f:X \to X'$, $\mathcal{P}(f):\mathcal{P}(X) \to \mathcal{P}(X')$ is defined by $\mathcal{P}(f)(Y) = \{f(y) \mid y \in Y\}$. □

Example 3.4.7 (Contravariant power set functor). $\mathcal{P}_{-1}: \mathbf{Set}^{op} \to \mathbf{Set}$ is a functor, where $\mathcal{P}_{-1}(X) = \{Y \mid Y \subseteq X\}$ for any set X, and for any morphism $f: X \to X'$ in $\mathbf{Set}^{op}$ (i.e. any function $f: X' \to X$), $\mathcal{P}_{-1}(f): \mathcal{P}_{-1}(X) \to \mathcal{P}_{-1}(X')$ is defined by $\mathcal{P}_{-1}(f)(Y) = \{x' \in X' \mid f(x') \in Y\}$. □

Example 3.4.8 (Sequence functor). $\mathbf{Seq}: \mathbf{Set} \to \mathbf{Mon}$ is a functor, where $\mathbf{Mon}$ is the category of monoids with monoid homomorphisms as morphisms. For any set $X \in |\mathbf{Set}|$, $\mathbf{Seq}(X) = \langle X^*, \hat{}, \varepsilon \rangle$, where X^* is the set of all finite sequences of elements from X, $\hat{}$ is sequence concatenation, and ε is the empty sequence. Then, for any function $f: X \to Y$, $\mathbf{Seq}(f): \mathbf{Seq}(X) \to \mathbf{Seq}(Y)$ is the homomorphism defined by $\mathbf{Seq}(f)(x_1 \dots x_n) = f(x_1) \dots f(x_n)$. □

Example 3.4.9 (Reduct functor). Given an algebraic signature morphism $\sigma: \Sigma \to \Sigma'$, $\_|_\sigma: \mathbf{Alg}(\Sigma') \to \mathbf{Alg}(\Sigma)$ is a functor that takes each Σ'-algebra A' to its σ-reduct $A'|_\sigma \in |\mathbf{Alg}(\Sigma)|$ and each Σ'-homomorphism h' to its σ-reduct $h'|_\sigma$ (cf. Definitions 1.5.4 and 1.5.8). □

Example 3.4.10 (Forgetful functor). For any signature $\Sigma = \langle S, \Omega \rangle$, $|\_|: \mathbf{Alg}(\Sigma) \to \mathbf{Set}^S$ is the functor that takes each Σ-algebra $A \in |\mathbf{Alg}(\Sigma)|$ to its S-sorted carrier set $|A| \in |\mathbf{Set}^S|$ and each Σ-homomorphism to its underlying S-sorted function. (The functor $|\_|$ should really be decorated with a subscript identifying the signature Σ — we hope that leaving it out will not confuse the reader.) These special reduct functors $|\_|$ will be referred to as *forgetful functors*.

More generally, the term "forgetful functor" is used to refer to any functor that, intuitively, forgets the structure of objects in a category, mapping any structured object to its underlying unstructured set of elements. Thus, in addition to covering examples that exactly fit the above definition (like the functor mapping any monoid to the set of its elements), this also covers examples like the functor that maps any topological space to the set of its points and the functor that forgets the metric of a metric space. □

Example 3.4.11 (Term algebra). For any signature $\Sigma = \langle S, \Omega \rangle$, there is a functor $T_\Sigma: \mathbf{Set}^S \to \mathbf{Alg}(\Sigma)$ that maps any S-sorted set X to the term algebra $T_\Sigma(X)$, and any S-sorted function $f: X \to Y$ to the unique Σ-homomorphism $f^\#: T_\Sigma(X) \to T_\Sigma(Y)$ that extends f. □

Exercise 3.4.12. For any signature Σ and set $\mathscr{E}$ of Σ-equations, define the *quotient functor* $\_/\mathscr{E}: \mathbf{Alg}(\Sigma) \to \mathbf{Alg}(\Sigma)$ such that for any Σ-algebra A, $A/\mathscr{E}$ is the quotient of A by the least congruence $\simeq$ on A generated by $\mathscr{E}$, that is, such that $t_A(v) \simeq t'_A(v)$ for each Σ-equation $\forall X \bullet t = t'$ in $\mathscr{E}$ and valuation $v: X \to |A|$. Check that what you define is a functor! □

Exercise 3.4.13. For any signature Σ, define the *restriction functor* $\mathbf{R}_\Sigma: \mathbf{Alg}(\Sigma) \to \mathbf{Alg}(\Sigma)$ such that for any Σ-algebra A, $\mathbf{R}_\Sigma(A)$ is the reachable subalgebra of A.

More generally: let $\mathbf{K}$ be an arbitrary category with an initial object and a factorisation system, and let $\mathbf{K}_R$ be the full subcategory of $\mathbf{K}$ determined by the collection of all reachable objects in $\mathbf{K}$ (cf. Section 3.3). Define a functor $\mathbf{R}_\mathbf{K}: \mathbf{K} \to \mathbf{K}_R$ that maps any $A \in |\mathbf{K}|$ to the (unique up to isomorphism) reachable subobject of A. □

Example 3.4.14 (Projection functor). For any two categories **K1** and **K2**, the projection functors $\Pi_{\mathbf{K1}}\colon \mathbf{K1} \times \mathbf{K2} \to \mathbf{K1}$ and $\Pi_{\mathbf{K2}}\colon \mathbf{K1} \times \mathbf{K2} \to \mathbf{K2}$ are defined by $\Pi_{\mathbf{K1}}(\langle A,B\rangle) = A$ and $\Pi_{\mathbf{K1}}(\langle f,g\rangle) = f$, and $\Pi_{\mathbf{K2}}(\langle A,B\rangle) = B$ and $\Pi_{\mathbf{K2}}(\langle f,g\rangle) = g$. □

Example 3.4.15 (Hom-functor). Let **K** be a locally small category. $\mathbf{Hom}\colon \mathbf{K}^{op} \times \mathbf{K} \to \mathbf{Set}$ is a functor, where $\mathbf{Hom}(\langle A,B\rangle) = \mathbf{K}(A,B)$ and

$$\mathbf{Hom}(\underbrace{\langle f\colon A' \to A, g\colon B \to B'\rangle}_{\in \mathbf{K}^{op}\times\mathbf{K}(\langle A,B\rangle,\langle A',B'\rangle)})(\underbrace{h\colon A \to B}_{\in \mathbf{Hom}(\langle A,B\rangle)}) = \underbrace{f;h;g}_{\in \mathbf{Hom}(\langle A',B'\rangle)}.$$

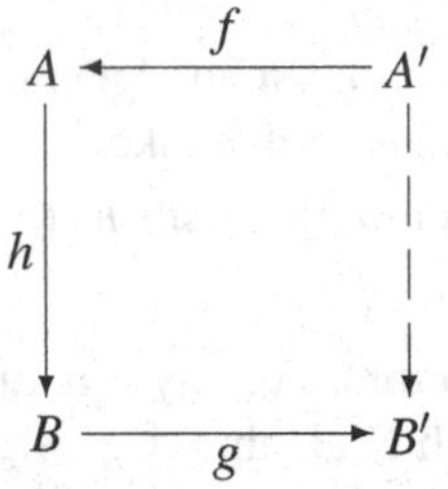

□

Exercise 3.4.16 (Exponent functor). For any set X define a functor $[\_\!\to\!X]\colon \mathbf{Set}^{op} \to \mathbf{Set}$ mapping any set to the set of all functions from it to X. That is, for any set $Y \in |\mathbf{Set}|$, $[Y\!\to\!X]$ is the set of all functions from Y to X and then for any morphism $f\colon Y \to Y'$ in $\mathbf{Set}^{op}$, which is a function $f\colon Y' \to Y$ in **Set**, $[f\!\to\!X]\colon [Y\!\to\!X] \to [Y'\!\to\!X]$ is defined by pre-composition with f as follows: $[f\!\to\!X](g) = f;g$. □

Example 3.4.17 (Converting partial function to total function). Recall the category **Pfn** of sets with partial functions (Example 3.1.8) and let $\mathbf{Set}_\perp$ be the subcategory of **Set** having sets containing a distinguished element $\perp$ as objects and $\perp$-preserving functions as morphisms. Then $\mathbf{Tot}\colon \mathbf{Pfn} \to \mathbf{Set}_\perp$ converts partial functions to total functions by using $\perp$ to represent "undefined" as follows:

- $\mathbf{Tot}(X) = X \uplus \{\perp\}$
- $\mathbf{Tot}(f)(x) = \begin{cases} f(x) & \text{if } f(x) \text{ is defined} \\ \perp & \text{otherwise} \end{cases}$

Exercise. Strictly speaking, the above definition is not well formed: according to the definition of disjoint union, if X is non-empty then $X \not\subseteq X \uplus \{\perp\}$; thus, given a partial function $f\colon X \to Y$, $\mathbf{Tot}(f)$ as defined above need not be a function from $\mathbf{Tot}(X)$ to $\mathbf{Tot}(Y)$. Restate this definition formally, using explicit injections $\iota_1\colon X \to X \uplus \{\perp\}$ and $\iota_2\colon \{\perp\} \to X \uplus \{\perp\}$ for each set X. □

Example 3.4.18 (Converting partial algebra to total algebra). The same "totalisation" idea as used in the above Example 3.4.17 yields a totalisation functor $\mathbf{Tot}_\Sigma\colon \mathbf{PAlg}_{\mathbf{str}}(\Sigma) \to \mathbf{Alg}(\Sigma)$, for each signature Σ, mapping partial Σ-algebras and their strong homomorphisms to total Σ-algebras and their homomorphisms (cf. Definitions 2.7.30 and 2.7.31, and Example 3.3.13).

Let $\Sigma = \langle S, \Omega\rangle \in |\mathbf{AlgSig}|$. $\mathbf{Tot}_\Sigma\colon \mathbf{PAlg}_{\mathbf{str}}(\Sigma) \to \mathbf{Alg}(\Sigma)$ is defined as follows:

- For any partial Σ-algebra $A \in |\mathbf{PAlg_{str}}(\Sigma)|$, $\mathbf{Tot}_\Sigma(A) = A_\perp \in |\mathbf{Alg}(\Sigma)|$ is the (total) Σ-algebra underlying A; see Definition 2.7.30.
- For any strong Σ-homomorphism $h: A \to B$ (which is a family of *total* functions between the corresponding carriers of A and B), $\mathbf{Tot}_\Sigma(h): \mathbf{Tot}_\Sigma(A) \to \mathbf{Tot}_\Sigma(B)$ is (the family of functions in) h extended to map $\perp$ to $\perp$.

Exercise. Check that $\mathbf{Tot}_\Sigma(h): \mathbf{Tot}_\Sigma(A) \to \mathbf{Tot}_\Sigma(B)$ is indeed a Σ-homomorphism for any strong Σ-homomorphism $h: A \to B$. Can you extend $\mathbf{Tot}_\Sigma$ to *weak* Σ-homomorphisms between partial algebras? □

Exercise 3.4.19. Do the above functors map monomorphisms to monomorphisms? What about epimorphisms? Isomorphisms? (Co)cones? (Co)limits? Anything else you can think of? □

Definition 3.4.20 (Diagram translation). Given a functor $\mathbf{F}: \mathbf{K1} \to \mathbf{K2}$ and a diagram D in $\mathbf{K1}$, the *translation of D by* $\mathbf{F}$ is defined as the diagram $\mathbf{F}(D)$ in $\mathbf{K2}$ with the same underlying graph as D and with the labels of D translated by $\mathbf{F}$:

- $G(\mathbf{F}(D)) = G(D)$;
- for each $n \in |G(D)|_{Node}$, $\mathbf{F}(D)_n = \mathbf{F}(D_n)$; and
- for each $e \in |G(D)|_{Edge}$, $\mathbf{F}(D)_e = \mathbf{F}(D_e)$. □

Exercise 3.4.21 (Diagram as functor). A diagram D in $\mathbf{K}$ corresponds to a functor from the category $\mathbf{Path}(G(D))$ of paths in the underlying graph of D to $\mathbf{K}$. Formalise this. HINT: Given a diagram D, define a functor that maps each path $e_1 \ldots e_n$ in $G(D)$ to $D_{e_1};\ldots;D_{e_n}$. Do not forget the case where $n = 0$.

Conversely, any functor into $\mathbf{K}$ with small source category may be considered as a diagram in $\mathbf{K}$: given a small category $\mathbf{J}$, a diagram in $\mathbf{K}$ on the underlying graph $G(\mathbf{J})$ of $\mathbf{J}$ may be obtained from any functor $\mathbf{D}: \mathbf{J} \to \mathbf{K}$ in the obvious way. In general though, not every diagram with the underlying graph $G(\mathbf{J})$ is so given by a functor from $\mathbf{J}$. Show an example of a category $\mathbf{K}$ and a small category $\mathbf{J}$ such that all (diagrams given by) functors $\mathbf{D}: \mathbf{J} \to \mathbf{K}$ have limits in $\mathbf{K}$ but not all diagrams with the underlying graph $G(\mathbf{J})$ have limits in $\mathbf{K}$. Prove though that $\mathbf{K}$ is (co)complete if and only if (co)limits for all functors $\mathbf{D}: \mathbf{J} \to \mathbf{K}$ with small $\mathbf{J}$ exist in $\mathbf{K}$.

Then, anticipating Definition 3.4.27, define the translation of a diagram by a functor in terms of functor composition. □

Definition 3.4.22 (Functor continuity and cocontinuity). A functor $\mathbf{F}: \mathbf{K1} \to \mathbf{K2}$ is (*finitely*) *continuous* if it preserves the existing limits of all (finite) diagrams in $\mathbf{K1}$, that is, if for any (finite) diagram D in $\mathbf{K1}$, $\mathbf{F}$ maps any limiting cone over D to a limiting cone over $\mathbf{F}(D)$.

A functor $\mathbf{F}: \mathbf{K1} \to \mathbf{K2}$ is (*finitely*) *cocontinuous* if it preserves the existing colimits of all (finite) diagrams in $\mathbf{K1}$, that is, if for any (finite) diagram D in $\mathbf{K1}$, $\mathbf{F}$ maps any colimiting cocone over D to a colimiting cocone over $\mathbf{F}(D)$. □

Exercise 3.4.23. Assuming that $\mathbf{K1}$ is (finitely) complete, use Exercise 3.2.49 to show that a functor $\mathbf{F}: \mathbf{K1} \to \mathbf{K2}$ is (finitely) continuous if and only if it preserves (finite) products and equalisers.

Similarly, show that $\mathbf{F}: \mathbf{K1} \to \mathbf{K2}$ is finitely continuous if and only if it preserves terminal objects and all pullbacks, and it is continuous if and only if it preserves terminal objects and all wide pullbacks. HINT: Exercises 3.2.48 and 3.2.50.

Dually, give similar characterisation of (finitely) cocontinuous functors, for instance as those that preserve (finite) coproducts and coequalisers. □

Exercise 3.4.24. Given a set X, show that the functor $[\_\to X]: \mathbf{Set}^{op} \to \mathbf{Set}$ from Exercise 3.4.16 is continuous. HINT: Use Exercise 3.4.23: relying on the explicit constructions of (co)products and (co)equalisers in **Set**, show that the functor maps any coproduct (disjoint union) of sets $\langle X_n \rangle_{n \in N}$ to a product of sets of functions $[X_n \to X]$, $n \in N$, and a coequaliser of functions $f, g: X_1 \to X_2$ to an equaliser of (precomposition) functions $(f;\_), (g;\_): [X_2 \to X] \to [X_1 \to X]$.

You may also want to similarly check which of the examples of functors given above are (finitely) (co)continuous. □

Exercise 3.4.25. Consider a category $\mathbf{K}$ with a terminal object $1 \in |\mathbf{K}|$. Given any functor $\mathbf{F}: \mathbf{K} \to \mathbf{K}'$, check that $\mathbf{F}$ determines a functor $\mathbf{F}_{\downarrow 1}: \mathbf{K} \to \mathbf{K}' \downarrow \mathbf{F}(1)$ from $\mathbf{K}$ to the slice category of $\mathbf{K}'$-objects over $\mathbf{F}(1)$ (Definition 3.1.29), where for any object $A \in |\mathbf{K}|$, $\mathbf{F}_{\downarrow 1}(A) = \mathbf{F}(!_A)$, with $!_A: A \to 1$ being the unique morphism from A to 1, and $\mathbf{F}_{\downarrow 1}$ coincides with $\mathbf{F}$ on morphisms.

Suppose now that $\mathbf{K}$ has all pullbacks (so that it is finitely complete) and $\mathbf{F}$ preserves them (but we do not require $\mathbf{F}$ to preserve the terminal object, so it does not have to be finitely continuous). Show that $\mathbf{F}_{\downarrow 1}: \mathbf{K} \to \mathbf{K}' \downarrow \mathbf{F}(1)$ is finitely continuous. HINT: Recall Exercise 3.2.51. By the discussion there, since $\mathbf{F}$ preserves pullbacks, $\mathbf{F}$ maps products in $\mathbf{K}$, which are pullbacks of morphisms to 1, to pullbacks in $\mathbf{K}'$ of morphisms to $\mathbf{F}(1)$ — and these are essentially products in $\mathbf{K}' \downarrow \mathbf{F}(1)$. Moreover, by the construction, $\mathbf{F}_{\downarrow 1}$ preserves the terminal object, and the conclusion follows by Exercise 3.4.23.

Similarly, show that if $\mathbf{K}$ has all wide pullbacks (so that it is complete) and $\mathbf{F}$ preserves them then $\mathbf{F}_{\downarrow 1}: \mathbf{K} \to \mathbf{K}' \downarrow \mathbf{F}(1)$ is continuous. □

Exercise 3.4.26. Recall the definition of the category $\mathbf{T}_{\Sigma,\mathscr{E}}$, the algebraic theory generated by a set $\mathscr{E}$ of equations over a signature Σ (cf. Exercise 3.1.16). Show that those functors from $\mathbf{T}^{op}_{\Sigma,\varnothing}$ to **Set** that preserve finite products (where products in $\mathbf{T}^{op}_{\Sigma,\mathscr{E}}$, that is coproducts in $\mathbf{T}_{\Sigma,\mathscr{E}}$, are given by concatenation of sequences of sort names — cf. Exercise 3.2.19 — and products in **Set** are given by the Cartesian product) are in a bijective correspondence with Σ-algebras in $|\mathbf{Alg}(\Sigma)|$. Generalise this correspondence further to product-preserving functors from $\mathbf{T}^{op}_{\Sigma,\mathscr{E}}$ to **Set** and Σ-algebras in $\mathit{Mod}_\Sigma(\mathscr{E})$. □

Definition 3.4.27 (Functor composition). The category **Cat** (the category of all categories) is defined as follows:

Objects of **Cat**: categories;[5]
Morphisms of **Cat**: functors;

[5] To be cautious about the set-theoretic foundations here, we should rather say *small* categories.

Composition in **Cat**: If $\mathbf{F}{:}\,\mathbf{K1} \to \mathbf{K2}$ and $\mathbf{G}{:}\,\mathbf{K2} \to \mathbf{K3}$ are functors, then $\mathbf{F};\mathbf{G}{:}\,\mathbf{K1} \to \mathbf{K3}$ is a functor defined as follows: $(\mathbf{F};\mathbf{G})_{Obj} = \mathbf{F}_{Obj};\mathbf{G}_{Obj}$ and $(\mathbf{F};\mathbf{G})_{A,B} = \mathbf{F}_{A,B};\mathbf{G}_{\mathbf{F}(A),\mathbf{F}(B)}$ for all $A,B \in \mathbf{K1}$. □

Example 3.4.28. In the following we will often use the functor $|\_|{:}\,\mathbf{Cat} \to \mathbf{Set}$[6] which for any category $\mathbf{K} \in |\mathbf{Cat}|$ yields the collection $|\mathbf{K}|$ of the objects of this category and for each functor $\mathbf{F}{:}\,\mathbf{K} \to \mathbf{K}'$ yields its object part $|\mathbf{F}| = \mathbf{F}_{Obj}{:}\,|\mathbf{K}| \to |\mathbf{K}'|$. □

Example 3.4.29. $\mathbf{Alg}{:}\,\mathbf{AlgSig}^{op} \to \mathbf{Cat}$ **is a functor, where:**

- for any $\Sigma \in |\mathbf{AlgSig}|$, $\mathbf{Alg}(\Sigma)$ is the category of Σ-algebras; and
- for any signature morphism $\sigma{:}\,\Sigma \to \Sigma'$ in **AlgSig**, $\mathbf{Alg}(\sigma)$ is the reduct functor $\_|_\sigma{:}\,\mathbf{Alg}(\Sigma') \to \mathbf{Alg}(\Sigma)$. □

Exercise 3.4.30. Define a functor $\mathbf{Alg}^{der}{:}\,(\mathbf{AlgSig}^{der})^{op} \to \mathbf{Cat}$ so that $\mathbf{Alg}^{der}(\Sigma) = \mathbf{Alg}(\Sigma)$ for any signature $\Sigma \in |\mathbf{AlgSig}^{der}|$, and for any derived signature morphism δ, $\mathbf{Alg}^{der}(\delta)$ is the δ-reduct as sketched in Definition 1.5.16 and Exercise 1.5.17. □

Exercise 3.4.31. Define the category **Poset** (objects: partially ordered sets; morphisms: order-preserving functions). Define the functor from **Poset** to **Cat** that maps a partially ordered set to the corresponding (preorder) category (cf. Example 3.1.3) and an order-preserving function to the corresponding functor. □

Exercise 3.4.32. Characterise isomorphisms in **Cat**. Show that product categories are products in **Cat**. What are terminal objects, pullbacks and equalisers in **Cat**? Conclude that **Cat** is complete. HINT: Use constructions analogous to those in **Set**, as summarized in Exercise 3.2.53. □

Exercise 3.4.33. Prove that $\mathbf{Alg}{:}\,\mathbf{AlgSig}^{op} \to \mathbf{Cat}$ (cf. Example 3.4.29) is continuous, that is, that it maps colimits in the category **AlgSig** of signatures to limits in the category **Cat** of all categories.

HINT: By Exercise 3.4.23 it is enough to show that **Alg** maps coproducts of signatures to products of the corresponding categories of algebras and coequalisers of signature morphisms to equalisers of the corresponding reduct functors.

(*Coproducts*): Recall that by Exercise 3.2.17, a coproduct of signatures is in fact their disjoint union. Now, it is easy to see that an algebra over a disjoint union of a family of signatures may be identified with a tuple of algebras over the signatures in the family. Since a similar fact holds for homomorphisms, the rest of the proof in this case is straightforward (cf. Exercise 3.4.32). Notice that this argument covers the coproduct of the empty family of signatures as well.

---

[6] Again, we should restrict attention to small categories here. Alternatively, in place of **Set** we could use the category of all discrete categories, inheriting all of the foundational problems of **Cat**.

(*Coequalisers*): Recall (cf. Exercise 3.2.25) that a coequaliser of two signature morphisms $\sigma, \sigma' : \Sigma \to \Sigma'$ is the natural projection $p : \Sigma' \to (\Sigma'/{\equiv})$, where $\equiv$ is the least equivalence relation on Σ' such that $\sigma(x) \equiv \sigma'(x)$ for all sort and operation names x in Σ (this is just a sketch of the construction). Notice now that $(\Sigma'/{\equiv})$-algebras correspond exactly to those Σ'-algebras that have identical components $\sigma(x)$ and $\sigma'(x)$ for all sort and operation names x in Σ, or equivalently, to those algebras $A' \in |\mathbf{Alg}(\Sigma')|$ for which $A'|_{\sigma} = A'|_{\sigma'}$. Moreover, the correspondence is given by the functor $\_|_p : \mathbf{Alg}(\Sigma'/{\equiv}) \to \mathbf{Alg}(\Sigma')$. Since a similar fact holds for homomorphisms, it is straightforward now to prove that $\_|_p = \mathbf{Alg}(p)$ is an equaliser of $\_|_{\sigma} = \mathbf{Alg}(\sigma)$ and $\_|_{\sigma'} = \mathbf{Alg}(\sigma')$ (cf. Exercises 3.4.32 and 3.2.23). □

Exercise 3.4.34 (Amalgamation Lemma for algebras). Consider a pushout in the category **AlgSig** of signatures:

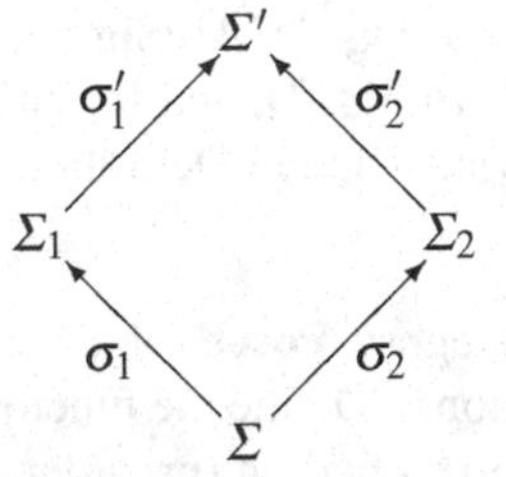

Conclude from Exercise 3.4.33 above that for any Σ_1-algebra A_1 and Σ_2-algebra A_2 such that $A_1|_{\sigma_1} = A_2|_{\sigma_2}$, there exists a unique Σ'-algebra A' such that $A'|_{\sigma'_1} = A_1$ and $A'|_{\sigma'_2} = A_2$.

Similarly, for any two homomorphisms $h_1 : A_{11} \to A_{12}$ in $\mathbf{Alg}(\Sigma_1)$ and $h_2 : A_{21} \to A_{22}$ in $\mathbf{Alg}(\Sigma_2)$ such that $h_1|_{\sigma_1} = h_2|_{\sigma_2}$, there exists a unique Σ'-homomorphism $h' : A'_1 \to A'_2$ such that $h'|_{\sigma'_1} = h_1$ and $h'|_{\sigma'_2} = h_2$. □

Example 3.4.35. Recall Example 3.2.36 of a simple pushout of algebraic signatures. Let $N \in |\mathbf{Alg}(\Sigma_{\mathrm{NAT}})|$ be the standard model of natural numbers. Build $N_1 \in |\mathbf{Alg}(\Sigma_{\mathrm{NAT}_{\mathit{fib}}})|$ by adding to N the interpretation of the operation *fib* as the standard Fibonacci function, and $N_2 \in |\mathbf{Alg}(\Sigma_{\mathrm{NAT}_{\mathit{mult}}})|$ by adding to N the interpretation of the operation *mult* as multiplication. By construction we have $N_1|_{\Sigma_{\mathrm{NAT}}} = N = N_2|_{\Sigma_{\mathrm{NAT}}}$ and so N_1 and N_2 amalgamate to a unique algebra $N' \in |\mathbf{Alg}(\Sigma_{\mathrm{NAT}_{\mathit{fib},\mathit{mult}}})|$ such that $N'|_{\Sigma_{\mathrm{NAT}_{\mathit{fib}}}} = N_1$ and $N'|_{\Sigma_{\mathrm{NAT}_{\mathit{mult}}}} = N_2$. Clearly, N' is the only expansion of N that defines *fib* as the Fibonacci function (as N_1 does) and *mult* as multiplication (as N_2 does). □

Exercise 3.4.36. Define initial objects and coproducts in **Cat**. (HINT: This is easy.) Try to define coequalisers and then pushouts in **Cat**. (HINT: This is difficult.) □

3.4.2 Natural transformations

Let $\mathbf{F}:\mathbf{K1} \to \mathbf{K2}$ and $\mathbf{G}:\mathbf{K1} \to \mathbf{K2}$ be two functors with common source and target categories.

A transformation from $\mathbf{F}$ to $\mathbf{G}$ should map the results of $\mathbf{F}$ to the results of $\mathbf{G}$. This means that it should consist of a family of morphisms in $\mathbf{K2}$, one $\mathbf{K2}$-morphism from $\mathbf{F}(A)$ to $\mathbf{G}(A)$ for each $\mathbf{K1}$-object A. An extra requirement is that this family should be compatible with the application of $\mathbf{F}$ and $\mathbf{G}$ to $\mathbf{K1}$-morphisms, as formalised by the following definition:

Definition 3.4.37 (Natural transformation). A *natural transformation* from $\mathbf{F}$ to $\mathbf{G}$, $\tau:\mathbf{F} \to \mathbf{G}$,[7] is a family $\langle \tau_A:\mathbf{F}(A) \to \mathbf{G}(A)\rangle_{A \in |\mathbf{K1}|}$ of $\mathbf{K2}$-morphisms such that for any $A, B \in |\mathbf{K1}|$ and $\mathbf{K1}$-morphism $f:A \to B$ the following diagram commutes:

$$\begin{array}{ccccc}
\mathbf{K1}: & \qquad & \mathbf{K2}: & & \\
A & & \mathbf{F}(A) & \xrightarrow{\ \tau_A\ } & \mathbf{G}(A) \\
\downarrow f & & \downarrow \mathbf{F}(f) & & \downarrow \mathbf{G}(f) \\
B & & \mathbf{F}(B) & \xrightarrow[\ \tau_B\ ]{} & \mathbf{G}(B)
\end{array}$$

(this property is often referred to as the *naturality* of the family τ).

Furthermore, τ is a *natural isomorphism* if for all $A \in |\mathbf{K1}|$, τ_A is iso (in $\mathbf{K2}$). □

Example 3.4.38. The identity transformation $id_{\mathbf{F}}:\mathbf{F} \to \mathbf{F}$, where $(id_{\mathbf{F}})_A = id_{\mathbf{F}(A)}$, is a natural isomorphism.

For any morphism $f:A \to B$ in a category $\mathbf{K2}$ and for any category $\mathbf{K1}$, there is a constant natural transformation $\mathbf{c}_f:\mathbf{C}_A \to \mathbf{C}_B$ between the constant functors $\mathbf{C}_A, \mathbf{C}_B:\mathbf{K1} \to \mathbf{K2}$ (cf. Example 3.4.4) defined by $(\mathbf{c}_f)_o = f$ for all objects $o \in |\mathbf{K1}|$. □

Example 3.4.39. The family of singleton functions $sing\_set: \mathbf{Id}_{\mathbf{Set}} \to \mathcal{P}$, where for any set X, $sing\_set_X:X \to \mathcal{P}(X)$ is defined by $sing\_set_X(a) = \{a\}$, is a natural transformation.

Let $(\_)^* = \mathbf{Seq};|\_|:\mathbf{Set} \to \mathbf{Set}$ be the composition of $\mathbf{Seq}:\mathbf{Set} \to \mathbf{Mon}$ (Example 3.4.8) with the forgetful functor $|\_|:\mathbf{Mon} \to \mathbf{Set}$ mapping any monoid to its underlying carrier set. The family of singleton functions $sing\_seq: \mathbf{Id}_{\mathbf{Set}} \to (\_)^*$, where for any set X, $sing\_seq_X:X \to X^*$ is defined by $sing\_seq_X(a) = a$ ($sing\_seq$ maps any element to the singleton sequence consisting of this element only), is a natural transformation. □

---

[7] Some authors would use a dotted or double arrow here, writing $\tau:\mathbf{F} \dot{\to} \mathbf{G}$ or $\tau:\mathbf{F} \Rightarrow \mathbf{G}$, respectively. We prefer to use the same symbol for all morphisms, and also for natural transformations, since they are morphisms in certain categories; see Definition 3.4.61 below.

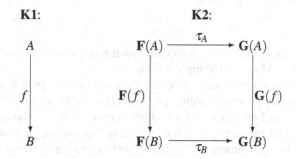

Exercise 3.4.40. Consider the functor $(\_)^*\colon \mathbf{Set} \to \mathbf{Set}$ mapping any set X to the set X^* of sequences over X (cf. Example 3.4.39 above). Show that the following families of functions (indexed by sets $X \in |\mathbf{Set}|$) yield natural transformations from $(\_)^*$ to $(\_)^*$:

- for each $k \geq 0$, for $n \geq 0$ and $x_1, \ldots, x_n \in X$,
  $stutter^k_X(x_1 \ldots x_n) = \underbrace{x_1 \ldots x_1}_{k\text{ times}} \ldots \underbrace{x_n \ldots x_n}_{k\text{ times}}$;
- for each $k \geq 0$, for $n \geq 0$ and $x_1, \ldots, x_n \in X$,
  $repeat^k_X(x_1 \ldots x_n) = \underbrace{\underbrace{x_1 \ldots x_n} \ldots \underbrace{x_1 \ldots x_n}}_{k\text{ times}}$;
- for $n \geq 0$ and $x_1, \ldots, x_n \in X$,
  $reverse_X(x_1 \ldots x_n) = x_n \ldots x_1$;
- for $n \geq 0$ and $x_1, \ldots, x_{2n+1} \in X$,
  $odds_X(x_1 x_2 x_3 \ldots x_{2n}) = x_1 x_3 \ldots x_{2n-1}$ and
  $odds_X(x_1 x_2 x_3 \ldots x_{2n+1}) = x_1 x_3 \ldots x_{2n+1}$.

Check which of these functions also yield natural transformations from **Seq** to **Seq** (where $\mathbf{Seq}\colon \mathbf{Set} \to \mathbf{Mon}$; cf. Example 3.4.8).

The above examples indicate a close link between polymorphic functions as encountered in functional programming languages (like Standard ML [MTHM97] or Haskell [Pey03]) and natural transformations between functors representing polymorphic types. This property, often referred to as "parametric polymorphism" (as opposed to "ad hoc polymorphism") can be explored to derive some properties of polymorphic functions directly from their types [Wad89]. □

Exercise 3.4.41. Recall (Exercise 3.4.26) the correspondence between product-preserving functors from $\mathbf{T}^{op}_{\Sigma,\mathscr{E}}$ to **Set** and Σ-algebras in $|\mathbf{Mod}_\Sigma(\mathscr{E})|$. Show that this correspondence extends to morphisms: each Σ-homomorphism between algebras gives rise to a natural transformation between the corresponding functors, and vice versa, each natural transformation between such functors determines a homomorphism between the corresponding algebras. HINT: To prove that this yields a bijective correspondence, first use the naturality condition for product projections to show that for any natural transformation $\tau\colon \mathbf{F} \to \mathbf{G}$ between product-preserving functors $\mathbf{F}, \mathbf{G}\colon \mathbf{T}^{op}_{\Sigma,\mathscr{E}} \to \mathbf{Set}$, any sequence $s_1 \ldots s_n$ of sort names (an object in $\mathbf{T}_{\Sigma,\mathscr{E}}$) and any $\langle a_1, \ldots, a_n \rangle \in \mathbf{F}(s_1 \ldots s_n)$, $\tau_{s_1 \ldots s_n}(\langle a_1, \ldots, a_n \rangle) = \langle \tau_{s_1}(a_1), \ldots, \tau_{s_n}(a_n) \rangle$. □

Natural transformations have been introduced as morphisms between functors. The obvious thing to do next is to define composition of natural transformations. Traditionally, two different composition operations for natural transformations are introduced: *vertical* and *horizontal* composition. The former is a straightforward composition of natural transformations between parallel functors. The latter is somewhat more involved; in a sense, it shows how natural transformations "accumulate" when functors are composed.

Definition 3.4.42 (Vertical composition). Let $\mathbf{F1}, \mathbf{F2}, \mathbf{F3}\colon \mathbf{K1} \to \mathbf{K2}$ be three functors with common source and target categories. Let $\tau\colon \mathbf{F1} \to \mathbf{F2}$ and $\sigma\colon \mathbf{F2} \to \mathbf{F3}$ be natural transformations:

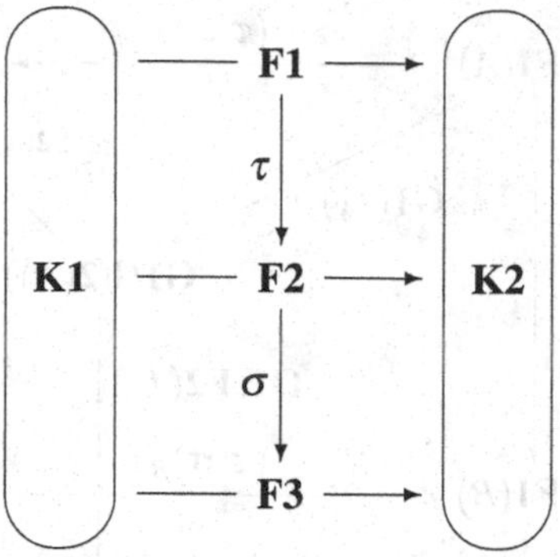

Then the *vertical composition* of τ and σ, $\tau;\sigma\colon \mathbf{F1} \to \mathbf{F3}$, is defined by $(\tau;\sigma)_A = \tau_A;\sigma_A$ (in **K2**) for all $A \in |\mathbf{K1}|$. □

Exercise 3.4.43. Prove that $\tau;\sigma$ is indeed a natural transformation. □

Definition 3.4.44 (Horizontal composition). Given two pairs of parallel functors, $\mathbf{F1},\mathbf{F2}\colon \mathbf{K1} \to \mathbf{K2}$ and $\mathbf{G1},\mathbf{G2}\colon \mathbf{K2} \to \mathbf{K3}$, let $\tau\colon \mathbf{F1} \to \mathbf{F2}$ and $\sigma\colon \mathbf{G1} \to \mathbf{G2}$ be natural transformations:

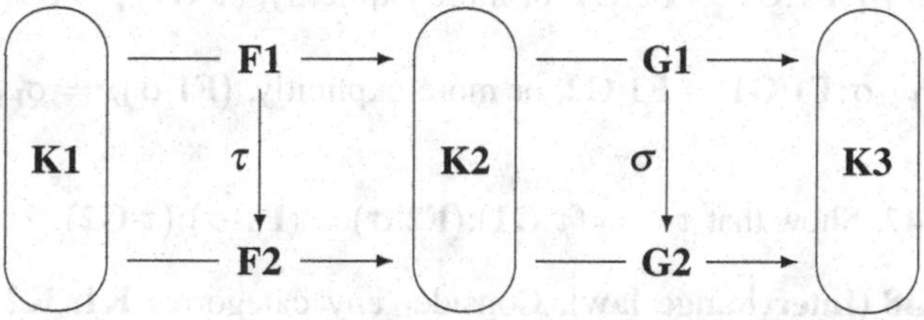

Then the *horizontal composition* of τ and σ, $\tau{\cdot}\sigma\colon \mathbf{F1};\mathbf{G1} \to \mathbf{F2};\mathbf{G2}$, is defined by $(\tau{\cdot}\sigma)_A = \mathbf{G1}(\tau_A);\sigma_{\mathbf{F2}(A)} = \sigma_{\mathbf{F1}(A)};\mathbf{G2}(\tau_A)$ (in **K3**) for all $A \in |\mathbf{K1}|$:

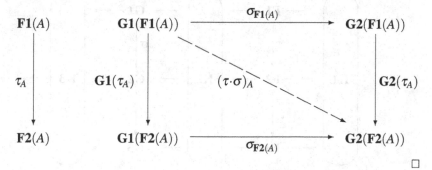

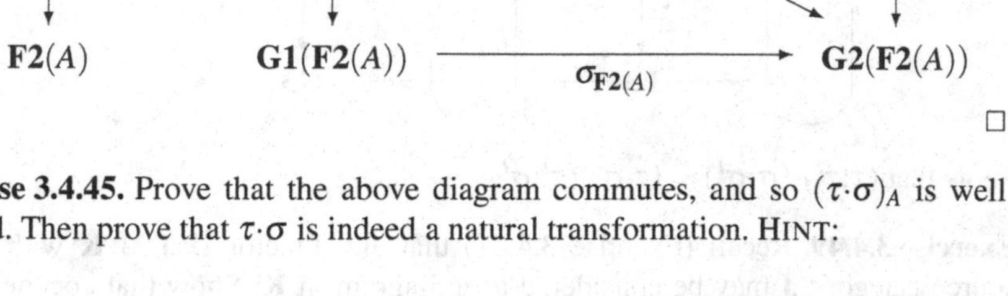

□

Exercise 3.4.45. Prove that the above diagram commutes, and so $(\tau{\cdot}\sigma)_A$ is well defined. Then prove that $\tau{\cdot}\sigma$ is indeed a natural transformation. HINT:

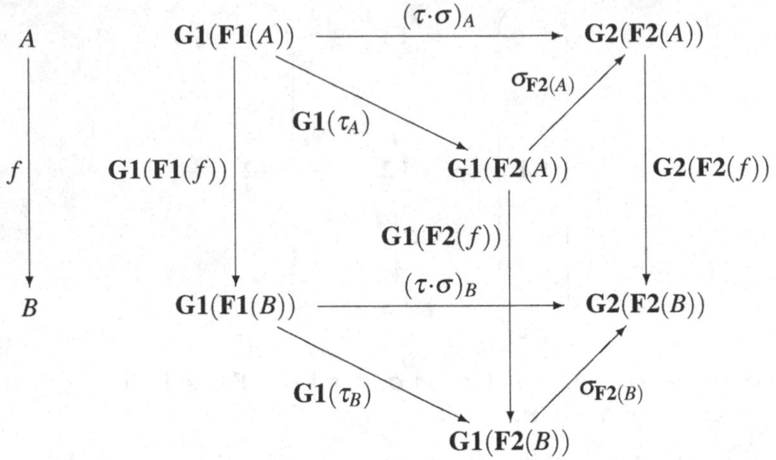

□

Definition 3.4.46 (Multiplication by a functor). A special case of the horizontal composition of natural transformations is the *multiplication* of a natural transformation by a functor. Under the assumptions of Definition 3.4.44, we define:

- $\tau\cdot\mathbf{G1} = \tau\cdot id_{\mathbf{G1}}:\mathbf{F1};\mathbf{G1} \to \mathbf{F2};\mathbf{G1}$, or more explicitly, $(\tau\cdot\mathbf{G1})_A = \mathbf{G1}(\tau_A)$ for $A \in |\mathbf{K1}|$;
- $\mathbf{F1}\cdot\sigma = id_{\mathbf{F1}}\cdot\sigma:\mathbf{F1};\mathbf{G1} \to \mathbf{F1};\mathbf{G2}$, or more explicitly, $(\mathbf{F1}\cdot\sigma)_A = \sigma_{\mathbf{F1}(A)}$ for $A \in |\mathbf{K1}|$. □

Exercise 3.4.47. Show that $\tau\cdot\sigma = (\tau\cdot\mathbf{G1});(\mathbf{F2}\cdot\sigma) = (\mathbf{F1}\cdot\sigma);(\tau\cdot\mathbf{G2})$. □

Exercise 3.4.48 (Interchange law). Consider any categories **K1**, **K2**, **K3**, functors $\mathbf{F1},\mathbf{F2},\mathbf{F3}:\mathbf{K1} \to \mathbf{K2}$ and $\mathbf{G1},\mathbf{G2},\mathbf{G3}:\mathbf{K2} \to \mathbf{K3}$, and natural transformations $\tau:\mathbf{F1} \to \mathbf{F2}$, $\tau':\mathbf{F2} \to \mathbf{F3}$, $\sigma:\mathbf{G1} \to \mathbf{G2}$, and $\sigma':\mathbf{G2} \to \mathbf{G3}$:

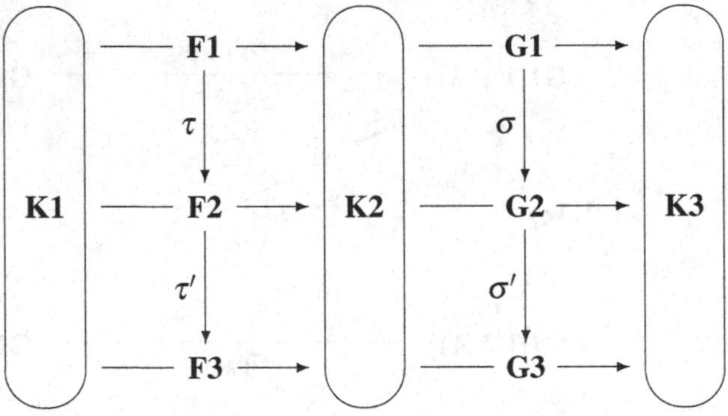

Show that $(\tau;\tau')\cdot(\sigma;\sigma') = (\tau\cdot\sigma);(\tau'\cdot\sigma')$. □

Exercise 3.4.49. Recall (Exercise 3.4.21) that any functor $\mathbf{D}:\mathbf{J} \to \mathbf{K}$ with small source category **J** may be considered as a diagram in **K**. Show that cocones over the diagram (given by) $\mathbf{D}:\mathbf{J} \to \mathbf{K}$ are natural transformations from **D** to constant functors. Use (vertical) composition of such cocones with constant natural transformations (Example 3.4.38) to rephrase the definition of a colimit. Characterise cones and limits in a similar way. □

3.4.3 Constructing categories, revisited

3.4.3.1 Comma categories

Definition 3.4.50 (Comma category). Let $\mathbf{F}:\mathbf{K1} \to \mathbf{K}$ and $\mathbf{G}:\mathbf{K2} \to \mathbf{K}$ be two functors with a common target category. The *comma category* $(\mathbf{F},\mathbf{G})$ is defined by:

Objects of $(\mathbf{F},\mathbf{G})$: triples $\langle A1,f,A2\rangle$, where $A1 \in |\mathbf{K1}|$, $A2 \in |\mathbf{K2}|$ and $f:\mathbf{F}(A1) \to \mathbf{G}(A2)$ is a morphism in $\mathbf{K}$;

Morphisms of $(\mathbf{F},\mathbf{G})$: a morphism from $\langle A1,f,A2\rangle$ to $\langle B1,g,B2\rangle$ is a pair $\langle h1,h2\rangle$ of morphisms where $h1:A1 \to B1$ (in $\mathbf{K1}$) and $h2:A2 \to B2$ (in $\mathbf{K2}$) such that (the middle part of) the following diagram commutes:

$$\begin{array}{cccccc}
A1 & \mathbf{F}(A1) & \xrightarrow{\ f\ } & \mathbf{G}(A2) & A2 \\
{\scriptstyle h1}\downarrow & {\scriptstyle \mathbf{F}(h1)}\downarrow & & \downarrow{\scriptstyle \mathbf{G}(h2)} & \downarrow{\scriptstyle h2} \\
B1 & \mathbf{F}(B1) & \xrightarrow[\ g\ ]{} & \mathbf{G}(B2) & B2
\end{array}$$

Composition in $(\mathbf{F},\mathbf{G})$: $\langle h1,h2\rangle;\langle h1',h2'\rangle = \langle h1;h1',h2;h2'\rangle$. □

Exercise 3.4.51. Construct the category $\mathbf{K}^{\to}$ of $\mathbf{K}$-morphisms and the category $\mathbf{K}\downarrow A$ of $\mathbf{K}$-objects over $A \in |\mathbf{K}|$ as comma categories (cf. Definitions 3.1.28 and 3.1.29). HINT: Consider categories $(\mathbf{Id_K},\mathbf{Id_K})$ and $(\mathbf{Id}_K,\mathbf{C}^{\mathbf{1}}_A)$, where $\mathbf{Id_K}$ is the identity functor on $\mathbf{K}$ and $\mathbf{C}^{\mathbf{1}}_A:\mathbf{1} \to \mathbf{K}$ is a constant functor from the terminal category $\mathbf{1}$. □

Example 3.4.52. Another way of presenting the category **Graph** is as the comma category $(\mathbf{Id_{Set}},\mathbf{CP})$, where $\mathbf{CP}:\mathbf{Set} \to \mathbf{Set}$ is the Cartesian product functor defined by $\mathbf{CP}(X) = X \times X$ and $\mathbf{CP}(f:X \to Y)\langle x1,x2\rangle = \langle f(x1),f(x2)\rangle$. To see this, write an object in $|(\mathbf{Id_{Set}},\mathbf{CP})|$ as $\langle E,\langle source:E \to N,target:E \to N\rangle,N\rangle$. □

Exercise 3.4.53. Another way to present the category of signatures **AlgSig** is as the comma category $(\mathbf{Id_{Set}},(\_)^+)$, where $(\_)^+:\mathbf{Set} \to \mathbf{Set}$ is the functor which for any set $X \in |\mathbf{Set}|$ yields the set X^+ of all finite non-empty sequences of elements from X.

First, complete the definition of the functor $(\_)^+$. Then, since $X^+ = X^* \times X$, any object in $|(\mathbf{Id_{Set}},(\_)^+)|$ may be written as $\langle \Omega,\langle arity:\Omega \to S^*,sort:\Omega \to S\rangle,S\rangle$. Indicate now why the category defined is almost, but not quite, the same as the category **AlgSig** of signatures (cf. Section 1.2 and Exercise 3.4.76 below). □

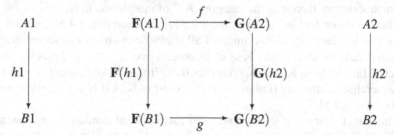

Exercise 3.4.54. Prove that if **K1** and **K2** are (finitely) complete categories, $\mathbf{F}\colon \mathbf{K1} \to \mathbf{K}$ is a functor, and $\mathbf{G}\colon \mathbf{K2} \to \mathbf{K}$ is a (finitely) continuous functor, then the comma category $(\mathbf{F},\mathbf{G})$ is (finitely) complete. Moreover, the obvious projections from $(\mathbf{F},\mathbf{G})$ to **K1** and **K2**, respectively, are (finitely) continuous. HINT: To construct a limit of a diagram in $(\mathbf{F},\mathbf{G})$, start by building limits of the projections of the diagram to **K1** and **K2**, respectively, and then use the continuity property of **G** to complete the construction of the limit object in $(\mathbf{F},\mathbf{G})$. If the notation in the proof gets too heavy, use Exercise 3.2.49 and spell the details out for the construction of products and equalisers.

Check that this construction of limits in $(\mathbf{F},\mathbf{G})$ works for diagrams of any given shape (given by the graph, or the class of graphs, underlying the diagrams): if **K1** and **K2** have limits of diagrams of a given shape, and **G** preserves them, then $(\mathbf{F},\mathbf{G})$ has limits of diagrams of this shape, and the projection functors preserve them.

State and prove the analogous facts about cocompleteness of $(\mathbf{F},\mathbf{G})$. HINT: Clearly, appropriate colimits must exist in **K1** and **K2**, but unlike with limits, it is **F** that must preserve them. □

Exercise 3.4.55. Use Exercises 3.4.51 and 3.4.54 to show that if **K** is a (finitely) complete category then so is the category $\mathbf{K}^{\to}$ of morphisms in **K**.

Then, without looking at Exercise 3.2.51, use Exercises 3.4.51 and 3.4.54 to prove that if a category **K** has limits of all (finite) non-empty connected diagrams then so does the slice category $\mathbf{K}{\downarrow}A$ of its objects over $A \in |\mathbf{K}|$, and that the obvious forgetful functor from $\mathbf{K}{\downarrow}A$ to **K** preserves these limits. Notice though that this does not generalise to arbitrary (finite) limits that exist in $\mathbf{K}{\downarrow}A$ if **K** is (finitely) complete by Exercise 3.2.51.

Check that your proof shows a stronger fact: without assuming the existence of any limits in **K**, the forgetful functor from $\mathbf{K}{\downarrow}A$ to **K** *creates* limits of all non-empty connected diagrams, that is, for any such diagram $D_{\downarrow A}$ in $\mathbf{K}{\downarrow}A$, if its projection D to **K** has a limit in **K** then there is a unique cocone on $D_{\downarrow A}$ in $\mathbf{K}{\downarrow}A$ that projects to this limit, and this cocone is a limit of $D_{\downarrow A}$ in $\mathbf{K}{\downarrow}A$. □

Exercise 3.4.56. Show that if **K** has all pullbacks and a terminal object (so it is finitely complete) and a functor $\mathbf{F}\colon \mathbf{K} \to \mathbf{K}'$ preserves pullbacks, then **F** also preserves the limits of all finite non-empty connected diagrams. HINT: Put together Exercises 3.4.25 and 3.4.55.

Similarly, show that if **K** has all wide pullbacks and a terminal object (so it is complete) and a functor $\mathbf{F}\colon \mathbf{K} \to \mathbf{K}'$ preserves wide pullbacks, then **F** also preserves the limits of all non-empty connected diagrams. □

3.4.3.2 Indexed categories

We frequently need to deal not just with a single category, but rather with a family of categories, "parameterised" by a certain collection of indices. The categories of S-sorted sets (one for each set S) and the categories of Σ-algebras (one for each signature Σ) are typical examples. A crucial property here is that all the categories in such a family are defined in a uniform way, and consequently any change of an index induces a smooth translation between the corresponding component categories. In typical examples, the translation goes in the opposite direction than the change of index, which leads to the following definition:

Definition 3.4.57 (Indexed category). An *indexed category* (over an *index category* **Ind**) is a functor $\mathbf{C}\colon \mathbf{Ind}^{op} \to \mathbf{Cat}$. □

Example 3.4.58. $\mathbf{Alg}\colon \mathbf{AlgSig}^{op} \to \mathbf{Cat}$ is an indexed category (cf. Example 3.4.29). □

Definition 3.4.59 (Grothendieck construction). Any indexed category $\mathbf{C}\colon \mathbf{Ind}^{op} \to \mathbf{Cat}$ gives rise to a "flattened" *Grothendieck* category $\mathbf{Groth}(\mathbf{C})$ defined as follows:

Objects of $\mathbf{Groth}(\mathbf{C})$: pairs $\langle i,A\rangle$ for all $i \in |\mathbf{Ind}|$ and $A \in |\mathbf{C}(i)|$;
Morphisms of $\mathbf{Groth}(\mathbf{C})$: a morphism from $\langle i,A\rangle$ to $\langle j,B\rangle$ is a pair $\langle \sigma,f\rangle\colon \langle i,A\rangle \to \langle j,B\rangle$, where $\sigma\colon i \to j$ is an **Ind**-morphism and $f\colon A \to \mathbf{C}(\sigma)(B)$ is a $\mathbf{C}(i)$-morphism;
Composition in $\mathbf{Groth}(\mathbf{C})$: $\langle \sigma,f\rangle;\langle \sigma',f'\rangle = \langle \sigma;\sigma', f;\mathbf{C}(\sigma)(f')\rangle$. □

Exercise 3.4.60. Show that if **Ind** is complete, $\mathbf{C}(i)$ is complete for all $i \in |\mathbf{Ind}|$, and $\mathbf{C}(\sigma)$ is continuous for all $\sigma \in \mathbf{Ind}$, then $\mathbf{Groth}(\mathbf{C})$ is complete.

HINT: Given a diagram in the Grothendieck category $\mathbf{Groth}(\mathbf{C})$, first consider its obvious projection on the index category **Ind**. Since **Ind** is complete, this has a limit $l \in |\mathbf{Ind}|$. Using the functors assigned by **C** to the projection morphism of the limit, "translate" all the nodes and edges of the diagram to the category $\mathbf{C}(l)$, thus obtaining a diagram in $\mathbf{C}(l)$. Since $\mathbf{C}(l)$ is complete, it has a limit. Check that the projection morphisms of the limit of the diagram constructed in **Ind** when paired with the corresponding projection morphisms of the limit of the diagram in $\mathbf{C}(l)$ form the limit of the original diagram in $\mathbf{Groth}(\mathbf{C})$.

To make the construction manageable, consider only products and equalisers: this is sufficient by Exercise 3.2.49. □

3.4.3.3 Functor categories

Definition 3.4.61 (Functor category). Let **K1** and **K2** be categories.[8] The *functor category* $[\mathbf{K1}\to\mathbf{K2}]$ is defined by:

Objects of $[\mathbf{K1}\to\mathbf{K2}]$: functors from **K1** to **K2**;
Morphisms of $[\mathbf{K1}\to\mathbf{K2}]$: natural transformations;
Composition in $[\mathbf{K1}\to\mathbf{K2}]$: vertical composition. □

Exercise 3.4.62. Define the category $\mathbf{Set}^S$ of S-sorted sets as a functor category. □

Exercise 3.4.63. For any category **K**, define its morphism category $\mathbf{K}^{\to}$ as the category of functors $[\mathbf{2}\to\mathbf{K}]$. □

[8] To be cautious about set-theoretic foundations, one may want to assume that **K1** is small.

Exercise 3.4.64. Let **K1** and **K2** be categories. Show that if **K2** is (finitely) complete then so is the functor category $[\mathbf{K1}{\to}\mathbf{K2}]$. State and show the dual fact as well. HINT: The limit of any diagram in $[\mathbf{K1}{\to}\mathbf{K2}]$ may be constructed "pointwise", for each object in $|\mathbf{K1}|$ separately. More precisely, using Exercise 3.2.49 to simplify the notational burden: consider any family of functors $\langle \mathbf{F}_n : \mathbf{K1} \to \mathbf{K2} \rangle_{n \in N}$. For each $X \in |\mathbf{K1}|$, let $\mathbf{Q}(X) \in |\mathbf{K2}|$ with projections $(\pi_n)_X : \mathbf{Q}(X) \to \mathbf{F}_n(X)$, $n \in N$, be a product of $\langle \mathbf{F}_n(X) \rangle_{n \in N}$ in **K2**. Check that there is a unique way to extend **Q** to a functor $\mathbf{Q} : \mathbf{K1} \to \mathbf{K2}$ so that all $\pi_n : \mathbf{Q} \to \mathbf{F}_n$, $n \in N$, become natural transformations. Show that **Q** with projections $\langle \pi_n \rangle_{n \in N}$ is a product of $\langle \mathbf{F}_n : \mathbf{K1} \to \mathbf{K2} \rangle_{n \in N}$ in $[\mathbf{K1}{\to}\mathbf{K2}]$. Then proceed similarly with equalisers: consider functors $\mathbf{F}, \mathbf{F}' : \mathbf{K1} \to \mathbf{K2}$ and natural transformations $\tau_1, \tau_2 : \mathbf{F} \to \mathbf{F}'$. For each $X \in |\mathbf{K1}|$, let $\tau_X : \mathbf{Q}'(X) \to \mathbf{F}(X)$ be an equaliser of $(\tau_1)_X, (\tau_2)_X : \mathbf{F}(X) \to \mathbf{F}'(X)$ in **K2**. This yields a unique functor $\mathbf{Q}' : \mathbf{K1} \to \mathbf{K2}$ such that $\tau : \mathbf{Q}' \to \mathbf{F}$ is a natural transformation, which is an equaliser of τ_1, τ_2 in $[\mathbf{K1}{\to}\mathbf{K2}]$. □

Exercise 3.4.65. Let **K1**, $\mathbf{K1}'$ and **K2** be categories. Show how any functor $\mathbf{F} : \mathbf{K1} \to \mathbf{K1}'$ induces a functor $(\mathbf{F};\_) : [\mathbf{K1}'{\to}\mathbf{K2}] \to [\mathbf{K1}{\to}\mathbf{K2}]$. Relying on the construction outlined in Exercise 3.4.64 and assuming that **K2** is (finitely) complete, show that this functor is (finitely) continuous.

Prove also that this yields a functor $[\_{\to}\mathbf{K2}] : \mathbf{Cat}^{op} \to \mathbf{Cat}$[9] (cf. Exercise 3.4.16). □

Exercise 3.4.66. For any category **K**, define a category **Funct**(**K**) of *functors into* **K** as follows:

Objects of **Funct**(**K**): functors $\mathbf{F} : \mathbf{K}' \to \mathbf{K}$ into **K**;
Morphisms of **Funct**(**K**): a morphism from $\mathbf{F} : \mathbf{K1} \to \mathbf{K}$ to $\mathbf{G} : \mathbf{K2} \to \mathbf{K}$ is a pair $\langle \boldsymbol{\Phi}, \rho \rangle$, where $\boldsymbol{\Phi} : \mathbf{K1} \to \mathbf{K2}$ is a functor and $\rho : \mathbf{F} \to \boldsymbol{\Phi};\mathbf{G}$ is a natural transformation (between functors from **K1** to **K**);
Composition in **Funct**(**K**): $\langle \boldsymbol{\Phi}, \rho \rangle ; \langle \boldsymbol{\Phi}', \rho' \rangle = \langle \boldsymbol{\Phi};\boldsymbol{\Phi}', \rho;(\boldsymbol{\Phi}\cdot\rho') \rangle$.

Show how the category **Funct**(**K**) arises from the Grothendieck construction of Definition 3.4.59 for the functor $[\_{\to}\mathbf{K}]$ as defined in the previous exercise.[10] □

Exercise 3.4.67. Show that if **K** is a (finitely) complete category then the category **Funct**(**K**) of functors into **K** is (finitely) complete as well. HINT: You may construct the limits in **Funct**(**K**) directly, perhaps using Exercise 3.2.49. Alternatively, rely on the construction of **Funct**(**K**) via the Grothendieck construction (Definition 3.4.59) for the functor $[\_{\to}\mathbf{K}] : \mathbf{Cat}^{op} \to \mathbf{Cat}$ and on Exercise 3.4.60; recall that **Cat** is complete by Exercise 3.4.32, for any category **K1**, $[\mathbf{K1}{\to}\mathbf{K}]$ is (finitely) complete by Exercise 3.4.64, and for every functor $\mathbf{F} : \mathbf{K1} \to \mathbf{K2}$, $(\mathbf{F};\_) : [\mathbf{K2}{\to}\mathbf{K}] \to [\mathbf{K1}{\to}\mathbf{K}]$ is (finitely) continuous by Exercise 3.4.65. □

Exercise 3.4.68. Show that if a category **K1** has a factorisation system (cf. Section 3.3) then for any category **K2**, the functor category $[\mathbf{K2}{\to}\mathbf{K1}]$ has a factorisation system as well.

---

[9] Requiring **K2** to be small would help to resolve potential foundational problems here.

[10] So, for foundational reasons, one may prefer to keep all categories small around here as well.

HINT: Let $\langle \mathbf{E1}, \mathbf{M1} \rangle$ be a factorisation system for $\mathbf{K1}$. Put $\mathbf{E} = \{\varepsilon \in [\mathbf{K2} \to \mathbf{K1}] \mid \varepsilon_A \in \mathbf{E1} \text{ for } A \in |\mathbf{K2}|\}$ and $\mathbf{M} = \{\eta \in [\mathbf{K2} \to \mathbf{K1}] \mid \eta_A \in \mathbf{M1} \text{ for } A \in |\mathbf{K2}|\}$. Now, to construct an $\langle \mathbf{E}, \mathbf{M} \rangle$-factorisation of a natural transformation $\tau : \mathbf{F} \to \mathbf{G}$ between functors $\mathbf{F}, \mathbf{G} : \mathbf{K2} \to \mathbf{K1}$, first for each object $A \in |\mathbf{K2}|$ obtain an $\langle \mathbf{E1}, \mathbf{M1} \rangle$-factorisation of τ_A, say $\tau_A = \varepsilon_A ; \eta_A$ with $\varepsilon_A \in \mathbf{E1}$ and $\eta_A \in \mathbf{M1}$, and $\varepsilon_A : \mathbf{F}(A) \to \mathbf{H}(A)$, $\eta_A : \mathbf{H}(A) \to \mathbf{G}(A)$ for some $\mathbf{H}(A) \in |\mathbf{K1}|$. Then use the diagonal fill-in lemma (Lemma 3.3.4) to extend the mapping $\mathbf{H} : |\mathbf{K2}| \to |\mathbf{K1}|$ to a functor $\mathbf{H} : \mathbf{K2} \to \mathbf{K1}$ such that $\varepsilon : \mathbf{F} \to \mathbf{H}$ and $\eta : \mathbf{H} \to \mathbf{G}$ are natural transformations. □

3.4.3.4 Equivalence of categories

Definition 3.4.69 (Isomorphic categories). Two categories **K1** and **K2** are *isomorphic* if there are functors $\mathbf{F} : \mathbf{K1} \to \mathbf{K2}$ and $\mathbf{F}^{-1} : \mathbf{K2} \to \mathbf{K1}$ such that $\mathbf{F} ; \mathbf{F}^{-1} = \mathbf{Id}_{\mathbf{K1}}$ and $\mathbf{F}^{-1} ; \mathbf{F} = \mathbf{Id}_{\mathbf{K2}}$. □

In other words, we say that two categories are isomorphic if they are isomorphic as objects of **Cat**. As with isomorphic objects of other kinds, we will view isomorphic categories as abstractly the same. It turns out, however, that in this case there is a coarser relation which allows us to identify categories that have all the same categorical properties, even though they may not be isomorphic.

Definition 3.4.70 (Equivalent categories). **K1** and **K2** are *equivalent* if there are functors $\mathbf{F} : \mathbf{K1} \to \mathbf{K2}$ and $\mathbf{G} : \mathbf{K2} \to \mathbf{K1}$ and natural isomorphisms $\tau : \mathbf{Id}_{\mathbf{K1}} \to \mathbf{F} ; \mathbf{G}$ and $\sigma : \mathbf{G} ; \mathbf{F} \to \mathbf{Id}_{\mathbf{K2}}$. □

To characterise equivalent categories, we need one more concept:

Definition 3.4.71 (Skeletal category). A category **K** is *skeletal* iff any two isomorphic **K**-objects are identical. A *skeleton of* **K** is any maximal skeletal subcategory of **K**. □

Exercise 3.4.72. Prove that two categories are equivalent iff they have isomorphic skeletons. □

Thus, intuitively, two categories are equivalent if and only if they differ only in the number of isomorphic copies of corresponding objects.

Example 3.4.73. The category **FinSet** of all finite sets is equivalent to its full subcategory of all natural numbers, where any natural number n is defined as the set $\{0, \ldots, n-1\}$ of all natural numbers smaller than n. In fact, the latter is a skeleton of **FinSet**. Similarly, the category **Set** of all sets is equivalent to its full subcategory of all ordinals. □

Exercise 3.4.74. Show that for any signature Σ and set $\mathscr{E}$ of Σ-equations, the full subcategory of $\mathbf{T}_\Sigma / \mathscr{E}$ given by the finite sets of variables is equivalent to the category $\mathbf{T}_{\Sigma, \mathscr{E}}$ (cf. Exercises 3.1.15 and 3.1.16). □

Exercise 3.4.75. Let **K1** and **K2** be equivalent categories. Show that if **K1** is (finitely) (co)complete then so is **K2**. □

Exercise 3.4.76. Recall Exercise 3.4.53. As indicated there, categories **AlgSig** and $(\mathbf{Id}_{\mathbf{Set}}, (\_)^{+})$ are not isomorphic. Show that they are equivalent. Then, using Exercises 3.4.75 and 3.4.54, conclude from this that **AlgSig** is complete and cocomplete. □

3.5 Adjoints

Recall Facts 1.4.4 and 1.4.10:

Fact 1.4.4. *For any Σ-algebra A and S-sorted function $v{:}X \to |A|$ there is exactly one Σ-homomorphism $v^{\#}{:}T_\Sigma(X) \to A$ which extends v, i.e. such that $v^{\#}_s(\iota_X(x)) = v_s(x)$ for all $s \in S$, $x \in X_s$, where $\iota_X{:}X \to |T_\Sigma(X)|$ is the embedding that maps each variable in X to the corresponding term.* □

Fact 1.4.10. *This property defines $T_\Sigma(X)$ up to isomorphism: if B is a Σ-algebra and $\eta{:}X \to |B|$ is an S-sorted function such that for any Σ-algebra A and S-sorted function $v{:}X \to |A|$ there is a unique Σ-homomorphism $v^{\$}{:}B \to A$ such that $\eta;|v^{\$}| = v$, then B is isomorphic to $T_\Sigma(X)$.* □

The construction of the algebra of Σ-terms is one example of an *adjoint functor* (it is *left adjoint* to the functor $|\_|{:}\mathbf{Alg}(\Sigma) \to \mathbf{Set}^{sorts(\Sigma)}$). The general concept of an adjoint functor, to which this section is devoted, has many other important instances. In fact, [Gog91a] goes so far as to say:

> *Any canonical construction from widgets to whatsits is an adjoint of another functor, from whatsits to widgets.*

3.5.1 Free objects

Let **K1** and **K2** be categories, $\mathbf{G}{:}\mathbf{K2} \to \mathbf{K1}$ be a functor, and $A1$ be an object of **K1**.

Definition 3.5.1 (Free object). A *free object over $A1$ w.r.t.* **G** is a **K2**-object $A2$ together with a **K1**-morphism $\eta_{A1}{:}A1 \to \mathbf{G}(A2)$ such that for any **K2**-object $B2$ and **K1**-morphism $f{:}A1 \to \mathbf{G}(B2)$ there is a unique **K2**-morphism $f^{\#}{:}A2 \to B2$ such that $\eta_{A1};\mathbf{G}(f^{\#}) = f$.

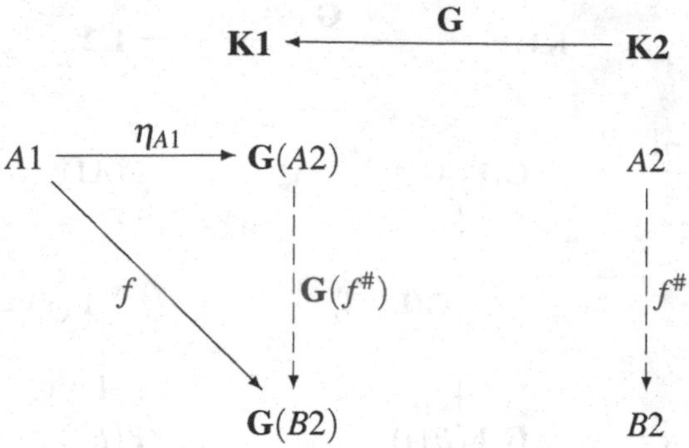

η_{A1} is called the *unit morphism*. □

Example 3.5.2. Let $\Sigma = \langle S, \Omega \rangle$ be an arbitrary signature. Consider the forgetful functor $|\_|: \mathbf{Alg}(\Sigma) \to \mathbf{Set}^S$. Fact 1.4.4 asserts that for any S-sorted set X, the term algebra $T_\Sigma(X)$ with the inclusion $\eta_X: X \hookrightarrow |T_\Sigma(X)|$ is a free object over X w.r.t. $|\_|$. □

Exercise 3.5.3. Define free monoids and the path categories $\mathbf{Path}(G)$ as free objects w.r.t. some obvious functors. Then, look around at the areas of mathematics with which you are familiar for more examples. For instance, check that free groups and discrete topologies, (ideal) completion of partial orders, of ordered algebras, and so on, may be defined as free objects w.r.t. some straightforward functors. □

Exercise 3.5.4. Prove that any free object over $A1$ w.r.t. $\mathbf{G}$ is an initial object in the comma category $(\mathbf{C}_{A1}, \mathbf{G})$, where $\mathbf{C}_{A1}: \mathbf{1} \to \mathbf{K1}$ is the constant functor. Conclude that a free object over $A1$ w.r.t. $\mathbf{G}$ is unique up to isomorphism. □

Exercise 3.5.5. Prove that if $A2 \in |\mathbf{K2}|$ is a free object over $A1 \in |\mathbf{K1}|$ w.r.t. $\mathbf{G}: \mathbf{K2} \to \mathbf{K1}$, then for any $B2 \in |\mathbf{K2}|$, the function $(\_)^\#: \mathbf{K1}(A1, \mathbf{G}(B2)) \to \mathbf{K2}(A2, B2)$ is a bijection.

Check that one consequence of this is that two morphisms $g, h: A2 \to B2$ coincide (in $\mathbf{K2}$) whenever $\eta_{A1}; \mathbf{G}(g) = \eta_{A1}; \mathbf{G}(h)$ in $\mathbf{K1}$. □

3.5.2 Left adjoints

Let $\mathbf{K1}$ and $\mathbf{K2}$ be categories and $\mathbf{G}: \mathbf{K2} \to \mathbf{K1}$ be a functor. So far we have considered free objects w.r.t. $\mathbf{G}$ one by one, without relating them with each other. One crucial property is that the construction of free objects, if they exist, is functorial.

Proposition 3.5.6. *If for any $A1 \in |\mathbf{K1}|$ there is a free object over $A1$ w.r.t. $\mathbf{G}$, say $F(A1) \in |\mathbf{K2}|$ with unit morphism $\eta_{A1}: A1 \to \mathbf{G}(F(A1))$ (in $\mathbf{K1}$), then $A1 \mapsto F(A1)$ and $f \in \mathbf{K1}(A1, B1) \mapsto (f;\eta_{B1})^\# \in \mathbf{K2}(F(A1), F(B1))$ determine a functor $\mathbf{F}: \mathbf{K1} \to \mathbf{K2}$.*

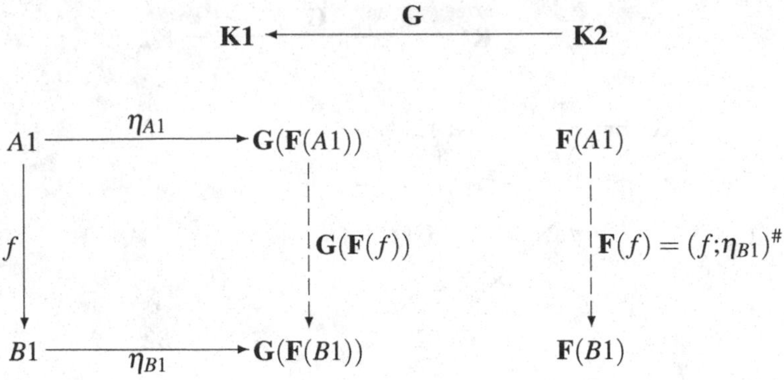

Proof. **F** *preserves identities*: $\mathbf{F}(id_{A1}) = (id_{A1};\eta_{A1})^{\#} = id_{\mathbf{F}(A1)}$ follows from the fact that the following diagram commutes:

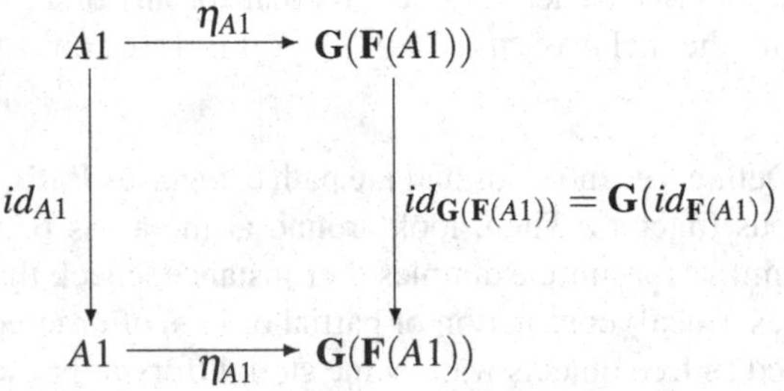

F *preserves composition*: Since the following diagram commutes,

$$
\begin{array}{ccc}
A1 & \xrightarrow{\eta_{A1}} & G(F(A1)) \\
f\downarrow & & \downarrow G(F(f)) \\
B1 & \xrightarrow{\eta_{B1}} & G(F(B1)) \\
g\downarrow & & \downarrow G(F(g)) \\
C1 & \xrightarrow{\eta_{C1}} & G(F(C1))
\end{array}
\qquad \mathbf{G}(\mathbf{F}(f));\mathbf{G}(\mathbf{F}(g)) = \mathbf{G}(\mathbf{F}(f);\mathbf{F}(g))
$$

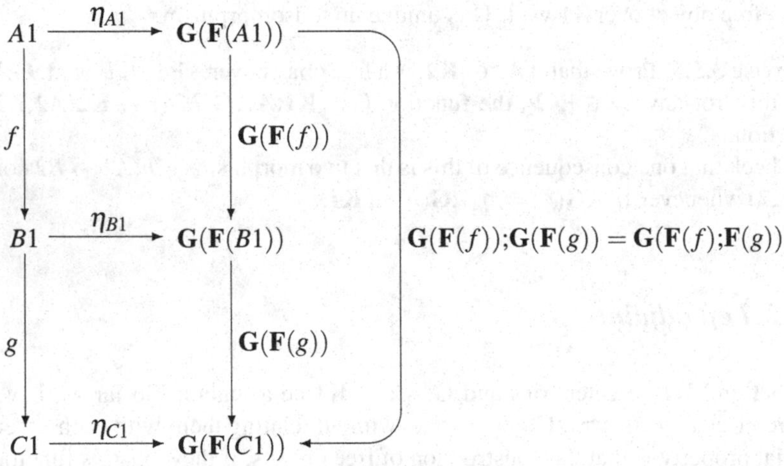

it follows that $\mathbf{F}(f;g) = (f;g;\eta_{C1})^{\#} = \mathbf{F}(f);\mathbf{F}(g)$. $\square$

Exercise 3.5.7. Check that $\eta: \mathbf{Id}_{\mathbf{K1}} \to \mathbf{F};\mathbf{G}$ in Proposition 3.5.6 is a natural transformation. $\square$

Definition 3.5.8 (Left adjoint). Let $\mathbf{F}:\mathbf{K1} \to \mathbf{K2}$ and $\mathbf{G}:\mathbf{K2} \to \mathbf{K1}$ be functors and $\eta:\mathbf{Id_{K1}} \to \mathbf{F};\mathbf{G}$ be a natural transformation. $\mathbf{F}$ is *left adjoint to* $\mathbf{G}$ *with unit* η if for any $A1 \in |\mathbf{K1}|$, $\mathbf{F}(A1)$ with unit morphism $\eta_{A1}:A1 \to \mathbf{G}(\mathbf{F}(A1))$ is a free object over $A1$ w.r.t. $\mathbf{G}$. □

Before we give any examples, let us prove a very important property of left adjoints.

Proposition 3.5.9. *A left adjoint to* $\mathbf{G}$ *is unique up to (natural) isomorphism: if* $\mathbf{F}$ *and* $\mathbf{F}'$ *are left adjoints of* $\mathbf{G}$ *with units* η *and* η' *respectively, then there is a natural isomorphism* $\tau:\mathbf{F} \to \mathbf{F}'$ *such that* $\eta;(\tau\cdot\mathbf{G}) = \eta'$.

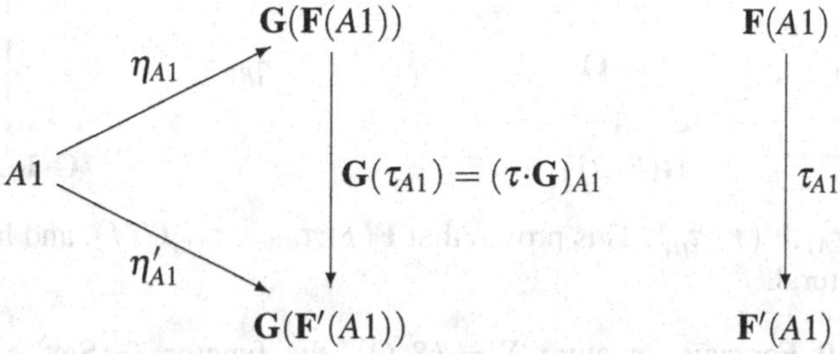

Proof. First notice that for any $f \in \mathbf{K1}(A1,B1)$, $\mathbf{F}(f) = (f;\eta_{B1})^{\#}$ and $\mathbf{F}'(f) = (f;\eta'_{B1})^{\#'}$.

Then, for $A1 \in |\mathbf{K1}|$, define $\tau_{A1} = (\eta'_{A1})^{\#}$ and $\tau^{-1}_{A1} = (\eta_{A1})^{\#'}$. Then $\tau_{A1};\tau^{-1}_{A1} = id_{\mathbf{F}(A1)}$ since the following diagrams commute,

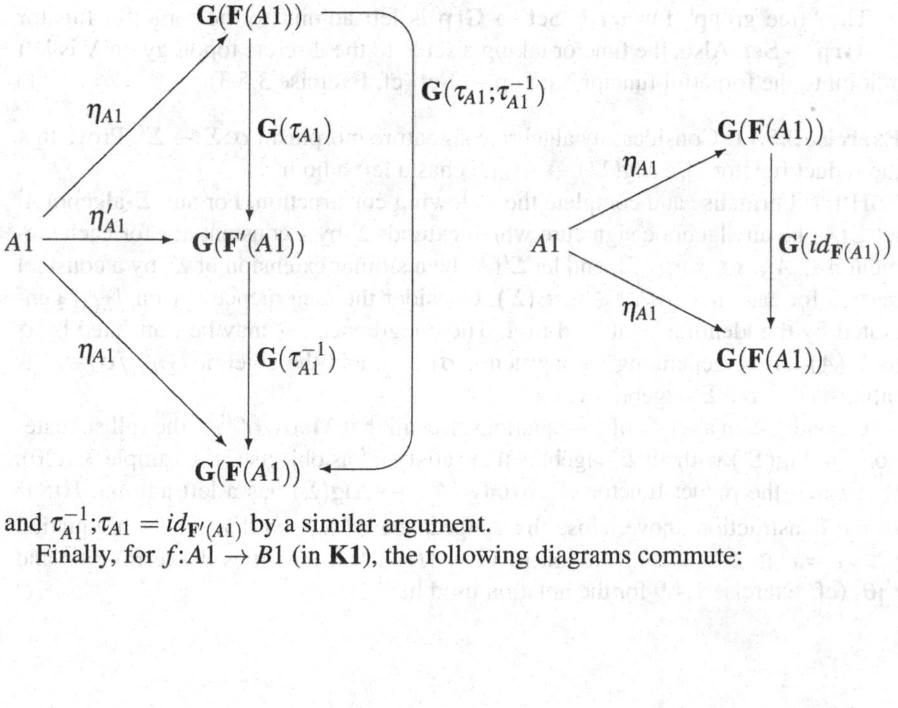

and $\tau^{-1}_{A1};\tau_{A1} = id_{\mathbf{F}'(A1)}$ by a similar argument.

Finally, for $f:A1 \to B1$ (in $\mathbf{K1}$), the following diagrams commute:

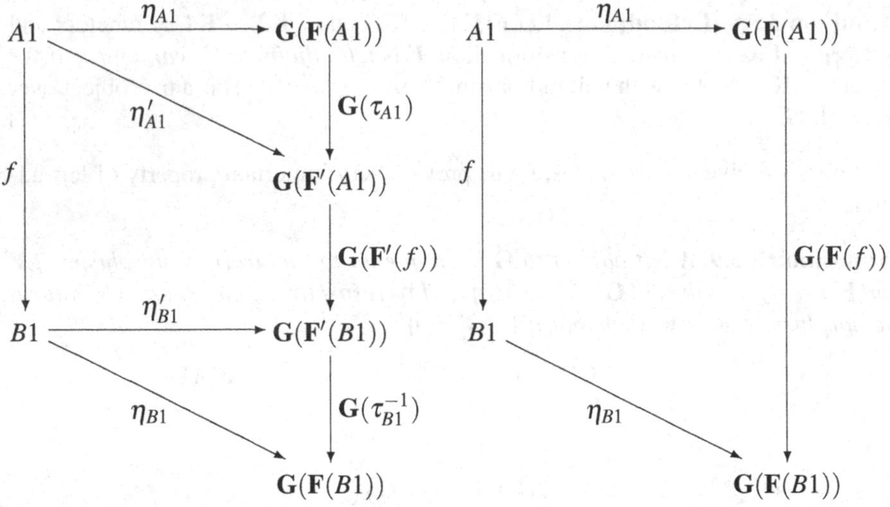

Thus, $\mathbf{F}(f) = \tau_{A1};\mathbf{F}'(f);\tau_{B1}^{-1}$. This proves that $\mathbf{F}(f);\tau_{B1} = \tau_{A1};\mathbf{F}'(f)$, and hence that $\tau:\mathbf{F} \to \mathbf{F}'$ is natural. □

Example 3.5.10. For any signature $\Sigma = \langle S, \Omega \rangle$, the functor $T_\Sigma:\mathbf{Set}^S \to \mathbf{Alg}(\Sigma)$ is left adjoint to the forgetful functor $|\_|:\mathbf{Alg}(\Sigma) \to \mathbf{Set}^S$ (cf. Examples 3.4.11 and 3.4.9).

The functor $\mathbf{Seq}:\mathbf{Set} \to \mathbf{Mon}$ is left adjoint to the forgetful functor $|\_|:\mathbf{Mon} \to \mathbf{Set}$ which takes a monoid to its underlying set of elements. The unit is the natural transformation $sing\_seq:\mathbf{Id_{Set}} \to \mathbf{Seq};|\_|$ (cf. Examples 3.4.8 and 3.4.39).

The "free group" functor $\mathbf{F}:\mathbf{Set} \to \mathbf{Grp}$ is left adjoint to the forgetful functor $|\_|:\mathbf{Grp} \to \mathbf{Set}$. Also, the functor taking a set X to the discrete topology on X is left adjoint to the forgetful functor $|\_|:\mathbf{Top} \to \mathbf{Set}$ (cf. Exercise 3.5.3). □

Exercise 3.5.11. Consider any algebraic signature morphism $\sigma:\Sigma \to \Sigma'$. Prove that the reduct functor $\_|_\sigma:\mathbf{Alg}(\Sigma') \to \mathbf{Alg}(\Sigma)$ has a left adjoint.

HINT: Formalise and complete the following construction. For any Σ-algebra A, let $\Sigma(A)$ be an algebraic signature which extends Σ by a constant $a:s$ for each element $a \in |A|_s$, $s \in sorts(\Sigma)$, and let $\Sigma'(A)$ be a similar extension of Σ' by a constant $a:\sigma(s)$ for each $a \in |A|_s$, $s \in sorts(\Sigma)$. Consider the congruence $\equiv_A$ on $T_{\Sigma(A)}$ generated by the identities that hold in A. The congruence $\equiv_A$ may be translated by σ to $\Sigma'(A)$-terms, generating a congruence $\sigma(\equiv_A)$, and the algebra $T_{\Sigma'(A)}/\sigma(\equiv_A)$ is (almost) the free Σ'-algebra over A.

Consider then a set $\mathscr{E}'$ of Σ'-equations. Recall that $\mathbf{Mod}_{\Sigma'}(\mathscr{E}')$ is the full subcategory of $\mathbf{Alg}(\Sigma')$ with all Σ'-algebras that satisfy $\mathscr{E}'$ as objects (cf. Example 3.1.20). Prove that the reduct functor $\_|_\sigma:\mathbf{Mod}_{\Sigma'}(\mathscr{E}') \to \mathbf{Alg}(\Sigma)$ has a left adjoint. HINT: In the construction above, close the congruence $\sigma(\equiv_A)$ so that for each equation $\forall X' \bullet t = t'$ in $\mathscr{E}'$ and substitution $\theta:X' \to |T_{\Sigma'(A)}|$, it identifies the terms $t[\theta]$ and $t'[\theta]$ (cf. Exercise 1.4.9 for the notation used here).

Finally, for any set $\mathscr{E}$ of Σ-equations such that $\mathscr{E}' \models_{\Sigma'} \sigma(\mathscr{E})$, prove that the reduct functor $\_|_\sigma : \mathbf{Mod}_{\Sigma'}(\mathscr{E}') \to \mathbf{Mod}_\Sigma(\mathscr{E})$, which is well defined by Proposition 2.3.13, has a left adjoint. HINT: This is easy now. □

Exercise 3.5.12. Generalise Exercise 3.5.11 to derived signature morphisms, with reduct functors as introduced in Exercise 3.4.30. □

Example 3.5.13. Let $\mathbf{K}$ be a category, and recall that $\mathbf{1}$ is a category containing a single object, say a. Let $\mathbf{F}: \mathbf{1} \to \mathbf{K}$ be left adjoint to $\mathbf{C}_a: \mathbf{K} \to \mathbf{1}$ (note that such a functor $\mathbf{F}$ may not exist). Then $\mathbf{F}(a)$ is an initial object in $\mathbf{K}$. □

Exercise 3.5.14. Let $\Delta: \mathbf{K} \to \mathbf{K} \times \mathbf{K}$ be the "diagonal" functor such that $\Delta(A) = \langle A, A\rangle$ and $\Delta(f: A \to B) = \langle f, f\rangle : \Delta(A) \to \Delta(B)$. Prove that $\mathbf{K}$ has all coproducts iff Δ has a left adjoint. What is the unit? □

Exercise 3.5.15. Formulate analogous theorems for coequalisers and pushouts and prove them. Show how this may be done for any colimit. □

Exercise 3.5.16. Let $\mathbf{K}$ be a category with an initial object and a factorisation system and let $\mathbf{K}_R$ be its full subcategory of reachable objects. Recall that $\mathbf{R}_\mathbf{K}: \mathbf{K} \to \mathbf{K}_R$ is a functor that maps any object to its reachable subobject (cf. Exercise 3.4.13). Show that the inclusion functor $\mathbf{I}: \mathbf{K}_R \to \mathbf{K}$ is left adjoint to $\mathbf{R}_K$. □

Exercise 3.5.17. Show that left adjoints preserve colimits of diagrams. Do they preserve limits as well? □

Exercise 3.5.18. Let $\mathbf{F}: \mathbf{K2} \to \mathbf{K1}$ be left adjoint to $\mathbf{G}: \mathbf{K1} \to \mathbf{K2}$ with unit $\eta: \mathbf{Id}_{\mathbf{K1}} \to \mathbf{F};\mathbf{G}$. Consider two objects $A, B \in |\mathbf{K1}|$ and suppose that for some epimorphism $e: A \to B$ there exists a morphism $h: B \to \mathbf{G}(\mathbf{F}(A))$ such that $e;h = \eta_A$. Prove that $\mathbf{F}(e): \mathbf{F}(A) \to \mathbf{F}(B)$ is an isomorphism.

HINT:

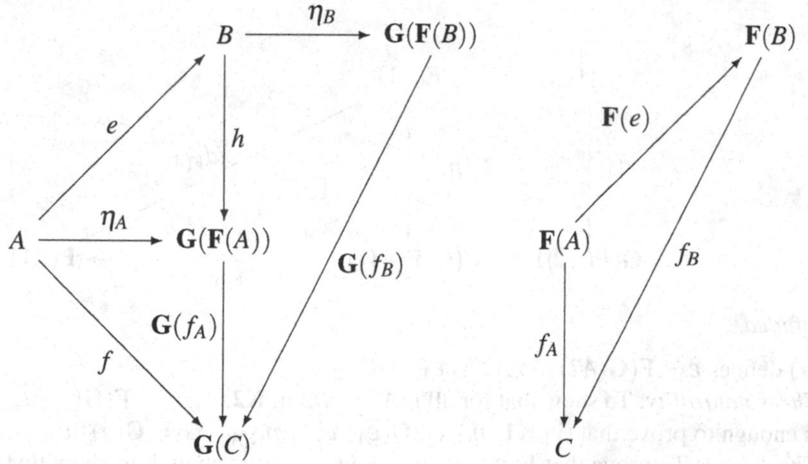

First show that $\mathbf{F}(B)$ with $e;\eta_B:A \to \mathbf{G}(\mathbf{F}(B))$ as the unit morphism is a free object over A w.r.t. $\mathbf{G}$. For this, use the following construction: for any $C \in |\mathbf{K2}|$ and $f:A \to \mathbf{G}(C)$, let $f_B:\mathbf{F}(B) \to C$ be the unique morphism such that $\eta_B;\mathbf{G}(f_B) = h;\mathbf{G}(f_A)$, where in turn $f_A:\mathbf{F}(A) \to C$ is the unique morphism such that $\eta_A;\mathbf{G}(f_A) = f$. Now, f_B satisfies $(e;\eta_B);\mathbf{G}(f_B) = f$, and moreover, it is the only morphism from $\mathbf{F}(B)$ to C with this property (use the fact that e is an epimorphism and the freeness of $\mathbf{F}(B)$ to prove the latter). Then, show that the conclusion holds by following the proof of the uniqueness of left adjoints; cf. Proposition 3.5.9. □

3.5.3 Adjunctions

Consider two categories **K1** and **K2** and functors $\mathbf{F}:\mathbf{K1} \to \mathbf{K2}$ and $\mathbf{G}:\mathbf{K2} \to \mathbf{K1}$ such that $\mathbf{F}$ is left adjoint to $\mathbf{G}$ with unit $\eta:\mathbf{Id_{K1}} \to \mathbf{F};\mathbf{G}$.

Proposition 3.5.19. *There is a natural transformation $\varepsilon:\mathbf{G};\mathbf{F} \to \mathbf{Id_{K2}}$ such that*

$$(*): \quad (\mathbf{G}\cdot\eta);(\varepsilon\cdot\mathbf{G}) = id_{\mathbf{G}}$$
$$(**): \quad (\eta\cdot\mathbf{F});(\mathbf{F}\cdot\varepsilon) = id_{\mathbf{F}}$$

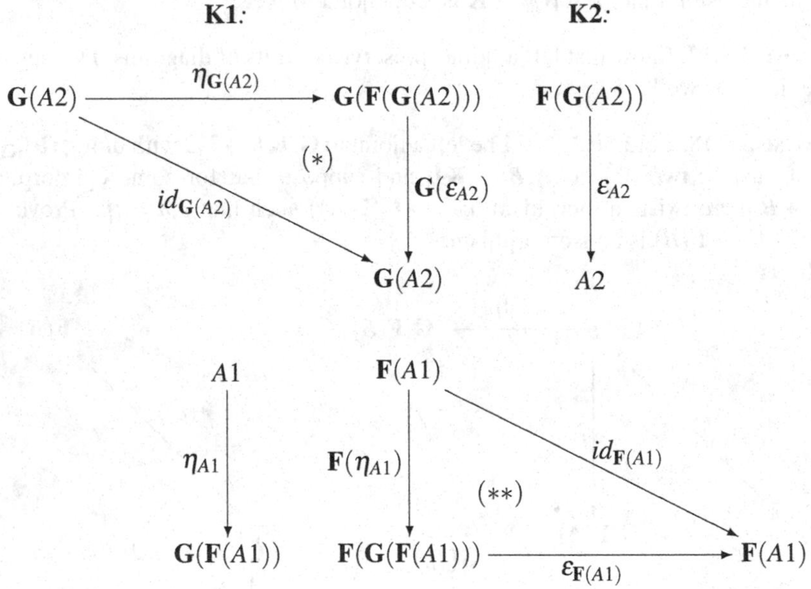

Proof idea.

- $(*)$ defines $\varepsilon_{A2}:\mathbf{F}(\mathbf{G}(A2)) \to A2$ as $\varepsilon_{A2} = (id_{\mathbf{G}(A2)})^{\#}$.
- *Check naturality*: To show that for all $g:A2 \to B2$ in **K2**, $\varepsilon_{A2};g = \mathbf{F}(\mathbf{G}(g));\varepsilon_{B2}$, it is enough to prove that (in **K1**) $\eta_{\mathbf{G}(A2)};\mathbf{G}(\varepsilon_{A2};g) = \eta_{\mathbf{G}(A2)};\mathbf{G}(\mathbf{F}(\mathbf{G}(g));\varepsilon_{B2})$.
- *Check* $(**)$: To prove that $\mathbf{F}(\eta_{A1});\varepsilon_{\mathbf{F}(A1)} = id_{\mathbf{F}(A1)}$, it is enough to show that (in **K1**) $\eta_{A1};\mathbf{G}(\mathbf{F}(\eta_{A1});\varepsilon_{\mathbf{F}(A1)}) = \eta_{A1};\mathbf{G}(id_{\mathbf{F}(A1)})$. □

Proposition 3.5.20. *Consider functors* $\mathbf{F}: \mathbf{K1} \to \mathbf{K2}$ *and* $\mathbf{G}: \mathbf{K2} \to \mathbf{K1}$ *and natural transformations* $\eta: \mathbf{Id_{K1}} \to \mathbf{F};\mathbf{G}$ *and* $\varepsilon: \mathbf{G};\mathbf{F} \to \mathbf{Id_{K2}}$ *such that*

$$(*): \quad (\mathbf{G}\cdot\eta);(\varepsilon\cdot\mathbf{G}) = id_{\mathbf{G}}$$
$$(**): \quad (\eta\cdot\mathbf{F});(\mathbf{F}\cdot\varepsilon) = id_{\mathbf{F}}$$

Then **F** *is left adjoint to* **G** *with unit* η.

Proof. For $A1 \in |\mathbf{K1}|$, $B2 \in |\mathbf{K2}|$, $f: A1 \to \mathbf{G}(B2)$, let $f^{\#} = \mathbf{F}(f);\varepsilon_{B2}: \mathbf{F}(A1) \to B2$.

- $\eta_{A1};\mathbf{G}(f^{\#}) = \eta_{A1};\mathbf{G}(\mathbf{F}(f));\mathbf{G}(\varepsilon_{B2}) = f;\eta_{\mathbf{G}(B2)};\mathbf{G}(\varepsilon_{B2}) = f;id_{\mathbf{G}(B2)} = f$.
- Suppose that for some $g: \mathbf{F}(A1) \to B2$, $\eta_{A1};\mathbf{G}(g) = f$. Then, $f^{\#} = \mathbf{F}(f);\varepsilon_{B2} = \mathbf{F}(\eta_{A1};\mathbf{G}(g));\varepsilon_{B2} = \mathbf{F}(\eta_{A1});\mathbf{F}(\mathbf{G}(g));\varepsilon_{B2} = \mathbf{F}(\eta_{A1});\varepsilon_{\mathbf{F}(A1)};g = g$. □

Definition 3.5.21 (Adjunction). Let **K1** and **K2** be categories. An *adjunction from* **K1** *to* **K2** is a quadruple $\langle \mathbf{F}, \mathbf{G}, \eta, \varepsilon \rangle$ where $\mathbf{F}: \mathbf{K1} \to \mathbf{K2}$ and $\mathbf{G}: \mathbf{K2} \to \mathbf{K1}$ are functors and $\eta: \mathbf{Id_{K1}} \to \mathbf{F};\mathbf{G}$ and $\varepsilon: \mathbf{G};\mathbf{F} \to \mathbf{Id_{K2}}$ are natural transformations such that

$$(*): \quad (\mathbf{G}\cdot\eta);(\varepsilon\cdot\mathbf{G}) = id_{\mathbf{G}}$$
$$(**): \quad (\eta\cdot\mathbf{F});(\mathbf{F}\cdot\varepsilon) = id_{\mathbf{F}}$$

□

Fact 3.5.22. *Equivalently, an adjunction may be given as either of the following:*

- *A functor* $\mathbf{G}: \mathbf{K2} \to \mathbf{K1}$ *and, for each* $A1 \in |\mathbf{K1}|$*, a free object over* $A1$ *w.r.t.* **G***;*
- *A functor* $\mathbf{G}: \mathbf{K2} \to \mathbf{K1}$ *and its left adjoint.* □

Exercise 3.5.23 (Galois connection). Recall that any partial order gives rise to a corresponding preorder category (cf. Example 3.1.3). Galois connections (Definition 2.3.3) arise as adjunctions between preorder categories:

Consider two partially ordered sets $\langle A, \leq_A \rangle$ and $\langle B, \leq_B \rangle$ and two order-preserving functions $f: A \to B$ and $g: B \to A$ (i.e. for $a, a' \in A$, if $a \leq_A a'$ then $f(a) \leq_B f(a')$ and for $b, b' \in B$, if $b \leq_B b'$ then $g(b) \leq_A g(b')$).

Show that f and g (viewed as functors) form an adjunction between $\langle A, \leq_A \rangle$ and $\langle B, \leq_B \rangle$ (viewed as categories) if and only if for all $a \in A$ and $b \in B$

$$a \leq_A g(b) \quad \text{iff} \quad f(a) \leq_B b.$$

Then show that this is further equivalent to the requirement that

- $a \leq_A g(f(a))$ for all $a \in A$; and
- $f(g(b)) \leq_B b$ for all $b \in B$.

View the Galois connection between sets of equations and classes of algebras on a given signature defined in Section 2.3 (cf. Proposition 2.3.2) as a special case of the above definition. That is, check that for any signature Σ, the function mapping any set of Σ-equations to the class of all Σ-algebras that satisfy this set of equations and the function mapping any class of Σ-algebras to the set of all Σ-equations that hold in this class form an adjunction between the power set of the set of Σ-equations (ordered by inclusion) and the power class of the class of Σ-algebras (ordered by containment).

Then check that the above definition of Galois connection coincides with the more explicit Definition 2.3.3 of a Galois connection between $\langle A, \leq_A \rangle$ and $\langle B, \geq_B \rangle$ (note the opposite order for B). □

Exercise 3.5.24. Dualise the development in this section. Begin with the following definition, dual to Definition 3.5.1:

Definition. Let $\mathbf{F}: \mathbf{K1} \to \mathbf{K2}$ be a functor and let $A2 \in |\mathbf{K2}|$. A *cofree object over* $A2$ *w.r.t.* $\mathbf{F}$ is a **K1**-object $A1$ together with a **K2**-morphism $\varepsilon_{A2}: \mathbf{F}(A1) \to A2$ (the *counit*) such that for any **K1**-object $B1$ and **K2**-morphism $f: \mathbf{F}(B1) \to A2$ there is a unique **K1**-morphism $f^{\#}: B1 \to A1$ such that $\mathbf{F}(f^{\#}); \varepsilon_{A2} = f$.

Then dually to Section 3.5.2 show how cofree objects induce *right adjoints*. Finally, prove facts dual to Propositions 3.5.19 and 3.5.20, thus proving that right adjoints and cofree objects give another equivalent definition of adjunction. □

Exercise 3.5.25. Develop yet another equivalent definition of an adjunction between locally small categories, centering around the bijection $\#: \mathbf{K1}(A1, \mathbf{G}(A2)) \to \mathbf{K2}(\mathbf{F}(A1), A2)$ using a generalised version of Hom-functors (cf. Example 3.4.15). HINT:

- For any locally small category $\mathbf{K}$ and two functors $\mathbf{F1}: \mathbf{K1} \to \mathbf{K}$ and $\mathbf{F2}: \mathbf{K2} \to \mathbf{K}$, define a functor $\mathbf{Hom}_{\mathbf{F1},\mathbf{F2}}: \mathbf{K1}^{op} \times \mathbf{K2} \to \mathbf{Set}$ by $\mathbf{Hom}_{\mathbf{F1},\mathbf{F2}}(\langle A1, A2 \rangle) = \mathbf{K}(\mathbf{F1}(A1), \mathbf{F2}(A2))$ and $\mathbf{Hom}_{\mathbf{F1},\mathbf{F2}}(\langle f1, f2 \rangle)(h) = \mathbf{F1}(f1); h; \mathbf{F2}(f2)$.
- Show that if $\mathbf{F}: \mathbf{K1} \to \mathbf{K2}$ is left adjoint to $\mathbf{G}: \mathbf{K2} \to \mathbf{K1}$ then $\#: \mathbf{Hom}_{\mathbf{Id_{K1}},\mathbf{G}} \to \mathbf{Hom}_{\mathbf{F},\mathbf{Id_{K2}}}$ is a natural isomorphism.
- Finally, prove that for any two functors $\mathbf{F}: \mathbf{K1} \to \mathbf{K2}$ and $\mathbf{G}: \mathbf{K2} \to \mathbf{K1}$, a natural isomorphism $\#: \mathbf{Hom}_{\mathbf{Id_{K1}},\mathbf{G}} \to \mathbf{Hom}_{\mathbf{F},\mathbf{Id_{K2}}}$ shows that $\mathbf{F}$ is left adjoint to $\mathbf{G}$. □

Exercise 3.5.26. Show that adjunctions compose: given any categories **K1**, **K2** and **K3**, and adjunctions $\langle \mathbf{F}, \mathbf{G}, \eta, \varepsilon \rangle$ from **K1** to **K2** and $\langle \mathbf{F}', \mathbf{G}', \eta', \varepsilon' \rangle$ from **K2** to **K3**, we have an adjunction of the form $\langle \mathbf{F};\mathbf{F}', \mathbf{G}';\mathbf{G}, \_, \_ \rangle$ from **K1** to **K3**. Fill in the holes! □

3.6 Bibliographical remarks

Category theory has found very many applications in computer science, and the material presented here covers just those fragments that we will require in later chapters. Books on category theory for mathematicians include the classic [Mac71] as well as the encyclopedic [HS73], with [AHS90] a more recent favourite, the three-volume handbook [Bor94], the modestly sized textbook [Awo06], and many more. An early book on category theory directed towards computer scientists is [AM75], followed by [Pie91], [Poi92] and [BW95]. An interesting angle is in [RB88], where categorical concepts are presented by coding them in ML.

Our terminology is mainly based on [Mac71], although we prefer to write composition in diagrammatic order, denoted by semicolon. The reader should be warned that the terminology and notation in category theory is not completely standardized, and differ from one author to another.

We have decided to keep to the basics, and have not ventured into many more advanced topics, some of which are quite important for computer science applications. In particular, Cartesian closed categories [BW95], [Mit96] and the Curry-Howard isomorphism [SU06], categorical logic [LS86], monads [Man76], [Mog91], [Pho92], fibrations [Jac99], and topoi [Joh02], [Gol06] all deserve attention.

We have presented somewhat more material than usual on certain topics that will find application in some of the subsequent chapters. For example, in the material on factorisation systems (with Section 3.3 taken from [Tar85]) and on indexed categories (with Section 3.4.3.2 based on [TBG91]), we include some exercises which formulate facts that we will rely on later. We will work with indexed categories throughout the book, sometimes implicitly, since we find them more natural for these applications than equivalent formulations in terms of fibrations [Jac99].

We have deliberately chosen to use a notion of factorisation system based on [HS73]. The later book [AHS90] uses a somewhat more general concept, where factorisation morphisms are not required to be epi and mono, respectively, and therefore the uniqueness of the isomorphism between different factorisations of the same morphisms — or equivalently, of the diagonal in Lemma 3.3.4 — must be required explicitly. Although much of the material carries over, some results are simpler under our assumptions: for instance, we rely on Exercise 3.3.5, which does not hold in this form in the framework of [AHS90].

Our presentation of signatures, terms and algebras in Chapter 1 was elementary and set-theoretic, and we retain this style throughout the book. But category theory offers a whole spectrum of possibilities for doing universal algebra fruitfully in a different style. Exercises 3.4.26 and 3.4.41 relate to a categorical "Lawvere-style" presentation of some of the same concepts; see [Law63], [Man76], [BW85]. This was used in some early papers on algebraic specification, e.g. [GTWW75], but as it abstracts away from the choice of operation names in the signature, it seems less useful for applications to program specification. (This argument was put forward in [BG80], with the notion of "signed theory" from [GB78] called to the rescue.) An alternative approach to specifications in this framework is given by sketches; see [BW95], which presents specifications as graphs with indicated diagrams, cones and cocones that in a functorial model of the graph are mapped to commutative diagrams, limits and colimits, respectively. Commutative diagrams capture equational requirements here, with (co)limiting (co)cones offering additional specification power. Another related approach takes the general notion of a T-algebra for a functor $T\colon \mathbf{K} \to \mathbf{K}$ as its starting point, where a T-algebra on an object $A \in |\mathbf{K}|$ is a morphism from $T(A)$ to A. This works smoothly if T is a monad; see [Man76]. Such abstract approaches offer natural generalisations based on semantic interpretation in categories other than **Set**, but again, in our view, abstraction from familiar concepts and syntactic presentations makes them less convenient for our purposes here.

Chapter 4
Working within an arbitrary logical system

Several approaches to specification were discussed in Chapter 2. Each approach involved a different *logical system* as a part of its mathematical underpinnings. We encountered different definitions of:

Signatures: "ordinary" many-sorted signatures, signatures containing *Bool*, *true* and *false* (for final and reachable semantics), error signatures, order-sorted signatures;

Algebras (on a signature Σ): "ordinary" Σ-algebras, error Σ-algebras, partial Σ-algebras, order-sorted Σ-algebras;

Logical sentences (on a signature Σ): Σ-equations, conditional Σ-equations, error Σ-equations (with safe and unsafe variables), Σ-definedness formulae, order-sorted Σ-equations; and

Satisfaction (of a Σ-sentence by a Σ-algebra): of a Σ-equation by a Σ-algebra, of an error Σ-equation by an error Σ-algebra, of a Σ-equation by a partial Σ-algebra, of a Σ-definedness formula by a partial Σ-algebra, of an order-sorted Σ-equation by an order-sorted Σ-algebra.

All of these choices can be combined to obtain many different logical systems and hence different approaches to specification, e.g. partial error specifications with conditional axioms. There are also several alternative approaches to the specification of partial algebras and likewise for the specification of error handling. Furthermore, there are many other variations that have not been considered, including the following (some of them briefly mentioned in Section 2.7.6):

- polymorphic signatures which permit polymorphic type constructors (rather than just sorts) and operations having polymorphic types;
- continuous algebras to handle infinite data objects such as streams;
- higher-order algebras to handle higher-order functions (i.e. functions taking functions as arguments and/or yielding functions as results);
- relational structures to model specifications containing predicates;
- inequations and conditional inequations;
- first-order formulae, with and without equality;

- various modal logics, including algorithmic, dynamic, and temporal logics, for formulating properties of (possibly non-functional) programs.

Some of these variations depart quite considerably from the usual algebraic framework presented in Chapters 1 and 2. But none of them (and very few of the others considered in the literature) are artificial, resulting merely from a theoretician's toying with formal definitions. All of them arise from the practical need to specify different aspects of software systems, often reflected by diverse features of different programming languages.

The resulting wealth of choice of definitions of the basic concepts is not a bad thing. None of the logical systems used in specifications is clearly better than all the others — and we should not expect that such a "best" system will ever be developed. In theory, we can imagine putting all of the above concepts together, producing a single logical system where signatures, algebras, sentences and the satisfaction relation would cover as special cases all we have considered up to now. But the result would be so huge and complex as to make it unmanageable. Moreover, what would we do if one day somebody pointed out that yet another view of software is important and should be reflected in specifications, and hence included in the logical system we use? Scrap everything and start again?

Different specification tasks may call for different systems to express most conveniently the properties required. Moreover, different logical systems may be appropriate for describing different aspects of the same software system, and so a number of logical systems may be useful in a single specification task. It is thus important that the designer of a software system be able to choose which logical system(s) to use.

An unfortunate effect of this necessary wealth of choice is that different researchers have tended to adopt different combinations of basic definitions in their work, sometimes varying their choices from one paper to the next. This makes it difficult to build on the work of others, to compare the results obtained for different logical systems, and to transfer results from one system to another. Such results include not only mathematical definitions and theorems, but also practically useful tools supporting software specification, development and verification produced at great expense of effort, time and money.

In fact, much of the work done turns out to be independent of the particular choice of the basic definitions, although this is often not obvious. The main objective of this chapter, and one of the main objectives of this book, is to lay out the mathematical foundations necessary to make this independence explicit. We achieve this using the notion of an *institution*, which formalises the informal concept of a logical system devised to fit the purposes of specification theory; see Section 4.1 below for the definition. Our thesis is that building as much as possible on the notion of an institution brings important benefits to both the theory and the practice of software specification and development. On the one hand, this allows much work on theories, results, and practical tools to be done just once for many different specific logical systems; on the other hand it forces, via abstraction, a better understanding of and deeper insight into the essence of the concepts and results.

A first example of this general approach is given in Section 4.2, where we recast the fundamental ideas of the standard approach to specification from Chapter 2 in the framework of an arbitrary institution.

It should be stressed that the notion of an institution captures only certain aspects of the informal concept of a logical system. In particular, it takes a model-theoretic view of logical systems, and no direct attempt is made to accommodate proof-theoretic concepts. See Section 9.1 for a discussion of how proof fits into the picture.

When discussing different approaches to specification in Chapter 2, apart from considering various basic notions of signature, algebra, sentence and satisfaction, we also considered different kinds of models (algebras satisfying a set of axioms) as particularly interesting:

- the initial models;
- the reachable models satisfying $true \neq false$;
- the final models in the category of reachable models satisfying $true \neq false$.

These options, although important for the overall style of specification, are of a different nature than the choice of the basic definitions embodied in the particular institution used. We show in Section 4.3 how such "interesting models" may be singled out in an arbitrary institution, thus suggesting that the choice here is in a sense orthogonal to the choice of the underlying institution.

Our general programme is to strive to work in an arbitrary institution as much as possible. However, the concepts involved in the basic theory of institutions are often too general, and hence too weak, to express all that is necessary. When this happens, it would be premature to give up, and switch to working in a particular institution. The "game" is then to identify a (hopefully) minimal set of additional assumptions under which the job can be done, covering most or all of the logical systems of interest. This gives rise to an enriched notion of institution with some additional structure that is relevant to the particular purpose we have in mind. A few examples of this are given in Sections 4.4 and 4.5.

Before proceeding we should warn the reader that although working in an arbitrary institution is very important, it is only one side of the story. The other side is to define an institution appropriate for the needs of the particular task at hand, and quite often this is far from trivial. Indeed, in many areas of computer science, the fundamental problem yet to be satisfactorily solved is the development of a logical system appropriate for the aspects of computing addressed. A number of examples are given in Section 4.1, and references to many others are in Section 4.6. An example of an area for which a satisfactory, commonly accepted solution still seems to be outstanding (despite numerous proposals and active research) is the theory of concurrency.

4.1 Institutions

Following Goguen and Burstall [GB92], we introduce the notion of an *institution*, capturing some essential aspects of the informal concept of a "logical system". The basic ingredients of an institution are a notion of a signature in the system, and then for each signature, notions of an algebra with this signature and of a logical sentence over this signature, and finally a satisfaction relation between algebras and sentences.

In contrast to classical logic and model theory, we are not content with considering logical systems "pointwise" for an "arbitrary but fixed" signature. To capture the process of building a specification and designing a software system, some means of moving from one signature to another is required, that is, some notion of signature morphism. These morphisms typically enable signatures to be extended by new components, renaming and/or identifying others, as well as hiding some components used "internally" but not intended to be visible "externally". Any signature morphism should give rise to a translation of sentences and a translation of algebras determined by the change of names involved. Furthermore, these translations must be consistent with one another, preserving the satisfaction relation. As usual, when we switch from syntax (signatures, sentences) to semantics (algebras), the direction of translation is reversed.

The language of category theory is used in the definition to express the above ideas. This concisely and elegantly captures structure arising from signature morphisms, and forces an appropriate level of generality and abstraction.

Definition 4.1.1 (Institution). An *institution* **INS** consists of

- a category $\mathbf{Sign}_{\mathbf{INS}}$ of *signatures*;
- a functor $\mathbf{Sen}_{\mathbf{INS}}\colon \mathbf{Sign}_{\mathbf{INS}} \to \mathbf{Set}$, giving a set $\mathbf{Sen}_{\mathbf{INS}}(\Sigma)$ of Σ*-sentences* for each signature $\Sigma \in |\mathbf{Sign}_{\mathbf{INS}}|$ and a function $\mathbf{Sen}_{\mathbf{INS}}(\sigma)\colon \mathbf{Sen}_{\mathbf{INS}}(\Sigma) \to \mathbf{Sen}_{\mathbf{INS}}(\Sigma')$ which translates Σ-sentences to Σ'-sentences for each signature morphism $\sigma\colon\Sigma \to \Sigma'$;
- a functor $\mathbf{Mod}_{\mathbf{INS}}\colon \mathbf{Sign}_{\mathbf{INS}}^{op} \to \mathbf{Cat}$, giving a category $\mathbf{Mod}_{\mathbf{INS}}(\Sigma)$ of Σ*-models* for each signature $\Sigma \in |\mathbf{Sign}_{\mathbf{INS}}|$ and a functor $\mathbf{Mod}_{\mathbf{INS}}(\sigma)\colon \mathbf{Mod}_{\mathbf{INS}}(\Sigma') \to \mathbf{Mod}_{\mathbf{INS}}(\Sigma)$ which translates Σ'-models to Σ-models (and Σ'-morphisms to Σ-morphisms) for each signature morphism $\sigma\colon \Sigma \to \Sigma'$; and
- a family $\langle \models_{\mathbf{INS},\Sigma} \subseteq |\mathbf{Mod}_{\mathbf{INS}}(\Sigma)| \times \mathbf{Sen}_{\mathbf{INS}}(\Sigma)\rangle_{\Sigma\in|\mathbf{Sign}_{\mathbf{INS}}|}$ of *satisfaction relations*, determining satisfaction of Σ-sentences by Σ-models for each signature $\Sigma \in |\mathbf{Sign}_{\mathbf{INS}}|$

such that for any signature morphism $\sigma\colon \Sigma \to \Sigma'$ the translations $\mathbf{Mod}_{\mathbf{INS}}(\sigma)$ of models and $\mathbf{Sen}_{\mathbf{INS}}(\sigma)$ of sentences preserve the satisfaction relation, that is, for any $\varphi \in \mathbf{Sen}_{\mathbf{INS}}(\Sigma)$ and $M' \in |\mathbf{Mod}_{\mathbf{INS}}(\Sigma')|$

$$M' \models_{\mathbf{INS},\Sigma'} \mathbf{Sen}_{\mathbf{INS}}(\sigma)(\varphi) \quad\text{iff}\quad \mathbf{Mod}_{\mathbf{INS}}(\sigma)(M') \models_{\mathbf{INS},\Sigma} \varphi$$

[*Satisfaction condition*].

□

The following picture may help us visualise the structure:

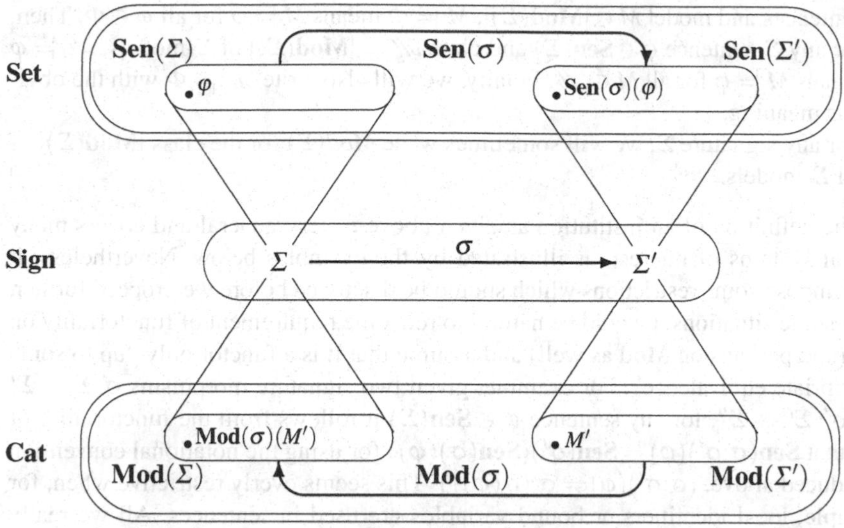

We will freely use standard terminology, and, for example, say that a Σ-model M *satisfies* a Σ-sentence φ, or that φ *holds* in M, whenever $M \models_{\mathbf{INS},\Sigma} \varphi$.

The term "model" (which we use following [GB92]) thereby becomes overloaded: it is used to refer both to objects in the category $\mathbf{Mod}_{\mathbf{INS}}(\Sigma)$ and to the algebras which satisfy a given set of axioms (we will soon extend the latter terminology to an arbitrary institution in Section 4.2, and then to an arbitrary structured specification in Chapter 5). Hopefully, this will not lead to confusion as the context will always determine which of the two meanings is meant. If in doubt, we will use "a Σ-model" (where Σ is a signature) for the former, and "a model of Φ" (where Φ is a set of sentences) for the latter meaning of the word.

Notation.

- When there is no danger of confusion, as in the picture above, we omit the subscript **INS** when referring to the components of an institution, writing $\mathbf{INS} = \langle \mathbf{Sign}, \mathbf{Sen}, \mathbf{Mod}, \langle \models_\Sigma \rangle_{\Sigma \in |\mathbf{Sign}|} \rangle$. Similarly, the subscript Σ on the satisfaction relations will often be omitted.
- For any signature morphism $\sigma : \Sigma \to \Sigma'$, the function $\mathbf{Sen}(\sigma) : \mathbf{Sen}(\Sigma) \to \mathbf{Sen}(\Sigma')$ will be denoted by $\sigma : \mathbf{Sen}(\Sigma) \to \mathbf{Sen}(\Sigma')$ and the functor $\mathbf{Mod}(\sigma) : \mathbf{Mod}(\Sigma') \to \mathbf{Mod}(\Sigma)$ by $\_|_\sigma : \mathbf{Mod}(\Sigma') \to \mathbf{Mod}(\Sigma)$. Thus for any Σ-sentence $\varphi \in \mathbf{Sen}(\Sigma)$, $\sigma(\varphi) \in \mathbf{Sen}(\Sigma')$ is its σ*-translation* to a Σ'-sentence, and for any Σ'-model $M' \in |\mathbf{Mod}(\Sigma')|$, $\mathbf{M'}|_\sigma \in |\mathbf{Mod}(\Sigma)|$ is its σ*-reduct* to a Σ-model. We will also refer to M' as a σ*-expansion* of $M'|_\sigma$. Using this notation, the satisfaction condition of Definition 4.1.1 may be expressed as follows: $M' \models \sigma(\varphi) \iff M'|_\sigma \models \varphi$.

- For any signature Σ, the satisfaction relation extends naturally to sets of Σ-sentences and classes[1] of Σ-models. Namely, for any set $\Phi \subseteq \mathbf{Sen}(\Sigma)$ of Σ-sentences and model $M \in |\mathbf{Mod}(\Sigma)|$, $M \models \Phi$ means $M \models \varphi$ for all $\varphi \in \Phi$. Then, for any Σ-sentence $\varphi \in \mathbf{Sen}(\Sigma)$ and class $\mathcal{M} \subseteq |\mathbf{Mod}(\Sigma)|$ of Σ-models, $\mathcal{M} \models \varphi$ means $M \models \varphi$ for all $M \in \mathcal{M}$. Finally, we will also write $\mathcal{M} \models \Phi$ with the obvious meaning.
- For any signature Σ, we will sometimes write $Mod(\Sigma)$ for the class $|\mathbf{Mod}(\Sigma)|$ of all Σ-models. □

The definition of an institution as given above is very general and covers many logical systems of interest, as illustrated by the examples below. Nevertheless, it does impose some restrictions which should be discussed before we proceed further.

In some situations it would be natural to relax the requirement of functoriality on **Sen** (and perhaps on **Mod** as well) and assume that it is a functor only "up to some appropriate equivalence". For example, given two signature morphisms $\sigma: \Sigma \to \Sigma'$ and $\sigma': \Sigma' \to \Sigma''$, for any sentence $\varphi \in \mathbf{Sen}(\Sigma)$ it follows from the functoriality of **Sen** that $\mathbf{Sen}(\sigma;\sigma')(\varphi) = \mathbf{Sen}(\sigma')(\mathbf{Sen}(\sigma)(\varphi))$ (or using the notational convention introduced above, $(\sigma;\sigma')(\varphi) = \sigma'(\sigma(\varphi))$). This seems overly restrictive when, for example, local identifiers or bound variables are used in sentences. All we really care about here is that the two translations of φ to a Σ''-sentence are *semantically equivalent*: $(\sigma;\sigma')(\varphi)$ and $\sigma'(\sigma(\varphi))$ hold in the same Σ''-models. A solution is to consider sentences up to this semantic equivalence, and work in an institution where sentences simply *are* the corresponding equivalence classes. This solution would resemble the usual practice in λ-calculi, where terms are considered "up to α-conversion" (renaming of bound variables), meaning that terms are really classes of mutually α-convertible syntactic terms.

The only explicit requirement in the definition of an institution is that the satisfaction condition holds. Speaking informally, this deals with the situation where a "small" signature Σ and a "big" signature Σ' are related by a signature morphism $\sigma: \Sigma \to \Sigma'$, and we have a model $M' \in |\mathbf{Mod}(\Sigma')|$ over the "big" signature, and a sentence $\varphi \in \mathbf{Sen}(\Sigma)$ over the "small" signature. There are then two ways to check whether M' "satisfies" φ: we can either reduce the model M' to the "small" signature and check whether the reduct satisfies the sentence φ, or translate the sentence φ to the "big" signature and check whether the translated sentence holds in M'.

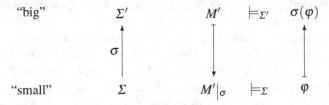

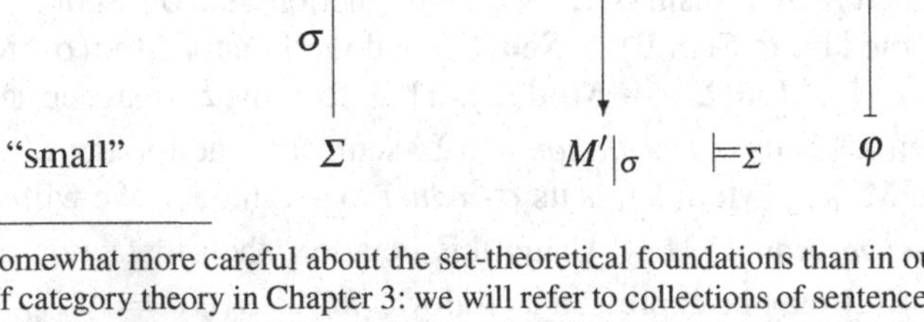

[1] We will be somewhat more careful about the set-theoretical foundations than in our presentation of the basics of category theory in Chapter 3: we will refer to collections of sentences as "sets" and to collections of models as "classes", as in Chapter 2. This is consistent with the formal definition of an institution above, and satisfactory for the logical systems formalised as institutions given as examples (but see Example 4.1.46, footnote 17).

The satisfaction condition states that these two alternatives are equivalent. This embodies two fundamental assumptions. One is that the meaning of a sentence depends only on the components used in the sentence, and does not depend on the context in which the sentence is considered. The other is that the meaning of a sentence is preserved under translation; as [GB92] states:

Truth is invariant under change of notation.

The latter requirement does not raise much doubt — we are not aware of any natural system in which it would not hold. The former, however, is sometimes violated. There are natural logical systems where the meaning of a sentence depends on the context in which it is used, or in other words on the signature over which the sentence is considered. For instance, in logical systems involving quantifiers, the range of quantification may implicitly depend on the signature, e.g. with quantified variables ranging only over reachable values, so that "$\exists x \bullet \ldots$" is interpreted as "there exists an element x which is the value of a ground term, such that ... ", and similarly for universal quantification. For such a logic the satisfaction condition does not hold unless very strong restrictions are placed on signature morphisms.

Exercise 4.1.2. Give a concrete counterexample to the satisfaction condition for a logical system similar to equational logic, but with the universally quantified variables in equations ranging only over reachable values. Show how the logical system you give may be modified to make the satisfaction condition hold. HINT: The satisfaction condition failed because the interpretation of universal quantification over reachable values implicitly depends on the signature; try to make this dependence explicit! □

4.1.1 Examples of institutions

Example 4.1.3 (Ground equational logic GEQ). The institution **GEQ** of ground equational logic is defined as follows:

- The category $\mathbf{Sign_{GEQ}}$ is just **AlgSig**, the usual category of algebraic signatures.
- The functor $\mathbf{Sen_{GEQ}}\colon \mathbf{AlgSig} \to \mathbf{Set}$ gives:
  - the set of ground Σ-equations for each $\Sigma \in |\mathbf{AlgSig}|$; and
  - the σ-translation function taking ground Σ-equations to ground Σ'-equations for each signature morphism $\sigma\colon \Sigma \to \Sigma'$.
- The functor $\mathbf{Mod_{GEQ}}\colon \mathbf{AlgSig}^{op} \to \mathbf{Cat}$ is the functor $\mathbf{Alg}\colon \mathbf{AlgSig}^{op} \to \mathbf{Cat}$ as defined in Example 3.4.29, that is, $\mathbf{Mod_{GEQ}}$ gives:
  - the category $\mathbf{Alg}(\Sigma)$ of Σ-algebras and Σ-homomorphisms for each $\Sigma \in |\mathbf{AlgSig}|$; and
  - the reduct functor $\_|_\sigma\colon \mathbf{Alg}(\Sigma') \to \mathbf{Alg}(\Sigma)$ which maps Σ'-algebras and Σ'-homomorphisms to Σ-algebras and Σ-homomorphisms for each signature morphism $\sigma\colon \Sigma \to \Sigma'$.

- For $\Sigma \in |\mathbf{AlgSig}|$, the satisfaction relation $\models_{\mathbf{GEQ},\Sigma} \subseteq |\mathbf{Alg}(\Sigma)| \times \mathbf{Sen}_{\mathbf{GEQ}}(\Sigma)$ is the usual relation of satisfaction of a ground Σ-equation by a Σ-algebra.

The Satisfaction Lemma (Lemma 2.1.8) ensures that the required satisfaction condition holds and so the above definition indeed yields an institution. □

Example 4.1.4 (Equational logic EQ). The institution **EQ** of (ordinary) equational logic is defined as follows:

- The category $\mathbf{Sign}_{\mathbf{EQ}}$ is just **AlgSig**.
- The functor $\mathbf{Sen}_{\mathbf{EQ}}\colon \mathbf{AlgSig} \to \mathbf{Set}$ gives:
  - the set of Σ-equations for each $\Sigma \in |\mathbf{AlgSig}|$; and
  - the σ-translation function taking Σ-equations to Σ'-equations for each signature morphism $\sigma\colon \Sigma \to \Sigma'$.[2]
- The functor $\mathbf{Mod}_{\mathbf{EQ}}$ is $\mathbf{Alg}\colon \mathbf{AlgSig}^{op} \to \mathbf{Cat}$, just like $\mathbf{Mod}_{\mathbf{GEQ}}$ for ground equational logic.
- For each $\Sigma \in |\mathbf{AlgSig}|$, the satisfaction relation $\models_{\mathbf{EQ},\Sigma} \subseteq |\mathbf{Alg}(\Sigma)| \times \mathbf{Sen}_{\mathbf{EQ}}(\Sigma)$ is the usual relation of satisfaction of a Σ-equation by a Σ-algebra.

The Satisfaction Lemma (Lemma 2.1.8) again ensures that the required satisfaction condition holds and so the above definition indeed yields an institution. □

There is an obvious sense in which **GEQ** can be regarded as a "subinstitution" of **EQ**. We will encounter further such cases below. We refrain from formulating a notion of subinstitution because the concept turns out to be more subtle than it might appear at first. We postpone a proper treatment of relationships between institutions to Chapter 10 (in particular, see Exercise 10.4.9).

Exercise 4.1.5 (Reachable ground equational logic RGEQ). Define an institution **RGEQ** of ground equational logic on reachable algebras by modifying the definition of **GEQ** so that only reachable algebras are considered as models. Do not forget to adjust the definition of reduct functors!

Try to extend this to an institution **REQ** of equational logic on reachable algebras — and notice that the satisfaction condition cannot be ensured without modifying the notion of an equation to include "data constructors" to determine the reachable values for which the equation is to be considered, as already hinted at in Exercise 4.1.2. □

Example 4.1.6 (Partial equational logic PEQ). The institution **PEQ** of partial equational logic is defined as follows (cf. Section 2.7.4):

- $\mathbf{Sign}_{\mathbf{PEQ}}$ is **AlgSig** again.

---

[2] The exact treatment of variables in equations requires special care to ensure that the translation of equations along possibly non-injective signature morphisms is indeed functorial. The use of disjoint union in the translation of many-sorted sets of variables in Definition 1.5.10 causes problems here. The simplest way to make this work is to assume that, in each equation, variables of different sorts are distinct. See [GB92] for details.

- $\mathbf{Sen}_{\mathbf{PEQ}}: \mathbf{AlgSig} \to \mathbf{Set}$ gives:
  - the set of Σ-equations and Σ-definedness formulae for each $\Sigma \in |\mathbf{AlgSig}|$; and
  - the σ-translation function taking Σ-equations and Σ-definedness formulae to Σ'-equations and Σ'-definedness formulae for each signature morphism $\sigma: \Sigma \to \Sigma'$.[3]
- $\mathbf{Mod}_{\mathbf{PEQ}}: \mathbf{AlgSig}^{op} \to \mathbf{Cat}$ gives:
  - the category $\mathbf{PAlg}(\Sigma)$ of partial Σ-algebras and weak Σ-homomorphisms for each $\Sigma \in |\mathbf{AlgSig}|$ (cf. Example 3.3.13); and
  - the reduct functor $\_|_\sigma: \mathbf{PAlg}(\Sigma') \to \mathbf{PAlg}(\Sigma)$ defined as in the total case for each signature morphism $\sigma: \Sigma \to \Sigma'$.
- For $\Sigma \in |\mathbf{AlgSig}|$, the satisfaction relation $\models_{\mathbf{PEQ},\Sigma} \subseteq |\mathbf{PAlg}(\Sigma)| \times \mathbf{Sen}_{\mathbf{PEQ}}(\Sigma)$ is the satisfaction of Σ-equations (with strong equality) and Σ-definedness formulae by partial Σ-algebras.

Exercise. Following the lines of the proof of the Satisfaction Lemma (Lemma 2.1.8), show that the satisfaction condition holds for **PEQ**. □

Example 4.1.7 (Ground partial equational logic PGEQ). The institution **PGEQ** of ground partial equational logic is defined just like the institution **PEQ** of partial equational logic above, except that only ground equations and ground definedness formulae are considered. □

Exercise 4.1.8. Recalling the notion of existential equality for partial algebras from Section 2.7.4, define institutions $\mathbf{PEQ}^e$ and $\mathbf{PGEQ}^e$ of partial existence equational logic and ground partial existence equational logic, respectively, modifying the definitions in Examples 4.1.6 and 4.1.7 by using existential equations of the form $\forall X.t \stackrel{e}{=} t'$ and their ground versions only. □

Example 4.1.9 (Propositional logic PROP). The institution **PROP** of propositional logic is defined as follows:

- $\mathbf{Sign}_{\mathbf{PROP}}$ is **Set**, the usual category of sets. In this context, for each "signature" $P \in |\mathbf{Set}|$, we call elements of P *propositional variables*.
- $\mathbf{Sen}_{\mathbf{PROP}}: \mathbf{Set} \to \mathbf{Set}$ gives
  - For each $P \in |\mathbf{Set}|$, $\mathbf{Sen}_{\mathbf{PROP}}(P)$ is the least set that contains P, sentences true and false, and is closed under the usual propositional connectives, that is, if $\varphi, \varphi' \in \mathbf{Sen}_{\mathbf{PROP}}(\Sigma)$ then also $\varphi \vee \varphi' \in \mathbf{Sen}_{\mathbf{PROP}}(\Sigma)$, $\neg\varphi \in \mathbf{Sen}_{\mathbf{PROP}}(\Sigma)$, $\varphi \wedge \varphi' \in \mathbf{Sen}_{\mathbf{PROP}}(\Sigma)$, and $\varphi \Rightarrow \varphi' \in \mathbf{Sen}_{\mathbf{PROP}}(\Sigma)$.[4]

[3] As in Example 4.1.4, care is needed with the treatment of variables and their translation under signature morphisms; see footnote 2.

[4] We tacitly assume here that true, false, $\vee$, $\wedge$, $\Rightarrow$, $\neg$ are new symbols (not in P), and rely on the usual precedence rules and parentheses to make sure that no ambiguities in the "parsing" of propositional sentences arise.

  - For each function $\sigma: P \to P'$, $\mathbf{Sen_{PROP}}(\sigma)$ extends σ to the translation of arbitrary propositional sentences with propositional variables in P to propositional sentences with propositional variables in P', preserving the propositional connectives in the obvious way.
- $\mathbf{Mod_{PROP}}: \mathbf{Set}^{op} \to \mathbf{Cat}$ gives:
  - For each set of propositional variables $P \in |\mathbf{Set}|$, P-models are all functions from P to $\{\mathit{ff}, \mathit{tt}\}$. These functions can be identified with subsets of P, where $M: P \to \{\mathit{ff}, \mathit{tt}\}$ yields $\{p \in P \mid M(p) = \mathit{tt}\}$). Model morphisms are just inclusions of these sets, i.e. given two P-models $M_1, M_2: P \to \{\mathit{ff}, \mathit{tt}\}$, we have a (unique) morphism from M_1 to M_2 if for all $p \in P$, $M_2(p) = \mathit{tt}$ whenever $M_1(p) = \mathit{tt}$.
  - For $\sigma: P \to P'$, the functor $\mathbf{Mod_{PROP}}(\sigma): \mathbf{Mod_{PROP}}(P') \to \mathbf{Mod_{PROP}}(P)$ maps any model $M': P' \to \{\mathit{ff}, \mathit{tt}\}$ to $\sigma;M': P \to \{\mathit{ff}, \mathit{tt}\}$.
- For $P \in |\mathbf{Set}|$, the satisfaction relation $\models_{\mathbf{PROP},P} \subseteq |\mathbf{Mod_{PROP}}(P)| \times \mathbf{Sen_{PROP}}(P)$ is the usual relation of satisfaction of propositional sentences, that is, for any P-model $M: P \to \{\mathit{ff}, \mathit{tt}\}$, $p \in P$ and $\varphi, \varphi' \in \mathbf{Sen_{PROP}}(P)$:
  - $M \models_{\mathbf{PROP},P} p$ if and only if $M(p) = \mathit{tt}$,
  - $M \models_{\mathbf{PROP},P} \varphi \vee \varphi'$ if and only if $M \models_{\mathbf{PROP},P} \varphi$ or $M \models_{\mathbf{PROP},P} \varphi'$,
  - $M \models_{\mathbf{PROP},P} \neg\varphi$ if and only if $M \not\models_{\mathbf{PROP},P} \varphi$,
  - $M \models_{\mathbf{PROP},P} \varphi \wedge \varphi'$ if and only if $M \models_{\mathbf{PROP},P} \varphi$ and $M \models_{\mathbf{PROP},P} \varphi'$.
  - $M \models_{\mathbf{PROP},P} \varphi \Rightarrow \varphi'$ if and only if $M \models_{\mathbf{PROP},P} \varphi'$ or $M \not\models_{\mathbf{PROP},P} \varphi$
  - $M \models_{\mathbf{PROP},P}$ true, and
  - $M \not\models_{\mathbf{PROP},P}$ false. □

Exercise 4.1.10. Recall the specification of Boolean algebras in Example 2.2.4.

Note that one way to view the definitions in Example 4.1.9 is to define the set of P-sentences as terms of sort *Bool* with variables from P. Then, one can consider the two-element Boolean algebra $\mathbb{B}$ with the carrier $\{\mathit{ff}, \mathit{tt}\}$ (with $\mathsf{true}_{\mathbb{B}} = \mathit{tt}$ and $\mathsf{false}_{\mathbb{B}} = \mathit{ff}$). Furthermore, any propositional model $M: P \to \{\mathit{ff}, \mathit{tt}\}$ induces evaluation of terms $M^\sharp: \mathbf{Sen_{PROP}}(P) \to |\mathbb{B}|$, with $M^\sharp(\varphi) = \mathit{tt}$ if and only if $M \models_{\mathbf{PROP},P} \varphi$ as defined above.

Define another institution of propositional logic, $\mathbf{PROP^{BA}}$, where signatures and sentences are as in **PROP**, but models use arbitrary Boolean algebras rather than just $\mathbb{B}$. That is, for any set $P \in |\mathbf{Set}|$ of propositional variables, a P-model in $\mathbf{PROP^{BA}}$ consists of a Boolean algebra B together with valuation $M: P \to |B|$, where we define $\langle B, M\rangle \models_{\mathbf{PROP^{BA}},P} \varphi$ if and only if $\varphi_B(M) = \mathsf{true}_B$ (where $\varphi_B(M)$ is the value of term φ in B under valuation M).

Prove now that the semantic consequence relations (Definition 2.3.6, cf. Definition 4.2.5 below) in **PROP** and $\mathbf{PROP^{BA}}$ coincide.

HINT: Clearly, if $\Psi \models_{\mathbf{PROP^{BA}},P} \varphi$ then also $\Psi \models_{\mathbf{PROP},P} \varphi$ for any set P of propositional variables, $\Psi \subseteq \mathbf{Sen_{PROP}}(P)$ and $\varphi \in \mathbf{Sen_{PROP}}(P)$. Suppose now that $\Psi \not\models_{\mathbf{PROP^{BA}},P} \varphi$. Use the following lemma:[5]

[5] The proof of this lemma is beyond the scope of this book, but see [RS63], I,8.5 and II,5.2,(a)⇒(e).

Lemma. *Given any Boolean algebra B and element $b \in |B|$ such that $b \neq \mathsf{true}_B$, there exists a homomorphism $h: B \to \mathbb{B}$ from B to the two-element Boolean algebra $\mathbb{B}$ such that $h(b) = \mathsf{false}_{\mathbb{B}}$.*

Now, given any Boolean algebra B and valuation $M: P \to |B|$ such that for all $\psi \in \Psi$, $\psi_B(M) = \mathsf{true}_B$ and $\varphi_B(M) \neq \mathsf{true}_B$, conclude that $(M;h)^\sharp(\psi) = tt$ for all $\psi \in \Psi$, while $(M;h)^\sharp(\varphi) = ff$, where the homomorphism $h: B \to \mathbb{B}$ such that $h(\varphi_B(M)) = ff$ is given by the above lemma. □

Exercise 4.1.11. Define the institution of intuitionistic propositional logic, $\mathbf{PROP^I}$, following the pattern of $\mathbf{PROP^{BA}}$ in Exercise 4.1.10, but using arbitrary Heyting algebras (see Example 2.7.6) rather than just Boolean algebras.

Show that if $\Psi \models_{\mathbf{PROP^I},P} \varphi$ then also $\Psi \models_{\mathbf{PROP},P} \varphi$ for any set P of propositional variables, $\Psi \subseteq \mathbf{Sen_{PROP}}(P)$ and $\varphi \in \mathbf{Sen_{PROP}}(P)$, and give a counterexample to show that the opposite implication fails in general. □

Example 4.1.12 (First-order predicate logic with equality FOPEQ). The institution **FOPEQ** of first-order predicate logic with equality is defined as follows:

- $\mathbf{Sign_{FOPEQ}}$, from now on denoted by **FOSig**, is the category of *first-order signatures*, where we define:
  - A *first-order signature* Θ is a triple $\langle S, \Omega, \Pi \rangle$, where S is a set (of *sort names*), $\Omega = \langle \Omega_{w,s} \rangle_{w \in S^*, s \in S}$ is a family of sets (of *operation names* with their arities and result sorts indicated — just as in algebraic signatures) and $\Pi = \langle \Pi_w \rangle_{w \in S^*}$ is a family of sets (of *predicate* or *relation names* with their arities indicated).
  - A *first-order signature morphism* $\theta: \langle S, \Omega, \Pi \rangle \to \langle S', \Omega', \Pi' \rangle$ consists again of three components: a function $\theta_{sorts}: S \to S'$, an $(S^* \times S)$-indexed family of functions $\theta_{ops} = \langle (\theta_{ops})_{w,s}: \Omega_{w,s} \to \Omega'_{\theta^*_{sorts}(w), \theta_{sorts}(s)} \rangle_{w \in S^*, s \in S}$ (these are as in algebraic signature morphisms) and $\theta_{preds} = \langle (\theta_{preds})_w: \Pi_w \to \Pi'_{\theta^*_{sorts}(w)} \rangle_{w \in S^*}$. (As with algebraic signature morphisms, all the components of a first-order signature morphism θ will be denoted by θ when there is no danger of ambiguity.)
- $\mathbf{Sen_{FOPEQ}}: \mathbf{FOSig} \to \mathbf{Set}$ gives:
  - For each first-order signature $\Theta = \langle S, \Omega, \Pi \rangle$, $\mathbf{Sen_{FOPEQ}}(\Theta)$ is the set of all closed (i.e. without unbound occurrences of variables) *first-order formulae* built out of atomic formulae using the standard propositional connectives ($\vee$, $\wedge$, $\Rightarrow$, $\Leftrightarrow$, $\neg$) and quantifiers ($\forall$, $\exists$). The *atomic formulae* are: equalities of the form $t = t'$, where t and t' are $\langle S, \Omega \rangle$-terms (possibly with variables) of the same sort; atomic predicate formulae of the form $p(t_1, \ldots, t_n)$, where $p \in \Pi_{s_1 \ldots s_n}$ and $t_1, \ldots, t_n$ are terms (possibly with variables) of sorts $s_1, \ldots, s_n$, respectively; and the logical constants true and false.
  - For each first-order signature morphism $\theta: \Theta \to \Theta'$, $\mathbf{Sen_{FOPEQ}}(\theta)$ is the translation of first-order Θ-sentences to first-order Θ'-sentences determined in the

obvious way by the renaming θ of sort, operation and predicate names in Θ to the corresponding names in Θ'.[6]

- $\mathbf{Mod_{FOPEQ}}$: $\mathbf{FOSig}^{op} \to \mathbf{Cat}$, from now on denoted by **FOStr**, gives:
  - For each first-order signature $\Theta = \langle S, \Omega, \Pi \rangle$, the category $\mathbf{FOStr}(\Theta)$ of *first-order Θ-structures* is defined as follows:
    - A *first-order Θ-structure* $A \in |\mathbf{FOStr}(\Theta)|$ consists of a carrier set $|A|_s$ for each sort name $s \in S$, a function $f_A : |A|_{s_1} \times \ldots \times |A|_{s_n} \to |A|_s$ for each operation name $f \in \Omega_{s_1 \ldots s_n, s}$ (these are the same as in $\langle S, \Omega \rangle$-algebras) and a relation $p_A \subseteq |A|_{s_1} \times \ldots \times |A|_{s_n}$ for each predicate name $p \in \Pi_{s_1 \ldots s_n}$. In the following we write $p_A(a_1, \ldots, a_n)$ for $\langle a_1, \ldots, a_n \rangle \in p_A$.
    - For any first-order Θ-structures A and B, a *first-order Θ-morphism* between them, $h : A \to B$, is a family of functions $h = \langle h_s : |A|_s \to |B|_s \rangle_{s \in S}$ which preserves the operations (as ordinary $\langle S, \Omega \rangle$-homomorphisms do) and predicates (i.e. for $p \in \Pi_{s_1 \ldots s_n}$ and $a_1 \in |A|_{s_1}$, ..., $a_n \in |A|_{s_n}$, if $p_A(a_1, \ldots, a_n)$ then $p_B(h_{s_1}(a_1), \ldots, h_{s_n}(a_n))$ as well). A Θ-morphism is *strong* if it reflects predicates as well, so that for $p \in \Pi_{s_1 \ldots s_n}$ and $a_1 \in |A|_{s_1}$, ..., $a_n \in |A|_{s_n}$, $p_A(a_1, \ldots, a_n)$ if and only if $p_B(h_{s_1}(a_1), \ldots, h_{s_n}(a_n))$.
  - For each first-order signature morphism $\theta : \Theta \to \Theta'$, we have the *$\theta$-reduct functor* $\mathbf{FOStr}(\theta) : \mathbf{FOStr}(\Theta') \to \mathbf{FOStr}(\Theta)$ defined similarly to reduct functors corresponding to algebraic signature morphisms.
- For each $\Theta \in |\mathbf{FOSig}|$, the satisfaction relation $\models_{\mathbf{FOPEQ},\Theta} \subseteq |\mathbf{FOStr}(\Theta)| \times \mathbf{Sen_{FOPEQ}}(\Theta)$ is the usual relation of satisfaction of first-order sentences in first-order structures, determined by the usual interpretation of $\vee$, $\wedge$, $\Rightarrow$ and $\neg$ as disjunction, conjunction, implication and negation, respectively, of $\forall$ and $\exists$ as universal and existential quantifiers, respectively, of equalities $t = t'$ as identity of the values of t and t', of atomic predicate formulae $p(t_1, \ldots, t_n)$ as the value of the predicate named p in the structure on the values of the terms t_1, ..., t_n, and of true and false.

Exercise. Work out all the details omitted from the above definition. Then, generalising the proof of the Satisfaction Lemma, show that the satisfaction condition holds for **FOPEQ**. □

Exercise 4.1.13 (First-order predicate logic FOP, first-order logic with equality FOEQ). First-order predicate logic with equality contains some standard "sublogics". Define the institution **FOP** of first-order predicate logic (without equality) by referring to the same signatures and models as in **FOPEQ**, but limiting the sentences to those that do not contain equality.

Define also the institution **FOEQ** with signatures and models as in the institution **EQ** of equational logic, but with first-order sentences (without predicates). □

[6] As in Example 4.1.4, some care is needed with the exact treatment of quantified variables and their translation under signature morphisms (cf. footnote 2) — again, the simplest solution is to assume that, in each formula, variables of different sorts are distinct. See [GB92] for a careful presentation.

Exercise 4.1.14 (Infinitary logics). Define an institution $\mathbf{L}_{\omega_1\omega}$ of so-called $L_{\omega_1\omega}$ logic, which extends first-order predicate logic with equality by allowing conjunctions and disjunctions of *countable* families of formulae (but still only finitary quantification). Extend this further by allowing quantification over countable sets of variables, obtaining an institution of $L_{\omega_1\omega_1}$ logic. You may also want to define institutions of $L_{\alpha\beta}$ logics, for any infinite cardinal numbers α and β such that $\beta \leq \alpha$, with conjunctions and disjunctions of sets of formulae of cardinality smaller than α and quantification over sets of variables of cardinality smaller than β. □

Exercise 4.1.15 (Higher-order logics). Define an institution **SOL** of *second-order logic*, which extends first-order logic by introducing variables ranging over predicates (which in a model denote subsets of a product of the carrier sets) and quantification over such (first-order) predicates. Then generalise this further to an institution **HOL** of *higher-order logic*, which introduces variables that range over (second-order) predicates with arities that may include arities of first-order predicates, and predicates with arities that may include arities of second-order predicates, and so on, and allows for quantification over such higher-order predicates. Without much additional effort, you may want to extend this further, by allowing variables that range over functions of an arbitrary higher-order type, and quantification over such functions. Note though that this will be different from first-order logic for higher-order algebras as sketched in Example 2.7.56, where quantification over higher-order function types does not necessarily coincide with quantification over *all* functions of this type. □

Exercise 4.1.16 (First-order equational logic with boolean values FOEQBool). Define an institution **FOEQBool** which differs from **FOEQ** by considering only signatures that contain a subsignature Σ_{Bool} of *truth values* (Σ_{Bool} has a special, distinguished sort *Bool* and two constants *true*,*false*: *Bool*) and assuming that signature morphisms preserve and reflect symbols in Σ_{Bool} and that algebras interpret them in the standard way (the carrier of sort *Bool* has exactly two distinct elements that are values of *true* and *false*, respectively).

There is now an obvious equivalence between the categories of signatures of **FOPEQ** and **FOEQBool** obtained by mapping each first-order signature to the algebraic signature it contains with the sort *Bool* and constants *true*,*false*: *Bool* added, and with new operation name $f_p{:}s_1 \times \ldots \times s_n \to Bool$ for each predicate $p{:}s_1 \times \ldots \times s_n$. First-order structures give rise to algebras with the standard interpretation of Σ_{Bool} and with functions f_p that yield the value of *true* exactly on those arguments for which the predicate p holds. Clearly, this yields a one-to-one correspondence between first-order structures and algebras over the corresponding signatures. However, this does not extend to model morphisms in general. (**Exercise:** Find a counterexample. Notice though that every *strong* morphism between first-order structures extends to a homomorphism between their corresponding algebras.) We then consider translation of atomic sentences $p(t_1,\ldots,t_n)$ to equalities $p(t_1,\ldots,t_n) = true$, and extend it further to arbitrary first-order sentences with predicates and equality in the obvious way.

Prove that such translations of sentences and models preserve and reflect satisfaction. □

It is not much more difficult to define, for example, the institution **PFOPEQ** of partial first-order predicate logic with equality, or any other institution formalising one of the many standard variants of the classical notions.

Exercise 4.1.17 (Partial first-order predicate logic with equality PFOPEQ). Define the institution **PFOPEQ** of partial first-order predicate logic with equality according to the following sketch:

- $\mathbf{Sign}_{\mathbf{PFOPEQ}} = \mathbf{FOSig}$.
- For each $\Theta \in |\mathbf{FOSig}|$, partial first-order Θ-sentences are defined in the same way as usual first-order Θ-sentences on atomic formulae, which here include *atomic definedness formulae* $def(t)$ for any Θ-term t, in addition to equalities and atomic predicate formulae. The translation of sentences along signature morphisms is defined in the obvious way.
- For each $\Theta \in |\mathbf{FOSig}|$, the models in $\mathbf{Mod}_{\mathbf{PFOPEQ}}(\Theta)$ are like first-order Θ-structures except that the operations may be partial. Morphisms between Θ-models in $\mathbf{Mod}_{\mathbf{PFOPEQ}}(\Theta)$ are like first-order Θ-morphisms but are required to preserve definedness of operations, as weak homomorphisms of partial algebras do. The reduct functors are defined similarly as for first-order structures.
- For each signature $\Theta \in |\mathbf{FOSig}|$, the satisfaction relation $\models_{\mathbf{PFOPEQ},\Theta}$ is defined like the usual first-order satisfaction relation, building on the interpretation of atomic equalities and definedness formulae, which follows the interpretation of (strong) equations and definedness formulae in partial algebras as defined in the institution **PEQ** of partial equational logic, and on the usual interpretation of atomic predicate formulae $p(t_1, \ldots, t_n)$, which yields *false* when any of $t_1, \ldots, t_n$ is undefined. □

Exercise 4.1.18 (Partial first-order logic with equality PFOEQ). Following Exercise 4.1.13, define the institution **PFOEQ** of partial first-order logic with equality with signatures and models inherited from the institution **PEQ** of partial equational logic, but with first-order sentences (without predicates). Similarly, define the institution **PFOP** of partial first-order predicate logic (without equality). □

Exercise 4.1.19 (Partial first-order equational logic with truth PFOEQTruth). As in Exercise 4.1.16, define now an institution **PFOEQBool** of partial first-order logic with equality and built-in boolean values.

However, using partial functions predicates may be modelled differently (and more faithfully when model morphisms are considered). Namely, define an institution **PFOEQTruth** which differs from **PFOEQ** by assuming that the signatures contain a subsignature Σ_{Truth} (which has a special, distinguished sort *Truth* with a single constant *true*: *Truth*), that signature morphisms preserve and reflect symbols in Σ_{Truth}, and that algebras interpret them in the standard way: the carrier of sort *Truth* has exactly one element that is the value of *true*.

The equivalence of categories of signatures and the translation of sentences between **PFOPEQ** and **PFOEQTruth** can now be given in essentially the same way as in Exercise 4.1.16. Moreover, first-order partial structures are in one-to-one correspondence with algebras over the corresponding algebraic signature, and this correspondence may be described exactly as in Exercise 4.1.16 as well. The difference is that now for arguments for which predicates do not hold, their corresponding operations are undefined instead of yielding a non-*true* value. This allows us to extend this correspondence to model morphisms as well.

Prove that such translations of sentences and models preserve and reflect satisfaction. □

Exercise 4.1.20. Recall the notion of a strong homomorphism between partial algebras (Definition 2.7.31) and between first-order structures (given in Example 4.1.12). For each of the institutions above with models that involve partial operations or predicates (**FOPEQ**, **FOP**, **PFOPEQ**, **PEQ**, and so on) define a variant in which all morphisms are strong. We will refer to these institutions as $\mathbf{FOPEQ_{str}}$, $\mathbf{FOP_{str}}$, $\mathbf{PFOPEQ_{str}}$, $\mathbf{PEQ_{str}}$, and so on. In particular, model morphisms in $\mathbf{PFOPEQ_{str}}$ preserve and reflect predicates as well as definedness of operations. □

Exercise 4.1.21. Using the material in Sections 2.7.1, 2.7.3 and 2.7.5, respectively, define institutions $\mathbf{EQ}^{\Rightarrow}$ of conditional equations with signatures and models as in **EQ**; **Horn** of Horn formulae built over signatures and models of **FOPEQ**, where sentences have the form $\forall X \bullet \varphi_1 \wedge \ldots \wedge \varphi_n \Rightarrow \varphi$ for atomic formulae $\varphi_1, \ldots, \varphi_n, \varphi$; **ErrEQ** of error equational logic; and **OrdEQ** of order-sorted equational logic. □

Example 4.1.22 (Equational logic for continuous algebras CEQ). We need some auxiliary definitions. Let $\Sigma = \langle S, \Omega \rangle$ be an algebraic signature.

Recall (cf. Example 3.3.14) that a continuous Σ-algebra $A \in |\mathbf{CAlg}(\Sigma)|$ consists of carriers, which are complete partial orders $\langle |A|_s, \leq_s \rangle$ for $s \in S$, and operations, which are continuous functions $f_A : |A|_{s_1} \times \ldots \times |A|_{s_n} \to |A|_s$ for $f : s_1 \times \ldots \times s_n \to s$ in Σ.

For any S-sorted set X (of variables), the (S-sorted) set $|T_\Sigma^\infty(X)|$ of *infinitary Σ-terms* is the least set such that:[7]

- $X \subseteq |T_\Sigma^\infty(X)|$;
- for each $f : s_1 \times \ldots \times s_n \to s$ in Σ, if $t_1 \in |T_\Sigma^\infty(X)|_{s_1}, \ldots, t_n \in |T_\Sigma^\infty(X)|_{s_n}$ then $f(t_1, \ldots, t_n) \in |T_\Sigma^\infty(X)|_s$; and
- for each $s \in S$, if for $k \geq 0$, $t_k \in |T_\Sigma^\infty(X)|_s$, then $\bigsqcup \langle t_k \rangle_{k \geq 0} \in |T_\Sigma^\infty(X)|_s$.

Intuitively, $|T_\Sigma^\infty(X)|$ contains all the usual finitary Σ-terms and in addition is closed under formal "least upper bounds" of countable sequences of terms. Notice, however, that we do not provide $|T_\Sigma^\infty(X)|$ with the structure of a continuous Σ-algebra; in particular, a term $\bigsqcup \langle t_k \rangle_{k \geq 0}$ is just a formal expression here, not a least upper bound.

Then, for any continuous Σ-algebra A and valuation of variables $v : X \to |A|$, we define a *partial* function $v^\# : |T_\Sigma^\infty(X)| \to |A|$ which for any term $t \in |T_\Sigma^\infty(X)|$ yields the *value $v^\#(t)$ of t* (if defined):

---

[7] For simplicity, we omit the decoration of terms by their target sorts. Formally, to avoid any potential ambiguities, the definition should follow the pattern of Definition 1.4.1.

- for $x \in X$, $v^{\#}(x) = v(x)$;
- for $f: s_1 \times \ldots \times s_n \to s$ and $t_1 \in |T^{\infty}_{\Sigma}(X)|_{s_1}, \ldots, t_n \in |T^{\infty}_{\Sigma}(X)|_{s_n}$, $v^{\#}(f(t_1,\ldots,t_n))$ is defined if and only if $v^{\#}(t_1),\ldots,v^{\#}(t_n)$ are all defined, and then $v^{\#}(f(t_1,\ldots,t_n)) = f_A(v^{\#}(t_1),\ldots,v^{\#}(t_n))$; and
- for $t_k \in T^{\infty}_{\Sigma}(X)_s$, $k \geq 0$, $v^{\#}(\bigsqcup\langle t_k\rangle_{k\geq 0})$ is defined if and only if all $v^{\#}(t_k)$, $k \geq 0$, are defined and form a chain $v^{\#}(t_0) \leq_s v^{\#}(t_1) \leq_s \ldots$, and then $v^{\#}(\bigsqcup\langle t_k\rangle_{k\geq 0}) = \bigsqcup_{k\geq 0} v^{\#}(t_k)$ (where $\bigsqcup$ on the right-hand side stands for the least upper bound in the cpo $\langle |A|_s, \leq_s\rangle$).

As usual, we write $t_A(v)$ for $v^{\#}(t)$.

Finally, an *infinitary Σ-equation* is a triple $\langle X,t,t'\rangle$, written $\forall X \bullet t = t'$, where X is an S-sorted set of variables[8] and $t,t' \in |T^{\infty}_{\Sigma}(X)|_s$ for some $s \in S$. A continuous Σ-algebra A *satisfies* an infinitary Σ-equation $\forall X \bullet t = t'$, written $A \models_{\mathbf{CEQ},\Sigma} \forall X \bullet t = t'$, if for all valuations $v: X \to |A|$, $t_A(v)$ and $t'_A(v)$ are both defined and equal.

We are now ready to define the institution **CEQ** of equational logic for continuous algebras:

- $\mathbf{Sign_{CEQ}}$ is **AlgSig** again.
- $\mathbf{Sen_{CEQ}}$: **AlgSig** $\to$ **Set** gives:
  - the set of infinitary Σ-equations for each $\Sigma \in |\mathbf{AlgSig}|$; and
  - the σ-translation function, mapping infinitary Σ-equations to infinitary Σ'-equations in the obvious way, for each signature morphism $\sigma: \Sigma \to \Sigma'$.
- $\mathbf{Mod_{CEQ}}$: $\mathbf{AlgSig}^{op} \to$ **Cat** gives:
  - the category $\mathbf{CAlg}(\Sigma)$ of continuous Σ-algebras and continuous Σ-homomorphisms for each $\Sigma \in |\mathbf{AlgSig}|$; and
  - the reduct functor $\_|_{\sigma}: \mathbf{CAlg}(\Sigma') \to \mathbf{CAlg}(\Sigma)$, defined similarly as in the case of usual (discrete) algebras for each signature morphism $\sigma: \Sigma \to \Sigma'$.
- For $\Sigma \in |\mathbf{AlgSig}|$, the satisfaction relation $\models_{\mathbf{CEQ},\Sigma} \subseteq |\mathbf{CAlg}(\Sigma)| \times \mathbf{Sen_{CEQ}}(\Sigma)$ is the relation of satisfaction of infinitary Σ-equations by continuous Σ-algebras.

Exercise. Proceeding similarly as in the proof of the Satisfaction Lemma, show that the satisfaction condition holds for **CEQ**.

Exercise. Show that even though we have introduced only infinitary equations as sentences in **CEQ**, infinitary inequalities of the form $\forall X \bullet t \leq t'$ are expressible here as well. HINT: $a \leq b$ iff $a \sqcup b = b$. □

Exercise 4.1.23. For each of the institutions **INS** defined above, define formally its version $\mathbf{INS}^{der}$ based on the category of signatures with derived signature morphisms as presented in Section 1.5.2 (cf. Exercises 3.1.13 and 3.4.30). □

---

[8] For $s \in S$, the sets $X_s \subseteq \mathcal{X}$ come from a fixed vocabulary of variables as in Definition 2.1.1 and are mutually disjoint as in footnote 2.

Example 4.1.24 (Three-valued first-order logic 3FOPEQ). We sketch here the institution **3FOPEQ** of three-valued first-order predicate logic with equality as an example of how the notion of an institution can cope with logical systems based on multiple truth values, where the interpretation of sentences may yield a number of values rather than just being true or false.

- $\mathbf{Sign}_{\mathbf{3FOPEQ}}$ is the category **FOSig** of first-order signatures.
- $\mathbf{Sen}_{\mathbf{3FOPEQ}}\colon \mathbf{Sign}_{\mathbf{3FOPEQ}} \to \mathbf{Set}$ gives:
  - For each $\Theta \in |\mathbf{FOSig}|$, $\mathbf{Sen}_{\mathbf{3FOPEQ}}(\Theta)$ is the set of sentences of the form φ **is** *tt*, φ **is** *ff*, or φ **is** *undef*, where φ is a Θ-sentence of partial first-order predicate logic with equality **PFOPEQ** (see Exercise 4.1.17).
  - For each first-order signature morphism $\theta\colon\Theta \to \Theta'$, we define the translation function $\mathbf{Sen}_{\mathbf{3FOPEQ}}(\theta)\colon\mathbf{Sen}_{\mathbf{3FOPEQ}}(\Theta) \to \mathbf{Sen}_{\mathbf{3FOPEQ}}(\Theta')$ in the obvious way, using the translation of first-order Θ-sentences to Θ'-sentences induced by the morphism θ.
- $\mathbf{Mod}_{\mathbf{3FOPEQ}}\colon \mathbf{Sign}^{op}_{\mathbf{3FOPEQ}} \to \mathbf{Cat}$ is defined as usual for first-order logic, except that operations in structures are partial functions and predicates are interpreted as *partial relations*, which for any tuple of arguments may yield one of three logical values: *tt* (for truth), *ff* (for falsity) and a "third truth value" *undef* (for undefinedness).
- Atomic formulae, propositional connectives and quantifiers may be interpreted over the three-element set of truth values $\{tt, ff, undef\}$ in a number of ways; see [KTB91] and references there for a discussion. Here, we adopt the following interpretation:
  - Atomic definedness formulae have the expected meaning: $def(t)$ is *tt* if the value of t is defined, and is *ff* otherwise.
  - Equalities are interpreted as *strict equalities*: $t = t'$ is *tt* if the values of t and t' are defined and equal, is *ff* if they are defined and different, and is *undef* otherwise.
  - The propositional connectives and quantifiers are interpreted as in Kleene's calculus (cf. [KTB91]). For example, $\varphi \vee \varphi'$ is *tt* if either φ or φ' is *tt*, is *ff* if both φ and φ' are *ff*, and is *undef* otherwise.

  For any $\varphi \in \mathbf{Sen}_{\mathbf{PFOPEQ}}(\Theta)$ and $M \in |\mathbf{Mod}_{\mathbf{3FOPEQ}}(\Theta)|$, this gives the *interpretation of* φ *in* M, $[\![\varphi]\!]_M \in \{tt, ff, undef\}$.

  For $\Theta \in |\mathbf{FOSig}|$, the satisfaction relation $\models_{\mathbf{3FOPEQ},\Theta} \subseteq |\mathbf{Mod}_{\mathbf{3FOPEQ}}(\Theta)| \times \mathbf{Sen}_{\mathbf{3FOPEQ}}(\Theta)$ is now defined in the obvious way: for any $M \in |\mathbf{Mod}_{\mathbf{3FOPEQ}}(\Theta)|$ and $\varphi \in \mathbf{Sen}_{\mathbf{3FOPEQ}}(\Theta)$:

  - $M \models_{\mathbf{3FOPEQ},\Theta} \varphi$ **is** *tt* holds if and only if $[\![\varphi]\!]_M = tt$;
  - $M \models_{\mathbf{3FOPEQ},\Theta} \varphi$ **is** *ff* holds if and only if $[\![\varphi]\!]_M = ff$; and
  - $M \models_{\mathbf{3FOPEQ},\Theta} \varphi$ **is** *undef* holds if and only if $[\![\varphi]\!]_M = undef$.

Exercise. Work out all the details omitted from the above definition; notice that, in particular, model morphisms may be defined in a number of sensible ways. Then show that the satisfaction condition holds. □

Example 4.1.25 (A logic for functional programs FPL). The institution **FPL** of a logic for a simple functional programming language with a first-order monomorphic type system is defined as follows:

- A signature $\mathsf{SIG} = \langle S, \Omega, D \rangle$ consists of a set S of sort names, a family of sets of operation names $\Omega = \langle \Omega_{w,s} \rangle_{w \in S^*, s \in S}$, and a set D of *sorts with value constructors*. Elements of D have the form $\langle d, \mathscr{F} \rangle$ with $d \in S$ and $\mathscr{F} = \langle F_{w,d} \rangle_{w \in S^*}$, where $F_{w,d} \subseteq \Omega_{w,d}$ for $w \in S^*$, with no sort given more than one set of value constructors, i.e. $\langle d, \mathscr{F} \rangle, \langle d, \mathscr{F}' \rangle \in D$ implies $\mathscr{F} = \mathscr{F}'$. So SIG consists of an ordinary algebraic signature $\langle S, \Omega \rangle$ together with a set of *value constructors* for some of the sorts. Sorts with value constructors correspond to algebraic datatypes in functional programming languages. In examples we use a CASL-like notation,[9] for instance,

  sort *Nat* **free with** 0 | *succ*(*Nat*)

  adds the sort name *Nat* to S, the operation names $0{:}Nat$ and $succ{:}Nat \to Nat$ to Ω, and $\langle Nat, \{0{:}Nat, succ{:}Nat \to Nat\} \rangle$ as a sort with value constructors to D. We assume for convenience that each **FPL**-signature SIG contains the sort *Bool* with value constructors *true* and *false*:

  sort *Bool* **free with** *true* | *false*.

- A model over a signature $\mathsf{SIG} = \langle S, \Omega, D \rangle$ is a partial $\langle S, \Omega \rangle$-algebra A such that for each set[10] of sorts with value constructors $\{\langle d_1, \mathscr{F}_1 \rangle, \ldots, \langle d_n, \mathscr{F}_n \rangle\} \subseteq D$, for $1 \le i \le n$, each value constructor in $\mathscr{F}_i$ is total and each element $a \in |A|_{d_i}$ is uniquely constructed from the values in $|A|$ of sorts other than $d_1, \ldots, d_n$ using the value constructors in $\mathscr{F}_1 \cup \cdots \cup \mathscr{F}_n$; that is, $\langle |A|_{d_i} \rangle_{1 \le i \le n}$ is freely generated by $\mathscr{F}_1 \cup \cdots \cup \mathscr{F}_n$ from the carriers of the other sorts in A.
  We assume that all **FPL**-models interpret the sort *Bool* and its constructors *true* and *false* in the same standard way.
  A SIG-morphism between SIG-models A and B is an $\langle S, \Omega \rangle$-homomorphism between A and B viewed as partial $\langle S, \Omega \rangle$-algebras. It is *strong* if it is strong when viewed as a homomorphism between partial algebras; see Definition 2.7.31.
- The set $|T_{\mathsf{SIG}}(X)|$ of **FPL**-terms over $\mathsf{SIG} = \langle S, \Omega, D \rangle$ with variables X and their interpretation in an **FPL**-model A are defined by extending the usual definition

---

[9] CASL notation: this would be written **free type** $Nat ::= 0 \mid succ(Nat)$ in CASL.

[10] This definition is complicated because of the possible presence of mutually dependent sorts with value constructors. **Exercise:** Check that imposing the same requirement for each sort with value constructors separately is more permissive and would not capture the intended meaning. Check also that it would be sufficient to consider only maximal sets of sorts with value constructors that are mutually dependent.

of terms over $\langle S, \Omega \rangle$ and their interpretation by the following additional functional programming constructs (local recursive function definitions and pattern-matching case analysis, respectively):[11]

- **let fun** $f(x_1{:}s_1,\ldots,x_n{:}s_n){:}s' = t'$ **in** t is an **FPL**-term of sort s with variables in X if:
  - $s_1,\ldots,s_n,s' \in S$;
  - t' is an **FPL**-term of sort s' over SIG extended by $f{:}s_1 \times \cdots \times s_n \to s'$ with variables in $X \cup \{x_1{:}s_1,\ldots,x_n{:}s_n\}$; and
  - t is an **FPL**-term of sort s over SIG extended by $f{:}s_1 \times \cdots \times s_n \to s'$ with variables in X.

  The value of such a term under a valuation $v{:}X \to |A|$ is determined as follows:
  - extend A to give an algebra $\widehat{A}$ by interpreting $f{:}s_1 \times \cdots \times s_n \to s'$ as the least-defined partial function $f_{\widehat{A}}$ such that for all $a_1 \in |A|_{s_1},\ldots,a_n \in |A|_{s_n}$, the value of $f_{\widehat{A}}(a_1,\ldots,a_n)$ is the same as the value of t' in $\widehat{A}$ under v modified by mapping x_1 to a_1 and ... and x_n to a_n, whenever the latter is defined.[12]
  - the resulting value is then the value of t in $\widehat{A}$ under v.
- **case** t **of** pat_1 => t_1 | $\cdots$ | pat_n => t_n is an **FPL**-term of sort s with variables in X if:
  - t is an **FPL**-term of some sort s' over SIG with variables in X;
  - for each $1 \le j \le n$, pat_j is a *pattern* over SIG of sort s', where a pattern is an $\langle S, \Omega \rangle$-term containing only variables and value constructors, with no repeated variable occurrences; and
  - for each $1 \le j \le n$, t_j is an **FPL**-term of sort s with variables in the set X extended by the variables of pat_j.

  The value of such a term under a valuation $v{:}X \to |A|$ is determined as follows:
  - obtain the value a of t in A under v;
  - find the least j such that a *matches* pat_j, yielding a valuation v' of the variables in pat_j, where matching a value against a pattern proceeds as follows:
    - a variable x is matched by any value a, yielding a valuation $\{x \mapsto a\}$;
    - a pattern $f(p_1,\ldots,p_m)$ is matched by a yielding a valuation v' iff[13] $a = f_A(a_1,\ldots,a_m)$ and each p_i $(1 \le i \le m)$ is matched by a_i yielding v'_i, with $v' = v'_1 \cup \cdots \cup v'_m$;
  - if such a j exists then the resulting value is that of t_j in A under the extension of v by v'; otherwise, the resulting value is undefined.

---

[11] We will use additional parentheses and indentation to disambiguate **FPL**-terms when the syntax leaves any doubt.

[12] The fact that this unambiguously defines $f_{\widehat{A}}$, and that $f_{\widehat{A}}$ can be equivalently given via the natural operational semantics of recursively defined functions, is a standard result of denotational semantics; see for instance [Sch86].

[13] This uniquely determines a result because non-variable patterns are of sorts that are freely generated by the value constructors and there are no repeated occurrences of variables in patterns.

- Sentences over SIG are first-order sentences built over atomic formulae that are equalities between **FPL**-terms over SIG of the same sort and definedness assertions for such terms. Interpretation of **FPL**-terms in a model determines satisfaction of such sentences, as in **PFOEQ**; see Exercises 4.1.17 and 4.1.18. (Recall that **PFOEQ** uses *strong* equality; see Section 2.7.4.)
  For convenience, we introduce *function definitions* of the form

  $$\textbf{fun}\ f(x_1{:}s_1,\ldots,x_n{:}s_n){:}s = t$$

  to abbreviate the formula

  $$\forall x_1{:}s_1,\ldots,x_n{:}s_n \bullet f(x_1,\ldots,x_n) = \textbf{let fun}\ f(x_1{:}s_1,\ldots,x_n{:}s_n){:}s = t\ \textbf{in}\ f(x_1,\ldots,x_n).$$

  To make the scopes of identifiers clearer, this can be rewritten using a new operation name g and new variables $x'_1,\ldots,x'_n$ as

  $$\forall x_1{:}s_1,\ldots,x_n{:}s_n \bullet f(x_1,\ldots,x_n) = \textbf{let fun}\ g(x'_1{:}s_1,\ldots,x'_n{:}s_n){:}s = t'\ \textbf{in}\ g(x_1,\ldots,x_n)$$

  where t' results from t by replacing f by g and x_i by x'_i, $i = 1,\ldots,n$. Such a recursive function definition is different from the equality $f(x_1,\ldots,x_n) = t$: for instance, $\textbf{fun}\ f(x_1{:}s_1,\ldots,x_n{:}s_n){:}s = f(x_1,\ldots,x_n)$ holds only when f is totally undefined while $f(x_1,\ldots,x_n) = f(x_1,\ldots,x_n)$ trivially always holds.
- Given $\mathsf{SIG} = \langle S,\Omega,D\rangle$ and $\mathsf{SIG}' = \langle S',\Omega',D'\rangle$, an **FPL**-signature morphism $\delta{:}\mathsf{SIG} \to \mathsf{SIG}'$ is a derived signature morphism $\delta{:}\langle S,\Omega\rangle \to \langle S',\Omega'\rangle$ (using **FPL**-terms in place of ordinary terms in Definition 1.5.13), such that for each $\langle d,\mathscr{F}\rangle \in D$, we have $\langle \delta(d),\mathscr{F}'\rangle \in D'$ such that δ restricted to $\mathscr{F}$ is determined by a bijection from $\mathscr{F}$ to $\mathscr{F}'$.
  We require all **FPL**-signature morphisms to preserve the sort *Bool* and its constructors *true* and *false*.
  Such signature morphisms go well beyond the usual renaming of sort and operation names; here we allow (non-constructor) operations to be mapped to complicated terms involving programming constructs like recursion and pattern-matching case analysis. This will be used in Chapters 6–9 to give examples, starting with Example 6.1.6, that suggest how programs fit into the overall specification and development framework.
  Such a signature morphism determines a translation of SIG-sentences to SIG′-sentences in the usual manner,[14] and the same for the reduct from SIG′-models to SIG-models.

[14] Care is required to avoid unintended clashes of **let**-bound operation names in SIG-terms with operation names in SIG′. To avoid consequent problems with functoriality of sentence translation, we can regard **FPL**-terms as being defined up to renaming of **let**-bound operation names.

Moreover, as in **FOPEQ** (see Example 4.1.12), care is needed with the treatment of bound variables (which now also include variables in patterns and formal parameters in **let**-bound operation definitions); cf. footnote 6.

Exercise. Prove that the requirements on **FPL**-signature morphisms concerning the mapping of sorts with value constructors ensure that the reduct of a SIG′-model is indeed a SIG-model.

Exercise. Complete the above definition and prove the satisfaction condition. □

Exercise 4.1.26. The functional programming constructs used above are inspired by those in Standard ML [Pau96]. Add more constructs from Standard ML to the definition of **FPL**. Try adding type definitions, polymorphism, higher-order functions, exceptions.

It is easy to add built-in types other than *Bool* by basing the definition of **FPL** on an arbitrary algebra *DT* as in **IMP** (Example 4.1.32 below). □

Exercise 4.1.27. Mutual recursion need not be added explicitly since it is already expressible using local definitions of recursive functions. Show how. HINT: It may be necessary to resort to copying function definitions, to make each function available for the definitions of the others. □

Exercise 4.1.28. Consider an **FPL**-signature SIG containing a sort *s* with value constructors having arities containing only such sorts. Show how an equality operation $eq_s : s \times s \to Bool$ may be defined using a recursive function definition with pattern-matching case analysis. Use this to view conditionals of the form

$$\textbf{if } t_1 = t_2 \textbf{ then } t \textbf{ else } t'$$

(where t_1, t_2 are SIG-terms of sort *s*, and t, t' have the same sort) as an abbreviation for

$$\textbf{let fun } eq_s(x{:}s, y{:}s){:}Bool = \ldots \textbf{ in case } eq_s(t_1, t_2) \textbf{ of } true \Rightarrow t \mid false \Rightarrow t'.$$ □

Exercise 4.1.29. We could also introduce a conditional of the form **if** φ **then** t **else** t' where φ is a formula. Spell out the details. This would be unusual as a programming construct because branching is controlled by an arbitrary logical formula, allowing terms that would be problematic from a programming point of view, such as **if** $def(t)$ **then** t' **else** t'' and **if** $\forall x{:}s \bullet t_1 = t_2$ **then** t' **else** t''. Note that the meaning of such a conditional would be different from the one introduced in Exercise 4.1.28 when the check for equality involves a term with no defined value. □

Exercise 4.1.30. While **FPL** involves constructs borrowed from functional programming languages, it puts them in a logical context involving equality, logical connectives and quantifiers, which results in sentences capable not only of defining functions, but also of specifying their properties. Identify the "programming part" of **FPL** by defining its "subinstitution" **FProg** with the same signatures and models, but with sets of sentences restricted to function definitions (with satisfaction relations inherited from **FPL** as well). As function definitions may not be closed under translation along arbitrary (derived) signature morphisms in **FPL**, restrict the class of signature morphisms in **FProg** to the standard morphisms, where operation names are mapped to operation names rather than to arbitrary terms.

The use of *strong* equality (rather than, say, *existential* equality; see Sect. 2.7.4) is essential here to ensure that any function definition has a model (in fact, it defines the function unambiguously). Check that under existential interpretation of equality in partial algebras, some function definitions would not be satisfiable. □

Exercise 4.1.31. **FPL** and its programming part **FProg** relate to eager functional programming languages like Standard ML because partial functions are required to be strict. Formulate an analogous institution for lazy functional programming as in Haskell [Pey03]. □

The institutions **FPL** and **FProg** will be used in the sequel to present examples that are meant to appeal to the reader's programming intuition. Later on, the connection with functional programming will be further enhanced by introducing notations for defining ML-style modules in **FPL** (see Example 6.1.9 and Exercise 7.3.5 below).

Example 4.1.32 (A simple imperative language IMP). The institution **IMP** of an imperative programming language with simple type definitions is parameterised by an algebra DT on a signature Σ_{DT} of primitive (built-in) data types and functions of the language. The components of $\mathbf{IMP}_{DT}$ are defined as follows:

- A signature $\Pi = \langle T, P \rangle$ consists of a set T of type names and a set P of functional procedure names with types of the form $s_1, \ldots, s_n \to s$, where each of $s_1, \ldots, s_n, s$ is either a sort in Σ_{DT} or a type name in T. The names in T and P are distinct from those in Σ_{DT}. Thus $\Pi \cup \Sigma_{DT}$ is an algebraic signature — we will denote it by Π_{DT}. Signature morphisms map type names to type names and procedure names to procedure names, preserving their types.
- There are two kinds of sentences over a signature $\Pi = \langle T, P \rangle$.
  First, sentences can be type definitions of the form

  $$\mathbf{type}\ s = \textit{type-expr}$$

  where $s \in T$ is a type name and *type-expr* is a type expression in a simple language of types built over the sorts in Σ_{DT} and a unit type `unit` using the operators $+$ (disjoint union) and $\times$ (Cartesian product). The type expression *type-expr* may contain the type name s as well, which provides for recursive type definitions.[15]
  Second, sentences can be procedure definitions of the form

  $$\mathbf{proc}\ p(x_1{:}s_1, \ldots, x_n{:}s_n) = \textit{while-program};\mathbf{result}\ \textit{expr}{:}s$$

  where $p{:}s_1, \ldots, s_n \to s$ is a procedure name in P, *expr* is a Π_{DT}-term (with variables) of sort s, and *while-program* is a statement in a deterministic programming language over the built-in data types and functions given in DT (*while-program* may be empty, and so the program part of a procedure body may be omitted). We

---

[15] Other type names from T are excluded, to prevent mutual recursion in type definitions — with some extra work this restriction can be removed.

assume that the usual iterative program constructions are provided: sequential statements, conditionals and while loops. This requires that Σ_{DT} contain the sort *Bool* with $|DT|_{Bool} = \{tt, ff\}$. The basic statements are well-typed assignments (of expression values to formal parameters or variables scoped within each procedure body).

Expressions may use projections $\mathtt{proj}_1(v)$ and $\mathtt{proj}_2(v)$ for values v of product types of the form $s_1 \times s_2$ and pairing $\langle v_1, v_2 \rangle$ to build values of product types, as well as boolean tests $\mathtt{is\text{-}in}_1(v)$ and $\mathtt{is\text{-}in}_2(v)$ for values v of union types of the form $s_1 + s_2$ and the constant $\langle\rangle$ of type `unit` denoting the only element of this type. The usual coercions between union types and their component types may also be used. With a bit of additional complication we can also allow expressions to contain (recursive) procedure calls.

- A model M over a signature $\Pi = \langle T, P \rangle$ has a carrier set $|M|_s$ for each $s \in T$. We write $|M|_s$ for $|DT|_s$ if s is a sort name in Σ_{DT}.

  We have the usual notion of *state*, where each state maps formal parameters and variables to values of their sorts in M, or marks them as undefined. An obvious operational semantics may be given that determines, for each statement and state, a sequence of states that formally captures the execution of that statement starting in that state.

  Then, M assigns to each procedure name $p: s_1, \ldots, s_n \to s$ in P and every sequence $v_1 \in |M|_{s_1}, \ldots, v_n \in |M|_{s_n}$ of (actual parameter) values a formal execution which has one of the following forms:

  (*Successful termination*): a finite sequence of states and a result value $v \in |M|_s$;
  (*Unsuccessful termination*): a finite sequence of states; or
  (*Divergence*): an infinite sequence of states.

  Given any such model M, for any procedure name $p: s_1, \ldots, s_n \to s$ in P we get a partial function $p_M : |M|_{s_1} \times \cdots \times |M|_{s_n} \to |M|_s$, where $p_M(v_1, \ldots, v_n) = v$ if $M(P)(v_1, \ldots, v_n)$ is a (finite) formal execution with the result value v.

  The models defined in this way form a discrete category.
- For any signature $\Pi = \langle T, P \rangle$ and Π-model M:
  - M satisfies a Π-sentence of the form

    type $s = \textit{type-expr}$

    if $|M|_s$ is the least set $\mathbb{D}$ such that $\mathbb{D}$ is the value of the type expression *type-expr* in which the type name s is interpreted as $\mathbb{D}$ and sort names s' in Σ_{DT} are interpreted as $|DT|_{s'}$.
  - M satisfies a Π-sentence of the form

    proc $p(x_1 : s_1, \ldots, x_n : s_n) = \textit{while-program}$; **result** $\textit{expr} : s$

    if for all $v_1 \in |M|_{s_1}, \ldots, v_n \in |M|_{s_n}$, $M(p)(v_1, \ldots, v_n)$ is the formal execution of the statement *while-program* starting in the state $\{x_1 \mapsto v_1, \ldots, x_n \mapsto v_n\}$, and if the execution terminates successfully in a state in which *expr* has value v then $M(p)(v_1, \ldots, v_n)$ contains v as the result value.

Exercise. Complete the above definition and prove the satisfaction condition. □

Exercise 4.1.33. Sentences in **IMP** are essentially programs; they provide no means of writing loose specifications. Add sentences of **PFOPEQ** for specifying properties of the procedures of **IMP** viewed as partial functions. A different way of achieving a similar effect will be presented in Examples 10.1.9, 10.1.14 and 10.1.17. □

Example 4.1.34 (Commutative diagrams CDIAG). The following example is of a rather non-standard character. We present a simple logical system for stating that certain diagrams in a category with named objects and morphisms commute. Sentences of the logical system allow one to require that morphisms produced by composition of series of (named) morphisms coincide.

- The category of signatures in **CDIAG** is the category **Graph** of graphs (see Definition 3.2.37).
- A *path equation* in a graph G is a pair of paths in G with the same sources and targets, respectively. For any graph G (a signature in $\mathbf{Sign}_{\mathbf{CDIAG}}$), G-sentences in **CDIAG** are sets of path equations in G.
- A model over a graph G is a (small) category $\mathbf{C}$ with a diagram D of "shape" G, i.e. (via Exercise 3.4.21) a functor $D\colon \mathbf{Path}(G) \to \mathbf{C}$. For any two G-models $D1\colon \mathbf{Path}(G) \to \mathbf{C1}$ and $D2\colon \mathbf{Path}(G) \to \mathbf{C2}$, a *$G$-morphism* in $\mathbf{Mod}_{\mathbf{CDIAG}}(G)$ from $D1$ to $D2$ is a functor $\mathbf{F}\colon \mathbf{C1} \to \mathbf{C2}$ such that $D1;\mathbf{F} = D2$.
- For any G-model $D\colon \mathbf{Path}(G) \to \mathbf{C}$, a path p from s to t in G determines a morphism $D(p)\colon D(s) \to D(t)$ in $\mathbf{C}$. We say that a G-model $D\colon \mathbf{Path}(G) \to \mathbf{C}$ *satisfies* a path equation $\langle p,q\rangle$ if $D(p) = D(q)$. A G-model satisfies a G-sentence Φ if it satisfies all path equations $\varphi \in \Phi$.

Exercise. Complete the definition and prove the satisfaction condition for **CDIAG**.

Exercise. Reformulate the above definitions so that a sentence over a graph G would be a subdiagram of G used to denote the set of path equations in G which make the subdiagram commute. □

The last few examples show that the notion of institution covers much more than what one usually associates with the concept of a logical system.

The next two examples are perhaps even more unusual: we show that the definition of an institution does not restrict the sentences of a logic to be syntactic objects, and does not force models to provide semantic domains and operations used to determine the meanings of the syntactic objects. Thus, the notion of an institution covers systems in which such a distinction is entirely blurred.

Example 4.1.35. Consider an arbitrary category **Sign** and functor $\mathbf{Mod}\colon \mathbf{Sign}^{op} \to \mathbf{Cat}$. We think of **Sign** as a category of signatures and of **Mod** as yielding categories of models and reduct functors. To be cautious about foundations, we should make sure that **Mod** yields only small categories.

We can now define an institution $\mathbf{INS}^{\mathrm{Sen}(\mathbf{Mod})}$ where "sentences" are classes of models:

- The category of signatures of $\mathbf{INS}^{\mathrm{Sen}(\mathbf{Mod})}$ is $\mathbf{Sign}$.
- The "sentence" functor of $\mathbf{INS}^{\mathrm{Sen}(\mathbf{Mod})}$ is defined as follows:
  - For any signature $\Sigma \in |\mathbf{Sign}|$, a Σ-"sentence" of $\mathbf{INS}^{\mathrm{Sen}(\mathbf{Mod})}$ is a collection $\mathscr{M} \subseteq |\mathbf{Mod}(\Sigma)|$ of Σ-models.
  - For any signature morphism $\sigma: \Sigma \to \Sigma'$, the σ-translation of any Σ-"sentence" $\mathscr{M} \subseteq |\mathbf{Mod}(\Sigma)|$ to a Σ'-"sentence" $\sigma(\mathscr{M}) \subseteq |\mathbf{Mod}(\Sigma')|$ is defined as the coimage of $\mathscr{M}$ w.r.t. the σ-reduct functor, i.e. $\sigma(\mathscr{M}) = \{M' \in |\mathbf{Mod}(\Sigma')| \mid \mathbf{Mod}(\sigma)(M') \in \mathscr{M}\}$.
- The model functor of $\mathbf{INS}^{\mathrm{Sen}(\mathbf{Mod})}$ is $\mathbf{Mod}$.
- For each signature Σ, the Σ-satisfaction relation of $\mathbf{INS}^{\mathrm{Sen}(\mathbf{Mod})}$ is just the membership relation: for any Σ-model $M \in |\mathbf{Mod}(\Sigma)|$ and Σ-"sentence" $\mathscr{M} \subseteq |\mathbf{Mod}(\Sigma)|$, $M \models_{\mathbf{INS}^{\mathrm{Sen}(\mathbf{Mod})},\Sigma} \mathscr{M}$ if and only if $M \in \mathscr{M}$.

Exercise. Complete the definition and check the satisfaction condition. □

Example 4.1.36. Consider an arbitrary category **Sign** and functor $\mathbf{Sen}: \mathbf{Sign} \to \mathbf{Set}$. We think of **Sign** as a category of signatures and of **Sen** as yielding sets of sentences and their translations.

We can now define an institution $\mathbf{INS}^{\mathrm{Mod}(\mathbf{Sen})}$ where "models" are sets of sentences:

- The category of signatures of $\mathbf{INS}^{\mathrm{Mod}(\mathbf{Sen})}$ is $\mathbf{Sign}$.
- The sentence functor of $\mathbf{INS}^{\mathrm{Mod}(\mathbf{Sen})}$ is $\mathbf{Sen}$.
- The "model" functor of $\mathbf{INS}^{\mathrm{Mod}(\mathbf{Sen})}$ is defined as follows:
  - For any signature $\Sigma \in |\mathbf{Sign}|$, a Σ-"model" of $\mathbf{INS}^{\mathrm{Mod}(\mathbf{Sen})}$ is a set $\Phi \subseteq \mathbf{Sen}(\Sigma)$ of Σ-sentences. The category of Σ-"models" is just the preorder category where the set of all such subsets is ordered by inclusion.
  - For any signature morphism $\sigma: \Sigma \to \Sigma'$, the σ-reduct functor of $\mathbf{INS}^{\mathrm{Mod}(\mathbf{Sen})}$ from the category of Σ'-"models" to the category of Σ-"models" maps any Σ'-"model" $\Phi' \subseteq \mathbf{Sen}(\Sigma')$ to its coimage $\{\varphi \in \mathbf{Sen}(\Sigma) \mid \mathbf{Sen}(\sigma)(\varphi) \in \Phi'\} \subseteq \mathbf{Sen}(\Sigma)$; this obviously extends to a functor between the preorder categories of Σ'- and Σ-"models".
- For each signature Σ, the Σ-satisfaction relation of $\mathbf{INS}^{\mathrm{Mod}(\mathbf{Sen})}$ is (the inverse of) the membership relation: for any Σ-"model" $\Phi \subseteq \mathbf{Sen}(\Sigma)$ and Σ-sentence $\varphi \in \mathbf{Sen}(\Sigma)$, $\Phi \models_{\mathbf{INS}^{\mathrm{Mod}(\mathbf{Sen})},\Sigma} \varphi$ if and only if $\varphi \in \Phi$.

Exercise. Complete the definition and check the satisfaction condition. □

Let us complete this list of examples by pointing out that the definition of institution admits a number of trivial situations:

Example 4.1.37 (Trivial institutions).

- Recall that $\mathbf{0}$ is the empty category. Hence, there is a unique (empty) functor from $\mathbf{0}$ to **Set** and a unique (empty) functor from $\mathbf{0}^{op} = \mathbf{0}$ to **Cat**. Together with the empty family of relations, they form an empty institution (no signatures, and hence no sentences and no models).

- Given any category **Sign** and functor $\mathbf{Mod}: \mathbf{Sign}^{op} \to \mathbf{Cat}$, a trivial institution with signatures **Sign**, with models given by **Mod**, and with no sentences may be constructed. Formally, the sentences of this institution are given by the functor $\mathbf{Sen}_{\varnothing}: \mathbf{Sign} \to \mathbf{Set}$ which yields the empty set for each signature.
- Given any category **Sign** and functor $\mathbf{Sen}: \mathbf{Sign} \to \mathbf{Set}$, a trivial institution with signatures **Sign**, with sentences given by **Sen**, and with no models may be constructed. Formally, the models of this institution are given by the functor $\mathbf{Mod}_{\mathbf{0}}: \mathbf{Sign}^{op} \to \mathbf{Cat}$ which yields the empty category for each signature.
- Given any category **Sign** and functors $\mathbf{Sen}: \mathbf{Sign} \to \mathbf{Set}$ and $\mathbf{Mod}: \mathbf{Sign}^{op} \to \mathbf{Cat}$, two trivial institutions with signatures **Sign**, with sentences given by **Sen**, and with models given by **Mod** may be constructed. One is obtained by making all sentences false in all models, that is by defining each satisfaction relation to be empty. The other is obtained by making all sentences hold in all models, that is by defining each satisfaction relation to be total (i.e. for each $\Sigma \in |\mathbf{Sign}|$, $\models_\Sigma = |\mathbf{Mod}(\Sigma)| \times \mathbf{Sen}(\Sigma)$). □

4.1.2 Constructing institutions

In the examples of the previous subsection, each of the institutions was constructed "from scratch" by explicitly defining its signatures, sentences, models and satisfaction relations. This is often a rather tedious task (we have simplified it in many cases by referring to the standard definitions), and then checking the satisfaction condition is not always easy. In this subsection we will give some examples of constructions leading from an institution to a more complex one. The complexity added by the construction does not necessarily imply that the institution so obtained has any extra "expressive power". We start with some examples of "formal juggling" with institution components, very much in the spirit of Examples 4.1.35 and 4.1.36, and only then show how adding propositional connectives to a logic may be viewed as a construction of a new institution from an existing one.

Example 4.1.38. Sets of sentences of any institution may be regarded as single sentences (with the obvious "conjunctive" interpretation).

For any institution **INS** define the institution $\mathbf{INS}^\wedge$ of sets of **INS**-sentences as follows:

- The category of $\mathbf{INS}^\wedge$-signatures is the same as the category **Sign** of **INS**-signatures.
- The sentence functor $\mathbf{Sen}_{\mathbf{INS}^\wedge}$ is defined as follows:
  - For any signature $\Sigma \in |\mathbf{Sign}|$, $\mathbf{Sen}_{\mathbf{INS}^\wedge}(\Sigma)$ is the set of all sets $\Phi \subseteq \mathbf{Sen}_{\mathbf{INS}}(\Sigma)$ of Σ-sentences in **INS**.

- For any signature morphism $\sigma: \Sigma \to \Sigma'$, the translation of a Σ-sentence Φ in $\mathbf{INS}^\wedge$ is its image w.r.t. the σ-translation function in **INS**: $\mathbf{Sen}_{\mathbf{INS}^\wedge}(\sigma)(\Phi) = \{\mathbf{Sen}_{\mathbf{INS}}(\sigma)(\varphi) \mid \varphi \in \Phi\} \subseteq \mathbf{Sen}_{\mathbf{INS}}(\Sigma')$.

- The model functor of $\mathbf{INS}^\wedge$ is the same as the model functor $\mathbf{Mod}: \mathbf{Sign}^{op} \to \mathbf{Cat}$ of **INS**.
- For any signature $\Sigma \in |\mathbf{Sign}|$, the satisfaction relation of $\mathbf{INS}^\wedge$ gives the conjunctive interpretation of (sets of) sentences: for any Σ-model $M \in |\mathbf{Mod}(\Sigma)|$ and Σ-sentence $\Phi \subseteq \mathbf{Sen}_{\mathbf{INS}}(\Sigma)$, $M \models_{\mathbf{INS}^\wedge,\Sigma} \Phi$ if and only if for all $\varphi \in \Phi$, $M \models_{\mathbf{INS},\Sigma} \varphi$. □

Example 4.1.39. Signatures of any institution may be enriched to incorporate sentences which restrict the class of models considered over the given signature.

For any institution **INS** define the institution $\mathbf{INS}^{\mathbf{Sign}^+}$ with signatures enriched by sentences as follows:

- Signatures of $\mathbf{INS}^{\mathbf{Sign}^+}$ are pairs $\langle \Sigma, \Phi \rangle$, where $\Sigma \in |\mathbf{Sign}_{\mathbf{INS}}|$ is an **INS**-signature and $\Phi \subseteq \mathbf{Sen}_{\mathbf{INS}}(\Sigma)$ is a set of Σ-sentences. Then, an $\mathbf{INS}^{\mathbf{Sign}^+}$-signature morphism $\sigma: \langle \Sigma, \Phi \rangle \to \langle \Sigma', \Phi' \rangle$ is a signature morphism $\sigma: \Sigma \to \Sigma'$ in $\mathbf{Sign}_{\mathbf{INS}}$ such that for all $\varphi \in \Phi$, $\sigma(\varphi) \in \Phi'$. This defines a category $\mathbf{Sign}_{\mathbf{INS}^{\mathbf{Sign}^+}}$ of $\mathbf{INS}^{\mathbf{Sign}^+}$-signatures (with composition inherited from $\mathbf{Sign}_{\mathbf{INS}}$).
- Sentences of $\mathbf{INS}^{\mathbf{Sign}^+}$ are the same as **INS**-sentences: for any $\mathbf{INS}^{\mathbf{Sign}^+}$-signature $\langle \Sigma, \Phi \rangle$, $\mathbf{Sen}_{\mathbf{INS}^{\mathbf{Sign}^+}}(\langle \Sigma, \Phi \rangle) = \mathbf{Sen}_{\mathbf{INS}}(\Sigma)$, with the translation functions inherited from **INS** as well.
- Models of $\mathbf{INS}^{\mathbf{Sign}^+}$ are again the same as models of **INS**; we consider, however, only those models that satisfy the sentences in the given signature. For any $\mathbf{INS}^{\mathbf{Sign}^+}$-signature $\langle \Sigma, \Phi \rangle$, $\mathbf{Mod}_{\mathbf{INS}^{\mathbf{Sign}^+}}(\langle \Sigma, \Phi \rangle)$ is the full subcategory of $\mathbf{Mod}_{\mathbf{INS}}(\Sigma)$ consisting of all Σ-models (in **INS**) that satisfy (according to $\models_{\mathbf{INS},\Sigma}$) all the sentences in Φ. The reduct functors are again inherited from **INS**.
- The satisfaction relations of $\mathbf{INS}^{\mathbf{Sign}^+}$ are inherited from **INS**.

Exercise. Spell out all the details of the above definition. In particular, check that the reduct functors of the new institution $\mathbf{INS}^{\mathbf{Sign}^+}$ are well defined (cf. Fact 4.2.26 below). □

Example 4.1.40. For any institution, we can enlarge its categories of models by considering models over extended signatures.

For any institution **INS**, define the institution $\mathbf{INS}^{\mathbf{Mod}^+}$ with categories of models containing models over extended signatures as follows:

- The category of $\mathbf{INS}^{\mathbf{Mod}^+}$-signatures is the category **Sign** of **INS**-signatures.
- The sentence functor of $\mathbf{INS}^{\mathbf{Mod}^+}$ is the sentence functor $\mathbf{Sen}: \mathbf{Sign} \to \mathbf{Set}$ of **INS**.
- The model functor $\mathbf{Mod}_{\mathbf{INS}^{\mathbf{Mod}^+}}: \mathbf{Sign}^{op} \to \mathbf{Cat}$ is defined as follows:

– For any signature $\Sigma \in |\mathbf{Sign}|$, a Σ-model of $\mathbf{INS}^{\mathbf{Mod}^+}$ is an $\mathbf{INS}$-model over an extension of the signature Σ. Formally, a Σ-model in $\mathbf{INS}^{\mathbf{Mod}^+}$ is a pair $\langle \sigma : \Sigma \to \Sigma', M' \in |\mathbf{Mod}_{\mathbf{INS}}(\Sigma')| \rangle$. A Σ-model morphism between two such Σ-models is again a pair $\langle \sigma', f \rangle : \langle \sigma_1 : \Sigma \to \Sigma'_1, M'_1 \in |\mathbf{Mod}_{\mathbf{INS}}(\Sigma'_1)| \rangle \to \langle \sigma_2 : \Sigma \to \Sigma'_2, M'_2 \in |\mathbf{Mod}_{\mathbf{INS}}(\Sigma'_2)| \rangle$, which consists of an $\mathbf{INS}$-signature morphism $\sigma' : \Sigma'_1 \to \Sigma'_2$ such that $\sigma_1 ; \sigma' = \sigma_2$ and a model morphism $f : M'_1 \to \mathbf{Mod}_{\mathbf{INS}}(\sigma')(M'_2)$ in $\mathbf{Mod}_{\mathbf{INS}}(\Sigma'_1)$.
– For any signature morphism $\sigma : \Sigma_1 \to \Sigma_2$, the σ-reduct functor $\mathbf{Mod}_{\mathbf{INS}^{\mathbf{Mod}^+}}(\sigma)$ maps any Σ_2-model $\langle \sigma_2 : \Sigma_2 \to \Sigma'_2, M'_2 \in |\mathbf{Mod}_{\mathbf{INS}}(\Sigma'_2)| \rangle$ to the Σ_1-model given by pre-composition with σ, namely $\langle \sigma ; \sigma_2 : \Sigma_1 \to \Sigma'_2, M'_2 \in |\mathbf{Mod}_{\mathbf{INS}}(\Sigma'_2)| \rangle$, and $\mathbf{Mod}_{\mathbf{INS}^{\mathbf{Mod}^+}}(\sigma)$ preserves model morphisms.

- For $\Sigma \in |\mathbf{Sign}|$, the Σ-satisfaction relation of $\mathbf{INS}^{\mathbf{Mod}^+}$ is determined by the Σ-satisfaction relation of $\mathbf{INS}$: for any Σ-model $\langle \sigma : \Sigma \to \Sigma', M' \in |\mathbf{Mod}_{\mathbf{INS}}(\Sigma')| \rangle$ and Σ-sentence $\varphi \in \mathbf{Sen}(\Sigma)$, $\langle \sigma, M' \rangle \models_{\mathbf{INS}^{\mathbf{Mod}^+},\Sigma} \varphi$ if and only if $M' \models_{\mathbf{INS},\Sigma'} \mathbf{Sen}(\sigma)(\varphi)$, which is equivalent to $\mathbf{Mod}_{\mathbf{INS}}(\sigma)(M') \models_{\mathbf{INS},\Sigma} \varphi$ by the satisfaction condition for $\mathbf{INS}$.

Exercise. Complete the definition and check the satisfaction condition. Try to reformulate the construction of the categories of models of $\mathbf{INS}^{\mathbf{Mod}^+}$ using the Grothendieck construction for indexed categories (Definition 3.4.59) and the machinery of comma categories (Definition 3.4.50). □

Example 4.1.41. For any institution $\mathbf{INS}$ define the institution $\mathbf{INS}^{prop}$ by closing the sets of its sentences under propositional connectives (with the usual interpretation) as follows:

- The category of signatures of $\mathbf{INS}^{prop}$ is just the category $\mathbf{Sign}$ of $\mathbf{INS}$-signatures.
- The sentence functor $\mathbf{Sen}_{\mathbf{INS}^{prop}} : \mathbf{Sign} \to \mathbf{Set}$ is defined as follows:

  – For any signature $\Sigma \in |\mathbf{Sign}|$, $\mathbf{Sen}_{\mathbf{INS}^{prop}}(\Sigma)$ is the least set that contains all of the Σ-sentences of $\mathbf{INS}$ and two special sentences true and false, and is closed under the usual propositional connectives as introduced in Example 4.1.9, that is, if $\varphi, \varphi' \in \mathbf{Sen}_{\mathbf{INS}^{prop}}(\Sigma)$ then also $\varphi \vee \varphi' \in \mathbf{Sen}_{\mathbf{INS}^{prop}}(\Sigma)$, $\neg\varphi \in \mathbf{Sen}_{\mathbf{INS}^{prop}}(\Sigma)$, $\varphi \wedge \varphi' \in \mathbf{Sen}_{\mathbf{INS}^{prop}}(\Sigma)$, and $\varphi \Rightarrow \varphi' \in \mathbf{Sen}_{\mathbf{INS}^{prop}}(\Sigma)$.[16]
  – For any $\sigma : \Sigma \to \Sigma'$, the σ-translation function $\mathbf{Sen}_{\mathbf{INS}^{prop}}(\sigma)$ coincides with $\mathbf{Sen}_{\mathbf{INS}}(\sigma)$ on $\mathbf{Sen}_{\mathbf{INS}}(\Sigma)$ and preserves the propositional connectives in the new sentences in the obvious way.

- The model functor of $\mathbf{INS}^{prop}$ is the model functor $\mathbf{Mod} : \mathbf{Sign}^{op} \to \mathbf{Cat}$ of $\mathbf{INS}$.
- For each signature $\Sigma \in |\mathbf{Sign}|$, the Σ-satisfaction relation of $\mathbf{INS}^{prop}$ is the same as the Σ-satisfaction relation of $\mathbf{INS}$ for sentences in $\mathbf{Sen}_{\mathbf{INS}}(\Sigma)$, and then, for any Σ-model $M \in |\mathbf{Mod}(\Sigma)|$, for the sentences built using the propositional connectives, the satisfaction of such sentences in M is defined inductively as in Example 4.1.9.

[16] The remarks in footnote 4 apply as appropriate.

Exercise. Show how **PROP**, the institution of propositional logic (see Example 4.1.9) arises as the propositional closure of a simple institution with propositional variables as the only sentences. □

In Section 4.4.2 below we define yet another similar construction on institutions by showing how quantifiers may be introduced.

The constructions described in the examples above may naturally be viewed as extensions of the original institution — this will be made formal in Section 10.2; cf. Example 10.2.5. In Section 10.3 we will discuss how such extensions may be combined using the limit construction in a suitable category of institutions.

These examples are about adding new sentences built using logical connectives to an institution. The new sentences are added even if the connectives were already expressible in the following sense:

Definition 4.1.42. The institution **INS** *has negation* if for every signature $\Sigma \in |\mathbf{Sign}|$ and Σ-sentence φ, there exists a Σ-sentence ψ such that for every Σ-model M, $M \models_\Sigma \varphi$ iff $M \not\models_\Sigma \psi$. Any such ψ may be referred to as $\neg\varphi$.

The properties of *having conjunction, having disjunction* and *having implication* are defined in the analogous way, and similarly for *having truth, having falsity, having infinitary conjunction*, and so on. □

Exercise 4.1.43. Suppose that the institution **INS** has negation. Using the satisfaction condition, show that for every signature morphism $\sigma\colon\Sigma\to\Sigma'$ and Σ-sentence φ, $\neg\sigma(\varphi)$ may be taken to be $\sigma(\neg\varphi)$. Show a similar property for the other connectives. □

Example 4.1.44. For any institutions $\mathbf{INS}_1 = \langle \mathbf{Sign}_1, \mathbf{Sen}_1, \mathbf{Mod}_1, \langle \models_{1,\Sigma_1} \rangle_{\Sigma_1 \in |\mathbf{Sign}_1|} \rangle$ and $\mathbf{INS}_2 = \langle \mathbf{Sign}_2, \mathbf{Sen}_2, \mathbf{Mod}_2, \langle \models_{2,\Sigma_2} \rangle_{\Sigma_2 \in |\mathbf{Sign}_2|} \rangle$, their *sum* $\mathbf{INS}_1 + \mathbf{INS}_2$ puts $\mathbf{INS}_1$ and $\mathbf{INS}_2$ side by side without any "interaction". Formally, $\mathbf{INS}_1 + \mathbf{INS}_2$ is defined as follows:

- The category of signatures of $\mathbf{INS}_1 + \mathbf{INS}_2$ is the disjoint union $\mathbf{Sign}_1 + \mathbf{Sign}_2$ of the categories of signatures of $\mathbf{INS}_1$ and of $\mathbf{INS}_2$.
- The sentence functor $\mathbf{Sen}_{\mathbf{INS}_1+\mathbf{INS}_2}\colon \mathbf{Sign}_1 + \mathbf{Sign}_2 \to \mathbf{Set}$ acts as $\mathbf{Sen}_1$ on $\mathbf{Sign}_1$ and as $\mathbf{Sen}_2$ on $\mathbf{Sign}_2$ (that is, $\mathbf{Sen}_{\mathbf{INS}_1+\mathbf{INS}_2}$ is determined by $\mathbf{Sen}_1$ and $\mathbf{Sen}_2$ according to the coproduct property of $\mathbf{Sign}_1 + \mathbf{Sign}_2$).
- The model functor $\mathbf{Mod}_{\mathbf{INS}_1+\mathbf{INS}_2}\colon (\mathbf{Sign}_1 + \mathbf{Sign}_2)^{op} \to \mathbf{Cat}$ acts as $\mathbf{Mod}_1$ on $\mathbf{Sign}_1$ and as $\mathbf{Mod}_2$ on $\mathbf{Sign}_2$ (that is, $\mathbf{Mod}_{\mathbf{INS}_1+\mathbf{INS}_2}$ is determined by $\mathbf{Mod}_1$ and $\mathbf{Mod}_2$ according to the coproduct property of $\mathbf{Sign}_1 + \mathbf{Sign}_2$).
- The family of satisfaction relations of $\mathbf{INS}_1 + \mathbf{INS}_2$ is the union of the families of satisfaction relations of $\mathbf{INS}_1$ and of $\mathbf{INS}_2$ (that is, for $\Sigma_1 \in |\mathbf{Sign}_1|$, $\models_{\mathbf{INS}_1+\mathbf{INS}_2,\Sigma_1}$ is $\models_{1,\Sigma_1}$, and for $\Sigma_2 \in |\mathbf{Sign}_2|$, $\models_{\mathbf{INS}_1+\mathbf{INS}_2,\Sigma_2}$ is $\models_{2,\Sigma_2}$). □

Example 4.1.45. Given institutions $\mathbf{INS}_1 = \langle \mathbf{Sign}_1, \mathbf{Sen}_1, \mathbf{Mod}_1, \langle \models_{1,\Sigma_1} \rangle_{\Sigma_1 \in |\mathbf{Sign}_1|} \rangle$ and $\mathbf{INS}_2 = \langle \mathbf{Sign}_2, \mathbf{Sen}_2, \mathbf{Mod}_2, \langle \models_{2,\Sigma_2} \rangle_{\Sigma_2 \in |\mathbf{Sign}_2|} \rangle$, their *product* $\mathbf{INS}_1 \times \mathbf{INS}_2$ is defined as follows:

- The category of signatures of $\mathbf{INS}_1 \times \mathbf{INS}_2$ is the product $\mathbf{Sign}_1 \times \mathbf{Sign}_2$ of the categories of signatures of $\mathbf{INS}_1$ and of $\mathbf{INS}_2$; thus a signature in $\mathbf{INS}_1 \times \mathbf{INS}_2$ is a pair consisting of one signature from $\mathbf{INS}_1$ and one from $\mathbf{INS}_2$, and similarly for signature morphisms.
- The sentence functor $\mathbf{Sen}_{\mathbf{INS}_1 \times \mathbf{INS}_2} : \mathbf{Sign}_1 \times \mathbf{Sign}_2 \to \mathbf{Set}$ is defined as follows:
  - For any signature $\langle \Sigma_1, \Sigma_2 \rangle \in |\mathbf{Sign}_1 \times \mathbf{Sign}_2|$, the set $\mathbf{Sen}_{\mathbf{INS}_1 \times \mathbf{INS}_2}(\langle \Sigma_1, \Sigma_2 \rangle) = \mathbf{Sen}_1(\Sigma_1) + \mathbf{Sen}_2(\Sigma_2)$ of $\langle \Sigma_1, \Sigma_2 \rangle$-sentences is the disjoint union of the sets of $\mathbf{INS}_1$-sentences over Σ_1 and of $\mathbf{INS}_2$-sentences over Σ_2.
  - For any signature morphism $\langle \sigma_1, \sigma_2 \rangle : \langle \Sigma_1, \Sigma_2 \rangle \to \langle \Sigma'_1, \Sigma'_2 \rangle$, the translation of sentences $\mathbf{Sen}_{\mathbf{INS}_1 \times \mathbf{INS}_2}(\langle \sigma_1, \sigma_2 \rangle) = \mathbf{Sen}_1(\sigma_1) + \mathbf{Sen}_2(\sigma_2)$ acts as $\mathbf{Sen}_1(\sigma_1)$ on $\mathbf{INS}_1$-sentences and as $\mathbf{Sen}_2(\sigma_2)$ on $\mathbf{INS}_2$-sentences over the signature $\langle \Sigma_1, \Sigma_2 \rangle$.
- The model functor $\mathbf{Mod}_{\mathbf{INS}_1 \times \mathbf{INS}_2} : (\mathbf{Sign}_1 \times \mathbf{Sign}_2)^{op} \to \mathbf{Cat}$ is defined as follows:
  - For any $\langle \Sigma_1, \Sigma_2 \rangle \in |\mathbf{Sign}_1 \times \mathbf{Sign}_2|$, the category $\mathbf{Mod}_{\mathbf{INS}_1 \times \mathbf{INS}_2}(\langle \Sigma_1, \Sigma_2 \rangle) = \mathbf{Mod}_1(\Sigma_1) \times \mathbf{Mod}_2(\Sigma_2)$ of $\langle \Sigma_1, \Sigma_2 \rangle$-models is the product of the categories of $\mathbf{INS}_1$-models over Σ_1 and of $\mathbf{INS}_2$-models over Σ_2; thus a model in $\mathbf{INS}_1 \times \mathbf{INS}_2$ is a pair consisting of one model from $\mathbf{INS}_1$ and one from $\mathbf{INS}_2$, and similarly for model morphisms.
  - For any signature morphism $\langle \sigma_1, \sigma_2 \rangle : \langle \Sigma_1, \Sigma_2 \rangle \to \langle \Sigma'_1, \Sigma'_2 \rangle$, the reduct functor $\mathbf{Mod}_{\mathbf{INS}_1 \times \mathbf{INS}_2}(\langle \sigma_1, \sigma_2 \rangle) = \mathbf{Mod}_1(\sigma_1) \times \mathbf{Mod}_2(\sigma_2)$ acts as $\mathbf{Mod}_1(\sigma_1)$ on the $\mathbf{INS}_1$-components of $\langle \Sigma'_1, \Sigma'_2 \rangle$-models and model morphisms and as $\mathbf{Mod}_2(\sigma_2)$ on the $\mathbf{INS}_2$-components of $\langle \Sigma'_1, \Sigma'_2 \rangle$-models and model morphisms.
- For any signature $\langle \Sigma_1, \Sigma_2 \rangle \in |\mathbf{Sign}_1 \times \mathbf{Sign}_2|$, sentences $\varphi_1 \in \mathbf{Sen}_1(\Sigma_1)$ and $\varphi_2 \in \mathbf{Sen}_2(\Sigma_2)$, and model $\langle M_1, M_2 \rangle \in |\mathbf{Mod}_{\mathbf{INS}_1 \times \mathbf{INS}_2}(\langle \Sigma_1, \Sigma_2 \rangle)|$, we define $\langle M_1, M_2 \rangle \models_{\mathbf{INS}_1 \times \mathbf{INS}_2, \langle \Sigma_1, \Sigma_2 \rangle} \varphi_1$ to hold if and only if $M_1 \models_{1,\Sigma_1} \varphi_1$, and similarly, $\langle M_1, M_2 \rangle \models_{\mathbf{INS}_1 \times \mathbf{INS}_2, \langle \Sigma_1, \Sigma_2 \rangle} \varphi_2$ if and only if $M_2 \models_{2,\Sigma_2} \varphi_2$. That is, satisfaction in $\mathbf{INS}_1 \times \mathbf{INS}_2$ is defined as $\mathbf{INS}_1$-satisfaction for $\mathbf{INS}_1$-sentences (extracting the $\mathbf{INS}_1$-components of $(\mathbf{INS}_1 \times \mathbf{INS}_2)$-models) and as $\mathbf{INS}_2$-satisfaction for $\mathbf{INS}_2$-sentences (extracting the $\mathbf{INS}_2$-components of $(\mathbf{INS}_1 \times \mathbf{INS}_2)$-models). □

The next example indicates a technically correct but intuitively somewhat artificial way of dealing with the translation of sentences along signature morphisms. The simple idea is that instead of actually translating sentences from one signature to another, we can always keep the original sentence over its original signature together with a morphism "fitting" it to another signature.

Example 4.1.46. Let $\mathbf{INS} = \langle \mathbf{Sign}, \mathbf{Sen}, \mathbf{Mod}, \langle \models_\Sigma \rangle_{\Sigma \in |\mathbf{Sign}|} \rangle$ be an institution. Consider a function $\mathit{NewSen} : |\mathbf{Sign}| \to |\mathbf{Set}|$ coming together with a family of relations $\langle \models_{\mathit{NewSen},\Sigma} \subseteq |\mathbf{Mod}(\Sigma)| \times \mathit{NewSen}(\Sigma) \rangle_{\Sigma \in |\mathbf{Sign}|}$. Intuitively, for any signature Σ, $\mathit{NewSen}(\Sigma)$ is a set of new sentences over Σ with the satisfaction relation $\models_{\mathit{NewSen},\Sigma}$. We define an institution $\mathbf{INS} + \mathit{NewSen}$ by adding these new sentences fitted to other signatures via signature morphisms:

- The category of signatures of $\mathbf{INS} + \mathit{NewSen}$ is just the category **Sign** of **INS**-signatures.
- The sentence functor $\mathbf{Sen}_{\mathbf{INS}+\mathit{NewSen}}: \mathbf{Sign} \to \mathbf{Set}$ is defined as follows:
  - For any signature $\Sigma \in |\mathbf{Sign}|$, $\mathbf{Sen}_{\mathbf{INS}+\mathit{NewSen}}(\Sigma)$ is the (disjoint) union of the "old" sentences $\mathbf{Sen}(\Sigma)$ and the set[17] of "new" sentences fitted to the signature Σ by a signature morphism. The latter are pairs $\langle \varphi', \theta \rangle$, written as φ' **through** θ, with $\theta: \Sigma' \to \Sigma$ and $\varphi' \in \mathit{NewSen}(\Sigma')$ for an arbitrary signature Σ'.
  - For any $\sigma: \Sigma \to \Sigma_1$, $\mathbf{Sen}_{\mathbf{INS}+\mathit{NewSen}}(\sigma)$ works as $\mathbf{Sen}(\sigma)$ on the **INS**-sentences; for $\theta: \Sigma' \to \Sigma$ and $\varphi' \in \mathit{NewSen}(\Sigma')$, $\mathbf{Sen}_{\mathbf{INS}+\mathit{NewSen}}(\sigma)(\varphi'$ **through** $\theta) = \varphi'$ **through** $\theta;\sigma$.
- The model functor of $\mathbf{INS} + \mathit{NewSen}$ is the model functor $\mathbf{Mod}: \mathbf{Sign}^{op} \to \mathbf{Cat}$ of **INS**.
- For each signature $\Sigma \in |\mathbf{Sign}|$, the Σ-satisfaction relation of $\mathbf{INS} + \mathit{NewSen}$ is the same as the Σ-satisfaction relation of **INS** for the "old" Σ-sentences, and then, for any Σ-model $M \in |\mathbf{Mod}(\Sigma)|$, $\theta: \Sigma' \to \Sigma$ and $\varphi' \in \mathit{NewSen}(\Sigma')$, $M \models_{\mathbf{INS}+\mathit{NewSen}} \varphi'$ **through** θ if and only if $M|_\theta \models_{\mathit{NewSen},\Sigma'} \varphi'$.

Exercise. Check the satisfaction condition. □

We conclude this list of constructions on institutions with a sketch of how various modal (and temporal) logics may be built over an arbitrary institution.

Example 4.1.47 (Linear-time temporal logic $\mathbf{LTL}_{\mathbf{INS}}$). For any institution $\mathbf{INS} = \langle \mathbf{Sign}, \mathbf{Sen}, \mathbf{Mod}, \langle \models_\Sigma \rangle_{\Sigma \in |\mathbf{Sign}|} \rangle$, we define the institution $\mathbf{LTL}_{\mathbf{INS}}$ of linear-time temporal logic over **INS**, using sequences of models from **INS** as models and sentences from **INS** as "state sentences", that is:

- The category of signatures of $\mathbf{LTL}_{\mathbf{INS}}$ is **Sign**, the same as in **INS**.
- For each signature Σ, a Σ-model in $\mathbf{LTL}_{\mathbf{INS}}$ is a countably infinite sequence $\overline{M} = \langle M_n \rangle_{n \geq 0}$ of models $M_n \in |\mathbf{Mod}(\Sigma)|$ for $n \geq 0$. Reducts of such models w.r.t. a signature morphism σ are defined componentwise, using the reduct w.r.t. σ as defined in **INS**. (We disregard model morphisms here, taking $\mathbf{Mod}_{\mathbf{LTL}_{\mathbf{INS}}}(\Sigma)$ to be the discrete category.)
- For each signature Σ, the set of Σ-sentences in $\mathbf{LTL}_{\mathbf{INS}}$ is the least set that contains true and all the sentences in $\mathbf{Sen}(\Sigma)$ (called *state sentences* in this context) and is closed under negation, written $\neg\varphi$, conjunction, $\varphi \wedge \psi$, and two modal operators: *next time*, $\mathsf{X}\varphi$, and *until*, $\varphi \mathbin{\mathsf{U}} \psi$.
- For each signature Σ, satisfaction is defined in terms of an auxiliary relation of satisfaction at a given position in the temporal sequence; for each model $\overline{M} = \langle M_n \rangle_{n \geq 0}$, and $j \geq 0$ we define:

---

[17] This may lead to some foundational difficulties, since the collection of all signature morphisms into Σ, and hence the collection of all new Σ-sentences, need not form a set. One argument for ignoring these problems here is that we can typically limit the size of the category of signatures of the institution we start with, for example assuming that the category **Sign** is small.

– for any state sentence φ, $\overline{M} \models^j \varphi$ if $M_j \models \varphi$ (in **INS**);
– $\overline{M} \models^j \neg\varphi$ if it is not the case that $\overline{M} \models^j \varphi$;
– $\overline{M} \models^j \varphi \wedge \psi$ if $\overline{M} \models^j \varphi$ and $\overline{M} \models^j \psi$;
– $\overline{M} \models^j \mathsf{X}\varphi$ if $\overline{M} \models^{j+1} \varphi$; and
– $\overline{M} \models^j \varphi \,\mathsf{U}\, \psi$ if for some $k \geq j$, $\overline{M} \models^k \psi$ and for all $j \leq i < k$, $\overline{M} \models^i \varphi$.

We put now $\overline{M} \models_{\mathbf{LTL_{INS}},\Sigma} \varphi$ if $\overline{M} \models^0 \varphi$.

Exercise. Complete the definition and check the satisfaction condition.

Exercise. Add other temporal modalities, like "eventually/finally" and "henceforth/globally", either by defining them explicitly, or as abbreviations, for instance: $\mathsf{F}\varphi \equiv \mathsf{true}\,\mathsf{U}\,\varphi$, $\mathsf{G}\varphi \equiv \neg(\mathsf{F}(\neg\varphi))$, and so on.

Also, add "past" temporal modalities (previous, since, sometimes in the past, always in the past, and so on). □

Exercise 4.1.48 (Modal logic $\mathbf{MDL_{INS}}$). Proceeding similarly as in Example 4.1.47, given an institution **INS**, define the institution $\mathbf{MDL_{INS}}$ of modal logic:

- The category of signatures of $\mathbf{MDL_{INS}}$ is **Sign**, the same as in **INS**.
- For each signature Σ, a Σ-model in $\mathbf{MDL_{INS}}$ is a Kripke structure, i.e., a triple $\langle W, \rightsquigarrow, \overline{M}\rangle$, which consists of a set W (of "possible worlds" or "state names") and a relation $\rightsquigarrow \subseteq W \times W$ ("transition relation"), together with a family $\overline{M} = \langle M_w \rangle_{w \in W}$ of Σ-models in **INS**, $M_w \in |\mathbf{Mod}(\Sigma)|$ for $w \in W$. Again, we disregard model morphisms.
- For each signature Σ, the set of Σ-sentences in $\mathbf{MDL_{INS}}$ is the least set that contains true and all the sentences in $\mathbf{Sen}(\Sigma)$ and is closed under negation $\neg\varphi$, conjunction $\varphi \wedge \psi$, and the modal operator $\Box\varphi$.
- For each signature Σ, satisfaction is defined in terms of an auxiliary relation of satisfaction at a given world in a Kripke structure; here is the crucial clause:
  – $\langle W, \rightsquigarrow, \overline{M}\rangle \models^w \Box\varphi$ if for all $v \in W$ such that $w \rightsquigarrow v$, $\langle W, \rightsquigarrow, \overline{M}\rangle \models^v \varphi$.

  Then a model satisfies a sentence in $\mathbf{MDL_{INS}}$ if the sentence holds in the above sense at each of its possible worlds.

Complete the definition and check the satisfaction condition.

To keep the definition closer to $\mathbf{LTL_{INS}}$, you may want to define a somewhat different version of modal logic, where Kripke structures in addition indicate an initial world, and then the satisfaction of a sentence in a model is determined by its satisfaction at this initial world. You may also want to impose requirements on the transition relation (for instance, that it is transitive, or that all possible worlds can be reached from the initial world).

Combining the ideas behind $\mathbf{MDL_{INS}}$ and $\mathbf{LTL_{INS}}$, define the institution $\mathbf{CTL^*_{INS}}$ of branching-time temporal logic, where signatures and models are as in $\mathbf{MDL_{INS}}$, but sentences are closed under a variety of temporal operators used to quantify (separately) over paths in the Kripke structure and over worlds in these paths. HINT: Distinguish between two kinds of sentences: path sentences that are evaluated for a given path in the Kripke structure, and state sentences that are evaluated for a given world in the Kripke structure — or see [Eme90].

You may also start by defining a simpler institution $\mathbf{CTL_{INS}}$ where the use of temporal operators is limited by requiring that quantification over paths and over worlds in these paths always happen together, so in fact we have only bundled path/state temporal operators, as in "for some path, always in this path", "for some path, eventually in this path", and so on. □

Exercise 4.1.49. Consider the institution $\mathbf{MDL_{FOPEQ}}$ of modal logic built over first-order predicate logic with equality. Note that this is *not* the institution of first-order modal logic, since quantification is internal to state sentences only and cannot be interleaved with the modal operator. Define an institution **FOMDL** of first-order modal logic in which such an arbitrary interleaving of quantifiers, propositional connectives and the modal operator is allowed. HINT: The trouble here is with moving valuations of variables from one world to another in the definition of satisfaction. At least, define such an institution assuming that the carriers of all models in any Kripke structure coincide. Discuss possible generalisations.

Similarly, try to define institutions of first-order temporal logics that extend $\mathbf{LTL_{FOPEQ}}$, $\mathbf{CTL^*_{FOPEQ}}$ and $\mathbf{CTL_{FOPEQ}}$, respectively. □

4.2 Flat specifications in an arbitrary institution

Throughout this section we will deal with an arbitrary but fixed institution. This means that we will be working with a logical system about which we know nothing beyond what is given in the definition of an institution. For example, we will not be able to refer to any particular component of signatures, any particular syntax of sentences, any particular internal structure of models, or any particular definition of satisfaction. Indeed, we cannot even be sure that signatures have components, that sentences are syntactic entities in any sense, or that models have any internal structure at all.

Given these limitations, one may think that there is very little that can be done. However, the structure of an institution is rich enough to allow us to recast in these terms the material on simple equational specifications presented in Sections 2.2 and 2.3 (this will be done in the present section, without repeating the discussion and motivation) and then to proceed further into the theory of specifications and software development.

Let us then fix an arbitrary institution $\mathbf{INS} = \langle \mathbf{Sign}, \mathbf{Sen}, \mathbf{Mod}, \langle \models_\Sigma \rangle_{\Sigma \in |\mathbf{Sign}|} \rangle$. We start with the basic concepts built around the notion of satisfaction.

Definition 4.2.1 ($Mod_\Sigma(\Phi)$, $Th_\Sigma(\mathscr{M})$, $Cl_\Sigma(\Phi)$ and $Cl_\Sigma(\mathscr{M})$). Let Σ be a signature.

- For any set $\Phi \subseteq \mathbf{Sen}(\Sigma)$ of Σ-sentences, the class $Mod_\Sigma(\Phi) \subseteq |\mathbf{Mod}(\Sigma)|$ of *models of* Φ is defined as the class of all Σ-models that satisfy all the sentences in Φ.[18]
- For any class $\mathcal{M} \subseteq |\mathbf{Mod}(\Sigma)|$ of Σ-models, the *theory of* $\mathcal{M}$ is the set $Th_\Sigma(\mathcal{M}) \subseteq \mathbf{Sen}(\Sigma)$ of all the Σ-sentences that are satisfied by all the models in $\mathcal{M}$.
- A set $\Phi \subseteq \mathbf{Sen}(\Sigma)$ of Σ-sentences is *closed* if $\Phi = Th_\Sigma(Mod_\Sigma(\Phi))$. We will write $Cl_\Sigma(\Phi)$ for $Th_\Sigma(Mod_\Sigma(\Phi))$ and refer to $Cl_\Sigma(\Phi)$ as the *closure of* Φ.
- A class $\mathcal{M} \subseteq |\mathbf{Mod}(\Sigma)|$ of Σ-models is *closed* if $\mathcal{M} = Mod_\Sigma(Th_\Sigma(\mathcal{M}))$. Closed classes of models will be called *definable*. The *closure* of $\mathcal{M}$, written $Cl_\Sigma(\mathcal{M})$, is the class $Mod_\Sigma(Th_\Sigma(\mathcal{M}))$. □

The basic properties of the above notions follow from the fact that Th_Σ and Mod_Σ form a Galois connection:

Proposition 4.2.2. *For any signature Σ, the mappings Th_Σ and Mod_Σ form a Galois connection between sets of Σ-sentences and classes of Σ-models ordered by inclusion.*

Proof. The proof is the same (and just as easy) as in the equational case; cf. Proposition 2.3.2. □

Corollary 4.2.3. *For any signature Σ, set $\Phi \subseteq \mathbf{Sen}(\Sigma)$ of Σ-sentences, and class $\mathcal{M} \subseteq |\mathbf{Mod}(\Sigma)|$ of Σ-models,*

$$\Phi \subseteq Th_\Sigma(\mathcal{M}) \quad \textit{iff} \quad Mod_\Sigma(\Phi) \supseteq \mathcal{M}. \qquad \square$$

Exercise 4.2.4. Construct counterexamples that show that under the assumptions of Corollary 4.2.3 neither of the following implications holds:

$$Mod_\Sigma(\Phi) \subseteq \mathcal{M} \text{ implies } Th_\Sigma(\mathcal{M}) \subseteq \Phi$$
$$Th_\Sigma(\mathcal{M}) \subseteq \Phi \text{ implies } Mod_\Sigma(\Phi) \subseteq \mathcal{M}.$$

Prove that the former implication holds if Φ is closed (i.e. if Φ is the theory of a class of models) and the latter if $\mathcal{M}$ is closed (i.e. if $\mathcal{M}$ is definable). □

The satisfaction relation determines in the obvious way a consequence relation between sentences of the institution:

Definition 4.2.5 (Semantic consequence). Let Σ be an arbitrary signature. A Σ-sentence $\varphi \in \mathbf{Sen}(\Sigma)$ is a *semantic consequence* of a set $\Phi \subseteq \mathbf{Sen}(\Sigma)$ of Σ-sentences, written $\Phi \models_\Sigma \varphi$, if $\varphi \in Cl_\Sigma(\Phi)$, or equivalently, if $Mod_\Sigma(\Phi) \models_\Sigma \varphi$. □

As usual, the subscript Σ will often be omitted.

In the following we will often implicitly rely on three basic properties of semantic consequence, namely that it is reflexive, closed under weakening, and transitive, in the following sense:

---

[18] Note the overloading of the term "model" as discussed after Definition 4.1.1. We continue to follow the terminology of [GB92], hoping that this will not lead to any confusion.

Proposition 4.2.6. *Let Σ be a signature. Consider any Σ-sentences $\varphi, \psi \in \mathbf{Sen}(\Sigma)$, sets of Σ-sentences $\Phi, \Psi \subseteq \mathbf{Sen}(\Sigma)$, and $\Psi_\varphi \subseteq \mathbf{Sen}(\Sigma)$ for each $\varphi \in \Phi$. Then:*

1. $\{\varphi\} \models_\Sigma \varphi$.
2. *If* $\Phi \models_\Sigma \varphi$ *then* $\Phi \cup \Psi \models_\Sigma \varphi$.
3. *If* $\Phi \models_\Sigma \psi$ *and* $\Psi_\varphi \models_\Sigma \varphi$ *for each* $\varphi \in \Phi$ *then* $\bigcup_{\varphi\in\Phi} \Psi_\varphi \models_\Sigma \psi$.

Proof. Directly from the definition. □

We consider semantic consequence for arbitrary, possibly infinite, sets of sentences. For some (but not all!) standard logical systems it is sufficient to restrict attention to consequences of finite sets of sentences:

Definition 4.2.7 (Compactness). INS is *compact* if for every signature Σ and all $\Phi \subseteq \mathbf{Sen}(\Sigma)$ and $\varphi \in \mathbf{Sen}(\Sigma)$, whenever $\Phi \models_\Sigma \varphi$ then $\Phi_{fin} \models_\Sigma \varphi$ for some finite $\Phi_{fin} \subseteq \Phi$. □

Exercise 4.2.8. It is well known that classical single-sorted first-order logic and equational logic are compact; this carries over to the institutions **FOPEQ** and **EQ** of, respectively, first-order and equational logic (Examples 4.1.12 and 4.1.4). Show, however, that infinitary logics (Exercise 4.1.14) and higher-order logics (Exercise 4.1.15) are not compact. Check which of the other institutions introduced in Section 4.1.1 are compact. HINT: You may want to return to this exercise after reading through Section 9.1. □

Another important property of semantic consequence is that it is preserved by translation along signature morphisms:

Proposition 4.2.9. *For any signature morphism $\sigma\colon\Sigma \to \Sigma'$, set $\Phi \subseteq \mathbf{Sen}(\Sigma)$ of Σ-sentences, and Σ-sentence $\varphi \in \mathbf{Sen}(\Sigma)$,*

$$\Phi \models_\Sigma \varphi \textit{ implies } \sigma(\Phi) \models_{\Sigma'} \sigma(\varphi).$$

In other words, $\sigma(Cl_\Sigma(\Phi)) \subseteq Cl_{\Sigma'}(\sigma(\Phi))$.

Proof. Let $M' \in Mod_{\Sigma'}(\sigma(\Phi))$. Then by the satisfaction condition $M'|_\sigma \in Mod_\Sigma(\Phi)$, and so by the definition of the consequence relation $M'|_\sigma \models \varphi$. Thus, by the satisfaction condition again, $M' \models \sigma(\varphi)$, which shows that indeed $\sigma(\Phi) \models \sigma(\varphi)$. □

In general, the reverse implication does not hold, that is, the consequence relation is not reflected by translation along signature morphisms.

Exercise 4.2.10. Try to prove the opposite implication, and notice where the proof breaks down. Then construct a counterexample showing that $\sigma(\Phi) \models \sigma(\varphi)$ does not imply that $\Phi \models \varphi$ even in the standard equational institution **EQ**. HINT: See Proposition 4.2.17 below. □

Corollary 4.2.11. *Under the assumptions of Proposition 4.2.9, $Cl_{\Sigma'}(\sigma(Cl_\Sigma(\Phi))) = Cl_{\Sigma'}(\sigma(\Phi))$.* □

The above corollary implies that when we want to "move" the closure of a set of sentences from one signature to another, it is enough to move only the set itself; all its consequences can be derived over the target signature as well.

Another consequence of Proposition 4.2.9 is that closure of a set of sentences is reflected by translation along signature morphisms:

Corollary 4.2.12. *For any signature morphism $\sigma : \Sigma \to \Sigma'$ and set $\Phi' \subseteq \mathbf{Sen}(\Sigma')$ of Σ'-sentences, if Φ' is closed then so is $\sigma^{-1}(\Phi')$.*

Proof. Suppose Φ' is closed and let $\varphi \in Cl_\Sigma(\sigma^{-1}(\Phi'))$. First, notice that since $\sigma(\sigma^{-1}(\Phi')) \subseteq \Phi'$, $Cl_{\Sigma'}(\sigma(\sigma^{-1}(\Phi'))) \subseteq Cl_{\Sigma'}(\Phi')$. Now, by Proposition 4.2.9, $\sigma(\varphi) \in Cl_{\Sigma'}(\sigma(\sigma^{-1}(\Phi'))) \subseteq Cl_{\Sigma'}(\Phi') = \Phi'$. Thus, $\varphi \in \sigma^{-1}(\Phi')$. □

Notice that the above does not imply that "closure commutes with inverse image" in general; only one of the required inclusions holds:

Corollary 4.2.13. *For any signature morphism $\sigma : \Sigma \to \Sigma'$, set $\Phi' \subseteq \mathbf{Sen}(\Sigma')$ of Σ'-sentences, and Σ-sentence $\varphi \in \mathbf{Sen}(\Sigma)$, if $\sigma^{-1}(\Phi') \models \varphi$ then $\Phi' \models \sigma(\varphi)$. In other words, $Cl_\Sigma(\sigma^{-1}(\Phi')) \subseteq \sigma^{-1}(Cl_{\Sigma'}(\Phi'))$.* □

Exercise 4.2.14. Show that the reverse inclusion does not hold in the standard equational institution **EQ**. □

Forming the closure of a set of sentences consists of two phases: first taking the class of models the set defines, and then taking the theory of this class. Separation of these two phases by translation along a signature morphism preserves the closure to some extent only:

Proposition 4.2.15. *For any signature morphism $\sigma : \Sigma \to \Sigma'$ and set $\Phi' \subseteq \mathbf{Sen}(\Sigma')$ of Σ'-sentences,*

$$Cl_\Sigma(\sigma^{-1}(\Phi')) \subseteq Th_\Sigma(Mod_{\Sigma'}(\Phi')|_\sigma) = \sigma^{-1}(Cl_{\Sigma'}(\Phi'))$$

where for any class $\mathcal{M} \subseteq |\mathbf{Mod}(\Sigma')|$, $\mathcal{M}|_\sigma = \{M'|_\sigma \mid M' \in \mathcal{M}\}$.

Proof. For the first part, let $\varphi \in Cl_\Sigma(\sigma^{-1}(\Phi'))$. By Corollary 4.2.13, $\Phi' \models_{\Sigma'} \sigma(\varphi)$. By the satisfaction condition, $Mod_{\Sigma'}(\Phi')|_\sigma \models_\Sigma \varphi$, and so $\varphi \in Th_\Sigma(Mod_{\Sigma'}(\Phi')|_\sigma)$.

We have $Mod_{\Sigma'}(\Phi') = Mod_{\Sigma'}(Cl_{\Sigma'}(\Phi'))$, which shows $Th_\Sigma(Mod_{\Sigma'}(\Phi')|_\sigma) = Th_\Sigma(Mod_{\Sigma'}(Cl_{\Sigma'}(\Phi'))|_\sigma) \supseteq Cl_\Sigma(\sigma^{-1}(Cl_{\Sigma'}(\Phi'))) \supseteq \sigma^{-1}(Cl_{\Sigma'}(\Phi'))$, and hence also proves one inclusion ("$\supseteq$") of the second part. For the opposite inclusion, consider $\varphi \in Th_\Sigma(Mod_{\Sigma'}(\Phi')|_\sigma)$, that is $Mod_{\Sigma'}(\Phi')|_\sigma \models_\Sigma \varphi$. By the satisfaction condition, $Mod_{\Sigma'}(\Phi') \models_{\Sigma'} \sigma(\varphi)$, which means $\sigma(\varphi) \in Cl_{\Sigma'}(\Phi')$, and so indeed $\varphi \in \sigma^{-1}(Cl_{\Sigma'}(\Phi'))$. □

Corollary 4.2.16. *For any signature morphism $\sigma : \Sigma \to \Sigma'$ and set $\Phi \subseteq \mathbf{Sen}(\Sigma)$ of Σ-sentences, $Cl_\Sigma(\Phi) \subseteq \sigma^{-1}(Cl_{\Sigma'}(\sigma(\Phi)))$.* □

Just as the implication opposite to the one stated in Proposition 4.2.9 does not hold in general, the inclusion opposite to the one above does not hold in general either. This changes for signature morphisms that induce *surjective* reduct functors.

Proposition 4.2.17. *Consider a signature morphism $\sigma : \Sigma \to \Sigma'$ such that the reduct functor $\_|_\sigma : \mathbf{Mod}(\Sigma') \to \mathbf{Mod}(\Sigma)$ is surjective on models. For any set $\Phi \subseteq \mathbf{Sen}(\Sigma)$ of Σ-sentences and Σ-sentence $\varphi \in \mathbf{Sen}(\Sigma)$,*

$$\Phi \models_\Sigma \varphi \quad \textit{iff} \quad \sigma(\Phi) \models_{\Sigma'} \sigma(\varphi).$$

Proof. We prove only the implication opposite to that of Proposition 4.2.9. Let $M \in |\mathbf{Mod}(\Sigma)|$ be an arbitrary Σ-model, and let $M' \in |\mathbf{Mod}(\Sigma')|$ be a σ-expansion of M, i.e. $M'|_\sigma = M$ (such an M' exists since $\_|_\sigma$ is surjective on models). If $M \models_\Sigma \Phi$ then by the satisfaction condition $M' \models_{\Sigma'} \sigma(\Phi)$, and so $M' \models_{\Sigma'} \sigma(\varphi)$. Thus, by the satisfaction condition again, $M \models_\Sigma \varphi$. □

Corollary 4.2.18. *Under the assumptions of Proposition 4.2.17, we have $Cl_\Sigma(\Phi) = \sigma^{-1}(Cl_{\Sigma'}(\sigma(\Phi)))$.* □

This shows that the surjectivity of the reduct functor ensures that moving along a signature morphism is "sound" and "complete" as a strategy for deciding if $\Phi \models_\Sigma \varphi$ by checking whether or not $\sigma(\Phi) \models_{\Sigma'} \sigma(\varphi)$ — without this property, such a strategy is still "complete" (the satisfaction condition ensures that no consequences are lost) but is not always "sound" (new consequences between "old" sentences may be added).

Exercise 4.2.19. Provide an example showing that surjectivity of $\_|_\sigma : \mathbf{Mod}(\Sigma') \to \mathbf{Mod}(\Sigma)$ is not a necessary condition for the conclusions of Proposition 4.2.17 and Corollary 4.2.18. □

Exercise 4.2.20. Show that the inclusion $Cl_\Sigma(\Phi) \subseteq \sigma^{-1}(Cl_{\Sigma'}(\sigma(\Phi)))$, for any $\sigma : \Sigma \to \Sigma'$ and $\Phi \subseteq \mathbf{Sen}(\Sigma)$, directly implies (and, in fact, is equivalent to) Corollary 4.2.13. However, the opposite inclusion $Cl_\Sigma(\Phi) \supseteq \sigma^{-1}(Cl_{\Sigma'}(\sigma(\Phi)))$ does not imply the opposite to the inclusion there: even under the assumptions of Proposition 4.2.17 and Corollary 4.2.18, the inclusion $Cl_\Sigma(\sigma^{-1}(\Phi')) \supseteq \sigma^{-1}(Cl_{\Sigma'}(\Phi'))$ may fail for a set $\Phi' \subseteq \mathbf{Sen}(\Sigma')$ of Σ'-sentences. HINT: One way to construct a counterexample is to add *false* to the set of sentences of **EQ** for some, but not all signatures.

Show, however, that under the assumptions of Proposition 4.2.17, we have $Cl_\Sigma(\sigma^{-1}(\Phi')) = Th_\Sigma(Mod_{\Sigma'}(\Phi')|_\sigma)$ and $Cl_\Sigma(\sigma^{-1}(\Phi')) = \sigma^{-1}(Cl_{\Sigma'}(\Phi'))$ provided that in addition $\sigma : \mathbf{Sen}(\Sigma) \to \mathbf{Sen}(\Sigma')$ is surjective. Discuss why this fact does not seem very interesting. □

The following generalisation of Proposition 4.2.17 underlies the key corollary below.

Proposition 4.2.21. *Let $\sigma : \Sigma \to \Sigma'$ be a signature morphism. Suppose that a set $\Gamma \subseteq \mathbf{Sen}(\Sigma)$ of Σ-sentences exactly characterises the σ-reducts of Σ'-models that satisfy a set $\Gamma' \subseteq \mathbf{Sen}(\Sigma')$ of Σ'-sentences, that is, $Mod_\Sigma(\Gamma) = Mod_{\Sigma'}(\Gamma')|_\sigma$. Then for any set $\Phi \subseteq \mathbf{Sen}(\Sigma)$ of Σ-sentences and Σ-sentence $\varphi \in \mathbf{Sen}(\Sigma)$, $\Phi \cup \Gamma \models_\Sigma \varphi$ if and only if $\sigma(\Phi) \cup \Gamma' \models_{\Sigma'} \sigma(\varphi)$.*

Proof. For the "if" part, assume that $\sigma(\Phi) \cup \Gamma' \models_{\Sigma'} \sigma(\varphi)$ and let $M \models_{\Sigma} \Phi \cup \Gamma$. Then, since $M \in Mod_{\Sigma}(\Gamma)$, there exists $M' \in Mod_{\Sigma'}(\Gamma')$ with $M'|_{\sigma} = M$. By the satisfaction condition, $M' \models_{\Sigma'} \sigma(\Phi)$, hence $M' \models_{\Sigma'} \sigma(\Phi) \cup \Gamma'$ and so $M' \models_{\Sigma'} \sigma(\varphi)$ as well. Thus, by the satisfaction condition again, $M \models_{\Sigma} \varphi$.

For the "only if" part, assume that $\Phi \cup \Gamma \models_{\Sigma} \varphi$ and let $M' \models_{\Sigma'} \sigma(\Phi) \cup \Gamma'$. Then by the satisfaction condition, $M'|_{\sigma} \models_{\Sigma} \Phi$ and moreover, by the assumption, $M'|_{\sigma} \models_{\Sigma} \Gamma$. Hence, $M'|_{\sigma} \models_{\Sigma} \Phi \cup \Gamma$, and so $M'|_{\sigma} \models_{\Sigma} \varphi$ as well, which by the satisfaction condition again proves that $M' \models_{\Sigma'} \sigma(\varphi)$. □

Corollary 4.2.22. *Let $\sigma: \Sigma \to \Sigma'$ be a signature morphism. Suppose that a set $\Gamma \subseteq$ $\mathbf{Sen}(\Sigma)$ of Σ-sentences exactly characterises the σ-reducts of Σ'-models, that is, $Mod_{\Sigma}(\Gamma) = (|\mathbf{Mod}(\Sigma')|)|_{\sigma}$. Then for any set $\Phi \subseteq \mathbf{Sen}(\Sigma)$ of Σ-sentences and Σ-sentence $\varphi \in \mathbf{Sen}(\Sigma)$, $\Phi \cup \Gamma \models_{\Sigma} \varphi$ if and only if $\sigma(\Phi) \models_{\Sigma'} \sigma(\varphi)$.* □

Exercise 4.2.23. Show that Proposition 4.2.17 follows from Proposition 4.2.21 (or Corollary 4.2.22). Generalise Corollary 4.2.18 in a similar way. □

Definition 4.2.24 (Presentation). For any signature Σ, a *Σ-presentation* (or *flat specification*) is a pair $\langle \Sigma, \Phi \rangle$ where $\Phi \subseteq \mathbf{Sen}(\Sigma)$. $M \in |\mathbf{Mod}(\Sigma)|$ is a *model* of a Σ-presentation $\langle \Sigma, \Phi \rangle$ if $M \models \Phi$. $Mod[\langle \Sigma, \Phi \rangle]$ denotes the class of all models of the presentation $\langle \Sigma, \Phi \rangle$, and $\mathbf{Mod}[\langle \Sigma, \Phi \rangle]$ the full subcategory of $\mathbf{Mod}(\Sigma)$ with objects in $Mod[\langle \Sigma, \Phi \rangle]$. □

Definition 4.2.25 (Category of theories). For any signature Σ, a *Σ-theory T* is a Σ-presentation $\langle \Sigma, \Phi \rangle$ where Φ is closed. A Σ-presentation $\langle \Sigma, \Psi \rangle$ *presents* the Σ-theory $\langle \Sigma, Cl_{\Sigma}(\Psi) \rangle$.

For any theories $T = \langle \Sigma, \Phi \rangle$ and $T' = \langle \Sigma', \Phi' \rangle$, a *theory morphism $\sigma: T \to T'$* is a signature morphism $\sigma: \Sigma \to \Sigma'$ such that $\sigma(\varphi) \in \Phi'$ for every $\varphi \in \Phi$.

The category $\mathbf{Th}_{\mathbf{INS}}$ of theories in **INS** has theories as objects and theory morphisms as morphisms, with identities and composition inherited from the category $\mathbf{Sign}_{\mathbf{INS}}$ of signatures of **INS**. □

The satisfaction condition implies the following important characterisation of theory morphisms, analogous to that given for equational theory morphisms in Proposition 2.3.13.

Proposition 4.2.26. *For any signature morphism $\sigma: \Sigma \to \Sigma'$ and sets $\Phi \subseteq \mathbf{Sen}(\Sigma)$ and $\Phi' \subseteq \mathbf{Sen}(\Sigma')$ of sentences, the following conditions are equivalent:*

1. *σ is a theory morphism $\sigma: \langle \Sigma, Cl_{\Sigma}(\Phi) \rangle \to \langle \Sigma', Cl_{\Sigma'}(\Phi') \rangle$.*
2. *$\sigma(\Phi) \subseteq Cl_{\Sigma'}(\Phi')$.*
3. *For every $M' \in Mod_{\Sigma'}(\Phi')$, $M'|_{\sigma} \in Mod_{\Sigma}(\Phi)$.*

Proof.

$1 \Rightarrow 2$: Obvious, since $\Phi \subseteq Cl_{\Sigma}(\Phi)$.

$2 \Rightarrow 3$: Consider $M' \in Mod_{\Sigma'}(\Phi')$. Then also $M' \in Mod_{\Sigma'}(Cl_{\Sigma'}(\Phi'))$, and so for all $\varphi \in \Phi$, $M' \models \sigma(\varphi)$ (since $\sigma(\varphi) \in Cl_{\Sigma'}(\Phi')$). Hence, by the satisfaction condition, $M'|_{\sigma} \models \varphi$, and thus indeed $M'|_{\sigma} \in Mod_{\Sigma}(\Phi)$.

3 ⇒ 1: Consider any $\varphi \in Cl_\Sigma(\Phi)$. We have to show that $\sigma(\varphi) \in Cl_{\Sigma'}(\Phi')$, that is that for all $M' \in Mod_{\Sigma'}(\Phi')$, $M' \models \sigma(\varphi)$. However, if $M' \in Mod_{\Sigma'}(\Phi')$ then $M'|_\sigma \in Mod_\Sigma(\Phi)$. Hence, $M'|_\sigma \models \varphi$, and the conclusion follows from the satisfaction condition. □

Exercise 4.2.27. Define the category $\mathbf{Pres_{INS}}$ of presentations in **INS**, with morphisms $\sigma: \langle\Sigma, \Phi\rangle \to \langle\Sigma', \Phi'\rangle$ that are signature morphisms $\sigma: \Sigma \to \Sigma'$ such that $\Phi' \models \sigma(\varphi)$ for all $\varphi \in \Phi$. Check that $\mathbf{Th_{INS}}$ is a full subcategory of $\mathbf{Pres_{INS}}$ and that the two categories are equivalent. □

Exercise 4.2.28. Show that by Proposition 4.2.26 above, the mapping which to any theory assigns the category of its models extends to a functor $\mathbf{Mod_{Th}}: \mathbf{Th}_{\mathbf{INS}}^{op} \to \mathbf{Cat}$, where:

- for any theory $T = \langle\Sigma, \Phi\rangle$, $\mathbf{Mod}[T]$ is the full subcategory of $\mathbf{Mod}(\Sigma)$ with objects in $Mod[T]$, as in Definition 4.2.24; and
- for any theory morphism $\sigma: T \to T'$, $\mathbf{Mod}(\sigma)$ is $\_|_\sigma: \mathbf{Mod}[T'] \to \mathbf{Mod}[T]$, the σ-reduct functor restricted to the subcategory $\mathbf{Mod}[T']$ of $\mathbf{Mod}(\Sigma')$, where $T' = \langle\Sigma', \Phi'\rangle$. □

Many standard properties of theories (and presentations) investigated in the realm of classical model theory may be formulated in the framework of an arbitrary institution. For example:

Definition 4.2.29 (Consistency and completeness of a presentation). A presentation $\langle\Sigma, \Phi\rangle$ is *consistent* if it has a model, i.e. if $Mod[\langle\Sigma, \Phi\rangle] \neq \varnothing$.

A presentation $\langle\Sigma, \Phi\rangle$ is *complete* if it is a maximal consistent presentation, i.e. if it is consistent and no presentation $\langle\Sigma, \Phi'\rangle$ such that Φ' properly contains Φ is consistent. □

Proposition 4.2.30. *A presentation $\langle\Sigma, \Phi\rangle$ is consistent if and only if the theory $\langle\Sigma, Cl_\Sigma(\Phi)\rangle$ is consistent. Any complete presentation is a (consistent) theory.* □

Definition 4.2.31 (Conservative theory morphism). For any theories $T = \langle\Sigma, \Phi\rangle$ and $T' = \langle\Sigma', \Phi'\rangle$, a theory morphism $\sigma: T \to T'$ is *conservative* if for every Σ-sentence φ, $\varphi \in \Phi$ whenever $\sigma(\varphi) \in \Phi'$.

A theory morphism $\sigma: T \to T'$ *admits model expansion* if the corresponding reduct function $\_|_\sigma: Mod_{\Sigma'}(\Phi') \to Mod_\Sigma(\Phi)$ is surjective, that is, for every Σ-model M such that $M \models_\Sigma \Phi$, there exists a Σ'-model M' such that $M' \models_{\Sigma'} \Phi'$ and $M'|_\sigma = M$. □

Exercise 4.2.32. As in Proposition 4.2.17, show that a theory morphism $\sigma: T \to T'$ is conservative if it admits model expansion. Note that the opposite implication does not hold by Exercise 4.2.19. □

The careful reader has probably realised that in this section we have not even mentioned model morphisms. Indeed, everything above works equally well if we forget about the category structure provided on the collections of models in an institution. But this proves inadequate for some purposes; see for example the next section, where the category structure on models is exploited.

4.3 Constraints

As discussed in Section 2.5, the class of all models that satisfy a given presentation often contains some models that intuitively are undesirable realisations of the presentation. Different methods are used to constrain the semantics of presentations so that from among all its models only the ones that are "desirable" are selected: for example, one may take its initial semantics, reachable semantics, or final semantics (cf. Sections 2.5 and 2.7.2). How do these fit into the institutional framework introduced above? Let us consider initiality constraints[19] first.

There is clearly no problem with expressing the basic concept of initial model in an arbitrary institution: models over any signature form a category; hence the class of models satisfying a given presentation determines a full subcategory of this category — and we know what initiality means in any category (cf. Section 3.2.1).

Let $\mathbf{INS} = \langle \mathbf{Sign}, \mathbf{Sen}, \mathbf{Mod}, \langle \models_\Sigma \rangle_{\Sigma \in |\mathbf{Sign}|} \rangle$ be an institution, fixed throughout this section.

Definition 4.3.1 (Initial model of a presentation). For any signature $\Sigma \in |\mathbf{Sign}|$ and set $\Phi \subseteq \mathbf{Sen}(\Sigma)$ of sentences, the *initial model* of the presentation $\langle \Sigma, \Phi \rangle$ is the (unique up to isomorphism) initial object in $\mathbf{Mod}[\langle \Sigma, \Phi \rangle]$. □

We might feel tempted to pursue any of a number of ways to incorporate the idea of initiality into the institutional framework:

- We may hope to be able to modify all institutions of interest so that they yield initial semantics directly, by changing the model functor **Mod** to yield only the initial models as models over any signature. Clearly, this fails: requiring initiality only makes sense relative to a presentation. If sentences are not taken into account then, for example, the only initial models in the institution **EQ** of equational logic are ground term algebras.
- We can attempt to modify the satisfaction relation so that only the initial models of a sentence will be defined to satisfy it. Quite obviously, this does not work, since it would then be impossible to adequately define models of presentations involving more than one sentence. Without modifying the satisfaction relation, we could modify Definitions 4.2.1 and 4.2.24 and consider only initial models of presentations by defining $Mod_\Sigma(\Phi)$ to consist only of the initial models in $\{M \mid M \models \Phi\}$ considered as a full subcategory of $\mathbf{Mod}(\Sigma)$. But this would make the whole theory rather clumsy, and the various definitions would not fit together as neatly as they do now. For example, Propositions 4.2.9 and 4.2.26 would no longer hold. Worse, this would not allow the user to write axioms that are to be interpreted in a loose, non-initial fashion, indicating that only certain parts of a specification are to be interpreted in an initial way. See Example 4.3.2 below.

---

[19] We use the term "constraint" here following the terminology of [BG80], [GB92]. Initiality and data constraints as discussed and formally defined below have nothing to do with constraints as used in "constraint logic programming" [JL87].

- We can view the requirement of initiality with respect to a presentation as just another *sentence*. This would be a rather complicated sentence, as it has to contain other sentences within it, but in view of examples like 4.1.38 (not to mention 4.1.35) there is no reason why this should bother us. This is the approach we will take.

It is not sufficient to define initiality constraints simply as sets of sentences over a given signature, and then to define their satisfaction via the notion of an initial model. The problem is that we do not always want to constrain the entire model of a presentation. As the following example illustrates, we need to be able to constrain only a certain part of this model, that is, to impose initiality constraints on its reduct to a certain subsignature.

Example 4.3.2. Recall Exercise 2.5.21, which concerned the specification of a function $ch{:}Nat \to Nat$ that for each natural number n chooses an arbitrary number greater than n. As argued there, we certainly do not want to take the initial model of the entire specification: the initial model would generate "artificial elements" of sort *Nat* (as the results of the function *ch*), and then artificial elements of sort *Bool* as well (as results of comparisons by $<$ that involve the artificial elements of sort *Nat*). What one would like is to first interpret the original specification NAT of natural numbers in an initial way, do the same for the specification BOOL, add the operation $\_\_<\_\_{:}Nat \times Nat \to Bool$ (which is defined by its axioms in a sufficiently complete way) — it so happens that this would be the same as taking an initial model of these specifications put together — and only then add an operation $ch{:}Nat \to Nat$ with the corresponding axiom interpreted in the underlying logic, with no initiality restrictions intervening in any way at this stage. □

By allowing initiality requirements to be "fitted" to larger signatures by signature morphisms, along the lines of the construction presented in Example 4.1.46, we can impose the initiality requirement on parts of models.

Definition 4.3.3 (Initiality constraint). Let $\Sigma \in |\mathbf{Sign}|$ be a signature. A *Σ-initiality constraint* is a pair $\langle \Phi', \theta \rangle$, written as **initial** Φ' **through** θ, where $\theta{:}\Sigma' \to \Sigma$ is a signature morphism and $\Phi' \subseteq \mathbf{Sen}(\Sigma')$ is a set of Σ'-sentences. A Σ-model $M \in |\mathbf{Mod}(\Sigma)|$ *satisfies* a Σ-initiality constraint **initial** Φ' **through** θ if its reduct $M|_\theta \in |\mathbf{Mod}(\Sigma')|$ is an initial model of $\langle \Sigma', \Phi' \rangle$. □

Now, such an initiality constraint may be regarded as just another sentence in a presentation, and freely mixed with "ordinary" sentences.

Exercise 4.3.4. Redo Exercise 2.5.21 using initiality constraints. Discuss the possibility of achieving the same effect without the "fitting morphism" component in initiality constraints. □

The specification built in Exercise 4.3.4 is not a presentation in **FOEQ** — we have to extend this institution by adding initiality constraints first. Indeed, given an institution **INS** we can always form a new institution $\mathbf{INS}^{init}$ in which initiality constraints are allowed as additional sentences. Such a construction is implicitly involved whenever initiality constraints are used.

Definition 4.3.5 (Institution with initiality constraints). The institution $\mathbf{INS}^{init}$ with initiality constraints in **INS** is defined as follows:

- The category $\mathbf{Sign}_{\mathbf{INS}^{init}}$ of signatures is just **Sign**, the same as in **INS**.
- The functor $\mathbf{Sen}_{\mathbf{INS}^{init}}$ gives:
  - for each signature Σ, the (disjoint) union of the set $\mathbf{Sen}(\Sigma)$ of Σ-sentences in **INS** and of the set of Σ-initiality constraints;[20] and
  - for each signature morphism $\sigma : \Sigma \to \Sigma_1$, the translation function $\mathbf{Sen}_{\mathbf{INS}^{init}}(\sigma)$ that works as $\mathbf{Sen}(\sigma)$ on all the "old" Σ-sentences in **INS**, and for any Σ-initiality constraint **initial** Φ' **through** θ, where $\theta : \Sigma' \to \Sigma$ and $\Phi' \subseteq \mathbf{Sen}(\Sigma')$, is defined by $\mathbf{Sen}_{\mathbf{INS}^{init}}(\sigma)(\textbf{initial } \Phi' \textbf{ through } \theta) = \textbf{initial } \Phi' \textbf{ through } \theta;\sigma$.
- The functor $\mathbf{Mod}_{\mathbf{INS}^{init}}$ is just **Mod**, the same as in **INS**.
- For each signature $\Sigma \in |\mathbf{Sign}_{\mathbf{INS}^{init}}|$, the Σ-satisfaction relation $\models_{\mathbf{INS}^{init},\Sigma}$ is the same as the Σ-satisfaction relation in **INS** for the Σ-sentences from **INS**, and is given by Definition 4.3.3 for Σ-initiality constraints. □

Exercise 4.3.6. Present the above definition as an instance of the construction given in Example 4.1.46. Notice that this is sufficient to conclude that $\mathbf{INS}^{init}$ is indeed an institution.

Show (referring for example to Exercise 4.3.4) that in general the translation of an initiality constraint cannot be given without the "fitting morphism" component, and so we would not be able to define an institution where only initiality constraints with trivial (identity) fitting morphisms would be allowed. □

Exercise 4.3.7. Working in the institution **EQ**, follow Definition 4.3.3 and define *reachability constraints* that are satisfied only by algebras having an indicated reduct that is reachable. Note that axioms used in initiality constraints play no role here, so you can adopt a syntax like **reachable through** θ. Following Definition 4.3.5, define an institution $\mathbf{EQ}^{reach}$ extending **EQ** by reachability constraints.

Assuming that each category of models in **INS** comes equipped with a factorisation system (Section 3.3), introduce reachability constraints for **INS** using Definition 3.3.7 and extend **INS** correspondingly. □

The use of initiality constraints as introduced above is not always entirely satisfactory. Often, rather than requiring that a certain part of a model be initial, we want to require that it be a *free extension* of some other part. Natural examples arise when we want to specify data structures built on an arbitrary set of elements, like lists, sets or bags of arbitrary elements. This involves imposing the requirement that an algebra modelling the data structure is a free extension of its reduct to the sort of elements. To formalise this, the concept of a data constraint is introduced below.

Definition 4.3.8 (Data constraint). Let $\Sigma \in |\mathbf{Sign}|$ be a signature.

[20] As in Example 4.1.46, this may lead to some foundational difficulties which we disregard here; cf. footnote 17.

A Σ-*data constraint* is a triple $\langle\sigma,\Phi',\theta\rangle$, written as **data** Φ' **over** σ **through** θ, where $\sigma{:}\Sigma_1\to\Sigma'$ and $\theta{:}\Sigma'\to\Sigma$ are signature morphisms and $\Phi'\subseteq\mathbf{Sen}(\Sigma')$ is a set of Σ'-sentences.

A Σ-model $M\in|\mathbf{Mod}(\Sigma)|$ *satisfies* the data constraint **data** Φ' **over** σ **through** θ if its reduct $M|_\theta\in|\mathbf{Mod}(\Sigma')|$ to a Σ'-model is a free model of Φ' w.r.t. the reduct functor $\_|_\sigma{:}\mathbf{Mod}[\langle\Sigma,\Phi'\rangle]\to\mathbf{Mod}(\Sigma_1)$ over $(M|_\theta)|_\sigma$, with the identity as unit. That is, M satisfies **data** Φ' **over** σ **through** θ if:

- $M|_\theta\models_{\Sigma'}\Phi'$; and
- for any $M'\in Mod_{\Sigma'}(\Phi')$ and Σ_1-morphism $f{:}M|_{\sigma;\theta}\to M'|_\sigma$ there exists a unique Σ'-morphism $f^\#{:}M|_\theta\to M'$ such that $f^\#|_\sigma=f$. □

Exercise 4.3.9. Using data constraints, give a specification of finite bags of an arbitrary set of elements. □

Exercise 4.3.10. Following the pattern of Definition 4.3.5 (and of Example 4.1.46), define the institution $\mathbf{INS}^{data}$ by adding data constraints as additional sentences to **INS**. □

Note that nowhere in the above has it been assumed that initial models of presentations actually exist in general (nor that the reduct functor used in Definition 4.3.8 has a left adjoint). We do know that in some institutions (for example, in the institution **EQ** of equational logic and in the institution **PEQ** of partial equational logic) any set of sentences over a given signature has an initial model (see Theorem 2.5.14 for the case of **EQ**). On the other hand, there are institutions in which some sets of sentences do not have initial models; the institution **FOEQ** of first-order logic with equality is an example (see Example 2.7.11). Nevertheless, the above definitions work for an arbitrary institution. If a set $\Phi\subseteq\mathbf{Sen}(\Sigma)$ of Σ-sentences has no initial model, then an initiality constraint **initial** Φ **through** θ based on this set has no model, even if the class $Mod_\Sigma(\Phi)$ of models of this set of sentences is not empty.

Exercise 4.3.11. Any set of sentences in the equational institution **EQ** has a model, and moreover, it has an initial model. Show that neither of these properties carries over to the institution $\mathbf{EQ}^{init}$ of initiality constraints in **EQ**. That is, give a presentation in $\mathbf{EQ}^{init}$ that has no model. □

Exercise 4.3.12. Recall the institution **Horn** of Horn formulae from Exercise 4.1.21 and show that every set of sentences in **Horn** has an initial model. Discuss the interpretation of predicates in initial models: notice that they hold "minimally", meaning that only positive cases, where a predicate is required to hold, need to be explicitly specified. Extend this analysis to data constraints, and use this to specify the transitive and reflexive closure of an arbitrary binary predicate. □

Exercise 4.3.13. Working in the institution **EQ** as in Exercise 4.3.7, follow Definition 4.3.8 and define *generation constraints* **generated over** σ **through** θ that are satisfied by algebras A such that $A|_\theta$ is generated in a suitable sense by $A|_{\sigma;\theta}$. Define an institution $\mathbf{EQ}^{gen}$ extending **EQ** by generation constraints.

Assuming that each category of models in **INS** comes equipped with a factorisation system (Section 3.3), introduce generation constraints for **INS** anticipating Definition 4.5.1 and extend **INS** correspondingly. □

Exercise 4.3.14. Following Exercise 3.5.24, dualise the concept of data constraint. Given an institution **INS**, write **codata** Φ' **over** σ **through** θ, where Φ', σ and θ are as in Definition 4.3.8, for a *codata constraint* in **INS**. A Σ-model $M \in |\mathbf{Mod}(\Sigma)|$ *satisfies* **codata** Φ' **over** σ **through** θ if $M|_\theta$ is a cofree model of Φ' w.r.t. the reduct functor $\_|_\sigma : \mathbf{Mod}[\langle\Sigma',\Phi'\rangle] \to \mathbf{Mod}(\Sigma_1)$ over its σ-reduct, with the identity as counit, that is, if $M|_\theta \models_{\Sigma'} \Phi'$ and for any $M' \in Mod_{\Sigma'}(\Phi')$ and Σ_1-morphism $f: M'|_\sigma \to M|_{\sigma;\theta}$ there exists a unique Σ'-morphism $f^\#: M' \to M|_\theta$ such that $f^\#|_\sigma = f$. Extend this definition to build an institution $\mathbf{INS}^{codata}$ by adding codata constraints as additional sentences to **INS**.

Explore the use of codata constraints in **EQ** and **FOPEQ**. For instance, consider the following simple presentation:

$$
\begin{array}{lll}
\mathbf{spec}\ \text{STREAM} = & \mathbf{sorts} & Elem, Stream \\
& \mathbf{ops} & hd: Stream \to Elem \\
& & tl: Stream \to Stream \\
& & cons: Elem \times Stream \to Stream \\
& \multicolumn{2}{l}{\forall x: Elem, s: Stream} \\
& & \bullet\ hd(cons(x,s)) = x \\
& & \bullet\ tl(cons(x,s)) = s
\end{array}
$$

Check that any model M of STREAM that is cofree over $E = |M|_{Elem}$ (w.r.t. the reduct functor given by the obvious signature inclusion) is isomorphic to the algebra E^ω of (countably) infinite streams of elements from E, with the operations defined in the standard way.

Much the same effect is achieved even when we remove the operation *cons* and the two axioms from this presentation: check that if Σ is a signature with sorts *Elem*, *Stream* and operations $hd: Stream \to Elem$, $tl: Stream \to Stream$ then cofree Σ-models over their carrier E of sort *Elem* are (up to isomorphism) the algebras E^ω of (countably) infinite streams of elements from E, with *hd* and *tl* defined in the standard way. Check then that in any such algebra the two axioms in STREAM define the operation *cons* unambiguously. □

4.4 Exact institutions

As illustrated in Sections 4.2 and 4.3, institutions provide a sufficient basis for much of the standard machinery of specifications, without the need for further assumptions. Still, the structure and properties of a logical system exposed by the definition of an institution are very limited, and do not provide an adequate basis for many other aspects of the theory and practice of software specification and development. As discussed in the introduction to this chapter, this should not discourage us from working within the institutional framework. On the contrary, it is worth trying to find some adequately abstract additional assumptions that are sufficient for the purpose at hand. As always in mathematics, the main informal guideline to follow is to keep the additional assumptions to a minimum. Part of the payoff is that this forces us to work at a level of generality and abstraction that ensures a deeper understanding of the essence of the studied phenomena, while at the same time covering as many cases of potential interest as possible.

In this section and the next we will illustrate this strategy by presenting some extensions to the notion of an institution by additional structure or properties that are required to support study of more detailed properties of specifications.

The ways in which specifications (or programs, systems, or structures of any kind) are put together is the very essence of the theory and methodology of software specification and development. One of the basic tools for putting things together is the categorical notion of colimit (cf. Section 3.2), with pushouts as a particularly important special case; see for instance Section 6.3 below. Putting specifications together then involves taking colimits in the category of theories. It would be rather inconvenient to have to establish the existence of a colimit for each diagram of interest separately, so we normally require the category of theories to be cocomplete (or at least finitely cocomplete). Checking this directly would be tedious — and this is why the following general result is useful.

Theorem 4.4.1. *For any institution* **INS**, *if the category* $\mathbf{Sign}_{\mathbf{INS}}$ *of signatures in* **INS** *is cocomplete then so is the category* $\mathbf{Th}_{\mathbf{INS}}$ *of theories in* **INS**.

Proof. Let D be a diagram in $\mathbf{Th}_{\mathbf{INS}}$ with $|G(D)|_{Node} = N$ and $D_n = \langle \Sigma_n, \Phi_n \rangle$ for $n \in N$. Let D' be the corresponding diagram in $\mathbf{Sign}_{\mathbf{INS}}$; hence $D'_n = \Sigma_n$ for $n \in N$. By the assumption of the theorem, D' has a colimit, say $\langle \alpha_n : \Sigma_n \to \Sigma \rangle_{n \in N}$. Let $\Phi = Cl_\Sigma(\bigcup_{n \in N} \alpha_n(\Phi_n))$. Then for each $n \in N$, $\alpha_n : \langle \Sigma_n, \Phi_n \rangle \to \langle \Sigma, \Phi \rangle$ is a theory morphism (this is obvious) and $\langle \alpha_n \rangle_{n \in N}$ is a colimit of D in $\mathbf{Th}_{\mathbf{INS}}$: first notice that it is a cocone on D (since it is a cocone on D' in $\mathbf{Sign}_{\mathbf{INS}}$), and then consider another cocone on D, say $\langle \beta_n : \langle \Sigma_n, \Phi_n \rangle \to \langle \Sigma', \Phi' \rangle \rangle_{n \in N}$. By the construction, there exists a unique signature morphism $\sigma : \Sigma \to \Sigma'$ such that for each $n \in N$, $\alpha_n;\sigma = \beta_n$. To complete the proof, it is sufficient to show that $\sigma : \langle \Sigma, \Phi \rangle \to \langle \Sigma', \Phi' \rangle$ is a theory morphism. By Proposition 4.2.26, it is enough to show that $\sigma(\bigcup_{n \in N} \alpha_n(\Phi_n)) \subseteq \Phi'$. This easily follows from the fact that for each $n \in N$, β_n is a theory morphism, and hence $\sigma(\alpha_n(\Phi_n)) = (\alpha_n;\sigma)(\Phi_n) = \beta_n(\Phi_n) \subseteq \Phi'$. □

The above proof shows that in fact a stronger property holds: in any institution, the category of theories has all of the colimits that the category of signatures has: the forgetful functor mapping theories to their underlying signatures *lifts colimits*. So, for instance:

Corollary 4.4.2. *For any institution* **INS**, *if the category* $\mathbf{Sign}_{\mathbf{INS}}$ *of signatures in* **INS** *is finitely cocomplete then so is the category* $\mathbf{Th}_{\mathbf{INS}}$ *of theories in* **INS**. □

Notice that the above theorem applies to *any* institution, regardless of the means used to construct it. Hence, for example, if the category $\mathbf{Sign}_{\mathbf{INS}}$ of signatures in an institution **INS** is cocomplete, then not only is the category $\mathbf{Th}_{\mathbf{INS}}$ of theories in **INS** cocomplete, but so are the categories $\mathbf{Th}_{\mathbf{INS}^{init}}$, $\mathbf{Th}_{\mathbf{INS}^{data}}$ and $\mathbf{Th}_{\mathbf{INS}^{codata}}$ of theories in the corresponding institutions with initiality constraints, data constraints and codata constraints respectively (cf. Definition 4.3.5, Exercise 4.3.10 and Exercise 4.3.14).

Exercise 4.4.3. Assume that the category of signatures of a certain institution has an initial object. What is then an initial object in the category of theories? □

Example 4.4.4. Working in the institution **EQ** of equational logic, recall Example 3.2.36 of a simple pushout of algebraic signatures, and the set $\mathscr{E}_{\mathrm{NAT}}$ of equational axioms over the signature Σ_{NAT} given in Exercise 2.5.4. Let T_{NAT} be the Σ_{NAT}-theory presented by $\mathscr{E}_{\mathrm{NAT}}$. Let $T_{\mathrm{NAT}_{fib}}$ be the $\Sigma_{\mathrm{NAT}_{fib}}$-theory presented by the axioms $\mathscr{E}_{\mathrm{NAT}_{fib}}$ that include $\mathscr{E}_{\mathrm{NAT}}$ plus the following:

$$\begin{array}{l} fib(0) = succ(0) \\ fib(succ(0)) = succ(0) \\ \forall n{:}Nat \bullet fib(succ(succ(n))) = fib(succ(n)) + fib(n) \end{array}$$

Finally, let $T_{\mathrm{NAT}_{mult}}$ be the $\Sigma_{\mathrm{NAT}_{mult}}$-theory presented by the axioms $\mathscr{E}_{\mathrm{NAT}_{mult}}$ that include $\mathscr{E}_{\mathrm{NAT}}$ plus the following:

$$\begin{array}{l} \forall n{:}Nat \bullet mult(0,n) = 0 \\ \forall n,m{:}Nat \bullet mult(succ(n),m) = mult(n,m) + m \end{array}$$

Now, we have theory inclusions:

$$T_{\mathrm{NAT}_{fib}} \hookleftarrow T_{\mathrm{NAT}} \hookrightarrow T_{\mathrm{NAT}_{mult}}$$

with the corresponding signature inclusions given in Example 3.2.36. Their pushout is the $\Sigma_{\mathrm{NAT}_{fib,mult}}$-theory $T_{\mathrm{NAT}_{fib,mult}}$ presented by the union of $\mathscr{E}_{\mathrm{NAT}}$, $\mathscr{E}_{\mathrm{NAT}_{fib}}$ and $\mathscr{E}_{\mathrm{NAT}_{mult}}$.

As in Example 3.2.36, this is deceptively simple, as only single-sorted theory inclusions that introduce different operation names are involved.

Exercise. Give examples of pushouts in the category of equational theories with signatures involving more than one sort, extensions with overlapping sets of operation names, and theory morphisms that are not injective on sort and/or on operation names. Notice, however, that the extra complications come only from the construction of signature pushouts; the theories are defined in much the same way.

Exercise. Obviously, when giving the set of axioms for $T_{\mathrm{NAT}_{fib,mult}}$, $\mathscr{E}_{\mathrm{NAT}}$ may be omitted, as it is already included in the other sets of axioms. Try to generalise this remark to "optimise" the construction of the colimit in the category of theories given in the proof of Theorem 4.4.1. □

We have seen how the assumption that the category of signatures of an institution is (finitely) cocomplete ensures that the institution provides means for putting theories together. It is also interesting to investigate how this relates to putting models together, which is what structured programming in the large is all about. There is an important difference here: in the above, and in general when dealing with specifications, we were interested in combining theories (*sets* of sentences). In model-theoretic terms, this corresponds to combining classes of models. However, when the specified system is being built, we are interested in expanding and combining *individual* models.

Example 4.4.5. Recall Example 4.4.4 of a simple pushout in the category of theories of the institution **EQ** of equational logic. Consider an arbitrary model N of T_{NAT}, any $\Sigma_{\mathrm{NAT}_{mult}}$-algebra N_2 built by adding to N an interpretation of *fib* such that the axioms in $\mathscr{E}_{\mathrm{NAT}_{fib}}$ are satisfied, and any $\Sigma_{\mathrm{NAT}_{mult}}$-algebra N_2 built by adding to N an interpretation of *mult* such that the axioms in $\mathscr{E}_{\mathrm{NAT}_{mult}}$ are satisfied. Then, much as in Example 3.4.35 where specific such algebras were considered, N_1 and N_2 may be uniquely combined to a $\Sigma_{\mathrm{NAT}_{fib,mult}}$-algebra N' that expands them both. The key property now is that the algebras built in this way are models of the theory $T_{\mathrm{NAT}_{fib,mult}}$, and moreover, that all its models may be built in this way. □

It turns out that the crucial link which ensures that constructions to combine theories and to combine models work together smoothly, as in the above example, is the continuity of the model functor in the underlying institution.

Definition 4.4.6 (Exact institution). An institution **INS** is (*finitely*) *exact* when its category of signatures $\mathbf{Sign}_{\mathbf{INS}}$ is (finitely) cocomplete and its model functor $\mathbf{Mod}_{\mathbf{INS}} \colon \mathbf{Sign}_{\mathbf{INS}}^{op} \to \mathbf{Cat}$ is (finitely) continuous, mapping (finite) colimits in $\mathbf{Sign}_{\mathbf{INS}}$ to limits in $\mathbf{Cat}$. □

Example 4.4.7. All of the institutions defined in the examples and sketched in the exercises in Section 4.1.1, with the major exception of **FPL** (Example 4.1.25) and perhaps those given in Examples 4.1.35, 4.1.36 and 4.1.37, where we know nothing about the signature categories, are exact. See Exercises 3.2.53 and 3.4.33 for the standard algebraic case of the equational institution **EQ** — all of the other cases require a similar argument. □

Exercise 4.4.8. The abstract formulation of exactness above may somewhat hide the role of this property in putting models together. Consider an exact institution **INS** and a diagram D in $\mathbf{Sign}_{\mathbf{INS}}$ with colimit signature Σ'. Anticipating the crucial case of preservation of signature pushouts treated in Definition 4.4.12, show that (up to isomorphism of categories) $\mathbf{Mod}_{\mathbf{INS}}(\Sigma')$ can be defined as follows, where N is the set of nodes in D:

- Σ'-models are families $\langle M_n \in |\mathbf{Mod}_{\mathbf{INS}}(D_n)| \rangle_{n \in N}$ that are compatible with signature morphisms in D in the sense that $M_n = M_m|_{D_e}$ for each edge $e \colon n \to m$ in the graph of D; and
- Σ'-morphisms between any such Σ'-models $\langle M_n \rangle_{n \in N}$ and $\langle M'_n \rangle_{n \in N}$ are families $\langle h_n \colon M_n \to M'_n \rangle_{n \in N}$ of morphisms in $\mathbf{Mod}_{\mathbf{INS}}(D_n)$, $n \in N$, that are compatible with signature morphisms in D in the sense that $h_n = h_m|_{D_e}$ for each edge $e \colon n \to m$ in the graph of D.

Moreover, for each $n \in N$, the reduct functor w.r.t. the colimit injection from D_n to Σ' is just the projection of such families on the nth component.

HINT: Use Exercise 3.4.32 (and indirectly Exercise 3.2.53). □

Exercise 4.4.9. Consider a finitely exact institution. Present initiality constraints (Definition 4.3.3) as a special case of data constraints (Definition 4.3.8). Is the assumption that the institution is finitely exact essential? □

Exercise 4.4.10. An interesting standard institution with a cocomplete category of signatures and a model functor that preserves "nearly all" finite colimits of signatures is the institution **SSEQ** of single-sorted equational logic. Give a precise definition of this institution and indicate which colimits of signature diagrams are not preserved by the model functor. HINT: Consider the initial single-sorted signature. □

Definition 4.4.11 (Semi-exact institution). An institution **INS** is *semi-exact* if all pushouts exist in its category of signatures $\mathbf{Sign_{INS}}$ and its model functor $\mathbf{Mod_{INS}}: \mathbf{Sign}^{op}_{\mathbf{INS}} \to \mathbf{Cat}$ preserves pushouts, mapping them to pullbacks in **Cat**. □

A consequence of the assumption that the model functor of an institution preserves signature pushouts is the well-known *Amalgamation Lemma*.

Definition 4.4.12 (Amalgamation). Let $\mathbf{INS} = \langle \mathbf{Sign}, \mathbf{Sen}, \mathbf{Mod}, \langle \models_\Sigma \rangle_{\Sigma \in |\mathbf{Sign}|} \rangle$ be an institution and consider the following diagram in **Sign**:

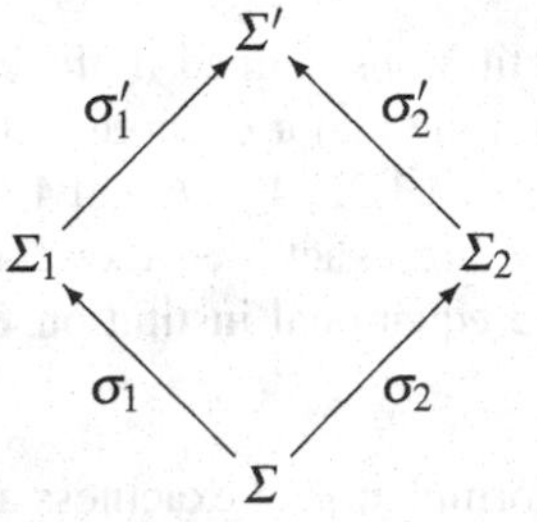

This diagram *admits amalgamation* if:

- for any two models $M_1 \in |\mathbf{Mod}(\Sigma_1)|$ and $M_2 \in |\mathbf{Mod}(\Sigma_2)|$ such that $M_1|_{\sigma_1} = M_2|_{\sigma_2}$, there exists a unique model $M' \in |\mathbf{Mod}(\Sigma')|$ such that $M'|_{\sigma'_1} = M_1$ and $M'|_{\sigma'_2} = M_2$ (we call such M' the *amalgamation* of M_1 and M_2); and
- for any two model morphisms $f_1: M_{11} \to M_{12}$ in $\mathbf{Mod}(\Sigma_1)$ and $f_2: M_{21} \to M_{22}$ in $\mathbf{Mod}(\Sigma_2)$ such that $f_1|_{\sigma_1} = f_2|_{\sigma_2}$, there exists a unique model morphism $f': M'_1 \to M'_2$ in $\mathbf{Mod}(\Sigma')$ such that $f'|_{\sigma'_1} = f_1$ and $f'|_{\sigma'_2} = f_2$ (we call such f' the *amalgamation* of f_1 and f_2).

The institution **INS** *has the amalgamation property* if all pushouts in **Sign** exist and every pushout diagram in **Sign** admits amalgamation. □

Exercise 4.4.13. Show that if a diagram as in Definition 4.4.12 admits amalgamation and is commutative then all models and morphisms in $\mathbf{Mod}(\Sigma')$ are amalgamations of pairs of (compatible) models and morphisms from $\mathbf{Mod}(\Sigma_1)$ and $\mathbf{Mod}(\Sigma_2)$, respectively. □

Lemma 4.4.14 (Amalgamation Lemma). *Any semi-exact institution has the amalgamation property.* □

The proof of the Amalgamation Lemma is based on the construction of pullbacks in **Cat**; cf. Exercise 3.4.32. See also Exercise 3.4.34, which is the same result in the standard algebraic framework. Note that the opposite implication also holds, so semi-exactness is equivalent to the amalgamation property.

Clearly, every exact institution is finitely exact, and every finitely exact institution is semi-exact. However, the last property is strictly weaker: for example, the institution **SSEQ** of single-sorted equational logic is semi-exact, but not finitely exact (see Exercise 4.4.10). In semi-exact institutions coproducts of signatures need not exist, or if they exist, need not be preserved by the model functor. However, if signature coproducts exist, the colimits for a large interesting class of signature diagrams (exist and) are preserved:

Proposition 4.4.15. *In any semi-exact institution, if the category of signatures has an initial object then it is finitely cocomplete and the model functor maps colimits of all finite non-empty connected diagrams of signatures to limits in* **Cat**.

Proof sketch. The first part (existence of colimits of finite signature diagrams) follows as usual, by dualising Exercise 3.2.48; the second part (preservation of limits of finite non-empty connected signature diagrams) follows by Exercise 3.4.56. □

Exercise 4.4.16. Define institutions: **SSFOPEQ** of single-sorted first-order predicate logic with equality, **SSPFOPEQ** of single-sorted partial first-order predicate logic with equality, **SSCEQ** of single-sorted equational logic for continuous algebras, and so on. Check that all of these institutions have cocomplete categories of signatures and are semi-exact. However, check that their model functors do not map coproducts of their signatures to products of the corresponding model categories, so these institutions are not (finitely) exact. □

Exercise 4.4.17. Let **INS** be a (finitely) exact institution. Recall that there is a functor $\mathbf{Mod_{Th}}\colon \mathbf{Th}_{\mathbf{INS}}^{op} \to \mathbf{Cat}$ mapping theories to their model categories and theory morphisms to the corresponding reduct functors (cf. Exercise 4.2.28). Prove that $\mathbf{Mod_{Th}}$ preserves (finite) limits.

HINT: First use the satisfaction condition for **INS** and the Amalgamation Lemma for signatures (Lemma 4.4.14) to prove the following generalisation of the Amalgamation Lemma:

Lemma (Amalgamation Lemma for theories). *Let* **INS** *be a semi-exact institution. Consider a pushout in the category* $\mathbf{Th}_{\mathbf{INS}}$ *of theories:*

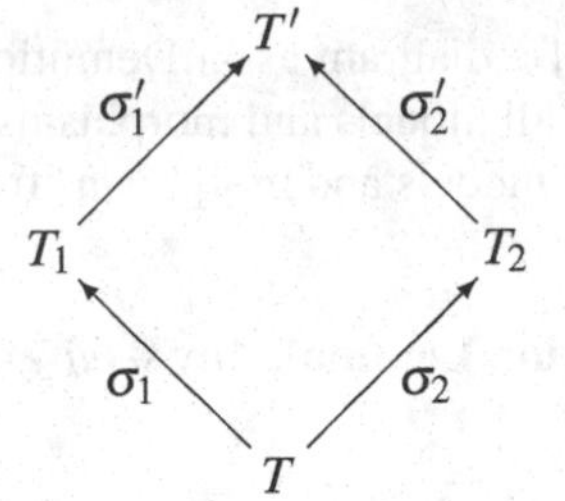

Then, for any two models $M_1 \in Mod[T_1]$ and $M_2 \in Mod[T_2]$ such that $M_1|_{\sigma_1} = M_2|_{\sigma_2}$, there exists a unique model $M' \in Mod[T']$ such that $M'|_{\sigma_1'} = M_1$ and $M'|_{\sigma_2'} = M_2$, and similarly for morphisms.

To complete the proof that $\mathbf{Mod_{Th}}$ is finitely continuous, by Exercise 3.2.48 it is enough to consider the initial theory and its category of models. To show that it is continuous, by Exercise 3.4.23 it is enough to consider coproducts of arbitrary families of theories and their categories of models. □

The trouble with **FPL** and with other institutions based on derived signature morphisms (see Exercise 4.1.23) is more severe than with single-sorted institutions: they are not semi-exact since not all pushouts exist in their signature categories; see Exercise 3.2.54. This motivates the following relaxation of semi-exactness, which is important for applications later on.

Definition 4.4.18 (I-semi-exact institution). For any institution **INS**, we say that a collection **I** of signature morphisms in **INS** is *closed under pushouts* if **I** contains all the identities, is closed under composition (so that **I** is a wide subcategory of $\mathbf{Sign_{INS}}$) and, for any signature morphism $\sigma\colon \Sigma \to \Sigma_1$ and "**I**-extension of Σ" $\iota\colon \Sigma \to \Sigma'$ in **I**, there is a pushout in **Sign**

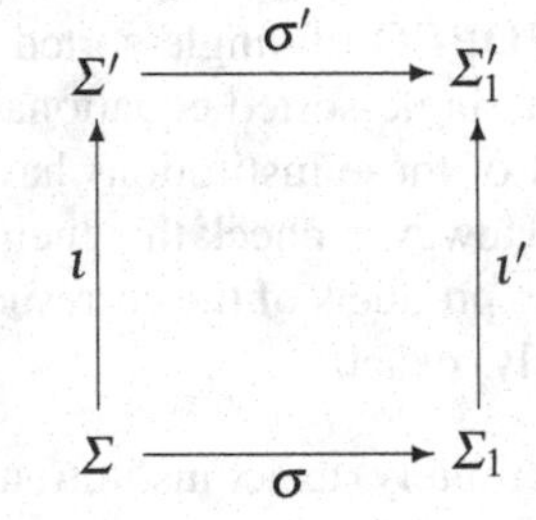

such that $\iota' \in \mathbf{I}$.

Moreover, if all such pushouts with $\iota, \iota' \in \mathbf{I}$ admit amalgamation (i.e. the model functor maps them to pullbacks in **Cat**) we say that **INS** is *semi-exact w.r.t.* **I** (or **I**-*semi-exact*). □

Exercise 4.4.19. As mentioned above, institutions with derived signature morphisms do not have cocomplete signature categories. Check, however, that for example the institution $\mathbf{GEQ}^{der}$ is semi-exact w.r.t. the class of all inclusions (where inclusions are derived signature morphisms that map any n-ary operation name f to the term $f(\boxed{1},\ldots,\boxed{n})$; cf. Definition 1.5.14). Similarly, check that $\mathbf{GEQ}^{der}$ is semi-exact w.r.t. the class of inclusions that introduce only new constants. (Notice that in general an institution may be **I**-semi-exact without being **I**$'$-semi-exact for some $\mathbf{I}' \subseteq \mathbf{I}$.)

For **FPL**, consider the class $\mathbf{I_{FPL}}$ of signature morphisms $\delta: \mathsf{SIG} \to \mathsf{SIG}'$ that are injective renamings of sort and operation names such that no new value constructors are added for "old" sorts (i.e. sorts in $\delta(\mathsf{SIG})$). Show that **FPL** is $\mathbf{I_{FPL}}$-semi-exact. Notice that both parts of the assumption on these morphisms are essential. Give an example of a non-injective renaming that does not have a pushout with another **FPL**-signature morphism. Give an example of an injective renaming that adds value constructors for an old sort and does not have a pushout with another **FPL**-signature morphism. Finally, give an example of a pushout in the category of **FPL**-signatures that is not mapped by the **FPL**-model functor to a pullback in **Cat**. HINT: Consider two morphisms that add a new sort and a new unary value constructor for a previously unconstrained sort, with the new sort as its argument sort. □

Exercise 4.4.20. To complete the formal picture, note that the category of theories in **FPL** is cocomplete even though its category of signatures is not. Discuss why this is not useful for combining models over different signatures. HINT: Consider a simple signature with one sort and one binary operation, and two morphisms which map this operation to the projections on the first and second argument respectively. Then these two morphisms do not have a coequaliser in $\mathbf{Sign_{FPL}}$ while in $\mathbf{Th_{FPL}}$ their coequaliser is obtained by adding an equation to assert that the two projections coincide. □

We have introduced and studied amalgamation, exactness and semi-exactness as purely technical properties of institutions. However, as hinted at by Example 4.4.5 and the examples it builds on, amalgamation, and hence semi-exactness and exactness, provide a fundamental tool for combining models over different signatures. The point is easiest to see in institutions with standard signatures, like **FOPEQ** or **EQ**, when all the morphisms are inclusions. In that case, generalising the simple example of natural numbers and their extensions by the Fibonacci function and multiplication in Example 3.2.36, given signatures Σ_1 and Σ_2 with $\Sigma = \Sigma_1 \cap \Sigma_2$, we get $\Sigma' = \Sigma_1 \cup \Sigma_2$ as the pushout signature. Now, the amalgamation property ensures that, given a Σ_1-model M_1 and a Σ_2-model M_2 which give the same interpretation to all of the common symbols (in Σ), we can put them together in the obvious way (generalising Example 4.4.5) to interpret all of the symbols in the combined signature Σ'. In the institutional context, this intuition applies as well, but the sharing requirement is expressed by insisting on a common reduct along the indicated signature morphisms, and the combined signature is obtained using the pushout.

4.4.1 Abstract model theory

One of the ideas behind the definition of institution is that it is important to indicate over which signature one is working. In classical logic, there are a number of theorems in which the signature (or *language*, as logicians would say) over which formulae are constructed must be considered. Here is an example (for this, and for a classical formulation of the Robinson consistency theorem mentioned below, see [CK90]):

Theorem (Craig interpolation theorem). *In first-order logic, for any two formulae φ_1 and φ_2, if $\varphi_1 \models \varphi_2$ then there exists a formula θ using only the common symbols of φ_1 and φ_2 — that is, those symbols that occur in both formulae — such that $\varphi_1 \models \theta$ and $\theta \models \varphi_2$.* □

In our view, this standard formulation is not very elegant: referring to "the common symbols of φ_1 and φ_2" feels rather clumsy, even though it is easy enough to make it precise in the case of first-order logic. In the institutional framework this can be expressed in a more general and abstract way using colimits in the category of signatures.

Definition 4.4.21 (Craig interpolation property). Let **INS** be an institution with a finitely cocomplete category **Sign** of signatures. **INS** satisfies the *Craig interpolation property* for the following commutative diagram in **Sign**

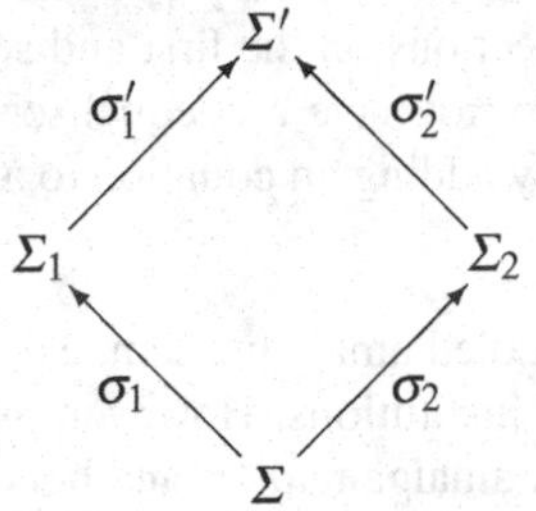

if for any Σ_1-sentence $\varphi_1 \in \mathbf{Sen}(\Sigma_1)$ and Σ_2-sentence $\varphi_2 \in \mathbf{Sen}(\Sigma_2)$, if $\sigma_1'(\varphi_1) \models_{\Sigma'} \sigma_2'(\varphi_2)$ then there exists a Σ-sentence $\theta \in \mathbf{Sen}(\Sigma)$ (called an *interpolant* for φ_1 and φ_2) such that $\varphi_1 \models_{\Sigma_1} \sigma_1(\theta)$ and $\sigma_2(\theta) \models_{\Sigma_2} \varphi_2$. **INS** satisfies the *Craig interpolation property* if it satisfies this property for all pushouts in **Sign**. □

Not only has "the common symbols of φ_1 and φ_2" been captured by the simple categorical concept of a pushout here, but we have also been forced to identify the signatures over which the individual consequence relations are considered. In our view, this is a much improved statement of the Craig interpolation property! Not only does it seem clearer — of course, any comparison should be made with a fully formal statement of the Craig interpolation theorem in the classical framework, not with the presentation given above — it is also more abstract and may be used for any logical system formalised as an institution, not just for first-order logic.

Here is another example, which states that consistent extensions of a complete theory (cf. Definition 4.2.29) combine safely:

Definition 4.4.22 (Robinson consistency property). Let **INS** be an institution with a finitely cocomplete category **Sign** of signatures. **INS** satisfies the *Robinson consistency property* for the following commutative diagram in **Sign**

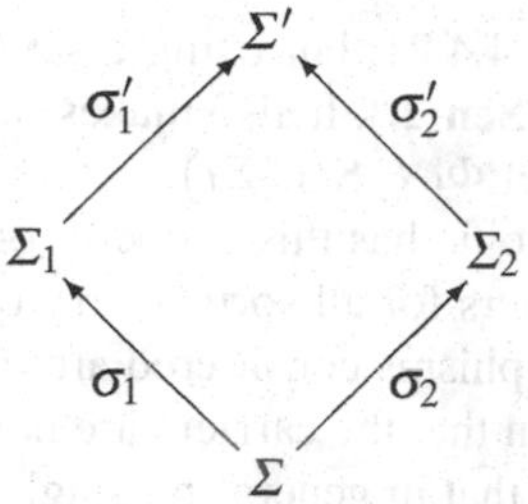

if for any complete Σ-theory $T = \langle \Sigma, \Phi \rangle$ and consistent theories $T_1 = \langle \Sigma_1, \Phi_1 \rangle$ and $T_2 = \langle \Sigma_2, \Phi_2 \rangle$ such that $\sigma_1 : T \to T_1$ and $\sigma_2 : T \to T_2$ are theory morphisms, the Σ'-presentation $\langle \Sigma', \sigma_1'(\Phi_1) \cup \sigma_2'(\Phi_2) \rangle$ is consistent. **INS** satisfies the *Robinson consistency property* if it satisfies this property for all pushouts in **Sign**. □

Exercise 4.4.23. Adapt any standard proof of the Craig interpolation theorem to show that **FOPEQ** has the Craig interpolation property for those pushouts where at least one of σ_1 or σ_2 is injective on sorts. Construct a counterexample which shows that the proof must break down if neither σ_1 nor σ_2 is injective on sort names (injectivity on operation and predicate names does not have to be required). HINT: See [Bor05].

Show also that the Craig interpolation theorem for **FOPEQ** implies the analogous result for some of the subinstitutions of **FOPEQ** (see Exercise 4.1.13), for instance for **FOEQ**. Note though that your argument will not work for **FOP**, first-order predicate logic without equality — in fact, Craig interpolation may fail in **FOP** when one of the morphisms involved is non-injective on operation names, even if all the morphisms are injective on sort names. Of course, the standard proofs of Craig interpolation easily adapt to **FOP** when the morphisms involved are injective (on sort names as well as on operation names). □

It is well known that equational logic does not have the interpolation property:

Counterexample 4.4.24. In **EQ**, consider the signature Σ with three sorts s, s_1 and s_2, and two constants $a, b{:}s$. Let Σ_1 and Σ_2 extend Σ by a constant $e{:}s_1$ and by a unary operation $f{:}s_1 \to s_2$ respectively. Let Σ' be the union of Σ_1 and Σ_2 (this is the pushout signature for the two signature inclusions). Consider the sentences $\forall x{:}s_2 \bullet a = b \in \mathbf{Sen}_{\mathbf{EQ}}(\Sigma_1)$ and $a = b \in \mathbf{Sen}_{\mathbf{EQ}}(\Sigma_2)$. Clearly, over Σ' we have $\forall x{:}s_2 \bullet a = b \models a = b$ (since all Σ'-algebras have non-empty carriers for all sorts).

Suppose that we have an interpolant $\theta \in \mathbf{Sen}_{\mathbf{EQ}}(\Sigma)$ for $\forall x{:}s_2 \bullet a = b$ and $a = b$, so that $\forall x{:}s_2 \bullet a = b \models \theta$ over Σ_1 and $\theta \models a = b$ over Σ_2. Consider a Σ_1-algebra A_1 with the carrier of sort s_2 empty and with $a_{A_1} \neq b_{A_1}$. Clearly, $A_1 \models_{\Sigma_1} \forall x{:}s_2 \bullet a = b$, and so also $A_1 \models_{\Sigma_1} \theta$. Hence, $A_1|_\Sigma \models_\Sigma \theta$. Take a subalgebra of $A_1|_\Sigma$ with the empty carrier of sort s_1, which satisfies θ, and consider its expansion A_2 to a Σ_2-algebra. Then $A_2 \models_{\Sigma_2} \theta$ but $A_2 \not\models_{\Sigma_2} a = b$. Contradiction. □

Exercise 4.4.25. It is often stated that equational logic has interpolation (at least for pushouts w.r.t. injective signature morphisms) if one admits a *set of interpolants*, rather than just a single interpolant sentence θ as in Definition 4.4.21. Spell out this property following Definition 4.4.21, but using a set of sentences $\Theta \subseteq \mathbf{Sen}(\Sigma)$ in place of a single sentence $\theta \in \mathbf{Sen}(\Sigma)$. It also makes sense then to replace the single sentence $\varphi_1 \in \mathbf{Sen}(\Sigma_1)$ by a set $\Phi_1 \subseteq \mathbf{Sen}(\Sigma_1)$.

Unfortunately, equational logic has this property only if we restrict attention to algebras with non-empty carriers for all sorts. Carry out the proof for this case assuming that the signature morphisms considered are injective (HINT: see [Rod91]) and note where the assumption that the carriers are non-empty is important. Give a counterexample which shows that in general no single interpolant can be sufficient here. Extend this proof to the case where only one of the signature morphisms (σ_2 in Definition 4.4.21) is injective, and give a counterexample to show that the injectivity requirement cannot be dropped. HINT: See [RG00], [PŞR09].

Check that Counterexample 4.4.24 shows that the institution **EQ** of equational logic (with models that admit empty carriers) does not have the interpolation property, not even when sets of interpolants are allowed (and the morphisms involved are signature inclusions).

Go through other examples of institutions in Section 4.1.1 and check which of them have the interpolation property, either with a single interpolant, or with a set of interpolants (at least for pushouts involving signature inclusions, where this notion makes sense). □

Of course, we cannot expect to be able to prove that either the Craig interpolation or Robinson consistency properties are satisfied by an arbitrary institution — they simply do not hold for some logics. However, one may attempt to identify other conditions on the underlying institution which imply the two properties. Along these lines, under some further technical assumptions, the two properties are equivalent: an institution satisfying certain technical assumptions satisfies the Craig interpolation property if and only if it satisfies the Robinson consistency property.

Exercise 4.4.26. Assume that an institution **INS** has falsity, implication and appropriate conjunction (in general, infinitary conjunction may be needed, but if we assume the institution to be compact — see Definition 4.2.7 — then binary conjunction suffices). Show that **INS** satisfies the Craig interpolation property for a given pushout in **Sign** if and only if **INS** satisfies the Robinson consistency property for that pushout. □

This reflects what is well known in classical model theory for first-order logic, where the Craig interpolation and Robinson consistency properties are indeed derivable from one another. However, for institutions in which some of the required connectives are not available, these properties need not be equivalent. Different versions of these properties, such as Craig interpolation with sets of interpolants as considered in Exercise 4.4.25, may bring the two concepts closer again:

Exercise 4.4.27. Given an institution **INS** with a finitely cocomplete category **Sign** of signatures, consider the following version of Craig interpolation with additional "parameters". **INS** satisfies the *Craig-Robinson interpolation property* for the following commutative diagram in **Sign**

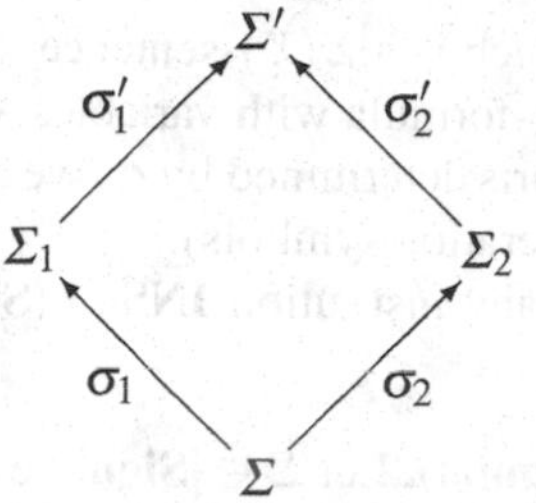

if for any $\Phi_1 \subseteq \mathbf{Sen}(\Sigma_1)$, $\Phi_2 \subseteq \mathbf{Sen}(\Sigma_2)$, $\varphi_2 \in \mathbf{Sen}(\Sigma_2)$ such that $\sigma_1'(\Phi_1) \cup \sigma_2'(\Phi_2) \models \sigma_2'(\varphi_2)$, there exists a set $\Theta \subseteq \mathbf{Sen}(\Sigma)$ such that $\Phi_1 \models \sigma_1(\theta)$ for all $\theta \in \Theta$ and $\sigma_2(\Theta) \cup \Phi_2 \models \varphi_2$.

To warm up, show that Craig-Robinson interpolation implies the Craig interpolation property with sets of interpolants. Show also that if **INS** has falsity then Craig-Robinson interpolation implies the Robinson consistency property. Finally, show that if **INS** has falsity, implication and appropriate conjunction (as in Exercise 4.4.26) then all three of these properties are equivalent.

Adapt your proof from Exercise 4.4.25 to show that equational logic has the Craig-Robinson interpolation property if we restrict attention to algebras with non-empty carriers and assume that σ_2 in the pushout diagram above is injective. □

4.4.2 Free variables and quantification

In logic, formulae may contain free variables; such formulae are called *open*, as opposed to *closed* formulae, which have no free variables. To interpret an open formula, one needs not only an interpretation for the symbols of the underlying signature (a model) but also an interpretation for the free variables (a valuation of variables in the model). This provides a natural way to deal with quantifiers. The need for open formulae also arises in the study of specification languages. In fact, we will use them to abstractly express the basic notion of behavioural equivalence in Section 8.5.3; see Exercise 8.5.61.

Fortunately we do not have to change the notion of an institution to cope with free variables — we can provide open formulae in the present framework. Note that we use here the term "formula" rather than "sentence", which is reserved for the sentences of the underlying institution, corresponding to closed formulae.

Consider the institution **GEQ** of ground equational logic (Example 4.1.3). Let $\Sigma = \langle S, \Omega \rangle$ be an algebraic signature. For any S-indexed family of sets, $X = \langle X_s \rangle_{s \in S}$, define $\Sigma(X)$ to be the extension of Σ by the elements of X as new constants of the appropriate sorts. Any sentence over $\Sigma(X)$ may be viewed as an open formula over Σ with free variables X. Given a Σ-algebra A, to determine whether an open Σ-formula with variables X holds in A we have to first fix a valuation of variables X in $|A|$. Such a valuation corresponds exactly to an expansion of A to a $\Sigma(X)$-algebra.

Given a translation of sentences along an algebraic signature morphism $\sigma : \Sigma \to \Sigma'$, we can extend it to a translation of open formulae: we translate an open Σ-

formula with variables X, which is a $\Sigma(X)$-sentence, to the corresponding $\Sigma'(X')$-sentence, which is an open Σ'-formula with variables X'. Here X' results from X by an appropriate renaming of sorts determined by σ (we also have to avoid unintended "clashes" of variables and operation symbols).

The above generalises to any institution $\mathbf{INS} = \langle \mathbf{Sign}, \mathbf{Sen}, \mathbf{Mod}, \langle \models_\Sigma \rangle_{\Sigma \in |\mathbf{Sign}|} \rangle$ that is semi-exact.

Definition 4.4.28 (Open formula). Let $\Sigma \in |\mathbf{Sign}|$ be a signature in **INS**. Any pair $\langle \varphi, \theta \rangle$, where $\theta : \Sigma \to \Sigma'$ is a signature morphism and $\varphi \in \mathbf{Sen}(\Sigma')$, is an *open Σ-formula* with variables "$\Sigma' \setminus \theta(\Sigma)$". For any Σ-model $M \in |\mathbf{Mod}(\Sigma)|$, a *valuation* of variables "$\Sigma' \setminus \theta(\Sigma)$" in M is a Σ'-model $M' \in |\mathbf{Mod}(\Sigma')|$ which is a θ-expansion of M, i.e. $M'|_\theta = M$. We say that *$\langle \varphi, \theta \rangle$ holds in M under valuation M'* iff $M' \models_{\Sigma'} \varphi$. If $\sigma : \Sigma \to \Sigma_1$ is a signature morphism then we define the translation of $\langle \varphi, \theta \rangle$ along σ as $\langle \sigma'(\varphi), \theta' \rangle$, where

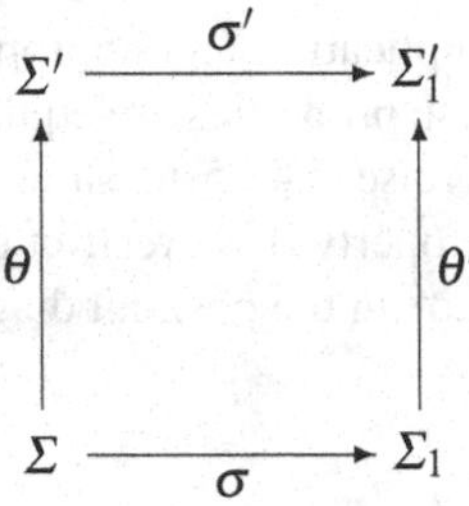

is a pushout in **Sign**. □

Note the quotation marks around the "set of variables" $\Sigma' \setminus \theta(\Sigma)$ in the above definition: since $\Sigma' \setminus \theta(\Sigma)$ makes no sense in an arbitrary institution, it is only meaningful as an aid to our intuition.

In the standard logical framework there may be no valuation of a set of variables in a model containing an empty carrier. Similarly here, a valuation need not always exist. For example, in **GEQ** if a signature morphism $\theta : \Sigma \to \Sigma'$ is not injective then some Σ-models have no θ-expansion.

There is a rather subtle problem with the above definition: pushouts are defined only up to isomorphism, so strictly speaking the translation of open formulae is not well defined. The following exercise shows that (at least for semantic analysis) an arbitrary pushout may be selected and so we may safely accept the above definition of translation.

Exercise 4.4.29. Consider an isomorphism $\iota : \Sigma'_1 \to \Sigma''_1$ in **Sign**, with inverse ι^{-1}. Since functors preserve isomorphisms, $\mathbf{Sen}(\iota) : \mathbf{Sen}(\Sigma'_1) \to \mathbf{Sen}(\Sigma''_1)$ is a bijection and $\mathbf{Mod}(\iota) : \mathbf{Mod}(\Sigma''_1) \to \mathbf{Mod}(\Sigma'_1)$ is an isomorphism in **Cat**. Show that moreover, for any $\psi \in \mathbf{Sen}(\Sigma'_1)$ and $M'_1 \in |\mathbf{Mod}(\Sigma'_1)|$, $M'_1 \models_{\Sigma'_1} \psi \iff M'_1|_{\iota^{-1}} \models_{\Sigma''_1} \iota(\psi)$. □

Sometimes we want to restrict the class of signature morphisms that may be used to construct open formulae. In fact, in the above remarks sketching how free variables may be introduced into **GEQ** we used just those algebraic signature inclusions $\iota : \Sigma \hookrightarrow \Sigma'$ where the only new symbols in Σ' were constants. To guarantee that the translation of open formulae is defined under such a restriction, we consider only restrictions to a collection **I** of signature morphisms that is closed under pushouts (see Definition 4.4.18).

Examples of such collections **I** in **AlgSig** include the collection of all algebraic signature inclusions, the restriction of this to inclusions $\theta: \Sigma \hookrightarrow \Sigma'$ such that Σ' contains no new sorts, the further restriction of this by the requirement that Σ' contain new constants only (as above), the collection of all algebraic signature morphisms which are surjective on sorts, the collection of all identities, and the collection of all morphisms. Note that most of these permit variables denoting operations or even sorts.

4.4.2.1 Universal quantification

In the rest of this section we briefly sketch how to universally close the open formulae introduced above.

Let **I** be a collection of signature morphisms that is closed under pushouts. Let Σ be a signature and let $\langle \varphi, \theta \rangle$ be an open Σ-formula such that $\theta \in \mathbf{I}$. Consider the universal closure of $\langle \varphi, \theta \rangle$, written $\forall \theta \bullet \varphi$, as a new Σ-sentence. The satisfaction relation and the translation of a sentence $\forall \theta \bullet \varphi$ along a signature morphism are defined in the expected way:

- A Σ-model satisfies the Σ-sentence $\forall \theta \bullet \varphi$ if $\langle \varphi, \theta \rangle$ holds in this model under any valuation of the variables "$\Sigma' \setminus \theta(\Sigma)$", that is, for any $M \in |\mathbf{Mod}(\Sigma)|$, $M \models_\Sigma \forall \theta \bullet \varphi$ if for all $M' \in |\mathbf{Mod}(\Sigma')|$ such that $M'|_\theta = M$, $M' \models_{\Sigma'} \varphi$.
- For any signature morphism $\sigma: \Sigma \to \Sigma_1$, $\sigma(\forall \theta \bullet \varphi)$ is $\forall \theta' \bullet \sigma'(\varphi)$, where

$$\begin{array}{ccc} \Sigma' & \xrightarrow{\sigma'} & \Sigma'_1 \\ \uparrow \theta & & \uparrow \theta' \\ \Sigma & \xrightarrow[\sigma]{} & \Sigma_1 \end{array}$$

  is a pushout in **Sign** such that $\theta' \in \mathbf{I}$.

Note that in the above we have extended our underlying institution **INS**. Formally:

Definition 4.4.30 (Institution with universally closed formulae). Let **INS** be an institution, and let **I** be a collection of signature morphisms in **INS** that is closed under pushouts such that **INS** is **I**-semi-exact. The *extension of* **INS** *by universal closure w.r.t.* **I** is the following institution $\mathbf{INS}^{\forall(\mathbf{I})}$:

- $\mathbf{Sign}_{\mathbf{INS}^{\forall(\mathbf{I})}}$ is $\mathbf{Sign}_{\mathbf{INS}}$.

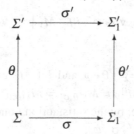

- For any signature Σ, $\mathbf{Sen}_{\mathbf{INS}^{\forall(\mathbf{I})}}(\Sigma)$ is the disjoint union of $\mathbf{Sen}_{\mathbf{INS}}(\Sigma)$ with the collection[21] of all universal closures $\forall\theta\bullet\varphi$ of open Σ-formulae, where $\theta\in\mathbf{I}$; for any signature morphism $\sigma:\Sigma\to\Sigma_1$, $\mathbf{Sen}_{\mathbf{INS}^{\forall(\mathbf{I})}}(\sigma)$ is the function induced by $\mathbf{Sen}_{\mathbf{INS}}(\sigma)$ on $\mathbf{Sen}_{\mathbf{INS}}(\Sigma)$ and by the notion of translation defined above on universally closed open Σ-formulae.
- $\mathbf{Mod}_{\mathbf{INS}^{\forall(\mathbf{I})}}$ is $\mathbf{Mod}_{\mathbf{INS}}$.
- The satisfaction relation in $\mathbf{INS}^{\forall(\mathbf{I})}$ is induced by the satisfaction relation of **INS** for **INS**-sentences and the notion of satisfaction for universally closed open formulae as defined above. □

The following theorem guarantees that $\mathbf{INS}^{\forall(\mathbf{I})}$ is in fact an institution, modulo the above remark about the definition of the translation of open formulae.

Theorem 4.4.31 (Satisfaction condition for $\mathbf{INS}^{\forall(\mathbf{I})}$). *Let* **INS** *and* **I** *be as in Definition 4.4.30. For any signature morphism $\sigma:\Sigma\to\Sigma_1$, open Σ-formula $\langle\varphi,\theta\rangle$ (where $\theta\in\mathbf{I}$), Σ_1-model $M_1\in|\mathbf{Mod}(\Sigma_1)|$, and pushout*

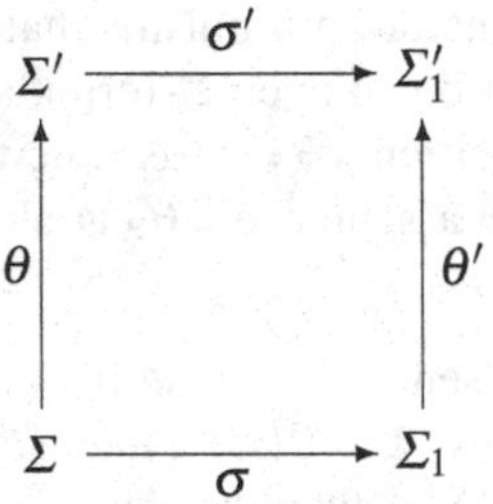

in **Sign** *such that $\theta'\in\mathbf{I}$,*

$$M_1|_\sigma \models_\Sigma \forall\theta\bullet\varphi \quad \textit{iff} \quad M_1\models_{\Sigma_1}\forall\theta'\bullet\sigma'(\varphi).$$

Proof.

($\Rightarrow$): Assume that $M_1|_\sigma\models_\Sigma\forall\theta\bullet\varphi$ and let M_1' be a θ'-expansion of M_1. Put $M'=M_1'|_{\sigma'}$. Obviously, $M'|_\theta=M_1'|_{\theta;\sigma'}=M_1'|_{\sigma;\theta'}=M_1|_\sigma$. Thus, since $M_1|_\sigma\models_\Sigma\forall\theta\bullet\varphi$, $M'\models_{\Sigma'}\varphi$. Hence, by the satisfaction condition of **INS**, $M_1'\models_{\Sigma_1'}\sigma'(\varphi)$, which proves $M_1\models_{\Sigma_1}\forall\theta'\bullet\sigma'(\varphi)$.

($\Leftarrow$): Assume that $M_1\models_{\Sigma_1}\forall\theta'\bullet\sigma'(\varphi)$ and let M' be a θ-expansion of $M_1|_\sigma$. Since **INS** is **I**-semi-exact, there exists a θ'-expansion M_1' of M_1 such that $M_1'|_{\sigma'}=M'$. Then, since $M_1\models_{\Sigma_1}\forall\theta'\bullet\sigma'(\varphi)$, $M_1'\models_{\Sigma_1'}\sigma'(\varphi)$. Thus, by the satisfaction condition, $M'\models_{\Sigma'}\varphi$, which proves $M_1|_\sigma\models_\Sigma\forall\theta\bullet\varphi$. □

Example 4.4.32. Let **I** be the collection of algebraic signature inclusions $\iota:\Sigma\hookrightarrow\Sigma'$ in **AlgSig** such that $\Sigma'\setminus\Sigma$ contains a finite set of new constants only. The institution $\mathbf{GEQ}^{\forall(\mathbf{I})}$ essentially coincides with the institution **EQ** of equational logic (modulo the details of the notation used for sentences), as suggested already in Exercise 2.1.6.

[21] As usual, we disregard here the foundational problems which may arise if **I** is not a set.

If $\Sigma' \setminus \Sigma$ is allowed to contain new operation names (not just constants), then quantification along morphisms in **I** leads to a version of second-order logic. □

Other quantifiers (there exists, there exists a unique, there exist infinitely many, for almost all, and so on) may be introduced in the same manner in which we have just introduced universal quantifiers. Example 4.1.41 illustrates how one may introduce logical connectives. By iterating these constructions one can, for example, derive the institution of first-order logic from the institution of ground atomic formulae.

4.5 Institutions with reachability structure

An alternative to the standard initial algebra approach to specifications is to take the reachable semantics of presentations, as discussed in Section 2.7.2, where from among all the algebras satisfying a presentation only the *reachable* algebras are selected. In Section 4.3 we argued that it is important to consider not just initial algebras, but more generally, algebras that are free extensions of a specified part; similarly, it is important here to consider not just reachable algebras, but more generally, algebras that are generated by some specified part. Given an algebraic signature Σ and a subsignature $\Sigma' \subseteq \Sigma$, a Σ-algebra A is *reachable from* Σ' if it has no proper subalgebra with the same Σ'-reduct. (**Exercise:** Show that this is the same as requiring that the algebra is generated by the set of all its elements in the carriers of the sorts in Σ', as defined in Exercise 1.2.6.) To generalise this notion to the framework of an arbitrary institution we will proceed along the lines suggested by the "categorical theory of reachability" presented in Section 3.3, based on factorisation systems.

Definition 4.5.1 (Reachable model). Let $\langle \mathbf{Sign}, \mathbf{Sen}, \mathbf{Mod}, \langle \models_\Sigma \rangle_{\Sigma \in |\mathbf{Sign}|} \rangle$ be an institution. Assume that for each signature $\Sigma \in |\mathbf{Sign}|$, we have a factorisation system $\langle \mathbf{E}_\Sigma, \mathbf{M}_\Sigma \rangle$ for the category $\mathbf{Mod}(\Sigma)$ of Σ-models.

Let $\sigma: \Sigma' \to \Sigma$ be a signature morphism. A Σ-model $M \in |\mathbf{Mod}(\Sigma)|$ is *σ-reachable* if M has no proper submodel with an isomorphic σ-reduct, that is, if any factorisation monomorphism $m: N \to M$ in $\mathbf{M}_\Sigma$ such that $m|_\sigma$ is an isomorphism in $\mathbf{Mod}(\Sigma')$ is in fact an isomorphism in $\mathbf{Mod}(\Sigma)$. □

Example 4.5.2. Recall that for any algebraic signature $\Sigma \in \mathbf{AlgSig}$, the categories $\mathbf{Alg}(\Sigma)$, $\mathbf{PAlg}(\Sigma)$ and $\mathbf{CAlg}(\Sigma)$ of total, partial and continuous algebras come equipped with factorisation systems (Examples 3.3.3, 3.3.13 and 3.3.14, respectively). Hence, the above definition makes sense in the institutions **EQ** of equational logic, **PEQ** of partial equational logic and **CEQ** of equational logic for continuous algebras, yielding the expected notions. □

Exercise 4.5.3. Recall that by Definition 3.3.7 a Σ-model is reachable if it has no proper submodel. Show that if **INS** is finitely exact then reachability is a special case of σ-reachability as defined above. HINT: Use the fact that there is an initial signature with the singleton category **1** of models. □

In Section 3.3 it was shown how the notion of reachability introduced there may be related to an equivalent definition stated in terms of quotients of initial models (Theorem 3.3.8(1)). In the standard algebraic case, an algebra is reachable if and only if it is isomorphic to a quotient of the algebra of ground terms (Exercise 1.4.14). To give an analogous result for σ-reachability we have to be able to build terms over a specified reduct of the given algebra (cf. Exercise 3.5.11). Given such a construction, a Σ-algebra A is reachable from $\Sigma' \subseteq \Sigma$ if and only if evaluation in A of Σ-terms over the Σ'-reduct of A is surjective, or equivalently, if A is a natural quotient of the algebra of Σ-terms built over $A|_{\Sigma'}$. We introduce a generalisation of the construction of term algebras to an arbitrary institution by requiring that reduct functors induced by signature morphisms have left adjoints. Notice that only signatures are involved in this definition, and no sentences, and so this requirement indeed corresponds to the mild assumption that free models (term algebras) may be built along arbitrary signature morphisms.

Definition 4.5.4 (Institution with reachability structure). An *institution with reachability structure* is an institution $\langle \mathbf{Sign}, \mathbf{Sen}, \mathbf{Mod}, \langle \models_\Sigma \rangle_{\Sigma \in |\mathbf{Sign}|} \rangle$ together with:

- for each signature $\Sigma \in |\mathbf{Sign}|$, a factorisation system $\langle \mathbf{E}_\Sigma, \mathbf{M}_\Sigma \rangle$ for the category $\mathbf{Mod}(\Sigma)$ of Σ-models; and
- for each signature morphism $\sigma: \Sigma' \to \Sigma$, a *$\sigma$-free functor* $\mathbf{F}_\sigma: \mathbf{Mod}(\Sigma') \to \mathbf{Mod}(\Sigma)$ which is left adjoint to the σ-reduct functor $\_|_\sigma: \mathbf{Mod}(\Sigma) \to \mathbf{Mod}(\Sigma')$ with unit $\eta^\sigma: \mathbf{Id}_{\mathbf{Mod}(\Sigma')} \to \mathbf{F}_\sigma(\_)|_\sigma$.

(As usual, sub- and superscripts will be omitted when convenient.) □

Example 4.5.5. The institution **EQ** of equational logic equipped with factorisation systems for categories of algebras (cf. Example 3.3.3) has reachability structure — the free functors are given by Exercise 3.5.11. □

Exercise 4.5.6. Show that the institution **PEQ** of partial equational logic with the factorisation systems given by Example 3.3.13 for categories of partial algebras forms an institution with reachability structure. HINT: Free functors are rather trivial here.

Similarly, show that the institution **CEQ** of equational logic for continuous algebras with the factorisation systems given by Example 3.3.14 for categories of continuous algebras forms an institution with reachability structure. HINT: The construction of free functors is much more difficult here — follow the construction for ordinary algebras in Exercise 3.5.11, but when defining the new operations in a free way remember that you have to extend the complete partial order to cover the new values as well, ensuring continuity of the operations. □

Exercise 4.5.7. Let **INS** be a finitely exact institution. Prove that if every reduct functor in **INS** has a left adjoint, then for every signature Σ the category $\mathbf{Mod}_{\mathbf{INS}}(\Sigma)$ of Σ-models has an initial object. HINT: Use the fact that there is an initial signature with the singleton category **1** of models. □

The following theorem generalises well-known facts from the standard algebraic setting. Like its "predecessor" Theorem 3.3.8, it confirms our confidence in the abstract definitions by showing how their different versions "click together" nicely.

Theorem 4.5.8. *Let* $\mathbf{INS} = \langle \mathbf{Sign}, \mathbf{Sen}, \mathbf{Mod}, \langle \models_\Sigma \rangle_{\Sigma \in |\mathbf{Sign}|} \rangle$ *be an institution with reachability structure. Consider a signature morphism* $\sigma : \Sigma' \to \Sigma$.

1. *A* Σ*-model* $M \in |\mathbf{Mod}(\Sigma)|$ *is* σ*-reachable if and only if it is a natural quotient of the free object over its* σ*-reduct, that is, the counit morphism* $\varepsilon_M = (id_{M|_\sigma})^\# : \mathbf{F}_\sigma(M|_\sigma) \to M$ *belongs to* $\mathbf{E}_\Sigma$ *(cf. Exercise 3.5.24).*
2. *For any* σ*-reachable model* $M \in |\mathbf{Mod}(\Sigma)|$*, any model* $N \in |\mathbf{Mod}(\Sigma)|$ *and* Σ'*-model morphism* $f' : M|_\sigma \to N|_\sigma$*, there exists at most one* Σ*-model morphism* $f : M \to N$ *that extends* f' *(i.e.* $f|_\sigma = f'$*).*
3. *Every* Σ*-model has a unique (up to isomorphism)* σ*-reachable submodel with an isomorphic* σ*-reduct.*
4. *If* $M \in |\mathbf{Mod}(\Sigma)|$ *is* σ*-reachable then for any* Σ*-model morphism* $f : N \to M$ *such that* $f|_\sigma$ *is an isomorphism,* f *is a factorisation epimorphism (i.e.* $f \in \mathbf{E}_\Sigma$*).*

Proof.

1. ($\Rightarrow$): Let $\mathbf{F}_\sigma(M|_\sigma) \xrightarrow{e} N \xrightarrow{m} M$ be a factorisation of $\varepsilon_M : \mathbf{F}_\sigma(M|_\sigma) \to M$. Arguing dually to Exercise 3.5.18 we can show that $m|_\sigma : N|_\sigma \to M|_\sigma$ is an isomorphism. Hence, by the σ-reachability of M, m is an isomorphism, which proves that $\varepsilon_M \in \mathbf{E}_\Sigma$.
   ($\Leftarrow$): Let $m : N \to M$, $m \in \mathbf{M}_\Sigma$, with $m|_\sigma$ being an isomorphism. Let $f : \mathbf{F}_\sigma(M|_\sigma) \to N$ be defined by $f = ((m|_\sigma)^{-1})^\#$. Then $\eta_{M|_\sigma} ; (f;m)|_\sigma = id_{M|_\sigma}$. By the freeness of $\mathbf{F}_\sigma(M|_\sigma)$, this implies that $f;m = \varepsilon_M$. Thus, by the assumption that $\varepsilon_M \in \mathbf{E}_\Sigma$ and by Exercise 3.3.5, m is an isomorphism.
2. Suppose that $f_1, f_2 : M \to N$ are such that $f_1|_\sigma = f_2|_\sigma = f'$. Then $\eta_{M|_\sigma} ; (\varepsilon_M ; f_1)|_\sigma = f' = \eta_{M|_\sigma} ; (\varepsilon_M ; f_2)|_\sigma$, and so $\varepsilon_M ; f_1 = \varepsilon_M ; f_2$. Thus, we also have $f_1 = f_2$, since by (1) above ε_M is an epimorphism.
3. Consider an arbitrary Σ-model M. Let $\mathbf{F}_\sigma(M|_\sigma) \xrightarrow{e} N \xrightarrow{m} M$ be a factorisation of $\varepsilon_M : \mathbf{F}_\sigma(M|_\sigma) \to M$. Again, arguing dually to Exercise 3.5.18 we can show that $m|_\sigma : N|_\sigma \to M|_\sigma$ is an isomorphism. Moreover, by the naturality of ε, $\mathbf{F}_\sigma(m|_\sigma) ; \varepsilon_M = \varepsilon_N ; m$, that is $\mathbf{F}_\sigma(m|_\sigma) ; e ; m = \varepsilon_N ; m$, and so (since m is a monomorphism) $\varepsilon_N = \mathbf{F}_\sigma(m|_\sigma) ; e \in \mathbf{E}_\Sigma$. Thus, by (1) again, N is a σ-reachable submodel of M.
   To prove uniqueness up to isomorphism, consider a subobject $m_1 : N_1 \to M$ with $m_1|_\sigma$ being an isomorphism and $\varepsilon_{N_1} : F_\sigma(N_1|_\sigma) \to N_1$ in $\mathbf{E}_\Sigma$. Then $\mathbf{F}_\sigma(m_1|_\sigma) ; \varepsilon_M = \varepsilon_{N_1} ; m_1$, and since $\mathbf{F}_\sigma(m_1|_\sigma)$ is an isomorphism, we have two factorisations of $\varepsilon_M : \mathbf{F}_\sigma(M|_\sigma) \to M$, $\langle \mathbf{F}_\sigma(m_1|_\sigma)^{-1} ; \varepsilon_{N_1}, m_1 \rangle$ and $\langle e, m \rangle$, which by the uniqueness of factorisations implies that N and N_1 are isomorphic.
4. Let $N \xrightarrow{e} \cdot \xrightarrow{m} M$ be a factorisation of $f : N \to M$. Then, by naturality of ε, $\varepsilon_N ; e ; m = \mathbf{F}_\sigma(f|_\sigma) ; \varepsilon_M$. Now, since $f|_\sigma$ (and hence $\mathbf{F}_\sigma(f|_\sigma)$) is an isomorphism,

by σ-reachability of M and (1) above, $\varepsilon_N;e;m \in \mathbf{E}_\Sigma$. Thus, by Exercise 3.3.5, m is an isomorphism, and so $f \in \mathbf{E}_\Sigma$. □

4.5.1 The method of diagrams

In the standard algebraic framework, reachable algebras enjoy a number of useful properties which make them especially easy to deal with. As a consequence of the fact that we are able to "name" (using ground terms) all their elements, reachable algebras are easy to describe using the most elementary logical sentences, ground equations. To be more precise, for any algebraic signature Σ and reachable Σ-algebra A, the class

$$Ext(A) = \{B \in |\mathbf{Alg}(\Sigma)| \mid \text{there exists a } \Sigma\text{-homomorphism } h{:}A \to B\}$$

is the class of models of the ground Σ-equations that hold in A, that is, $Ext(A) = Mod_{\mathbf{GEQ}}(Th_{\mathbf{GEQ}}(\{A\}))$, and moreover, A is initial in $Ext(A)$. (We will refer to classes of algebras of the form $Ext(A)$ for a reachable algebra A as *ground varieties*.) This gives a one-to-one correspondence between ground equational theories and isomorphism classes of reachable algebras (and furthermore, congruences on ground term algebras by Exercise 1.4.14).

Unfortunately, not all algebras are reachable, and it is clear that this correspondence does not carry over to arbitrary algebras: there are algebras that cannot be characterised as initial models of equational theories. But there is a technical trick that may help: if a Σ-algebra A is not reachable, then consider the signature $\Sigma(A)$ obtained by adding to Σ the elements of $|A|$ as constants of the appropriate sorts. Now, the algebra A has an obvious expansion to a reachable $\Sigma(A)$-algebra $E(A)$, where the new constants are interpreted as the elements they correspond to. This expansion has a number of useful properties:

- Any Σ-homomorphism $h{:}A \to B$ determines unambiguously an expansion of B to a $\Sigma(A)$-algebra $E_h(B)$ where each new constant in $\Sigma(A)$ is interpreted as the value of h on the corresponding element of $|A|$. Moreover, this expansion is independent of any decomposition of h: for any Σ-homomorphisms $h_1{:}A \to C$ and $h_2{:}C \to B$ such that $h = h_1;h_2$, the homomorphism h_2 (or more precisely, its underlying map) is a $\Sigma(A)$-homomorphism from $E_{h_1}(C)$ to $E_h(B)$.
- Intuitively, the expansion does not introduce more structure than necessary to make A reachable; in particular, no new elements are added.

Putting all these together, any Σ-algebra A may be characterised by the set of ground equations on the signature $\Sigma(A)$ that hold in $E(A)$. This technique, known as *the method of diagrams*, is one of the basic tools of classical model theory (cf. [CK90]). We have already suggested its use in the construction of the free functor corresponding to a signature morphism in Exercise 3.5.11.

In the following the method of diagrams is formulated in the context of an arbitrary institution with reachability structure. We will assume that the institution is finitely exact in order to be able to deal with reachability (not just reachability relative to signature morphisms; cf. Exercises 4.5.3 and 4.5.7).

Definition 4.5.9 (Method of diagrams). A finitely exact institution with reachability structure $\mathbf{INS} = \langle \mathbf{Sign}, \mathbf{Sen}, \mathbf{Mod}, \langle \models_\Sigma \rangle_{\Sigma \in |\mathbf{Sign}|} \rangle$ *admits the method of diagrams* if:

- (*Definability of ground varieties*)
  for every signature $\Sigma \in |\mathbf{Sign}|$ and reachable Σ-model $M \in |\mathbf{Mod}(\Sigma)|$, the class
  $$Ext(M) = \{N \in |\mathbf{Mod}(\Sigma)| \mid \text{there exists a } \Sigma\text{-model morphism } h\colon M \to N\}$$
  of extensions of M is definable, that is, $Ext(M) = Mod_\Sigma(\Phi)$ for some set $\Phi \subseteq \mathbf{Sen}(\Sigma)$.
- (*Existence of diagrams*)
  for every signature $\Sigma \in |\mathbf{Sign}|$ and Σ-model $M \in |\mathbf{Mod}(\Sigma)|$, there exists a signature $\Sigma(M) \in |\mathbf{Sign}|$ and signature morphism $\iota\colon \Sigma \to \Sigma(M)$ such that:
  - M has a reachable ι-expansion $E(M)$: there exists $E(M)$ which is a reachable $\Sigma(M)$-model such that $E(M)|_\iota = M$;
  - ι-reduct is an isomorphism of the slice categories $\mathbf{Mod}(\Sigma(M)){\uparrow}E(M)$ and $\mathbf{Mod}(\Sigma){\uparrow}M$ (see Exercise 3.1.31), that is, for any Σ-model morphism $f\colon M \to N$, there exists a unique ι-expansion of N, $E_f(N)$, such that f has an ι-expansion $E(f)\colon E(M) \to E_f(N)$ and such that any Σ-model morphism $h\colon N \to N_1$ has a unique ι-expansion $E(h)\colon E_f(N) \to E_{f;h}(N_1)$; and
  - ι-reduct preserves the factorisation system on $\mathbf{Mod}(\Sigma(M)){\uparrow}E(M)$ as inherited from $\mathbf{Mod}(\Sigma(M))$, that is, for any $f\colon E(M) \to N'$ and $h\colon N' \to N''$, if $h \in \mathbf{E}_{\Sigma(M)}$ then $h|_\iota \in \mathbf{E}_\Sigma$ and if $h \in \mathbf{M}_{\Sigma(M)}$ then $h|_\iota \in \mathbf{M}_\Sigma$.

Then, $\Sigma(M)$ is called the *diagram signature for* M (with *signature inclusion* ι), $E(M)$ is called the *diagram expansion of* M, and finally the theory $\Delta^+(M) = Th_{\Sigma(M)}(Ext(E(M)))$ is called the (*positive*) *diagram of* M. □

Example 4.5.10. The institutions **EQ** of equational logic, **PEQ** of partial equational logic, and **CEQ** of equational logic for continuous algebras admit the method of diagrams. Ground varieties in **EQ** are definable by sets of ground equations; ground varieties of **PEQ** are definable by sets of ground equations and ground definedness formulae; ground varieties in **CEQ** are definable by sets of ground infinitary equations. For any (total, partial, or continuous) Σ-algebra A, the diagram signature for A is formed by adding constants corresponding to all the elements of $|A|$. The diagram expansion of a partial algebra is formed by requiring that the new constants are defined and have the expected values. □

Exercise 4.5.11. Show that in any institution that admits the method of diagrams, and for any model M, the class of models of the positive diagram of M is the class of all extensions of the diagram expansion of M: $Mod_{\Sigma(M)}(\Delta^+(M)) = Ext(E(M))$. □

4.5.2 Abstract algebraic institutions

In Exercise 3.5.11 we suggested the use of the method of diagrams to prove that in the standard algebraic framework, the reduct functor induced by a signature morphism has a left adjoint. With some more effort, one can generalise this result and prove that in the standard equational institution the reduct functor induced by a *theory* morphism has a left adjoint:

Exercise 4.5.12. Prove that in the equational institution **EQ**, for any theory morphism $\sigma: T \to T'$, the reduct functor $\_|_\sigma: \mathbf{Mod}[T'] \to \mathbf{Mod}[T]$ has a left adjoint.

HINT: Formalise and complete the following construction: Let $T = \langle \Sigma, \Phi \rangle$ and $T' = \langle \Sigma', \Phi' \rangle$. For any Σ-algebra $A \in Mod[T]$, let $\Sigma(A)$ be its diagram signature, and let

$$\begin{array}{ccc} \Sigma(A) & \xrightarrow{\ \sigma'\ } & \Sigma'(A) \\ \uparrow{\scriptstyle \iota} & & \uparrow{\scriptstyle \iota'} \\ \Sigma & \xrightarrow[\ \sigma\ ]{} & \Sigma' \end{array}$$

be a pushout in the category of signatures. Then, let $\Delta^+(A) \subseteq \mathbf{Sen}_{\mathbf{EQ}}(\Sigma(A))$ be the positive diagram of A. Consider the presentation $\langle \Sigma'(A), \sigma'(\Delta^+(A)) \cup \iota'(\Phi') \rangle$. By Theorem 2.5.14, this has an initial model. Its ι'-reduct is a free object over A. (See also Exercise 3.5.11 for a slightly different line of reasoning.) □

We will come back to a careful, more abstract analysis of this construction later (cf. Theorem 4.5.18 below). For now, notice that the construction not only uses the fact that the equational institution admits the method of diagrams, but also relies (directly or indirectly) on a number of simple facts about the reachability structure of the equational institution. We capture some of these additional properties in the following abstract definition:

Definition 4.5.13 (Abstract algebraic institution). An *abstract algebraic institution* is a finitely exact institution $\mathbf{INS} = \langle \mathbf{Sign}, \mathbf{Sen}, \mathbf{Mod}, \langle \models_\Sigma \rangle_{\Sigma \in |\mathbf{Sign}|} \rangle$ with reachability structure that admits the method of diagrams for which the following additional conditions hold:

- For any signature $\Sigma \in |\mathbf{Sign}|$, the category $\mathbf{Mod}(\Sigma)$ has all products (of sets of models) and is $\mathbf{E}_\Sigma$-co-well-powered (Definition 3.3.10).
- For any signature morphism $\sigma: \Sigma \to \Sigma'$, the σ-reduct functor preserves submodels (i.e. for all $m' \in \mathbf{M}_{\Sigma'}$, $m'|_\sigma \in \mathbf{M}_\Sigma$) and products.
- (*Abstraction condition*) For any signature Σ and Σ-models $M, N \in |\mathbf{Mod}(\Sigma)|$, if M and N are isomorphic then they satisfy exactly the same Σ-sentences. □

Example 4.5.14. The institutions **EQ** of equational logic, **PEQ** of partial equational logic, and **CEQ** of equational logic for continuous algebras are abstract algebraic institutions. □

Exercise 4.5.15. There is a certain asymmetry in the above definition: reduct functors in abstract algebraic institutions are required to preserve submodels but are not required to preserve quotients. Prove that in **EQ**, reduct functors preserve quotients as well: for all $\sigma:\Sigma \to \Sigma'$ and $e' \in \mathbf{E}_{\Sigma'}$, $e'|_{\sigma} \in \mathbf{E}_{\Sigma}$. Show, however, that this is not true in general in **PEQ**. □

4.5.3 Liberal abstract algebraic institutions

In Section 4.3 we have shown that it is possible to restrict attention to initial models of specifications written in an arbitrary institution, even if theories in the institution are not guaranteed to have initial models in general. Similarly, data constraints make sense in an arbitrary institution even if reduct functors induced by theory morphisms are not guaranteed to have left adjoints. This flexibility is useful, but nevertheless it may be important to know whether or not a theory used in an initiality constraint has an initial model, or whether a theory morphism used in a data constraint has a corresponding free functor. In some institutions this is always the case: the equational institution **EQ** is one example (cf. Theorem 2.5.14 and Exercise 4.5.12). In the rest of this section we present a characterisation of institutions that have this property. Of course, very little can be done in the framework of an arbitrary institution: however, abstract algebraic institutions as introduced above provide a sufficiently rich background.

Definition 4.5.16 (Liberal institution). An institution **INS** *admits initial models* if every theory in **INS** has an initial model. **INS** is *liberal* if for every theory morphism $\sigma:T \to T'$ in **INS**, the σ-reduct functor $\_|_{\sigma}:\mathbf{Mod}[T'] \to \mathbf{Mod}[T]$ has a left adjoint.

Then, an abstract algebraic institution **INS** *admits reachable initial models* if every theory in **INS** has an initial model which is reachable. **INS** is *strongly liberal* if for every theory morphism $\sigma:T \to T'$ in **INS**, the σ-reduct functor $\_|_{\sigma}:\mathbf{Mod}[T'] \to \mathbf{Mod}[T]$ has a left adjoint $\mathbf{F}_{\sigma}:\mathbf{Mod}[T] \to \mathbf{Mod}[T']$ such that for any $M \in Mod[T]$, $\mathbf{F}_{\sigma}(M) \in Mod[T']$ is σ-reachable. □

In the last part of the definition we have slightly abused notation by using σ as both a *theory* morphism and a *signature* morphism (which in fact it is). It is important that the notion of σ-reachability used here be taken with respect to signature morphisms (cf. Definition 4.5.1), without taking into account the theory context.

Exercise 4.5.17. Find an institution that admits initial models but does not admit reachable initial models. HINT: Consider an algebraic signature Σ with a unary operation symbol $f:s \to s$. Show that the class of Σ-algebras satisfying the axiom

$\exists! x{:}s \bullet f(x) = x$ has an initial model which is not reachable, where $\exists!$ reads "there exists a unique", that is, $\exists! x{:}s \bullet f(x) = x$ stands for $\exists x{:}s \bullet f(x) = x \wedge \forall x_1, x_2{:}s \bullet f(x_1) = x_1 \wedge f(x_2) = x_2 \Rightarrow x_1 = x_2$. □

For abstract algebraic institutions, the conditions introduced in Definition 4.5.16 are pairwise equivalent.

Theorem 4.5.18. *Let* **INS** *be an abstract algebraic institution.* **INS** *is liberal if and only if it admits initial models.*

Proof.

($\Rightarrow$): Let $T = \langle \Sigma, \Phi \rangle$ be a theory. Let $\iota_\Sigma{:}\Sigma_\varnothing \to \Sigma$ be the only signature morphism from the initial signature $\Sigma_\varnothing$ to Σ. Then $\iota_\Sigma{:}T_\varnothing \to T$ is a theory morphism, where $T_\varnothing = \langle \Sigma_\varnothing, Cl_{\Sigma_\varnothing}(\varnothing) \rangle$ is the initial theory, and so the reduct functor $\_|_{\iota_\Sigma}{:}\mathbf{Mod}[T] \to \mathbf{Mod}[T_\varnothing]$ has a left adjoint $\mathbf{F}_{\iota_\Sigma}{:}\mathbf{Mod}[T_\varnothing] \to \mathbf{Mod}[T]$. Now, there is exactly one $\Sigma_\varnothing$-model, say $M_\varnothing \in |\mathbf{Mod}[T_\varnothing]|$, and moreover, $\mathbf{F}_{\iota_\Sigma}(M_\varnothing)$ is an initial model of T.

($\Leftarrow$): We follow the proof for the equational institution **EQ** sketched in Exercise 4.5.12. For any theory morphism $\sigma{:}T \to T'$, where $T = \langle \Sigma, \Phi \rangle$ and $T' = \langle \Sigma', \Phi' \rangle$, and model $M \in Mod[T]$, we construct a model $\mathbf{F}_\sigma(M) \in Mod[T']$ with unit $\eta_M{:}M \to \mathbf{F}_\sigma(M)|_\sigma$ that is free over M w.r.t. $\_|_\sigma{:}\mathbf{Mod}[T'] \to \mathbf{Mod}[T]$.
Let $\Sigma(M)$ be the diagram signature for M with signature inclusion $\iota{:}\Sigma \to \Sigma(M)$, and let

$$\begin{array}{ccc} \Sigma(M) & \xrightarrow{\sigma'} & \Sigma'(M) \\ \uparrow{\scriptstyle \iota} & & \uparrow{\scriptstyle \iota'} \\ \Sigma & \xrightarrow[\sigma]{} & \Sigma' \end{array}$$

be a pushout in the category of signatures. Then, let $\Delta^+(M) \subseteq \mathbf{Sen}(\Sigma(M))$ be the positive diagram of M. Consider the presentation $\langle \Sigma'(M), \sigma'(\Delta^+(M)) \cup \iota'(\Phi') \rangle$. By the assumption, it has an initial model, say I. Put $\mathbf{F}_\sigma(M) = I|_{\iota'}$. Then, since by the satisfaction condition $I|_{\sigma'} \models_{\Sigma(M)} \Delta^+(M)$, $I|_{\sigma'} \in Ext(E(M))$ (cf. Exercise 4.5.11). Hence, there exists a (unique, since $E(M)$ is reachable) $\Sigma(M)$-model morphism $\widehat{\eta_M}{:}E(M) \to I|_{\sigma'}$. Put $\eta_M = \widehat{\eta_M}|_\iota{:}M \to \mathbf{F}_\sigma(M)|_\sigma = I|_{\iota;\sigma'}$.
First, notice that since $I \models_{\Sigma'(M)} \iota'(\Phi')$, $\mathbf{F}_\sigma(M) \in Mod[T']$. Then, consider an arbitrary model $N \in Mod[T']$ and a Σ-model morphism $f{:}M \to N|_\sigma$.
By the definition of the diagram signature for M, $N|_\sigma$ has a unique ι-expansion to a $\Sigma(M)$-model $E_f(N|_\sigma)$ such that there exists a $\Sigma(M)$-model morphism $E(f){:}E(M) \to E_f(N|_\sigma)$ with $E(f)|_\iota = f$. Amalgamation yields a unique $\Sigma'(M)$-model $E_f^\sigma(N|_\sigma) \in |\mathbf{Mod}(\Sigma'(M))|$ with $E_f^\sigma(N|_\sigma)|_{\sigma'} = E_f(N|_\sigma)$ and $E_f^\sigma(N|_\sigma)|_{\iota'} = N$. Since $N \models_{\Sigma'} \Phi'$, $E_f^\sigma(N|_\sigma) \models_{\Sigma'(M)} \iota'(\Phi')$. Then, since $E_f(N|_\sigma) \in Ext(E(M))$,

$E_f(N|_\sigma) \models_{\Sigma(M)} \Delta^+(M)$, and so $E_f^\sigma(N|_\sigma) \models_{\Sigma'(M)} \sigma'(\Delta^+(M))$. Consequently, we get a unique $\Sigma'(M)$-model morphism $\widehat{f'}{:}I \to E_f^\sigma(N|_\sigma)$. Put $f' = \widehat{f'}|_{\iota'}{:}\mathbf{F}_\sigma(M) \to N$. Notice that $\widehat{\eta_M};\widehat{f'}|_{\sigma'}{:}E(M) \to E_f(N|_\sigma)$. Hence, since $E(M)$ is reachable, $\widehat{\eta_M};\widehat{f'}|_{\sigma'} = E(f)$, and so we obtain $\eta_M;f'|_\sigma = f$. Moreover, f' is the only morphism with this property. To see this, suppose that for some $f''{:}\mathbf{F}_\sigma(M) \to N$, $\eta_M;f''|_\sigma = f$. Then, by the amalgamation property (this time for model morphisms) there exists a $\Sigma'(M)$-model morphism $\widehat{f''}{:}I \to E_f^\sigma(N|_\sigma)$ such that $\widehat{f''}|_{\iota'} = f''$ (and $\widehat{f''}|_{\sigma'} = E(f''|_\sigma){:}I|_{\sigma'} \to E_f(N|_\sigma)$). By initiality of I, $\widehat{f''} = \widehat{f'}$, and so $f'' = f'$, which completes the proof. □

Theorem 4.5.19. *Let* **INS** *be an abstract algebraic institution.* **INS** *is strongly liberal if and only if it admits reachable initial models.*

Proof. We extend the proof of the previous theorem, relying on the notation introduced there.

($\Rightarrow$): The only additional remark needed is that $\mathbf{F}_{\iota_\Sigma}(M_\varnothing)$ is reachable if it is ι_Σ-reachable (cf. Exercise 4.5.3).

($\Leftarrow$): We have to additionally prove that $\mathbf{F}_\sigma(M) = I|_{\iota'}$ is σ-reachable whenever I is reachable. To see this, consider an arbitrary submodel of $I|_{\iota'}$ with an isomorphic σ-reduct, say $m{:}N \to I|_{\iota'}$, where $m \in \mathbf{M}_{\Sigma'}$ and $m|_\sigma{:}N|_\sigma \to I|_{\sigma;\iota'}$ is an isomorphism. Put $f = \eta_M;(m|_\sigma)^{-1}{:}M \to N|_\sigma$. Then $f;m|_\sigma = \eta_M$, and so $m|_\sigma$ has an expansion to a $\Sigma(M)$-model morphism $E(m|_\sigma){:}E_f(N|_\sigma) \to E_{\eta_M}(I|_{\sigma;\iota'}) = I|_{\sigma'}$. Then, as in the corresponding part of the proof of Theorem 4.5.18, we get a unique $\Sigma'(M)$-model $E_f^\sigma(N|_\sigma) \in |\mathbf{Mod}(\Sigma'(M))|$, such that $E_f^\sigma(N|_\sigma)|_{\sigma'} = E_f(N|_\sigma)$ and $E_f^\sigma(N|_\sigma)|_{\iota'} = N$, and a $\Sigma'(M)$-model morphism $\widehat{f'}{:}I \to E_f^\sigma(N|_\sigma)$. On the other hand, by the amalgamation property again, there exists a unique $\Sigma'(M)$-model morphism $\widehat{m}{:}E_f^\sigma(N|_\sigma) \to I$ such that $\widehat{m}|_{\sigma'} = E(m|_\sigma)$ and $\widehat{m}|_{\iota'} = m$. By the initiality of I, $\widehat{f'};\widehat{m}$ is the identity, and so is $(\widehat{f'};\widehat{m})|_{\iota'} = \widehat{f'}|_{\iota'};m$. Thus, by Exercise 3.3.5, m is an isomorphism — which completes the proof. □

4.5.4 *Characterising abstract algebraic institutions that admit reachable initial models*

From the very beginning of work on algebraic specifications it has been known that the standard equational institution **EQ** admits reachable initial models (cf. Theorem 2.5.14). Moreover, the proof of this property generalises readily to the situation where conditional equations (even with infinite sets of premises) are permitted as axioms. On the other hand, Example 2.7.11 shows that if disjunction is permitted, the property is lost. Indeed, in the standard algebraic framework the infinitary conditional axioms, which define all non-empty quasi-varieties, form in some sense a border beyond which one cannot be sure of the existence of reachable initial models. We generalise this result to the framework of abstract algebraic institutions.

Theorem 4.5.20. *Let* **INS** *be an abstract algebraic institution.* **INS** *admits reachable initial models if and only if every class of models definable in* **INS** *is closed under products (of sets of models) and under submodels.*

Proof.

($\Leftarrow$): This follows directly by Lemma 3.3.12; just notice that any class of models closed under products and submodels is a *non-empty* quasi-variety (cf. Definition 3.3.11).

($\Rightarrow$): Let $\langle \Sigma, \Phi \rangle$ be a presentation in **INS**. We show the required closure properties of $Mod_\Sigma(\Phi)$.

(*Submodels*): Consider a model $M \in Mod_\Sigma(\Phi)$ and its submodel $m{:}N \to M$, $m \in \mathbf{M}_\Sigma$. Let $\Sigma(N)$ be a diagram signature for N with signature inclusion $\iota{:}\Sigma \to \Sigma(N)$, and let $\Delta^+(N) \subseteq \mathbf{Sen}(\Sigma(N))$ be the positive diagram of N. Recall that $Mod_{\Sigma(N)}(\Delta^+(N)) = Ext(E(N))$, where $E(N) \in |\mathbf{Mod}(\Sigma(N))|$ is the diagram expansion of N. The presentation $\langle \Sigma(N), \Delta^+(N) \cup \iota(\Phi) \rangle$ has a reachable initial model, say I. We show that $I|_\iota$ is isomorphic to N, which in particular implies $N \in Mod_\Sigma(\Phi)$.
Since $I \models_{\Sigma(N)} \Delta^+(N)$, there exists a $\Sigma(N)$-model morphism $f{:}E(N) \to I$. Moreover, since I is reachable, $f \in \mathbf{E}_{\Sigma(N)}$ (by Theorem 3.3.8(4)) and hence also $f|_\iota \in \mathbf{E}_\Sigma$. Then, let $E_m(M)$ be the unique expansion of M to a $\Sigma(N)$-model with $E(m){:}E(N) \to E_m(M)$ such that $E(m)|_\iota = m$. Since $M \models \Phi$, $E_m(M) \models_{\Sigma(N)} \iota(\Phi)$, and, since $E_m(M) \in Ext(E(N))$, $E_m(M) \models_{\Sigma(N)} \Delta^+(N)$. Hence, there is a (unique) morphism $g{:}I \to E_m(M)$. Now, since $E(N)$ is reachable, there exists at most one morphism from $E(N)$ to $E_m(M)$, and so we have $f;g = E(m)$, which implies $f|_\iota;g|_\iota = m \in \mathbf{M}_\Sigma$. Since $f|_\iota \in \mathbf{E}_\Sigma$, it follows from Exercise 3.3.5 that $f|_\iota{:}N \to I|_\iota$ is indeed an isomorphism.

(*Products*): Consider any family $M_i \in Mod_\Sigma(\Phi)$, $i \in J$, where J is any set (of indices). Let N with projections $\pi_i{:}N \to M_i$, $i \in J$, be the product of the family $\langle M_i \rangle_{i \in J}$. We proceed similarly as in the previous case: let $\Sigma(N)$ be a diagram signature for N with signature inclusion $\iota{:}\Sigma \to \Sigma(N)$, and let $\Delta^+(N) \subseteq \mathbf{Sen}(\Sigma(N))$ be the positive diagram of N. The presentation $\langle \Sigma(N), \Delta^+(N) \cup \iota(\Phi) \rangle$ has a reachable initial model, say I. We show that $I|_\iota$ is isomorphic to N, which implies that $N \in Mod_\Sigma(\Phi)$.
As in the previous case, there exists $f{:}E(N) \to I$ with $f|_\iota \in \mathbf{E}_\Sigma$. Then, for $i \in J$, let $E_{\pi_i}(M_i)$ be the unique $\Sigma(N)$-model such that there is an expansion of π_i to a $\Sigma(N)$-model morphism $E(\pi_i){:}E(N) \to E_{\pi_i}(M_i)$. $E_{\pi_i}(M_i)$ satisfies both $\Delta^+(N)$ and $\iota(\Phi)$, and so there exists a morphism $h_i{:}I \to E_{\pi_i}(M_i)$. Hence, by the definition of a product, there exists a (unique) Σ-model morphism $g{:}I|_\iota \to N$ such that for $i \in J$, $h_i|_\iota = g;\pi_i$. Moreover, for $i \in J$, since $E(N)$ is reachable and so there is at most one morphism from $E(N)$ to $E_{\pi_i}(M_i)$, $f;h_i = E(\pi_i)$. Consequently, $(f|_\iota;g);\pi_i = f|_\iota;h_i|_\iota = (f;h_i)|_\iota = E(\pi_i)|_\iota = \pi_i$. It follows that

$f|_\iota;g$ is an isomorphism, and thus $f|_\iota \in \mathbf{E}_\Sigma$ implies that $f|_\iota : N \to I|_\iota$ is an isomorphism as well. □

Exercise 4.5.21. As we have mentioned earlier, institutions of single-sorted logics, like those in Exercises 4.4.10 and 4.4.16, are only semi-exact, rather than finitely exact.

Call an institution **INS** *almost abstract algebraic* if it satisfies all the assumptions imposed on abstract algebraic institutions except for the requirement of finite exactness, instead of which we require that:

- **INS** is semi-exact; and
- for each signature $\Sigma \in |\mathbf{Sign_{INS}}|$, the category $\mathbf{Mod_{INS}}(\Sigma)$ of Σ-models has an initial object.

The above characterisation theorems nearly hold for almost abstract algebraic institutions:

- By direct inspection of their proofs, check that Theorem 4.5.20 as well as the "if" parts of Theorems 4.5.18 and 4.5.19 hold for almost abstract algebraic institutions.
- Prove that the "only if" part of Theorem 4.5.18 holds for almost abstract algebraic institutions. HINT: To show that a Σ-theory T has an initial model, consider the identity signature morphism as a morphism from the empty Σ-theory to T. Then use Exercise 3.5.17.
- Show that the "only if" part of Theorem 4.5.19 does not hold for almost abstract algebraic institutions. HINT: In **SSEQ**, the requirement of σ-reachability is trivial for any signature morphism σ. Consider the extension of **SSEQ** by sentences involving the quantifier "there exists a unique". □

4.6 Bibliographical remarks

This chapter has its origins in the seminal work of Goguen and Burstall on institutions. The reader may have noticed that the main paper on institutions [GB92] appeared later than many of its applications. The first appearance of institutions was in the semantics of Clear [BG80], under the name "language", and early versions of [GB92] were widely circulated, with [GB84a] an early published version. Most of our terminology (signature, sentence, model, liberal institution, and so on) comes from [GB92]. There is a minor technical difference with respect to the definition given in [GB92]: we take the contravariant functor $\mathbf{Mod_{INS}}$ to be $\mathbf{Mod_{INS}}: \mathbf{Sign}^{op}_{\mathbf{INS}} \to \mathbf{Cat}$ rather than $\mathbf{Mod_{INS}}: \mathbf{Sign_{INS}} \to \mathbf{Cat}^{op}$. This is consistent with the further refinement of this definition in Chapter 10 as well as with the notion of an indexed category (cf. Section 3.4.3 and [TBG91]).

A large number of variants, generalisations and extensions of the notion of institution have been considered. In some work where model morphisms are not important, institutions were considered with classes (rather than categories) of models, e.g. [BG80]. Somewhat dually, one way to bring deduction into the realm of institutions is by considering categories (rather than sets) of sentences, where morphisms capture proofs. These variants were present in some unpublished versions of [GB92]; see also [MGDT07] for some elaboration on these possibilities.

One line of generalisation is to allow a space of truth values other than just the standard two-valued set, leading to proposals like galleries [May85] or generalised institutions [GB86]. General logics [Mes89] add an explicit notion of entailment and proof to institutions; see Chapter 9 for developments in this direction. Foundations [Poi88] include a similar idea, in addition imposing a rich indexed category structure on sentences. Context institutions [Paw96] offer an explicit notion of context and hence of open formulae and valuation as a part of the institution structure. There have also been attempts to relax the satisfaction condition, with for instance pre-institutions ([SS93], [SS96]), where the equivalence in the satisfaction conditions is split into two separately imposed implications. This captures logical systems in which one or both of the directions of the satisfaction condition fail, as discussed before Exercise 4.1.2. This applies to the so-called ultra-loose approach to algebraic specification [WB89], Extended ML [KST97] and various notions of behavioural satisfaction; see Chapter 8. (In [Gog91b], the satisfaction condition holds for behavioural satisfaction but at the cost of restricting the notion of signature morphism.) Overall though, in spite of all these proposed variants and generalisations, most research has been based on the original notion, as we present it here.

The theory of institutions adopts a primarily model-theoretic view of logical systems. This does not preclude proof-theoretic investigation — see Chapter 9 — but it does exclude logical systems that are inherently not based on the Tarskian notion of satisfaction of a sentence in a model. Typically such systems are centred around a notion of logical consequence that is defined via deduction, in contrast to our Definition 4.2.5. One such example would be non-monotonic logics [MT93], where increasing the set of premises can render consequences invalid. Other examples include substructural logics such as linear logic [Gir87], where changing the number of occurrences of premises, or their order, may affect deduction and change the set of valid consequences. Clearly, such logics cannot be directly represented as institutions, but see [CM97], which indicates how an institution for linear logic can be defined by taking linear logic sequents (statements about consequence) as individual sentences. A view of logic based on proof rules and deduction underlies so-called "general logical frameworks", with Edinburgh LF [HHP93] as a prime example. For proposals in this direction related to institutions, see π-institutions [FS88] and also entailment systems [Mes89], [HST94], which re-emerge in Definition 9.1.2 below.

Section 4.1.1 gives only the beginning of the long list of examples of logical systems that have been formalised as institutions. Standard examples of institutions (**EQ**, **FOP**, **Horn**, **Horn** without equality, $\mathbf{EQ}^{\Rightarrow}$) were presented in [GB92], with further standard algebraic variants in [Mos96b], and **CEQ** is from [Tar86b].

Dozens of other logical systems have been formalised as institutions. Some examples: [Bor00] defines an institution of higher-order logic based on HOL; [SML05] defines an institution with type class polymorphism; [Roş94] defines an institution of order-sorted equational logic; [ACEGG91] defines a family of institutions of multiple-valued logics, including logical systems arising from fuzzy set theory; [Dia00] defines an institution of constraint logic; [Cîr02] defines an institution with models that have both coalgebraic and algebraic components, and sentences involving modal formulae; [FC96] defines an institution of temporal logic; [LS00] defines an institution of hybrid systems based on the specification language of HYTECH [HHWT97]; and [BH06a] defines the COL constructor-based observational logic institution based on viewing reachability and observability as dual concepts. The semantics of basic specifications in CASL [BCH$^+$04] defines an institution, the rest of the semantics being defined in an institution-independent fashion. Alternatives to the standard CASL institution include the institution underlying CO-CASL, which includes cogeneration constraints, cofreeness constraints, and modal formulae [MSRR06]; the institution underlying HASCASL, with partial higher-order functions, higher-order subtyping, shallow polymorphism, and type classes, designed for specifying functional programs [SM09]; an institution of labelled transition logic for specifying dynamic reactive systems [RAC99]; and the institution underlying CSP-CASL for describing systems of processes [Rog06]. The eight institutions involved in CafeOBJ [DF98] are defined in [DF02], with their combination leading to an institution via a version of the Grothendieck construction (Definition 3.4.59) that is applicable here [Dia02], and the Maude language [CDE$^+$02] is based on rewriting logic [Mes92] and on the institution of membership equational logic [Mes98] (with some technical nuances of their relationship pointed out in [CMRM10]). Institutions for three different UML diagram types are defined in [CK08a], [CK08b], [CK08c], with the relationships between them given by institution comorphisms (see Section 10.4 below). A spectrum of institutions capturing some aspects of Semantic Web languages are defined and linked with each other in [LLD06]. Different approaches to the specification of objects have led to the definition of a number of institutions; they include [SCS94], which defines an institution of temporal logic for specifying object behaviour, [GD94b], which argues that an institution based on hidden-sorted algebra is relevant, and [Zuc99], which shows how to construct an institution with features for specifying dynamic aspects of systems using so-called "d-oids" from an institution for specifying static data. Finally, some slightly non-standard examples include two institutions for graph colouring in [Sco04], a way of viewing a database as an institution [Gog11], and a framework based on institutions for typed object-oriented, XML and other data models [Ala02].

Some of the examples of constructions on institutions in Section 4.1.2 were independently introduced by others. For instance, [Mes89] constructs an institution "out of thin air" starting with theories in an entailment system, the idea of which is presented in Examples 4.1.36 and 4.1.40. Incidentally, a very interesting exercise is to use the method of diagrams (Definition 4.5.9) to show how the construction of models from theories recovers the institution for which the entailment system that generates the theories was built.

Overall though, Section 4.1.2 only hints at the issue of how institutions should be defined. In particular, we do not discuss here the notion of a *parchment* [GB86], which offers one convenient way to present institutions in a concise and uniform style, at the same time ensuring that the satisfaction condition holds. See also

[MTP97], [MTP98] and comments in Section 10.5 for variants of this notion and its use for combining presentations of logical systems.

The idea of data constraints originates in [BG80], but was independently introduced earlier by Reichel [Rei80]; cf. [KR71]. Our treatment in Section 4.3 follows [GB92]. Definition 4.3.8 is essentially equivalent to the definition there, although the technicalities are somewhat different; in particular, as in [ST88a], we do not require the institution to be liberal. Hierarchy constraints [SW82], also known as generating constraints [EWT83], are like data constraints but require that some carriers be generated from other carriers rather than freeness; see Exercise 4.3.13. Exercise 4.3.14 introduces a way to specify so-called coinductive data types involving infinitary data. This has been mixed with algebraic techniques both in specification — see CoCASL [MSRR06] — and in experimental programming languages — see [Hag87] and Charity [CS92], [CF92]. See [Rut00] for an introduction to a comprehensive coalgebraic approach to specification which provides an alternative perspective to the material on behavioural specifications in Chapter 8 below.

Colimits of signatures and theories built over them have been used as tools for combining theories and specifications at least since [BG77], [GB78]. This follows the general ideas of [Gog73] and underlies the semantics of Clear [BG80] and the commercial Specware system [Smi06]; support for the use of colimits to combine theories in a number of institutions is also offered by the HETS system [MML07]. A category-theoretic approach to software engineering which makes extensive use of these ideas is [Fia05]. Theorem 4.4.1 originates with [GB92], generalising a non-institutional version in [GB84b], and Corollary 4.4.2 is from [BG80].

The idea of amalgamation in model theory [CK90] refers to a subtler and deeper property of certain theories than does the notion defined here. The use of amalgamation in algebraic specification, in connection with pushout-style parameterisation mechanisms, originates with [EM85], following its introduction in [BPP85]; see also the Extension Lemma in [EKT$^+$83]. In the context of an arbitrary institution, it was first imposed as a requirement and linked with continuity of the model functor in [ST88a]; cf. [EWT83].

Limiting the amalgamation property to pushouts along a chosen collection of signature morphisms, as in Definition 4.4.18, is important not only because of examples like those in Exercise 4.4.19. The range of relevant cases includes systems emerging in practice. For instance, the institution of CASL [Mos04] admits amalgamation for pushouts along most, but not all, CASL signature morphisms, due to problems with the required unique interpretation of subsorting coercions; see [SMT$^+$05].

There has been some confusion with the terminology surrounding exactness of institutions in the literature. The term was first used in [Mes89], although for preservation of signature pushouts (the amalgamation property) only. It became widely used after [DGS93], where it meant that the model functor maps finite colimits of signatures to limits in **Cat**, so that neither infinite colimits nor existence of colimits were covered (the latter also applies to semi-exactness as introduced there). This was sometimes missed in the literature, leading to subtle mistakes in the presentation of some results. We decided to put all of these assumptions together under the single requirement of "exactness". The notion of an institution "with composable signatures" was used in early versions of this chapter and in [Tar99] to mean the same thing as exactness, and this terminology was adopted by other authors in a few papers. The notion of exactness as used in category theory is different, although for functors between so-called Abelian categories it implies preservation of finite colimits.

The consequences of semi-exactness for preservation of finite connected colimits of signature diagrams stated in Proposition 4.4.15 appear to be new in the literature concerning institutions; they had not been clear to us until we were pointed to [CJ95] and a result there which we give as Exercise 3.4.56.

Institutions with extra structure have been used as the basis for the definition of the semantics of a number of specification languages, beginning with ASL [ST88a], which required an exact institution. In [ST86], an institution-independent semantics for the Extended ML specification language is sketched in terms of an "institution with syntax"; this requires an additional functor which gives concrete syntactic representations of sentences, together with a natural transformation which associates these concrete objects with the "abstract" sentences they represent. In [BCH+04], the semantics of CASL is based on an "institution with qualified symbols" [Mos00], which requires considerable additional structure in order to support the operations on signatures used in the semantics; these include union of signatures and generation of signature morphisms from maps between symbols. Similar constructions on signatures are available when the category of signatures is equipped with a so-called inclusion system, which leads to the concept of an inclusive institution [DGS93], [GR04] (see also Exercise 5.2.1 below).

Although the theory of institutions emerged originally in the context of algebraic specification theory, it shares ideas and broad goals with abstract model theory as pursued within mathematical logic — see [Bar74], [BF85] — which concentrates on the study of definable classes of algebras (or rather first-order structures), abstracting away from the structure of sentences and from proof-theoretic mechanisms. The idea of developing an institutional version of abstract model theory, which also abstracts away from the nature of models, was first put forward in [Tar86a], where the equivalence of the Craig interpolation and Robinson consistency properties (Exercise 4.4.26) was shown.

The Craig interpolation property (Definition 4.4.21) will be used frequently in the sequel. In this formulation, it originates in [Tar86a]. Interpolation for first-order logic is a standard result in model theory [CK90] but the delicacy of its status in many-sorted first-order logic (see Exercise 4.4.23) was first pointed out in [Bor05]. There are several variants of the formulation of interpolation, including Craig-Robinson interpolation (Exercise 4.4.27), discussed extensively in [DM00]. The generalisation to arbitrary pushouts [Tar86a] and then to commutative squares of signature morphisms [Dia08] and sets of interpolants (see the discussion in [DGS93]) is especially important. In particular, sets of interpolants may always be found in the case of equational logic under the assumption that carriers are non-empty [Rod91], but the necessity of this assumption has been widely disregarded; see Exercise 4.4.25.

Our treatment of variables, open formulae and quantification in an arbitrary institution comes from [Tar86b], [ST88a]; see the concept of syntactic operator in [Bar74] for an earlier related idea. Section 4.5 is based on [Tar85], following [MM84], which is in an institutional style but based on the standard notion of logical structure. In [Tar86b], infinitary conditional "equations" were defined for an arbitrary abstract algebraic institution and it was shown that sets of these sentences define quasi-varieties — see [Mal71] — thus providing a "syntactic" version of Theorem 4.5.20. Further developments in institutional abstract model theory, with results and ideas that refine those in Sections 4.4 and 4.5 and reach much further into classical model theory than we have done here, are presented in [Dia08].

Chapter 5
Structured specifications

Chapter 2 appears to provide just what is needed to specify programs. A flat specification records the required sort and operation names along with the axioms that the operations must satisfy. Taking its initial interpretation is one way of excluding undesirable models, and the equational calculus (together with appropriate induction schemes, in the case of the initial interpretation) can be used to prove additional properties of models on the basis of the axioms. Most of this generalises smoothly to an arbitrary logical system, as discussed in Section 2.7 and Chapter 4.

In practice, it very quickly becomes apparent that these techniques are only suitable for writing and reasoning about small specifications containing at most a handful of sorts and perhaps two dozen operations. A flat specification of a large system would be impossible to understand or use — imagine trying to decide what a 200-page list of axioms really specifies!

What is needed is some means of building complex *structured* specifications by combining and extending simpler ones. Then, an understanding of a large specification may be achieved via an understanding of its components, and components of large specifications may be reused. Moreover, the structure of a specification conveys intangible but important aspects of the conceptual structure of the problem domain, such as the degree to which entities and concepts described in the specification are interrelated. As we will see in later chapters, the structure of specifications can often be exploited in their subsequent use, for instance both in structuring proof systems for specifications and in guiding proof search.

Building on the specification mechanisms in Chapter 2, this chapter introduces *specification-building operations* for use in constructing structured specifications. It is possible to give a treatment of the particular operations presented at the level of an arbitrary institution (Chapter 4), abstracting away from the particular details of the logical system in use. Although these operations are very simple, their expressive power is sufficient to capture most of the specification-building operations provided by existing specification languages. There are two exceptions: *parameterisation* mechanisms are treated in Chapter 6; and *behavioural abstraction* is treated in Chapter 8.

We deliberately do not define a specification language, or even fix a specific set of specification-building operations. This is partly because our aim is to study basic concepts that are common to all such languages, and partly because most of the sequel does not rely on the choice of language features. For results having such a dependence, the requirements will be stated explicitly. Examples will use a subset of the specification-building operations we define. Despite our desire to avoid a specific choice of language features, we use a notation — already in the preceding chapters — that is very close to CASL [Mos04]. Even for the part of our framework that overlaps with CASL, we go beyond what is in CASL and this has forced us to diverge from CASL notation and terminology in several respects. For the sake of CASL users, we record these differences in footnotes[1] when they first appear. The first such footnote appeared in Definition 4.1.25.

Some readers will be disturbed by the absence of examples of large and complex specifications in a chapter devoted to the provision of means for their construction. The examples that are provided are chosen for simplicity and to illustrate particular points rather than to motivate the need for mechanisms to deal with large specifications, which is regarded as self-evident. There is a somewhat more substantial example in Section 5.3 to give a taste of how the development of a structured specification might proceed. Larger examples, expressed using specification-building operations that are different from but closely related to those presented here, can be found in (for instance) [GH93], [BM04], [Mos04] and [MHST08].

5.1 Specification-building operations

Flat specifications were introduced in Chapter 2 and then generalised to the context of an arbitrary institution in Section 4.2. Here we will be dealing with structured specifications, built from flat specifications using the specification-building operations introduced below. At this stage, we will not dwell on the syntax of these operations, using a convenient but ad hoc notation. The main emphasis will be on the semantics of specifications: for each specification, we define its signature and its class of models. The signature of a specification defines an interface giving names to the required program components, while its models represent programs that are considered to be correct realisations. (There is a subtle issue involved in ensuring that *all* such programs are represented; see Chapter 8.) A specification with signature Σ is called a *Σ-specification*; it is called *consistent* if it has a non-empty class of models. We will write *Spec* for the class of all specifications, without defining what this class is, since all that matters is that specifications come with a semantics of the indicated form. We write $Spec(\Sigma)$ for the class of Σ-specifications.

While the meaning of a flat specification is given in one step, it is natural to define the meaning of a structured specification in a way that takes account of its structure. A *compositional* semantics is one in which the meaning of a phrase is de-

[1] CASL notation: such footnotes look like this.

termined from the meanings of its immediate constituents. A structured specification is built by consecutive application of specification-building operations; with a compositional semantics, the meaning of a structured specification is obtained by the consecutive application of functions on signatures and model classes corresponding to each specification-building operation. Provided these are relatively simple functions, understanding of a large specification can be achieved via an understanding of its individual components. Indeed, *any* manipulation of a large specification can be achieved via manipulation of its individual components, taking account of their compositional relationship.

In this chapter we focus on the following basic specification-building operations:

- Form the *union* of two Σ-specifications, combining the constraints imposed by each.
- *Translate* a Σ-specification to another signature Σ' with a signature morphism $\sigma: \Sigma \to \Sigma'$. This together with union allows large specifications to be built from smaller and more or less independent specifications.
- *Hide* components of a Σ-specification via a signature morphism $\sigma: \Sigma' \to \Sigma$ to obtain a view of the Σ-specification as a Σ'-specification. This allows auxiliary components of a specification to be hidden while essentially preserving its collection of models.

These are very representative operations that can be found in some form in practically all algebraic specification languages in existence, and are a sufficient basis for expressing the most common ways of combining and modifying specifications. Still, this choice of operations is somewhat arbitrary. It is easy to define further operations; some examples are given as exercises at the end of this section.

The semantics of these operations is defined in an arbitrary but fixed institution $\mathbf{INS} = \langle \mathbf{Sign}, \mathbf{Sen}, \mathbf{Mod}, \langle \models_\Sigma \rangle_{\Sigma \in |\mathbf{Sign}|} \rangle$. Results in later sections that refer to these operations are institution-independent as well. Expressing them at this level identifies the basic assumptions required to support them, making the whole theory applicable in a wide range of contexts and lifting our discourse to the appropriate level of generality.

Of course, any actual application requires the choice of an institution that models programs at the right level of abstraction, capturing all the features of interest while avoiding distracting details. Not all specification-building operations can be defined in an arbitrary institution; it is sometimes necessary to require additional structure; see, for instance, Exercise 5.1.10.

We now proceed to define the specification-building operations. Each definition gives the meaning of a specification built using the given operation in terms of the meanings of its component specifications. This style of presentation gives a compositional semantics, which for each specification SP defines its signature $Sig[SP]$ and its class of models $Mod[SP] \subseteq |\mathbf{Mod}(Sig[SP])|$. When we need to indicate **INS**, we refer to SP as a specification *over* **INS**.

Union

If both SP_1 and SP_2 are Σ-specifications (so $Sig[SP_1] = Sig[SP_2] = \Sigma$) then $SP_1 \cup SP_2$ is a specification with the following semantics:[2]

$$Sig[SP_1 \cup SP_2] = \Sigma$$
$$Mod[SP_1 \cup SP_2] = Mod[SP_1] \cap Mod[SP_2]$$

If both SP_1 and SP_2 are flat specifications, their union is expressible as a flat specification as well: $Mod[\langle \Sigma, \Phi_1 \rangle \cup \langle \Sigma, \Phi_2 \rangle] = Mod[\langle \Sigma, \Phi_1 \cup \Phi_2 \rangle]$. This motivates the use of the name "union" for this operation: intersection of model classes corresponds to union of requirements.

We give a simple example over the institution **EQ**. Consider the specification GROUP of groups over the signature with sort s and operations $\_\_+\_\_ : s \times s \to s$, $0 : s$ and $-\_\_ : s \to s$ and a specification COMM over the same signature requiring $+$ to be commutative. Then GROUP $\cup$ COMM specifies Abelian groups. Note that the signatures of the two specifications are required to match exactly even though commutativity can be expressed in a smaller signature without 0 or $-$.

Translation

If SP is a Σ-specification and $\sigma : \Sigma \to \Sigma'$ is a signature morphism then SP **with** σ is a specification with semantics defined as follows:[3]

$$Sig[SP \textbf{ with } \sigma] = \Sigma'$$
$$Mod[SP \textbf{ with } \sigma] = \{M' \in |\mathbf{Mod}(\Sigma')| \mid M'|_\sigma \in Mod[SP]\}$$

If SP is a flat specification $\langle \Sigma, \Phi \rangle$ then SP **with** σ has exactly the same models as $\langle \Sigma', \sigma(\Phi) \rangle$, where $\sigma(\Phi)$ is the image of Φ under σ (i.e. under $\mathbf{Sen}(\sigma) : \mathbf{Sen}(\Sigma) \to \mathbf{Sen}(\Sigma')$).

The specification COMM above can be given as follows:

spec SIMPLECOMM = **sorts** s
  ops $op : s \times s \to s$
  $\forall x, y : s \bullet op(x, y) = op(y, x)$

spec COMM = SIMPLECOMM **with** σ

where $\sigma : Sig[\text{SIMPLECOMM}] \to Sig[\text{GROUP}]$ maps s to s and op to $+$.

Using translation with a bijective signature morphism simply changes the names of symbols. If the signature morphism is not surjective, translation also adds new symbols without constraining their interpretation. Using translation with a non-injective signature morphism imposes a requirement that the interpretation of the symbols that the morphism identifies coincide.

[2] CASL notation: union is subsumed by sum (Section 5.2) which is written using **and** in CASL.

[3] CASL notation: σ is required to be surjective in CASL. The effect of translation along a non-surjective σ would be achieved there using a combination of translation and enrichment.

Exercise 5.1.1. Give an example of a consistent specification *SP* and signature morphism σ such that *SP* **with** σ is inconsistent. HINT: Use **FOEQ**. Can you do this in **EQ**? □

Hiding

If $\sigma: \Sigma' \to \Sigma$ is a signature morphism and *SP* is a Σ-specification then *SP* **hide via** σ is a specification with the following semantics:[4]

$$\begin{aligned} Sig[SP\ \mathbf{hide\ via}\ \sigma] &= \Sigma' \\ Mod[SP\ \mathbf{hide\ via}\ \sigma] &= \{M|_\sigma \mid M \in Mod[SP]\} \end{aligned}$$

$Mod[\langle\Sigma,\Phi\rangle\ \mathbf{hide\ via}\ \sigma] \subseteq Mod[\langle\Sigma',\sigma^{-1}(Cl_\Sigma(\Phi))\rangle] \subseteq Mod[\langle\Sigma',\sigma^{-1}(\Phi)\rangle]$, where $\sigma^{-1}(\Phi)$ is the coimage of Φ under σ (i.e. under $\mathbf{Sen}(\sigma)$). Note however that both inclusions may be proper: concerning the first, sometimes not all the properties of models of the specification obtained by hiding are expressible using just Σ'-sentences; concerning the second, see Exercise 4.2.14.

Let σ be the inclusion of the signature *Sig*[GROUP] without the operation "−" into *Sig*[GROUP]. Then GROUP **hide via** σ specifies groups having no explicit inverse operation. Note that it does *not* specify the class of all semi-groups: all models of GROUP **hide via** σ admit inverses, while this is of course not the case for some semi-groups.

Most uses of hiding are to hide auxiliary sorts and operations, and involve a signature morphism which is an inclusion. Such hiding adds expressive power: an example of a specification in the institution **EQ** of equational logic using hiding that cannot be finitely expressed without it can be found in [TWW82]. This extends to many institutions with more expressive sentences since second-order existential quantification may be needed to describe what otherwise can be captured by specifications with hidden operations. See Section 5.6 for related discussion. An injective signature morphism that is not an inclusion can be used to combine hiding of sorts and/or operations with renaming.

The use of hiding with a non-injective signature morphism is less common although this can be used to duplicate sorts and operations: if Σ' extends *Sig*[GROUP] by a function $inv: s \to s$ and $\sigma: \Sigma' \to Sig[\text{GROUP}]$ maps *inv* to "−" and is the identity otherwise, then GROUP **hide via** σ is just like GROUP, except with two names for the (same) inverse function.

Exercise 5.1.2. In most institutions, the satisfaction relation is preserved under isomorphism of models, so the class of models of a flat specification is always closed under isomorphism. This property (called the *abstraction condition* in Definition 4.5.13) is not guaranteed by the definition of an institution. Even in institutions like **EQ** that do satisfy this property, specifications produced by hiding may violate it: give a flat specification *SP* over **EQ** and signature morphism σ such that

[4] CASL notation: σ is required to be a signature inclusion in CASL, and then the notation is as for export in Section 5.2.

$Mod[SP$ **hide via** $\sigma]$ is not closed under isomorphism. HINT: Consider a signature morphism σ that is not injective on sorts. □

The above definitions enjoy a number of useful properties. All well-formedness conditions (for example, the requirement that $Sig[SP] = Sig[SP']$ in $SP \cup SP'$) are stated solely in terms of the signatures of constituent specifications. Moreover, the signature of a specification depends only on the signatures of its constituents:

Exercise 5.1.3. View the specification-building operations defined above as families of functions indexed by signatures and signature morphisms as appropriate. That is:

Union: $\_\_ \cup \_\_ : Spec(\Sigma) \times Spec(\Sigma) \rightarrow Spec(\Sigma)$ for each signature $\Sigma \in |\mathbf{Sign}|$;
Translation: $\_\_$ **with** $\sigma : Spec(\Sigma) \rightarrow Spec(\Sigma')$ for each signature morphism $\sigma : \Sigma \rightarrow \Sigma'$; and
Hiding: $\_\_$ **hide via** $\sigma : Spec(\Sigma') \rightarrow Spec(\Sigma)$ for each signature morphism $\sigma : \Sigma \rightarrow \Sigma'$;

Note that all of these functions are total. View all of the specification-building operations below in this style. □

Specification-building operations are monotone on model classes in the sense of the following exercise:

Exercise 5.1.4. Check that all of the specification-building operations above are *monotone* in the sense that for any specification-building operation $\mathbf{sbo} : Spec(\Sigma) \rightarrow Spec(\Sigma')$ and Σ-specifications SP, SP', if $Mod[SP] \subseteq Mod[SP']$ then $Mod[\mathbf{sbo}(SP)] \subseteq Mod[\mathbf{sbo}(SP')]$. □

In other words, building a specification from a more restrictive constituent specification yields a more restrictive result.

Exercise 5.1.5. It is straightforward to generalise union to an arbitrary indexed family of Σ-specifications for any signature Σ, written $\bigcup_{i \in I} SP_i$. (Note that if $I = \varnothing$, the signature Σ is still required and then $Mod[\bigcup_{i \in I} SP_i] = |\mathbf{Mod}(\Sigma)|$.)

For operations over families of specifications, two notions of monotonicity make sense: either fix all but one element in the family and compare the result for different choices of the remaining element; or let all elements in the family vary simultaneously. Show that union is monotone in both senses. Show that both notions coincide for operations which combine finite families of specifications. Define a specification-building operation (combining an infinite family of specifications) that is monotone in the former sense but not in the latter. Is the converse possible? □

The above specification-building operations are only illustrative examples. It is easy to think of further operations, and some possibilities are discussed in the following exercises.

Exercise 5.1.6. Define an operation $SP \cap SP'$ that is dual to $SP \cup SP'$. □

Exercise 5.1.7. Recall from Exercise 5.1.2 that the class of models of a specification need not be closed under isomorphism. Define an operation **iso-close** SP that can be used to force this by closing the class of models of SP under isomorphism. □

Exercise 5.1.8. The **iso-close** operation from Exercise 5.1.7 generalises in an obvious way to closure under any family of equivalences on models. Given $\simeq = \langle \simeq_\Sigma \rangle_{\Sigma \in |\mathbf{Sign}|}$, where $\simeq_\Sigma \subseteq |\mathbf{Mod}(\Sigma)| \times |\mathbf{Mod}(\Sigma)|$ is an equivalence for any $\Sigma \in |\mathbf{Sign}|$, define an operation $\mathbf{close}_\simeq$ *SP* which closes the class of models of *SP* under $\simeq_{Sig[SP]}$. Then prove that for any specifications *SP*, *SP'* over the same signature:

- Closure preserves the original models: $Mod[SP] \subseteq Mod[\mathbf{close}_\simeq SP]$.
- Closure is monotone: $Mod[\mathbf{close}_\simeq SP] \subseteq Mod[\mathbf{close}_\simeq SP']$ whenever $Mod[SP] \subseteq Mod[SP']$.
- Closure is idempotent: $Mod[\mathbf{close}_\simeq (\mathbf{close}_\simeq SP)] = Mod[\mathbf{close}_\simeq SP]$.
- Closure preserves and reflects consistency: $Mod[\mathbf{close}_\simeq SP] \neq \varnothing$ iff $Mod[SP] \neq \varnothing$. □

Exercise 5.1.9. Consider the following specification-building operation. If $\sigma\colon \Sigma' \to \Sigma$ is a signature morphism and *SP* is a Σ-specification then **free** *SP* **wrt** σ is a specification with the following semantics:[5]

$$\begin{aligned}
Sig[\mathbf{free}\ SP\ \mathbf{wrt}\ \sigma] &= \Sigma \\
Mod[\mathbf{free}\ SP\ \mathbf{wrt}\ \sigma] &= \\
\{M \in Mod[SP] \mid &\text{ for any } M1 \in Mod[SP] \text{ and } \Sigma'\text{-morphism } f\colon M|_\sigma \to M1|_\sigma, \\
&\text{there is a unique } \Sigma\text{-morphism } f^\sharp\colon M \to M1 \text{ with } f^\sharp|_\sigma = f\}
\end{aligned}$$

In other words, the models of **free** *SP* **wrt** σ are those models of *SP* that are free over their σ-reducts with respect to the functor $\_|_\sigma\colon \mathbf{Mod}[SP] \to \mathbf{Mod}(\Sigma')$ with the identity as unit, where $\mathbf{Mod}[SP]$ is the full subcategory of $\mathbf{Mod}(\Sigma)$ determined by the models of *SP*.

Show how the data constraints introduced in Section 4.3 may be captured using this operation together with translation of specifications. HINT: Compare the models of a data constraint **data** Φ' **over** σ **through** θ (Definition 4.3.8) and of the specification (**free** $\langle \Sigma', \Phi' \rangle$ **wrt** σ) **with** θ. You may also consider a generalisation of data constraints by allowing them to contain an arbitrary specification in place of a set of axioms.

Introduce an operation **initial** *SP* for selecting the initial models of a specification *SP* and relate it to the initiality constraints introduced in Definition 4.3.3.

Check that these two specification-building operations are not monotone, unlike the operations considered so far (cf. Exercise 5.1.4). □

Exercise 5.1.10. In **EQ**, introduce a specification-building operation which, given a specification *SP*, yields those models of *SP* that are reachable. Generalise this as follows.

First, define an operation that imposes reachability with respect to a signature morphism.[6] (A Σ-algebra is *reachable with respect to* $\sigma\colon \Sigma' \to \Sigma$ if it has no proper subalgebra having the same σ-reduct.) A special case could be written as follows:

[5] CASL notation: this is written **free** $\{SP\}$ in CASL, where σ is implicitly taken to be the signature inclusion from the signature of the current context to the signature of the context enriched by *SP*.

[6] CASL notation: this is called **generated** in CASL, with the same convention for the implicit signature inclusion as for **free**.

reachable SP **on** S

where S is a set of sorts in $Sig[SP]$, which has as models all those models of SP in which the carriers of sorts in S are generated from the other carriers.

Second, generalise these operations to an arbitrary institution (with additional structure) using the apparatus of Section 4.5; cf. Definition 4.5.1.

Finally, show how the reachability and generation constraints introduced in Exercises 4.3.7 and 4.3.13 may be expressed using this operation. □

The use of **reachable** and **free** in the context of other specification-building operations brings benefits similar to those of the use of data constraints with morphisms that "fit" them to larger signatures; see Section 4.3. In both cases, it is possible to constrain (require reachability or freeness for) selected parts of the specified models only. This allows specifications to be built up in layers, where the interpretation of each layer does not interfere with the interpretation of "deeper" layers. This is in contrast to the initial algebra framework of Section 2.5 where the initiality requirement applies to the whole specification.

Exercise 5.1.11. Redo Exercises 2.5.20 or 2.7.9, first using **reachable** or **free** to specify NAT and then adding subtraction. (You might want to redo it again for cosmetic reasons after enrichment is defined in Section 5.2 below.) What is the value of $0 - succ(0)$? □

Exercise 5.1.12. In **EQ**, define an operation **quotient** SP **preserving** S, where S is a subset of the sorts of $Sig[SP]$, such that each model of the resulting specification is a quotient of a model of SP by some congruence that is the identity on sorts in S. Define the "dual" operation **unquotient** SP **preserving** S, such that for each model M of the resulting specification, there is a congruence $\sim$ on M such that $\sim$ is the identity on sorts in S and $M/\!\sim$ is a model of SP.

Use the apparatus of Section 4.5 to generalise these operations to an arbitrary institution with reachability structure. HINT: First replace the set of sorts S by a signature morphism. □

Exercise 5.1.13. Prove that all of the above operations, except for **free**, are monotone in the sense of Exercise 5.1.4. □

Exercise 5.1.14. Given a class of specifications with semantics in terms of signatures and model classes as above, we can view specifications as sentences of a (more expressive) logical system developed over the underlying institution. The satisfaction relation for such sentences is obvious: for any specification SP and model $M \in |\mathbf{Mod}(Sig[SP])|$, $M \models_{Sig[SP]} SP$ iff $M \in Mod[SP]$. Use the construction of Example 4.1.46 to formally define an institution of specifications over **INS**. The morphisms that "fit" the new sentences to other signatures may be incorporated into specifications using translation. (Note that care is required to ensure that translation of these sentences along signature morphisms is functorial.) Check that semantic consequence $\{SP\} \models SP'$ is simply inclusion of model classes $Mod[SP] \subseteq Mod[SP']$. □

5.2 Towards specification languages

In the previous section we did not define a formal specification language but merely a small set of specification-building operations that might underlie such a language. Nor shall we do so elsewhere in this book. One ingredient that would be required to promote our operations to a specification language is a carefully designed concrete syntax. Although we provide a suggestive and unambiguous notation for our operations, this is not a complete syntax. Without fixing a particular institution, the syntax for writing down signatures, signature morphisms and (sets of) sentences cannot be chosen. Another missing ingredient concerns all the features of languages that are routine from a semantic point of view but are nevertheless required for convenient use, such as provisions for naming specifications and other entities, for local definitions with the usual rules for visibility and scoping, for comments, and so on. See CASL [BM04], [Mos04] for an example of a full-blown specification language that provides all of these. Defining such a language is not our aim. Our concern here is the basic underlying concepts and operations of specification languages. A well-developed theory of these basic aspects and their properties is fundamental to the effective use of any specification language.

The starting point is a minimal set of basic specification-building operations having a clear and simple semantics such as the ones defined in the last section. Although these are a little primitive for convenient use in examples, in combination they have considerable expressive power. Abbreviations can be introduced for some of their most common combinations and/or to instantiate them to frequently occurring special cases, perhaps in the context of a specific institution or a class of institutions equipped with additional structure. These bridge part of the gap to the specification-building mechanisms in realistic specification languages, allowing the theory of the basic specification-building operations to be applied in this richer context.

Export

A very common use of hiding is to "export" the constraints imposed by a specification on a subsignature, thus revealing only some of the sorts and operations specified and indicating that the others are auxiliary; it is convenient to provide syntax for this special case.

In any institution **INS** with algebraic signatures ($\mathbf{Sign}_{\mathbf{INS}} = \mathbf{AlgSig}$), if *SP* is a Σ-specification and Σ' is a subsignature of Σ then *SP* **reveal** Σ' is a specification defined by

$$SP\ \mathbf{reveal}\ \Sigma' = SP\ \mathbf{hide\ via}\ \iota$$

where $\iota{:}\Sigma' \hookrightarrow \Sigma$ is the signature inclusion. If $\Sigma = \langle S, \Omega \rangle$ and $\Sigma' = \langle S', \Omega' \rangle$ then the same specification will also be written as

$$SP\ \mathbf{hide\ sorts}\ S \setminus S'\ \mathbf{ops}\ \Omega \setminus \Omega'$$

with the obvious notational variants if either $S \setminus S'$ or $\Omega \setminus \Omega'$ is empty.

Exercise 5.2.1. Generalise revealing to an arbitrary institution with appropriate extra structure. Make sure that this covers **FPL** and other institutions with signatures that are similarly based on algebraic signatures. HINT: Use *inclusion systems* [DGS93] or *institutions with symbols* [Mos00]. The idea is to identify an order subcategory of the category of signatures having morphisms that play the role of inclusions, including all identities. This category might be required to enjoy further properties such as the existence of unions and intersections. □

Enrichment

In any institution **INS** with algebraic signatures, let *SP* be a Σ-specification with $\Sigma = \langle S, \Omega \rangle$, S' be a set of sort names, Ω' be a set of operation names with arities and result sorts over $S \cup S'$, and Φ' be a set of sentences over the signature $\Sigma' = \langle S \cup S', \Omega \cup \Omega' \rangle$. Then *SP* **then sorts** S' **ops** $\Omega' \bullet \Phi'$ is a specification that enriches *SP* by S', Ω' and Φ', defined as follows:

$$SP\ \textbf{then sorts}\ S'\ \textbf{ops}\ \Omega' \bullet \Phi' = (SP\ \textbf{with}\ \iota) \cup \langle \Sigma', \Phi' \rangle^{7}$$

where $\iota : \Sigma \hookrightarrow \Sigma'$ is the signature inclusion. Notational variants are used to enhance convenience, and to extend this operation to institutions where signatures have other components. For instance, if $S = \varnothing$ we simply omit the "**sorts**" keyword, and for institutions with first-order signatures we use "**preds** Π" to add new predicates. For **FPL** (Example 4.1.25), sorts introduced in the enrichment may come with value constructors, and we also allow the notation **fun** $f(x_1{:}s_1, \ldots, x_n{:}s_n){:}s = t^8$ to be used to introduce an operation together with its definition, rather than requiring its name, arity and result sort in the **ops** part to be separated from its definitional axiom.

Free enrichment

The enrichment operation above is used to add new symbols and axioms to an existing specification. When the new symbols are to be interpreted freely, we use the following operation instead.

In any institution **INS** with algebraic signatures, let *SP* be a Σ-specification with $\Sigma = \langle S, \Omega \rangle$, S' be a set of sort names, Ω' be a set of operation names with arities and result sorts over $S \cup S'$, and Φ' be a set of sentences over the signature $\Sigma' = \langle S \cup S', \Omega \cup \Omega' \rangle$. Then *SP* **then free sorts** S' **ops** $\Omega' \bullet \Phi'$ is a specification defined by

$$\begin{aligned} &SP\ \textbf{then free sorts}\ S'\ \textbf{ops}\ \Omega' \bullet \Phi' = \\ &\qquad \textbf{free}\ (SP\ \textbf{then sorts}\ S'\ \textbf{ops}\ \Omega' \bullet \Phi')\ \textbf{wrt}\ \iota \end{aligned}$$

---

7 CASL notation: curly braces are used in place of parentheses for grouping in CASL.

8 CASL notation: semicolons are used in CASL in place of commas to separate groups of differently-sorted variables in operation definition parameter lists, and the keyword **fun** is not used.

where $\iota:\Sigma \hookrightarrow \Sigma'$ is the signature inclusion and **free** is defined in Exercise 5.1.9.

Example 5.2.2. The following small example in **FOPEQ** illustrates the use of specification enrichments, assuming that a specification NAT of natural numbers is given (see Exercise 2.5.4, or better, Section 5.3 below).

spec SET = (NAT
  then free
    sorts *NatSet*
    ops $\varnothing$:*NatSet*
      add:*Nat* × *NatSet* → *NatSet*
    $\forall x,y{:}Nat, S{:}NatSet$[9]
      • $add(x,add(y,S)) = add(y,add(x,S))$
      • $add(x,add(x,S)) = add(x,S)$)
  then
    pred $\_\_\in\_\_$:*Nat* × *NatSet*
    $\forall x,y{:}Nat, S{:}NatSet$
      • $\neg(x \in \varnothing)$
      • $x \in add(x,S)$
      • $x \neq y \Rightarrow (x \in add(y,S) \Leftrightarrow x \in S)$

Exercise. Expand this example using the definitions above. Assuming that NAT has the usual interpretation, convince yourself that all models of SET are isomorphic to the usual model of finite sets of natural numbers. □

Exercise 5.2.3. Show that the following specification has the same models as SET (the discussion of free interpretation of predicates in Exercise 4.3.12 applies here as well).

spec SET′ = (NAT
  then free
    sorts *NatSet*
    ops $\varnothing$:*NatSet*
      add:*Nat* × *NatSet* → *NatSet*
    $\forall x,y{:}Nat, S{:}NatSet$
      • $add(x,add(y,S)) = add(y,add(x,S))$
      • $add(x,add(x,S)) = add(x,S)$)
  then free
    pred $\_\_\in\_\_$:*Nat* × *NatSet*
    $\forall x,y{:}Nat, S{:}NatSet$
      • $x \in add(x,S)$
      • $x \in S \Rightarrow x \in add(y,S)$ □

[9] CASL notation: semicolons are used in CASL in place of commas to separate groups of differently-sorted variables in lists of quantified variables, as in $\forall x,y{:}Nat; S{:}NatSet$.

Sum

In any institution **INS** with algebraic signatures, if SP is a Σ-specification and SP' is a Σ'-specification then SP **and** SP' is a specification defined by

$$SP \text{ and } SP' = (SP \text{ with } \iota) \cup (SP' \text{ with } \iota')$$

where $\iota : \Sigma \hookrightarrow \Sigma \cup \Sigma'$ and $\iota' : \Sigma' \hookrightarrow \Sigma \cup \Sigma'$ are the signature inclusions.

Exercise 5.2.4. Generalise sum to an arbitrary institution with the extra structure hinted at in Exercise 5.2.1, covering at least **FPL**. □

The sum operation above provides the simplest way to combine specifications over different signatures. When a symbol occurs in both SP_1 and SP_2, the specification SP_1 **and** SP_2 contains just one copy of that symbol ("same name, same thing"). Unintended name clashes can be avoided by applying translation before sum. An alternative is to use a more sophisticated version of sum in which the intended common subsignature is indicated explicitly.

Sum with explicit sharing

In any institution **INS** with finitely cocomplete signature category, if SP_1, SP_2 are two specifications and Σ is a signature with signature morphisms $\sigma_1 : \Sigma \to Sig[SP_1]$ and $\sigma_2 : \Sigma \to Sig[SP_2]$, then $SP_1 +_{\sigma_1,\sigma_2} SP_2$ is a specification defined by

$$SP_1 +_{\sigma_1,\sigma_2} SP_2 = (SP_1 \text{ with } \sigma_2') \cup (SP_2 \text{ with } \sigma_1')$$

where σ_1' and σ_2' are determined by the following pushout diagram:

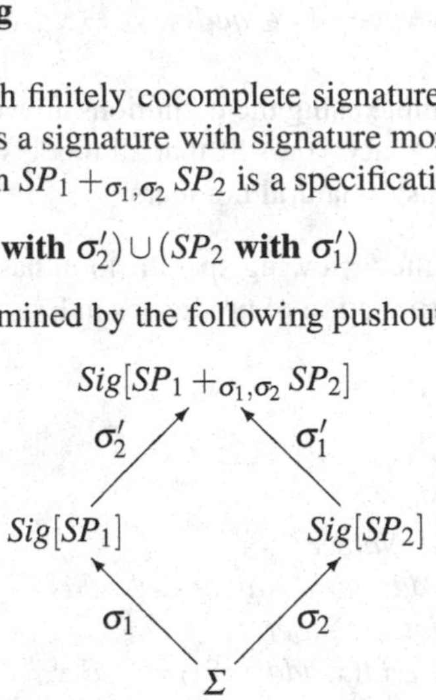

Note that the result signature of $SP_1 +_{\sigma_1,\sigma_2} SP_2$ is only defined to within isomorphism, and so strictly speaking some canonical pushout construction should be used. (See also Exercise 4.4.29.)

Exercise 5.2.5. Specialize the above operation to institutions with algebraic signatures where instead of giving σ_1 and σ_2, one would indicate a common subsignature of the two argument specifications, taking σ_1 and σ_2 to be signature inclusions. Then SP_1 **and** SP_2 would be the same as $SP_1 +_{Sig[SP_1] \cap Sig[SP_2]} SP_2$.

Notice also that this can be done in an arbitrary institution with the extra structure hinted at in Exercise 5.2.1. □

5.3 An example

The following sequence of specifications in **FOPEQ** illustrates the use of many of the specification-building operations introduced in the last two sections. The aim is a simplified specification of direct chaining hash tables, with the collision buckets again as hash tables (using another hash function) but with linear probing [Wik].

The items stored are strings and the only information we seek is whether a given string has been put into the table or not. Hash tables are built using arrays indexed by natural numbers.

spec NAT = **reachable**
    sorts *Nat*
    ops $0{:}\,Nat$
        $succ{:}\,Nat \to Nat$
        $\_\_+\_\_{:}\,Nat \times Nat \to Nat$
    preds $\_\_\leq\_\_{:}\,Nat \times Nat$
    $\forall n,m{:}\,Nat$
        • $0 \neq succ(n) \wedge (succ(m) = succ(n) \Rightarrow m = n)$
        • $0+n = n \wedge succ(m)+n = succ(m+n)$
        • $0 \leq n \wedge \neg(succ(n) \leq 0) \wedge (succ(m) \leq succ(n) \Leftrightarrow m \leq n)$
  on *Nat*

spec STRING = **initial**
    sorts *String*
    ops $\varepsilon{:}\,String$
        $\mathtt{a},\mathtt{b},\ldots,\mathtt{z}{:}\,String$
        $\_\_\hat{}\_\_{:}\,String \times String \to String$
    $\forall s,t,v{:}\,String$
        • $s\hat{}\varepsilon = s$
        • $\varepsilon\hat{}s = s$
        • $s\hat{}(t\hat{}v) = (s\hat{}t)\hat{}v$

Exercise 5.3.1. NAT and STRING above are expressed using different specification-building operations, **reachable** (see Exercise 5.1.10) and **initial** (see Exercise 5.1.9). Express NAT using **initial**. Without looking at Section 7.4 below, express NAT in the institution **FPL** (see Example 4.1.25) using a sort freely generated by value constructors.

Try to express STRING using **reachable**, adding axioms specifying when strings are different. How many additional axioms are required? Try to express STRING in **FPL** using a sort freely generated by value constructors; this will require auxiliary hidden operations. □

Here is the specification of the arrays used for the hash buckets. The use of "compound names" like *Array*[*String*] here and below bears no semantic significance and in particular has nothing to do with polymorphic types as in Example 2.7.57. They are just ordinary names, like *String*. But see Example 6.3.6 for a hint on how such structure can be exploited in full-blown specification languages.

spec STRINGARRAY =
  STRING **and** NAT
  then
    sorts $Array[String]$
    ops $empty{:}Array[String]$
      $put{:}Nat \times String \times Array[String] \to Array[String]$
      $get{:}Nat \times Array[String] \to String$
    preds $used{:}Nat \times Array[String]$
    $\forall i,j{:}Nat, s{:}String, a{:}Array[String]$
      • $\neg used(i, empty)$
      • $used(i, put(i,s,a))$
      • $i \neq j \Rightarrow (used(i, put(j,s,a)) \Leftrightarrow used(i,a))$
      • $get(i, put(i,s,a)) = s$
      • $i \neq j \Rightarrow get(i, put(j,s,a)) = get(i,a)$

The following is a specification of strings with the hashing function that will be used within the hash buckets. We require that hashing the empty string yields 0 as an example of a condition that may be imposed, but we make no use of this property below.

spec STRINGKEY0 =
  STRING **and** NAT
  then
    ops $hash0{:}String \to Nat$
    • $hash0(\varepsilon) = 0$

Here is the specification of the hash buckets. We begin by adjusting the sort name:

spec BUCKET = STRINGARRAY **with** $\sigma_{Array[String]\mapsto Bucket}$

where $\sigma_{Array[String]\mapsto Bucket}$ is a signature morphism from Sig[STRINGARRAY] that is surjective, renames $Array[String]$ to $Bucket$, and is the identity otherwise.[10]

spec STRINGHASHTABLE0 =
  STRINGKEY0 **and** BUCKET
  then
    ops $add{:}String \times Bucket \to Bucket$
      $putnear{:}Nat \times String \times Bucket \to Bucket$
    preds $present{:}String \times Bucket$
      $isnear{:}Nat \times String \times Bucket$
    $\forall i{:}Nat, s{:}String, a, a'{:}Bucket$
      • $add(s,a) = putnear(hash0(s), s, a)$
      • $putnear(i,s,a) = a' \Leftrightarrow$
        $\exists j{:}Nat \bullet (\neg used(i+j,a) \vee get(i+j,a) = s) \wedge$
          $(\forall k{:}Nat \bullet \neg(j \leq k) \Rightarrow used(i+k,a)) \wedge a' = put(i+j,s,a)$

[10] CASL notation: this signature morphism could be written $Array[String] \mapsto Bucket$ in CASL, since identity mappings in signature morphisms can be left implicit.

- $present(s,a) \Leftrightarrow isnear(hash0(s),s,a)$
- $isnear(i,s,a) \Leftrightarrow$
  $\exists j{:}Nat \bullet (\forall k{:}Nat \bullet k \leq j \Rightarrow used(i+k,a)) \wedge get(i+j,a) = s$

Perhaps unexpectedly, STRINGHASHTABLE0 excludes nearly all models having a value b of sort *Bucket* and an index n such that $used(i,b)$ whenever $i \geq n$. (**Exercise:** Why are such models excluded? Why aren't *all* such models excluded?) As a property in a requirements specification this is unproblematic; however the issue will re-emerge when the example is revisited in Section 6.5.

Here is the specification of the array used for the hash table as an array of hash buckets. (The similarity to STRINGARRAY is not accidental. We will address this in Section 6.3; see in particular Example 6.3.3.)

spec BUCKETARRAY =
  BUCKET
  then
    sorts $Array[Bucket]$
    ops $empty{:}Array[Bucket]$
      $put{:}Nat \times Bucket \times Array[Bucket] \rightarrow Array[Bucket]$
      $get{:}Nat \times Array[Bucket] \rightarrow Bucket$
    preds $used{:}Nat \times Array[Bucket]$
    $\forall i,j{:}Nat, b{:}Bucket, a{:}Array[Bucket]$
      - $\neg used(i, empty{:}Array[Bucket])$
      - $used(i, put(i, b{:}Bucket, a))$
      - $i \neq j \Rightarrow (used(i, put(j, b{:}Bucket, a)) \Leftrightarrow used(i,a))$
      - $get(i, put(i,b,a)) = b{:}Bucket$
      - $i \neq j \Rightarrow get(i, put(j, b{:}Bucket, a)) = get(i,a)$

In this specification, some of the operation names are overloaded, for instance, *get* (we have $get{:}Nat \times Bucket \rightarrow String$ from BUCKET and $get{:}Nat \times Array[Bucket] \rightarrow Bucket$ added here). The context of use of operations usually determines which arity and result sort is intended, but in this case it is not enough, so we have to attach some explicit sort information.

The specification of strings with the hashing function that will be used at the top level of the hash table is the same as STRINGKEY0, except for the name of the hash function.

spec STRINGKEY = STRINGKEY0 **with** $\sigma_{hash0 \mapsto hash}$

where $\sigma_{hash0 \mapsto hash}$ is a signature morphism from $Sig[\text{STRINGKEY0}]$ that is surjective, renames *hash0* to *hash*, and is the identity otherwise.

Finally, here is the specification of the hash table with all the auxiliary sorts and operations introduced so far. Again, we begin by adjusting the sort name:

spec TABLE = BUCKETARRAY **with** $\sigma_{Array[Bucket] \mapsto Table}$

where $\sigma_{Array[Bucket] \mapsto Table}$ is a signature morphism from $Sig[\text{BUCKETARRAY}]$ that is surjective, renames $Array[Bucket]$ to *Table*, and is the identity otherwise.

spec STRINGHASHTABLE =
  STRINGHASHTABLE0 **and** STRINGKEY **and** TABLE
  then
    ops $add\colon String \times Table \to Table$
    preds $present\colon String \times Table$
    $\forall i\colon Nat, s\colon String, a\colon Table$
      • $hash(s) = i \wedge used(i,a) \Rightarrow add(s,a) = put(i, add(s, get(i,a)), a)$
      • $hash(s) = i \wedge \neg used(i,a) \Rightarrow add(s,a) = put(i, add(s, empty), a)$
      • $hash(s) = i \wedge used(i,a) \Rightarrow (present(s,a) \Leftrightarrow present(s, get(i,a)))$
      • $hash(s) = i \wedge \neg used(i,a) \Rightarrow \neg present(s,a)$

Here, despite the overloading of names like *add* (we have $add\colon String \times Bucket \to Bucket$ and $add\colon String \times Table \to Table$) the context of use of operations is sufficient to resolve the ambiguity.

This concludes our specification. It contains many operations that are not relevant to the end user, for instance $putnear\colon Nat \times String \times Bucket \to Bucket$. As a final touch, we could hide all these, revealing only the sorts and operations of interest:

spec USERSTRINGHASHTABLE =
  STRINGHASHTABLE
  reveal sorts $String, Table$
      ops $empty\colon Table$
        $add\colon String \times Table \to Table$
        $\varepsilon\colon String$
        $\mathtt{a}, \mathtt{b}, \ldots, \mathtt{z}\colon String$
        $\_\_\text{\^{}}\_\_\colon String \times String \to String$
      pred $present\colon String \times Table$

Exercise 5.3.2. Redo the specifications above in the institution **FPL** (see Example 4.1.25) following Exercise 4.1.16 (that is, encoding predicates as boolean operations). Try to use pattern-matching case analysis to simplify some of the axioms. Don't forget to add definedness assertions to require that operations are total as they are in **FOPEQ**. However, note that in some cases, a better specification is obtained by allowing some operations to be partial. For instance, there is no need to require that *get* above be total. □

The overall meaning of this specification is a class of models over its signature. The structure of a specification facilitates its understanding and use, but there is no trace of such structure in the resulting models. The way that models are built is an orthogonal issue, which will be treated at length in Chapter 7; this need bear no relation to the structure of the specification.

The specification USERSTRINGHASHTABLE can be compared with the following one:

spec STRINGTABLE =
  STRING
  then

sorts *Table*
ops *empty*: *Table*
add: *String* × *Table* → *Table*
preds *present*: *String* × *Table*
$\forall s, s' : String, t : Table$
• $\neg present(s, empty)$
• $present(s, add(s,t))$
• $s \neq s' \Rightarrow (present(s, add(s',t)) \Leftrightarrow present(s,t))$

Exercise 5.3.3. The signatures of STRINGTABLE and USERSTRINGHASHTABLE are the same, and so their model classes may be directly compared. Show that they are *not* quite the same, so these two specifications are not equivalent. Try to modify STRINGTABLE by adding additional axioms to make them equivalent. □

Despite the fact that STRINGTABLE and USERSTRINGHASHTABLE are not equivalent, all the models of both specifications display the same external behaviour. See Chapter 8 for how this relationship between the two specifications may be captured.

5.4 A property-oriented semantics of specifications

In Section 5.1, the semantics of structured specifications was given by assigning to every specification a signature and a class of models over that signature. The need to assign signatures to specifications is self-evident for the "static" element of specification semantics. The use of model classes to assign interpretations to the symbols in the signature corresponds to a "model-oriented" view where models, representing programs, are taken as the primary objects of interest. This is in contrast to a "property-oriented" view where a specification denotes the set of properties that any realisation is required to satisfy, which ultimately determines a class of models as well, but in an indirect fashion via this intermediate stage. In the semantics of flat specifications in Chapter 2 (cf. Section 4.2), both views are present: a flat specification $\langle \Sigma, \Phi \rangle$ presents the Σ-theory $\langle \Sigma, Cl_\Sigma(\Phi) \rangle$ and has all the Σ-algebras in the class $Mod[\langle \Sigma, \Phi \rangle]$ as models. These are equivalent for flat specifications because of the Galois connection between presentations and model classes; see Propositions 2.3.2 and 4.2.2. The closed elements of this Galois connection are theories, which are in bijective correspondence with closed (i.e. definable) model classes. For any flat specification $\langle \Sigma, \Phi \rangle$ this gives an information-preserving translation from its theory to its class of models and vice versa: $Cl_\Sigma(\Phi) = Th_\Sigma(Mod[\langle \Sigma, \Phi \rangle])$ and $Mod[\langle \Sigma, \Phi \rangle] = Mod_\Sigma(Cl_\Sigma(\Phi))$.

The concepts involved generalise readily to arbitrary specifications:

Definition 5.4.1 (Semantic consequence and theory of specification). For any specification *SP* with signature $\Sigma = Sig[SP]$, we say that a Σ-sentence φ is a *semantic consequence* of *SP*, written $SP \models \varphi$, if $M \models_\Sigma \varphi$ for all $M \in Mod[SP]$.

The *theory* of *SP* is $Th[SP] = \langle \Sigma, Th_\Sigma(Mod[SP]) \rangle$. □

Note that we omit the subscript Σ in the notation for semantic consequence and for the theory of a specification, as it is unambiguously determined by the specification itself. This also formally resolves the potential confusion between Th (which maps specifications to theories) and Th_Σ (which maps classes of Σ-models to Σ-theories).

As for flat specifications, the following observation links the semantic consequences and the theory of a specification:

Proposition 5.4.2. *For any Σ-specification SP, $Th[SP] = \{\varphi \in \mathbf{Sen}(\Sigma) \mid SP \models \varphi\}$.* □

The mapping $\mathscr{T}_0$ of specifications to theories as defined below is an example of an alternative semantics for specifications, given in terms of properties (sentences of the institution) that the specification ensures. Any function that assigns a $Sig[SP]$-theory to every specification SP will be referred to as a *property-oriented semantics*. This raises the obvious question of whether a property-oriented semantics that for any specification SP exactly captures its theory $Th[SP]$ can be given to structured specifications directly, without reference to $Mod[SP]$, in contrast to the definition of semantic consequence and theory in Definition 5.4.1.

Definition 5.4.3. The following clauses inductively define a compositional property-oriented semantics that assigns a Σ-theory $\mathscr{T}_0[SP]$ to any well-formed structured Σ-specification SP built from flat specifications using union, translation and hiding.

$$\mathscr{T}_0[\langle\Sigma,\Phi\rangle] = Cl_\Sigma(\Phi)$$
$$\mathscr{T}_0[SP \cup SP'] = Cl_{Sig[SP]}(\mathscr{T}_0[SP] \cup \mathscr{T}_0[SP'])$$
$$\mathscr{T}_0[SP \textbf{ with } \sigma\colon Sig[SP] \to \Sigma] = Cl_\Sigma(\sigma(\mathscr{T}_0[SP]))$$
$$\mathscr{T}_0[SP \textbf{ hide via } \sigma\colon \Sigma \to Sig[SP]] = \sigma^{-1}(\mathscr{T}_0[SP])$$ □

The last clause of the above definition yields a theory by Corollary 4.2.12.

Exercise 5.4.4. Write out the definition of $\mathscr{T}_0$ for export, enrichment and sum by expanding the abbreviations.

An attempt to bring **free** or **reachable** into the picture may require the use of institutions with data constraints or generation constraints; see Section 4.3. □

The rest of this section is devoted to an analysis of the relationship between this compositional property-oriented semantics and the "reference" semantics Th.

Definition 5.4.5. Let $\mathscr{T}$ be a property-oriented semantics for specifications, assigning to any specification SP a $Sig[SP]$-theory $\mathscr{T}[SP]$.

- $\mathscr{T}$ is *monotone* if $\mathscr{T}[\mathbf{sbo}(SP_1,\ldots,SP_n)] \subseteq \mathscr{T}[\mathbf{sbo}(SP'_1,\ldots,SP'_n)]$ for all specifications $SP_1,\ldots,SP_n$ and $SP'_1,\ldots,SP'_n$ such that $Sig[SP_i] = Sig[SP'_i]$ and $\mathscr{T}[SP_i] \subseteq \mathscr{T}[SP'_i]$, for $i = 1,\ldots,n$, and specification-building operations **sbo** such that both $\mathbf{sbo}(SP_1,\ldots,SP_n)$ and $\mathbf{sbo}(SP'_1,\ldots,SP'_n)$ are well formed.
- $\mathscr{T}$ is *compositional* if $\mathscr{T}[\mathbf{sbo}(SP_1,\ldots,SP_n)] = \mathscr{T}[\mathbf{sbo}(SP'_1,\ldots,SP'_n)]$ for all specifications $SP_1,\ldots,SP_n$ and $SP'_1,\ldots,SP'_n$ such that $Sig[SP_i] = Sig[SP'_i]$ and $\mathscr{T}[SP_i] = \mathscr{T}[SP'_i]$, for $i = 1,\ldots,n$, and specification-building operations **sbo** such that both $\mathbf{sbo}(SP_1,\ldots,SP_n)$ and $\mathbf{sbo}(SP'_1,\ldots,SP'_n)$ are well formed.

- $\mathscr{T}$ is *sound* if $\mathscr{T}[SP] \subseteq Th[SP]$ for every specification SP.
- A sound $\mathscr{T}$ is *complete* if $\mathscr{T}[SP] = Th[SP]$ for every specification SP.
- Given a specification-building operation **sbo**, a sound $\mathscr{T}$ is *closed complete for* **sbo** if $\mathscr{T}[\mathbf{sbo}(SP_1,\ldots,SP_n)] = Th[\mathbf{sbo}(SP_1,\ldots,SP_n)]$ for all $SP_1,\ldots,SP_n$ such that $Mod_{Sig[SP_i]}(\mathscr{T}[SP_i]) = Mod[SP_i]$, $i = 1,\ldots,n$, and $\mathbf{sbo}(SP_1,\ldots,SP_n)$ is well formed. $\mathscr{T}$ is *closed complete* if it is closed complete for all specification-building operations.
- $\mathscr{T}$ is *non-absent-minded* if $\Phi \subseteq \mathscr{T}[\langle\Sigma,\Phi\rangle]$ for every signature Σ and set Φ of Σ-sentences.
- $\mathscr{T}$ is *flat complete* if $\mathscr{T}[\langle\Sigma,\Phi\rangle] = Cl_\Sigma(\Phi)$ for every signature Σ and set Φ of Σ-sentences. □

Exercise 5.4.6. Check that monotonicity implies compositionality but not vice versa. Check that flat completeness is equivalent to non-absent-mindedness for sound $\mathscr{T}$. Check that completeness implies flat completeness and closed completeness. Check that closed completeness for flat specifications, viewed as nullary specification-building operations, is the same as flat completeness. Try to use closed completeness and flat completeness to prove completeness and see where the proof breaks down. (The counterexample below shows that this implication doesn't hold.) □

Proposition 5.4.7. *For structured specifications built from flat specifications using union, translation and hiding, $\mathscr{T}_0$ is monotone, compositional, sound, closed complete, non-absent-minded and flat complete. It is complete for specifications built from flat specifications using union and translation, and for such specifications with hiding applied outermost.*

Proof sketch. Monotonicity and compositionality follow from the definitions, while soundness requires a simple inductive proof. Completeness for specifications built from flat specifications using union and translation follows from the fact that $Mod[SP] = Mod_{Sig[SP]}(\mathscr{T}_0[SP])$ for all such SP, which requires a simple inductive proof. Closed completeness for hiding remains to be checked; this follows from the definitions and the satisfaction condition, and can be used to extend completeness to specifications with hiding applied outermost. □

Counterexample 5.4.8. $\mathscr{T}_0$ is not complete for the following specification SP in **EQ**:

$$
\begin{array}{lll}
\mathbf{spec}\ SP_0 = & \mathbf{sorts} & s, s' \\
 & \mathbf{ops} & a{:}\, s \\
 & & b{:}\, s' \\
 & & c{:}\, s'
\end{array}
$$

$$\mathbf{spec}\ SP_1 = SP_0\ \mathbf{hide\ ops}\ a{:}\, s$$

$$\mathbf{spec}\ SP = SP_1\ \mathbf{then}\ \forall x{:}s \bullet b = c$$

This example relies on the fact that $\forall x{:}s \bullet b = c$ does not imply $b = c$, although it implies $b = c$ for $Sig[SP]$-algebras having a non-empty carrier of sort s; see Section 2.4.

Now, $Mod[SP_0]$ is the class of all algebras (over the indicated signature) and $Mod[SP_1]$ consists of all algebras that are reducts of $Sig[SP_0]$-algebras. Consequently, $Mod[SP_1]$ contains only those algebras having a non-empty carrier of sort s. Then, selecting from $Mod[SP_1]$ the algebras that satisfy $\forall x{:}s \bullet b = c$ yields the class $Mod[SP]$ — and all these algebras satisfy $b = c$ (since for the algebras in $Mod[SP_1]$, $b = c$ follows from $\forall x{:}s \bullet b = c$). This shows that $Th[SP]$ includes $b = c$.

On the other hand, $\mathscr{T}_0[SP_0]$ is the trivial theory containing only the equational tautologies, and so is $\mathscr{T}_0[SP_1]$ (equations cannot express the fact that a carrier is non-empty). Then, the additional axiom $\forall x{:}s \bullet b = c$ in the context of $\mathscr{T}_0[SP_1]$ does not imply the equation $b = c$. This shows that $b = c$ is *not* in $\mathscr{T}_0[SP]$. □

Exercise 5.4.9. Use Counterexample 4.4.24 to give another counterexample indicating the incompleteness of $\mathscr{T}_0$ in **EQ**. □

The above counterexample depends crucially on the use of **EQ**. In **FOEQ**, the class $Mod[SP_1]$ becomes definable (by the sentence $\exists x{:}s \bullet x = x$) and the discrepancy between $Th[SP]$ and $\mathscr{T}_0[SP]$ disappears. In Section 9.2 it will be shown that $\mathscr{T}_0$ is complete over any finitely exact institution that satisfies an interpolation property (Exercise 9.2.10, or Theorem 9.2.6 when propositional connectives are available). In general, however, completeness cannot be ensured since $\mathscr{T}_0$ is the strongest sound compositional non-absent-minded property-oriented semantics we can give:

Theorem 5.4.10. *Consider two property-oriented semantics $\mathscr{T}$ and $\mathscr{T}'$ for specifications constructed using a set of specification-building operations, including all flat specifications. Let $\mathscr{T}$ be sound, monotone, non-absent-minded and closed complete. Let $\mathscr{T}'$ be sound, compositional and non-absent-minded. Then $\mathscr{T}$ is at least as strong as $\mathscr{T}'$: for every SP, $\mathscr{T}'[SP] \subseteq \mathscr{T}[SP]$.*

Proof. By induction on the structure of SP. For flat specifications, $\mathscr{T}'[\langle \Sigma, \Phi \rangle] \subseteq Cl_\Sigma(\Phi) = \mathscr{T}[\langle \Sigma, \Phi \rangle]$ by soundness of $\mathscr{T}'$ and flat completeness of $\mathscr{T}$, which follows from its non-absent-mindedness (or closed completeness). Now, more generally, consider any well-formed specification $\mathbf{sbo}(SP_1, \ldots, SP_n)$ with $\Sigma_i = Sig[SP_i]$ (where $i = 1, \ldots, n$ here and below) and $\Sigma' = Sig[\mathbf{sbo}(SP_1, \ldots, SP_n)]$, and suppose $\mathscr{T}'[SP_i] \subseteq \mathscr{T}[SP_i]$. Then:

$\mathscr{T}'[\mathbf{sbo}(SP_1, \ldots, SP_n)]$

$= \mathscr{T}'[\mathbf{sbo}(\langle \Sigma_1, \mathscr{T}'[SP_1] \rangle, \ldots, \langle \Sigma_n, \mathscr{T}'[SP_n] \rangle)]$
by compositionality of $\mathscr{T}'$, since $\mathscr{T}'[SP_i] = \mathscr{T}'[\langle \Sigma_i, \mathscr{T}'[SP_i] \rangle]$ by flat completeness of $\mathscr{T}'$ which follows from its non-absent-mindedness

$\subseteq Th[\mathbf{sbo}(\langle \Sigma_1, \mathscr{T}'[SP_1] \rangle, \ldots, \langle \Sigma_n, \mathscr{T}'[SP_n] \rangle)]$
by soundness of $\mathscr{T}'$

$= \mathscr{T}[\mathbf{sbo}(\langle \Sigma_1, \mathscr{T}'[SP_1] \rangle, \ldots, \langle \Sigma_n, \mathscr{T}'[SP_n] \rangle)]$
by closed completeness (and flat completeness) of $\mathscr{T}$

$\subseteq \mathscr{T}[\mathbf{sbo}(SP_1, \ldots, SP_n)]$
by monotonicity of $\mathscr{T}$, since
$\mathscr{T}[\langle \Sigma_i, \mathscr{T}'[SP_i] \rangle] = \mathscr{T}'[SP_i]$ by flat completeness of $\mathscr{T}$
$\subseteq \mathscr{T}[SP_i]$ by the inductive hypothesis. □

Exercise 5.4.11. Prove another version of this theorem where $\mathscr{T}$ is required to be compositional and $\mathscr{T}'$ is required to be monotone rather than the other way around as above. □

Corollary 5.4.12. *$\mathscr{T}_0$ is stronger than any other sound, compositional and non-absent-minded property-oriented semantics for structured specifications built from flat specifications using union, translation and hiding.* □

Exercise 5.4.13. Show that in Theorem 5.4.10 the assumption that $\mathscr{T}'$ is non-absent-minded cannot be dropped.

HINT: Consider an institution with signatures Σ and Σ', and a signature morphism $\sigma:\Sigma \to \Sigma'$. Let $\mathbf{Sen}(\Sigma) = \{\alpha\}$, $\mathbf{Sen}(\Sigma') = \{\alpha,\beta\}$, with σ-translation preserving α, and let $|\mathbf{Mod}(\Sigma)| = |\mathbf{Mod}(\Sigma')| = \{M_1,M_2,M_3\}$, with the identity σ-reduct. Suppose $M_1 \models \alpha$, $M_2 \not\models \alpha$, $M_3 \models \alpha$, $M_1 \models \beta$, $M_2 \not\models \beta$, $M_3 \not\models \beta$, and that we have a Σ-specification SP_{bad} with $Mod[SP_{bad}] = \{M_1\}$. Then let $\mathscr{T}'$ be such that $\mathscr{T}'[SP_{bad}] = \{\alpha\}$ and $\mathscr{T}'[SP_{bad}\ \mathbf{with}\ \sigma] = \{\alpha,\beta\}$, and $\mathscr{T}'$ forgets the axiom α in all flat specifications. We can then ensure that for all Σ-specifications SP, if $\alpha \in \mathscr{T}'[SP]$ then $M_3 \notin Mod[SP]$. Then $\mathscr{T}'$ is sound and compositional, but for the Σ'-specification SP_{bad} **with** σ, it is stronger than the expected sound, monotone and closed complete property-oriented semantics $\mathscr{T}$, which yields $\mathscr{T}[SP_{bad}] = \{\alpha\}$ and $\mathscr{T}[SP_{bad}\ \mathbf{with}\ \sigma] = \{\alpha\}$.

Adapt this example to show that Corollary 5.4.12 does not hold for some institutions and some property-oriented semantics that are sound and compositional, but possibly absent-minded. □

These results show that there is a mismatch between compositional property-oriented semantics and model-oriented semantics for structured specifications, and that in general this is unavoidable, even for the small set of simple specification-building operations defined in Section 5.1. Since the objects of ultimate interest here are programs, represented as models, while axioms and theories are nothing more than logical means for describing them, the mismatch demonstrates that theories are not in general adequate as denotations of specifications.

5.5 The category of specifications

In Sections 2.3 and 4.2 we defined morphisms between theories. With the introduction of specification-building operations, it is natural to consider morphisms between structured specifications.

Recall that we write *Spec* for the class of specifications considered, with semantics given by *Sig* and *Mod* as discussed in Section 5.1. This could be the class of specifications built from flat specifications using union, translation and hiding, but the definitions and results below do not depend on this.

Definition 5.5.1 (Specification morphism). For any $SP, SP' \in Spec$, a *specification morphism* $\sigma: SP \to SP'$ is a signature morphism $\sigma: Sig[SP] \to Sig[SP']$ such that for each $M' \in Mod[SP']$, $M'|_\sigma \in Mod[SP]$. □

By Proposition 4.2.26, specification morphisms between flat specifications are exactly theory morphisms between the theories they present.

Exercise 5.5.2. Check that for any signature morphism $\sigma: Sig[SP] \to Sig[SP']$ the following properties are equivalent:

- $\sigma: SP \to SP'$ is a specification morphism.
- $Mod[SP' \textbf{ hide via } \sigma] \subseteq Mod[SP]$.
- $Mod[SP'] \subseteq Mod[SP \textbf{ with } \sigma]$.

Use this equivalence to conclude that hiding and translation w.r.t. any specification morphism $\sigma: \Sigma \to \Sigma'$ form a Galois connection (see Exercise 3.5.23) between classes of Σ-models and classes of Σ'-models.

Also note that the following two properties are independent in general:

- $Mod[SP] \subseteq Mod[SP' \textbf{ hide via } \sigma]$.
- $Mod[SP \textbf{ with } \sigma] \subseteq Mod[SP']$.

Provide counterexamples for the implications in both directions. □

Any specification generates a theory, by Definition 5.4.1; it is easy to check that this extends to morphisms in the obvious way:

Proposition 5.5.3. *If $\sigma: SP \to SP'$ is a specification morphism then for any $\varphi \in Th[SP]$, $\sigma(\varphi) \in Th[SP']$, that is, $\sigma: Th[SP] \to Th[SP']$ is a theory morphism.* □

Exercise 5.5.4. It is *not* the case that $\sigma: SP \to SP'$ is a specification morphism iff $\sigma: Th[SP] \to Th[SP']$ is a theory morphism. The forward implication holds by Proposition 5.5.3; provide a counterexample to the reverse implication. HINT: Consider a specification SP such that the theory of SP has more models than SP itself, as given in Counterexample 5.4.8. □

Exercise 5.5.5. Check that a specification morphism $\sigma: SP \to SP'$ need not be a theory morphism $\mathscr{T}[SP] \to \mathscr{T}[SP']$ for a property-oriented semantics $\mathscr{T}$ that is "weaker" than Th. In particular, give a counterexample to show that a specification morphism $\sigma: SP \to SP'$ between specifications built from flat specifications using union, translation and hiding need not be a theory morphism $\sigma: \mathscr{T}_0[SP] \to \mathscr{T}_0[SP']$ for the compositional property-oriented semantics $\mathscr{T}_0$ of specifications introduced in Definition 5.4.3. HINT: Such a counterexample may be built using the specification SP in Counterexample 5.4.8: the identity is a specification morphism to SP from SP **then** $\bullet\ b = c$.

Using the same idea, show that a sound property-oriented semantics is functorial if and only if it is complete, under the reasonable assumption that we can enrich any specification by an axiom over the same signature so that the semantics includes this axiom in the theory assigned to the enriched specification. □

Definition 5.5.6 (Conservative specification morphism). We say that a specification morphism $\sigma: SP \to SP'$ is *conservative* if for all $Sig[SP]$-sentences φ, $\varphi \in Th[SP]$ whenever $\sigma(\varphi) \in Th[SP']$.

A specification morphism $\sigma: SP \to SP'$ *admits model expansion* if the corresponding reduct function $\_|_\sigma: Mod[SP'] \to Mod[SP]$ is surjective, that is, for every $M \in Mod[SP]$, there exists an $M' \in Mod[SP']$ such that $M'|_\sigma = M$. □

Exercise 5.5.7. Following Exercise 4.2.32, show that a specification morphism $\sigma: SP \to SP'$ is conservative if it admits model expansion, and that the opposite implication does not hold. □

Definition 5.5.8 (Category of specifications). The category **Spec** has specifications in *Spec* as objects and specification morphisms as morphisms, with identities and composition inherited from the category of signatures. Let $\mathbf{Sig}: \mathbf{Spec} \to \mathbf{Sign}$ be the obvious functorial extension of *Sig*. □

Exercise 5.5.9. Check that identities are specification morphisms and that specification morphisms compose. □

The model class semantics for specifications extends to a contravariant functor:

Definition 5.5.10. For any $SP \in Spec$, let $\mathbf{Mod}[SP]$ be the full subcategory of $\mathbf{Mod}(Sig[SP])$ determined by $Mod[SP]$. For any specification morphism $\sigma: SP \to SP'$, let $\mathbf{Mod}(\sigma)$ be the σ-reduct functor $\_|_\sigma: \mathbf{Mod}[SP'] \to \mathbf{Mod}[SP]$. This defines a functor $\mathbf{Mod}: \mathbf{Spec}^{op} \to \mathbf{Cat}$. □

An important result about the category of theories was that it is cocomplete whenever the category of signatures is cocomplete (Theorem 4.4.1). This generalises to structured specifications as follows.

Theorem 5.5.11. *Suppose that Spec is closed under union of arbitrary families of specifications and under translation.*[11] *Then* **Spec** *is cocomplete provided that* **Sign** *is so.*

Proof. Let D be a diagram in **Spec** with $|G(D)|_{Node} = N$, $D_n = SP_n$ and $Sig[SP_n] = \Sigma_n$ for $n \in N$. Let D' be the corresponding diagram in **Sign**; hence $D'_n = \Sigma_n$ for $n \in N$. By assumption, D' has a colimit, say $\langle \alpha_n: \Sigma_n \to \Sigma \rangle_{n \in N}$. Let $SP = \bigcup_{n \in N} SP_n$ **with** α_n. Then for each $n \in N$, $\alpha_n: SP_n \to SP$ is a specification morphism (this is obvious) and $\langle \alpha_n \rangle_{n \in N}$ is a colimit of D in **Spec**: it is a cocone on D (since it is a cocone on D' in **Sign**); then consider another cocone on D, say $\langle \beta_n: SP_n \to SP' \rangle_{n \in N}$. By the construction, there exists a unique signature morphism $\sigma: \Sigma \to Sig[SP']$ such that for each $n \in N$, $\alpha_n;\sigma = \beta_n$. To complete the proof, we need to show that $\sigma: SP \to SP'$ is a specification morphism. This follows easily: if $M' \in Mod[SP']$ then $M'|_\sigma \in Mod[SP]$ since for each $n \in N$, $(M'|_\sigma)|_{\alpha_n} = M'|_{\alpha_n;\sigma} = M'|_{\beta_n} \in Mod[SP_n]$. □

The proof of the theorem shows that a stronger property holds: $\mathbf{Sig}: \mathbf{Spec} \to \mathbf{Sign}$ lifts colimits, and so in any institution, the category of specifications has all the colimits that the category of signatures has.

---

[11] That is, $\bigcup_{i \in I} SP_i \in Spec$ whenever for some signature Σ, $SP_i \in Spec(\Sigma)$ for $i \in I$, and SP **with** $\sigma \in Spec$ whenever $SP \in Spec$ and $\sigma: Sig[SP] \to \Sigma'$.

Exercise 5.5.12. Check that if *Spec* is closed under translation and binary union, and admits empty flat specifications (i.e. for each signature Σ, $\langle\Sigma,\varnothing\rangle \in Spec$) then **Spec** is *finitely* cocomplete provided that **Sign** is so. □

Exercise 5.5.13. Check that the above theorem still holds if union, translation and empty flat specifications are only *expressible* in *Spec* in the following sense: for any family of Σ-specifications $\langle SP_i \in Spec\rangle_{i\in I}$ there exists a Σ-specification $SP \in Spec$ such that $Mod[SP] = \bigcap_{i\in I} Mod[SP_i]$, and analogously for translation and empty flat specifications. □

Exercise 5.5.14. Show that the proof of Theorem 4.4.1 specializes the proof above, using the fact that $\bigcup_{i\in I}\langle\Sigma,\Phi_i\rangle$ and $\langle\Sigma,\Phi\rangle$ **with** $\sigma:\Sigma\to\Sigma'$ have the same models as $\langle\Sigma,\bigcup_{i\in I}\Phi_i\rangle$ and $\langle\Sigma',\sigma(\Phi)\rangle$ respectively. □

Theorem 5.5.11 shows that for each underlying signature diagram the colimit construction in the category of specifications may be viewed as an abbreviation for a particular combination of union and translation. The "sum with sharing" operation (Section 5.2) is what we obtain for pushout diagrams.

Exercise 5.5.15. Working in an exact institution with *Spec* closed under union and translation, generalise Exercise 4.4.17 to structured specifications, showing that $\mathbf{Mod}\colon \mathbf{Spec}^{op}\to\mathbf{Cat}$ preserves limits. Consequently, given a diagram D with nodes N and edges E in the category of specifications, the models of its colimit SP are in bijective correspondence with families of models $\langle M_m\rangle_{m\in N}$ that are compatible with D in the sense that $M_m \in Mod[D_m]$ for each $m\in N$, and $M_n = M_m|_{D_e}$ for each $e{:}n\to m$ in E. Use Exercise 5.5.12 to weaken the above requirements and still ensure that $\mathbf{Mod}\colon\mathbf{Spec}^{op}\to\mathbf{Cat}$ is finitely continuous. □

Exercise 5.5.16. Proposition 5.5.3 states that the mapping of specifications to their theories may be extended to morphisms in a natural way. Check that in any institution **INS** this yields a functor $Th\colon\mathbf{Spec}\to\mathbf{Th}_{\mathbf{INS}}$ from the category of specifications to the category of theories. Moreover, this functor preserves (commutes with) the obvious projection functors from the categories of specifications and theories, respectively, to the category of signatures.

Show that in general the functor $Th\colon\mathbf{Spec}\to\mathbf{Th}_{\mathbf{INS}}$ does not preserve colimits, not even if the institution is exact and the class of specifications is closed under union and translation. HINT: Consider a diagram D in the category of specifications. Check that the theory of its colimit as constructed in the proof of Theorem 5.5.11 may be larger than the colimit of the diagram $Th(D)$ of theories, as constructed in the proof of Theorem 4.4.1. In some institutions it may even happen that specifications in D are inconsistent (so that the colimit specification is inconsistent as well; see Exercise 5.5.15), their theories contain all the sentences over their respective signatures, but some sentences over the colimit signature are not consequences of the union of their translations. □

5.6 Algebraic laws for structured specifications

We have already given a few algebraic laws that might be used to transform specifications to equivalent ones having a different form. For example, $\langle \Sigma, \Phi \rangle$ **with** $\sigma: \Sigma \to \Sigma'$ is equivalent to $\langle \Sigma', \sigma(\Phi) \rangle$. To state such laws it is convenient to have a formal notion of equivalence of structured specifications, which is the obvious one:

Definition 5.6.1. Two specifications SP and SP' are *equivalent*, written $SP \equiv SP'$, if $Sig[SP] = Sig[SP']$ and $Mod[SP] = Mod[SP']$. □

We can now restate these equivalences formally, and add some others describing further interactions between the specification-building operations introduced in Section 5.1. Throughout this section, when we state an equivalence $SP \equiv SP'$ there is a tacit assumption that SP and SP' are both well-formed specifications.

Proposition 5.6.2.

1. $\langle \Sigma, \Phi_1 \rangle \cup \langle \Sigma, \Phi_2 \rangle \equiv \langle \Sigma, \Phi_1 \cup \Phi_2 \rangle$
2. $\langle \Sigma, \Phi \rangle$ **with** $\sigma: \Sigma \to \Sigma' \equiv \langle \Sigma', \sigma(\Phi) \rangle$
3. $SP \cup SP \equiv SP$, $SP_1 \cup SP_2 \equiv SP_2 \cup SP_1$ *and* $(SP_1 \cup SP_2) \cup SP_3 \equiv SP_1 \cup (SP_2 \cup SP_3)$
4. $SP \cup \langle Sig[SP], \varnothing \rangle \equiv SP$
5. SP **with** $id_{Sig[SP]} \equiv SP \equiv SP$ **hide via** $id_{Sig[SP]}$
6. *If* ι *is an isomorphism then* SP **with** $\iota \equiv SP$ **hide via** ι^{-1}
7. $(SP$ **with** $\sigma)$ **with** $\sigma' \equiv SP$ **with** $\sigma;\sigma'$
8. $(SP$ **hide via** $\sigma)$ **hide via** $\sigma' \equiv SP$ **hide via** $\sigma';\sigma$
9. $(SP \cup SP')$ **with** $\sigma \equiv (SP$ **with** $\sigma) \cup (SP'$ **with** $\sigma)$ □

Exercise 5.6.3. In general,

$$\langle \Sigma, \Phi \rangle \textbf{ hide via } \sigma: \Sigma' \to \Sigma \equiv \langle \Sigma', \sigma^{-1}(Cl_\Sigma(\Phi)) \rangle$$

does *not* hold, since the image of $Mod[\langle \Sigma, \Phi \rangle]$ under σ-reduct need not be definable even if $\Phi = \varnothing$, but there is an inclusion of model classes from left to right (this is a part of Proposition 4.2.15). Check that the same is true of the following:

1. $(SP \cup SP')$ **hide via** σ vs. $(SP$ **hide via** $\sigma) \cup (SP'$ **hide via** $\sigma)$
2. $(SP$ **with** $\sigma)$ **hide via** σ vs. SP
3. SP vs. $(SP$ **hide via** $\sigma)$ **with** σ

Link the model class inclusions indicated in the last two items with Exercise 5.5.2, noticing that for any signature morphism $\sigma: \Sigma \to \Sigma'$, __ **hide via** σ and __ **with** σ form a Galois connection between Σ-specifications and Σ'-specifications considered up to equivalence and ordered by the inclusion of model classes. □

Exercise 5.6.4. Justify the claim that for institutions with algebraic signatures, any use of hiding with an injective signature morphism can be replaced by the use of hiding with a signature inclusion and translation with a bijective signature morphism. HINT: Use Proposition 5.6.2(6,8) to show that if for some isomorphism ι, $\sigma = \iota;\sigma'$ then

$$SP \textbf{ hide via } \sigma \equiv (SP \textbf{ hide via } \sigma') \textbf{ with } \iota^{-1}. \qquad \square$$

The interaction between union, translation and hiding is best described in the context of a (finitely) exact institution; see Definitions 4.4.6, 4.4.12 and Lemma 4.4.14. The key result is the normal form theorem; see Theorem 5.6.10 below. Its proof will rely on the following two propositions.

Proposition 5.6.5. *In any institution* **INS**,

$$(SP \textbf{ hide via } \sigma) \textbf{ with } \tau \equiv (SP \textbf{ with } \sigma') \textbf{ hide via } \tau'$$

provided that the following pushout in the category of signatures admits amalgamation:

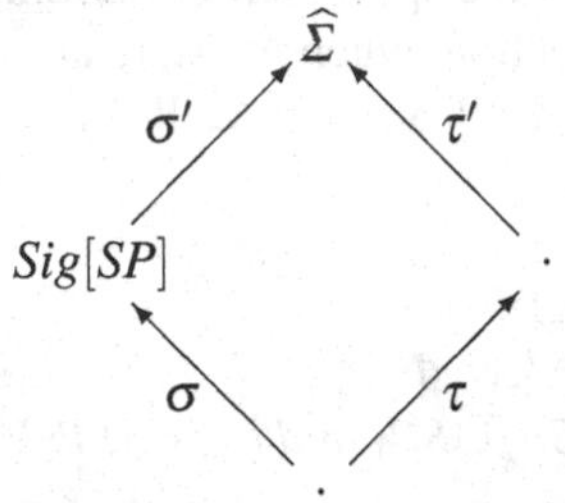

Proof. The fact that both specifications have the same signature is evident. We now show that their model classes coincide.

($\subseteq$): Let $M' \in Mod[(SP \textbf{ hide via } \sigma) \textbf{ with } \tau]$. Then $M'|_\tau \in Mod[SP \textbf{ hide via } \sigma]$; hence there exists $M \in Mod[SP]$ with $M|_\sigma = M'|_\tau$. Since the pushout considered admits amalgamation, there exists $\widehat{M} \in |\mathbf{Mod}(\widehat{\Sigma})|$ such that $\widehat{M}|_{\sigma'} = M$ and $\widehat{M}|_{\tau'} = M'$. Thus $\widehat{M} \in Mod[SP \textbf{ with } \sigma']$ and finally $M' \in Mod[(SP \textbf{ with } \sigma') \textbf{ hide via } \tau']$.

($\supseteq$): Suppose $M' \in Mod[(SP \textbf{ with } \sigma') \textbf{ hide via } \tau']$. Then $M' = \widehat{M}|_{\tau'}$ for some $\widehat{M} \in Mod[SP \textbf{ with } \sigma']$. We have $\widehat{M}|_{\sigma'} \in Mod[SP]$ and so $M'|_\tau = \widehat{M}|_{\tau;\tau'} = \widehat{M}|_{\sigma;\sigma'} = (\widehat{M}|_{\sigma'})|_\sigma \in Mod[SP \textbf{ hide via } \sigma]$. Thus $M' \in Mod[(SP \textbf{ hide via } \sigma) \textbf{ with } \tau]$. $\square$

Exercise 5.6.6. What does this law say in the case where τ is the identity? HINT: It does *not* follow that σ' is the identity as well. $\square$

Proposition 5.6.7. *In any institution* **INS**,

$$(SP_1 \textbf{ hide via } \sigma_1) \cup (SP_2 \textbf{ hide via } \sigma_2)$$
$$\equiv$$
$$((SP_1 \textbf{ with } \sigma_1') \cup (SP_2 \textbf{ with } \sigma_2')) \textbf{ hide via } \sigma_1;\sigma_1'$$

provided that the following pushout in the category of signatures admits amalgamation:

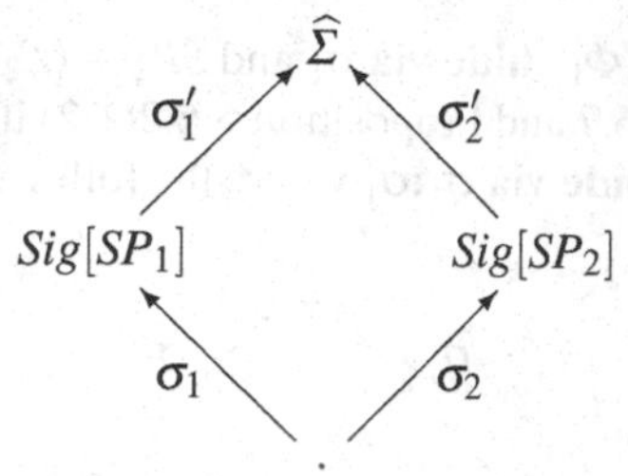

Proof.

($\subseteq$): Let $M \in Mod[(SP_1$ **hide via** $\sigma_1) \cup (SP_2$ **hide via** $\sigma_2)]$. Then there exist $M_1 \in Mod[SP_1]$ with $M_1|_{\sigma_1} = M$ and $M_2 \in Mod[SP_2]$ with $M_2|_{\sigma_2} = M$. By the assumption, there exists $\widehat{M} \in |\mathbf{Mod}(\widehat{\Sigma})|$ such that $\widehat{M}|_{\sigma_1'} = M_1$ and $\widehat{M}|_{\sigma_2'} = M_2$. Hence $\widehat{M} \in Mod[(SP_1$ **with** $\sigma_1') \cup (SP_2$ **with** $\sigma_2')]$ and therefore $M = \widehat{M}|_{\sigma_1;\sigma_1'} \in Mod[((SP_1$ **with** $\sigma_1') \cup (SP_2$ **with** $\sigma_2'))$ **hide via** $\sigma_1;\sigma_1']$.

($\supseteq$): Let $M \in Mod[(SP_1$ **with** $\sigma_1') \cup (SP_2$ **with** $\sigma_2')$ **hide via** $\sigma_1;\sigma_1']$. Hence there exists $\widehat{M} \in Mod[SP_1$ **with** $\sigma_1'] \cap Mod[SP_2$ **with** $\sigma_2']$ such that $\widehat{M}|_{\sigma_1;\sigma_1'} = \widehat{M}|_{\sigma_2;\sigma_2'} = M$. Therefore $\widehat{M}|_{\sigma_1'} \in Mod[SP_1]$ and $\widehat{M}|_{\sigma_2'} \in Mod[SP_2]$, which finally yields $M \in Mod[(SP_1$ **hide via** $\sigma_1) \cup (SP_2$ **hide via** $\sigma_2)]$. □

Exercise 5.6.8. Working in **EQ**, check that Proposition 5.6.7 implies

$$(SP_1 \textbf{ reveal } \Sigma) \textbf{ and } (SP_2 \textbf{ reveal } \Sigma) \equiv (SP_1 \textbf{ and } SP_2) \textbf{ reveal } \Sigma$$

where $\Sigma = Sig[SP_1] \cap Sig[SP_2]$. HINT: First check that the union of two signatures is a pushout of the inclusions of their intersection.

Generalise this to an institution with the appropriate extra structure using Exercises 5.2.1 and 5.2.4. □

These equivalences can be combined to give a procedure which reduces ("normalises") any structured specification built from flat specifications using union, translation and hiding to an equivalent one that is "almost" a flat specification. First, an easy result about specifications that do not involve hiding:

Proposition 5.6.9. *For every specification built from flat specifications using union and translation, there is an equivalent flat specification.*

Proof. By induction on the structure of specifications, using Proposition 5.6.2(1,2). □

Theorem 5.6.10 (Normal form theorem). *Let* **INS** *be a semi-exact institution. For every specification SP built from flat specifications using union, translation and hiding, there is an equivalent specification of the form* $\langle \Sigma, \Phi \rangle$ **hide via** σ *(where* Φ *is finite provided the flat specifications involved in SP are finite).*

Proof. By induction on the structure of *SP*.

- For flat specifications, $\langle \Sigma, \Phi \rangle \equiv \langle \Sigma, \Phi \rangle$ **hide via** id_Σ.

- For union, if $SP_1 \equiv \langle \Sigma_1, \Phi_1 \rangle$ **hide via** σ_1 and $SP_2 \equiv \langle \Sigma_2, \Phi_2 \rangle$ **hide via** σ_2 then directly by Proposition 5.6.7 and Proposition 5.6.2(1,2) it follows that $SP_1 \cup SP_2 \equiv \langle \widehat{\Sigma}, \sigma_1'(\Phi_1) \cup \sigma_2'(\Phi_2) \rangle$ **hide via** $\sigma_1;\sigma_1'$ where the following is a pushout in the category of signatures:

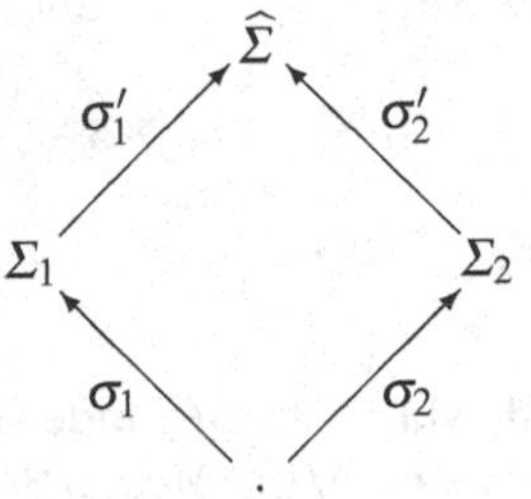

- For translation, if $SP \equiv \langle \Sigma, \Phi \rangle$ **hide via** σ then by Proposition 5.6.5 and Proposition 5.6.2(2), SP **with** $\tau \equiv \langle \widehat{\Sigma}, \sigma'(\Phi) \rangle$ **hide via** τ' where the following is a pushout in the category of signatures:

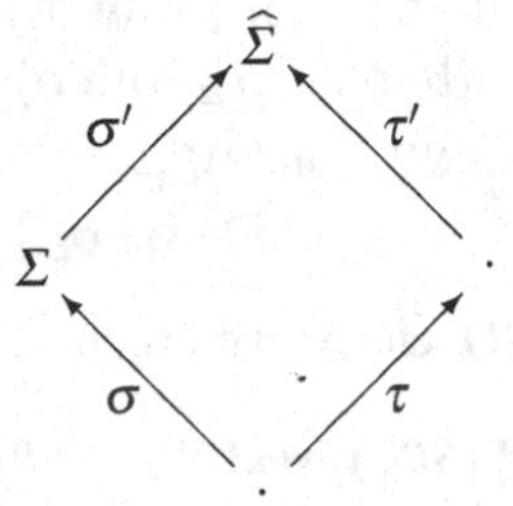

- Finally, for hiding, use Proposition 5.6.2(8). □

Counterexample 5.4.8 shows that the use of hiding in the normal form is unavoidable in general: the structured specifications SP and SP_1 there have classes of models that are not definable by flat specifications in **EQ**.

Exercise 5.6.11. Working in an exact institution, generalise Theorem 5.6.10 to permit unions of arbitrary families of specifications. HINT: The existence of arbitrary signature coproducts and their preservation by the model functor are needed, although a weaker requirement — that colimits of arbitrary non-empty families of signature morphisms with a common source (*wide pushouts*) exist and be preserved by the model functor — would be sufficient as well. □

To use Theorem 5.6.10 to obtain a normal form for a given specification SP, it is sufficient to require that the pushouts that are needed in the course of applying the normalisation procedure to SP exist and admit amalgamation. For example, even though an institution like $\mathbf{EQ}^{der}$ with algebraic signatures and derived signature morphisms is not (finitely) exact (see Example 4.4.7), the normal form theorem can be applied to specifications in $\mathbf{EQ}^{der}$ that involve hiding only with respect to signature inclusions, since the necessary pushouts exist and admit amalgamation:

Exercise 5.6.12. Consider an institution **INS** and a collection **I** of signature morphisms in **INS** such that **INS** is **I**-semi-exact; see Definition 4.4.18. Prove that for any structured specification built from flat specifications using union, translation and hiding which involves hiding only with respect to morphisms in **I**, there exists an equivalent specification of the form $\langle \Sigma, \Phi \rangle$ **hide via** σ where $\sigma \in \mathbf{I}$. □

Exercise 5.6.13. Working in the institution **EQ**, develop a normal form result for specifications built from flat specifications using union, translation, hiding and **reachable** with respect to a signature morphism (Exercise 5.1.10). HINT: The normal form will require one use of **reachable** constraining a subsignature of the signature of the specification inside the hiding. See [EWT83].

Repeat the exercise for **free** in place of **reachable**. HINT: As above, but see [WE87]. □

The normal form theorem can be used to prove meta-results about specifications; see, for instance, Section 9.2. It might also be regarded as a generally useful tool for reducing all structured specifications to a convenient standard form. However, if the result is intended for human consumption then this is usually inappropriate: normalising a specification obliterates all trace of its structure, tending to make it much more difficult to understand. Moreover, the structure of specifications can provide useful guidance in proving theorems about specifications and in developing programs from specifications by refinement, so it should not be lightly disregarded.

5.7 Bibliographical remarks

The idea of taking specification-building operations seriously is from the first algebraic specification language, Clear [BG77], [BG80], which is also where our hiding operation originated. Most of the other operations presented in Section 5.1 originated in the ASL "kernel" specification language [Wir82], [SW83], [Wir86], which is also where semantics of structured specifications in terms of model classes was first used. Their institution-independent versions are from [ST88a].

In the work mentioned above, a different notation for translation and hiding was used, based on the keywords **translate** and **derive**, respectively. Some authors also write $\sigma(SP)$ and $SP|_\sigma$ (or even $\sigma^{-1}(SP)$) for our SP **with** σ and SP **hide via** σ, respectively. Albeit pleasantly compact, the latter choice makes it too easy to confuse the specification-building operations with the corresponding semantic-level concepts, and moreover, cannot be maintained when larger examples are to be written and read. Instead, we decided to use a notation close to what was chosen in CASL [BM04], [Mos04]. Our notation for other operations (sum, enrichment, export, free enrichment, and so on) adheres to the spirit, if not quite always the details, of CASL syntax as well, giving up for instance the keyword **enrich** for enrichment.

Many algebraic specification languages have been developed; see [SW99] for an overview. CASL [BM04], [Mos04] is an attempt to design a common algebraic specification language that integrates all of their best features into a coherent whole. There are many aspects of the design of specification languages that are not discussed here, ranging from the notation used to express signature morphisms (see

[Wir86], where a small programming language is provided for such purposes) to the treatment of specification libraries (see CASL, which supports libraries distributed over the Internet).

The enrichment and sum operations appear in all specification languages in some form. The need for "hidden operations", as introduced by hiding, was first noticed in [Maj77] and studied in [TWW82] in the context of equational specifications. The issue of shared subspecifications touched on at the end of Section 5.2 and further developed in Theorem 5.5.11 was given serious attention in the design of Clear [BG80] via the use of so-called "based theories", which record the relationships between any specification and the environment. Although the design of the module system for Standard ML [Mac84] and so Extended ML [ST85] was influenced by this, it was disregarded by most later specification languages, with the exception of Specware [Smi06] and CASL. A related view, in which constructions apply to components taken from a background environment and produce a result which contributes to that environment, is in [BST08]; see also Section 8.4.4 below.

Section 5.4 analyses the discrepancy, noticed already in [San82], between the semantics for structured specifications given in terms of model classes versus one given in terms of theories. The view we take is that the interpretation in terms of model classes should be regarded as primary. This issue has been somewhat controversial, with the opposite point of view taken in (for instance) [DGS93]. Compatibility between the first and second levels of the semantics of ACT ONE [EM85] shows that no such discrepancies arises there, which is due to the fact that ACT ONE contains no specification-building operation corresponding to hiding. The subtleties of the relationship between theory-level and model-level semantics do not appear to have been analysed to the depth of Section 5.4 before, but see [HST94] for Counterexample 5.4.8 and a related discussion. An analysis of similar issues in a different setting is carried out in [GR04], leading to seemingly different conclusions.

The generalisation from theory morphisms and the cocompleteness of the category of theories (Definition 4.2.25 and Theorem 4.4.1) to specification morphisms and the cocompleteness of the category of structured specifications (Definition 5.5.8 and Theorem 5.5.11) originates with [ST88a].

The systematic study of algebraic laws for specification-building operations originated in [BHK90], although some algebraic laws were given earlier in [Wir82] and [SW83]. The earliest version of the normal form theorem (Theorem 5.6.10), corresponding to Exercise 5.6.13, appeared in [EWT83], while Theorem 5.6.10 is from [BHK90]. One question raised by [BHK90] concerned the distributive law for hiding over union, which does not hold in general (see Exercise 5.6.3). This was taken up in [DGS93], where several variants were considered, with Proposition 5.6.7 capturing the "middle distributive law"; cf. Exercise 5.6.8.

This chapter studies mechanisms for structuring specifications that are not targeted at any particular programming language or paradigm. Our view is that structure at specification level and structure at program level serve different purposes and need not correspond. See Chapters 6 and 7 for more on this topic. Other views of the connection between specification and programming formalisms have been taken in Larch [GH93] and Extended ML [ST89].

Chapter 6
Parameterisation

In the previous chapter, devoted to mechanisms for structuring specifications, discussion of the most important structuring mechanism of all — parameterisation, or genericity — was deferred to this chapter.

Parameterisation allows a definition or expression to be abstracted partly or fully from its context, the result of such abstraction being treated as an entity in its own right. Selected dependencies on the context are made explicit via an interface which describes the range of permissible parameters. The exercise of spelling out such dependencies often suggests ways in which they can be minimised, leading to increased generality. Subsequent instantiation, by supplying an appropriate parameter, then allows the parameterised entity to be (re)used as required. An example of parameterisation in mathematics is the way in which one defines linear equations, and the Gauss-Jordan method for solving such equations, using coefficients taken from an arbitrary field, rather than specifically from the rationals or the real numbers. In programming we build compilers using lexical analysers constructed by plugging a set of regular expressions describing the lexical structure of the language at hand into a tool like `lex`. In logic, inference rules are schematic (given $\Gamma \vdash \varphi \Rightarrow \psi$ and $\Gamma' \vdash \varphi$, conclude $\Gamma \cup \Gamma' \vdash \psi$). And the theory presented in Chapters 5–9 of this book is parameterised by an arbitrary institution.

This book is concerned both with algebras, modelling programs, and with specifications, and this chapter will focus on the use of parameterisation at both of these levels. Parameterised specifications, which take specifications as parameters and deliver specifications as results, correspond to (user-defined) specification-building operations. Parameterised algebras, which take algebras as parameters and deliver algebras as results, correspond to generic program modules as found in languages like Standard ML and Modula-3. Specifying a parameterised algebra involves describing its admissible parameters and the properties of the algebra that is produced when it is instantiated; this is not the same thing as a parameterised specification. It is possible and useful to extend parameterisation to higher order, where parameterised entities are themselves used as parameters.

For the sake of generality (via parameterisation!) we work in an arbitrary but fixed institution $\mathbf{INS} = \langle \mathbf{Sign}, \mathbf{Sen}, \mathbf{Mod}, \langle \models_\Sigma \rangle_{\Sigma \in |\mathbf{Sign}|} \rangle$ as in the last chapter, requiring additional structure only when necessary.

6.1 Modelling generic modules

The use of algebras to model programs was explained in Section 0.1 (and generalised to models in the underlying institution in Chapter 4). This view is adequate for complete programs or self-contained program modules. However, it does not cover *generic* modules [Nel91] (*functors* in Standard ML [Pau96], *parameterised programs* in [Gog96]), which take a module as a parameter and produce a module as a result, where the definition of components in the result module may rely on the components of the parameter. To use the components provided by a generic module, it must first be *instantiated* by *applying* it to an argument module; this yields a (non-generic) module.

Such generic modules correspond to *functions* taking models over one signature (for the parameter) to models over another (for the result), with instantiation corresponding to function application. In general such a function will be *partial* — we will consider cases in which the result is not defined for all models over the parameter signature, but only for those that satisfy additional constraints. This matches Standard ML: instantiation of a functor involves computation which may diverge, although it rarely does in practice. We will call such partial functions *constructors*.[1]

Definition 6.1.1 (Constructor). Given signatures Σ and Σ', $\Sigma \Rightarrow \Sigma'$ is a *constructor signature*, with $Mod(\Sigma \Rightarrow \Sigma')$ standing for the class of all partial functions mapping Σ-models to Σ'-models. A partial function $F \in Mod(\Sigma \Rightarrow \Sigma')$ is called a *constructor*. □

An obvious additional requirement to impose on constructors is that they be *effective* functions in a suitable formal sense. Just as we did not take computability into account when modelling programs as algebras, we do not impose such a requirement on constructors either. It does not seem to add anything significant (except complexity) to the theory, and furthermore, in the context of an arbitrary institution the concept of effective function seems problematic.

The use of the word "functor" in Standard ML for a generic module as well as the examples of constructors below might suggest that constructors should be functors between model categories rather than functions between model classes. This additional structure will not be needed here although it will re-emerge in Section 8.4.4.

[1] Constructors should not be confused with *value constructors* in Standard ML and similar programming languages and in the institution **FPL**; see Example 4.1.25.

Exercise 6.1.2. Show that all of the constructors given below, except for amalgamation, can be extended to functors on the full subcategory determined by their domain. □

Example 6.1.3 (Reduct). For any signature morphism $\sigma{:}\Sigma \to \Sigma'$, the object part of the σ-reduct functor is a function $\_|_\sigma{:}Mod(\Sigma') \to Mod(\Sigma)$. This is a (total) constructor $\_|_\sigma \in Mod(\Sigma' \Rightarrow \Sigma)$.[2]

In institutions like **EQ**, this constructs Σ-models from Σ'-models by forgetting and/or renaming their components. In institutions like $\mathbf{EQ}^{der}$ or $\mathbf{PEQ}^{der}$ (see Exercise 4.1.23), where *derived* signature morphisms are used (see Section 1.5.2), this constructor can be used as a definitional mechanism, where operations in Σ are defined via terms over Σ'.

This may be taken even further by generalising derived signature morphisms to allow conditional definitions and recursion (see Exercise 1.5.22), and to define carrier sets for sorts in Σ in terms of those in Σ' using product, disjoint union, recursion, and so on (see Exercise 1.5.21). Enriching this with appropriate syntactic sugar gives a language that approaches the expressive power of Standard ML functors. This is done below for the institution **FPL** (Example 4.1.25); see Examples 6.1.6, 6.1.8 and 6.1.9 below. □

Example 6.1.4. Let Σ' be a signature with a sort *Bool* and two boolean connectives, conjunction ($\wedge$) and negation ($\neg$), and let Σ extend Σ' by adding disjunction ($\vee$). Then the derived signature morphism $\delta{:}\Sigma \to \Sigma'$ given by (recall the notation for derived terms introduced in Definition 1.5.13) $\delta(Bool) = Bool$, $\delta(\wedge) = (\boxed{1}{:}Bool) \wedge (\boxed{2}{:}Bool)$, $\delta(\neg) = \neg(\boxed{1}{:}Bool)$, $\delta(\vee) = \neg(\neg(\boxed{1}{:}Bool) \wedge \neg(\boxed{2}{:}Bool))$ determines the constructor $\_|_\delta \in Mod(\Sigma' \Rightarrow \Sigma)$ which adds to any Σ'-model a disjunction operation defined as usual in terms of conjunction and negation. □

Exercise 6.1.5. Observe that the use of reducts w.r.t. derived signature morphisms to define constructors may be essentially replaced by the use of the corresponding "definitional" equations (and vice versa).

For instance, in the standard algebraic framework with derived signature morphisms introduced in Section 1.5.2, consider an algebraic signature Σ and let Σ_f be its extension by a new operation name $f{:}s_1 \times \cdots \times s_n \to s$ (for some sorts $s_1,\ldots,s_n,s$ in Σ). Let $\delta{:}\Sigma_f \to \Sigma$ be a derived signature morphism which is the identity on the symbols in Σ and maps f to a Σ-term, $\delta_f(f) = t \in |T_\Sigma(\boxed{1}{:}s_1,\ldots,\boxed{n}{:}s_n)|_s$. Then the "definitional" axiom $\forall \boxed{1}{:}s_1,\ldots,\boxed{n}{:}s_n \bullet f(\boxed{1},\ldots,\boxed{n}) = t$ captures the construction given by the reduct w.r.t. δ_f: check that for any Σ-algebra A, $A|_{\delta_f}$ is the unique expansion of A to a Σ_f-algebra that satisfies this definitional equation.

Extend this characterisation to any derived algebraic signature morphism $\delta{:}\Sigma' \to \Sigma$ such that $\iota;\delta = id_\Sigma$ for some signature morphism $\iota{:}\Sigma \to \Sigma'$. □

---

[2] CASL notation: reduct is written in CASL using the same **reveal/hide** notation as export for specifications (Section 5.2), with the same restriction on σ.

Example 6.1.6. Consider the following signatures in the institution **FPL** (see Example 4.1.25).

TREE =
  sorts *Nat* **free with** 0 | *succ*(*Nat*)
    Tree
  ops *empty*: *Tree*
    node: *Tree* × *Nat* × *Tree* → *Tree*
    sum: *Tree* → *Nat*

RAWTREEWITHOP =
  sorts *Nat* **free with** 0 | *succ*(*Nat*)
    Tree **free with** *empty* | *node*(*Tree*,*Nat*,*Tree*)[3]
  ops *op*: *Nat* → *Nat*

Consider the signature morphism δ: TREE → RAWTREEWITHOP such that

δ(*sum*) =
  let fun *plus*(*x*:*Nat*,*n*:*Nat*):*Nat* =
    case *x* **of** 0 => *n* | *succ*(*m*) => *succ*(*plus*(*m*,*n*))
  in let fun *sum*(*t*:*Tree*):*Nat* =
      case *t* **of** *empty* => 0
        | *node*(*t1*,*n*,*t2*) => *plus*(*plus*(*op*(*n*),*sum*(*t1*)),*sum*(*t2*))
    in *sum*($\boxed{1}$: *Tree*)

and δ is the identity on the remaining names. This determines the constructor $\_|_\delta \in Mod(\text{RAWTREEWITHOP} \Rightarrow \text{TREE})$, which implements the *sum* operation (sum of the results of applying the operation *op* to the node labels in a tree) over a representation of trees generated by *empty* and *node* as value constructors. □

The intuition behind the reduct constructor is that it builds a model by providing an explicit definition for each of its components in terms of the components of the argument model. This is only intuition, which applies to institutions like those mentioned above; it is possible to concoct institutions in which this intuition does not apply. (**Exercise:** Find one.) An alternative is to select a "standard" model from a specified range of possibilities. Taking the initial model of a specification or the free extension of the argument model to a model of a specification are examples of this.

Example 6.1.7 (Free extension). For any signature morphism $\sigma: \Sigma \to \Sigma'$ and Σ'-specification SP' and for $M \in Mod(\Sigma)$, let $F_{\sigma,SP'}(M)$ be the free object over M with respect to the reduct functor $\_|_\sigma: \mathbf{Mod}[SP'] \to \mathbf{Mod}(\Sigma)$, if it exists, where $\mathbf{Mod}[SP']$ is the full subcategory of $\mathbf{Mod}(\Sigma')$ determined by the models of SP'. This yields a constructor $F_{\sigma,SP'} \in Mod(\Sigma \Rightarrow \Sigma')$. (Strictly speaking, the above definition requires a canonical choice of the free object since otherwise $F_{\sigma,SP'}(M)$ is only defined up to

[3] CASL notation: semicolons are used in place of commas to separate arguments of value constructors in CASL.

isomorphism. Consequently, when we compare such constructors in the sequel the use of equality actually refers to natural isomorphism.) In general this constructor is partial since the free object need not exist, but, for instance, in **EQ** or any other liberal institution (see Definition 4.5.16 and Exercise 3.5.11 or 4.5.12) it is total if SP' is a flat specification.

An important special case of free extension $F_{\sigma,SP'}$ is when $SP' = \langle \Sigma', \varnothing \rangle$; then we use the abbreviation F_σ. This corresponds to algebraic datatype definitions in functional languages like Standard ML, sometimes referred to as the *anarchic* or *absolutely free* extension; cf. sorts freely generated by value constructors in the institution **FPL**, as in Example 6.1.6. In **FPL**, the constructor F_δ for $\delta\colon \mathsf{SIG} \to \mathsf{SIG}'$ is total provided that δ is an injective renaming of sort and operation names such that no operations are added that yield results of "old" sorts (i.e. sorts in $\delta(\mathsf{SIG})$) and no old operations are promoted to the status of value constructors. (This is a proper subclass of $\mathbf{I_{FPL}}$ as defined in Exercise 4.4.19.) This covers the case of **FPL**-signature morphisms that simply add new sorts with new value constructors, which will be of particular interest below. □

Example 6.1.8. Consider the following version of the signature RAWTREEWITHOP in the institution **FPL** from Example 6.1.6

SIMPLERAWTREEWITHOP =
  sorts *Nat* **free with** 0 | *succ*(*Nat*)
    Tree
  ops *empty*: *Tree*
    node: *Tree* × *Nat* × *Tree* → *Tree*
    op: *Nat* → *Nat*

and its subsignature

NATWITHOP =
  sorts *Nat* **free with** 0 | *succ*(*Nat*)
  ops *op*: *Nat* → *Nat*

with the signature inclusion j: NATWITHOP → SIMPLERAWTREEWITHOP. The free extension $F_j \in Mod$(NATWITHOP ⇒ SIMPLERAWTREEWITHOP) is a (total) constructor which adds a realisation of trees freely generated by *empty* and *node* to any model over NATWITHOP.

Alternatively, consider the signature RAWTREEWITHOP as given in Example 6.1.6 and the inclusion ι: NATWITHOP → RAWTREEWITHOP. Declaring *Tree* as a sort with value constructors *empty* and *node* "internalizes" the free extension requirement: $F_\iota \in Mod$(NATWITHOP ⇒ RAWTREEWITHOP) essentially coincides with F_j and is, up to isomorphism, the *only* constructor in *Mod*(NATWITHOP ⇒ RAWTREEWITHOP) that extends its argument models without modification. The latter property can be expressed without reference to freeness by requiring persistency; see Definition 6.1.16 below. □

Example 6.1.9 (FPL-constructor notation). Recall from Example 6.1.8 above the constructor $F_\iota \in Mod(\text{NATWITHOP} \Rightarrow \text{RAWTREEWITHOP})$ and from Example 6.1.6 the reduct constructor $\_|_\delta \in Mod(\text{RAWTREEWITHOP} \Rightarrow \text{TREE})$. The composition $F_\iota;(\_|_\delta) \in Mod(\text{NATWITHOP} \Rightarrow \text{TREE})$ is an instance of a pattern that is typical for constructors over the institution **FPL**, whereby one introduces new sorts freely generated by value constructors and then new operations defined by **FPL**-terms over those sorts. The composed constructor can be written using a notation that suggests a connection with functors in Standard ML, as follows:

> **constructor** K : NATWITHOP $\Rightarrow$ TREE =
>   **sorts** *Tree* **free with** *empty* | *node*(*Tree*, *Nat*, *Tree*)
>   **ops** **fun** *sum*(*t*:*Tree*):*Nat* =
>   **let fun** *plus*(*x*:*Nat*, *n*:*Nat*):*Nat* =
>   **case** *x* **of** 0 => *n* | *succ*(*m*) => *succ*(*plus*(*m*, *n*))
>   **in case** *t* **of** *empty* => 0
> | *node*(*t1*, *n*, *t2*) => *plus*(*plus*(*op*(*n*), *sum*(*t1*)), *sum*(*t2*))

This defines a total constructor with the given name (here K) and the given parameter signature (here NATWITHOP) and result signature (here TREE) as the composition of two steps. The first step introduces the new sorts, each freely generated by a set of (new) value constructors (here, *Tree* with *empty* and *node*), giving a free extension from the parameter signature to an intermediate signature consisting of the parameter signature enlarged by the new sorts and value constructors. Some of the sort and operation names in the result signature (here, *Nat*, 0, *succ*, *Tree*, *empty* and *node*) are already in the intermediate signature. The second step introduces definitions for the additional sorts (here, none) and operations (here, *sum*) of the result signature in terms of those in the intermediate signature. The definitions must be well formed with compatible argument and result sorts so as to determine an **FPL**-signature morphism from the result signature to the intermediate signature, and hence a reduct constructor.

Abstracting away from the details of this particular example, we have the following diagram:

$$\mathsf{SIG}_{parameter} \xrightarrow{\iota} \mathsf{SIG}_{intermediate} \xleftarrow{\delta} \mathsf{SIG}_{result}$$

where ι is an **FPL**-signature inclusion which introduces only new sorts with value constructors, and δ is an **FPL**-signature morphism which preserves the sort and operation names in $\mathsf{SIG}_{parameter} \cap \mathsf{SIG}_{result}$. Then the constructor defined is $F_\iota;(\_|_\delta) \in Mod(\mathsf{SIG}_{parameter} \Rightarrow \mathsf{SIG}_{result})$.

In the examples below, we will allow ourselves to go beyond the term constructs that are explicitly introduced for **FPL** in Example 4.1.25, making use of conditional terms as in Exercise 4.1.28, and occasionally using local definitions of values (not just functions).

Exercise. Add a new construct **let** $x = t$ **in** t' to the syntax of terms of **FPL**. This captures local definitions of values of arbitrary sorts. Define its semantics. □

Exercise 6.1.10. Redo Exercise 6.1.5 for derived signature morphisms in the institution **FPL**. Notice that the "definitional" equations needed are exactly function definitions, that is, axioms available in **FProg**, the subinstitution of **FPL** that captures its "programming" fragment (introduced in Exercise 4.1.30).

Now, referring to the constructor notation introduced in Example 6.1.9 above, assume that we have an inclusion $\iota'\colon \mathsf{SIG}_{intermediate} \to \mathsf{SIG}_{result}$ such that $\iota';\delta$ is the identity on $\mathsf{SIG}_{intermediate}$. Construct then a SIG_{result}-specification $\mathrm{CODE}_{\iota,\delta}$ in **FProg** which captures the construction described in Example 6.1.9. That is, for any $\mathsf{SIG}_{parameter}$-model A, $F_\iota(A)|_\delta$ is the unique (up to isomorphism) model of $\mathrm{CODE}_{\iota,\delta}$ that is a $(\iota;\iota')$-expansion of A. (The fact that it is unique only up to isomorphism is related to the possible choice of representation of the new constrained sorts introduced by F_ι.) Note that the specification $\mathrm{CODE}_{\iota,\delta}$ required is in fact just what we actually write as the "body" of the constructor in this notation (perhaps modulo some syntactic details).

In case not all the sorts and operations of $\mathsf{SIG}_{intermediate}$ are included in SIG_{result}, so that ι' needed above cannot be given directly, show that the required specification in **FProg** can be given over a signature that comprises the symbols from both $\mathsf{SIG}_{intermediate}$ and SIG_{result}, and then that the overall construction is captured as a unique expansion given by the "code", as above, followed by a simple "hiding" step (reduct to a subsignature). □

Example 6.1.11 (Quotient). Another special case of free extension $F_{\sigma,SP}$ is when $\sigma\colon \Sigma \to Sig(SP)$ is the identity signature morphism. This yields a constructor that we will write as $\_/SP \in Mod(\Sigma \Rightarrow \Sigma)$. In the institution **EQ** this is a total constructor when SP is a flat specification and yields the usual quotient by the least congruence generated by the equations in SP; see Exercise 3.4.12. □

Exercise 6.1.12. Work out the conditions under which $F_{\sigma,SP'}$ can be defined as the composition of the absolutely free extension F_σ and the quotient $\_/SP'$. □

Example 6.1.13 (Restriction to sort-generated subalgebra). In **EQ** and similar institutions, for any signature Σ, the object part of the restrict functor (cf. Exercise 3.4.13) is a (total) constructor $R_\Sigma \in Mod(\Sigma \Rightarrow \Sigma)$ which returns the reachable subalgebra of any Σ-algebra. For any set $S \subseteq sorts(\Sigma)$ of sorts, there is a constructor $R_S \in Mod(\Sigma \Rightarrow \Sigma)$ such that for $A \in |\mathbf{Alg}(\Sigma)|$, $R_S(A)$ is the subalgebra of A that is generated by its carriers of sorts not in S. Obviously, $R_\Sigma = R_{sorts(\Sigma)}$. □

Exercise 6.1.14. Generalise R_Σ and R_S to an arbitrary institution (with additional structure) using the apparatus of Section 4.5; see also Exercise 5.1.10. □

Exercise 6.1.15. Any constructor $F \in Mod(\Sigma \Rightarrow \Sigma')$ gives rise to a specification-building operation which takes a Σ-specification SP to a Σ'-specification $F(SP)$ having the image of $Mod[SP]$ under F as its models.

Link the examples of constructors above with the specification-building operations of Section 5.1. Be careful though: for instance, $R_S(SP)$ is in general not equivalent to **reachable** SP **on** S (but the equivalence *does* hold if $Mod[SP]$ is closed under subalgebras).

Find an example of a specification-building operation in Section 5.1 that cannot arise in this way from a constructor. In particular, consider $\_$ **with** σ: show that in typical institutions (like **FOEQ**) it does arise from a constructor if σ is surjective and that this constructor is total if σ is an isomorphism.

Consider constructors given as reducts w.r.t. (derived) signature morphisms. Show that in typical institutions, under the circumstances given in Exercise 6.1.5, the specification-building operation determined by such a constructor may be expressed as an enrichment. In particular, by Exercise 6.1.10, for constructors in the institution **FPL** that are expressible in the notation of Example 6.1.9, the corresponding specification-building operation may be given as an enrichment by "code" — function definitions from **FProg**, the subinstitution of **FPL** which captures its programming fragment — followed by a simple "hiding" step (in case not all sorts and operations of the intermediate signature constructed in Example 6.1.9 are included in the result signature). □

Constructors are often considered "along" a signature morphism which connects the source signature to the target signature; typically this signature morphism is an inclusion. Intuitively, such a constructor extends the model to which it is applied with some additional components to build the result. This intuition is not accurate unless the constructor is required to preserve the part of the result that corresponds to the argument.

Definition 6.1.16 (Persistent constructor). For any signature morphism $\iota\colon\Sigma \to \Sigma'$ and class of models $\mathcal{M} \subseteq Mod(\Sigma)$, a constructor $F \in Mod(\Sigma \Rightarrow \Sigma')$ is *persistent on* $\mathcal{M}$ *along* ι if for every Σ-model $M \in \mathcal{M}$ we have $M \in dom(F)$ and $F(M)|_{\iota} = M$; we write $Mod(\Sigma \overset{\iota}{\Rightarrow}_{\mathcal{M}} \Sigma')$ for the class of all such constructors. F is *persistent along* ι if it is persistent on $dom(F)$; we write $Mod(\Sigma \overset{\iota}{\Rrightarrow} \Sigma')$ for the class of all such constructors. □

It is trivial to see that the composition of persistent constructors is a persistent constructor.

A typical example of a persistent constructor is given in Example 6.1.4 where disjunction is defined in terms of conjunction and negation. This is an instance of the following general situation.

Proposition 6.1.17. *If* $\iota\colon\Sigma \to \Sigma'$ *and* $\sigma\colon\Sigma' \to \Sigma$ *are such that* $\iota;\sigma = id_{\Sigma}$ *then* $\_|_{\sigma} \in Mod(\Sigma \overset{\iota}{\Rrightarrow} \Sigma')$. □

Exercise 6.1.18. Observe that the above proposition applies to the constructor in Example 6.1.6. Prove the following more general observation: in the institution **FPL**, for any constructor

constructor $K : \mathsf{SIG} \Rightarrow \mathsf{SIG}' = \ldots$

defined in the notation of Example 6.1.9 such that $\mathsf{SIG}' \supseteq \mathsf{SIG}$, K is persistent along the inclusion $\iota\colon\mathsf{SIG} \to \mathsf{SIG}'$. □

The following easy fact concerns the common situation, arising in the above exercise, where a persistent constructor introduces auxiliary components, some of which are subsequently forgotten.

Fact 6.1.19. *Consider a persistent constructor $F \in Mod(\Sigma \stackrel{\sigma}{\Rightarrow} \Sigma')$ where $\sigma\colon\Sigma \to \Sigma'$ is a signature morphism. Let $\iota\colon\Sigma'' \to \Sigma'$ and $\sigma'\colon\Sigma \to \Sigma''$ be signature morphisms such that $\sigma = \sigma';\iota$. Then $F;\_|_\iota \in Mod(\Sigma \stackrel{\sigma'}{\Longrightarrow} \Sigma'')$ is a persistent constructor.* □

Persistency of free extensions is a more complicated issue. In fact, mixing persistency and freeness properties naturally leads to a stronger concept:

Definition 6.1.20 (Naturally persistent free extension). Given a signature morphism $\sigma\colon\Sigma \to \Sigma'$ and Σ'-specification SP', the free extension $F_{\sigma,SP'}$ (Example 6.1.7) is *naturally persistent* if for each $M \in dom(F_{\sigma,SP'})$, $F_{\sigma,SP'}(M)$ is the free object over M with respect to the reduct functor $\_|_\sigma\colon\mathbf{Mod}[SP'] \to \mathbf{Mod}(\Sigma)$ with the unit $\eta_M\colon M \to F_{\sigma,SP'}(M)|_\sigma$ being the identity. □

Note that, trivially, natural persistency implies persistency of the free extension. The opposite implication does not hold in general:

Exercise 6.1.21. Let Σ be an algebraic signature with the single sort s and no operations, and let Σ' be its extension by a constant $a\colon s$ with the inclusion σ. Show that the absolutely free extension $F_\sigma \in Mod(\Sigma \Rightarrow \Sigma')$ is persistent (strictly speaking, only up to isomorphism, but it can be chosen to be persistent) on algebras having an infinite carrier of sort s. Show that it is not naturally persistent on any of these algebras. □

Persistency of free extensions strongly depends on the details of the underlying institution. In **EQ**, natural persistency is ensured by the following two conditions:

Definition 6.1.22 (Sufficient completeness and hierarchy consistency). In **EQ**, consider $\sigma\colon\Sigma \to \Sigma'$ and a set $\mathscr{E}'$ of Σ'-equations. $\langle\Sigma', \mathscr{E}'\rangle$ is *sufficiently complete w.r.t.* σ if for any set X of variables of sorts in Σ, sort $s \in sorts(\Sigma)$ and term $t' \in |T_{\Sigma'}(X')|_{\sigma(s)}$, where X' is the translation of X along σ as given in Definition 1.5.10, there exists a term $t \in |T_\Sigma(X)|_s$ such that $\mathscr{E}' \models_{\Sigma'} \forall X' \bullet \sigma(t) = t'$. $\langle\Sigma', \mathscr{E}'\rangle$ is *hierarchically consistent w.r.t.* σ if there is no Σ-equation $\forall X \bullet t_1 = t_2$ with distinct t_1 and t_2 such that $\mathscr{E}' \models_{\Sigma'} \forall X' \bullet \sigma(t_1) = \sigma(t_2)$. □

Exercise 6.1.23. Look at Exercises 2.5.20, 2.7.9 and 2.7.10 in the light of the above definition of sufficient completeness. □

Proposition 6.1.24. *In* **EQ**, *if $\langle\Sigma', \mathscr{E}'\rangle$ is sufficiently complete and hierarchically consistent w.r.t. $\sigma\colon\Sigma \to \Sigma'$ then the free extension $F_{\sigma,\langle\Sigma',\mathscr{E}'\rangle}$ is persistent up to natural isomorphism. Moreover, if σ is injective on sorts then $F_{\sigma,\langle\Sigma',\mathscr{E}'\rangle}$ may be chosen to be naturally persistent, and so $F_{\sigma,\langle\Sigma',\mathscr{E}'\rangle} \in Mod(\Sigma \stackrel{\sigma}{\Rightarrow} \Sigma')$.* □

Exercise 6.1.25. Prove this proposition. HINT: See Exercise 3.5.11 for the construction of the free extension.

Then consider a Σ-specification SP. For σ injective on sorts, natural persistency of the constructor $F_{\sigma,\langle\Sigma',\mathscr{E}'\rangle} \in Mod(\Sigma \Rightarrow \Sigma')$ restricted to $Mod[SP]$ follows from weaker versions of the sufficient completeness and hierarchical consistency conditions. Formulate these requirements and show that they are sufficient. HINT: Sufficient completeness may be limited to models of the equational consequences of SP and hierarchical consistency need only consider equations that are not such consequences. Try to construct a counterexample to show that limiting sufficient completeness to the models of SP would be too weak.

Show that all the above facts remain valid for the institution $\mathbf{EQ}^{\Rightarrow}$ of conditional equations. □

Exercise 6.1.26. In some institution (e.g. in **FOEQ**), give an example of a signature morphism $\sigma\colon\Sigma \to \Sigma'$ and Σ'-specification SP' such that SP' is sufficiently complete and hierarchically consistent w.r.t. σ (under the obvious generalisation of Definition 6.1.22) and $F_{\sigma,SP'}$ is a total constructor, but Proposition 6.1.24 fails for SP'. Can you produce a similar example in **EQ**? HINT: See Exercise 4.5.17. In **EQ**, SP' cannot be equivalent to a flat specification. Hiding may be used to simulate existential quantification, and free extension to ensure uniqueness. □

Exercise 6.1.27. In **FPL**, let $\iota\colon \mathsf{SIG} \to \mathsf{SIG}'$ be an injective renaming of sort and operation names such that no operations are added that yield results of old sorts and no old operations are promoted to the status of value constructors, as at the end of Example 6.1.7. Check that the free extension F_ι is total and naturally persistent. □

Exercise 6.1.25 hinted at one way of building a constructor from another one, namely by restricting to a smaller domain. Another obvious way is to compose two constructors as we did in Example 6.1.9. Both ways preserve persistency. There is also a natural way of "lifting" a persistent constructor along a signature morphism that fits its source signature into a larger context.

Example 6.1.28 (Translation of a constructor). Suppose that the following pushout diagram in **Sign** admits amalgamation

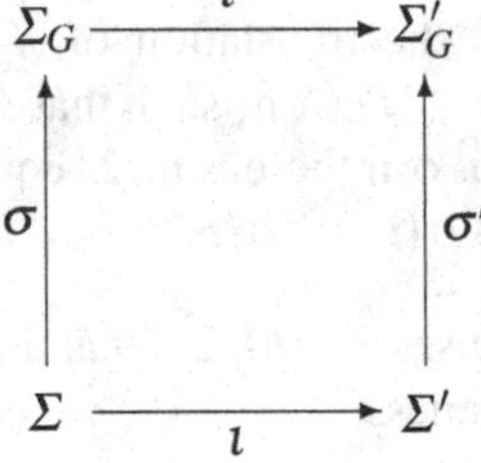

and that $F \in Mod(\Sigma \overset{\iota}{\Rightarrow} \Sigma')$ is a persistent constructor. Then for any $M_G \in Mod(\Sigma_G)$ such that $M_G|_\sigma \in dom(F)$, define $\sigma(F)(M_G)$ to be the unique Σ'_G-model such that $\sigma(F)(M_G)|_{\sigma'} = F(M_G|_\sigma)$ and $\sigma(F)(M_G)|_{\iota'} = M_G$.[4] Thus we have defined a persistent constructor $\sigma(F) \in Mod(\Sigma_G \overset{\iota'}{\Rightarrow} \Sigma'_G)$ which we call the *translation of F along*

[4] CASL notation: $\sigma(F)(M_G)$ would be written $F[M_G \textbf{ fit } \sigma]$ in CASL, and σ can sometimes be left implicit.

σ. Intuitively, $\sigma(F)$ performs the "local" construction F on the "Σ part" of a "global context" given as a Σ_G-model, and leaves the rest unchanged. □

Exercise 6.1.29. Show that translation of constructors preserves composition in the following sense. Suppose that the following two pushout diagrams in **Sign** admit amalgamation:

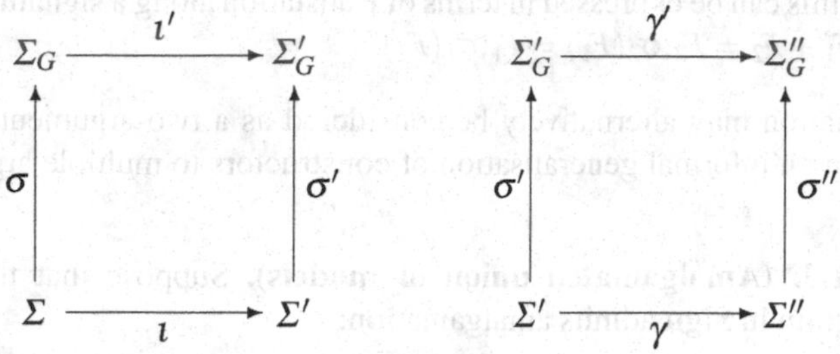

Consider constructors $F \in Mod(\Sigma \stackrel{\iota}{\Rightarrow} \Sigma')$ and $F' \in Mod(\Sigma' \stackrel{\gamma}{\Rightarrow} \Sigma'')$. Then prove that $\sigma(F;F') = \sigma(F);\sigma'(F') \in Mod(\Sigma_G \stackrel{\iota';\gamma'}{\Longrightarrow} \Sigma''_G)$. □

Exercise 6.1.30. Suppose that the following pushout diagram in **Sign** admits amalgamation

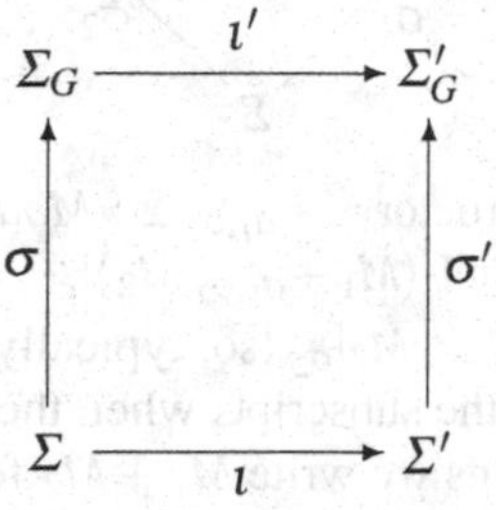

and let SP' be a Σ'-specification such that the free extension $F_{\iota,SP'}$ is naturally persistent and total. Show that $F_{\iota',SP'\,\mathbf{with}\,\sigma'}$ is naturally persistent and total as well, and moreover coincides with $\sigma(F_{\iota,SP'})$. HINT: Amalgamation of morphisms is crucial.

Conclude that in **EQ**, $\sigma(F_{\iota,\langle\Sigma',\mathscr{E}'\rangle}) = F_{\iota',\langle\Sigma'_G,\sigma'(\mathscr{E}')\rangle}$ for any set $\mathscr{E}'$ of Σ'-equations such that $F_{\iota,\langle\Sigma',\mathscr{E}'\rangle}$ is naturally persistent.

Generalise this property to free extensions restricted to a class of Σ-models. □

Example 6.1.31 (Amalgamated union of constructors). Suppose that the following pushout diagram in **Sign** admits amalgamation:

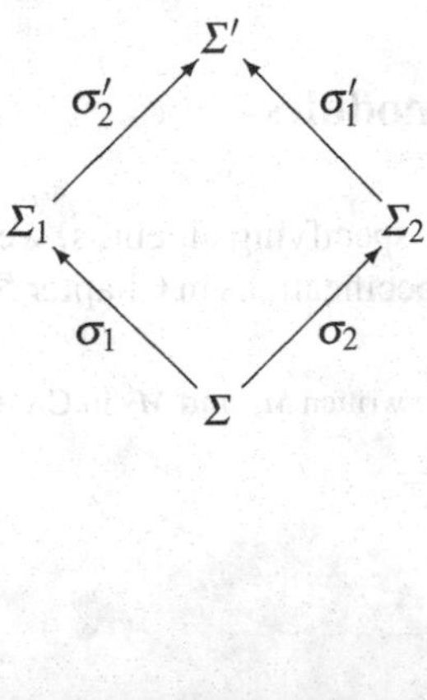

Given two persistent constructors $F_1 \in Mod(\Sigma \overset{\sigma_1}{\Longrightarrow} \Sigma_1)$ and $F_2 \in Mod(\Sigma \overset{\sigma_2}{\Longrightarrow} \Sigma_2)$, for any Σ-model $M \in dom(F_1) \cap dom(F_2)$, define $(F_1 + F_2)(M)$ to be the unique Σ'-model such that $(F_1 + F_2)(M)|_{\sigma'_1} = F_2(M)$ and $(F_1 + F_2)(M)|_{\sigma'_2} = F_1(M)$. We have thus defined a persistent constructor $F_1 + F_2 \in Mod(\Sigma \overset{\sigma}{\Longrightarrow} \Sigma')$, where $\sigma = \sigma_1;\sigma'_2 = \sigma_2;\sigma'_1$.

Note that this can be expressed in terms of translation along a signature morphism as follows: $F_1 + F_2 = F_2;\sigma_2(F_1) = F_1;\sigma_1(F_2)$. □

Amalgamation may alternatively be considered as a two-argument constructor. This relies on an informal generalisation of constructors to multiple arguments; see Section 7.3.

Example 6.1.32 (Amalgamated union of models). Suppose that the following pushout diagram in **Sign** admits amalgamation:

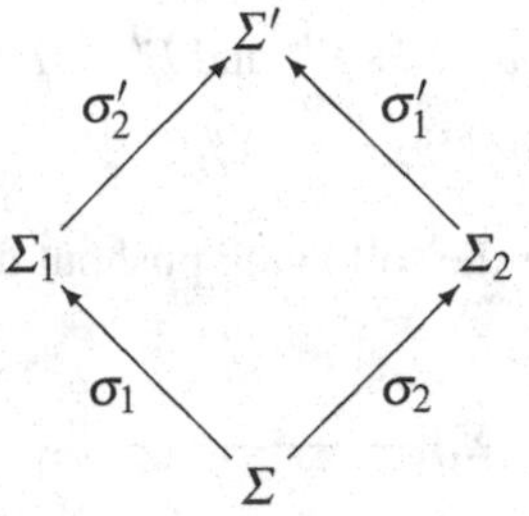

Define a two-argument constructor $\_ +_{\sigma_1,\sigma_2} \_ : Mod(\Sigma_1) \times Mod(\Sigma_2) \to Mod(\Sigma')$ by $(M_1 +_{\sigma_1,\sigma_2} M_2)|_{\sigma'_1} = M_2$ and $(M_1 +_{\sigma_1,\sigma_2} M_2)|_{\sigma'_2} = M_1$ for $M_1 \in Mod(\Sigma_1)$ and $M_2 \in Mod(\Sigma_2)$ such that $M_1|_{\sigma_1} = M_2|_{\sigma_2}$ (so, typically this constructor is not total).

In examples, we will omit the subscripts when they are evident. For instance, in **FOPEQ** and similar institutions we write $M_1 + M_2$ for $M_1 +_{\iota_1,\iota_2} M_2$ where $\iota_1 : \Sigma_1 \cap \Sigma_2 \to \Sigma_1$ and $\iota_2 : \Sigma_1 \cap \Sigma_2 \to \Sigma_2$ are the signature inclusions.[5] □

Exercise 6.1.33. Generalise this to an arbitrary diagram in **Sign**: define a constructor that combines a family of models over signatures in the nodes of the diagram to give a model over its colimit signature. □

Exercise 6.1.34. Generalise Exercise 6.1.15 to multi-argument constructors and show that union arises from amalgamated union of models (over a diagram of identities). □

6.2 Specifying generic modules

Just as we provided means for specifying algebras, we will now provide means for specifying constructors. The specifications in Chapter 5 described classes of models,

[5] CASL notation: $M_1 + M_2$ would be written M_1 **and** M_2 in CASL.

so analogously these specifications will describe classes of constructors. The basic idea is that we need to specify separately the conditions that the argument is required to satisfy and those that the result should then guarantee. This leads to the following definition.

Definition 6.2.1 (Constructor specification, first version). Given specifications SP and SP', $SP \Rightarrow SP'$ is a constructor specification with the following semantics:

$$Sig[SP \Rightarrow SP'] = Sig[SP] \Rightarrow Sig[SP']$$
$$Mod[SP \Rightarrow SP'] = \{F \in Mod(Sig[SP] \Rightarrow Sig[SP']) \mid \text{for all } M \in Mod[SP], M \in dom(F) \text{ and } F(M) \in Mod[SP']\}$$

□

Example 6.2.2. A signature morphism $\sigma: Sig[SP] \to Sig[SP']$ is a specification morphism $\sigma: SP \to SP'$ iff $\_|_\sigma \in Mod[SP' \Rightarrow SP]$.

Let $\sigma: Sig[SP] \to Sig[SP']$. Suppose that for every model $M \in Mod[SP]$ there exists a free object over M with respect to the reduct functor $\_|_\sigma: \mathbf{Mod}[SP'] \to \mathbf{Mod}(Sig[SP])$. Then $F_{\sigma,SP'} \in Mod[SP \Rightarrow SP']$. In particular, for liberal institutions and flat specifications SP', $F_{\sigma,SP'} \in Mod[SP \Rightarrow SP']$ for every specification SP. □

Exercise 6.2.3. Using Exercise 5.1.10, show that in **EQ**, for any Σ-specification SP such that $Mod[SP]$ is closed under subalgebras, we have $R_\Sigma \in Mod[SP \Rightarrow$ **reachable** SP **on** $sorts(\Sigma)]$, where R_Σ is the constructor mapping any algebra to its reachable subalgebra; see Example 6.1.13.

Generalise this to any institution with appropriate additional structure — cf. Section 4.5 — as in Exercises 5.1.10 and 6.1.14. □

Fact 6.2.4. *Suppose $Sig[SP] = Sig[SP_1]$ and $Sig[SP'] = Sig[SP'_1]$. If $Mod[SP_1] \subseteq Mod[SP]$ and $Mod[SP'] \subseteq Mod[SP'_1]$ then $Mod[SP \Rightarrow SP'] \subseteq Mod[SP_1 \Rightarrow SP'_1]$.* □

Fact 6.2.5. *If $F \in Mod[SP \Rightarrow SP']$ and $F' \in Mod[SP' \Rightarrow SP'']$ then $F;F' \in Mod[SP \Rightarrow SP'']$.* □

Constructor specifications of the form $SP \Rightarrow SP'$ do not provide any way of requiring dependency between the argument of a constructor and its result. In particular, even if SP' extends SP via an inclusion $\iota: SP \to SP'$, constructors in $Mod[SP \Rightarrow SP']$ need not be persistent in general.

Example 6.2.6. Recall Example 6.1.4, which defined a constructor $\_|_\delta$ extending models of booleans with conjunction and negation by the addition of disjunction. If SP is a Σ-specification of booleans without disjunction and SP' is a Σ'-specification of booleans with disjunction, then $\_|_\delta \in Mod[SP \Rightarrow SP']$. However, $Mod[SP \Rightarrow SP']$ also admits the constructor that disregards its argument, mapping all models of SP to the *same* model of SP'.

A related example would be a constructor specification for a sorting module where SP would specify a type with an order relation and SP' would specify a function to sort lists of elements with respect to an order relation; $SP \Rightarrow SP'$ does not require that the lists being sorted contain elements of the type provided by the argument, or that the order relation used for sorting be the one provided by the argument. □

This problem can be circumvented by imposing a persistency requirement, for example using the notation $SP \stackrel{\iota}{\Rightarrow} SP'$ to specify constructors in $Mod[SP \Rightarrow SP']$ that are persistent along $\iota : Sig[SP] \to Sig[SP']$. Unfortunately there appears to be no way of extending this idea to higher-order constructors — see Section 6.4 — so we take a different, more flexible, solution. Namely, we allow the specification of the result to explicitly depend on the argument to which the constructor is applied. But so far we have no construct which allows a specification to refer to a model and thus we need to add this as well.

Definition 6.2.7 (Singleton specification). Given a signature Σ and a model $M \in Mod(\Sigma)$, $\{M\}$ is a specification with the following semantics:[6]

$$\begin{aligned} Sig[\{M\}] &= \Sigma \\ Mod[\{M\}] &= \{M\} \end{aligned}$$

□

Definition 6.2.8 (Constructor specification). Given a specification SP and a specification $SP'[X]$ that may contain a free variable[7] X ranging over $Sig[SP]$-models, $\Pi X{:}SP \bullet SP'[X]$ is a *constructor specification* with the following semantics:

$$\begin{aligned} Sig[\Pi X{:}SP \bullet SP'[X]] &= Sig[SP] \Rightarrow Sig[SP'[X]] \\ Mod[\Pi X{:}SP \bullet SP'[X]] &= \{F \in Mod(Sig[SP] \Rightarrow Sig[SP'[X]]) \mid \\ &\qquad \text{for each } M \in Mod[SP], \\ &\qquad M \in dom(F) \text{ and } F(M) \in Mod[SP'[M/X]]\} \end{aligned}$$

A constructor specification $\Pi X{:}SP \bullet SP'[X]$ is *consistent* if $Mod[\Pi X{:}SP \bullet SP'[X]] \neq \varnothing$. □

Some of the concepts and notations used in this definition have not been formally introduced. Π is a binding construct and in order to give a completely formal definition of its syntax we would need to give a typing system for *specifications in context* in which specifications are expressions built using specification-building operations, perhaps involving variables that range over models. (In Section 6.3, we will need variables ranging over specifications as well.) Then $SP'[M/X]$ stands for the specification SP' in which all occurrences of the variable X are replaced by the model M; to do this more formally one needs to interpret specification expressions in an environment. See Section 6.4 below for the details; in the meantime the meaning should be clear enough.

The above definition properly generalises Definition 6.2.1: $SP \Rightarrow SP'$ is equivalent to $\Pi X{:}SP \bullet SP'$, where X does not occur in SP'.

Example 6.2.9 (Persistent constructor specification). The requirement of persistency can be captured as promised above. Let SP and SP' be specifications with a signature morphism $\iota : Sig[SP] \to Sig[SP']$. Then $SP \stackrel{\iota}{\Rightarrow} SP'$ is a constructor specification defined by

---

6 CASL notation: curly braces are only used for grouping in CASL.

7 So far we have systematically used the meta-variable X for sets of variables as may occur in terms and formulae in institutions like **EQ**; the use of X also as a variable in specification expressions here and below should cause no confusion.

$$SP \stackrel{\iota}{\Rightarrow} SP' = \Pi X{:}SP\bullet (\{X\} \textbf{ with } \iota) \cup SP'.$$

It is easy to check that

$$Mod[SP \stackrel{\iota}{\Rightarrow} SP'] = Mod[SP \Rightarrow SP'] \cap Mod(Sig[SP] \stackrel{\iota}{\Rightarrow}_{Mod[SP]} Sig[SP']). \qquad \square$$

Example 6.2.10. The problem indicated in Example 6.2.6 can now be resolved by using the constructor specification $SP \stackrel{\iota}{\Rightarrow} SP'$ where SP and SP' are as sketched there and $\iota{:}Sig[SP] \to Sig[SP']$ is the signature inclusion.

In this case, a tight specification of the constructor $\_|_\delta$ can be given, namely

$$\Pi X{:}SP\bullet \{X\} \textbf{ hide via } \delta,$$

and the same applies to any reduct constructor. (However, there is little point in writing a specification that admits a single constructor and is so close to a program via Exercise 6.1.5; one normally starts from loose requirements; see Chapter 7.) □

Example 6.2.11. Recall the specifications STRINGARRAY and BUCKETARRAY in Section 5.3. Constructors that implement such arrays in a generic fashion can be specified as follows.

spec ELEM$_\exists$ =
  sorts *Elem*
  $\exists x{:}Elem\bullet x = x$

spec CONSARRAY =
  ΠX:ELEM$_\exists\bullet \{X\}$ **and** NAT
    then
      sorts *Array*[*Elem*]
      ops $empty{:}Array[Elem]$
        $put{:}Nat \times Elem \times Array[Elem] \to Array[Elem]$
        $get{:}Nat \times Array[Elem] \to Elem$
      preds $used{:}Nat \times Array[Elem]$
      $\forall i,j{:}Nat, s{:}Elem, a{:}Array[Elem]$
        • $\neg used(i,empty)$
        • $used(i,put(i,s,a))$
        • $i \neq j \Rightarrow (used(i,put(j,s,a)) \Leftrightarrow used(i,a))$
        • $get(i,put(i,s,a)) = s$
        • $i \neq j \Rightarrow get(i,put(j,s,a)) = get(i,a)$

This specifies constructors which, when given a model of ELEM, combine it with a model of NAT and expand the result by some realisation of *Nat*-indexed arrays storing elements of the sort *Elem*.[8]

Exercise. The axiom in ELEM$_\exists$ guarantees that there is at least one value of sort *Elem*. Otherwise CONSARRAY would be inconsistent. Explain why.

---

[8] CASL notation: this would be written **unit spec** CONSARRAY = ELEM$_\exists \to \ldots$ in CASL, where the result is implicitly a persistent extension of the argument; see the notation ELEM$_\exists \stackrel{\iota}{\Rightarrow}$ ELEMARRAY$_\exists$ below.

The explicit dependency of the specification of the result on the model of ELEM given as argument ensures persistency of the specified constructors. So, exactly the same thing may be written as follows:

$$\text{CONSARRAY} \equiv \text{ELEM}_{\exists} \stackrel{\iota}{\Rightarrow} \text{ELEMARRAY}_{\exists}$$

where ELEMARRAY is defined in the obvious way

spec ELEMARRAY$_{\exists}$ =
 ELEM$_{\exists}$ **and** NAT
 then
  sorts *Array*[*Elem*]
  ops *empty*: *Array*[*Elem*]
   ...
  preds *used*: *Nat* × *Array*[*Elem*]
  $\forall i, j{:}Nat, s{:}Elem, a{:}Array[Elem]$
   • $\neg used(i, empty)$
   • ...

and ι is the signature inclusion. □

Exercise 6.2.12. Give a version of CONSARRAY in the institution **FPL** along the lines of Example 5.3.2. Then the following is a constructor in **FPL** that realises it (see Example 6.1.9):

constructor *A* : CONSARRAY =
 sorts *Nat* **free with** 0 | *succ*(*Nat*)
  Array[*Elem*] **free with** *empty* | *put*(*Nat*, *Elem*, *Array*[*Elem*])
 ops **fun** *used*(*i*:*Nat*, *a*:*Array*[*Elem*]):*Bool* =
   case *a* **of** *empty* => *false*
    | *put*(*n*, *e*, *a*′) => **if** *n* = *i* **then** *true* **else** *used*(*i*, *a*′)
  fun *get*(*i*:*Nat*, *a*:*Array*[*Elem*]):*Elem* =
   case *a* **of** *put*(*n*, *e*, *a*′) => **if** *n* = *i* **then** *e* **else** *get*(*i*, *a*′)

Check that this definition indeed yields a constructor in *Mod*[CONSARRAY]. Notice that the requirement on the argument model to contain at least one element of sort *Elem* may be dropped here.

Use the extensions to **FPL** suggested in Exercise 4.1.26 to provide another such constructor, for instance one in which arrays are represented as functions from *Nat* to *Elem*. □

Fact 6.2.13. $Mod[SP \stackrel{\iota}{\Rightarrow} SP'] \subseteq Mod[SP \stackrel{\iota}{\Rightarrow} SP' \cup (SP \textbf{ with } \iota)]$. □

Exercise 6.2.14. It follows that $Mod[SP \stackrel{\iota}{\Rightarrow} SP'] = Mod[SP \stackrel{\iota}{\Rightarrow} SP' \cup (SP \textbf{ with } \iota)]$ since the opposite inclusion to the one in the fact is trivial; cf. Fact 6.2.4. So we can always assume that in a persistent constructor specification $SP \stackrel{\iota}{\Rightarrow} SP'$, $\iota{:}SP \to SP'$ is a specification morphism. Show that $SP \stackrel{\iota}{\Rightarrow} SP'$ is consistent iff $\iota{:}SP \to SP'$ admits model expansion, and then $\iota{:}SP \to SP'$ is conservative by Exercise 5.5.7. □

Fact 6.2.15. *Given $F \in Mod[SP \overset{\iota}{\Rightarrow} SP']$ and $F' \in Mod[SP' \overset{\iota'}{\Rightarrow} SP'']$, we also have $F;F' \in Mod[SP \overset{\iota;\iota'}{\Longrightarrow} SP'']$.* □

Exercise 6.2.16. Prove the following generalisation of Fact 6.2.5 to arbitrary constructor specifications: if $F \in Mod[\Pi X{:}SP\bullet SP'_1[X]]$, $F' \in Mod[\Pi Y{:}SP'_2\bullet SP''[Y]]$ and for each $M \in Mod[SP]$, $Mod[SP'_1[M/X]] \subseteq Mod[SP'_2]$ then we also have $F;F' \in Mod[\Pi X{:}SP\bullet SP''[F(X)/Y]]$. (See Section 6.4 for clarification of the notation used in case it is not self-evident.) Use this to prove Fact 6.2.15. □

Fact 6.2.17. *Consider a persistent constructor $F \in Mod(SP \overset{\sigma}{\Rightarrow} SP')$ with a signature morphism $\sigma{:}Sig[SP] \to Sig[SP']$. Let $\iota{:}SP'' \to SP'$ be a specification morphism and $\sigma'{:}Sig[SP] \to Sig[SP'']$ be a signature morphism such that $\sigma = \sigma';\iota$. Then $F;\_|_\iota \in Mod(SP \overset{\sigma'}{\Longrightarrow} SP'')$ is a persistent constructor.* □

Fact 6.2.18. *Suppose that the following pushout diagram in* **Sign** *admits amalgamation*

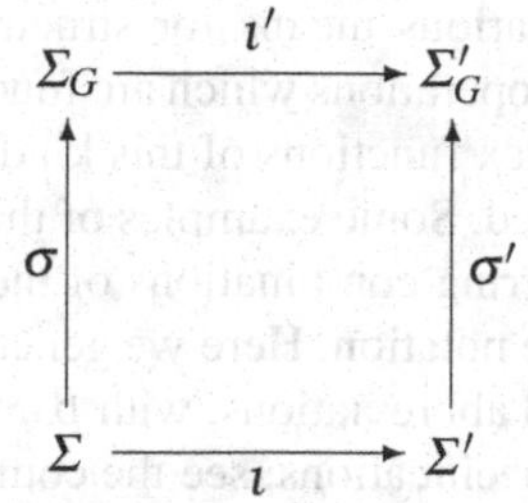

*and that $F \in Mod[SP \overset{\iota}{\Rightarrow} SP']$ where $Sig[SP] = \Sigma$ and $Sig[SP'] = \Sigma'$. Furthermore, let SP_G be a specification such that $Sig[SP_G] = \Sigma_G$ and $\sigma{:}SP \to SP_G$ is a specification morphism. Then $\sigma(F) \in Mod[SP_G \overset{\iota'}{\Longrightarrow} (SP'$ **with** $\sigma') \cup (SP_G$ **with** $\iota')]$.* □

Fact 6.2.19. *Suppose that the following pushout diagram in* **Sign** *admits amalgamation:*

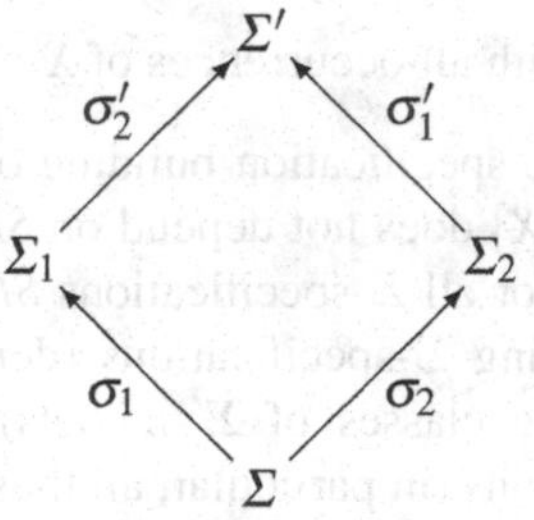

*Given two persistent constructors $F_1 \in Mod[SP \overset{\sigma_1}{\Longrightarrow} SP_1]$ and $F_2 \in Mod[SP \overset{\sigma_2}{\Longrightarrow} SP_2]$, $F_1 + F_2 \in Mod[SP \overset{\sigma}{\Rightarrow} (SP_1$ **with** $\sigma'_2) \cup (SP_2$ **with** $\sigma'_1)]$ where $\sigma = \sigma_1;\sigma'_2 = \sigma_2;\sigma'_1$.* □

Fact 6.2.20. *$\Pi X{:}SP\bullet SP'[X]$ is inconsistent iff for some $M \in Mod[SP]$, $SP'[M/X]$ is inconsistent. In particular, $SP \Rightarrow SP'$ is inconsistent iff SP is consistent and SP' is inconsistent.* □

Fact 6.2.21. *$SP \overset{\iota}{\Rrightarrow} SP'$ is consistent iff $Mod[SP] \subseteq Mod[SP'$ **hide via** $\iota]$.* □

Exercise 6.2.22. Give an example of specifications *SP*, *SP′* in **FOEQ** and a signature morphism $\iota{:}Sig[SP] \to Sig[SP']$ such that $SP \Rrightarrow SP'$ is consistent while $SP \overset{\iota}{\Rrightarrow} SP'$ is inconsistent. □

Exercise 6.2.23. Give an example of specifications *SP*, *SP′* in **FOEQ** and a signature morphism $\iota{:}Sig[SP] \to Sig[SP']$ such that $SP \overset{\iota}{\Rrightarrow} SP'$ is consistent but there is no persistent *functor* from **Mod**[*SP*] to **Mod**[*SP′*]. HINT: Consider an axiom of the form $a = b \Leftrightarrow c \neq d$ where a, b are constants in *SP*, and c, d are new constants in *SP′*. □

6.3 Parameterised specifications

In Chapter 5 we presented various means for structuring specifications. This involved specification-building operations which are functions mapping specifications to specifications. More complex functions of this kind may be defined as combinations of the operations provided. Some examples of these were given in Section 5.2 where a few commonly occurring combinations of the basic specification-building operations were given concise notation. Here we generalise this further and provide λ-abstraction for user-defined abbreviations, with β-reduction for application, with variables now ranging over specifications; see the comments after Definition 6.2.8.

Definition 6.3.1 (Parameterised specification). Given a signature Σ and a specification $SP'[X]$ that may contain a free variable X ranging over Σ-specifications, $\lambda X{:}Spec(\Sigma)\bullet SP'[X]$ is a *parameterised specification.*[9] The result of *applying* such a parameterised specification to a Σ-specification *SP* is defined as follows:

$$(\lambda X{:}Spec(\Sigma)\bullet SP'[X])(SP) = SP'[SP/X]$$

where $SP'[SP/X]$ is $SP'[X]$ with all occurrences of X replaced by *SP*. □

A characteristic feature of the specification-building operations we have defined is that the signature of $SP'[SP/X]$ does not depend on *SP*, so there is a common signature $\Sigma' = Sig[SP'[SP/X]]$ for all Σ-specifications *SP*. Thus $\lambda X{:}Spec(\Sigma)\bullet SP'[X]$ determines a function mapping Σ-specifications (denoting classes of Σ-models) to Σ'-specifications (denoting classes of Σ'-models). Another property of most specification-building operations (in particular, all those in Section 5.1 except **free**) is *monotonicity* with respect to model class inclusion; see Exercises 5.1.4 and 5.1.13. This carries over to parameterised specifications that involve only such specification-building operations.

[9] CASL notation: *generic specifications* in CASL have the same motivation but they are done differently; see Definition 6.3.5.

Exercise 6.3.2. Generalise the above definition of parameterised specifications to allow one to narrow the class of admissible arguments by requiring that all the models of the argument satisfy a given specification. □

Example 6.3.3. The example in Section 5.3 introduced two closely related versions of arrays: STRINGARRAY and BUCKETARRAY. Their similarity can be made explicit by introducing a specification of arrays that is parameterised by a specification of the elements to be stored in arrays.

spec ELEM =
  sorts *Elem*

spec ARRAY =
  $\lambda X{:}Spec(Sig[\text{ELEM}])\bullet$
    X **and** NAT
    then
      sorts $Array[Elem]$
      ops $empty{:}Array[Elem]$
        $put{:}Nat\times Elem\times Array[Elem]\to Array[Elem]$
        $get{:}Nat\times Array[Elem]\to Elem$
      preds $used{:}Nat\times Array[Elem]$
      $\forall i,j{:}Nat, s{:}Elem, a{:}Array[Elem]$
        • $\neg used(i,empty)$
        • $used(i,put(i,s,a))$
        • $i\neq j\Rightarrow(used(i,put(j,s,a))\Leftrightarrow used(i,a))$
        • $get(i,put(i,s,a))=s$
        • $i\neq j\Rightarrow get(i,put(j,s,a))=get(i,a)$

Now

$$\text{STRINGARRAY}\equiv\text{STRING}\ \textbf{and}\ (\text{ARRAY}(\text{STRING}\ \textbf{hide via}\ \sigma_{Elem\mapsto String})\ \textbf{with}\ \sigma_{Elem\mapsto String, Array[Elem]\mapsto Array[String]})$$

where $\sigma_{Elem\mapsto String}$ is the unique morphism from $Sig[\text{ELEM}]$ to $Sig[\text{STRING}]$ (mapping *Elem* to *String*) and $\sigma_{Elem\mapsto String, Array[Elem]\mapsto Array[String]}$ is the surjective morphism from $Sig[\text{ARRAY}(\text{ELEM})]$ that maps *Elem* to *String* and $Array[Elem]$ to $Array[String]$ and is the identity otherwise.

BUCKETARRAY may be expressed in a similar fashion. □

A parameterised specification often extends its argument as a constructor does by adding some additional components and axioms. Then it can be expressed in the form $\lambda X{:}Spec(\Sigma)\bullet X$ **then** $(\cdots)$ with no use of X in "$(\cdots)$"; see ARRAY above. When P is in this form there is an inclusion between Σ and the result signature, and then for any Σ-specification SP, the application $P(SP)$ gives SP **then** $(\cdots)$.

Exercise 6.3.4. Check that if P is of the form $\lambda X{:}Spec(\Sigma)\bullet X$ **then** $(\cdots)$ then for any Σ-specification SP, $P(SP)\equiv P(\langle\Sigma,\varnothing\rangle)$ **and** SP.

Use this fact to show that

$\text{STRINGARRAY} \equiv \text{STRING}$ **and** $(\text{ARRAY}(\text{ELEM})$
with $\sigma_{Elem \mapsto String, Array[Elem] \mapsto Array[String]})$.

The equivalence above does not hold when P does not merely enrich its parameter. An example is a specification of ordered lists where the order relation, given as a parameter, is used in the specification of insertion but is not included in the result. Work out the details of this example. □

It is often the case that the argument SP arises from a larger specification SP_G by "cutting it down" to fit the signature Σ. A common idiom is then to restore what has been cut out by adding it back to the result, as in the equivalent presentation of STRINGARRAY in Example 6.3.3 above. Generalising this to allow for an arbitrary signature morphism between parameter and result rather than only an inclusion, while constraining the choice of argument specification as suggested in Exercise 6.3.2, we obtain the following.

Definition 6.3.5 (Pushout-style parameterised specification). In an institution with a cocomplete category of signatures, a specification morphism $p: SP \to SP'$ may be considered as a *parameterised specification. Application* of p to an argument specification SP_G via a specification morphism $\sigma: SP \to SP_G$ results in the specification $p(SP_G[\sigma])$[10] defined by the following pushout diagram in the category of specifications:

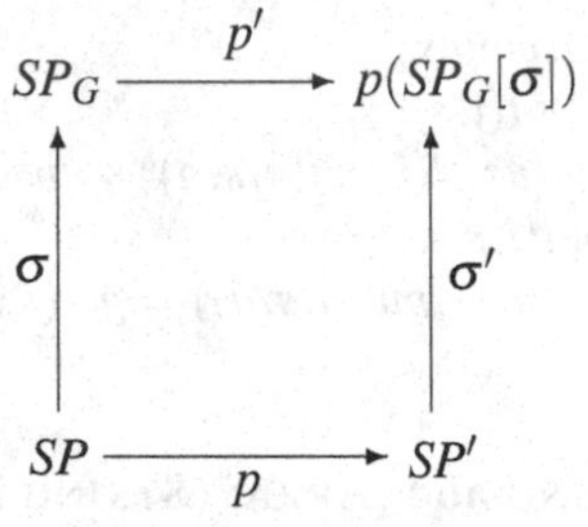

Here, the specification morphism σ is called the *fitting morphism*. □

Note that $p(SP_G[\sigma])$ is $(SP_G$ **with** $p') \cup (SP'$ **with** $\sigma')$ where the following is the pushout in the category of signatures with p and σ viewed as signature morphisms (cf. Theorem 5.5.11):

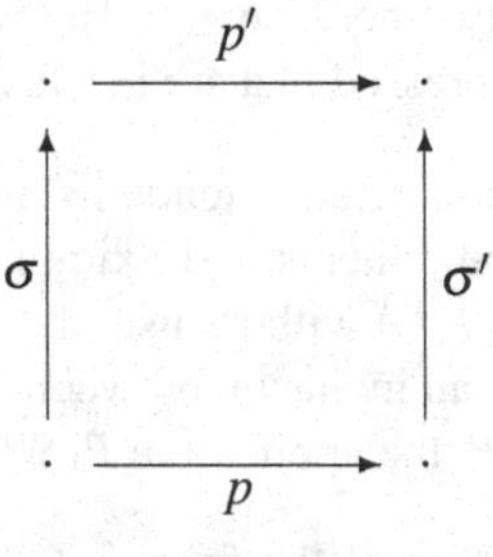

10 CASL notation: if p is an inclusion than this would be written $P[SP_G$ **fit** $\sigma]$ in CASL for an appropriate generic specification P corresponding to p. The fitting morphism σ can sometimes be left implicit, and CASL has special treatment of compound names like *Array*[*Elem*]; see Example 6.3.6.

Example 6.3.6. Example 6.3.3 may be simplified by the use of pushout-style parameterisation. Let PARRAY: ELEM $\to$ ARRAY(ELEM) be the inclusion. Then

$$\text{STRINGARRAY} \equiv \text{PARRAY}(\text{STRING}[\sigma_{Elem \mapsto String}])$$

where $\sigma_{Elem \mapsto String}$ is the unique morphism from Sig[ELEM] to Sig[STRING] (mapping *Elem* to *String*). To be precise, this equivalence depends on the choice of names in the signature of the instantiation, since the pushout is defined up to isomorphism. It holds when we follow the convention that names like *Array*[*Elem*] are mapped to *Array*[*String*] when the fitting morphism maps *Elem* to *String*; see [Mos04].

BUCKETARRAY may be expressed in a similar fashion. □

A specification morphism $p: SP \to SP'$ may also be viewed as a persistent constructor specification $SP \stackrel{p}{\Rightarrow} SP'$.

Exercise 6.3.7. Check that in any semi-exact institution, using the notation of Definition 6.3.5, if $F \in Mod[SP \stackrel{p}{\Rightarrow} SP']$ then $\sigma(F) \in Mod[SP_G \stackrel{p'}{\Longrightarrow} p(SP_G[\sigma])]$ (cf. Fact 6.2.18). Then it easily follows that for any specification morphism $q: SP_0 \to SP_G$ and $G \in Mod[SP_0 \stackrel{q}{\Rightarrow} SP_G]$ we get $G;F \in Mod[SP_0 \stackrel{q;p'}{\Longrightarrow} p(SP_G[\sigma])]$. □

The point of this exercise is that lifting of constructors specified in this way is compatible with pushout-style instantiation. This must *not* be taken as a directive to identify parameterised specifications with specifications of constructors. The two concepts serve inherently different methodological purposes, and semantically they give rise to objects of different types: a parameterised specification determines a function from specifications (denoting classes of models) to specifications (denoting classes of models), while a constructor specification denotes a class of constructors, that is, a class of functions from models to models.

If Σ and Σ' are the parameter and result signatures respectively, then a parameterised specification determines a function from specifications (denoting classes of models) to specifications (denoting classes of models), that is, $P: \mathcal{P}(Mod(\Sigma)) \to \mathcal{P}(Mod(\Sigma'))$ (where $\mathcal{P}(Mod(\Sigma))$ is the class of all classes of Σ-models). In contrast, a constructor specification denotes a class of constructors, that is, a class of functions from models to models, $Q \subseteq Mod(\Sigma \Rightarrow \Sigma')$.

The following exercise sheds some light on the relationship between the two concepts.

Exercise 6.3.8. Given signatures Σ and Σ', consider the following two semantic domains

- **ConstSpec**: all classes of *total* constructors in $Mod(\Sigma \Rightarrow \Sigma')$, ordered by inclusion
- **ParSpec**: all *monotone* functions from classes of Σ-models to classes of Σ'-models, ordered by pointwise containment

and define the following functions:

- For $\mathcal{Q} \in$ **ConstSpec**, define $\mathcal{Q}^\dagger \in$ **ParSpec** by $\mathcal{Q}^\dagger(C) = \{F(A) \mid F \in \mathcal{Q}, A \in C\}$

- For $\mathscr{P} \in$ **ParSpec**, define $\mathscr{P}^\# \in$ **ConstSpec** by $\mathscr{P}^\# = \{F \in Mod(\Sigma) \rightarrow Mod(\Sigma') \mid F(A) \in \mathscr{P}(\{A\})$ for all $A \in Mod(\Sigma)\}$.

Show that $\_^\dagger$ and $\_^\#$ form a Galois connection (see Exercise 3.5.23) between **ConstSpec** and **ParSpec**. Characterise closed elements of the Galois connection (HINT: If you get stuck, have a look at [SST92]) and relate them to (total) functions mapping Σ-models to classes of Σ'-models.

Give examples of objects in **ConstSpec** and **ParSpec** that are not closed in this sense. HINT: For the former, consider objects that are not expressible using a constructor specification. For the latter, try using the parameter variable more than once. □

Exercise 6.3.9. Recall the constructor specification CONSARRAY and the parameterised specification ARRAY from Examples 6.2.11 and 6.3.3, respectively. Check that the class of constructors that CONSARRAY denotes and the function on classes of models that ARRAY determines arise from a function mapping models of ELEM to classes of ARRAY(ELEM)-models. But note that they are not related by the Galois connection in Exercise 6.3.8. (HINT: The trouble is that CONSARRAY specifies partial constructors which are not defined on algebras with the empty carrier on sort *Elem*.) A simple remedy is to add a constant *noelem*: *Elem* to the signature of $\text{ELEM}_\exists$ in CONSARRAY and to the signature of ELEM in ARRAY. Check that then the expected relationship holds. Alternatively, work out a Galois connection as in Exercise 6.3.8 for classes of partial constructors that are defined at least on models in a pre-specified class. □

6.4 Higher-order parameterisation

The material in the preceding sections extends naturally to higher-order parameterisation, both for constructors and for parameterised specifications. Higher-order constructors, which take constructors as arguments and produce constructors as results, correspond to higher-order generic modules in programming languages, and constructor specifications as introduced in Definition 6.2.8 extend easily to this case.

However, so far (see Definitions 6.2.8 and 6.3.1) we have been rather informal about the exact syntax of specifications involving variables. Such informality is potentially dangerous for higher-order languages with binding operators, so we give a system of formal typing rules for the syntactic constructs involved. We start by generalising the notion of constructor signature.

Definition 6.4.1 (Constructor signature and context). A *constructor signature* $\mathscr{S}$ is either a signature $\Sigma \in |\mathbf{Sign}|$ or has the form $\mathscr{S}_1 \Rightarrow \mathscr{S}_2$, where $\mathscr{S}_1$ and $\mathscr{S}_2$ are constructor signatures. We generalise the notation used so far by writing $Mod(\mathscr{S}_1 \Rightarrow \mathscr{S}_2)$ for the class of all partial functions from $Mod(\mathscr{S}_1)$ to $Mod(\mathscr{S}_2)$. A *context* Γ is a sequence of the form $X_1{:}\mathscr{S}_1, \ldots, X_n{:}\mathscr{S}_n$ where $X_1, \ldots, X_n$ are distinct variables and $\mathscr{S}_1, \ldots, \mathscr{S}_n$ are constructor signatures. We write $dom(\Gamma)$ for $\{X_1, \ldots, X_n\}$ and identify contexts with the obvious mappings, writing $\Gamma(X_i)$ for $\mathscr{S}_i$. □

Definition 6.4.2 (Constructor and constructor specification). For any constructor signature $\mathscr{S}$, elements in $Mod(\mathscr{S})$ are called *constructors*. This extends Definition 6.1.1 and also covers models in $Mod(\Sigma)$ which we view here as nullary constructors.

The following typing rules introduce syntax for expressions that are well formed in a given context. We use the judgement form $\Gamma \rhd W : \mathscr{G}$ where Γ is a context (which may be empty, as in $\rhd W : \mathscr{G}$), W is an expression, and $\mathscr{G}$ is either a constructor signature $\mathscr{S}$ or is a *specification type* of the form $Spec(\mathscr{S})$ where $\mathscr{S}$ is a constructor signature; this will be generalised in Definition 6.4.5 below. *Constructor expressions*, that is, expressions E of type $\mathscr{S}$, will also be called *constructors*, and *specification expressions*, that is, expressions SP of type $Spec(\mathscr{S})$, will be called *constructor specifications*. As usual, we regard α-convertible[11] expressions as equal, where the binding constructs are λ and Π.

$$\frac{}{\Gamma \rhd \langle \Sigma, \Phi \rangle : Spec(\Sigma)} \quad \Sigma \in |\mathbf{Sign}|, \Phi \subseteq \mathbf{Sen}(\Sigma)$$

$$\frac{\Gamma \rhd SP : Spec(\Sigma)}{\Gamma \rhd SP \ \mathbf{with}\ \sigma : Spec(\Sigma')} \quad \sigma{:}\Sigma \to \Sigma'$$

$$\frac{\Gamma \rhd SP' : Spec(\Sigma')}{\Gamma \rhd SP' \ \mathbf{hide\ via}\ \sigma : Spec(\Sigma)} \quad \sigma{:}\Sigma \to \Sigma'$$

. . . and similarly for other specification-building operations . . .

$$\frac{\Gamma \rhd SP_1 : Spec(\mathscr{S}) \qquad \Gamma \rhd SP_2 : Spec(\mathscr{S})}{\Gamma \rhd SP_1 \cup SP_2 : Spec(\mathscr{S})} \qquad \frac{\Gamma \rhd E : \mathscr{S}}{\Gamma \rhd \{E\} : Spec(\mathscr{S})}$$

$$\frac{\Gamma \rhd SP : Spec(\mathscr{S}) \qquad \Gamma, X{:}\mathscr{S} \rhd SP' : Spec(\mathscr{S}')}{\Gamma \rhd \Pi X{:}SP \bullet SP' : Spec(\mathscr{S} \Rightarrow \mathscr{S}')}$$

$$\frac{}{\Gamma \rhd X : \Gamma(X)} \quad X \in dom(\Gamma)$$

$$\frac{\Gamma \rhd SP : Spec(\mathscr{S}) \quad \Gamma, X{:}\mathscr{S} \rhd E' : \mathscr{S}'}{\Gamma \rhd \lambda X{:}SP \bullet E' : \mathscr{S} \Rightarrow \mathscr{S}'} \qquad \frac{\Gamma \rhd E : \mathscr{S}_1 \Rightarrow \mathscr{S}_2 \quad \Gamma \rhd E_1 : \mathscr{S}_1}{\Gamma \rhd E(E_1) : \mathscr{S}_2}$$

$$\frac{}{\Gamma \rhd [F]^{\mathscr{S}} : \mathscr{S}} \quad F \in Mod(\mathscr{S})$$

As before, $SP \Rightarrow SP'$ stands for $\Pi X{:}SP \bullet SP'$ where X does not occur in SP'. We adopt the usual notational convention that $\Rightarrow$ associates to the right, so that $SP \Rightarrow SP' \Rightarrow SP''$ stands for $SP \Rightarrow (SP' \Rightarrow SP'')$. □

[11] Two expressions are *α-convertible* if they are the same up to capture-avoiding renaming of bound variables.

The final rule above is of a different character from the rest since F in its side condition is a semantic object used as an expression in the formal typing judgement in its conclusion. This is a deliberate choice in order to avoid prescribing any syntax for defining individual models and constructors. We tag these objects with their constructor signatures in order to avoid any ambiguity. A reasonable option is to limit this rule to models only,

$$\frac{}{\Gamma \rhd [M]^{\Sigma} : \Sigma} \quad M \in Mod(\Sigma),$$

and introduce a particular set of constructors either as constants, with their typing defined by rules like

$$\frac{}{\Gamma \rhd \_|_{\sigma} : \Sigma' \Rightarrow \Sigma} \quad \sigma : \Sigma \to \Sigma'$$

or as additional syntax, with rules like

$$\frac{\Gamma \rhd E : \Sigma'}{\Gamma \rhd E|_{\sigma} : \Sigma} \quad \sigma : \Sigma \to \Sigma'.$$

Exercise 6.4.3. Devise similar rules for the free extension constructor as defined in Example 6.1.7. For instance, the second rule might be

$$\frac{\Gamma \rhd E : \Sigma \qquad \Gamma \rhd SP' : Spec(\Sigma')}{\Gamma \rhd F_{\sigma,SP'}(E) : \Sigma'} \quad \sigma : \Sigma \to \Sigma'.$$

□

Exercise 6.4.4. Suppose that the following pushout diagram in **Sign** admits amalgamation:

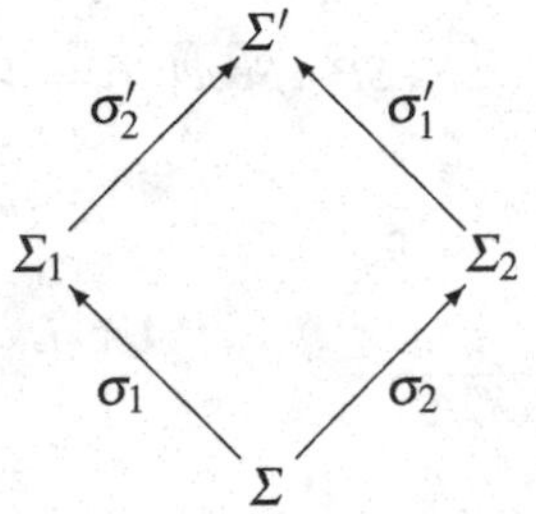

Example 6.1.32 defines a two-argument constructor of amalgamated union which we will write in curried form, taking its arguments one at a time rather than as a pair, $\_ +_{\sigma_1,\sigma_2} \_ : \Sigma_1 \Rightarrow (\Sigma_2 \Rightarrow \Sigma')$; cf. Exercise 6.4.16 below. Devise similar rules for this constructor. For instance, the second rule might be

$$\frac{\Gamma \rhd E_1 : \Sigma_1 \qquad \Gamma \rhd E_2 : \Sigma_2}{\Gamma \rhd E_1 +_{\sigma_1,\sigma_2} E_2 : \Sigma'} \quad \text{given the pushout diagram above.}$$

□

Higher-order parameterised specifications are harder to motivate but their syntax and semantics present no difficulties. We extend the form of specification types and contexts and use these to accommodate syntax for parameterised specifications.

Definition 6.4.5 (Parameterised specification). A *specification type* $\mathscr{T}$ is either of the form $Spec(\mathscr{S})$ for a constructor signature $\mathscr{S}$, as above, or has the form $\mathscr{T}_1 \to \mathscr{T}_2$. We extend the notion of context to allow variables typings to be specification types $\mathscr{T}$ as well as constructor signatures $\mathscr{S}$, and judgements to allow expressions to be typed to arbitrary specification types. Expressions P of type $\mathscr{T}$ are referred to as *$\mathscr{T}$-specifications*. When $\mathscr{T}$ has the form $Spec(\mathscr{S})$, these are constructor specifications, as before; when $\mathscr{T}$ has the form $\mathscr{T}_1 \to \mathscr{T}_2$, they are *parameterised specifications*. We add the following typing rules to those in Definition 6.4.2.

$$\frac{\Gamma, X{:}\mathscr{T} \rhd P' : \mathscr{T}'}{\Gamma \rhd \lambda X{:}\mathscr{T} \bullet P' : \mathscr{T} \to \mathscr{T}'} \qquad \frac{\Gamma \rhd P : \mathscr{T}_1 \to \mathscr{T}_2 \quad \Gamma \rhd P_1 : \mathscr{T}_1}{\Gamma \rhd P(P_1) : \mathscr{T}_2}$$

We write $[\![\mathscr{T}]\!]$ for the class of all subclasses of $Mod(\mathscr{S})$ when $\mathscr{T}$ has the form $Spec(\mathscr{S})$, and for the class of all functions from $[\![\mathscr{T}_1]\!]$ to $[\![\mathscr{T}_2]\!]$ when $\mathscr{T}$ has the form $\mathscr{T}_1 \to \mathscr{T}_2$. □

Exercise 6.4.6. Replace the rules for specification-building operations in Definition 6.4.2 by introducing these operations as constants as in

$$\frac{}{\Gamma \rhd \text{\_ } \mathbf{hide\ via}\ \sigma : Spec(\Sigma') \to Spec(\Sigma)} \ \sigma{:}\Sigma \to \Sigma'.$$ □

Exercise 6.4.7. Check that any typing judgement has at most one derivation in the system above, and that, in any context, every expression has at most one typing. Check that the type of an expression depends only on the types that the context assigns to its free variables. In particular, a closed expression (i.e. one with no free variables) has at most one type, which does not depend on the choice of the context. □

Exercise 6.4.8. Check that the following *substitutivity* properties hold in the system above:

- If $\Gamma, X{:}\mathscr{S}' \rhd E : \mathscr{S}$ and $\Gamma \rhd E' : \mathscr{S}'$ then $\Gamma \rhd E[E'/X] : \mathscr{S}$.
- If $\Gamma, X{:}\mathscr{S}' \rhd P : \mathscr{T}$ and $\Gamma \rhd E' : \mathscr{S}'$ then $\Gamma \rhd P[E'/X] : \mathscr{T}$.
- If $\Gamma, X{:}\mathscr{T}' \rhd E : \mathscr{S}$ and $\Gamma \rhd P' : \mathscr{T}'$ then $\Gamma \rhd E[P'/X] : \mathscr{S}$.
- If $\Gamma, X{:}\mathscr{T}' \rhd P : \mathscr{T}$ and $\Gamma \rhd P' : \mathscr{T}'$ then $\Gamma \rhd P[P'/X] : \mathscr{T}$.

For each of these cases, find expressions to which it is applicable. Check that the following property captures all four cases, where $\mathscr{G}$ and $\mathscr{G}'$ range over both constructor signatures and specification types:

- If $\Gamma, X{:}\mathscr{G}' \rhd W : \mathscr{G}$ and $\Gamma \rhd W' : \mathscr{G}'$ then $\Gamma \rhd W[W'/X] : \mathscr{G}$. □

Exercise 6.4.9. λ-abstraction and application are subject to *β-reduction*, defined by the rewrite rule (see Section 2.6)

$$(\lambda X{:}\mathscr{G}\bullet W)(W') \to_\beta W[W'/X],$$

which may be applied to reduce any occurrence of the redex within an expression; the resulting relation on expressions is also written $\to_\beta$.

Using Exercise 6.4.8, show that the following *subject reduction* properties hold in the system above:

- If $\Gamma \rhd E : \mathscr{S}$ and $E \to_\beta E'$ then $\Gamma \rhd E' : \mathscr{S}$.
- If $\Gamma \rhd P : \mathscr{T}$ and $P \to_\beta P'$ then $\Gamma \rhd P' : \mathscr{T}$.

The relation of η-reduction is defined in a similar way by the rewrite rule

$$\lambda X{:}\mathscr{G}\bullet W(X) \to_\eta W$$

whenever X is not free in W. Show that:

- If $\Gamma \rhd E : \mathscr{S}$ and $E \to_\eta E'$ then $\Gamma \rhd E' : \mathscr{S}$.
- If $\Gamma \rhd P : \mathscr{T}$ and $P \to_\eta P'$ then $\Gamma \rhd P' : \mathscr{T}$. □

We now give semantics to the expressions introduced in Definitions 6.4.2 and 6.4.5.

- Nullary constructors $E : \Sigma$ will denote models in $Mod(\Sigma)$.
- Constructors $E : \mathscr{S}_1 \Rightarrow \mathscr{S}_2$ will denote partial functions from $Mod(\mathscr{S}_1)$ to $Mod(\mathscr{S}_2)$.
- Constructor specifications $SP : Spec(\mathscr{S})$ will denote classes of $\mathscr{S}$-constructors.
- Parameterised specifications $P : \mathscr{T}_1 \to \mathscr{T}_2$ will denote functions mapping denotations of $\mathscr{T}_1$-specifications to denotations of $\mathscr{T}_2$-specifications.

Even well-typed expressions may fail to denote. This is due solely to the partiality of constructors as functions.

Definition 6.4.10. A *Γ-environment* ρ is a function that assigns values to variables in $dom(\Gamma)$ in a way that is consistent with their types. That is, for $X \in dom(\Gamma)$, if $\Gamma(X)$ is a constructor signature $\mathscr{S}$ then $\rho(X) \in Mod(\mathscr{S})$; and if $\Gamma(X)$ is a specification type $\mathscr{T}$ then $\rho(X) \in [\![\mathscr{T}]\!]$. □

Definition 6.4.11. For any context Γ and Γ-environment ρ, the following semantics gives a meaning to each expression that is well typed in Γ.

$[\![\Gamma \rhd \langle\Sigma,\Phi\rangle : Spec(\Sigma)]\!]_\rho = Mod[\langle\Sigma,\Phi\rangle]$
$[\![\Gamma \rhd SP \textbf{ with } \sigma : Spec(\Sigma')]\!]_\rho = \{M \in Mod(\Sigma') \mid M|_\sigma \in [\![\Gamma \rhd SP : Spec(\Sigma)]\!]_\rho\}$
    where $\sigma{:}\Sigma \to \Sigma'$
$[\![\Gamma \rhd SP' \textbf{ hide via } \sigma : Spec(\Sigma)]\!]_\rho = \{M|_\sigma \mid M \in [\![\Gamma \rhd SP' : Spec(\Sigma')]\!]_\rho\}$
    where $\sigma{:}\Sigma \to \Sigma'$
... and similarly for other specification-building operations ...

$[\![\Gamma \rhd SP_1 \cup SP_2 : Spec(\mathscr{S})]\!]_\rho = [\![\Gamma \rhd SP_1 : Spec(\mathscr{S})]\!]_\rho \cap [\![\Gamma \rhd SP_2 : Spec(\mathscr{S})]\!]_\rho$
$[\![\Gamma \rhd \{E\} : Spec(\mathscr{S})]\!]_\rho = \{[\![\Gamma \rhd E : \mathscr{S}]\!]_\rho\}$
$[\![\Gamma \rhd \Pi X{:}SP\bullet SP' : Spec(\mathscr{S} \Rightarrow \mathscr{S}')]\!]_\rho =$
    $\{F \in Mod(\mathscr{S} \Rightarrow \mathscr{S}') \mid \text{for all } G \in [\![\Gamma \rhd SP : \mathscr{S}]\!]_\rho, G \in dom(F)$
        $\text{and } F(G) \in [\![\Gamma, X{:}\mathscr{S} \rhd SP' : \mathscr{S}']\!]_{\rho[X \mapsto G]}\}$

$$
\begin{aligned}
&[\![\Gamma \rhd X : \Gamma(X)]\!]_\rho = \rho(X)\\
&[\![\Gamma \rhd \lambda X{:}SP\bullet E' : \mathscr{S} \Rightarrow \mathscr{S}']\!]_\rho =\\
&\qquad\qquad \{F \mapsto [\![\Gamma, X{:}\mathscr{S} \rhd E' : \mathscr{S}']\!]_{\rho[X\mapsto F]} \mid F \in [\![\Gamma \rhd SP : Spec(\mathscr{S})]\!]_\rho\}\\
&[\![\Gamma \rhd E(E_1) : \mathscr{S}_2]\!]_\rho = [\![\Gamma \rhd E : \mathscr{S}_1 \Rightarrow \mathscr{S}_2]\!]_\rho([\![\Gamma \rhd E_1 : \mathscr{S}_1]\!]_\rho)\\
&\quad \text{if } [\![\Gamma \rhd E_1 : \mathscr{S}_1]\!]_\rho \in dom([\![\Gamma \rhd E : \mathscr{S}_1 \Rightarrow \mathscr{S}_2]\!]_\rho)\\
&[\![\Gamma \rhd [F]^{\mathscr{S}} : \mathscr{S}]\!]_\rho = F\\
&\qquad \text{where } F \in Mod(\mathscr{S})
\end{aligned}
$$

$$
\begin{aligned}
&[\![\Gamma \rhd \lambda X{:}\mathscr{T}\bullet P' : \mathscr{T} \to \mathscr{T}']\!]_\rho = \{F \mapsto [\![\Gamma, X{:}\mathscr{T} \rhd P' : \mathscr{T}']\!]_{\rho[X\mapsto F]} \mid F \in [\![\mathscr{T}]\!]\}\\
&[\![\Gamma \rhd P(P_1) : \mathscr{T}_2]\!]_\rho = [\![\Gamma \rhd P : \mathscr{T}_1 \to \mathscr{T}_2]\!]_\rho([\![\Gamma \rhd P_1 : \mathscr{T}_1]\!]_\rho)
\end{aligned}
$$

Here, $\rho[Z \mapsto v]$ is the environment that results from ρ by assigning v to the variable Z and leaving the values of other variables unchanged. □

We will sometimes leave the context Γ and type $\mathscr{G}$ implicit and then abbreviate $[\![\Gamma \rhd W : \mathscr{G}]\!]_\rho$ as $[\![W]\!]_\rho$, as justified by Exercise 6.4.7. We use the notation $Mod[SP]$ to stand for $[\![SP]\!]_\varnothing$, where $\varnothing$ is the empty environment, when $\rhd SP : Spec(\mathscr{S})$. This properly extends the notation used for specifications SP as introduced before this section, for which we have $\rhd SP : Spec(\Sigma)$ or $\rhd SP : Spec(\Sigma \Rightarrow \Sigma')$.

Exercise 6.4.12. Check that the following *substitution property* holds: if $\Gamma, X{:}\mathscr{G}' \rhd W : \mathscr{G}$ and $\Gamma \rhd W' : \mathscr{G}'$, then $[\![W[W'/X]]\!]_\rho = [\![W]\!]_{\rho[X\mapsto[\![W']\!]_\rho]}$ for any Γ-environment ρ, provided that X has a free occurrence in W. The final proviso is needed because $[\![W]\!]_{\rho[X\mapsto[\![W']\!]_\rho]}$ is undefined when $[\![W']\!]_\rho$ is undefined, whether X is free in W or not. □

Exercise 6.4.13. Check that the typing system in Definitions 6.4.2 and 6.4.5 is sound with respect to the semantics in Definition 6.4.11. That is, show that for any context Γ and any Γ-environment ρ, whenever $\Gamma \rhd E : \mathscr{S}$ and $[\![\Gamma \rhd E : \mathscr{S}]\!]_\rho$ is defined then $[\![\Gamma \rhd E : \mathscr{S}]\!]_\rho \in Mod(\mathscr{S})$, and similarly, whenever $\Gamma \rhd P : \mathscr{T}$ and $[\![\Gamma \rhd P : \mathscr{T}]\!]_\rho$ is defined then $[\![\Gamma \rhd P : \mathscr{T}]\!]_\rho \in [\![\mathscr{T}]\!]$. Proceed by induction on the structure of the derivation, first showing that each rule is sound in the obvious sense. Note that the definedness proviso cannot be dropped, as it is not in general ensured by the typing rules. □

Exercise 6.4.14. Lift the Galois connection in Exercise 6.3.8 to the higher-order case. You may want to take partiality of constructors into account, as suggested at the end of Exercise 6.3.9. □

Exercise 6.4.15. Show that, for any constructor signature $\mathscr{S}$, there exists a specification $lift(\mathscr{S})$ such that $\rhd lift(\mathscr{S}) : Spec(\mathscr{S})$ and $[\![lift(\mathscr{S})]\!] = Mod(\mathscr{S})$. The base case is easy: $lift(\Sigma) = \langle \Sigma, \varnothing \rangle$. But the induction step also requires a specification with *no* models over the argument signature, because $Mod(\mathscr{S} \Rightarrow \mathscr{S}')$ is the *partial* function space. This requires an assumption on the underlying institution. □

Exercise 6.4.16. The extension to higher order admits constructors and parameterised specifications with multiple arguments via the usual currying convention: if $\Gamma \rhd E : \Sigma_1 \Rightarrow \Sigma_2 \Rightarrow \Sigma$, $\Gamma \rhd E_1 : \Sigma_1$ and $\Gamma \rhd E_2 : \Sigma_2$, then $\Gamma \rhd E(E_1)(E_2) : \Sigma$, so E

may be regarded as a two-argument constructor. Extend the notions of constructor signature, specification type and specification to cover explicit multi-argument functions taking tuples of arguments, with constructor signatures of the form $\mathscr{S}_1 \times \cdots \times \mathscr{S}_n \Rightarrow \mathscr{S}$, and modify the syntax, typing rules and semantics accordingly. Show, however, that the natural embedding that maps any two-argument constructor in $Mod(\mathscr{S}_1 \times \mathscr{S}_2 \Rightarrow \mathscr{S})$ to its curried form in $Mod(\mathscr{S}_1 \Rightarrow \mathscr{S}_2 \Rightarrow \mathscr{S})$ is not surjective in general. HINT: Constructors are partial functions. □

Exercise 6.4.17. Present amalgamation of persistent constructors having a common source signature (Example 6.1.31), application of a constructor to an argument, composition of constructors and lifting of a persistent constructor along a signature morphism (Example 6.1.28) as higher-order constructors. □

Exercise 6.4.18. Given an institution **INS**, define an institution with constructor signatures as signatures, constructors as models, and other components as appropriate. HINT: A morphism from $\mathscr{S}_1 \Rightarrow \mathscr{S}'_1$ to $\mathscr{S}_2 \Rightarrow \mathscr{S}'_2$ is a pair of morphisms $\langle \sigma : \mathscr{S}_2 \to \mathscr{S}_1, \sigma' : \mathscr{S}'_1 \to \mathscr{S}'_2 \rangle$. Use discrete model categories if you can't think of a good notion of morphism between constructors. Try using constructor specifications as sentences. If you have trouble with translation, see Example 4.1.46. □

Exercise 6.4.19. For "first-order" constructor signatures $\Sigma_1 \Rightarrow \Sigma'_1$ and $\Sigma_2 \Rightarrow \Sigma'_2$, Exercise 6.4.18 suggests that a morphism $\langle \sigma, \sigma' \rangle : (\Sigma_1 \Rightarrow \Sigma'_1) \to (\Sigma_2 \Rightarrow \Sigma'_2)$ consists of **INS**-signature morphisms $\sigma : \Sigma_2 \to \Sigma_1$ and $\sigma' : \Sigma'_1 \to \Sigma'_2$. Try to give natural examples in **FOEQ**. It is likely that σ will turn out to be an isomorphism or at least surjective in all of these examples. In typical examples, there are inclusions $\iota_1 : \Sigma_1 \to \Sigma'_1$ and $\iota_2 : \Sigma_2 \to \Sigma'_2$. Then, when Σ_1 is larger than Σ_2, the additional components of Σ_1, which are preserved in Σ'_1, cannot easily be mapped to components in Σ'_2 (in such a way that $\sigma;\iota_1;\sigma' = \iota_2$).

For semi-exact **INS**, consider a generalisation of the concept of first-order constructor signature morphism $\langle \sigma, \sigma' \rangle : (\Sigma_1 \Rightarrow \Sigma'_1) \to (\Sigma_2 \Rightarrow \Sigma'_2)$ in which σ is as before but $\sigma' : \Sigma'_1 \to \Sigma_{po}$ where Σ_{po} is given by the pushout of σ and ι_2

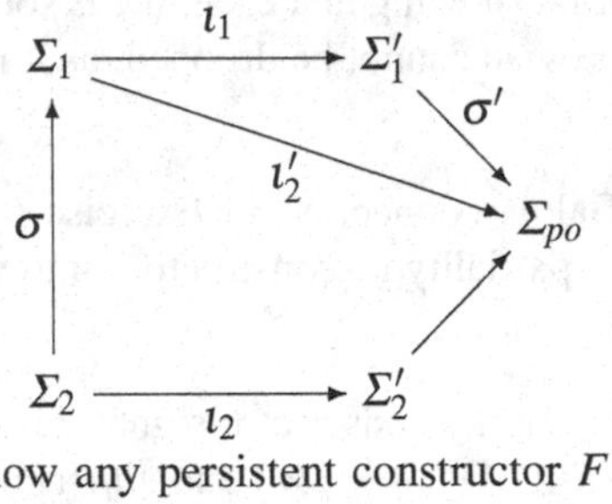

with $\iota_1;\sigma' = \iota'_2$. Show how any persistent constructor $F \in Mod(\Sigma_2 \overset{\iota_2}{\Rightarrow} \Sigma'_2)$ determines a persistent constructor $F|_{\langle \sigma, \sigma' \rangle} \in Mod(\Sigma_1 \overset{\iota_1}{\Rightarrow} \Sigma'_1)$.

Try to lift this idea to give an improved notion of morphism between higher-order constructor signatures. □

Exercise 6.4.20. The system above is *stratified*, with a clear separation between constructor signatures and specification types. Combine the two notions into a single more general notion of type: a generalised type $\mathscr{G}$ is either a signature Σ, or of the form $Spec(\mathscr{G}')$, or of the form $\mathscr{G}_1 \to \mathscr{G}_2$. Constructor types correspond to generalised types with no occurrences of *Spec*; what do specification types correspond to? Generalised types include new forms like:

- $\Sigma \to Spec(\Sigma')$, for functions from Σ-models to classes of Σ'-models (functions from Σ-models to Σ'-specifications);
- $Spec(\Sigma \to Spec(\Sigma'))$, for classes (specifications) of such functions;
- $Spec(Spec(\Sigma))$, for classes of classes of Σ-models (specifications of Σ-specifications).

Extend the syntax, typing rules and semantics correspondingly. Write down some examples of expressions having types like those above. □

6.5 An example

Recall the example of a structured specification in Section 5.3. Here we provide specifications for constructors that might be used to build models for the specifications introduced there. We begin by rewriting the specification CONSARRAY from Example 6.2.11, which specified a generic constructor for arrays of arbitrary elements, to make explicit its dependency on the choice of a model for NAT.

spec CONSARRAY =
  ΠN:NAT • ΠX:ELEM$_\exists$ •
    $\{N+X\}$
    then
      sorts $Array[Elem]$
      ops $empty: Array[Elem]$
        $put: Nat \times Elem \times Array[Elem] \to Array[Elem]$
        $get: Nat \times Array[Elem] \to Elem$
      preds $used: Nat \times Array[Elem]$
      $\forall i,j: Nat, s: Elem, a: Array[Elem]$
        • $\neg used(i, empty)$
        • $used(i, put(i,s,a))$
        • $i \neq j \Rightarrow (used(i, put(j,s,a)) \Leftrightarrow used(i,a))$
        • $get(i, put(i,s,a)) = s$
        • $i \neq j \Rightarrow get(i, put(j,s,a)) = get(i,a)$

Note that $\{N+X\}$, employing the amalgamated union constructor from Example 6.1.32, is equivalent to $\{N\}$ **and** $\{X\}$, employing the sum specification-building operation from Section 5.2 (assuming as usual the obvious choice of pushout signature). In this case amalgamation and sum are especially simple since the signatures of N and X are disjoint.

spec CONSBUCKET =
  ΠN:NAT • ΠS:STRING • ΠA:CONSARRAY •
    $\{(A(N)(S|_{\sigma_{Elem \mapsto String}})+S)|_{\sigma^{-1}_{Array[String] \mapsto Bucket}}\}$

where $\sigma_{Elem \mapsto String}: Sig[\text{ELEM}] \to Sig[\text{STRING}]$ is as introduced in Example 6.3.3 and $\sigma^{-1}_{Array[String] \mapsto Bucket}: Sig[\text{BUCKET}] \to Sig[\text{STRINGARRAY}]$ is the inverse of the morphism $\sigma_{Array[String] \mapsto Bucket}$ introduced in Section 5.3. We use the $+$ constructor, omitting the signature morphism subscripts. (Their common source would be $Sig[\text{ELEM}]$, and the morphisms would be the signature inclusion and $\sigma_{Elem \mapsto String}$ respectively.) Proposition 5.6.2(6) implies that $\{(\cdots)|_{\sigma^{-1}_{Array[String] \mapsto Bucket}}\}$ is equivalent to $\{(\cdots)\}$ **with** $\sigma_{Array[String] \mapsto Bucket}$. This is a tight specification which specifies the single constructor

$$
\begin{array}{l}
F_{\text{BUCKET}} = \\
\quad \lambda N{:}\text{NAT}\bullet \lambda S{:}\text{STRING}\bullet \lambda A{:}\text{CONSARRAY}\bullet \\
\qquad (A(N)(S|_{\sigma_{Elem \mapsto String}}) + S)|_{\sigma^{-1}_{Array[String] \mapsto Bucket}}
\end{array}
$$

We will treat this as a constructor definition, making F_{BUCKET} available for use in subsequent definitions.

Exercise 6.5.1. Express $A(N)(S|_{\sigma_{Elem \mapsto String}}) + S$ in F_{BUCKET} by means of the constructor $\sigma_{Elem \mapsto String}(A(N))$. Do the same for F_{TABLE} below. □

$$
\begin{array}{l}
\textbf{spec } \text{CONSSTRINGKEY0} = \\
\quad \Pi N{:}\text{NAT}\bullet \Pi S{:}\text{STRING}\bullet \\
\qquad \{N + S\} \\
\qquad \textbf{then} \\
\qquad\quad \textbf{ops } hash0{:}\, String \to Nat \\
\qquad\quad \bullet\ hash0(\varepsilon) = 0
\end{array}
$$

$$
\begin{array}{l}
\textbf{spec } \text{CONSSTRINGHASHTABLE0} = \\
\quad \Pi N{:}\text{NAT}\bullet \Pi S{:}\text{STRING}\bullet \Pi A{:}\text{CONSARRAY}\bullet \Pi K0{:}\text{CONSSTRINGKEY0}\bullet \\
\qquad \{K0(N)(S) + F_{\text{BUCKET}}(N)(S)(A)\} \\
\qquad \textbf{then} \\
\qquad\quad \textbf{ops} \quad add{:}\, String \times Bucket \to Bucket \\
\qquad\quad \phantom{\textbf{ops}} \quad putnear{:}\, Nat \times String \times Bucket \to Bucket \\
\qquad\quad \textbf{preds } present{:}\, String \times Bucket \\
\qquad\quad \phantom{\textbf{preds }} isnear{:}\, Nat \times String \times Bucket \\
\qquad\quad \forall i{:}Nat, s{:}Elem, a, a'{:}Bucket \\
\qquad\qquad \bullet\ add(s,a) = putnear(hash0(s), s, a) \\
\qquad\qquad \bullet\ putnear(i,s,a) = a' \Leftrightarrow \\
\qquad\qquad\qquad \exists j{:}Nat\bullet (\neg used(i+j,a) \vee get(i+j,a) = s) \wedge \\
\qquad\qquad\qquad\quad (\forall k{:}Nat\bullet \neg(j \le k) \Rightarrow used(i+k,a)) \wedge a' = put(i+j,s,a) \\
\qquad\qquad \bullet\ present(s,a) \Leftrightarrow isnear(hash0(s), s, a) \\
\qquad\qquad \bullet\ isnear(i,s,a) \Leftrightarrow \\
\qquad\qquad\qquad \exists j{:}Nat\bullet (\forall k{:}Nat\bullet k \le j \Rightarrow used(i+k,a)) \wedge get(i+j,a) = s
\end{array}
$$

Exercise 6.5.2. Perhaps unexpectedly, CONSSTRINGHASHTABLE0 is an inconsistent specification. There is an implicit requirement, as remarked following the specification STRINGHASHTABLE0 in Section 5.3, that excludes nearly all models having a value b of sort *Bucket* and an index n such that $used(i,b)$ whenever $i \ge n$.

In Section 5.3 this was not a problem, even though it amounts to an additional requirement on buckets that is not in the specification BUCKET but rather in the later specification STRINGHASHTABLE0. The problem here is that constructors satisfying CONSSTRINGHASHTABLE0 must be able to extend an *arbitrary* model of BUCKET, which does not exclude those that violate this additional requirement.

Repair the problem by adjusting the specification CONSARRAY, for example to exclude arrays that are *used* at infinitely many positions. □

spec CONSSTRINGKEY =
  ΠN:NAT• ΠS:STRING•
    $\{N+S\}$
    then
      ops $hash: String \to Nat$
      • $hash(\varepsilon) = 0$

A specification CONSTABLE for constructors that builds models of TABLE would be similarly tight as CONSBUCKET above. We give only the definition of the corresponding constructor. The definition allows the use of different constructors satisfying CONSARRAY, one for building the arrays realising buckets and another for arrays realising tables. This makes sense for efficiency reasons because different operations on these arrays will have to be provided.

$$F_{\text{TABLE}} = \lambda N{:}\text{NAT}\bullet \lambda S{:}\text{STRING}\bullet \lambda A{:}\text{CONSARRAY}\bullet \lambda A'{:}\text{CONSARRAY}\bullet (A(N)(F_{\text{BUCKET}}(N)(S)(A')|_{\sigma_{Elem\mapsto Bucket}}) + F_{\text{BUCKET}}(N)(S)(A'))|_{\sigma^{-1}_{Array[Bucket]\mapsto Table}}$$

where $\sigma_{Elem\mapsto Bucket}: Sig[\text{ELEM}] \to Sig[\text{BUCKET}]$ is the morphism mapping *Elem* to *Bucket* and $\sigma^{-1}_{Array[Bucket]\mapsto Table}: Sig[\text{TABLE}] \to Sig[\text{BUCKETARRAY}]$ is the inverse of $\sigma_{Array[Bucket]\mapsto Table}$ introduced in Section 5.3.

spec CONSSTRINGHASHTABLE =
  ΠN:NAT• ΠS:STRING• ΠA:CONSARRAY• $\Pi K0$:CONSSTRINGKEY0•
    $\Pi T0$:CONSSTRINGHASHTABLE0• ΠK:CONSSTRINGKEY•
      $\Pi A'$:CONSARRAY•
        $\{T0(N)(S)(A)(K0) + K(N)(S) + F_{\text{TABLE}}(N)(S)(A)(A')\}$
        then
          ops $add: String \times Table \to Table$
          preds $present: String \times Table$
          $\forall s: String, i: Nat, a: Table$
            • $hash(s) = i \wedge used(i,a) \Rightarrow add(s,a) = put(i, add(s, get(i,a)), a)$
            • $hash(s) = i \wedge \neg used(i,a) \Rightarrow add(s,a) = put(i, add(s, empty), a)$
            • $hash(s) = i \wedge used(i,a) \Rightarrow (present(s,a) \Leftrightarrow present(s, get(i,a)))$
            • $hash(s) = i \wedge \neg used(i,a) \Rightarrow \neg present(s,a)$

Finally, we define a constructor that builds models of USERSTRINGHASHTABLE.

$$
\begin{array}{l}
F_{\text{USERSTRINGHASHTABLE}} = \\
\quad \lambda N{:}\text{NAT}\bullet\ \lambda S{:}\text{STRING}\bullet\ \lambda A{:}\text{CONSARRAY}\bullet\ \lambda K0{:}\text{CONSSTRINGKEY0}\bullet \\
\quad\quad \lambda T0{:}\text{CONSSTRINGHASHTABLE0}\bullet\ \lambda K{:}\text{CONSSTRINGKEY}\bullet \\
\quad\quad\quad \lambda A'{:}\text{CONSARRAY}\bullet \\
\quad\quad\quad\quad \lambda T{:}\text{CONSSTRINGHASHTABLE}\bullet \\
\quad\quad\quad\quad\quad T(N)(S)(A)(K0)(T0)(K)(A')|_{\iota}
\end{array}
$$

where ι is the signature inclusion corresponding to the final hiding (or rather, revealing) step in the specification USERSTRINGHASHTABLE in Section 5.3.

The above presentation may seem overly complex. This is largely due to the need to handle the flow of dependencies explicitly, rather than to a conceptual fault. In Section 7.4, a style of presentation will be used that makes the management of dependencies less intrusive. The conventions used there to handle dependencies behind the scenes can be built into a formal notation for presenting developments; see [BST02].

6.6 Bibliographical remarks

Language features giving the power of generic modules have been available in programming languages for quite some time, in support of what has come to be known as "generic programming" [GJL+03]. The first language to provide generic modules in full generality as functions from modules to modules was Standard ML [Mac84]. Generic modules modelled as total functions on algebras appear in [ST88b], where they were used to capture a general notion of specification implementation; see Chapter 7. Constructors as partial functions, as they are defined here, appear in [ST89] and [Asp97].

Constructor specifications correspond to Extended ML functors [ST89], [KST97] and generic package specifications in Goguen's approach to parameterised programming [Gog96]. They first appear in the form of Π-specifications in [SST92] using notation from the theory of dependent types [AH05], with similar ideas appearing earlier in a programming language context in Pebble [LB88].

For specifying persistent constructors as discussed in Section 6.2, there appear to be two alternatives to the use of singleton specifications in constructor specifications, both of which can be viewed as special cases of the approach taken here. The first alternative is to use $SP \overset{\iota}{\Rrightarrow} SP'$ directly without considering this as an abbreviation, as in CASL [Mos04]; unfortunately it does not appear to be possible to extend this smoothly to the specification of higher-order constructors. The second alternative is to allow specifications to refer to the *components* of parameters using a form of "dot notation" as in Extended ML and [ST97]. Our use of singleton specifications is from [SST92]; singleton types also appear in [Hay94] and were studied in depth in [Asp95], and singleton kinds later found use in the context of typed intermediate languages [SH00].

The issue of functoriality of constructors (cf. Exercise 6.2.23) was perhaps first raised in [Ber87]; it will come up again in Section 8.4.4.

Parameterised specifications in the λ-calculus style of Section 6.3 originate in [Wir82] and [SW83]; see also [Wir86]. For the sake of simplicity we take $\lambda X{:}Spec(\Sigma) \bullet SP'[X]$ with the argument required to have a given signature, as in [ST88a], rather than $\lambda X{:}Spec(SP) \bullet SP'[X]$ where it is required to match a given specification, as in [SST92]. Further generalisations have been considered, for example passing the fitting morphism itself as an additional parameter [SW83], which may even result in dependency of the result signature on the actual argument; see also [Wir86], where anything is allowed as a parameter, including e.g. formulae.

Extensions of Standard ML [MT94], [Rus98], [RRS00] and Objective Caml [LDF$^+$10] provide higher-order generic modules. For a practical example, see Chapter 10 of [Oka98] where they are used for "data-structural bootstrapping". Specifications of higher-order constructors as in Section 6.4 first appeared in [SST92], while higher-order parameterised specifications, including recursive ones, are available in ASL [Wir82], [SW83], [Wir86].

The system of rules in Section 6.4 amounts to a stratified version of a system in [SST92]. The calculus there includes rules for *satisfaction* of specifications which generalise these *typing* rules. The *Spec* operator on types is available there as an operator on specifications and is used to eliminate the need for an explicit subtyping relation (which there amounts to model class inclusion rather than structural typing); this device was also used in [Car88]; see [Asp00]. This calculus was investigated in depth in [Asp97], where it was extended by a simple functional language that is similar to what we use here to define constructors in the institution **FPL**, and related to systems of dependent types with subtyping; see [AC01].

Pushout-style parameterised specifications originate in Clear [BG77], with other early papers being [TWW82] and [Ehr82]; this line has been pursued in a series of papers culminating in a monograph presentation [EM85]. In this approach, constructor specifications and parameterised specifications are regarded as two sides of the same coin. The main results concern the relationship between translation of free constructions along signature morphisms and pushout-style parameter passing, under the rubric "correctness of parameter passing"; see the Extension Lemma of [EM85] and results based on it. The essence of these results is captured by a combination of Example 6.1.28, Exercise 6.1.30, and Exercise 6.3.7. The problem of ensuring persistency naturally arises here; the requirements of sufficient completeness and hierarchy consistency (Proposition 6.1.24) are due to [Gan83] and [Pad85], based on [GH78]. "Parameterised parameter passing" amounts to composition of parameterised specifications, not to be confused with higher-order parameterised specifications as treated in Section 6.4. A generalisation of the pushout approach to the situation where the parameter occurs more than once in the body is [JOE95]; this can be regarded as an intermediate step towards parameterised specifications in the λ-calculus style. A first step towards generalisation of pushout-style parameterisation to higher order, where the basic notion of a morphism between constructor signatures is as given in Exercise 6.4.19, is in [Lin03].

Our view, which is built into the presentation of this chapter, is that it is conceptually and methodologically important to keep a clear distinction between constructor specifications and parameterised specifications, despite the fact that there is an intimate semantical correspondence between subclasses of these, which include most common examples (Exercises 6.3.8 and 6.4.14, from [SST92]). This point of view was put forward in [SST92] with the slogan

parameterised (program specification) $\neq$ *(parameterised program) specification.*

Chapter 7
Formal program development

The previous chapters have described a powerful and flexible specification framework. The point of constructing a specification is so that it may be used to define a programming task by precisely delimiting the range of program behaviours that are to be regarded as permissible. This presupposes an appropriate choice of institution providing models that adequately reflect the computational phenomena of interest for the problem at hand. Then, as we have seen, a specification describes a class of models in this institution, and programs correspond to individual models. Thus, the goal of producing a program that satisfies a specification *SP* has been achieved when a program *P* in the desired target programming language has been obtained such that the model corresponding to *P* is in the class of models of *SP*. This chapter describes a framework which supports the gradual step-by-step development of a program from a specification of requirements.

In keeping with the previous chapters, this is a model-theoretic framework which takes no account of the details of the target programming language: we assume the existence of programs (like *P* above) and a semantics that maps programs to models of the underlying institution, but refrain from making explicit reference to the syntax of the programming language or its semantics. This view entirely ignores many issues involved in the design of algorithms, not to mention efficiency aspects. Its advantage is that issues concerned with the overall modular design of programs and with the representation of data are addressed on a level at which the features of particular programming languages do not intrude. On the other hand, it is possible to put the programming language explicitly into the picture by using an institution that incorporates programming constructs; see Examples 4.1.25 and 6.1.9 and Exercise 6.2.12. Another way of achieving the same effect is to appropriately link an institution used for specifying with an institution used for programming, as will be described in Section 10.1; see Example 10.1.9 and Exercise 10.1.10, followed by Examples 10.1.17 and 10.1.18.

One way or another, there will always be a formal gap between the target programming language and the specification language, even when the latter admits specifications that describe individual models, i.e. denote classes that consist of a single model. The size of the gap, and how easy it is to make the transition from a sufficiently detailed specification to a program, depends on the choice of institution and the relationship between the syntax of axioms and the phrases of the target programming language. The gap is very small in cases like Example 4.1.25, where the axioms essentially incorporate programming constructs, and large in cases where an abstract logical formalism is used for specification and a low-level language for programming. The benefits of using high levels of abstraction in specification are conciseness and more powerful reasoning tools, while the benefit of using low levels of abstraction in programming is increased efficiency. Keeping the gap small means sacrificing one or more of these benefits to some extent.

The framework presented here formalises the basic notion of refinement step as an *implementation* of one specification by another, and then proceeds to define what it means for such a refinement step to be correct, and under what conditions correct individual refinement steps may be composed to lead ultimately to a correct realisation of an entire requirements specification. The focus is on correctness, compositionality and generality. This leads to a methodology of stepwise refinement involving modular decomposition into components connected via explicit interfaces. A different topic, and one that is not addressed here at all, is that of providing techniques and heuristics that assist in coming up with individual refinement steps; these will often be oriented towards specific problem areas and programming technologies. The soundness of such techniques should ultimately be justified by reference to the framework we present. In any case, since the power of our specification framework is such that there is no way to proceed automatically from an arbitrary specification to a correct program, such techniques can only offer a partial solution.

7.1 Simple implementations

There is a wide gulf between the realm of high-level user-oriented requirements specifications and that of programs full of technical decisions and algorithmic details. The programming discipline of stepwise refinement suggests that a program be evolved from a high-level specification by working gradually via a series of successively more detailed lower-level intermediate specifications. Each successive specification is called an *implementation* of the previous specification in the sequence. The process is complete when a specification is obtained which involves only types and functions that are already available in the target programming language. This gives rise to a program that is guaranteed to be correct (that is, to satisfy the original specification) provided all the implementation steps involved in its creation are correct.

Development of a program from a specification thus proceeds via a sequence of small, easy to understand and easy to verify steps:

$$SP_0 \rightsquigarrow SP_1 \rightsquigarrow \cdots \rightsquigarrow SP_n.$$

In such a chain, SP_0 is the original requirements and SP_n is a specification that is detailed enough that its conversion into a program is relatively straightforward.

A formalisation of this approach requires a precise definition of the concept of implementation, and of what it means for an implementation to be correct. We will use the notation $SP \rightsquigarrow SP'$ as above to mean that the specification SP' is a correct implementation of the specification SP; the wiggly arrow points from the higher-level specification SP to the lower-level specification SP'. The specification SP' is obtained from SP by incorporating additional design decisions, typically in the form of local modifications, each of which restricts the class of acceptable realisations. These will include decisions concerning the concrete representation of abstractly defined data types, decisions about how to compute abstractly specified functions, and decisions about options of behaviour left open by the specification.

The following very simple formal notion of implementation captures this general idea in its most basic form.

Definition 7.1.1 (Simple implementation). Given specifications SP and SP' with $Sig[SP] = Sig[SP']$, we say that SP' is a *simple implementation* of SP, written $SP \rightsquigarrow SP'$, if $Mod[SP] \supseteq Mod[SP']$. □

Note that, for simplicity, the definition of simple implementation requires the signatures of both specifications to be the same. The hiding operation may be used to adjust the signatures (for example, by removing auxiliary functions from the signature of the implementing specification — see Example 7.1.4 below) if this is not the case.

The definition of simple implementation ensures that the correctness of the final outcome of stepwise development may be inferred from the correctness of the individual implementation steps:

Proposition 7.1.2. *Given a chain $SP_0 \rightsquigarrow SP_1 \rightsquigarrow \cdots \rightsquigarrow SP_n$ of simple implementation steps and a model $M \in Mod[SP_n]$, we have $M \in Mod[SP_0]$.* □

This is what we seek: if SP_n is detailed enough that its conversion into a program P determining a model $M \in Mod[SP_n]$ is straightforward, then $M \in Mod[SP_0]$ and thus P is a solution to the original program development task.

An indirect way to prove the correctness of the final outcome is to notice that the simple implementation relation is (obviously) transitive; this is referred to as *vertical composability*.

Proposition 7.1.3 (Vertical composition). *If $SP \rightsquigarrow SP'$ and $SP' \rightsquigarrow SP''$ then $SP \rightsquigarrow SP''$.* □

Example 7.1.4. The following specifications are given in the institution **FPL**; see Example 4.1.25.

spec NAT =
  sorts *Nat* **free with** 0 | *succ*(*Nat*)

spec NATORD =
  NAT
  then
    ops $le{:}\,Nat \times Nat \rightarrow? Bool$
    $\forall m,n,p{:}\,Nat$
      • $def(le(m,n))$
      • $le(m,m) = true$
      • $le(m,n) = true \wedge le(n,m) = true \Rightarrow m = n$
      • $le(m,n) = true \wedge le(n,p) = true \Rightarrow le(m,p) = true$
      • $le(m,n) = true \vee le(n,m) = true$

spec NATLIST =
  NATORD
  then
    sorts *NatList* **free with** *nil* | *cons*(*Nat*,*NatList*)
    ops **fun** *append*(*l*:*NatList*, *l′*:*NatList*):*NatList* =
        case *l* **of** *nil* => *l′* | *cons*(*n*,*l″*) => *cons*(*n*,*append*(*l″*,*l′*))
      fun *is_in*(*n*:*Nat*, *l*:*NatList*):*Bool* =
        case *l* **of** *nil* => *false*
            | *cons*(*m*,*l′*) => **if** *n* = *m* **then** *true* **else** *is_in*(*n*,*l′*)

spec SORT1 =
  NATLIST
  then
    ops $sort{:}\,NatList \rightarrow NatList$
      $is\_sorted{:}\,NatList \rightarrow? Bool$
    $\forall n{:}\,Nat, l{:}\,NatList$
      • $def(is\_sorted(l))$
      • $is\_sorted(nil) = true$
      • $is\_sorted(cons(n,l)) = true \Leftrightarrow$
          $(\forall m{:}Nat \bullet is\_in(m,l) \Rightarrow le(n,m) = true) \wedge is\_sorted(l) = true$
      • $is\_sorted(sort(l)) = true$
      • $is\_in(n,l) = true \Leftrightarrow is\_in(n,sort(l)) = true$

spec SORT =
  SORT1 **hide ops** $is\_sorted{:}\,NatList \rightarrow? Bool$

Note that SORT deliberately does not require *sort* to preserve repetitions in its input, leaving this decision to the implementor. Any choice that is consistent with the rest of SORT would be deemed acceptable. Apart from the obvious possibilities (preserving repetitions, removing repetitions) it would be possible, for instance, to require all prime numbers in the input to occur exactly twice in the output. The following adds to SORT the requirement that *sort* preserve repetitions in its input, delivering a permutation of its argument:

```
spec SortCount =
  Sort
  then
    ops fun count(n:Nat, l:NatList):Nat =
          case l of nil => 0
                 | cons(m, l') => if n = m then succ(count(n, l'))
                                           else count(n, l')
    ∀n:Nat, l:NatList
       • count(n, l) = count(n, sort(l))

spec SortPerm =
  SortCount hide ops count: Nat × NatList → Nat
```

Then we choose the algorithm (insertion sort) and "code" *sort* but, for illustrative purposes, we refrain at this stage from giving the "code" for the additional operation *insert* and leave it specified only.

```
spec Ins =
  NatList
  then
    ops insert: Nat × NatList → NatList
    ∀n:Nat, l:NatList
      • def(insert(n, l))
      • ∃l1, l2:NatList •
          insert(n, l) = append(l1, cons(n, l2)) ∧ l = append(l1, l2)
          ∧ (∀l1':NatList • ∀m:Nat • l1 = append(l1', cons(m, nil)) ⇒
                                                    le(m, n) = true)
          ∧ (∀l2':NatList • ∀m:Nat • l2 = cons(m, l2') ⇒ le(n, m) = true)

spec SortByInsert =
  Ins
  then
    ops fun sort(l:NatList):NatList =
          case l of nil => nil | cons(n, l') => insert(n, sort(l'))

spec SortIns =
  SortByInsert hide ops insert: Nat × NatList → NatList
```

Finally, we "code" *insert*, preserving the "code" for *sort*:

```
spec InsDone =
  NatList
  then
    ops fun insert(n:Nat, l:NatList):NatList =
          case l of nil => cons(n, nil)
                 | cons(m, l') => if le(n, m) = true then cons(n, l)
                                          else cons(m, insert(n, l'))
```

spec SortByInsertDone =
  InsDone
  then
    ops **fun** *sort*(*l*:*NatList*):*NatList* =
        case *l* **of** *nil* => *nil* | *cons*(*n*, *l*′) => *insert*(*n*, *sort*(*l*′))

spec SortDone =
  SortByInsertDone **hide ops** *insert*: *Nat* × *NatList* → *NatList*

The above constitutes a sequence of simple implementation steps:

$$\text{SORT} \rightsquigarrow \text{SORTPERM} \rightsquigarrow \text{SORTINS} \rightsquigarrow \text{SORTDONE}$$

With SortDone, development may be considered complete since the axioms in NatList, InsDone and SortByInsertDone amount to code in **FProg** (Exercise 4.1.30), the "programming part" of **FPL**; this disregards the fact that *le* is only specified as a linear order, rather than being coded as a specific order relation. We will make this more explicit in the next section; see Example 7.2.8.

The simple implementation steps above satisfy the correctness criteria stated in the definition. This is trivial for Sort ⇝ SortPerm, since SortPerm just adds a constraint on the class of models of Sort. For SortPerm ⇝ SortIns, it is necessary to prove that to each model of SortIns, we can add *count* and *is_sorted* so that the axioms of SortPerm are satisfied. Since *count* and *is_sorted* are determined by the corresponding axioms in SortPerm, this amounts to proving that the axioms of SortPerm follow from the "code" for *sort* in SortIns, assuming that *insert* satisfies the axioms in Ins and that *count* and *is_sorted* satisfy their axioms. Finally, SortIns ⇝ SortDone requires a proof that the axiom in Ins follows from the "code" for *insert* in InsDone. (**Exercise:** Check the details.) Issues of formal proof of correctness of implementation steps will be addressed in full in Section 9.3, where we revisit this example; see Examples 9.3.8, 9.3.10 and 9.3.13. □

In Example 7.1.4, *Mod*[SortPerm] = *Mod*[SortIns] = *Mod*[SortDone] (even though *Mod*[Ins] ≠ *Mod*[InsDone], and *count*, hidden in SortPerm, is not even mentioned in SortIns and SortDone); this means that the last two implementation steps are semantically trivial (but does not mean that justifying their correctness is trivial). It then follows that, for example, SortDone ⇝ SortPerm. The notion of simple implementation is not fine enough to capture the sense in which SortDone is "closer" to a program than SortPerm is. A more elaborate notion of implementation, which provides a place to record "progress" towards a program, will be presented in Section 7.2.

An issue which may seem worrying is that the definition of implementation does not guarantee preservation of consistency: any specification is implemented by an inconsistent one. This can be seen as a problem, since any inconsistent specification opens a blind alley in the development process. From this point of view, it would be worthwhile to be able to check consistency of every specification as soon as it is formulated. Unfortunately, in general (for any sufficiently powerful specification framework) this is an undecidable property. Fortunately, inconsistency of specifications cannot lead to incorrect programs: if at any point in the development process we are able to convert the specification at hand into a program, then by definition this specification is consistent, and therefore all the specifications leading to it must have been consistent as well.

Apart from inconsistencies, there are many other sources of blind alleys and failures in the development process: there might be no computable realisation of a specification, there might be no "computationally feasible" realisation, we might not be clever enough to find a realisation, we might run out of money to finish the project, and so on.

Example 7.1.5. Consider a specification of the natural numbers with a pre-order $\prec$ characterised as follows:

$$m \prec n \Leftrightarrow \forall x{:}Nat \bullet M_m \downarrow x \Rightarrow M_n \downarrow x$$

where for all natural numbers k and x, the predicate $M_k \downarrow x$ is specified to mean that the Turing machine with Gödel number k terminates on input x. This specification is consistent but it has no computable models since the halting problem is undecidable. □

The main feature of the development framework we really do ensure is its *safety*: if we arrive at a program, then it is a correct realisation of the original specification.

Some implementation steps are more or less routine. For instance, there are standard ways of implementing many data abstractions (e.g. sets, queues) and standard ways of decomposing problems into simpler subproblems (e.g. "divide and conquer"). Such implementation steps can sometimes be described schematically by means of so-called *transformation rules* such that any instance is guaranteed to be correct provided certain conditions are met.

One way to make development more routine is to retain the structure of the original requirements specification and concentrate on refining the representation of data and providing algorithmic details. Then correctness of implementation steps can often be established easily using the following fact, which is referred to as "horizontal composability":

Proposition 7.1.6 (Horizontal composition). *For any n-argument monotone specification-building operation* **sbo** *(see Exercises 5.1.4 and 5.1.5), if* $SP_1 \rightsquigarrow SP'_1$ *and … and* $SP_n \rightsquigarrow SP'_n$ *then* $\mathbf{sbo}(SP_1,\ldots,SP_n) \rightsquigarrow \mathbf{sbo}(SP'_1,\ldots,SP'_n)$. □

Even though most of the specification-building operations we use are monotone (see Exercise 5.1.13), horizontal composability should not be misread as a directive to retain the structure of the original requirements specification throughout the development process, thus effectively decomposing the task of realising any specification $SP = \mathbf{sbo}(SP_1,\ldots,SP_n)$ into separate tasks to realise each of $SP_1,\ldots,SP_n$. Requiring the structure of the initial specification to be preserved in its implementation would be unrealistic and unreasonable. The aims of structuring requirements specifications are often contradictory with the aims of structuring software.

Example 7.1.7. The implementations in Example 7.1.4 illustrate the point that the structure of the final implementation differs from that of the original specification, even though the difference is only that different auxiliary operations are used (*is_sorted* versus *insert*). The essential change takes place in the step SORTPERM $\rightsquigarrow$ SORTINS. There is no way to build this implementation using the horizontal composability rule: stripping off the hiding operations from both SORTPERM and SORTINS (naively disregarding the fact that different things are hidden) yields two incomparable specifications. □

Another problem is that when attempting to retain the structure of $\mathbf{sbo}(SP_1,\ldots,SP_n)$ and build its implementation from implementations of $SP_1,\ldots,SP_n$, it is possible for the design decisions taken in the solutions of these separate tasks to be in conflict with each other so that their combination is inconsistent.

Example 7.1.8. Consider the following specification (taking mild liberties with what NAT provides):

$$\begin{array}{l}\textbf{spec } \mathrm{NAT}_c = (\,\mathrm{NAT}\\ \qquad\qquad \textbf{then}\\ \qquad\qquad\quad \textbf{ops } c{:}\,\mathit{Nat}\\ \qquad\qquad\quad \bullet\ 10 < c < 15\,)\\ \qquad\qquad \textbf{then}\\ \qquad\qquad\quad \bullet\ 12 < c < 27\end{array}$$

Since

$$\left(\begin{array}{l}\mathrm{NAT}\\ \textbf{then}\\ \quad \textbf{ops } c{:}\,\mathit{Nat}\\ \quad \bullet\ 10 < c < 15\end{array}\right) \rightsquigarrow \left(\begin{array}{l}\mathrm{NAT}\\ \textbf{then}\\ \quad \textbf{ops } c{:}\,\mathit{Nat}\\ \quad \bullet\ 10 < c < 12\end{array}\right)$$

and

$$\left(\begin{array}{l}\mathrm{NAT}\\ \textbf{then}\\ \quad \textbf{ops } c{:}\,\mathit{Nat}\\ \quad \bullet\ 12 < c < 27\end{array}\right) \rightsquigarrow \left(\begin{array}{l}\mathrm{NAT}\\ \textbf{then}\\ \quad \textbf{ops } c{:}\,\mathit{Nat}\\ \quad \bullet\ 14 < c < 20\end{array}\right)$$

we also have

$$\mathrm{NAT}_c \rightsquigarrow \left(\begin{array}{l}(\,\mathrm{NAT}\\ \quad \textbf{then}\\ \qquad \textbf{ops } c{:}\,\mathit{Nat}\\ \qquad \bullet\ 10 < c < 12\,)\\ \textbf{then}\\ \quad \bullet\ 14 < c < 20\end{array}\right)$$

However, even though NAT_c is consistent, and both of the resulting component specifications are consistent as well, the resulting composed specification to which NAT_c is refined is inconsistent!

This happens because the two specification arguments to the enrichment operation implicitly share a loosely specified part ($c{:}Nat$ in the example). If the decisions constraining this common part in separate developments of the two specifications are different, as above, then putting the resulting specifications together may yield inconsistency. This is of course a contrived example but the same phenomenon arises in more realistic situations.

Exercise. Expand NAT_c using the definition of enrichment in Section 5.2 and apply horizontal composability to derive the implementation steps sketched above. □

The material in this section applies just as well to constructor specifications as introduced in Section 6.2 and their extension to higher-order constructors as introduced in Definition 6.4.2. All that is required is a trivial change to the definition of simple implementation and to Proposition 7.1.2, and then the rest carries over without alteration.

Definition 7.1.9 (Simple implementation for constructor specifications). Given specifications SP and SP' with $SP, SP' : Spec(\mathscr{S})$ for some constructor signature $\mathscr{S}$, we say that SP' is a *simple implementation* of SP, written $SP \rightsquigarrow SP'$, if $Mod[SP] \supseteq Mod[SP']$. □

Proposition 7.1.10. *Given a chain $SP_0 \rightsquigarrow SP_1 \rightsquigarrow \cdots \rightsquigarrow SP_n$ of simple implementation steps and a constructor $F \in Mod[SP_n]$, we have $F \in Mod[SP_0]$.* □

Example 7.1.11. Recall the specification CONSSTRINGKEY from Section 6.5. We might want to add further constraints on the hash function such as the requirement that it is injective on strings of length 3:

spec BETTERCONSSTRINGKEY =
  $\Pi N{:}\mathrm{NAT}\bullet\ \Pi S{:}\mathrm{STRING}\bullet$
    $\{N+S\}$
    then
      ops $hash\colon String \to Nat$
        $length\colon String \to Nat$
      $\forall s, s'\colon String$
        • $length(\varepsilon) = 0$
        • $length(\mathtt{a}\,\hat{}\,s) = succ(length(s))$
        • $\cdots$
        • $length(\mathtt{z}\,\hat{}\,s) = succ(length(s))$
        • $hash(\varepsilon) = 0$
        • $length(s) = length(s') = succ(succ(succ(0))) \wedge s \neq s' \Rightarrow$
          $hash(s) \neq hash(s')$
    hide ops $length\colon String \to Nat$

Then clearly CONSSTRINGKEY $\rightsquigarrow$ BETTERCONSSTRINGKEY. □

This is a very typical example which can be derived from the following general fact:

Proposition 7.1.12. *If $SP_1 \rightsquigarrow SP$ and $SP' \rightsquigarrow SP'_1$ then $(SP \Rightarrow SP') \rightsquigarrow (SP_1 \Rightarrow SP'_1)$.* □

Exercise 7.1.13. Generalise the above proposition to the forms of constructor specifications in Example 6.2.9 and then in Definitions 6.2.8 and 6.4.2. Use this to justify the simple implementation in Example 7.1.11. Remember to take into account the fact that constructors satisfying $SP_1 \overset{\iota}{\Rightarrow} SP_1'$ yield models that extend models of SP_1, which allows one of the premises in the proposition to be weakened. □

7.2 Constructor implementations

The notion of simple implementation is powerful enough (in the context of a sufficiently rich specification language) to handle all concrete examples of interest. However, it is not very convenient to use in practice. During the process of developing a program, the successive specifications incorporate more and more details arising from successive design decisions. Thereby, some parts become fully determined, and remain unchanged as a part of the specification until the final program is obtained. For instance, in Example 7.1.4 we had consecutive simple implementations SORTPERM $\rightsquigarrow$ SORTINS $\rightsquigarrow$ SORTDONE, where the "code" for *sort* introduced in SORTINS and the hiding of *insert* are still present in the same form in SORTDONE. The following diagram is a visual representation of this situation, where $\kappa_1, \ldots, \kappa_n$ label the parts that become determined at consecutive steps:

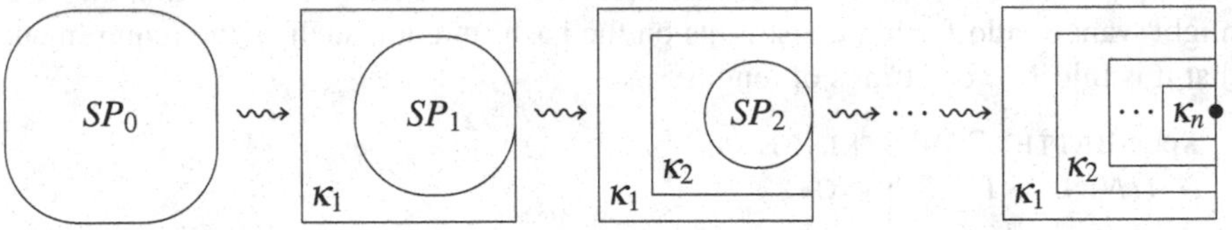

It is more convenient to avoid such clutter by separating the finished parts from the specification, putting them aside, and proceeding with the development of the unresolved parts only:

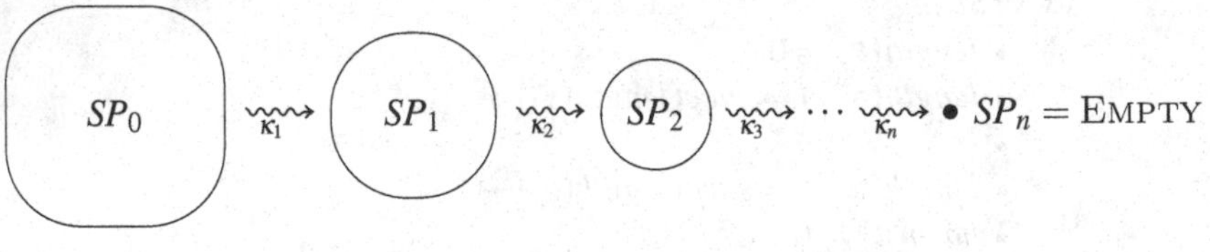

where EMPTY is a specification for which a standard model empty is available. For instance, in finitely exact institutions the obvious choice for EMPTY is the flat specification over the initial signature $\Sigma_\varnothing$ having no axioms, which has a unique model by the continuity of the model functor. In an institution like **FOEQ**, the initial signature $\Sigma_\varnothing$ is empty and empty is its unique (trivial) model, while in $\mathbf{IMP}_{DT}$ (Example 4.1.32) the unique model empty over the initial signature provides a standard realisation of some primitive (built-in) data types and functions. Similarly for **FPL** (Example 4.1.25), where the only built-in data type is *Bool* with constructors *true* and *false*. One way or another, we can consider that EMPTY and empty are given.

It is important for the finished parts $\kappa_1, \ldots, \kappa_n$ to be independent of the particular choice of realisation for what is left: they should extend any realisation of the unresolved part to a realisation of what is being implemented. Thus, they are simply *constructors* in the sense of Definition 6.1.1, and may be specified as in Definitions 6.2.1 and 6.2.8. This motivates a more elaborate version of the notion of implementation:

Definition 7.2.1 (Constructor implementation). Given specifications *SP* and *SP′*, we say that *SP′* is a *constructor implementation of SP via* κ, written $SP \underset{\kappa}{\rightsquigarrow} SP'$, if $\kappa \in Mod[SP' \Rightarrow SP]$. □

In the development diagram above, we have $\kappa_i{:}Mod(Sig[SP_i]) \to Mod(Sig[SP_{i-1}])$ with $\kappa_i \in Mod[SP_i \Rightarrow SP_{i-1}]$ for $1 \le i \le n$; that is, each κ_i corresponds to a generic module with input interface SP_i and output interface SP_{i-1}. Given a model M of SP_i, κ_i may be applied to yield a model $\kappa_i(M)$ of SP_{i-1}.

Exercise 7.2.2. Check that constructor implementations are a special case of simple implementations, since each constructor gives rise to a specification-building operation, see Exercise 6.1.15: $SP \underset{\kappa}{\rightsquigarrow} SP'$ iff $SP \rightsquigarrow \kappa(SP')$ and $Mod[SP'] \subseteq dom(\kappa)$. A trivial consequence of this is that $\kappa(SP') \underset{\kappa}{\rightsquigarrow} SP'$ whenever $Mod[SP'] \subseteq dom(\kappa)$. □

Once the development process is finally complete (that is, when nothing is left unresolved, as above) we can successively apply the constructors to obtain a correct realisation of the original specification. The correctness of the final outcome follows from the correctness of the individual constructor implementation steps:

Proposition 7.2.3. *Given a chain of constructor implementation steps*

$$SP_0 \underset{\kappa_1}{\rightsquigarrow} SP_1 \underset{\kappa_2}{\rightsquigarrow} \cdots \underset{\kappa_n}{\rightsquigarrow} SP_n = \text{EMPTY}$$

we have $\kappa_1(\kappa_2(\ldots\kappa_n(\mathsf{empty})\ldots)) \in Mod[SP_0]$, *where* $\mathsf{empty} \in Mod[\text{EMPTY}]$. □

The proof is by induction on n using the simple observation that constructor implementations reflect realisations: if $SP \underset{\kappa}{\rightsquigarrow} SP'$ and $M' \in Mod[SP']$ then $\kappa(M') \in Mod[SP]$.

As in the case of simple implementations, a slightly stronger but still easy fact is that constructor implementations vertically compose:

Proposition 7.2.4 (Vertical composition). *If* $SP \underset{\kappa}{\rightsquigarrow} SP'$ *and* $SP' \underset{\kappa'}{\rightsquigarrow} SP''$ *then* $SP \underset{\kappa';\kappa}{\rightsquigarrow} SP''$. □

Then given a chain $SP_0 \underset{\kappa_1}{\rightsquigarrow} SP_1 \underset{\kappa_2}{\rightsquigarrow} \cdots \underset{\kappa_n}{\rightsquigarrow} SP_n = \text{EMPTY}$ of constructor implementations, we obtain $(\kappa_n;\cdots;\kappa_2;\kappa_1)(\mathsf{empty}) \in Mod[SP_0]$. Semantically, this is the same as Proposition 7.2.3 but its use in connection with a programming language offering generic modules to code constructors is potentially problematic since it requires a composition operation on such modules. In Standard ML, there is no explicit functor composition operation but the composite of two functors may easily be defined using functor instantiation and abstraction. In the notation we have been using for defining constructors in the institution **FPL** — see Example 6.1.9 — a stronger property holds: it is possible to replace the composition of two constructors by a single constructor defined in the same notation, without using constructor application.

Lemma 7.2.5. *Consider the following signature diagram*

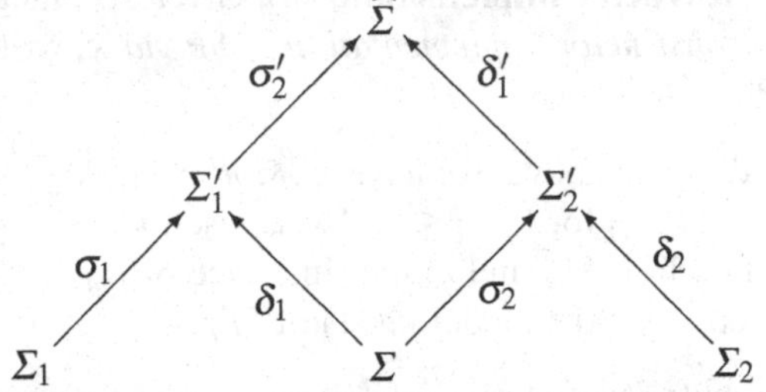

where the central square is a pushout which admits amalgamation, and the free extension $F_{\sigma_2} \in Mod(\Sigma \Rightarrow \Sigma_2')$ — see Example 6.1.7 — is a total constructor that is naturally persistent along σ_2. Then the following equality between constructors in $Mod(\Sigma_1 \Rightarrow \Sigma_2)$ holds:

$$F_{\sigma_1};(\_|_{\delta_1});F_{\sigma_2};(\_|_{\delta_2}) = F_{\sigma_1;\sigma_2'};(\_|_{\delta_2;\delta_1'}).$$

Proof. First note that $F_{\sigma_2'} = \delta_1(F_{\sigma_2})$, as in Exercise 6.1.30, and so $(\_|_{\delta_1});F_{\sigma_2} = F_{\sigma_2'};(\_|_{\delta_1'})$. Then the conclusion follows easily, since $F_{\sigma_1;\sigma_2'} = F_{\sigma_1};F_{\sigma_2'}$ (see Exercise 3.5.26) and $(\_|_{\delta_2;\delta_1'}) = (\_|_{\delta_1'});(\_|_{\delta_2})$. □

Exercise 7.2.6. Use Lemma 7.2.5 to show that in the institution **EQ**, any composition of reduct and naturally persistent free extension constructors can be presented as a composition of a single free extension followed by a single reduct constructor. □

Exercise 7.2.7. Check that the institution **FPL** is semi-exact w.r.t. **FPL**-signature inclusions that only add new sorts with value constructors; this is a proper subclass of $\mathbf{I_{FPL}}$, defined in Exercise 4.4.19. The notation introduced in Example 6.1.9 defines constructors in **FPL** that involve free extensions along such signature inclusions, which are total and naturally persistent (see Exercise 6.1.27). Use Lemma 7.2.5 to show that any composition of constructors defined using this notation can be expressed in the same notation as a single constructor. □

Example 7.2.8. Recall the consecutive simple implementation steps

$$\text{SORT} \leadsto \text{SORTPERM} \leadsto \text{SORTINS} \leadsto \text{SORTDONE}$$

in Example 7.1.4.

Consider the simple implementation SORTPERM ⇝ SORTINS in which "code" for *sort* is first introduced. Using the notation of Example 6.1.9, the constructor corresponding to this step can be expressed as follows:

constructor *K1* : *Sig*[INS] ⇒ *Sig*[SORTPERM] =
  ops fun *sort*(*l*:*NatList*):*NatList* =
    case *l* **of** *nil* => *nil* | *cons*(*n*, *l′*) => *insert*(*n*, *sort*(*l′*))

Recall that INS is the part of SORTINS that remains after "peeling off" *sort* and the operation of hiding *insert*, the part of the specification whose implementation is fixed in this step. The relationship between the specification SORTPERM (as built on INS) and the constructor *K1* introduced above is exactly as explained in Exercises 6.1.10 and 6.1.15. By moving

fun *sort*(*l*:*NatList*):*NatList* = ...

from SORTINS to the body of *K1*, its status changes from an axiom that resembles code, which *constrains* the interpretation of *sort*, to actual code that *defines sort*. Notice that the constructor definition provides not only code for *sort* but also (implicitly) realises the hiding of *insert* since *insert* is not present in the result signature. Similar remarks apply to other constructors below.

The next simple implementation step, SORTINS ⇝ SORTDONE, which introduces code for *insert*, corresponds to the following constructor:

constructor *K2* : *Sig*[NATLIST] ⇒ *Sig*[INS] =
  ops fun *insert*(*n*:*Nat*, *l*:*NatList*):*NatList* =
    case *l* **of** *nil* => *cons*(*n*, *nil*)
      | *cons*(*m*, *l′*) => **if** *le*(*n*, *m*) = *true* **then** *cons*(*n*, *l*)
                  else *cons*(*m*, *insert*(*n*, *l′*))

The code for *sort*, which in the original simple implementation step was still present in SORTDONE, has been dealt with in the previous step. Thus in this step we are able to focus on what remains, namely the *insert* operation, without the distraction of the surrounding context.

The definitional axioms in NATLIST may be translated directly into code, and we can choose a particular realisation to implement *le*, giving the following:

constructor *K3* : $\Sigma_\varnothing$ ⇒ *Sig*[NATLIST] =
  sorts *Nat* **free with** 0 | *succ*(*Nat*)
    NatList **free with** *nil* | *cons*(*Nat*, *NatList*)
  ops **fun** *le*(*m*:*Nat*, *n*:*Nat*):*Bool* =
      case *m* **of** 0 => *true*
        | *succ*(*m′*) => **case** *n* **of** 0 => *false*
                | *succ*(*n′*) => *le*(*m′*, *n′*)
    fun *append*(*l*:*NatList*, *l′*:*NatList*):*NatList* =
      case *l* **of** *nil* => *l′* | *cons*(*n*, *l″*) => *cons*(*n*, *append*(*l″*, *l′*))
    fun *is_in*(*n*:*Nat*, *l*:*NatList*):*Bool* =
      case *l* **of** *nil* => *false*
        | *cons*(*m*, *l′*) => **if** *n* = *m* **then** *true* **else** *is_in*(*n*, *l′*)

Here, $\Sigma_\varnothing$ is the "empty" **FPL**-signature (containing only the built-in sort *Bool* with its constructors).

To finish the example, we need to provide a constructor corresponding to the simple implementation step SORT $\rightsquigarrow$ SORTPERM. Since all that is done in this step is to impose a (non-constructive) restriction on the class of permissible realisations of *sort*, this is trivial:

constructor $K0 : Sig[\textsc{SortPerm}] \Rightarrow Sig[\textsc{Sort}] =$

(This is not a typo: there is nothing in the body!)

To sum up, we have the following constructor implementations:

$$\textsc{Sort} \underset{K0}{\rightsquigarrow} \textsc{SortPerm} \underset{K1}{\rightsquigarrow} \textsc{Ins} \underset{K2}{\rightsquigarrow} \textsc{NatList} \underset{K3}{\rightsquigarrow} \textsc{Empty}.$$

The justification requires proofs similar to those sketched in Example 7.1.4 for the corresponding simple implementation steps. Then

$K0(K1(K2(K3(\mathsf{empty}))))$

yields an **FPL**-model satisfying SORT, where empty is the unique **FPL**-model over $\Sigma_\varnothing$.

Alternatively, as suggested by Exercise 7.2.7, we can put together all the constructors involved, defining their composition explicitly as follows

constructor *K3_K2_K1_K0* : $\Sigma_\varnothing \Rightarrow Sig[\textsc{Sort}] =$
  sorts *Nat* **free with** 0 | *succ*(*Nat*)
    NatList **free with** *nil* | *cons*(*Nat*, *NatList*)
  ops **fun** *le*(*m*:*Nat*, *n*:*Nat*):*Bool* = ...
    fun *append*(*l*:*NatList*, *l'*:*NatList*):*NatList* = ...
    fun *is_in*(*n*:*Nat*, *l*:*NatList*):*Bool* = ...
    fun *insert*(*n*:*Nat*, *l*:*NatList*):*NatList* = ...
    fun *sort*(*l*:*NatList*):*NatList* = ...

and then *K3_K2_K1_K0*(empty) $\in Mod[\textsc{Sort}]$. □

Section 6.1 introduced a number of constructors corresponding to standard constructions on algebras used throughout earlier chapters. Among those, free extension and reduct have been used in the examples of constructor implementation above. Other constructors introduced there include quotient and restriction to the reachable subalgebra. These constructors typically arise in examples involving *data refinement* in which a high-level abstract representation of data (for instance sets or dictionaries) is implemented in terms of a lower-level "machine-oriented" representation (for instance lists or arrays).

Example 7.2.9 (Data refinement). We present an implementation of sets of natural numbers by lists of natural numbers, working in the institution **FPL**. We refer to the specifications NAT and NATLIST from Example 7.1.4. We begin by specifying sets of natural numbers:

spec NATSET =
  NAT
  then
    sorts *NatSet*
    ops *empty*: *NatSet*
      add: $Nat \times NatSet \rightarrow NatSet$
      $\_\_ \in \_\_ : Nat \times NatSet \rightarrow Bool$
    $\forall n, m: Nat, S: NatSet$
      • $(n \in empty) = false$
      • $(n \in add(n,S)) = true$
      • $n \neq m \Rightarrow (n \in add(m,S)) = (n \in S)$
      • $add(n, add(m,S)) = add(m, add(n,S))$
      • $add(n, add(n,S)) = add(n,S)$

The implementation of NATSET by NATLIST proceeds via three constructors. The first adds an idempotent addition operation to lists, adjusts the names of *NatList*, *nil* and *is_in*, and hides *cons*, *append* and *le*. The operations *append* and *le* are entirely superfluous for sets, while *cons* can be hidden once it has been used to define addition.

constructor *ListSet* : *Sig*[NATLIST] $\Rightarrow$ *Sig*[NATSET] =
  sorts *NatSet* = *NatList*
  ops *empty* = *nil*
    fun *add*(*n*:*Nat*, *l*:*NatList*):*NatList* =
      if *is_in*(*n*, *l*) = *true* **then** *l* **else** *cons*(*n*, *l*)
    $\in$ = *is_in*

The next step is to remove the unreachable values of sort *set*, which correspond to lists having multiple occurrences of one or more elements. Note that such values are reachable using *cons* and *nil*, but not using the operations available in NATSET.

$R_{Sig[\text{NATSET}]} : Sig[\text{NATSET}] \Rightarrow Sig[\text{NATSET}]$

See Example 6.1.13 and Exercise 6.1.14.

Finally, we quotient (see Exercise 2.7.32) the resulting **FPL**-models, viewed as partial algebras, by a congruence which identifies lists that differ only in the order of their elements, noting that a quotient by this congruence yields an **FPL**-model:

$\_/SP : Sig[\text{NATSET}] \Rightarrow Sig[\text{NATSET}]$

where *SP* is a flat *Sig*[NATSET]-specification with the axiom $add(n, add(m,S)) = add(m, add(n,S))$; see Example 6.1.11.

The overall outcome is a constructor implementation

$$\text{NATSET} \underset{\kappa}{\rightsquigarrow} \text{NATLIST}$$

using a constructor formed by composition: $\kappa = ListSet;R_{Sig[\text{NATSET}]};(\_/SP)$. This is an implementation of sets as equivalence classes of lists. The lists have no repeated elements and the equivalence identifies lists having the same elements but in a different order.

This is one possible implementation of sets in terms of lists. There are many other possibilities. One is a simpler variant of the above implementation with just two steps, the first renaming *cons* to *add* instead of coding it and adjusting the other names as above, and the second quotienting by both idempotence and commutativity of *add*. Another is to use ordered lists without duplicates, coding a version of *add* that inserts new elements into the appropriate position; this does not require a quotient step. □

A complete development chain that connects the initial requirements specification to EMPTY does not reflect the *process* of developing a system from a specification, which normally involves false starts, blind alleys and backtracking. It documents only the final outcome of this process. An incomplete development chain may be used to record a stage in the development process, so the development process corresponds to a sequence of such chains which culminates in a complete chain. Ideally, each chain in the sequence is an expansion of the previous one, but backtracking corresponds to deletion or alteration of parts of the chain that have already been filled in.

Related to this is the issue of whether to proceed in a completely top-down manner (i.e. left to right, in our development chains), by attempting to gradually reduce the problem stated in the requirements specification to simpler and simpler problems, or in a bottom-up manner, building up more and more complex programs in an incremental fashion in an attempt to obtain a solution to the original problem. The definitions and results do not actually prescribe either of these. What matters is that at the end a chain of correct implementation steps emerges which links the requirements specification with EMPTY. In reality a mixture of top-down and bottom-up, or even "middle-out", is common. But even if top-down refinement were to be prescribed, it is possible to represent bottom-up and middle-out development using an obvious generalisation of constructor implementations to constructor specifications.

Definition 7.2.10 (Constructor implementation for constructor specifications). Given specifications SP and SP' with $SP : Spec(\mathscr{S})$ and $SP' : Spec(\mathscr{S}')$ for constructor signatures $\mathscr{S}$ and $\mathscr{S}'$, we say that SP' is a *constructor implementation of SP via* κ, written $SP \underset{\kappa}{\rightsquigarrow} SP'$, if $\kappa \in Mod[SP' \Rightarrow SP]$. □

Example 7.2.11. Recall the specifications and constructors of Example 7.2.8. We then have, for instance:

$$\text{SORT} \underset{F1}{\rightsquigarrow} (\text{NATLIST} \Rightarrow \text{SORTPERM}) \underset{F2}{\rightsquigarrow} (\text{INS} \Rightarrow \text{SORTPERM})$$

where

$$F1(K{:}\,\text{NATLIST} \Rightarrow \text{SORTPERM}) = K0(K(K3(\mathsf{empty})))$$
$$F2(K{:}\,\text{INS} \Rightarrow \text{SORTPERM}) = K2;K$$

and then realising INS $\Rightarrow$ SORTPERM by $K1$ yields $F1(F2(K1)) \in Mod[\text{SORT}]$.

This amounts to an "outside-in" step, providing

$$\textsc{Sort} \underset{K0}{\rightsquigarrow} \textsc{SortPerm} \underset{???}{\rightsquigarrow} \textsc{NatList} \underset{K3}{\rightsquigarrow} \textsc{Empty}$$

as a partial solution to $\textsc{Sort} \underset{???}{\rightsquigarrow} \textsc{Empty}$, followed by a bottom-up step, providing

$$\textsc{SortPerm} \underset{???}{\rightsquigarrow} \textsc{Ins} \underset{K2}{\rightsquigarrow} \textsc{NatList}$$

as a partial solution to $\textsc{SortPerm} \underset{???}{\rightsquigarrow} \textsc{NatList}$.

Exercise. Capture a "middle-out" step that provides

$$\textsc{Sort} \underset{???}{\rightsquigarrow} \textsc{Ins} \underset{K2}{\rightsquigarrow} \textsc{NatList} \underset{???}{\rightsquigarrow} \textsc{Empty}$$

as a partial solution to $\textsc{Sort} \underset{???}{\rightsquigarrow} \textsc{Empty}$. HINT: The most obvious way is to use a two-argument constructor — see the next section — but try to do without it. □

7.3 Modular decomposition

When developing a large program it is crucial to progressively decompose the job into smaller tasks that can be handled separately. Each task is defined by a specification, and solving a task means producing a program component that satisfies this specification. Once all tasks are solved, producing the final system is a simple matter of appropriately assembling these components.

A development step involving the decomposition of a programming task into separate subtasks is modelled using a constructor implementation with a multi-argument constructor.

Definition 7.3.1 (Constructor implementation with decomposition). Given specifications SP and $SP'_1, \ldots, SP'_n$, we say that the tuple $\langle SP'_1, \ldots, SP'_n \rangle$ is a *constructor implementation of SP via* κ, written $SP \underset{\kappa}{\rightsquigarrow} \langle SP'_1, \ldots, SP'_n \rangle$, if $\kappa \in Mod[SP'_1 \Rightarrow \cdots \Rightarrow SP'_n \Rightarrow SP]$. □

Constructor implementations as introduced earlier are obviously a special case of these, identifying $\langle SP' \rangle$ with SP'. Notice that κ takes its arguments in curried form, so $\kappa(M'_1) \cdots (M'_n) \in Mod[SP]$ if $M'_1 \in Mod[SP'_1], \ldots, M'_n \in Mod[SP'_n]$, and recall that $\Rightarrow$ is right-associative.

Now the development takes on a tree-like shape. It is complete once a tree is obtained that has EMPTY as its leaves:

$$SP \underset{\kappa}{\rightsquigarrow} \begin{cases} SP_1 \underset{\kappa_1}{\rightsquigarrow} \textsc{Empty} \\ \vdots \\ SP_n \underset{\kappa_n}{\rightsquigarrow} \begin{cases} SP_{n1} \underset{\kappa_{n1}}{\rightsquigarrow} \{ SP_{n11} \underset{\kappa_{n11}}{\rightsquigarrow} \textsc{Empty} \\ \ldots \\ SP_{nm} \underset{\kappa_{nm}}{\rightsquigarrow} \textsc{Empty} \end{cases} \end{cases}$$

Then an appropriate application of the constructors in the tree yields a realisation of the original requirements specification. The above development tree yields

$$\kappa(\kappa_1(\mathsf{empty}))\cdots(\kappa_n(\kappa_{n1}(\kappa_{n11}(\mathsf{empty})))\cdots(\kappa_{nm}(\mathsf{empty}))) \in Mod[SP].$$

Our discussion following Proposition 7.2.4 concerning the modularisation facilities of the programming language at hand and whether they are sufficient to express such a composition directly still applies here.

The structure of the final realisation is determined by the shape of the development tree, which is in turn determined by the decomposition steps. Each such step corresponds to what software engineers call a *design* specification, or sometimes an *organizational* specification: it defines the structure of the system by specifying its components and describing how they fit together. This style of development leads to modular programs, built from fully specified, correct and reusable components.

Decisions concerning the structure of the final realisation, recorded in the development tree, need have no relation to the structure of the requirements specification. Nevertheless, given a structured specification $\kappa(SP_1,\ldots,SP_n)$ where κ is the specification-building operation corresponding to a constructor κ — see Exercise 6.1.15 — we *may* reuse its structure in the design of its final realisation as follows

$$\kappa(SP_1,\ldots,SP_n) \underset{\kappa}{\leadsto} \langle SP_1,\ldots,SP_n\rangle$$

provided that for all $M_1 \in Mod[SP_1],\ldots,M_n \in Mod[SP_n]$, $\kappa(M_1)\cdots(M_n)$ is defined.

Exercise 7.3.2. Recall the horizontal composability property of simple implementations in Proposition 7.1.6. Reformulate it (and prove it) for constructor implementations, where **sbo** arises from a constructor as above. Discuss the case of arbitrary monotone specification-building operations, referring to Example 7.1.8. □

Example 7.3.3. Recall Example 7.1.4, which defined a specification SORT for sorting lists of natural numbers, and Example 7.2.8, which gave a chain of successive constructor implementations leading to a realisation of the sorting function that preserves the number of occurrences of elements in its argument list.

Let us try to construct another realisation, this time with a sorting function that removes multiple occurrences of elements. We begin by giving a simple implementation of SORT that captures this requirement.

spec SORTONCE =
  SORT
  then
    ops $all\_once{:}\, NatList \to Bool$
    $\forall x{:}\,Nat, l{:}\,NatList$
      • $def(all\_once(l))$
      • $all\_once(nil) = true$
      • $all\_once(cons(x,l)) = true \Leftrightarrow (all\_once(l) = true \wedge is\_in(x,l) = false)$
      • $all\_once(sort(l)) = true$
  hide ops $all\_once{:}\, NatList \to Bool$

Clearly, SORT $\leadsto$ SORTONCE.

Consider an additional specification that introduces a function that is specified to build a maximal strictly increasing sublist of its argument (note that there may be many such sublists and we do not require that the result be the longest among them):

spec MAXINCSUBLIST =
  NATLIST
  then
    ops *increasing*: *NatList* → *Bool*
      sublist: *NatList* × *NatList* → *Bool*
      max_inc_sublist: *NatList* → *NatList*
    $\forall n{:}Nat, l, l'{:}NatList$
      • $def(increasing(l)) \wedge def(sublist(l,l')) \wedge def(max\_inc\_sublist(l))$
      • $increasing(nil) = true$
      • $increasing(cons(n,l)) = true \Leftrightarrow$
          $(\forall m{:}Nat \bullet is\_in(m,l) = true \Rightarrow le(m,n) = false) \wedge$
          $increasing(l) = true$
      • $sublist(nil,l') = true$
      • $sublist(cons(n,l),l') = true \Leftrightarrow$
          $\exists l1, l1'{:}NatList \bullet l' = append(l1, cons(n, l1')) \wedge$
            $sublist(l, l1') = true$
      • $increasing(max\_inc\_sublist(l)) = true$
      • $sublist(max\_inc\_sublist(l), l) = true$
      • $(increasing(l') = true \wedge sublist(l',l) = true \wedge$
          $sublist(max\_inc\_sublist(l), l') = true) \Rightarrow max\_inc\_sublist(l) = l'$
  hide ops *increasing*: *NatList* → *Bool*
      sublist: *NatList* × *NatList* → *Bool*

Now, the problem of implementing the specification SORTONCE may be decomposed (not very efficiently!) into the problems of implementing SORTPERM and MAXINCSUBLIST,

$$\text{SORTONCE} \underset{K4}{\rightsquigarrow} \langle \text{SORTPERM}, \text{MAXINCSUBLIST} \rangle,$$

where given models M_1 of SORTPERM and M_2 of MAXINCSUBLIST, the constructor *K4* : *Sig*[SORTPERM] ⇒ *Sig*[MAXINCSUBLIST] ⇒ *Sig*[SORTONCE] builds a model M of SORTONCE by adding a sort function that first performs sorting as given in M_1 and then selects its maximal strictly increasing sublist by applying *max_inc_sublist* from M_2. In essence, M is given by the reduct with respect to the signature morphism $\delta{:}Sig[\text{SORTONCE}] \to Sig[\text{SORTPERM} \textbf{ and } \text{MAXINCSUBLIST}]$, where $\delta(sort) = max\_inc\_sublist(sort(\boxed{1}{:}NatList))$ and δ is the identity on other names.[1]

Since we already have a realisation for SORTPERM (Example 7.2.8), it remains to provide one for MAXINCSUBLIST. We will do this using an auxiliary function

[1] This glosses over some of the details for expository reasons; see below for another version that is fully correct.

that removes initial elements from the list that are smaller than a given element, as captured by the following specification:

spec REMLIST =
  NATLIST
  then
    ops $rem: Nat \times NatList \rightarrow NatList$
    $\forall n: Nat, l, l': NatList$
      • $def(rem(n,l))$
      • $rem(n,l) = l' \Leftrightarrow$
        $\exists l1: NatList \bullet\ l = append(l1, l') \wedge$
          $(\forall p: Nat \bullet\ is\_in(p, l1) \Rightarrow le(p,n) = true) \wedge$
          $(l' = nil \vee (l' = cons(m, l2) \wedge le(m,n) = false))$

A realisation for MAXINCSUBLIST may come from the following constructor implementation steps:

$$\text{MAXINCSUBLIST} \underset{K5}{\rightsquigarrow} \text{REMLIST} \underset{K6}{\rightsquigarrow} \text{NATLIST} \underset{K7}{\rightsquigarrow} \text{EMPTY}$$

where

constructor *K5* : *Sig*[REMLIST] ⇒ *Sig*[MAXINCSUBLIST] =
  ops fun *max_inc_sublist*(*l*:*NatList*):*NatList* =
    case *l* **of** *nil* => *nil*
      | *cons*(*m*, *l'*) =>
        case *rem*(*m*, *l'*) **of** *nil* => *cons*(*m*, *nil*)
          | *ll'* => *cons*(*m*, *max_inc_sublist*(*ll'*))

constructor *K6* : *Sig*[NATLIST] ⇒ *Sig*[REMLIST] =
  ops fun *rem*(*n*:*Nat*, *ll*:*NatList*):*NatList* =
    case *ll* **of** *nil* => *nil*
      | *cons*(*m*, *l'*) => **if** *le*(*m*, *n*) = *true* **then** *rem*(*n*, *l'*) **else** *ll*

constructor *K7* : $\Sigma_\varnothing$ ⇒ *Sig*[NATLIST] =
  sorts *Nat* **free with** 0 | *succ*(*Nat*)
    NatList **free with** *nil* | *cons*(*Nat*, *NatList*)
  ops **fun** *le*(*m*:*Nat*, *n*:*Nat*):*Bool* =
      case *n* **of** 0 => *true*
        | *succ*(*n'*) => **case** *m* **of** 0 => *false*
          | *succ*(*m'*) => *le*(*m'*, *n'*)
    fun *append*(*l*:*NatList*, *l'*:*NatList*):*NatList* =
      case *l* **of** *nil* => *l'* | *cons*(*n*, *l''*) => *cons*(*n*, *append*(*l''*, *l'*))
    fun *is_in*(*n*:*Nat*, *l*:*NatList*):*Bool* =
      case *l* **of** *nil* => *false*
        | *cons*(*m*, *l'*) => **if** *n* = *m* **then** *true* **else** *is_in*(*n*, *l'*)

So we should have

K4(*K1*(*K2*(*K3*(empty))))(*K5*(*K6*(*K7*(empty)))) ∈ *Mod*[SORTONCE], but this doesn't work! (**Exercise:** Describe *sort* in this model, and check which axioms of SORTONCE are not satisfied.) □

The trouble with the above example is that the decomposition step

$$\text{SORTONCE} \underset{K4}{\rightsquigarrow} \langle \text{SORTPERM}, \text{MAXINCSUBLIST} \rangle$$

is incorrect. Informally, SORTPERM and MAXINCSUBLIST share the part that is specified by NATLIST, and this decomposition allows it to be realised differently in the models of SORTPERM and MAXINCSUBLIST that are constructed by the subsequent independent developments, as in *K3* and *K7* above (where *le* is defined differently).[2] This means that SORTPERM and MAXINCSUBLIST cannot be developed independently. This is a general problem that will arise whenever a task is decomposed into subtasks that are not entirely independent. The only time that decomposition is entirely unproblematic is when the axioms of both subtask specifications force a unique implementation of the common part; when the signatures of the subtask specifications are disjoint, this requirement holds trivially.

One way to capture the required sharing in our framework is to make the relationship between subtasks explicit using parameterisation. A typical schema for expressing a decomposition of a task *SP* into subtasks SP_1 and SP_2 is

$$SP \underset{\kappa}{\rightsquigarrow} \langle SP_0, SP_0 \stackrel{\iota_1}{\Rightarrow} SP_1, SP_0 \stackrel{\iota_2}{\Rightarrow} SP_2 \rangle$$

where $Sig[SP_1] \cap Sig[SP_2] = Sig[SP_0]$, ι_1 and ι_2 are the inclusions, $\kappa(M)(F_1)(F_2) = F_1(M) + F_2(M)$ and $SP \rightsquigarrow (SP_1 \textbf{ and } SP_2)$. The persistency of F_1 and F_2 ensures that κ is well defined and the last condition ensures that it yields a result in $Mod[SP]$.

Example 7.3.4. We replace the decomposition step in Example 7.3.3 by:

$$\text{SORTONCE} \underset{K4'}{\rightsquigarrow} \langle \text{NATLIST}, \text{NATLIST} \stackrel{\iota}{\Rightarrow} \text{SORTPERM}, \\ \text{NATLIST} \stackrel{\iota'}{\Rightarrow} \text{MAXINCSUBLIST} \rangle$$

where ι and ι' are the obvious signature inclusions and the constructor *K4'* builds a model *M* of SORTONCE from a model *N* of NATLIST and persistent constructors $F \in Mod[\text{NATLIST} \stackrel{\iota}{\Rightarrow} \text{SORTPERM}]$, $G \in Mod[\text{NATLIST} \stackrel{\iota'}{\Rightarrow} \text{MAXINCSUBLIST}]$ by adding to $F(N)+G(N)$ a sort function that first performs sorting as given in $F(N)$ and then applies *max_inc_sublist* from $G(N)$. Then, rearranging the development steps in Example 7.3.3 above to make it explicit that $K2;K1 \in Mod[\text{NATLIST} \stackrel{\iota}{\Rightarrow} \text{SORTPERM}]$ and $K6;K5 \in Mod[\text{NATLIST} \stackrel{\iota'}{\Rightarrow} \text{MAXINCSUBLIST}]$, we obtain

$$K4'(K3(\mathsf{empty}))(K2;K1)(K6;K5) \in Mod[\text{SORTONCE}].$$

We also have

$$K4'(K7(\mathsf{empty}))(K2;K1)(K6;K5) \in Mod[\text{SORTONCE}].$$

[2] That was the source of our trouble in describing *K4* above, where the models M_1 and M_2 need not amalgamate to give a model of SORTPERM **and** MAXINCSUBLIST.

This builds another model with a function to sort lists of natural numbers w.r.t. *le* as defined in *K7*. □

Constructors like *K4'* in Example 7.3.4 cannot be expressed in the notation we have been using to write constructors in the institution **FPL** without extending it by adding names for parameters.

Exercise 7.3.5 (FPL-constructor extended notation). Modify and extend the notation of Example 6.1.9 by allowing multiple named parameters; first-order constructor parameters; use of parameter names to refer to parameters; use of constructor application, giving the result a name; use of "dot notation" like $N.f$ to refer to components of named models; and replacing the implicit import of the parameter into the intermediate model by explicit inclusion of all the components of the named models.

The resulting notation should allow *K4'* from Example 7.3.4 to be defined as follows:

constructor $K4'(NL : \textsc{NatList})$
$(F : \textsc{NatList} \stackrel{\iota}{\Rightarrow} \textsc{SortPerm})$
$(G : \textsc{NatList} \stackrel{\iota'}{\Rightarrow} \textsc{MaxIncSubList}) : \textsc{SortOnce} =$
let $S = F(NL)$
let $M = G(NL)$
include M
ops fun *sort*(*l*:*NatList*):*NatList* = *max_inc_sublist*(*S.sort*(*l*))

A subtle point is that the persistency of the constructor parameters is needed in order to justify the well-formedness of the definitions in the body. These issues become considerably more involved when higher-order constructors are allowed as parameters. Extend this notation to cater for higher-order constructor parameters, but without attempting to give a fully adequate treatment of persistency for higher-order parameters. □

Exercise 7.3.6. Use the extended notation to provide realisations of all the specifications in Section 6.5. You may want to compare the outcome with the development in Section 7.4 below. □

Once we consider decomposition steps into subtasks involving development of (higher-order) constructors, it is possible to limit the constructors κ used in the implementation relation $\underset{\kappa}{\rightsquigarrow}$ to be of a particularly simple form, merely combining the subtask realisations using a few standard combinators, where those listed in Exercise 6.4.17 seem to be all that are required. For instance, the very decision of implementing a specification *SP* by *SP'* via an as-yet-unknown constructor may now be recorded by the decomposition step $SP \underset{app}{\rightsquigarrow} \langle SP' \Rightarrow SP, SP' \rangle$ where *app* is the application combinator, turning the choice of a specific constructor into a separate development task. See also Example 7.2.11.

Example 7.3.7. We can redo Example 7.3.4 in this style as follows.

$$\begin{aligned}\mathrm{SORTONCE} \underset{C}{\rightsquigarrow} \langle\ & \mathrm{NATLIST}, \\ & \mathrm{NATLIST} \overset{\iota}{\Rightarrow} \mathrm{SORTPERM}, \\ & \mathrm{NATLIST} \overset{\iota'}{\Rightarrow} \mathrm{MAXINCSUBLIST}, \\ & \mathrm{NATLIST} \Rightarrow (\mathrm{NATLIST} \overset{\iota}{\Rightarrow} \mathrm{SORTPERM}) \Rightarrow \\ & \qquad (\mathrm{NATLIST} \overset{\iota'}{\Rightarrow} \mathrm{MAXINCSUBLIST}) \Rightarrow \mathrm{SORTONCE}\ \rangle\end{aligned}$$

where $C(NL)(F)(G)(H) = H(NL)(F)(G)$ and then

$$C(K3(\mathsf{empty}))(K2;K1)(K6;K5)(K4') \in Mod[\mathrm{SORTONCE}]. \qquad \square$$

Exercise 7.3.8. Continue the above example by the following constructor implementation step:

$$\begin{pmatrix}\mathrm{NATLIST} \Rightarrow (\mathrm{NATLIST} \overset{\iota}{\Rightarrow} \mathrm{SORTPERM}) \Rightarrow \\ (\mathrm{NATLIST} \overset{\iota'}{\Rightarrow} \mathrm{MAXINCSUBLIST}) \Rightarrow \mathrm{SORTONCE}\end{pmatrix} \underset{C'}{\rightsquigarrow}$$
$$(\mathrm{SORTPERM}\ \textbf{and}\ \mathrm{MAXINCSUBLIST}) \Rightarrow \mathrm{SORTONCE}$$

where for $H' \in Mod[(\mathrm{SORTPERM}\ \textbf{and}\ \mathrm{MAXINCSUBLIST}) \Rightarrow \mathrm{SORTONCE}]$

$$\begin{aligned}C'(H') = \lambda NL{:}\mathrm{NATLIST}\bullet \\ \lambda F{:}\mathrm{NATLIST} \overset{\iota}{\Rightarrow} \mathrm{SORTPERM}\bullet \\ \lambda G{:}\mathrm{NATLIST} \overset{\iota'}{\Rightarrow} \mathrm{MAXINCSUBLIST}\bullet \\ H'(F(NL) + G(NL)).\end{aligned}$$

Adapt vertical composability (Proposition 7.2.4) to constructor implementations with decomposition, and use this to justify

$$\begin{aligned}\mathrm{SORTONCE} \underset{C''}{\rightsquigarrow} \langle\ & \mathrm{NATLIST}, \\ & \mathrm{NATLIST} \overset{\iota}{\Rightarrow} \mathrm{SORTPERM}, \\ & \mathrm{NATLIST} \overset{\iota'}{\Rightarrow} \mathrm{MAXINCSUBLIST}, \\ & (\mathrm{SORTPERM}\ \textbf{and}\ \mathrm{MAXINCSUBLIST}) \Rightarrow \mathrm{SORTONCE}\ \rangle\end{aligned}$$

where $C''(NL)(F)(G)(H') = H'(F(NL) + G(NL))$. Finally, conclude

$$C''(K3(\mathsf{empty}))(K2;K1)(K6;K5)(K4'') \in Mod[\mathrm{SORTONCE}]$$

where[3]

constructor *K4''*(*SM* : SORTPERM **and** MAXINCSUBLIST) : SORTONCE =
  include *SM*
  ops fun *sort*(*l*:*NatList*):*NatList* = *max_inc_sublist*(*SM*.*sort*(*l*)). □

[3] We assume that a new definition of a name (function, sort) overwrites the previous meaning, if any, for this name — as with *sort* here. You may need to adjust your definition for Exercise 7.3.5 to take this into account.

7.4 Example

We now present a development of a realisation of the specification STRINGTABLE from Section 5.3. This is a standard specification of sets of strings. The strategy of implementing these by two-level hash tables will be gradually "discovered" in the course of the development. Constructors like those specified in Section 6.5 will arise in the process.

As we develop a realisation of STRINGTABLE, we will also reuse other specifications from Section 5.3 without further reference. All these specifications are in the institution **FOPEQ**. Here we will work in **FPL** in order to facilitate coding constructor definitions, making use of the notations of Example 6.1.9 and Exercise 7.3.5. To translate specifications in **FOPEQ** into specifications in **FPL**, predicates need to be replaced by total boolean functions and definedness requirements need to be added for other operations; see Exercise 5.3.2. We will refer to specifications from Section 5.3 as if this transformation had been performed. In addition, we rewrite NAT and STRING to make the generated nature of natural numbers and strings explicit:

spec NAT =
  sorts *Nat* **free with** $0 \mid succ(Nat)$
  ops $\_\_+\_\_ : Nat \times Nat \to Nat$
    $\_\_\leq\_\_ : Nat \times Nat \to bool$
  $\forall m,n{:}Nat$
    • $m+n =$ **case** m **of** $0 \Rightarrow n \mid succ(m') \Rightarrow succ(m'+n)$
    • $m \leq n =$ **case** m **of** $0 \Rightarrow true$
      $\mid succ(m') \Rightarrow$ **case** n **of** $0 \Rightarrow false$
        $\mid succ(n') \Rightarrow m' \leq n'$

spec STRING =
  sorts *String* **free with** $\varepsilon \mid$ a^(*String*) $\mid \cdots \mid$ z^(*String*)

A more principled way to handle the switch from one institution to another will be presented in Section 10.1 below.

The development is presented as a sequence of tasks, with later tasks resulting from decomposition in earlier steps; see the end of the section for a summary. The components that emerge are given names for ease of reference in later development steps. While describing subsequent development steps, we indicate which of the components specified earlier are to be considered given. Such "imports" do not introduce a dependency on the specific realisation of the specification of the imported component; correctness is required for an arbitrary realisation. This is illustrated in the fully explicit version of the first step below.

Task 1: Build *HT* : STRINGTABLE

The first design decision is to use direct chaining hash tables to represent sets. This involves introducing hash functions, and arrays that store sets of strings at the position determined by the application of the hash function. At this stage, all decisions concerning the treatment of these sets are left open; we simply require arrays to store "buckets" of strings having the same hash value.

This leads to the following subtasks:

- N : NAT
- S : STRING
- SK : STRINGKEY **given** N, S
- B : SIMPLEBUCKET **given** S
  where

  spec SIMPLEBUCKET = STRINGTABLE **with** $\sigma_{Table \mapsto Bucket}$

  where $\sigma_{Table \mapsto Bucket}$ is a surjective signature morphism from Sig[STRINGTABLE] that renames *Table* to *Bucket* and is the identity otherwise.
- BA : SIMPLEBUCKETARRAY **given** N, B
  where (see Example 6.3.3 for the definition of ARRAY, translated here to a specification in **FPL**, but exceptionally without the requirement on *get* to be total, as hinted at in Exercise 5.3.2)

  spec SIMPLEBUCKETARRAY =
  SIMPLEBUCKET **and** (ARRAY(ELEM)
  with $\sigma_{Elem \mapsto Bucket, Array[Elem] \mapsto Array[Bucket]}$)

  where $\sigma_{Elem \mapsto Bucket, Array[Elem] \mapsto Array[Bucket]}$ is a surjective signature morphism from Sig[ARRAY(ELEM)] that maps *Elem* to *Bucket* and *Array*[*Elem*] to *Array*[*Bucket*] and is the identity otherwise.

Now, the implementation of *HT* may use all the components specified above. We present it as follows:

HT = **sorts** *Table* = *BA.Array*[*Bucket*]
        String = *S.String*
    ops *empty* = *BA.empty*
        fun *add*(*s*:*String*, *t*:*Table*):*Table* =
            let $i = SK.hash(s)$
            in if $BA.used(i,t) = true$ **then** $BA.put(i, B.add(s, BA.get(i,t)), t)$
                else $BA.put(i, B.add(s, B.empty), t)$
        fun *present*(*s*:*String*, *t*:*Table*):*Bool* =
            let $i = SK.hash(s)$
            in if $BA.used(i,t) = true$ **then** $B.present(s, BA.get(i,t))$
                else *false*

The "**given**" notation for imports used in the specifications above can be translated to the higher-order constructor notation of Section 6.4 (in essence, by adding an additional parameter for the specified constructor and immediately instantiating it with the given component, as illustrated below). Presenting lists of subtasks as above allows us to avoid making all of the parameters explicit, which would also involve repeating their specifications. The informal notation used in the definition of HT is like the one introduced in Example 6.1.9 and Exercise 7.3.5, but again omits explicit mention of parameters corresponding to the subtasks and their specifications. Making everything fully explicit gives the following implementation step:

$$\begin{array}{ll}\textsc{StringTable} \underset{K_{HT}}{\rightsquigarrow} \langle & \textsc{Nat}, \\ & \textsc{String}, \\ & \Pi N{:}\textsc{Nat}\bullet \Pi S{:}\textsc{String}\bullet \{N+S\} \textbf{ and } \textsc{StringKey}, \\ & \Pi S{:}\textsc{String}\bullet \{S\} \textbf{ and } \textsc{SimpleBucket}, \\ & \Pi N{:}\textsc{Nat}\bullet \Pi B{:}\textsc{SimpleBucket}\bullet \\ & \qquad \{N+B\} \textbf{ and } \textsc{SimpleBucketArray} \rangle \end{array}$$

where

$$\begin{array}{l}\textbf{constructor } K_{HT}(N : \textsc{Nat}) \\ \qquad (S : \textsc{String}) \\ \qquad (SK_{???} : \Pi N{:}\textsc{Nat}\bullet \Pi S{:}\textsc{String}\bullet \{N+S\} \textbf{ and } \textsc{StringKey}) \\ \qquad (B_{???} : \Pi S{:}\textsc{String}\bullet \{S\} \textbf{ and } \textsc{SimpleBucket}) \\ \qquad (BA_{???} : \Pi N{:}\textsc{Nat}\bullet \Pi B{:}\textsc{SimpleBucket}\bullet \\ \qquad\qquad \{N+B\} \textbf{ and } \textsc{SimpleBucketArray}) \\ \qquad : \textsc{StringTable} = \\ \quad \textbf{let } SK = SK_{???}(N)(S) \\ \quad \textbf{let } B = B_{???}(S) \\ \quad \textbf{let } BA = BA_{???}(N)(B) \\ \quad \textbf{sorts } Table = BA.Array[Bucket] \\ \quad \ldots \end{array}$$

Notice that, as one would expect, we require constructors to provide persistent extensions of their parameters. This is expressed in the explicit form of the subtask specifications above by the use of singleton specifications as in Example 6.2.9.

Imports that are generic will arise later in this section. For such imports, the persistency requirement will not be imposed, in accordance with the comments at the end of Exercise 7.3.5.

Exercise 7.4.1. Expand the "**given**" notation as explained above for all of the steps below. Some care will be needed in the case of specifications of generic modules with imports, where persistency is required over both imports and parameters. □

We can now proceed with the subtasks above independently, and in any order we choose. Realisations of NAT and STRING can be given easily; we will not pursue this here. The development of a hash function for given realisations of STRING and NAT to give a realisation of STRINGKEY is left to the reader.

Task 2: Build B : SIMPLEBUCKET given S

We have various options. One would be to simply implement buckets as lists; this corresponds to HT implementing standard direct chaining hash tables. A more elaborate solution is to use hash tables again to implement buckets, using a different hash function. Of course, the second hash function may again yield collisions, and we must decide how to resolve them. One possibility would be to repeat the design from the top level, and defer the problem to the sub-buckets. But unless we can find a finite collection of hash functions that collectively give perfect hashing — that is, determine an injection from strings to tuples of natural numbers — some other technique must be used sooner or later. In Section 5.3, linear probing was used at this second level to resolve collisions. We make the same decision here, modulo an adjustment to the specification of arrays hinted at in Exercise 6.5.2. This leads to the following subtasks:

- N' : NAT
- SK' : STRINGKEY **given** N', S
- SA : STRINGFINITEARRAY **given** N', S
  where

  spec STRINGFINITEARRAY =
  STRINGARRAY
  then
  $\forall a{:}Array[String] \bullet \exists n{:}Nat \bullet \forall j{:}Nat \bullet (n \leq j) = true \Rightarrow used(j,a) = false$

- LP : LINEARPROBING **given** SA
  where

  spec LINEARPROBING =
  STRINGFINITEARRAY
  then
  ops $putnear : Nat \times String \times Array[String] \rightarrow Array[String]$
  $isnear : Nat \times String \times Array[String] \rightarrow Bool$
  $\forall i{:}Nat, s{:}String, a, a'{:}Array[String]$
  $\bullet\ def(putnear(i,s,a)) \wedge def(isnear(i,s,a))$
  $\bullet\ putnear(i,s,a) = a' \Leftrightarrow$
  $\exists j{:}Nat \bullet (used(i+j,a) = false \vee get(i+j,a) = s) \wedge$
  $(\forall k{:}Nat \bullet (j \leq k) = false \Rightarrow used(i+k,a) = true) \wedge$
  $a' = put(i+j,s,a)$
  $\bullet\ isnear(i,s,a) = true \Leftrightarrow$
  $\exists j{:}Nat \bullet (\forall k{:}Nat \bullet (k \leq j) = true \Rightarrow used(i+k,a) = true) \wedge$
  $get(i+j,a) = s$

We then implement B as follows:

$B =$ **include** S
sorts $Bucket = SA.Array[String]$
ops $empty = SA.empty$
fun $add(s{:}String, b{:}Bucket){:}Bucket = LP.putnear(SK'.hash(s), s, b)$
fun $present(s{:}String, b{:}Bucket){:}Bool = LP.isnear(SK'.hash(s), s, b)$

As before, we do not pursue the realisation of NAT except to note that this need not be the same as in Task 1 above. The realisation of SK' is again left to the reader.

(Note that using the same function as in *SK* is not a good idea, although correctness of the development presented here does not depend on this.)

Exercise 7.4.2. The specification LINEARPROBING above requires *putnear* and *isnear* to probe successive positions in the bucket. Modify it to require probing at fixed intervals, with the length of the "hop" depending on the string in question rather than always being 1. (This is so-called "double hashing".) Adjust the development in Task 4 below accordingly. □

Task 3: Build *SA* : STRINGFINITEARRAY given N', S

This requires implementation of arrays of strings in which only a finite number of positions are used. We recognize that arrays are a general data structure and implement them for arbitrary kinds of data values. This leads to the following subtask:

- A : ELEM $\stackrel{\iota}{\Rightarrow}$ ELEMFINITEARRAY **given** N'
  where

  spec ELEMFINITEARRAY =
  ARRAY(ELEM)
  then
  $\forall a{:}Array[Elem]$
  $\bullet\ \exists n{:}Nat\bullet \forall j{:}Nat\bullet (n \leq j) = true \Rightarrow used(j,a) = false$

  and ι is the signature inclusion.

We then implement *SA* as follows:

$SA =$ **let** $SA' = A(S|_{\sigma})$
sorts $Array[String] = SA'.Array[Elem]$
include S
include SA'

where $\sigma = \{Elem \mapsto String\}$.

Using Example 6.1.28, *SA* can alternatively be written as $SA = \sigma(A)(S)$, with the choice of names in the result (pushout) signature as hinted at in Example 6.3.6; see Exercise 6.5.1. The same applies to the implementation *BA* in Task 5 below.

A may be realised in a number of ways; see for instance Exercise 6.2.12. In reality, an implementation of arrays would of course be available from a library. So this completes Task 3.

Task 4: Build *LP* : LINEARPROBING given *SA*

This can be implemented directly:

$$
\begin{array}{l}
LP = \textbf{include } SA \\
\quad\quad \textbf{ops fun } putnear(n{:}Nat, s{:}String, a{:}Array[String]){:}Array[String] = \\
\quad\quad\quad\quad \textbf{if } used(n,a) = false \textbf{ then } put(n,s,a) \\
\quad\quad\quad\quad \textbf{else if } get(n,a) = s \textbf{ then } a \textbf{ else } putnear(succ(n),s,a) \\
\quad\quad\quad \textbf{fun } isnear(n{:}Nat, s{:}String, a{:}Array[String]){:}Bool = \\
\quad\quad\quad\quad \textbf{if } used(n,a) = false \textbf{ then } false \\
\quad\quad\quad\quad \textbf{else if } get(n,a) = s \textbf{ then } true \textbf{ else } isnear(succ(n),s,a)
\end{array}
$$

This completes Task 4, which discharges the last remaining subtask of Task 2. We are left with one more subtask from Task 1.

Task 5: Build BA **:** SIMPLEBUCKETARRAY **given** N,B

Here we could again give an implementation from scratch. However, an obvious alternative is to simply reuse the generic implementation of arrays from Task 3, even though SIMPLEBUCKETARRAY does not require arrays to satisfy the extra finiteness condition imposed there. Then BA can be implemented using the same pattern as SA in Task 3:

$$
\begin{array}{l}
BA = \textbf{let } BA' = A[B|_{\sigma'}] \\
\quad\quad \textbf{sorts } Array[Bucket] = BA'.Array[Elem] \\
\quad\quad \textbf{include } B \\
\quad\quad \textbf{include } BA'
\end{array}
$$

where $\sigma' = \{Elem \mapsto Bucket\}$.

Note that this decision actually requires the current design to be restructured to make the generic module A available both here, for implementing BA, and in Task 3, for implementing SA. Consequently, a common realisation of NAT must be used in both Task 1 and Task 2.

This gives the following in place of the list of subtasks in Task 1:

- N : NAT
- S : STRING
- SK : STRINGKEY **given** N,S
- A : ELEM $\stackrel{\iota}{\Rightarrow}$ ELEMFINITEARRAY **given** N
- B : SIMPLEBUCKET **given** N,S,A
- BA : SIMPLEBUCKETARRAY **given** N,B,A

Then the subtask

- A : ELEM $\stackrel{\iota}{\Rightarrow}$ ELEMFINITEARRAY **given** N

disappears from Task 3, and the subtasks of Task 2 become

- SK' : STRINGKEY **given** N,S
- SA : STRINGFINITEARRAY **given** N,S,A
- LP : LINEARPROBING **given** SA

Exercise 7.4.3. Revisit the developments affected by this restructuring, and check that all the realisations of subtasks remain valid. □

This completes the development. The resulting development tree is as follows:

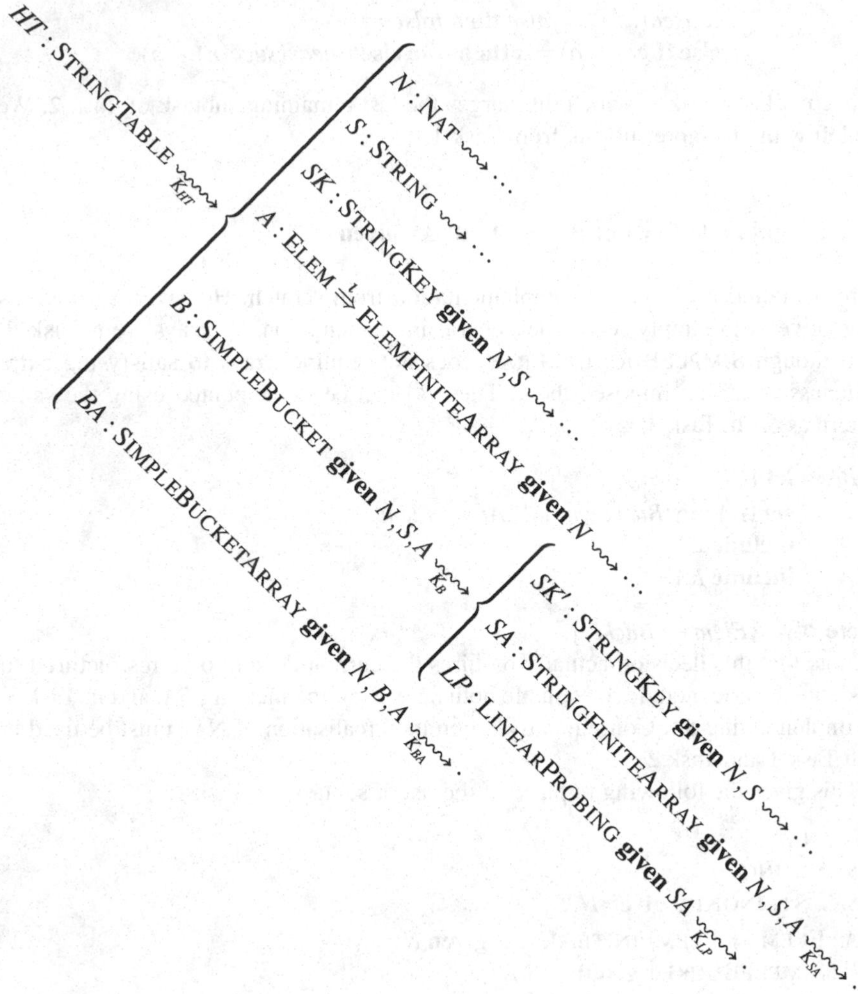

7.5 Bibliographical remarks

Most of the material in this chapter is taken from [ST88b] where simple implementation was called *refinement*. Simple implementation has been used in [SW83], [ST85] and [ST87] among other places. Constructor implementation is related to the notion of implementation in [Ehr81] — although the latter dealt with implementations of algebras rather than of specifications — and [Lip83]. A type-theoretic formulation of [ST88b] is presented in [Luo93].

The question of the "right" definition of implementation received a great deal of attention in the late 1970s and early 1980s, starting with the seminal paper [Hoa72], on which the approach in VDM [Jon80] is based. This corresponds to implementations using the pattern of Example 7.2.9, consisting of a free construction (to define new sorts and operations), followed by a reduct, a restriction to the reachable subalgebra (imposing a so-called "representation invariant"), and finally a quotient (where the natural quotient homomorphism corresponds to the so-called "retrieve function"). A representative sample of relevant papers at the time of this discussion is [GB80], [Ehr81], [Ehr82], [EKMP82], [GM82], [Wan82], [Gan83], [Lip83] and [Ore83], of which [EKMP82] is probably the most influential and well developed. From the perspective of our definition of constructor implementations, all of these approaches can be seen as special cases using specific patterns of constructors, typically (using 1980s terminology) forget-restrict-identify (i.e. a reduct to the signature of the implemented specification, then a restriction to the reachable subalgebra, and finally a quotient) as in [SW82], or forget-identify-restrict, as in [GTW76], or synthesize-restrict-identify, as in [EKMP82]. Vertical composition of two such implementations was required to yield an implementation of the same form, but this is not always possible except under additional assumptions about the specifications involved. Exercises 7.2.6 and 7.2.7 show that this can be achieved for certain patterns, and see [Ehr81] for some transformations between patterns of different forms. In the proof-oriented notion of implementation in [TM87], there is an interesting link between composability and Craig interpolation (see Definition 4.4.21).

In addition to these problems with vertical composability, a problem with each of these early definitions is that it fails to cover certain naturally arising examples, and most approaches are not able to cope with loose specifications. The approach described here is a unifying and generalising one which subsumes most of these early definitions; see the next chapter for a still more general version which subsumes the rest. Problems with the vertical composability of these early notions of implementation can be viewed as an artefact of the unnecessarily inflexible style of definition: the requirement that the composition of generic modules be forced into some given normal form corresponds to requiring programs to be written in a rather restricted programming language that provides no way of defining the composition of two generic modules without syntactically combining their actual code.

Probably the most influential unpublished paper in the algebraic specification literature is [GB80], which introduced the notions of vertical and horizontal composability as fundamental properties to be satisfied by any adequate definition of implementation. It observes that this gives rise to a two-dimensional structure and suggests that different ways of composing implementations should give rise to the same result, as in Exercise 3.4.48. Given the difficulty of achieving vertical composability, most early approaches to implementation did not tackle the related question of horizontal composability. Even [GB80] speculates that this property may not hold for some specification-building operations. Indeed, the horizontal structure of a specification may well be quite different from the modular structure of the final program that implements it; see [FJ90] for a convincing concrete example. Thus, although horizontal composition will sometimes be relevant, it cannot be the only way to implement structured specifications. We need separate means to explicitly mark design decisions that fix the final modular structure of the program under development. Once such a *design specification* [AG97] has been fixed, this top-level horizontal structure is to be preserved in programs resulting from the development process, and further development proceeds for each component specification separately. The final result is then obtained by applying the top-level constructors to the outcomes of these separate developments. Section 7.3 presents foundations for such an approach, with an example in Section 7.4. Explicit notation to support this style of modular development is provided by *architectural specifications* and their refinement in CASL [BST02], [Mos04], [MST04]. See [ST06] for an in-depth discussion of these issues.

So-called "wide-spectrum languages" [BW82a], which can be used to express specifications, programs, and intermediate stages in program development, include CIP-L [BBB$^+$85] and Extended ML [ST85]. One proposal for developing programs from specifications in such a language is by application of transformation rules which replace specification/program fragments by equivalent or more refined fragments [BD77], [PS83]; such rules correspond to pre-packaged implementation schemes where correctness of a given instantiation follows from given enabling conditions.

A programming language that supports stepwise development in the style suggested here needs to provide syntax and modularisation facilities for defining generic modules and their instantiations. One good example of such a language is Standard ML [MTHM97]. As discussed in Section 7.3 and illustrated by the example in Section 7.4, decomposition of a task into subtasks needs to take account of required sharing between the resulting modules. Our treatment here makes these dependencies explicit using higher-order parameterisation; this is sometimes referred to as *sharing by parameterisation* and was advocated in Pebble [LB88]. Unfortunately, this approach soon becomes notationally unwieldy. Standard ML deals with the same issues in a neater way, using so-called *sharing specifications*, where sharing requirements between arguments of a constructor are given explicitly in the form of equations. These can be regarded as a convenient shorthand for sharing via parameterisation, but the full details, especially in combination with higher-order parameterisation, are problematic; see for instance [MT94].

Chapter 8
Behavioural specifications

In this chapter we present a *behavioural* view of specifications and discuss the role of *behavioural equivalence* in software specification and development. In the context of algebraic specification the "behaviour" of a program is fully determined by the results of computations that the program may perform. When a program is modelled as an algebra over an algebraic signature Σ, a computation is described by a Σ-term and its result is the value of the term in the algebra. However, not all such results can be directly observed: values of only certain *observable sorts* may be directly returned to the user. Two Σ-algebras are *behaviourally equivalent* with respect to the observable sorts if any Σ-term of an observable sort has the same value in both algebras.

For instance, consider an algebraic signature Σ with sorts *Nat*, *Bool*, and *Bunch* and operations $empty{:}\,Bunch$, $add{:}\,Nat \times Bunch \to Bunch$ and $\_\_ \in \_\_{:}\,Nat \times Bunch \to Bool$ (as well as the usual operations on *Nat* and *Bool*), and suppose that A and B are Σ-algebras with carriers

$|A|_{Bunch} =$ the set of finite *sets* of natural numbers,

$|B|_{Bunch} =$ the set of finite *lists* of natural numbers,

and with the remaining carriers and operations defined in the expected way. (In particular, B implements $add(n,b)$ by adding the element n to the front of the list b, but it does *not* provide list operations like *hd* and *tl*.) Then A and B are behaviourally equivalent w.r.t. *Bool* as the only observable sort since every term of sort *Bool* has the same value in A as in B — the important terms here are those of the form $n \in add(n_1, \ldots add(n_k, empty) \ldots)$, and these have the same value in A and B. Indeed, from the user's point of view both algebras display identical behaviour if the user is unable to directly observe the actual values of "bunches" constructed in computations — all that can be observed, via the membership operation $\in$, is whether particular elements are contained in such bunches. This is the essential idea behind *data abstraction*: we encapsulate a data type together with operations to be used by a client, hiding the actual data representation and any representation-level operations. Then different representations that exhibit the same behaviour (in our terminology:

are behaviourally equivalent with respect to the unencapsulated sorts) can be used interchangeably.

As we have argued earlier, to adequately capture the user's initial requirements and to provide sufficient information barriers between the modules of a system to be developed, specifications should describe the *whats* of the system in a sufficiently abstract way, without constraining its *hows* at all. Even the above simple example indicates that the initial semantics of equational specifications (cf. Section 2.5), which always yields a single isomorphism class of algebras, is not good enough from this point of view: algebras A and B above are not isomorphic, so no specification can have both of them as initial models. As we will indicate in Section 8.1 below, taking the loose semantics of specifications (cf. Section 2.2) is not always good enough either.

What is needed is an interpretation of specifications which makes them sufficiently abstract, where abstraction means that the system is described only via its observable behaviour. Under such an interpretation, specifications should not distinguish between algebras that are behaviourally equivalent. As explained in Section 8.2 below, this may be ensured by explicitly closing the class of models of a specification under behavioural equivalence. Consequently, *abstract model specifications* are possible, where the required system behaviour is prescribed by giving a specific model (see Section 8.2.2). A viable alternative, presented in Section 8.3, is to relax the interpretation of equality in sentences so that *indistinguishable* values of non-observable sorts are regarded as equal. The behavioural view of specifications leads to a more permissive notion of implementation, as discussed in Section 8.4.

For most of this chapter we will abandon working in an arbitrary institution and instead consider the standard case of algebraic signatures and many-sorted total algebras. We leave open the choice of a particular institution having these signatures and models since sentences and their satisfaction are irrelevant for the developments here except in part of Section 8.3 and in examples. Partial algebras as used in **FPL** (Example 4.1.25) are considered in Section 8.5, with possible generalisations to the framework of an arbitrary institution outlined in Section 8.5.3.

8.1 Motivating example

Let us consider the following rather well-known example, where the standard loose interpretation of a specification turns out to be inadequate.

We start with a specification of stacks of strings. We build on the specification STRING from Section 5.3 and on a specification BOOL of booleans. In the example here and throughout this chapter (except for Section 8.5, where we switch to a different institution) we will assume for presentational convenience that standard realisations of STRING and of BOOL are given, and that all the algebras considered use these to interpret strings, booleans, and their operations in the same way. When we later use the specification NAT, we similarly assume a standard realisation of natural numbers.

spec STRINGSTACK =
 BOOL **and** STRING
 then
  sorts *Stack*
  ops $empty: Stack$
   $isempty: Stack \to Bool$
   $push: String \times Stack \to Stack$
   $top: Stack \to String$
   $pop: Stack \to Stack$
  $\forall x: String, s: Stack$
   • $isempty(empty) = true$
   • $isempty(push(x,s)) = false$
   • $top(push(x,s)) = x$
   • $pop(push(x,s)) = s$

Exercise 8.1.1. Rewrite the above specification of stacks of strings to obtain a specification of a generic module that builds stacks of arbitrary data elements. □

Exercise 8.1.2. STRINGSTACK leaves the result of applying *pop* and *top* to the empty stack totally unconstrained. When necessary, we will adopt in the following the simplest possible approach, defining the operations to yield a standard default value in this case. Try to reinterpret this specification and the rest of the example in this section in the more appropriate framework of partial algebras as introduced in Section 2.7.4; see Section 8.5 below. □

Let $\Sigma SS = Sig[\text{STACKSTRING}]$ be the signature of STACKSTRING. In one "standard" model of STACKSTRING — call it SS — stacks are lists of strings and the operations are defined as follows:

- $empty_{SS}$ is the empty list;
- $isempty_{SS}(s)$ checks whether the list s is empty;
- $push_{SS}(x,s)$ adds x to the front of the list s;
- $top_{SS}(s)$ yields the first element of the list s, and the empty string if the list s is empty; and
- $pop_{SS}(s)$ removes the first element from the list s, and does nothing if s is empty.

We will attempt to implement stacks of strings as specified above using infinite arrays of strings indexed by natural numbers — see STRINGARRAY in Section 5.3 — but with predicates replaced by boolean functions; see Exercise 4.1.16. Let $\Sigma SA = Sig[\text{STRINGARRAY}]$ be the signature of this specification.

Define a constructor $AwP: \Sigma SA \Rightarrow \Sigma SS$ (i.e. a function $AwP: |\mathbf{Alg}(\Sigma SA)| \to |\mathbf{Alg}(\Sigma SS)|$; see Definition 6.1.1), which for any ΣSA-algebra $SA \in |\mathbf{Alg}(\Sigma SA)|$ builds a ΣSS-algebra $AwP(SA) \in |\mathbf{Alg}(\Sigma SS)|$ as follows:

- $|AwP(SA)|_{Stack} = |SA|_{Array[String]} \times |SA|_{Nat}$;
- $empty_{AwP(SA)} = \langle empty_{SA}, 0_{SA} \rangle$;
- $isempty_{AwP(SA)}(\langle a,i \rangle) = true_{SA}$ if $i = 0_{SA}$ and $isempty_{AwP(SA)}(\langle a,i \rangle) = false_{SA}$ otherwise;

- $push_{AwP(SA)}(x, \langle a,i \rangle) = \langle put_{SA}(i,x,a), succ_{SA}(i) \rangle$;
- $pop_{AwP(SA)}(\langle a, succ_{SA}(i) \rangle) = \langle a,i \rangle$ and $pop_{AwP(SA)}(\langle a, 0_{SA} \rangle) = \langle a, 0_{SA} \rangle$; and
- $top_{AwP(SA)}(\langle a, succ_{SA}(i) \rangle) = get_{SA}(i,a)$ and $top_{AwP(SA)}(\langle a, 0_{SA} \rangle) = \varepsilon_{SA}$.

Informally, the constructor *AwP* provides the usual "array-with-pointer" realisation of stacks, where each stack is represented as an array storing in its initial segment the elements on the stack together with a pointer indicating the length of the relevant array segment. In other words, a pair $\langle a,i \rangle$ represents the stack $push(get(i-1,a), \ldots, push(get(0,a), empty) \ldots)$. This is a rather standard realisation of stacks, which works well and is commonly used in practice.

However, consider any model $SA \in Mod[\text{STRINGARRAY}]$. For $a \in |SA|_{Array[String]}$, $x \in |SA|_{String}$ and $i \in |SA|_{Nat}$, $pop_{AwP(SA)}(push_{AwP(SA)}(x, \langle a,i \rangle)) = \langle put_{SA}(i,x,a), i \rangle$, which of course in general is distinct from $\langle a,i \rangle$. Consequently,

$$AwP(SA) \not\models \forall x{:}String, s{:}Stack \bullet pop(push(x,s)) = s,$$

and so, $AwP(SA) \notin Mod[\text{STRINGSTACK}]$. Thus, according to Definition 7.2.1, the attempted constructor implementation of STRINGSTACK by STRINGARRAY via the constructor $AwP{:}\Sigma SA \Rightarrow \Sigma SS$ is incorrect!

The discrepancy between the formal definition of implementation, which renders the implementation of STRINGSTACK by STRINGARRAY via *AwP* incorrect, and practice, which in spite of this formal incorrectness views this implementation as quite satisfactory, has its origin in the way in which the models of STRINGSTACK are to be used. Intuitively, users will never be able to directly observe stack values. They can print out elements on a stack one by one, but not print its underlying representation; similarly, they can check whether two stacks contain the same elements in the same order, but not whether their underlying representations are identical. Therefore, even though the specification STRINGSTACK contains the axiom $pop(push(x,s)) = s$, it is not really relevant for the users whether the axiom holds or not. What is important is that all its *observable consequences* hold (quantification is omitted for simplicity):

$$\begin{aligned}
isempty(\underline{pop(push(x,s))}) &= isempty(\underline{s}) \\
top(\underline{pop(push(x,s))}) &= top(\underline{s}) \\
isempty(pop(\underline{pop(push(x,s))})) &= isempty(pop(\underline{s})) \\
top(pop(\underline{pop(push(x,s))})) &= top(pop(\underline{s})) \\
isempty(pop(pop(\underline{pop(push(x,s))}))) &= isempty(pop(pop(\underline{s}))) \\
top(pop(pop(\underline{pop(push(x,s))}))) &= top(pop(pop(\underline{s}))) \\
&\vdots
\end{aligned}$$

Let $\text{STRINGSTACK}^{\infty}$ be the infinitary specification that results from STRINGSTACK by replacing the single axiom $pop(push(x,s)) = s$ with the set of its observable consequences as listed above. Of course, any model of STRINGSTACK is a model of $\text{STRINGSTACK}^{\infty}$, but not vice versa: there are models of $\text{STRINGSTACK}^{\infty}$, for instance those produced by the constructor *AwP* when applied to models of

STRINGARRAY, that are not models of STRINGSTACK. In practice, the user will not be able to distinguish between the additional models of STRINGSTACK$^\infty$ and models of STRINGSTACK. For instance, all observable results are the same in models produced by the constructor *AwP* when applied to models of STRINGARRAY as in the standard stack model *SS* described above. In a sense, the original specification STRINGSTACK may be viewed as insufficiently abstract: it differentiates between algebras that cannot be distinguished by the user.

One may wonder therefore why the specification of stacks is usually given with the axiom $pop(push(x,s)) = s$, the satisfaction of which we have just argued is not the real issue, rather than directly with the above family of its observable consequences since these are what we really want to hold. There are at least two reasons: first, it is often very convenient to treat any implementation of stacks as if it satisfied this single axiom. We will discuss this in more detail in Section 8.4.2. Second, the above list of observable consequences of this single axiom is infinite, and in first-order logic it cannot be equivalently replaced by a finite presentation.

Exercise 8.1.3. Try to state and prove the theorem that there is no finite first-order presentation with the same class of models as STRINGSTACK$^\infty$. HINT: See [Sch92]. □

We may overcome the latter difficulty by resorting to a more powerful specification formalism. For instance, a finitary specification that is equivalent to STRINGSTACK$^\infty$ may be given in higher-order logic, or even in first-order logic via the use of an auxiliary function from natural numbers to enumerate all the contexts that appear in the list of observable consequences above. However, neither of these specifications would be as clear, simple and intuitively appealing as STRINGSTACK.

In the rest of this chapter we will describe an alternative interpretation of specifications like STRINGSTACK, under which all models of STRINGSTACK$^\infty$ are admitted as acceptable realisations of STRINGSTACK, mitigating in this way the discrepancy between practical usefulness and formal correctness of implementations that is illustrated by this example.

8.2 Behavioural equivalence and abstraction

Given a software system modelled as an algebra over an algebraic signature $\Sigma = \langle S, \Omega \rangle$, the sorts S name the sorts of data that can be constructed and manipulated when the system is used. Some of these sorts are rather special: they classify data that can be directly seen and exactly identified by the user of the system. Typically, these are just built-in elementary data types of the programming language used to code the system, with built-in procedures to input and output the data and with a built-in equality test, which can be used in programs. They will be shared by all systems coded in the given programming language. Integers, characters and boolean values are the most standard examples here. We will call these data, as well as the sorts that classify them, *observable*. Other data cannot be observed so directly. They can be constructed and manipulated using the operations provided by the system, of course, but they can be observed only by extracting their observable components. Stacks and arrays (as discussed in Section 8.1) may be used as examples here. All we can do is to traverse their structure and print one by one the observable elements they contain, or input these observable elements one by one and build a complex data structure out of them. Comparison, if needed, can again only be defined by traversing two data structures and directly comparing the observable data they contain. Consequently, two quite different systems, represented as non-isomorphic algebras, can still display the same behaviour.

These comments concerning non-observable data relate not only to structured data, but to any data type whose representation is hidden from the user, being accessible only by means of a number of pre-specified operations. An example would be stacks of natural numbers where each stack is represented using a single natural number, coding the stack as the product of consecutive primes to the power determined by the consecutive elements on the stack. These natural numbers representing stacks will not be directly observable if we allow the user to apply only the usual stack operations (which deal appropriately with the coded representation) to stacks.

8.2.1 Behavioural equivalence

Consider an algebraic signature $\Sigma = \langle S, \Omega \rangle$ together with a distinguished set of *observable* sorts $OBS \subseteq S$.

In the sequel we will use the following notational convention: given an S-sorted set X, by X_{OBS} we mean the S-sorted set which coincides with X on sorts $s \in OBS$ and is empty on all other sorts:

$$(X_{OBS})_s = \begin{cases} \varnothing & \text{for } s \in S \setminus OBS \\ X_s & \text{for } s \in OBS \end{cases}$$

In particular, for any S-sorted set of variables X, X_{OBS} is a set containing variables of observable sorts only, and given a Σ-algebra A, $|A|_{OBS}$ contains the carriers of A of observable sorts only (and is empty on the other sorts). Then, an S-sorted function $v: X_{OBS} \to |A|$ is determined by its observable components, and will be identified with the same function, $v: X_{OBS} \to |A|_{OBS}$.

Definition 8.2.1 (Compatible algebras). Two Σ-algebras $A, B \in |\mathbf{Alg}(\Sigma)|$ are *OBS-compatible* (or simply, *compatible*, if OBS is clear) if they have the same observable carriers, that is, $|A|_{OBS} = |B|_{OBS}$. □

Following the above motivation, given a Σ-algebra A modelling a system with observable sorts OBS, all the user of this system can observe are the values of terms of sorts in OBS under valuations of variables of the sorts in OBS (which correspond to supplying inputs of these sorts). Consequently, algebras where the corresponding values of all such terms coincide can be viewed as equivalent.

Definition 8.2.2 (Behavioural equivalence). *OBS*-compatible Σ-algebras $A, B \in |\mathbf{Alg}(\Sigma)|$ are *behaviourally equivalent w.r.t. OBS*, written $A \equiv_{OBS} B$ (or simply, *behaviourally equivalent*, written $A \equiv B$, if *OBS* is clear), if for every set of variables X_{OBS} (of observable sorts), term $t \in |T_\Sigma(X_{OBS})|_{OBS}$ of observable sort, and valuation $v: X_{OBS} \to |A|_{OBS}$ (so that $v: X_{OBS} \to |B|_{OBS}$ as well), $t_A(v) = t_B(v)$. □

Exercise 8.2.3. Check that the behavioural equivalence is indeed an equivalence relation on any class of Σ-algebras. □

Example 8.2.4. Recall the specifications and constructions from Section 8.1. For any $SA \in Mod[\text{STRINGARRAY}]$, SS and $AwP(SA)$ are behaviourally equivalent with respect to the observable sorts $\{String, Bool\}$. This holds even though they are not isomorphic, and the "internal" data of sort *Stack* are represented quite differently in the two algebras.

To see that $AwP(SA) \equiv_{\{String,Bool\}} SS$, consider the relation $\rho \subseteq |AwP(SA)| \times |SS|$ defined as the identity on (the carriers of sorts) *String* and *Bool*, and on *Stack* given by

$$\langle a, \overline{n}_{SA} \rangle \; \rho_{Stack} \; [get_{SA}(\overline{n}_{SA}, a), get_{SA}(\overline{n-1}_{SA}, a), \ldots, get_{SA}(\overline{0}_{SA}, a)]$$

where for any natural number n, $\overline{n}_{SA} = \underbrace{succ_{SA}(\ldots succ_{SA}}_{n \text{ times}}(0_{SA})\ldots)$. Now, for any set of variables $X_{\{String,Bool\}}$ and valuation $v: X_{\{String,Bool\}} \to |SS|_{\{String,Bool\}}$, by a simple induction on the structure of a term $t \in |T_{\Sigma SS}(X_{\{String,Bool\}})|$, we can show that $t_{AwP(SA)}(v) \; \rho \; t_{SS}(v)$; this implies that $t_{AwP(SA)}(v) = t_{SS}(v)$ for terms $t \in |T_{\Sigma SS}(X_{\{String,Bool\}})|_{\{String,Bool\}}$ of observable sorts. □

The technique used in the above example to show behavioural equivalence of the two models *SS* and $AwP(SA)$ works in the general case as well.

Definition 8.2.5 (Correspondence). Given two *OBS*-compatible Σ-algebras $A, B \in |\mathbf{Alg}(\Sigma)|$, an *OBS-correspondence* (or simply, a *correspondence*, if *OBS* is clear) between A and B, written $\rho: A \bowtie_{OBS} B$, is a relation $\rho \subseteq |A| \times |B|$ such that:

- ρ is the identity on observable carriers, that is, for each observable sort $o \in OBS$, $\rho_o = id_{|A|_o} = id_{|B|_o}$; and
- ρ is closed under the operations, that is, for each operation $f: s_1 \times \cdots \times s_n \to s$ in Σ and elements $a_1 \in |A|_{s_1}, \ldots, a_n \in |A|_{s_n}$ and $b_1 \in |B|_{s_1}, \ldots, b_n \in |B|_{s_n}$, if $a_1 \; \rho_{s_1} \; b_1, \ldots, a_n \; \rho_{s_n} \; b_n$ then also $f_A(a_1, \ldots, a_n) \; \rho_s \; f_B(b_1, \ldots, b_n)$. □

Exercise 8.2.6. Check that the identity relation on the carrier of a Σ-algebra is an *OBS*-correspondence, that the inverse of an *OBS*-correspondence is an *OBS*-correspondence, and that the composition of two *OBS*-correspondences is an *OBS*-correspondence. □

Exercise 8.2.7. Consider an algebraic signature morphism $\sigma: \Sigma \to \Sigma'$, with distinguished sets *OBS* and *OBS'* of observable sorts in Σ and Σ', respectively, such that $\sigma(OBS) \subseteq OBS'$. Show that the reduct $\_|_\sigma$ of an *OBS'*-correspondence (defined in the obvious way) is an *OBS*-correspondence. □

Theorem 8.2.8. *Two Σ-algebras $A, B \in |\mathbf{Alg}(\Sigma)|$ are behaviourally equivalent w.r.t. OBS if and only if there exists an OBS-correspondence between A and B.*

Proof.
($\Rightarrow$): Consider the relation $\rho \subseteq |A| \times |B|$ given by

$$\rho = \{\langle t_A(v), t_B(v)\rangle \mid t \in |T_\Sigma(X_{OBS})|, X_{OBS} \text{ is a set of variables}, v{:}X_{OBS} \to |A|_{OBS}\}.$$

It is easy to check that ρ is closed under the operations (directly from its definition) and that it is the identity on observable sorts (since $A \equiv_{OBS} B$).
($\Leftarrow$): Let $\rho{:}A \bowtie_{OBS} B$ be an *OBS*-correspondence. For any set of variables X_{OBS} and valuation $v{:}X_{OBS} \to |A|_{OBS}$, by an easy induction on the structure of a term $t \in |T_\Sigma(X_{OBS})|$ we can show that $t_A(v)\ \rho\ t_B(v)$, which implies $t_A(v) = t_B(v)$ for terms of observable sorts. □

A special case of a correspondence between compatible Σ-algebras is a Σ-homomorphism between them that is the identity on observable carriers. Such a homomorphism may even be partial, that is, defined only on a subalgebra.

Definition 8.2.9 (Behavioural subalgebra and homomorphism). A subalgebra A' of a Σ-algebra $A \in |\mathbf{Alg}(\Sigma)|$ is *behavioural* (*w.r.t. OBS*) if A' and A are *OBS*-compatible, that is, $|A'|_{OBS} = |A|_{OBS}$. A' is *OBS-generated* if it is the least behavioural subalgebra of A (or equivalently, of itself). The *OBS-generated subalgebra* of a Σ-algebra $A \in |\mathbf{Alg}(\Sigma)|$ will be denoted by $\langle A\rangle_{OBS}$.

A Σ-homomorphism $h{:}A \to B$ is *behavioural* (*w.r.t. OBS*) if it is the identity on carriers of observable sorts (it follows that A and B are *OBS*-compatible then). □

Corollary 8.2.10. *For any $A, B \in |\mathbf{Alg}(\Sigma)|$, if there exists a behavioural homomorphism $h{:}A' \to B$ where A' is a behavioural subalgebra of A, then A and B are behaviourally equivalent.*

Proof. Easy by Theorem 8.2.8, since h is a correspondence between A and B. □

The above corollary describes perhaps the most typical case of behaviourally equivalent algebras, corresponding to what is sometimes referred to as *data refinement*, where B is an "abstract" data model, and A provides a more concrete data representation; then A' is given by a *representation invariant* and each "concrete" data value $a \in |A'|$ represents $h(a) \in |B|$. Of course, any abstract data value $b \in |B|$ may have more than one concrete representation in $|A|$. Moreover, every data value in the *OBS*-generated subalgebra of B has at least one such concrete representation, and every data value in $|B|_{OBS}$ is represented by itself only.

Example 8.2.11. Recall Example 8.2.4. The correspondence ρ between $AwP(SA)$ and SS given there is in fact a behavioural homomorphism $\rho{:}AwP(SA) \to SS$. If we used integers rather than natural numbers as pointers to the top of arrays in the array-with-pointer representation of stacks, the requirement that the pointer be non-negative would provide an appropriate representation invariant to identify a behavioural subalgebra of the so-modified $AwP(SA)$ representing the standard stack model SS. □

Exercise 8.2.12. Recall Example 7.2.9. Give a behavioural homomorphism that captures the restrict and quotient steps in that data refinement example. □

Of course, not all correspondences are of such a simple form, and there exist behaviourally equivalent algebras with no behavioural homomorphisms between them.

Example 8.2.13. We rely again on specifications and constructions from Section 8.1. We modify the standard stack model *SS* by recording whether or not a *pop* operation was performed on an empty stack in the course of constructing each stack value, defining $SS^{err} \in Mod[\mathrm{STRINGSTACK}]$ as follows:

- $|SS^{err}|_{Stack}$ is the set of all pairs $\langle s, err\rangle$, where s is a list of strings and err is a boolean value (either tt or $f\!f$);
- all stack operations in SS^{err} act on the first component of such "stacks" as they do in the standard stack model *SS*, ignoring the *err* component; and
- the second component of the result of the stack operations is determined as follows:
  - $empty_{SS^{err}}$ sets err to $f\!f$;
  - $push_{SS^{err}}(\langle s, err\rangle)$ leaves err unchanged; and
  - $pop_{SS^{err}}(\langle s, err\rangle)$ sets err to tt if s is empty, and leaves it unchanged otherwise.

(**Exercise:** Check that indeed $SS^{err} \in Mod[\mathrm{STRINGSTACK}]$.)

For any $SA \in Mod[\mathrm{STRINGARRAY}]$, $AwP(SA)$ and SS^{err} are behaviourally equivalent w.r.t. the observable sorts $\{String, Bool\}$. To show this, a correspondence $\rho^{err} \subseteq |AwP(SA)| \times |SS^{err}|$ between $AwP(SA)$ and SS^{err} may be given by slightly modifying the definition from Example 8.2.4:

$$\langle a, \overline{n}_{SA}\rangle \ \rho^{err}_{Stack} \ \langle [get_{SA}(\overline{n}_{SA}, a), get_{SA}(\overline{n-1}_{SA}, a), \ldots, get_{SA}(\overline{0}_{SA}, a)], err\rangle$$

for $err \in \{tt, f\!f\}$ (and $\overline{n}_{SA}$ as before). The fact that there is no operation that extracts the "error flag" from a stack is essential here; ρ^{err} would not be closed under such an operation.

However, no behavioural homomorphism between (any subalgebras of) $AwP(SA)$ and SS^{err} exists. We have $AwP(SA) \models pop(empty) = empty$ and so there is no homomorphism from $AwP(SA)$ to SS^{err} since $SS^{err} \not\models pop(empty) = empty$. We also have $SS^{err} \models pop(push(t, empty)) = empty$ for all terms t of sort *String*, and so there is no homomorphism in the opposite direction since for some terms t of sort *String*, $AwP(SA) \not\models pop(push(t, empty)) = empty$. □

Even though there is no behavioural homomorphism between $AwP(SA)$ and SS^{err} in the above example, it is easy to see that there are behavioural homomorphisms from each of these algebras to the standard stack model *SS*. It turns out that this is an instance of a general fact:

Theorem 8.2.14. *Two Σ-algebras $A, B \in |\mathbf{Alg}(\Sigma)|$ are behaviourally equivalent w.r.t. OBS if and only if there exist behavioural subalgebras A' of A and B' of B, a Σ-algebra $C \in |\mathbf{Alg}(\Sigma)|$ and behavioural homomorphisms $h_A : A' \to C$ and $h_B : B' \to C$.*

$$A \hookleftarrow A' \xrightarrow{h_A} C \xleftarrow{h_B} B' \hookrightarrow B$$

Proof. The "if" part easily follows from Corollary 8.2.10 by Exercise 8.2.3.

For the "only if" part, consider a correspondence $\rho : A \bowtie_{OBS} B$ (such a correspondence exists by Theorem 8.2.8). Define a Σ-algebra C as follows:

- $|C|_{OBS} = |A|_{OBS}$ $(= |B|_{OBS})$;
- for each sort $s \in S \setminus OBS$, $|C|_s = \{[\langle a,b\rangle]_{\simeq_s} \mid a \rho_s b\}$, where $\simeq_s \subseteq \rho_s \times \rho_s$ is the least equivalence relation on the set of pairs ρ_s such that $\langle a,b\rangle \simeq_s \langle a',b\rangle$ and $\langle a,b\rangle \simeq_s \langle a,b'\rangle$ for all $\langle a,b\rangle, \langle a',b\rangle, \langle a,b'\rangle \in \rho_s$;
- for each operation $f : s_1 \times \cdots \times s_n \to s$ in Σ, for $p_1 \in |C|_{s_1}, \ldots, p_n \in |C|_{s_n}$, let $a_i = b_i = p_i$ if $s_i \in OBS$ and $p_i = [\langle a_i, b_i\rangle]_{\simeq_{s_i}}$ otherwise, $i = 1, \ldots, n$. Put $f_C(p_1, \ldots, p_n) = p$, where $p = f_A(a_1, \ldots, a_n) = f_B(b_1, \ldots, b_n)$ if $s \in OBS$ and $p = [\langle f_A(a_1, \ldots, a_n), f_B(b_1, \ldots, b_n)\rangle]_{\simeq_s}$ otherwise. (**Exercise:** Check that f_C is well defined.)

Note now that since ρ is closed under the operations, so are its domain and codomain (as subsets of $|A|$ and $|B|$, respectively). Therefore, there are subalgebras A' of A and B' of B such that $|A'| = \{a \in |A| \mid a \rho b \text{ for some } b \in |B|\}$, and $|B'| = \{b \in |B| \mid a \rho b \text{ for some } a \in |A|\}$. Moreover, since ρ is the identity on observable carriers, A' and B' are behavioural subalgebras of A and B, respectively.

Finally, for each $s \in S \setminus OBS$ define $|h_A|_s : |A'|_s \to |C|_s$ and $|h_B|_s : |B'|_s \to |C|_s$ as follows:

- for $a \in |A'|_s$, $|h_A|_s(a) = [\langle a,b\rangle]_{\simeq_s}$ for any $b \in |B|$ such that $a \rho_s b$, and
- for $b \in |B'|_s$, $|h_B|_s(b) = [\langle a,b\rangle]_{\simeq_s}$ for any $a \in |A|$ such that $a \rho_s b$.

It is easy to check now that for $s \in S \setminus OBS$, $|h_A|_s$ and $|h_B|_s$ are well defined and together with the identities on observable carriers form behavioural homomorphisms $h_A : A' \to C$ and $h_B : B' \to C$. □

Exercise 8.2.15. A dual characterisation of behavioural equivalence follows from the fact that any correspondence may be presented as a "span" of two behavioural homomorphisms: two Σ-algebras $A, B \in |\mathbf{Alg}(\Sigma)|$ are behaviourally equivalent w.r.t. OBS if and only if there exists a Σ-algebra C with behavioural homomorphisms $h_A : C \to A$ and $h_B : C \to B$. The "if" direction follows by Corollary 8.2.10. Complete the following proof sketch of the "only if" direction: for behaviourally equivalent $A, B \in |\mathbf{Alg}(\Sigma)|$, by Theorem 8.2.8 there is a correspondence $\rho : A \bowtie_{OBS} B$; define C to have the set of pairs ρ as its carrier and its operations defined componentwise (for sorts $s \in OBS$, identify pairs $\langle a,a\rangle \in \rho_s$ with a), and h_A and h_B to be the projections.

Remark: Note that $\rho = (h_A)^{-1}; h_B$. □

Exercise 8.2.16. We have presented the notion of behavioural equivalence with respect to an arbitrary but fixed set of observable sorts. Investigate how behavioural equivalence varies when the set of observable sorts changes. Given an algebraic signature $\Sigma = \langle S, \Omega\rangle$, characterise the behavioural equivalences $\equiv_\varnothing$, where no sorts are observable, and $\equiv_S$, where all sorts are observable. Show that if $OBS \subseteq OBS' \subseteq S$ then $\equiv_{OBS'} \subseteq \equiv_{OBS}$. Show also that for $OBS_1 \subseteq S$ and $OBS_2 \subseteq S$, $\equiv_{OBS_1 \cup OBS_2} \subseteq \equiv_{OBS_1} \cap \equiv_{OBS_2}$ but in general the inclusion may be proper. □

Exercise 8.2.17. The definition of behavioural equivalence above tacitly assumes that the observable computations take inputs from and produce outputs in the same set *OBS* of observable sorts. It is possible to use two different sets of sorts instead, say, *IN* and *OUT*, respectively. Define behavioural equivalence w.r.t. IN, OUT. Define (IN, OUT)-correspondences and prove a counterpart of Theorem 8.2.8. Make sure that when $IN = OUT = OBS$, the definitions coincide with those above. HINT: An (IN, OUT)-correspondence should include the identity on *IN*-carriers and be included in the identity on *OUT*-carriers.

Spell this out for the cases $IN = \varnothing$ (no inputs allowed) and $IN = S$ (arbitrary inputs allowed). □

8.2.2 Behavioural abstraction

We have argued earlier that it is desirable that specifications not distinguish between algebras that cannot be distinguished by the user. The notion of behavioural equivalence was introduced to capture what it means exactly for two algebras to be indistinguishable from the user's point of view. Unfortunately, typical specifications that can be built using specification techniques provided so far (Chapter 5) determine classes of models that need not be closed under behavioural equivalence, as shown by the example of STRINGSTACK in Section 8.1.

To alleviate this problem, we provide a new specification-building operation that closes the class of models of a specification under behavioural equivalence. This amounts to abstracting away from the details of the specifications that cannot be observed by the user of the specified system.

As before, let $\Sigma = \langle S, \Omega \rangle$ be an algebraic signature with a distinguished set of observable sorts $OBS \subseteq S$.

Definition 8.2.18 (Behavioural closure and abstraction). For any class $\mathscr{A} \subseteq |\mathbf{Alg}(\Sigma)|$ of Σ-algebras, its *behavioural closure* (*w.r.t. OBS*) is

$$Abs_{OBS}(\mathscr{A}) = \{B \in |\mathbf{Alg}(\Sigma)| \mid B \equiv_{OBS} A \text{ for some } A \in \mathscr{A}\}.$$

If *SP* is a Σ-specification then **abstract** *SP* **wrt** *OBS* is a specification with the following semantics:

$$Sig[\textbf{abstract } SP \textbf{ wrt } OBS] = \Sigma$$
$$Mod[\textbf{abstract } SP \textbf{ wrt } OBS] = Abs_{OBS}(Mod[SP])$$ □

A few simple facts follow immediately from the definition; cf. Exercise 5.1.8:

Proposition 8.2.19. *Let* $\mathscr{A}, \mathscr{A}' \subseteq |\mathbf{Alg}(\Sigma)|$ *be classes of* Σ*-algebras.*

1. *Behavioural closure preserves the original models:* $\mathscr{A} \subseteq Abs_{OBS}(\mathscr{A})$.
2. *Behavioural closure is monotone: if* $\mathscr{A} \subseteq \mathscr{A}'$ *then* $Abs_{OBS}(\mathscr{A}) \subseteq Abs_{OBS}(\mathscr{A}')$.
3. *Behavioural closure is idempotent:* $Abs_{OBS}(Abs_{OBS}(\mathscr{A})) = Abs_{OBS}(\mathscr{A})$.

4. Behavioural closure preserves and reflects consistency: $Abs_{OBS}(\mathscr{A}) \neq \varnothing$ *if and only if* $\mathscr{A} \neq \varnothing$. □

Exercise 8.2.20. Derive from Proposition 8.2.19 the analogous properties of specifications built using behavioural abstraction. □

Example 8.2.21. When presenting the example specification STRINGSTACK in Section 8.1 we have argued that what we really want is a specification of stacks that prescribes only the required behaviour. We would like to disregard properties like $\forall x{:}String, s{:}Stack \bullet pop(push(x,s)) = s$ that cannot be directly observed, as long as the observable behaviour implied by them is guaranteed. Using the operation of behavioural abstraction, we can now define a specification that captures exactly what we want:

$$\textbf{spec}\ \textsc{StringStackBeh} = \textbf{abstract}\ \textsc{StringStack}\ \textbf{wrt}\ \{Bool, String\}.$$

We can now explain why the array-with-pointer realisation of stacks, as encoded in the constructor $AwP{:}\Sigma SA \Rightarrow \Sigma SS$ (see Section 8.1), provided a perfectly satisfactory implementation even though the attempted constructor implementation of STRINGSTACK by STRINGARRAY via AwP is not correct. Namely, the point is that the constructor $AwP{:}\Sigma SA \Rightarrow \Sigma SS$ provides a correct constructor implementation of STRINGSTACKBEH by STRINGARRAY:

$$\textsc{StringStackBeh} \underset{AwP}{\leadsto} \textsc{StringArray}.$$

The role of behavioural abstraction in implementation and development will be further discussed in Section 8.4. □

Behavioural abstraction can be used to express a rather different specification technique than those based on axiomatic presentations and specification-building operations to manipulate them. *Abstract model specifications* describe the required behaviour of a system by explicitly defining a particular reference model — an algebra — which displays the desired behaviour.[1] This is not meant to suggest that the final system will be built in the same way as the reference model — any algebra which is behaviourally equivalent to that model is accepted. We can capture this by a combination of behavioural abstraction and singleton specifications from Definition 6.2.7: for any Σ-algebra $M \in |\mathbf{Alg}(\Sigma)|$, **abstract** $\{M\}$ **wrt** OBS defines the class of models that are behaviourally equivalent to M.

Example 8.2.22. Instead of attempting to specify stacks of strings axiomatically, as in STRINGSTACK, we could simply give a particular model of stacks of strings. For instance, we could just define the standard stack model $SS \in |\mathbf{Alg}(\Sigma SS)|$, and then put

$$\textbf{spec}\ \textsc{StandardStringStack} = \textbf{abstract}\ \{SS\}\ \textbf{wrt}\ \{Bool, String\}.$$

[1] The word "abstract" refers to the fact that rather high-level constructions that are not found in most programming languages may be used to build the reference model.

However, the resulting specification is more restrictive than STRINGSTACKBEH, since it fixes the result of applying the *top* and *pop* operations to the empty stack — which is left unconstrained by STRINGSTACKBEH. □

Exercise 8.2.23. Following Exercise 8.2.16, for any class $\mathscr{A} \subseteq |\mathbf{Alg}(\Sigma)|$ of Σ-algebras, characterise $Abs_{\varnothing}(\mathscr{A})$ and $Abs_S(\mathscr{A})$, and prove that if $OBS \subseteq OBS' \subseteq S$ then $Abs_{OBS'}(\mathscr{A}) \subseteq Abs_{OBS}(\mathscr{A})$. State consequences of these facts for behavioural abstraction. □

8.2.3 Weak behavioural equivalence

In the above we discussed behavioural equivalence between compatible algebras, which share the interpretation of observable sorts. Even though there are good arguments for this — typically, observable sorts are built-in data types of the programming language used to code the programs — one may feel that it is overly restrictive for use in connection with loose specifications. In this subsection we briefly generalise the previous developments to the case where observable carriers may be distinct, and provide results which in fact indicate that such a generalisation may not be needed.

As before, let $\Sigma = \langle S, \Omega \rangle$ be an algebraic signature with a distinguished set of observable sorts $OBS \subseteq S$.

Definition 8.2.24 (Weak behavioural equivalence). Two Σ-algebras $A, B \in |\mathbf{Alg}(\Sigma)|$ are *weakly behaviourally equivalent w.r.t. OBS*, written $A \overset{w}{\equiv}_{OBS} B$ (or simply, *weakly behaviourally equivalent*, written $A \overset{w}{\equiv} B$, if OBS is clear), if there exist a set of variables X_{OBS} (of observable sorts) and two surjective valuations $v_A{:}X_{OBS} \to |A|_{OBS}$ and $v_B{:}X_{OBS} \to |B|_{OBS}$ such that for each pair of terms $t, t' \in |T_\Sigma(X_{OBS})|_o$ of a common observable sort $o \in OBS$, $t_A(v_A) = t'_A(v_A)$ if and only if $t_B(v_B) = t'_B(v_B)$. □

Exercise 8.2.25. Check that weak behavioural equivalence is an equivalence on $|\mathbf{Alg}(\Sigma)|$. □

Definition 8.2.26 (Weak correspondence). Given two Σ-algebras $A, B \in |\mathbf{Alg}(\Sigma)|$, a *weak OBS-correspondence* (or simply, a *weak correspondence*, if OBS is clear) between A and B, written $\rho{:}A \overset{w}{\bowtie}_{OBS} B$, is a relation $\rho \subseteq |A| \times |B|$ such that:

- ρ is bijective on observable carriers, that is, for each observable sort $o \in OBS$, $\rho_o \subseteq |A|_o \times |B|_o$ is a bijection; and
- ρ is closed under the operations, that is, for each operation $f{:}s_1 \times \cdots \times s_n \to s$ in Σ and elements $a_1 \in |A|_{s_1}, \ldots, a_n \in |A|_{s_n}$ and $b_1 \in |B|_{s_1}, \ldots, b_n \in |B|_{s_n}$, if $a_1 \mathrel{\rho_{s_1}} b_1, \ldots, a_n \mathrel{\rho_{s_n}} b_n$ then also $f_A(a_1, \ldots, a_n) \mathrel{\rho_s} f_B(b_1, \ldots, b_n)$. □

Exercise 8.2.27. Check that isomorphisms, inverses of weak correspondences and compositions of weak correspondences are weak correspondences. □

Theorem 8.2.28. *Two Σ-algebras $A,B \in |\mathbf{Alg}(\Sigma)|$ are weakly behaviourally equivalent w.r.t. OBS if and only if there exists a weak OBS-correspondence between A and B.*

Proof.

($\Rightarrow$): Let X_{OBS}, $v_A: X_{OBS} \to |A|_{OBS}$ and $v_B: X_{OBS} \to |B|_{OBS}$ be as in Definition 8.2.24 of weak behavioural equivalence. Consider a relation $\rho \subseteq |A| \times |B|$ given by

$$\rho = \{\langle t_A(v_A), t_B(v_B)\rangle \mid t \in |T_\Sigma(X_{OBS})|\}.$$

It is easy to check that ρ is closed under the operations (directly from its definition). Moreover, for each observable sort $o \in OBS$, $\rho_o \subseteq |A|_o \times |B|_o$ is a bijection: first, since $v_A: X_{OBS} \to |A|_{OBS}$ is surjective, ρ_o is total on $|A|_o$. Then suppose that for some $a \in |A|_o$ and $b, b' \in |B|_o$, $a\ \rho_o\ b$ and $a\ \rho_o\ b'$, that is, for terms $t, t' \in |T_\Sigma(X_{OBS})|_o$, $a = t_A(v_A)$ and $b = t_B(v_B)$, and $a = t'_A(v_A)$ and $b' = t'_B(v_B)$. But then $b = b'$, since $t_A(v_A) = t'_A(v_A)$ implies $t_B(v_B) = t'_B(v_B)$. Hence, ρ_o is a total function from $|A|_o$ to $|B|_o$. By symmetry, ρ_o^{-1} is a total function from $|B|_o$ to $|A|_o$, and it follows that ρ_o is indeed a bijection.

($\Leftarrow$): Let $\rho: A \overset{w}{\bowtie}_{OBS} B$ be a weak *OBS*-correspondence. Let $X_{OBS} = |A|_{OBS}$ and let the valuation $v_A: X_{OBS} \to |A|_{OBS}$ be the identity and the valuation $v_B: X_{OBS} \to |B|_{OBS}$ be ρ_{OBS}. By an easy induction on the structure of a term $t \in |T_\Sigma(X_{OBS})|$, we can show that $t_A(v_A)\ \rho\ t_B(v_B)$. Therefore, for any observable sort $o \in OBS$, since ρ_o is a bijection, for any terms $t', t'' \in |T_\Sigma(X_{OBS})|_o$, $t'_A(v_A) = t''_A(v_A)$ if and only if $t'_B(v_B) = t''_B(v_B)$. □

Corollary 8.2.29. *Behavioural equivalence of algebras implies their weak equivalence: for any Σ-algebras $A,B \in |\mathbf{Alg}(\Sigma)|$, if $A \equiv_{OBS} B$ then $A \overset{w}{\equiv}_{OBS} B$.*

Proof. Follows by Theorems 8.2.8 and 8.2.28, since any correspondence is a weak correspondence. □

Lemma 8.2.30. *For any Σ-algebras $A,B \in |\mathbf{Alg}(\Sigma)|$, if $A \overset{w}{\equiv}_{OBS} B$ then there exists a Σ-algebra $C \in |\mathbf{Alg}(\Sigma)|$ that is isomorphic to A and OBS-compatible with B such that $C \equiv_{OBS} B$.*

Proof. Assume $A \overset{w}{\equiv}_{OBS} B$ and let $\rho: A \overset{w}{\bowtie}_{OBS} B$ be a weak correspondence, which exists by Theorem 8.2.28. Define a Σ-algebra $C \in |\mathbf{Alg}(\Sigma)|$ as follows:

- for each sort $s \in S$, $|C|_s = |B|_s$ if $s \in OBS$ and $|C|_s = |A|_s$ otherwise; and
- for each operation $f: s_1 \times \cdots \times s_n \to s$ in Σ and elements $a_1 \in |A|_{s_1}, \ldots, a_n \in |A|_{s_n}$, put $f_C(|\iota|_{s_1}(a_1), \ldots, |\iota|_{s_n}(a_n)) = |\iota|_s(f_A(a_1, \ldots, a_n))$, where the function $|\iota|: |A| \to |C|$ is a bijection such that for each sort $s \in S$, $|\iota|_s = \rho_s$ if $s \in OBS$ and $|\iota|_s = id_{|A|_s}$ otherwise.

Then $\iota: A \to C$ is a Σ-isomorphism and the composition $\iota^{-1};\rho \subseteq |C| \times |B|$ is a weak correspondence (by Exercise 8.2.27) which is the identity on observable carriers (by the construction), and so it is a correspondence between *OBS*-compatible algebras. Thus, $C \equiv_{OBS} B$ by Theorem 8.2.8, and so indeed A is isomorphic to an algebra that is behaviourally equivalent to B. □

Corollary 8.2.31. *Two Σ-algebras are weakly behaviourally equivalent if and only if they are isomorphic to behaviourally equivalent Σ-algebras.*

Proof. The "if" part easily follows from Theorems 8.2.8 and 8.2.28, by Exercise 8.2.27; Lemma 8.2.30 directly implies the "only if" part. □

The above lemma and corollary seem to be of crucial importance: informally, in any context where algebras are only identified up to isomorphism, working with behavioural equivalence is the same as working with weak behavioural equivalence.

Corollary 8.2.32. *Two Σ-algebras $A, B \in |\mathbf{Alg}(\Sigma)|$ are weakly behaviourally equivalent w.r.t. OBS if and only if there exist behavioural subalgebras A' of A and B' of B, a Σ-algebra $C \in |\mathbf{Alg}(\Sigma)|$ and homomorphisms $h_A: A' \to C$ and $h_B: B' \to C$ that are bijective on carriers of observable sorts.*

Proof. Follows from Corollary 8.2.31 and Theorem 8.2.14. □

An alternative form of behavioural abstraction can be based on weak behavioural equivalence:

Definition 8.2.33 (Weak behavioural closure and abstraction). For any class $\mathscr{A} \subseteq |\mathbf{Alg}(\Sigma)|$ of Σ-algebras, its *weak behavioural closure* (*w.r.t. OBS*) is

$$WAbs_{OBS}(\mathscr{A}) = \{B \in |\mathbf{Alg}(\Sigma)| \mid B \overset{w}{\equiv}_{OBS} A \text{ for some } A \in \mathscr{A}\}.$$

If *SP* is a Σ-specification then **weak abstract** *SP* **wrt** *OBS* is a specification with the following semantics:

$$Sig[\textbf{weak abstract } SP \textbf{ wrt } OBS] = \Sigma$$
$$Mod[\textbf{weak abstract } SP \textbf{ wrt } OBS] = WAbs_{OBS}(Mod[SP])$$ □

Corollary 8.2.34. *For any class $\mathscr{A} \subseteq |\mathbf{Alg}(\Sigma)|$ of Σ-algebras, if $\mathscr{A}$ is closed under isomorphism then $Abs_{OBS}(\mathscr{A}) = WAbs_{OBS}(\mathscr{A})$.*

In particular, if the class Mod[SP] of models of a Σ-specification SP is closed under isomorphism then

$$Mod[\textbf{abstract } SP \textbf{ wrt } OBS] = Mod[\textbf{weak abstract } SP \textbf{ wrt } OBS].$$

Proof. The inclusion "$\subseteq$" always holds by Corollary 8.2.29. Under the assumption that $\mathscr{A}$ (respectively, *Mod*[*SP*]) is closed under isomorphism, Lemma 8.2.30 directly implies the inclusion "$\supseteq$". □

Exercise 8.2.35. Study Section 8.4 below and redo it, replacing behavioural equivalence, closure and abstraction by weak behavioural equivalence, closure and abstraction, respectively. Are there any important changes? HINT: Feel free to assume that classes of models of all the specifications involved are closed under isomorphism and that constructors preserve isomorphisms. But also, feel obliged to check that this is indeed the case for the specifications and constructions used in examples there. □

Exercise 8.2.36. Building on Exercise 8.2.17, define a weak version of behavioural equivalence and of correspondence w.r.t. possibly different sets *IN* and *OUT* of input and output sorts, respectively. Adapt the results in this section for these notions. □

8.3 Behavioural satisfaction

In this section we will discuss an alternative view of behavioural abstraction. The overall idea is that instead of introducing a behavioural equivalence between algebras, considering the usual semantics of specifications and then closing the class of admissible models of a specification under the behavioural equivalence, we may want to attempt to describe the resulting class of models more directly. However, as indicated in Section 8.1, this cannot typically be done within the usual institutions.

Recall again the specification STRINGSTACK from Section 8.1. One way of looking at the problematic equation $\forall x{:}String, s{:}Stack \bullet pop(push(x,s)) = s$ is that we do not really want to require the values of $pop(push(x,s))$ and s to be identical, but only to be *indistinguishable* in the models of the specification. We will formally introduce such an indistinguishability relation between algebra elements and use it to define a new notion of satisfaction of formulae in algebras.

As before, let $\Sigma = \langle S, \Omega \rangle$ be an algebraic signature with a distinguished set of observable sorts $OBS \subseteq S$.

The indistinguishability relation should surely be compatible with the operations of the algebra: applying operations to indistinguishable values yields indistinguishable results. Moreover, it should identify as many elements as possible, without identifying directly observable data.

Definition 8.3.1 (Behavioural congruence). A congruence $\simeq \subseteq |A| \times |A|$ on a Σ-algebra $A \in |\mathbf{Alg}(\Sigma)|$ is *behavioural* (*w.r.t. OBS*) if $\simeq_{OBS}$ is the identity. □

Exercise 8.3.2. Show that any relation $\simeq \subseteq |A| \times |A|$ on a Σ-algebra $A \in |\mathbf{Alg}(\Sigma)|$ is a behavioural congruence w.r.t. *OBS* iff it is a reflexive *OBS*-correspondence $\simeq : A \bowtie_{OBS} A$. □

Proposition 8.3.3. *For any Σ-algebra $A \in |\mathbf{Alg}(\Sigma)|$, there is a largest behavioural congruence on A.*

Proof. There is at least one behavioural congruence on A: the identity relation. Consider the family of all behavioural congruences on A. Let $\sim \subseteq |A| \times |A|$ be the transitive closure of the union of this family. Then $\sim$ is a congruence on A and it is the identity on the observable carriers. So, $\sim$ is a behavioural congruence on A, and it is the largest such congruence on A. □

The above proposition provides a good candidate for the indistinguishability relation in an algebra. However, the simple proof given does not really reflect the intuition underlying this concept, which should relate directly to the possible observations of an element in a Σ-algebra. Intuitively, such observations are computations starting from this element, leading to an observable result, and perhaps using additional inputs from observable sorts. This can be modelled by restricting the notion of a context; cf. Definition 2.6.1.

Definition 8.3.4 (Observable context). A Σ-context $C \in |T_\Sigma(X \uplus \{\Box{:}s\})|$ for sort s is *observable* (*w.r.t. OBS*) if it contains only variables in X_{OBS} and is of a sort in *OBS*.

For any Σ-algebra $A \in |\mathbf{Alg}(\Sigma)|$ and valuation $v: X \to |A|$, the *value of the context C on $a \in |A|_s$ under valuation v*, written $C^v_A[a]$, is $C_A(v_a) \in |A|$ where $v_a: (X \uplus \{\Box{:}s\}) \to |A|$ extends v by $v_a(\Box) = a$. $\Box$

Exercise 8.3.5. Show that for any Σ-context $C \in |T_\Sigma(X \uplus \{\Box{:}s\})|$, Σ-term $t \in |T_\Sigma(X)|_s$ and valuation $v: X \to |A|$, $C^v_A[t_A(v)] = (C[t])_A(v)$. $\Box$

Definition 8.3.6 (Behavioural indistinguishability). Consider any Σ-algebra $A \in |\mathbf{Alg}(\Sigma)|$. *Behavioural indistinguishability on A w.r.t. OBS* (or simply, *indistinguishability on A*) is the relation $\sim^{OBS}_A \subseteq |A| \times |A|$ defined as follows: for $s \in S$ and $a, b \in |A|_s$, $a \sim^{OBS}_A b$ if and only if for all observable contexts $C \in |T_\Sigma(X_{OBS} \uplus \{\Box{:}s\})|_{OBS}$ and valuations $v: X_{OBS} \to |A|_{OBS}$, $C^v_A[a] = C^v_A[b]$. $\Box$

We omit the set *OBS* of observable sorts if it is clear from the context, and write simply $\sim_A$, or even $\sim$ if A is clear as well, for behavioural indistinguishability on A.

Theorem 8.3.7. *For any OBS-generated Σ-algebra $A \in |\mathbf{Alg}(\Sigma)|$, the behavioural indistinguishability relation $\sim \subseteq |A| \times |A|$ on A is the largest behavioural congruence on A.*

Proof. First, note that since for any observable sort $o \in OBS$, $\Box{:}o$ is an observable context for o, so $\sim$ is the identity on $|A|_{OBS}$.

Then, let $\simeq \subseteq |A| \times |A|$ be a congruence on A, and let $a, b \in |A|_s$ for some sort $s \in S$ be such that $a \simeq_s b$. By an easy induction on the structure of a term $t \in |T_\Sigma(X_{OBS} \uplus \{\Box{:}s\})|$, for any valuation $v: X_{OBS} \to |A|_{OBS}$, we can show that $t_A(v_a) \simeq t_A(v_b)$, where $v_a, v_b: (X_{OBS} \uplus \{\Box{:}s\}) \to |A|$ extend v by $v_a(\Box) = a$ and $v_b(\Box) = b$, respectively. If $\simeq$ is behavioural, this implies that $C^v_A[a] = C^v_A[b]$ for any observable context $C \in |T_\Sigma(X_{OBS} \uplus \{\Box{:}s\})|_{OBS}$, and so $a \sim_s b$, which proves $\simeq \subseteq \sim$.

To show that $\sim$ is a congruence on A, consider an operation $f: s_1 \times \cdots \times s_n \to s$ in Σ and elements $a_1, a'_1 \in |A|_{s_1}, \ldots, a_n, a'_n \in |A|_{s_n}$ such that $a_1 \sim_{s_1} a'_1, \ldots, a_n \sim_{s_n} a'_n$. Since A is *OBS*-generated, all its elements are values of terms with variables of observable sorts under some valuation of these variables in $|A|_{OBS}$. Choosing the variables appropriately, we can assume that there are terms $t_1, t'_1 \in |T_\Sigma(Y_{OBS})|_{s_1}, \ldots, t_n, t'_n \in |T_\Sigma(Y_{OBS})|_{s_n}$ and a valuation $w: Y_{OBS} \to |A|_{OBS}$ such that $a_1 = (t_1)_A(w)$, $a'_1 = (t'_1)_A(w)$, …, $a_n = (t_n)_A(w)$, $a'_n = (t'_n)_A(w)$. Consider now an observable context $C \in |T_\Sigma(X_{OBS} \uplus \{\Box{:}s\})|_{OBS}$ and a valuation $v: X_{OBS} \to |A|_{OBS}$. We can assume that Y_{OBS} and X_{OBS} are disjoint. Let $(v+w): (X_{OBS} \cup Y_{OBS}) \to |A|_{OBS}$ be the obvious valuation that extends v and w. Then:

$$
\begin{array}{lll}
C^v_A[f_A(a_1, a_2, \ldots, a_n)] & & \\
\quad = C^{v+w}_A[f_A(a_1, t_2, \ldots, t_n)] & & \text{since } a_2 = (t_2)_A(w), \ldots, a_n = (t_n)_A(w) \\
\quad = C^{v+w}_A[f_A(a'_1, t_2, \ldots, t_n)] & & \text{since } a_1 \sim_{s_1} a'_1 \text{ and} \\
 & & C(\Box{:}s_1, t_2, \ldots, t_n) \text{ is an observable context} \\
\quad = C^{v+w}_A[f_A(t'_1, a_2, \ldots, t_n)] & & \text{since } a'_1 = (t'_1)_A(w) \text{ and } a_2 = (t_2)_A(w) \\
\quad = C^{v+w}_A[f_A(t'_1, a'_2, \ldots, t_n)] & & \text{since } a_2 \sim_{s_1} a'_2 \text{ and} \\
 & & C(t'_1, \Box{:}s_2, \ldots, t_n) \text{ is an observable context} \\
\quad \ldots & & \\
\quad = C^{v+w}_A[f_A(t'_1, t'_2, \ldots, a'_n)] & & \ldots \\
\quad = C^v_A[f_A(a'_1, a'_2, \ldots, a'_n)] & & \text{since } a'_1 = (t'_1)_A(w), \ldots, a'_{n-1} = (t'_{n-1})_A(w)
\end{array}
$$

Thus $f_A(a_1,a_2,\ldots,a_n) \sim_s f_A(a'_1,a'_2,\ldots,a'_n)$ and so $\sim$ is indeed a congruence on A. □

Exercise 8.3.8. Construct a Σ-algebra $A \in |\mathbf{Alg}(\Sigma)|$ such that the behavioural indistinguishability relation $\sim_A \subseteq |A| \times |A|$ is not a congruence on A (so, A cannot be OBS-generated). Modify the definition of $\sim$ so that a characterisation of the largest behavioural congruence on any algebra is obtained. (HINT: Admit computations with non-observable inputs; see Exercise 8.2.17.) Check that the two definitions coincide on OBS-generated algebras, but if a Σ-algebra A is not OBS-generated, they may yield different relations, even on the OBS-generated subalgebra $\langle A\rangle_{OBS}$ of A. □

The above exercise demonstrates that considering elements outside the OBS-generated subalgebra may influence properties of OBS-generated elements. Since only the latter may occur in observable computations performed in an algebra, we will disregard all such "non-observable junk". Consequently, we define a new notion of *behavioural satisfaction* of equations using the indistinguishability relation to interpret equality and restricting attention to OBS-generated data.

Definition 8.3.9 (Behavioural satisfaction). A Σ-algebra A *behaviourally satisfies* (*w.r.t. OBS*) a Σ-equation $\forall X \bullet t = t'$, written $A \models_{OBS} \forall X \bullet t = t'$, if for all valuations $v{:}X \to |\langle A\rangle_{OBS}|$ into the OBS-generated subalgebra of A, $(t)_A(v) \sim_A^{OBS} (t')_A(v)$. In this situation we will sometimes say that $\forall X \bullet t = t'$ *behaviourally holds* in A (*w.r.t. OBS*). □

Exercise 8.3.10. Extend the above definition to behavioural satisfaction of first-order formulae without predicates (cf. Exercise 4.1.13). Follow the usual inductive definition of satisfaction of a first-order formula with respect to a valuation of variables, but:

- restrict valuations, and hence the possible values of quantified variables, to the OBS-generated part of the algebra; and
- use behavioural indistinguishability to interpret equality.

Check that the resulting definition coincides with the one above for equations viewed as universally quantified first-order sentences. □

The new notion of satisfaction of equations (and first-order sentences) extends as usual to sets of sentences and classes of algebras. It may be used to reinterpret flat specifications as follows:

Definition 8.3.11 (Behavioural semantics of presentation). Given a presentation $\langle \Sigma, \Phi\rangle$ in equational logic (or first-order logic) and a set of observable sorts OBS in Σ, the *behavioural model class of* $\langle \Sigma, \Phi\rangle$ (*w.r.t. OBS*) is defined as follows:

$$Mod_{OBS}(\langle \Sigma, \Phi\rangle) = \{A \in |\mathbf{Alg}(\Sigma)| \mid A \models_{OBS} \Phi\}.$$ □

Example 8.3.12. Recalling again specifications and constructions from Section 8.1, for any model $SA \in Mod[\text{STRINGARRAY}]$ of the specification of arrays of natural numbers, $AwP(SA)$, the array-with-pointer realisation of stacks it determines, is a behavioural model of all the axioms in STRINGSTACK. In particular, $AwP(SA) \models_{OBS} \forall x{:}String, s{:}Stack \bullet pop(push(x,s)) = s$. □

Exercise 8.3.13. Check that ordinary satisfaction of equations implies behavioural satisfaction, and that the behavioural model class of an equational presentation is a variety and hence is equationally definable under ordinary satisfaction (but not necessarily finitely, even if the original presentation is finite; see Exercise 8.1.3). □

We have so far considered behavioural satisfaction w.r.t. a fixed signature $\Sigma = \langle S, \Omega \rangle$ and a fixed set $OBS \subseteq S$ of observable sorts.

Exercise 8.3.14. Investigate how behavioural satisfaction varies when the set of observable sorts changes. In particular, for any Σ-equation e and Σ-algebra $A \in |\mathbf{Alg}(\Sigma)|$, characterise satisfaction $A \models_S e$ and $A \models_\varnothing e$, and prove that if $OBS \subseteq OBS' \subseteq S$ then $A \models_{OBS'} e$ implies $A \models_{OBS} e$. Compare this with Exercise 8.2.16 and its consequences for behavioural closure and abstraction in Exercise 8.2.23. HINT: See Corollary 8.3.33 below. □

Proposition 8.3.15. *For any signature morphism $\sigma{:}\Sigma \to \Sigma'$, Σ-equation e and Σ'-algebra $A' \in |\mathbf{Alg}(\Sigma')|$, if $A' \models_{\sigma(OBS)} \sigma(e)$ then $A'|_\sigma \models_{OBS} e$.* □

Exercise 8.3.16. Give a counterexample to the converse of the implication in Proposition 8.3.15. Conclude that behavioural satisfaction (of equations by algebras over algebraic signatures with their usual morphisms) does not directly give rise to an institution. The situation is even worse for first-order logic: show that even the forward implication may fail for sentences of first-order logic. Show that Proposition 8.3.15 applies to conditional equations with premises of observable sorts. □

Exercise 8.3.17. Try to characterise (behavioural) theory morphisms $\sigma{:}\langle \Sigma, \Phi \rangle \to \langle \Sigma', \Phi' \rangle$ such that for any model $A' \in Mod_{\sigma(OBS)}(\langle \Sigma', \Phi' \rangle)$ and Σ-equation e, $A' \models_{\sigma(OBS)} \sigma(e)$ if and only if $A'|_\sigma \models_{OBS} e$. HINT: This holds when in Σ' there are no essentially new observable contexts for old sorts (that is, all observable contexts in Σ' for sorts in $\sigma(\Sigma)$ are observable contexts from Σ translated by σ). For signature morphisms this is difficult to ensure in typical examples. Here, however, this may be defined modulo the identification of contexts induced by the axioms in Φ', which allows for some non-trivial applications. You may find Exercise 8.3.35 below useful in justifying the characterisation you develop. □

8.3.1 Behavioural satisfaction vs. behavioural abstraction

In Section 8.2, behavioural interpretation of specifications was given by behavioural abstraction, based on behavioural equivalence of algebras. In the above, similar motivation led to a new notion of behavioural satisfaction, and consequently to a behavioural interpretation of flat specifications given by the class of their behavioural models. Are these two approaches equivalent? We will show here that this is indeed the case for many typical specifications.

We will need to generalise two standard concepts of universal algebra (cf. Definitions 1.3.13 and 1.3.15):

Definition 8.3.18 (Partial congruence and quotient). Let $A \in |\mathbf{Alg}(\Sigma)|$ be a Σ-algebra. A relation $\simeq \subseteq |A| \times |A|$ is a *partial congruence* on A if $dom(\simeq) = \{a \in |A| \mid a \simeq a' \text{ for some } a' \in |A|\}$ is (the carrier of) a subalgebra of A and $\simeq$ is a congruence on this subalgebra. That is, a partial congruence is a transitive and symmetric (but not necessarily reflexive) relation on $|A|$ that is closed under the operations of A.

A *quotient* of a Σ-algebra A by a partial congruence $\simeq$ on A, written $A/\simeq$, is the quotient of the subalgebra of A having carrier $dom(\simeq)$ by the congruence $\simeq$. □

As before, let $\Sigma = \langle S, \Omega \rangle$ be an algebraic signature with a distinguished set of observable sorts $OBS \subseteq S$.

Definition 8.3.19 (Partial behavioural indistinguishability congruence). Given a Σ-algebra $A \in |\mathbf{Alg}(\Sigma)|$, the *partial behavioural indistinguishability congruence on* A is $\approx_A^{OBS} = \sim_A^{OBS} \cap (|\langle A \rangle_{OBS}| \times |\langle A \rangle_{OBS}|)$, the restriction of the behavioural indistinguishability relation on A to the OBS-generated subalgebra $\langle A \rangle_{OBS}$ of A. □

Corollary 8.3.20. *For any Σ-algebra $A \in |\mathbf{Alg}(\Sigma)|$, the partial behavioural indistinguishability congruence $\approx_A^{OBS}$ on A is the largest behavioural congruence on the OBS-generated subalgebra $\langle A \rangle_{OBS}$.*

Proof. Follows directly from Theorem 8.3.7, since by Definition 8.3.6, $\sim_{\langle A \rangle_{OBS}}$ coincides with $\sim_A$ on $|\langle A \rangle_{OBS}|$. □

We will aim first at a direct characterisation of (weak) behavioural equivalence in terms of partial behavioural indistinguishability. A special role will be played here by algebras in which all elements are OBS-generated and distinguishable from each other.

Definition 8.3.21 (Fully abstract algebra). A Σ-algebra $A \in |\mathbf{Alg}(\Sigma)|$ is *fully abstract* (*w.r.t. OBS*) if the partial indistinguishability relation $\approx_A^{OBS}$ on A is the identity on $|A|$. □

The class of all fully abstract Σ-algebras will be denoted by $FAlg_{OBS}(\Sigma)$. For any class $\mathscr{A} \subseteq |\mathbf{Alg}(\Sigma)|$ of Σ-algebras, we will write $FA_{OBS}(\mathscr{A})$ for $\mathscr{A} \cap FAlg_{OBS}(\Sigma)$.

Lemma 8.3.22. *Fully abstract algebras are weakly behaviourally equivalent if and only if they are isomorphic.*

Proof. The "if" part is trivial. For the "only if" part, let $A, B \in |\mathbf{Alg}(\Sigma)|$ be such that $A \overset{w}{\equiv}_{OBS} B$. Since fully abstract algebras are OBS-generated, by Corollary 8.2.32, there exists an OBS-generated Σ-algebra $C \in |\mathbf{Alg}(\Sigma)|$ and surjective behavioural homomorphisms $h_A : A \to C$ and $h_B : B \to C$. Then, the kernels of h_A and h_B are behavioural congruences on A and B, respectively, and so, since A and B are fully abstract, the kernels are identities. Consequently, h_A and h_B are isomorphisms, and A and B are isomorphic. □

Lemma 8.3.23. *For any Σ-algebra $A \in |\mathbf{Alg}(\Sigma)|$, its quotient $A/{\approx_A}$ by partial behavioural indistinguishability is fully abstract.*

Proof. First, recall that $dom(\approx_A)$ coincides with (the carrier of) the OBS-generated behavioural subalgebra $\langle A\rangle_{OBS}$ of A. Therefore $A/{\approx_A}$ is OBS-generated as well, and so the partial behavioural indistinguishability congruence $\approx_{(A/\approx_A)}$ is in fact a total congruence on $A/{\approx_A}$. Then, notice that the relation $\simeq \subseteq |\langle A\rangle_{OBS}| \times |\langle A\rangle_{OBS}|$ given by $a \simeq a'$ if and only if $[a]_{\approx_A} \approx_{(A/\approx_A)} [a']_{\approx_A}$ is a congruence on $\langle A\rangle_{OBS}$, since it is the kernel of the composition of quotient homomorphisms (cf. Exercise 1.3.18) from $\langle A\rangle_{OBS}$ to $A/{\approx_A}$ and then to $(A/{\approx_A})/{\approx_{(A/\approx_A)}}$. Moreover, it is behavioural, since both $\approx_A$ and $\approx_{(A/\approx_A)}$ are behavioural. Consequently, by Corollary 8.3.20, $\simeq \subseteq \approx_A$, which proves that $\approx_{(A/\approx_A)}$ is indeed the identity. □

Theorem 8.3.24. *Two Σ-algebras $A, B \in |\mathbf{Alg}(\Sigma)|$ are weakly behaviourally equivalent if and only if $A/{\approx_A}$ and $B/{\approx_B}$ are isomorphic.*

Proof. The "if" part follows directly from Corollary 8.2.32. For the "only if" part notice that if A and B are weakly behaviourally equivalent then also $A/{\approx_A}$ and $B/{\approx_B}$ are weakly behaviourally equivalent. Thus, by Lemmas 8.3.23 and 8.3.22, they are isomorphic. □

We proceed now to link behavioural satisfaction with ordinary satisfaction.

Lemma 8.3.25. *In any fully abstract Σ-algebra $A \in |\mathbf{Alg}(\Sigma)|$, behavioural satisfaction coincides with ordinary satisfaction.*

Proof. Follows directly from the definitions. □

Theorem 8.3.26. *For any Σ-algebra $A \in |\mathbf{Alg}(\Sigma)|$, behavioural satisfaction in A coincides with ordinary satisfaction in $A/{\approx_A}$. That is, for any Σ-equation e,*

$$A \models_{OBS} e \quad \textit{iff} \quad A/{\approx_A} \models e.$$

Proof. For any set of variables X and valuation $v : X \to |\langle A\rangle_{OBS}|$, let $\overline{v} : X \to |A/{\approx_A}|$ be the valuation defined by $\overline{v}(x) = [v(x)]_{\approx_A}$ for all $x \in X$. Note that all valuations in $|A/{\approx_A}|$ are of this form. Then, since the quotient homomorphism preserves values of terms (cf. Exercise 1.4.8), for any term $t \in |T_\Sigma(X)|$ and valuation $v : X \to |\langle A\rangle_{OBS}|$, $t_{A/\approx_A}(\overline{v}) = [t_A(v)]_{\approx_A}$. It follows that for any terms $t', t'' \in |T_\Sigma(X)|_s$ of the same sort $s \in S$ and valuation $v : X \to |\langle A\rangle_{OBS}|$, $t'_A(v) \approx_A t''_A(v)$ if and only if $t'_{A/\approx_A}(\overline{v}) \approx_{(A/\approx_A)} t''_{A/\approx_A}(\overline{v})$, that is (by Lemma 8.3.23) $t'_{A/\approx_A}(\overline{v}) = t''_{A/\approx_A}(\overline{v})$. Consequently, $A \models_{OBS} \forall X \bullet t' = t''$ if and only if $A/{\approx_A} \models \forall X \bullet t' = t''$. □

Exercise 8.3.27. Generalise this theorem to first-order sentences. HINT: Prove that for any first-order formula φ with free variables X, φ behaviourally holds in an algebra A under a valuation $v{:}X \to |\langle A\rangle_{OBS}|$ if and only if φ holds in the usual sense in $A/{\approx_A}$ under the valuation $\overline{v}{:}X \to |A/{\approx_A}|$ as defined in the above proof. □

Exercise 8.3.28. Recall (Exercise 8.3.13) that if an algebra satisfies an equation in the usual sense, then it also satisfies this equation behaviourally (w.r.t. any set of observable sorts). Re-prove this using Theorem 8.3.26 and show that the proof carries over to conditional equations with premises of observable sorts.

Show, however, that for arbitrary first-order sentences this property may fail: give an example of a sentence and an algebra which satisfies it in the usual sense but does not satisfy it behaviourally. □

Corollary 8.3.29. *Weakly behaviourally equivalent algebras are logically equivalent w.r.t. behavioural satisfaction, that is, if Σ-algebras $A, B \in |\mathbf{Alg}(\Sigma)|$ are weakly behaviourally equivalent then for any first-order Σ-sentence φ,*

$$A \models_{OBS} \varphi \quad \textit{iff} \quad B \models_{OBS} \varphi.$$

Proof. Follows directly from Theorems 8.3.24 and 8.3.26 and Exercise 8.3.27, since satisfaction of sentences is preserved under isomorphism. □

Finally, we can directly characterise the relationship between the behavioural models of a flat specification and the behavioural closure of its class of models. Two auxiliary concepts first:

Definition 8.3.30 (Behavioural expansion and closedness). The *behavioural expansion* (*w.r.t. OBS*) of a class $\mathscr{A} \subseteq |\mathbf{Alg}(\Sigma)|$ of Σ-algebras is

$$Beh_{OBS}(\mathscr{A}) = \{A \in |\mathbf{Alg}(\Sigma)| \mid A/{\approx_A} \in \mathscr{A}\}.$$

A class $\mathscr{A} \subseteq |\mathbf{Alg}(\Sigma)|$ of Σ-algebras is *behaviourally closed* (*w.r.t. OBS*) if for all $A \in \mathscr{A}$, $A/{\approx_A} \in \mathscr{A}$ as well. □

Corollary 8.3.31. *For any set Φ of first-order Σ-sentences, $Mod_{OBS}(\langle\Sigma,\Phi\rangle) = Beh_{OBS}(Mod[\langle\Sigma,\Phi\rangle])$.*

Proof. Directly from Definition 8.3.30, by Theorem 8.3.26 and Exercise 8.3.27. □

Theorem 8.3.32. *Let $\mathscr{A} \subseteq |\mathbf{Alg}(\Sigma)|$ be a class of Σ-algebras closed under isomorphism. Then:*

1. *$Beh_{OBS}(\mathscr{A}) \subseteq Abs_{OBS}(\mathscr{A})$,*
2. *$Beh_{OBS}(\mathscr{A}) = Abs_{OBS}(FA_{OBS}(\mathscr{A}))$,*
3. *$Beh_{OBS}(\mathscr{A}) = Abs_{OBS}(\mathscr{A})$ if and only if $\mathscr{A}$ is behaviourally closed.*

Proof. First, note that since $\mathscr{A}$ is closed under isomorphism, so is $FA_{OBS}(\mathscr{A})$ and by Corollary 8.2.34, $WAbs_{OBS}(\mathscr{A}) = Abs_{OBS}(\mathscr{A})$ and $WAbs_{OBS}(FA_{OBS}(\mathscr{A})) = Abs_{OBS}(FA_{OBS}(\mathscr{A}))$. Also, by Corollary 8.2.32, $A \overset{w}{\equiv}_{OBS} A/{\approx_A}$ for all $A \in |\mathbf{Alg}(\Sigma)|$. Then:

1. Follows directly from the definitions by the above remark.
2. If $A/{\approx_A} \in \mathscr{A}$ then $A/{\approx_A} \in FA_{OBS}(\mathscr{A})$ by Lemma 8.3.23. Therefore, by the above remark, if $A \in Beh_{OBS}(\mathscr{A})$ then also $A \in Abs_{OBS}(FA_{OBS}(\mathscr{A}))$.
   On the other hand, if $A \in Abs_{OBS}(FA_{OBS}(\mathscr{A}))$ then $A/{\approx_A} \in Abs_{OBS}(FA_{OBS}(\mathscr{A}))$, which by Lemma 8.3.22 implies that $A/{\approx_A} \in FA_{OBS}(\mathscr{A}) \subseteq \mathscr{A}$, and so $A \in Beh_{OBS}(\mathscr{A})$.
3. By the previous items, we have to show that $Abs_{OBS}(\mathscr{A}) \subseteq Abs_{OBS}(FA_{OBS}(\mathscr{A}))$ if and only if $\mathscr{A}$ is behaviourally closed.
   For the "if" part, suppose that $A \in Abs_{OBS}(\mathscr{A})$, that is, by Theorem 8.3.24, $A/{\approx_A}$ is isomorphic to $B/{\approx_B}$ for some $B \in \mathscr{A}$. Since $\mathscr{A}$ is behaviourally closed, $B/{\approx_B} \in \mathscr{A}$, and so by Lemma 8.3.23, $B/{\approx_B} \in FA_{OBS}(\mathscr{A})$. Hence, since $A \equiv_{OBS} B/{\approx_B}$, $A \in Abs_{OBS}(FA_{OBS}(\mathscr{A}))$.
   For the "only if" part, suppose that $Abs_{OBS}(\mathscr{A}) \subseteq Abs_{OBS}(FA_{OBS}(\mathscr{A}))$. Then, for any $A \in \mathscr{A} \subseteq Abs_{OBS}(\mathscr{A})$, $A \in Abs_{OBS}(FA_{OBS}(\mathscr{A}))$ as well, and so, by Theorem 8.3.24 and Definition 8.3.21, $A/{\approx_A}$ is (isomorphic to) an algebra in $FA_{OBS}(\mathscr{A})$. Consequently, $A/{\approx_A} \in \mathscr{A}$, which proves that $\mathscr{A}$ is behaviourally closed. □

An important corollary of the above theorem, which sums up the developments of this section, is that behavioural satisfaction and behavioural abstraction result in the same behavioural interpretation of flat equational specifications.

Corollary 8.3.33. *For any set $\mathscr{E}$ of Σ-equations, $Mod[$***abstract*** $\langle\Sigma,\mathscr{E}\rangle$ ***wrt*** $OBS] = Mod_{OBS}(\langle\Sigma,\mathscr{E}\rangle)$.*

Proof. Follows from Corollary 8.3.31 and Theorem 8.3.32(3), since $Mod[\langle\Sigma,\mathscr{E}\rangle]$ is closed under isomorphism, and for equational axioms $\mathscr{E}$, it is behaviourally closed, since it is closed under arbitrary quotients by Proposition 2.2.8. □

Exercise 8.3.34. Try to generalise the above corollary to first-order logic. First, conclude from Theorem 8.3.32 that for any set Φ of first-order Σ-sentences, we have $Mod[$**abstract** $\langle\Sigma,\Phi\rangle$ **wrt** $OBS] = Mod_{OBS}(\langle\Sigma,\Phi\rangle)$ if (and only if) $Mod[\langle\Sigma,\Phi\rangle]$ is behaviourally closed. Give a first-order sentence with a class of models that is not behaviourally closed. Then find a class of first-order sentences, going beyond equations, such that presentations with axioms from that class are behaviourally closed. □

Exercise 8.3.35. Study the interaction between reducts along signature morphisms and quotients by the partial behavioural indistinguishability congruence. In particular, for any signature morphism $\sigma\colon\Sigma \to \Sigma'$, set OBS of observable sorts in Σ, and Σ'-algebra $A' \in |\mathbf{Alg}(\Sigma')|$, show that $\approx_{A'}^{\sigma(OBS)}|_\sigma \subseteq \approx_{A'|_\sigma}^{OBS}$. It follows that there is a natural Σ-homomorphism $h\colon (A'/{\approx_{A'}^{\sigma(OBS)}})|_\sigma \to (A'|_\sigma)/{\approx_{A'|_\sigma}^{OBS}}$ provided that for every sort s in Σ, for all $a' \in |\langle A'\rangle_{\sigma(OBS)}|_{\sigma(s)}$, there exists $a \in |\langle A'|_\sigma\rangle_{OBS}|_s$ such that $a' \approx_{A'}^{\sigma(OBS)} a$. Moreover, h is surjective. Then show that h is injective provided that for every sort s in Σ, for all $a,b \in |\langle A'|_\sigma\rangle_{OBS}|_s$, if $a \approx^{OBS}_{A'|_\sigma} b$ then $a \approx^{\sigma(OBS)}_{A'} b$ (and show that the opposite implication always holds). Conclude that if both of these conditions are satisfied then $(A'|_\sigma)/{\approx^{OBS}_{A'|_\sigma}}$ and $(A'/{\approx^{\sigma(OBS)}_{A'}})|_\sigma$ are isomorphic.

Re-examine the failure of the satisfaction condition for behavioural satisfaction (Exercise 8.3.16) in the light of these semantic conditions. □

8.4 Behavioural implementations

The behavioural abstraction operation plays a rather special role in the software specification and development process. In several ways it is quite different from the other specification-building operations presented earlier. For instance, unlike the others[2] it "enlarges" the class of models of the specification to which it is applied. One consequence is that the theory of the resulting specification is generally smaller than the theory of the specification to which it is applied: in a sense, properties are lost. Another is that the interaction of behavioural equivalence with other specification-building operations is unclear. This power of behavioural abstraction calls for some caution, and in practice turns it more into a tool for the designers of specification languages and development methodologies than into an operation directly offered to the user. Example 8.2.21 motivates and illustrates a typical situation in which behavioural abstraction should be used, namely to justify correctness of (constructor) implementations in the software development process. So instead of regarding it as a general-purpose specification-building operation, we will build it into a *behavioural* version of the implementation relation.

8.4.1 Implementing specifications up to behavioural equivalence

Perhaps the most naive approach to behavioural implementation would be to regard every specification in the development chain as implicitly surrounded by an appropriate application of the behavioural abstraction operation. Then a constructor implementation $SP \underset{\kappa}{\leadsto} SP'$ would really stand for

$$\textbf{abstract } SP \textbf{ wrt } OBS \underset{\kappa}{\leadsto} \textbf{abstract } SP' \textbf{ wrt } OBS'$$

where the sets of observable sorts OBS and OBS' are chosen appropriately. The corresponding correctness condition would require then that $\kappa(A') \in Abs_{OBS}(Mod[SP])$ for every $A' \in Abs_{OBS'}(Mod[SP'])$. A direct proof of this condition is often difficult because we have little information about the properties of A' from which to reason about $\kappa(A')$; we cannot rely on the axioms in SP' as they need not hold in

[2] The only exception is closure under isomorphism. But in typical institutions, where satisfaction is preserved under isomorphism, this is unproblematic.

$Abs_{OBS'}(Mod[SP'])$. As we will see, under certain conditions it turns out to be sound to *pretend* that $A' \in Mod[SP']$, which makes such proofs much easier.

Example 8.4.1. Consider the following trivial specification:

$$
\begin{array}{l}
\textbf{spec } ID = \\
\quad \textsc{String} \textbf{ and } \textsc{Nat} \\
\quad \textbf{then} \\
\qquad \textbf{ops } id{:}\, String \times Nat \to String \\
\qquad \forall n{:}Nat, x{:}String \bullet id(x,n) = x
\end{array}
$$

Let $\Sigma ID = Sig[ID]$. An odd but correct realisation of *ID* in terms of stacks of strings as specified in Section 8.1 is given by a constructor $SID{:}\,\Sigma SS \Rightarrow \Sigma ID$ that maps any algebra $SS \in |\mathbf{Alg}(\Sigma SS)|$ to the following algebra $SID(SS) \in |\mathbf{Alg}(\Sigma ID)|$:

- $id_{SID(SS)}(x,n) = top_{SS}(multipop(n, multipush(n, push_{SS}(x, empty_{SS}))))$, where
  - $multipop{:}\, |SS|_{Nat} \times |SS|_{Stack} \to |SS|_{Stack}$ is defined recursively[3] by:

    $$multipop(0_{SS}, s) = s$$
    $$multipop(succ_{SS}(n), s) = multipop(n, pop_{SS}(s)),$$

  - $multipush{:}\, |SS|_{Nat} \times |SS|_{Stack} \to |SS|_{Stack}$ is defined recursively by:

    $$multipush(0_{SS}, s) = s$$
    $$multipush(succ_{SS}(n), s) = push_{SS}(\mathtt{a\hat{}n\hat{}y\hat{}}\varepsilon, multipush(n, s)).$$

To verify that StringStack implements *ID* via the constructor $SID{:}\,\Sigma SS \Rightarrow \Sigma ID$, that is, that $SID(SS) \in Mod[ID]$ for any model $SS \in Mod[\textsc{StringStack}]$, it is crucial to know that $SS \models \forall x{:}String, s{:}Stack \bullet pop(push(x,s)) = s$. Given this, a simple proof by induction (on the second argument of *id*) justifies that indeed $SID(SS) \models \forall n{:}Nat, x{:}String \bullet id(x,n) = x$, and so $SID(SS) \in Mod[ID]$. The reasoning for the induction step goes as follows (the obvious algebra subscripts for operations are omitted):

$$
\begin{aligned}
id(x, succ(n)) &= top(multipop(succ(n), multipush(succ(n), push(x, empty)))) \\
&= top(multipop(n, pop(push(\mathtt{a\hat{}n\hat{}y\hat{}}\varepsilon, multipush(n, push(x, empty)))))) \\
&= top(multipop(n, multipush(n, push(x, empty)))) \\
&= id(x, n) \\
&= x
\end{aligned}
$$

where the final step follows by the induction hypothesis. However, verifying that $SID(SS) \in Mod[ID]$ for $SS \in Abs_{\{String,Stack\}}(Mod[\textsc{StringStack}])$ is much harder. Since now *SS* need not satisfy the axiom $\forall x{:}String, s{:}Stack \bullet pop(push(x,s)) = s$, the crucial step in the above inductive argument fails. (Instead, we would have to use the fact that *SS* behaviourally satisfies this axiom; see Sections 8.3 and 9.6.)

This is of course an extremely contrived example, but it is easy to come up with realistic programs using stacks where properties like this need to be proved. □

---

[3] Recall that $|SS|_{Nat}$ is freely generated by 0_{SS} and $succ_{SS}$.

These considerations lead to the following generalisation of the notion of constructor implementation (cf. Definition 7.2.1):

Definition 8.4.2 (Behavioural implementation). Given specifications SP and SP', a constructor $\kappa : Sig[SP'] \Rightarrow Sig[SP]$ and a set OBS of observable sorts in $Sig[SP]$, we say that *SP' behaviourally implements SP via κ w.r.t. OBS*, written $SP \overset{OBS}{\underset{\kappa}{\leadsto}} SP'$, if $Abs_{OBS}(Mod[SP]) \supseteq \kappa(Mod[SP'])$ and $dom(\kappa) \supseteq Mod[SP']$. □

In other words, $SP \overset{OBS}{\underset{\kappa}{\leadsto}} SP'$ provided that the constructor κ transforms every model $A' \in Mod[SP']$ to a model $\kappa(A') \in Mod[$**abstract** SP **wrt** $OBS]$, that is, $\kappa \in Mod[SP' \Rightarrow$ **abstract** SP **wrt** $OBS]$. For instance, from Example 8.2.21 we have STRINGSTACK $\overset{\{String,Bool\}}{\underset{AwP}{\leadsto}}$ STRINGARRAY.

Proposition 8.4.3. *Constructor implementations are behavioural implementations, that is, $SP \underset{\kappa}{\leadsto} SP'$ implies $SP \overset{OBS}{\underset{\kappa}{\leadsto}} SP'$ for any specifications SP and SP', constructor $\kappa : Sig[SP'] \Rightarrow Sig[SP]$ and set OBS of observable sorts in $Sig[SP]$.*

Proof. Follows directly from the definitions by Proposition 8.2.19(1). □

Hence, all the examples of constructor implementations from Chapter 7 are examples of behavioural implementations as well, and $ID \overset{\{String\}}{\underset{SID}{\leadsto}}$ STRINGSTACK (see Example 8.4.1).

8.4.2 Stepwise development and stability

The alert reader will have noticed that there is a problem with the application of the notion of behavioural implementation (Definition 8.4.2) in the process of stepwise development as presented in Section 7.1. Behavioural implementation $SP \overset{OBS}{\underset{\kappa}{\leadsto}} SP'$ ensures only that algebras in $Mod[SP']$ give rise to realisations of SP that are correct up to behavioural equivalence; this says nothing about the models in $Abs_{OBS}(Mod[SP'])$ that are not in $Mod[SP']$ but may arise from subsequent behavioural implementations of SP'.

Example 8.4.1 (continued)

Recall that $ID \overset{\{String\}}{\underset{SID}{\leadsto}}$ STRINGSTACK and STRINGSTACK $\overset{\{String,Bool\}}{\underset{AwP}{\leadsto}}$ STRINGARRAY. We might hope to deduce from this that for any model $SA \in Mod[$STRINGARRAY$]$, $SID(AwP(SA))$ is a realisation of ID that is correct up to behavioural equivalence. Unfortunately, this does not follow: neither the property embodied by $ID \overset{\{String\}}{\underset{SID}{\leadsto}}$ STRINGSTACK (that $SID(SS) \in Abs_{\{String\}}(ID)$ for $SS \in Mod[$STRINGSTACK$]$), nor its suggested proof, tells us anything about the application of SID to the algebra $AwP(SA) \notin Mod[$STRINGSTACK$]$, even though $AwP(SA) \in Abs_{\{String,Bool\}}(Mod[$STRINGSTACK$])$ and we have already identified $AwP(SA)$ as an acceptable realisation of STRINGSTACK.

In this case, it may be shown that $SID(AwP(SA)) \in Abs_{\{String\}}(Mod[ID])$. The most obvious proof involves the non-elementary fact that for any natural number n,

$$multipop(n,s) = \underbrace{pop_{AwP(SA)}(\dots(pop_{AwP(SA)}(s))\dots)}_{n \text{ times}}$$

and similarly for $multipush_{AwP(SA)}$, and then relies on the fact that since $AwP(SA) \in Abs_{\{String\}}(Mod[\text{STRINGSTACK}])$, all observable computations in $AwP(SA)$ yield the same results on s and on $pop(push(z,s))$, for any stack s and string z.

Consider, however, another trivial realisation of ID in terms of STRINGSTACK, given by a constructor $SID' : \Sigma SS \Rightarrow \Sigma ID$ defined as a function which to any algebra $SS \in |\mathbf{Alg}(\Sigma SS)|$ assigns the following algebra $SID'(SS) \in |\mathbf{Alg}(\Sigma ID)|$:

- $id_{SID'(SS)}(x,n) = x$ if $pop_{SS}(push_{SS}(x, empty_{SS})) = empty_{SS}$
- $id_{SID'(SS)}(x,n) = \texttt{n\^{}o\^{}t\^{}}x$ otherwise.

Then $SID'(SS) \in Mod[ID]$ for all $SS \in Mod[\text{STRINGSTACK}]$, and so we still have a (behavioural) implementation $ID \overset{\{String\}}{\underset{SID'}{\leadsto}} \text{STRINGSTACK}$. However, the new constructor may not yield expected results on algebras that are not models of STRINGSTACK, even on those in $Abs_{\{String\}}(Mod[\text{STRINGSTACK}])$, and in fact it is easy to see that $SID'(AwP(SA)) \notin Abs_{\{String\}}(Mod[ID])$. □

It might seem that all is lost: behavioural implementations do not vertically compose. But there is a way out, which relies on the observation that the constructor SID' in the example above could not be coded in any standard programming language, as it involves a test for identity of non-observable data values. As a consequence, it distinguishes between models that are not supposed to be distinguishable by the user: that is, it maps behaviourally equivalent models to models that are not behaviourally equivalent. The following definition captures the essential property of constructors that prevents this:

Definition 8.4.4 (Stable constructor). Let Σ and Σ' be algebraic signatures with distinguished sets OBS and OBS' of observable sorts in Σ and Σ', respectively. Let $\mathscr{A}' \subseteq |\mathbf{Alg}(\Sigma')|$ be a class of Σ'-algebras.

A constructor $\kappa : \Sigma' \Rightarrow \Sigma$ is *stable on* $\mathscr{A}'$ *w.r.t.* OBS' *and* OBS if for every algebra $A' \in \mathscr{A}'$ and $B' \in |\mathbf{Alg}(\Sigma')|$, whenever $A' \equiv_{OBS'} B'$ then also $\kappa(A') \equiv_{OBS} \kappa(B')$.

We say that κ is *stable* (*w.r.t.* OBS' *and* OBS) if it is stable on the class $dom(\kappa)$. □

In particular, if κ is stable on $\mathscr{A}'$ w.r.t. OBS' and OBS then $Abs_{OBS'}(\mathscr{A}') \subseteq dom(\kappa)$.

Example 8.4.1 (continued)

It may be shown that the constructor $SID : \Sigma SS \Rightarrow \Sigma ID$ is stable (w.r.t. the observable sorts $\{String, Bool\}$ and $\{String\}$) while $SID' : \Sigma SS \Rightarrow \Sigma ID$ is not. □

Under the assumption that the constructor is stable, behavioural implementation turns out to be equivalent to constructor implementation between abstracted specifications, which ensures vertical composability.

Proposition 8.4.5. *Assume that the constructor* $\kappa : Sig[SP'] \Rightarrow Sig[SP]$ *is stable on* $Mod[SP']$ *w.r.t. sets* OBS' *and* OBS *of observable sorts in* $Sig[SP']$ *and* $Sig[SP]$*, respectively. Then*

$$SP \overset{OBS}{\underset{\kappa}{\leadsto}} SP'$$

iff

$$\textbf{abstract } SP \textbf{ wrt } OBS \underset{\kappa}{\rightsquigarrow} \textbf{abstract } SP' \textbf{ wrt } OBS'.$$

Proof. The "if" part is obvious by Proposition 8.2.19(1). For the "only if" part, assume that $A' \in Abs_{OBS'}(Mod[SP'])$, that is, for some $B' \in Mod[SP']$, $A' \equiv_{OBS'} B'$. Then, by the definition of behavioural implementation, $\kappa(B') \in Abs_{OBS}(Mod[SP])$ and, since κ is stable, $\kappa(A') \equiv_{OBS} \kappa(B')$. Thus, $\kappa(A') \in Abs_{OBS}(Mod[SP])$. □

Theorem 8.4.6 (Vertical composition). *Let* $SP \overset{OBS}{\underset{\kappa}{\rightsquigarrow}} SP'$ *and* $SP' \overset{OBS'}{\underset{\kappa'}{\rightsquigarrow}} SP''$ *be behavioural implementations such that the constructor* $\kappa\colon Sig[SP'] \Rightarrow Sig[SP]$ *is stable on* $Mod[SP']$ *w.r.t.* OBS' *and* OBS*. Then* $SP \overset{OBS}{\underset{\kappa';\kappa}{\rightsquigarrow}} SP''$.

Proof. Follows directly from the definition of behavioural implementation by Proposition 8.4.5. □

Corollary 8.4.7. *Given a chain of behavioural implementations*

$$SP_0 \overset{OBS_0}{\underset{\kappa_1}{\rightsquigarrow}} SP_1 \overset{OBS_1}{\underset{\kappa_2}{\rightsquigarrow}} \cdots \overset{OBS_{n-1}}{\underset{\kappa_n}{\rightsquigarrow}} SP_n$$

such that the constructors $\kappa_i\colon Sig[SP_i] \Rightarrow Sig[SP_{i-1}]$ *are stable on* $Mod[SP_i]$ *w.r.t.* OBS_i *and* OBS_{i-1}*, for* $i = 1,\ldots,n-1$*, we have*

$$SP_0 \overset{OBS_0}{\underset{\kappa_n;\cdots;\kappa_1}{\rightsquigarrow}} SP_n.$$

In particular, if $SP_n =$ EMPTY *is the empty specification over the empty algebraic signature, and* empty *is its unique model, then* $\kappa_1(\kappa_2 \ldots (\kappa_n(\mathsf{empty}))\ldots) \in Abs_{OBS_0}(Mod[SP_0])$. □

8.4.3 Stable and behaviourally trivial constructors

The results of Section 8.4.2 state that behavioural implementations via stable constructors compose and can therefore be safely used in stepwise development of programs, just as illustrated in Chapter 7 for simple implementations and constructor implementations. Of course, the crucial issue is the stability of the constructors we have available, so we now discuss, one by one, the constructors introduced in Section 6.1. The following technical lemma will be used:

Lemma 8.4.8. *Consider algebraic signatures* Σ *and* Σ' *with sets* OBS *and* OBS' *of observable sorts in* Σ *and* Σ'*, respectively, and a class* $\mathscr{A} \subseteq |\mathbf{Alg}(\Sigma)|$ *of* Σ*-algebras. Let* $\mathbf{F}\colon \mathbf{Alg}(\Sigma) \to \mathbf{Alg}(\Sigma')$ *be a functor that maps* OBS*-behavioural* Σ*-homomorphisms between algebras in* $Abs_{OBS}(\mathscr{A})$ *to* Σ'*-homomorphisms that are* OBS'*-behavioural. Then the object part of* $\mathbf{F}$*,* $F\colon \Sigma \Rightarrow \Sigma'$*, is a constructor which is stable on* $\mathscr{A}$ *w.r.t.* OBS *and* OBS'.

Proof. Consider two behaviourally equivalent Σ-algebras $A \in \mathscr{A}$ and $B \in |\mathbf{Alg}(\Sigma)|$, $A \equiv_{OBS} B$. By Exercise 8.2.15, there exists a Σ-algebra $C \in |\mathbf{Alg}(\Sigma)|$ and OBS-

behavioural Σ-homomorphisms $h_A: C \to A$ and $h_B: C \to B$. Then, by the assumption, Σ'-homomorphisms $\mathbf{F}(h_A): F(C) \to F(A)$ and $\mathbf{F}(h_B): F(C) \to F(B)$ are OBS'-behavioural, which proves $F(A) \equiv_{OBS'} F(B)$. □

Example 8.4.9 (Reduct). As in Example 6.1.3 consider an algebraic signature morphism $\sigma: \Sigma \to \Sigma'$, now with sets *OBS* and *OBS'* of observable sorts in Σ and Σ', respectively, such that $\sigma(OBS) \subseteq OBS'$. Then, by Lemma 8.4.8 or by Exercise 8.2.7, the object part of the σ-reduct functor is a constructor $\_|_\sigma: \Sigma' \Rightarrow \Sigma$ which is stable w.r.t. *OBS'* and *OBS*. □

Quotients and restrictions to reachable subalgebras (Examples 6.1.11 and 6.1.13, respectively) are constructors that are not typically provided by programming languages. Therefore one might guess that they are not really necessary, even though they arose in Example 7.2.9. The point, however, is that not only do they not change the signature of models, they typically do not change their behaviour either; see Exercise 8.2.12.

Definition 8.4.10 (Behaviourally trivial constructor). Consider an algebraic signature Σ with a distinguished set *OBS* of observable sorts and a class $\mathscr{A} \subseteq |\mathbf{Alg}(\Sigma)|$ of Σ-algebras.

A constructor $\kappa: \Sigma \Rightarrow \Sigma$ is *behaviourally trivial w.r.t. OBS on* $\mathscr{A}$ if for every algebra $A \in \mathscr{A}$, $A \equiv_{OBS} \kappa(A)$. $\kappa: \Sigma \Rightarrow \Sigma$ is *behaviourally trivial* (*w.r.t. OBS*) if it is so on the class $|\mathbf{Alg}(\Sigma)|$. □

Proposition 8.4.11. *Let Σ be an algebraic signature with a distinguished set of observable sorts OBS, and let $\mathscr{A} \subseteq |\mathbf{Alg}(\Sigma)|$. Any constructor $\kappa: \Sigma \Rightarrow \Sigma$ that is behaviourally trivial on $Abs_{OBS}(\mathscr{A})$ w.r.t. OBS is stable on $\mathscr{A}$ w.r.t. OBS.*

Proof. Easily follows from the fact that behavioural equivalence is an equivalence (Exercise 8.2.3). □

However, behavioural triviality on $Abs_{OBS}(\mathscr{A})$ is in general a stronger requirement than behavioural triviality on $\mathscr{A}$, and a constructor that is behaviourally trivial on $\mathscr{A}$ need not be stable on $\mathscr{A}$.

Proposition 8.4.12. *Let SP and SP' be Σ-specifications, and $SP \overset{OBS}{\underset{\kappa}{\rightsquigarrow}} SP'$ be a behavioural implementation via the constructor $\kappa: \Sigma \Rightarrow \Sigma$ that is behaviourally trivial on $Mod[SP']$ w.r.t. OBS. Then we also have a behavioural implementation $SP \overset{OBS}{\underset{id}{\rightsquigarrow}} SP'$ via the identity constructor $id_{|\mathbf{Alg}(\Sigma)|}: \Sigma \Rightarrow \Sigma$.*

Proof. By the assumptions, $id_{|\mathbf{Alg}(\Sigma)|}(A) \equiv_{OBS} \kappa(A) \in Abs_{OBS}(Mod[SP])$, and so $id_{|\mathbf{Alg}(\Sigma)|}(A) \in Abs_{OBS}(Mod[SP])$ as well for any $A \in Mod[SP']$. □

The above proposition means that whenever a behaviourally trivial constructor is used in the development process, it may be omitted entirely, and a correct behavioural implementation is still obtained. Note that the stability of behaviourally trivial constructors is not required:

Proposition 8.4.13. *Given behavioural implementations* $SP_1 \overset{OBS}{\underset{\kappa}{\leadsto}} SP_2 \overset{OBS}{\underset{\kappa'}{\leadsto}} SP'$ *with* $Sig[SP_1] = Sig[SP_2]$ *and such that the constructor* $\kappa : Sig[SP_2] \Rightarrow Sig[SP_1]$ *is behaviourally trivial on* $Mod[SP_2]$ *w.r.t. OBS, we also have the behavioural implementation* $SP_1 \overset{OBS}{\underset{\kappa'}{\leadsto}} SP'$.

Proof. Trivial, for instance by Proposition 8.4.12 and Theorem 8.4.6. □

Consequently, given a chain of behavioural implementations

$$SP \overset{OBS}{\underset{\kappa}{\leadsto}} SP'_1 \overset{OBS'}{\underset{\kappa'}{\leadsto}} SP'_2 \overset{OBS'}{\underset{\kappa''}{\leadsto}} SP''$$

such that $Sig[SP'_1] = Sig[SP'_2]$ and $\kappa' : Sig[SP'_2] \Rightarrow Sig[SP'_1]$ is behaviourally trivial on $Mod[SP'_2]$ w.r.t. OBS', we can simplify it to

$$SP \overset{OBS}{\underset{\kappa}{\leadsto}} SP'_1 \overset{OBS'}{\underset{\kappa''}{\leadsto}} SP''.$$

If moreover $\kappa : Sig[SP'_1] \Rightarrow Sig[SP]$ is stable on $Mod[SP'_1]$ w.r.t. OBS' and OBS, we obtain a behavioural implementation of SP by SP'' via $\kappa'';\kappa : Sig[SP''] \Rightarrow Sig[SP]$. This explains why behaviourally trivial constructors may be dropped from any development chain. Note that this is possible and correct even if the constructor $\kappa' : Sig[SP'_2] \Rightarrow Sig[SP'_1]$ is not stable on $Mod[SP'_2]$, which might render incorrect the behavioural implementation of SP by SP'' via $\kappa'';\kappa';\kappa : Sig[SP''] \Rightarrow Sig[SP]$, suggested in the original chain of behavioural implementations.

Another view of this situation is that we may implicitly *insert* behaviourally trivial constructors in order to facilitate proofs of correctness of behavioural implementation steps. Namely, to prove correctness of

$$SP \overset{OBS}{\underset{\kappa}{\leadsto}} SP'$$

where κ is stable w.r.t. OBS' and OBS (here, OBS' is the set of observable sorts in $Sig[SP']$ to be used in the behavioural implementation of SP') we can introduce behaviourally trivial constructors as follows: let $\kappa_1 : Sig[SP] \Rightarrow Sig[SP]$ be behaviourally trivial on $Mod[\kappa(SP')]$ w.r.t. OBS, and $\kappa_2 : Sig[SP'] \Rightarrow Sig[SP']$ be behaviourally trivial on $Mod[SP']$ w.r.t. OBS'. Then the correctness of the above behavioural implementation step follows from the correctness of

$$SP \overset{OBS}{\underset{\kappa;\kappa_1}{\leadsto}} \kappa_2(SP').$$

(**Exercise:** Explain why this works.) The resulting task is often easier; see Exercise 8.4.15 below.

Example 8.4.14 (Restriction to sort-generated subalgebra). Consider an algebraic signature Σ with a distinguished set OBS of observable sorts, and let S be a set of sorts in Σ that is disjoint from OBS. Then the constructor $R_S : \Sigma \Rightarrow \Sigma$ which restricts any Σ-algebra to its subalgebra generated by its carriers of sorts not in S (see Example 6.1.13) is behaviourally trivial w.r.t. OBS (this easily follows from Corollary 8.2.10). □

Exercise 8.4.15. Consider a putative behavioural implementation step $SP \overset{OBS}{\underset{\kappa}{\rightsquigarrow}} SP'$ and a set OBS' of observable sorts in $Sig[SP']$, where κ is stable with respect to OBS' and OBS. Show that this step is correct provided that for any $A' \in Mod[SP']$, $\langle \kappa(\langle A' \rangle_{OBS'}) \rangle_{OBS} \in Abs_{OBS}(Mod[SP])$. Discuss how this allows the use of induction in the proof of correctness, even when models of SP' need not be OBS'-generated and the results of κ are not guaranteed to be OBS-generated, and link this with the use of representation invariants in data refinement proofs; see Examples 8.2.11 and 7.2.9. □

Example 8.4.16 (Quotient). Consider an algebraic signature Σ with a distinguished set OBS of observable sorts. An *observable Σ-equation* is a Σ-equation of the form $\forall X_{OBS} \bullet t = t'$ where $t, t' \in |T_\Sigma(X_{OBS})|_o$ for some $o \in OBS$. Consider a set $\mathscr{E}$ of Σ-equations and a Σ-algebra $A \in |\mathbf{Alg}(\Sigma)|$ which satisfies all of the observable consequences of $\mathscr{E}$, that is, for every observable Σ-equation e, if $\mathscr{E} \models e$ then $A \models e$. If A is OBS-generated then $A \equiv_{OBS} A/\mathscr{E}$.[4] To see this, recall from Exercise 3.4.12 that $A/\mathscr{E}$ is the quotient of A by the least congruence $\simeq$ such that $t_A(v) \simeq t'_A(v)$ for all equations $\forall X \bullet t = t'$ in $\mathscr{E}$ and valuations $v: X \to |A|$. It is then easy to check that $\simeq$ is included in the behavioural indistinguishability relation on A — see Definition 8.3.6 — and so $\simeq$ is behavioural.

It follows that if every model in $Mod[SP]$ satisfies all of the observable consequences of $\mathscr{E}$ then the quotient constructor $\_/\mathscr{E}: \Sigma \Rightarrow \Sigma$ (see Example 6.1.11) is behaviourally trivial w.r.t. OBS on OBS-generated subalgebras of the models of SP. Note that the required restriction to the OBS-generated subalgebras of the models of SP may be obtained by inserting a behaviourally trivial restrict constructor; see Example 8.4.14. □

Exercise 8.4.17. Give an example of a Σ-specification SP and a set $\mathscr{E}$ of Σ-equations such that every model of SP satisfies all of the observable consequences of $\mathscr{E}$ but the quotient constructor $\_/\mathscr{E}: \Sigma \Rightarrow \Sigma$ is not behaviourally trivial on $Mod[SP]$. (HINT: Junk of non-observable sorts might influence the congruence induced by a set of equations. See also Exercise 8.3.8.) Notice that, consequently, this quotient constructor is not stable on $Mod[SP]$.

Modify the definition of observable equation by allowing variables of arbitrary sorts. Check that then for any Σ-algebra A — not necessarily OBS-generated — $A \equiv_{OBS} A/\mathscr{E}$ provided that A satisfies all of the so-modified observable consequences of $\mathscr{E}$. But note that this requirement is stronger and therefore more restrictive than the one in Example 8.4.16. □

Exercise 8.4.18. Generalise Example 8.4.16 to quotient constructors as defined in Example 6.1.11 for any specification having a quasi-variety as its model class, and discuss behavioural triviality of the quotient constructor in this case. □

---

[4] Actually, since behavioural equivalence applies only to OBS-compatible algebras, we have to redefine the quotient construction so that on sorts where the congruence relation is the identity, singleton equivalence classes are replaced by their unique elements. Since on the one hand, quotients may be defined up to isomorphism anyway, and on the other hand, we could always switch to working with weak behavioural equivalence as in Section 8.2.3, this is an unimportant detail.

Exercise 8.4.19. Consider a putative behavioural implementation step $SP \overset{OBS}{\underset{\kappa}{\rightsquigarrow}} SP'$ and a set OBS' of observable sorts in $Sig[SP']$, where κ is stable with respect to OBS' and OBS. Show that this step is correct provided that for any $A' \in Mod[SP']$, $\kappa(A'/{\approx_{A'}})/{\approx_{\kappa(A'/\approx_{A'})}} \in Abs_{OBS}(Mod[SP])$. In other words, it is possible to restrict attention to fully abstract quotients of the models of SP' as arguments to κ and fully abstract quotients of its results, even when SP' does not ensure full abstraction and the results of κ are not guaranteed to be fully abstract either. □

In the light of the above examples and results, Propositions 8.4.12 and 8.4.13 explain why the quotient and restriction constructors are not necessary in programming languages. Assuming that the observable sorts are built-in data types of the programming language in use, we can neither restrict them nor quotient them, and so quotient and restriction constructors used in a correct chain of developments can be omitted since they are always behaviourally trivial — otherwise we would arrive at a specification that is inconsistent with the built-in interpretation of the observable data types. On the other hand, Exercises 8.4.15 and 8.4.19 show how the quotient and restriction constructions may be used as model-theoretic tools for proving correctness of behavioural implementations.

Example 8.4.20 (Free extension). Given a signature morphism $\sigma: \Sigma \to \Sigma'$ and set $\mathscr{E}'$ of Σ'-equations, the free extension constructor $F_{\sigma,\mathscr{E}'}: \Sigma \Rightarrow \Sigma'$, as defined in Example 6.1.7, need not be stable.

However, if the free extension $F_{\sigma,\mathscr{E}'}: \Sigma \Rightarrow \Sigma'$ is naturally persistent (see Definition 6.1.20 but note that $F_{\sigma,\mathscr{E}'}$ is total here), then for any set OBS of observable sorts in Σ, $F_{\sigma,\mathscr{E}'}$ is stable with respect to OBS and $\sigma(OBS)$. This follows by Lemma 8.4.8 since $F_{\sigma,\mathscr{E}'}$ is the object part of a functor $\mathbf{F}_{\sigma,\mathscr{E}'}$, which makes all diagrams of the following form commute:

$$\begin{array}{ccc} A & \xrightarrow{\;id_A\;} & \mathbf{F}_{\sigma,\mathscr{E}'}(A)|_\sigma \\ {\scriptstyle h}\downarrow & & \downarrow{\scriptstyle \mathbf{F}_{\sigma,\mathscr{E}'}(h)|_\sigma} \\ B & \xrightarrow[\;id_B\;]{} & \mathbf{F}_{\sigma,\mathscr{E}'}(B)|_\sigma \end{array}$$

Thus, sufficient completeness and hierarchy consistency may be used to ensure stability of free extensions via Proposition 6.1.24. □

Exercise 8.4.21. Give an example of a free extension that is not stable (and hence is not naturally persistent). □

Non-example 8.4.22 (Translation of a stable constructor). Consider the following pushout diagram in **AlgSig**

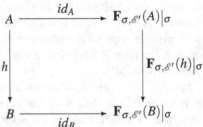

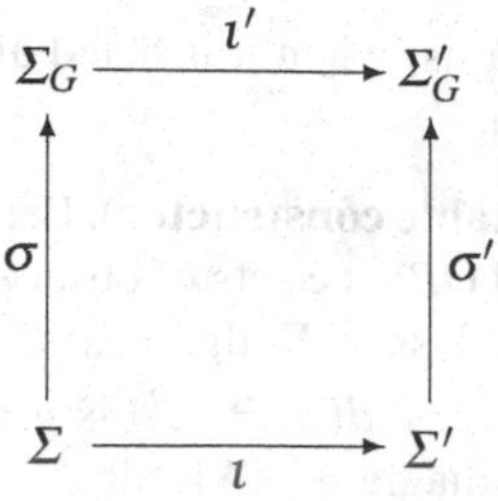

and a persistent constructor $F \in Mod(\Sigma \stackrel{\iota}{\Rightarrow} \Sigma')$ that is stable with respect to *OBS* and *OBS*′ of observable sorts in Σ and Σ' respectively. Recall from Example 6.1.28 that the translation of F along σ is a persistent constructor $\sigma(F) \in Mod(\Sigma_G \stackrel{\iota'}{\Rightarrow} \Sigma'_G)$. However, $\sigma(F)$ need not be stable with respect to $\sigma(OBS)$ and $\sigma'(OBS')$ since Σ_G may introduce new observable contexts for sorts from Σ. □

Exercise 8.4.23. Give an example of a pushout as above, a persistent constructor $F \in Mod(\Sigma \stackrel{\iota}{\Rightarrow} \Sigma')$ and a set *OBS* of observable sorts in Σ such that F is stable w.r.t. *OBS* and $\iota(OBS)$ but $\sigma(F)$ is not stable w.r.t. $\sigma(OBS)$ and $\iota'(\sigma(OBS))$. □

The fact that stability is not preserved under translation is a serious problem, since this construction is involved whenever a generic module is instantiated with an argument that originates from a richer context than is required by the parameter signature, as for example in Tasks 3 and 5 in Section 7.4. Conditions under which translation preserves stability may be given but they would exclude many of the situations of practical interest.

What is required is a more refined treatment of generic modules. In the non-behavioural case, studied in Chapter 7, constructors as used in implementation steps corresponded exactly to generic modules. In the behavioural case, there is a subtle difference. The constructor in a behavioural implementation of a Σ-specification by a Σ'-specification comes with a fixed context of use: it is always applied to a Σ'-algebra. In contrast, a generic module which is to be implemented as a subtask in the course of modular decomposition is used together with other modules, perhaps instantiated in a number of different contexts. Each of these contexts may include (different) extra operations, beyond those in its argument and result signatures, that allow observations of components in its argument and result. Therefore its implementation must be (behaviourally) correct in each of these contexts. This requires stronger correctness and stability conditions that are robust under change of application context.

8.4.4 Global stability and behavioural correctness

One conclusion from the way that generic modules are used is that the stability requirement in Section 8.4.2 needs to be strengthened to characterise constructors that are not merely stable in themselves but remain stable when instantiated in an arbitrary global context — that is, when translated along any signature morphism. Hence the following definition:

Definition 8.4.24 (Globally stable constructor). Let $\iota : \Sigma \to \Sigma'$ be an algebraic signature morphism and OBS and OBS' be sets of observable sorts in Σ and Σ', respectively. Let $\mathscr{A} \subseteq |\mathbf{Alg}(\Sigma)|$ be a class of Σ-algebras.

A persistent constructor $F \in Mod(\Sigma \overset{\iota}{\Rightarrow} \Sigma')$ is *globally stable on* $\mathscr{A}$ *w.r.t.* OBS *and* OBS' if for every pushout diagram in **AlgSig**

$$\begin{array}{ccc} \Sigma_G & \xrightarrow{\iota'} & \Sigma'_G \\ \uparrow{\scriptstyle \sigma} & & \uparrow{\scriptstyle \sigma'} \\ \Sigma & \xrightarrow[\iota]{} & \Sigma' \end{array}$$

and set OBS_G of observable sorts in Σ_G such that $\sigma(OBS) \subseteq OBS_G$, the translation of F along σ, $\sigma(F) \in Mod(\Sigma_G \overset{\iota'}{\Rightarrow} \Sigma'_G)$, is stable on $\mathscr{A}_G = \{A_G \in |\mathbf{Alg}(\Sigma_G)| \mid A_G|_\sigma \in \mathscr{A}\}$ w.r.t. OBS_G and OBS'_G, where $OBS'_G = \iota'(OBS_G) \cup \sigma'(OBS')$. We say that F is *globally stable* (*w.r.t.* OBS *and* OBS') if it is globally stable on the class $dom(F)$. □

Obviously, global stability implies stability: just take σ to be identity and $OBS = OBS_G$.

Exercise 8.4.25. Show that global stability is preserved by composition, translation of a constructor along a signature morphism (Example 6.1.28), and amalgamation of two constructors having a common source signature (Example 6.1.31). □

Establishing global stability of a constructor directly from the definition would be very difficult because of the quantification over all possible global contexts of use (that is, every signature morphism $\sigma : \Sigma \to \Sigma_G$ and every set OBS_G of observable sorts from Σ_G). Exercise 8.4.25 indicates how this task may be reduced to proving the global stability of "basic" constructors. For these there is fortunately a sufficient condition for global stability that is expressed entirely in *local* terms and that covers all the examples of interest (in fact, it covers all practical examples, by Exercise 8.4.33 below):

Theorem 8.4.26. *A constructor* $F \in Mod(\Sigma \overset{\iota}{\Rightarrow} \Sigma')$ *is globally stable w.r.t.* OBS *and* OBS' *on* $\mathscr{A} \subseteq |\mathbf{Alg}(\Sigma)|$ *if it* extends correspondences *on* $\mathscr{A}$, *that is, for any correspondence* $\rho : A \bowtie_{OBS} B$ *with* $A \in \mathscr{A}$ *there is a correspondence* $\rho' : F(A) \bowtie_{OBS'} F(B)$ *such that* $\rho = \rho'|_\iota$.

Proof. Consider the pushout diagram

$$\begin{array}{ccc} \Sigma_G & \xrightarrow{\iota'} & \Sigma'_G \\ \uparrow{\scriptstyle\sigma} & & \uparrow{\scriptstyle\sigma'} \\ \Sigma & \xrightarrow[\iota]{} & \Sigma' \end{array}$$

and set OBS_G of observable sorts in Σ_G such that $\sigma(OBS) \subseteq OBS_G$, with $OBS'_G = \iota'(OBS_G) \cup \sigma'(OBS')$, and two Σ_G-algebras A_G and B_G that are behaviourally equivalent w.r.t. OBS_G with $A_G|_\sigma \in \mathscr{A}$. By Theorem 8.2.8, there exists a correspondence $\rho_G{:}A_G \bowtie_{OBS_G} B_G$. Then we have $\rho_G|_\sigma{:}A_G|_\sigma \bowtie_{OBS} B_G|_\sigma$ and so by the assumption there exists a correspondence $\rho'{:}F(A_G|_\sigma) \bowtie_{OBS'} F(B_G|_\sigma)$ such that $\rho_G|_\sigma = \rho'|_\iota$. Amalgamation of correspondences ρ_G and ρ' yields a correspondence $\rho'_G{:}\sigma(F)(A_G) \bowtie_{OBS'_G} \sigma(F)(B_G)$. Thus $\sigma(F)(A_G)$ and $\sigma(F)(B_G)$ are behaviourally equivalent w.r.t. OBS'_G. □

Exercise 8.4.27. Check that the Amalgamation Lemma for homomorphisms (Exercise 3.4.34) extends to correspondences, as used in the proof above. □

Exercise 8.4.28. Show that the property of extending correspondences is preserved by composition, translation of a constructor along a signature morphism (Example 6.1.28), and amalgamation of two constructors having a common source signature (Example 6.1.31). □

The following lemma is a version of Lemma 8.4.8 for global stability, as required to prove global stability of constructors below.

Lemma 8.4.29. *Consider two algebraic signatures Σ and Σ' with sets OBS and OBS′ of observable sorts in Σ and Σ' respectively with a signature morphism $\iota{:}\Sigma \to \Sigma'$, and a class of Σ-algebras $\mathscr{A} \subseteq |\mathbf{Alg}(\Sigma)|$. Let $\mathbf{F}{:}\mathbf{Alg}(\Sigma) \to \mathbf{Alg}(\Sigma')$ be a functor that extends behavioural homomorphisms on $Abs_{OBS}(\mathscr{A})$, that is, for each OBS-behavioural Σ-homomorphism h between algebras in $Abs_{OBS}(\mathscr{A})$, $\mathbf{F}(h)$ is an OBS′-behavioural Σ'-homomorphism such that $\mathbf{F}(h)|_\iota = h$. Then the object part of $\mathbf{F}$, $F{:}\Sigma \Rightarrow \Sigma'$, is a persistent constructor which is globally stable on $\mathscr{A}$ w.r.t. OBS and OBS′.*

Proof. First note that F is persistent on $Abs_{OBS}(\mathscr{A})$ by applying the assumption to the identity homomorphism. Then, consider an OBS-correspondence $\rho{:}A \bowtie_{OBS} B$ on Σ-algebras $A \in \mathscr{A}$ and $B \in |\mathbf{Alg}(\Sigma)|$. By Exercise 8.2.15, there exists an algebra $C \in |\mathbf{Alg}(\Sigma)|$ and Σ-homomorphisms $h_A{:}C \to A$ and $h_B{:}C \to B$ that are behavioural w.r.t. OBS such that $\rho = (h_A)^{-1};h_B$. By the assumption, and since $A, B, C \in Abs_{OBS}(\mathscr{A})$, we have Σ'-homomorphisms $\mathbf{F}(h_A){:}F(C) \to F(A)$ and $\mathbf{F}(h_B){:}F(C) \to F(B)$ that are behavioural w.r.t. OBS' such that $\mathbf{F}(h_A)|_\iota = h_A$ and $\mathbf{F}(h_B)|_\iota = h_B$. Thus, $\rho' = (\mathbf{F}(h_A))^{-1};\mathbf{F}(h_B)$ is an OBS'-correspondence $\rho'{:}F(A) \bowtie_{OBS'} F(B)$, and moreover $\rho'|_\iota = \rho$. So by Theorem 8.4.26, F is globally stable on $\mathscr{A}$. □

Exercise 8.4.30. Check that from the lemma above, total naturally persistent free extensions (Example 8.4.20) are globally stable. □

Exercise 8.4.31. For reducts (Example 8.4.9), consider an algebraic signature morphism $\sigma: \Sigma' \to \Sigma$ and sets *OBS* and *OBS'* of observable sorts in Σ and Σ' respectively such that $\sigma(OBS') \subseteq OBS$. Then for any *OBS*-correspondence ρ, the σ-reduct $\rho|_\sigma$ is an *OBS'*-correspondence as well. Join this with Proposition 6.1.17 to check that given $\iota: \Sigma \to \Sigma'$ such that $\iota;\sigma = id_\Sigma$, the persistent constructor $\_|_\sigma \in Mod(\Sigma \overset{\iota}{\Rightarrow} \Sigma')$ is globally stable w.r.t. *OBS* and *OBS'*. □

The restrict and quotient constructors (Examples 8.4.16 and 8.4.16) are rarely persistent, let alone globally stable, but that makes no difference since they are superfluous in development chains as argued in Section 8.4.3.

Exercise 8.4.32. According to Theorem 8.4.26, the property of extending correspondences is a sufficient condition for global stability. Imposing an additional requirement on the correspondences involved yields a condition that is both sufficient and necessary.

Consider a signature Σ with a set *OBS* of observable sorts. A correspondence $\rho: A \bowtie_{OBS} B$ is *closed* if whenever $a \; \rho_s \; b$, $a \; \rho_s \; b'$ and $a' \; \rho_s \; b$, then also $a' \; \rho_s \; b'$ for each sort s in Σ, $a, a' \in |A|_s$ and $b, b' \in |B|_s$. Check that for any correspondence $\rho: A \bowtie_{OBS} B$ there is a least closed correspondence $\widehat{\rho}: A \bowtie_{OBS} B$ that contains ρ. Consequently, two Σ-algebras are behaviourally equivalent w.r.t. *OBS* iff there is a closed correspondence between them.

A constructor $F \in Mod(\Sigma \overset{\iota}{\Rightarrow} \Sigma')$ *extends closed correspondences on* $\mathscr{A} \subseteq |\mathbf{Alg}(\Sigma)|$ if for any closed correspondence $\rho: A \bowtie_{OBS} B$ with $A \in \mathscr{A}$ there exists a closed correspondence $\rho': F(A) \bowtie_{OBS'} F(B)$ such that $\rho = \rho'|_\iota$, where *OBS'* is a set of observable sorts in Σ'. Check that the proof of Theorem 8.4.26 may be reused to show that if F extends closed correspondences on $\mathscr{A}$ then it is globally stable on $\mathscr{A}$ with respect to *OBS* and *OBS'*.

Then check that given correspondences $\rho': F(A) \bowtie_{OBS'} F(B)$ and $\rho: A \bowtie_{OBS} B$, if $\rho'|_\iota = \rho$ then $\widehat{\rho'}|_\iota = \widehat{\rho}$ as well. One consequence of this is that if F extends correspondences (on $\mathscr{A}$), it extends closed correspondences (on $\mathscr{A}$) as well.

Consider now a closed correspondence $\widehat{\rho}: A \bowtie_{OBS} B$. Construct the extension Σ_G of Σ by a new sort *Bool* and the following new operations:

- $true, false: Bool$;
- for each sort s in Σ and $\langle a, b \rangle \in \widehat{\rho}_s$, $!^{a,b}: s$;
- for each sort s in Σ and $b \in |B|_s$, $?^b: s \to Bool$; and
- for each sort s in Σ, $?: s \to Bool$

and let $\sigma: \Sigma \to \Sigma_G$ be the signature inclusion. Let $OBS_G = OBS \cup \{Bool\}$. Construct the following expansions A_G and B_G of A and B respectively:

- $|A_G|_{Bool} = |B_G|_{Bool} = \{tt, ff\}$, $true_{A_G} = true_{B_G} = tt$ and $false_{A_G} = false_{B_G} = ff$;
- for each sort s in Σ and $a \; \widehat{\rho}_s \; b$, $!^{a,b}_{A_G} = a$ and $!^{a,b}_{B_G} = b$;

- for each sort s in Σ and $b \in |B|_s$, $?^b_{A_G}(a) = tt$ if $a\,\widehat{\rho}_s\,b$, and $?^b_{A_G}(a) = ff$ otherwise; $?^b_{B_G}(b') = tt$ if there exists $a \in |A|_s$ such that $a\,\widehat{\rho}_s\,b$ and $a\,\widehat{\rho}_s\,b'$, and $?^b_{B_G}(b') = ff$ otherwise;
- for each sort s in Σ and $a \in |A|_s$, $?_{A_G}(a) = tt$; and for each $b \in |B|_s$, $?_{B_G}(b) = tt$ if there exists $a \in |A|_s$ such that $a\,\widehat{\rho}_s\,b$ and $?_{B_G}(b) = ff$ otherwise.

Check that $\widehat{\rho}$ extended by the identity on $|A_G|_{Bool} = |B_G|_{Bool}$ is the only OBS_G-correspondence between A_G and B_G.

Use this to prove that if $F \in Mod(\Sigma \overset{\iota}{\Rightarrow} \Sigma')$ is globally stable on $\mathscr{A}$ w.r.t. OBS and OBS' then it extends closed correspondences on $\mathscr{A}$. □

Exercise 8.4.33. Prove that extending correspondences on a class of algebras closed under behavioural equivalence is a sufficient and necessary condition for a constructor to be globally stable on that class. HINT: It is a sufficient condition by Theorem 8.4.26. To prove that it is also necessary, start with Exercise 8.4.32. Then notice that behavioural homomorphisms and their inverses are closed correspondences, and use Exercise 8.2.15 to show that if a constructor extends closed correspondences on the behavioural closure of a class of algebras then it extends arbitrary correspondences on that class. □

Let us turn now to the issue of behavioural correctness of generic modules.

Definition 8.4.34 (Behaviourally correct constructor). Let $\iota: \Sigma \to \Sigma'$ be a signature morphism, let SP and SP' be Σ- and Σ'-specifications respectively, and let OBS' be a set of observable sorts in Σ'. A persistent constructor $F \in Mod(\Sigma \overset{\iota}{\Rightarrow} \Sigma')$ is *behaviourally correct along ι w.r.t. SP and SP' for OBS'* if for every $A \in Mod[SP]$ there exists a model $A' \in Mod[SP']$ and correspondence $\rho': A' \bowtie_{OBS'} F(A)$ such that $\rho'|_\iota = id_{|A|}$. □

The definition of behavioural correctness does not take into account which sorts of the argument specification SP are to be regarded as observable, requiring behavioural correctness of the result while taking *all* of the sorts in SP (in addition to those in OBS') as observable. Informally, this is because the application context may make arbitrary observations on the argument sorts and hence any of them may become fully observable.

This also restricts the use of behaviourally trivial constructors in verifying behavioural correctness as suggested in Exercises 8.4.15 and 8.4.19. While they can be meaningfully used to (for instance) restrict the result to an $\iota(sorts(\Sigma)) \cup OBS'$-generated subalgebra, trying to do the same on the argument brings no advantage since we take all the argument sorts as observable.

Exercise 8.4.35. Under the notation of Definition 8.4.34, consider a persistent constructor $F \in Mod(\Sigma \overset{\iota}{\Rightarrow} \Sigma')$. Show that the following conditions are equivalent:

- F is behaviourally correct along ι w.r.t. SP and SP' for OBS'.
- There exists $F' \in Mod[SP \overset{\iota}{\Rightarrow} SP']$ such that $F'(A) \equiv_{\iota(sorts(\Sigma)) \cup OBS'} F(A)$ for each $A \in Mod[SP]$.
- $F(Mod[SP]) \subseteq Abs_{\iota(sorts(\Sigma)) \cup OBS'}(SP')$ and $Mod[SP] \subseteq dom(F)$. □

A conclusion from the above is that "good" models of a constructor specification are those that are both behaviourally correct and globally stable:

Definition 8.4.36 (Behavioural model of a constructor specification). Let $\iota\colon\Sigma\to\Sigma'$ be a signature morphism, let SP and SP' be Σ- and Σ'-specifications respectively, and let OBS and OBS' be sets of observable sorts in Σ and Σ' respectively. We write $Mod[SP\,{}_{OBS}\!\overset{\iota}{\Rrightarrow}_{OBS'}\,SP']$ for the class of all *behavioural models of* $SP \overset{\iota}{\Rrightarrow} SP'$ *w.r.t.* OBS *and* OBS', that is, constructors that are globally stable w.r.t. OBS and OBS' and are behaviourally correct along ι w.r.t. SP and SP' for OBS'. □

It may happen that a consistent constructor specification has no behavioural models:

Exercise 8.4.37. Give an example of a consistent constructor specification of the form $SP \overset{\iota}{\Rrightarrow} SP'$ and sets OBS and OBS' of observable sorts in $Sig[SP]$ and $Sig[SP']$ respectively such that $Mod[SP\,{}_{OBS}\!\overset{\iota}{\Rrightarrow}_{OBS'}\,SP']$ is empty. See also Exercise 6.2.23. HINT: Consider a specification SP with non-observable sort s and constants $a,b\colon s$ and its extension SP' by a new observable sort s' with constants $c,d\colon s'$ and axiom $c\neq d \Leftrightarrow a=b$. □

Exercise 8.4.38. Consider a signature morphism $\iota\colon\Sigma\to\Sigma'$, Σ-specifications SP and SP_1, and Σ'-specifications SP' and SP'_1. Show that for any behavioural model $F\in Mod[SP\,{}_{OBS}\!\overset{\iota}{\Rrightarrow}_{OBS'}\,SP']$, we also have $F\in Mod[SP_1\,{}_{OBS}\!\overset{\iota}{\Rrightarrow}_{OBS'}\,SP'_1]$ provided that $Mod[SP_1]\subseteq Abs_{OBS}(Mod[SP])$ and $Mod[SP']\subseteq Abs_{\iota(sorts(\Sigma))\cup OBS'}(Mod[SP'_1])$. □

Exercise 8.4.39. Consider behavioural models $F_1\in Mod[SP_1\,{}_{OBS_1}\!\overset{\iota_1}{\Rrightarrow}_{OBS'_1}\,SP'_1]$ and $F_2\in Mod[SP_2\,{}_{OBS_2}\!\overset{\iota_2}{\Rrightarrow}_{OBS'_2}\,SP'_2]$. Check that if $SP'_1\equiv SP_2$ and $OBS'_1=OBS_2$ then $F_1;F_2\in Mod[SP_1\,{}_{OBS_1}\!\overset{\iota_1;\iota_2}{\Rrightarrow}_{OBS'_2}\,SP'_2]$. Check also that this holds under weaker assumptions, namely that $Sig[SP'_1]=Sig[SP_2]$, $Mod[SP'_1]\subseteq Abs_{OBS_2}(Mod[SP_2])$ and $OBS'_1\supseteq OBS_2$. □

Proposition 8.4.40. *Given specifications SP, SP′, signature morphism* $\sigma\colon Sig[SP]\to Sig[SP']$*, and sets OBS, OBS′ of observable sorts in* $Sig[SP]$ *and* $Sig[SP']$*, respectively, consider a behavioural model* $F\in Mod[SP\,{}_{OBS}\!\overset{\sigma}{\Rrightarrow}_{OBS'}\,SP']$*. Let* $\iota\colon SP''\to SP'$ *be a specification morphism and OBS″ be a set of observable sorts in* $Sig[SP'']$ *such that* $\iota(OBS'')\subseteq OBS'$*. Let* $\sigma'\colon Sig[SP]\to Sig[SP'']$ *be a signature morphism such that* $\sigma=\sigma';\iota$*, Then* $F;\_|_\iota\in Mod[SP\,{}_{OBS}\!\overset{\sigma'}{\Rrightarrow}_{OBS''}\,SP'']$. □

Exercise 8.4.41. Carry out the proof of Proposition 8.4.40, building on Fact 6.2.17. □

Lemma 8.4.42. *Let* $\iota\colon\Sigma\to\Sigma'$ *be a signature morphism, let SP and SP′ be* Σ*- and* Σ'*-specifications respectively, and let OBS and OBS′ be sets of observable sorts in* Σ *and* Σ' *respectively. Consider a behavioural model* $F\in Mod[SP\,{}_{OBS}\!\overset{\iota}{\Rrightarrow}_{OBS'}\,SP']$ *and a pushout diagram*

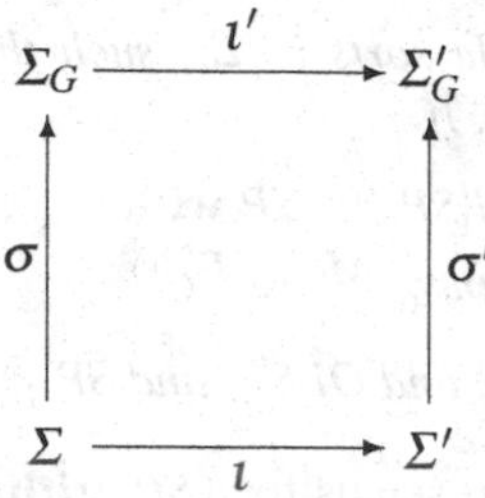

with a set OBS_G *of observable sorts in* Σ_G *such that* $\sigma(OBS) \subseteq OBS_G$, *and let* $OBS'_G = \iota'(OBS_G) \cup \sigma'(OBS')$. *We then have:*

$$\sigma(F) \in Mod[(SP\ \mathbf{with}\ \sigma)\ {}_{OBS_G}\!\overset{\iota'}{\Rightarrow}_{OBS'_G} (SP'\ \mathbf{with}\ \sigma')].$$

Proof. $\sigma(F)$ is globally stable by Exercise 8.4.25. To prove behavioural correctness, consider $A_G \in Mod[SP\ \mathbf{with}\ \sigma]$. Then $A_G|_\sigma \in Mod[SP]$, and so by behavioural correctness of F there exists $A' \in Mod[SP']$ and a correspondence $\rho' : A' \bowtie_{OBS'} F(A_G|_\sigma)$ with identity reduct $\rho'|_\iota = id_{A_G|_\sigma}$. Consider the unique Σ'_G-algebra A'_G such that $A'_G|_{\iota'} = A_G$ and $A'_G|_{\sigma'} = A'$. Then amalgamation of the identity $id_{A_G} : A_G \bowtie_{sorts(\Sigma_G)} A_G$ and $\rho' : A' \bowtie_{OBS'} F(A_G|_\sigma)$ yields a correspondence $\rho'_G : A'_G \bowtie_{OBS'_G} \sigma(F)(A_G)$, which completes the proof since $A'_G \in Mod[SP'\ \mathbf{with}\ \sigma']$. □

By the above lemma, whenever $F \in Mod[SP\ {}_{OBS}\!\overset{\iota}{\Rightarrow}_{OBS'} SP']$ and given a pushout diagram as above, $\sigma(F)$ may be used as a stable constructor to behaviourally implement Σ'_G-specifications by Σ_G-specifications. Namely, if SP_G and SP'_G are Σ_G- and Σ'_G-specifications respectively, then $SP'_G \overset{OBS'_G}{\underset{\sigma(F)}{\rightsquigarrow}} SP_G$ provided that:

(i) $Mod[SP_G] \subseteq Abs_{OBS_G}(Mod[SP\ \mathbf{with}\ \sigma])$; and
(ii) $Mod[SP'\ \mathbf{with}\ \sigma'] \subseteq Abs_{OBS'_G}(Mod[SP'_G])$.

But while condition (i) is acceptable, condition (ii) is too strong since it requires all the requirements in SP'_G to follow (up to behavioural equivalence) from the result specification SP' for the generic module. We need a condition that takes into account both SP' and the specification SP_G of the application context in ensuring SP'_G.

Theorem 8.4.43. *Let* $\iota : \Sigma \to \Sigma'$ *be a signature morphism, let* SP *and* SP' *be* Σ*- and* Σ'*-specifications respectively, and let* OBS *and* OBS' *be sets of observable sorts in* Σ *and* Σ' *respectively. Consider a behavioural model* $F \in Mod[SP\ {}_{OBS}\!\overset{\iota}{\Rightarrow}_{OBS'} SP']$ *and a pushout diagram*

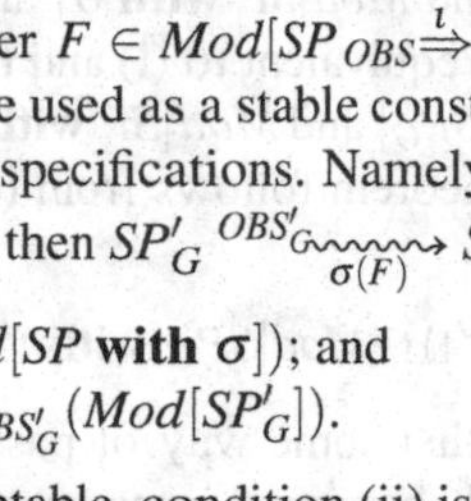

with a set OBS_G of observable sorts in Σ_G such that $\sigma(OBS) \subseteq OBS_G$, and let $OBS'_G = \iota'(OBS_G) \cup \sigma'(OBS')$. If

(i) $Mod[SP_G] \subseteq Abs_{OBS_G}(Mod[SP_G \cup (SP \textbf{ with } \sigma)])$*; and*
(ii) $Mod[SP' +_{\iota,\sigma} SP_G] \subseteq Abs_{OBS'_G}(Mod[SP'_G])$

then $\sigma(F)$ is stable w.r.t. OBS_G and OBS'_G, and $SP'_G \;{}^{OBS'_G}\!\!\underset{\sigma(F)}{\rightsquigarrow} SP_G$.

Proof. Recall that $SP' +_{\iota,\sigma} SP_G$ stands for $(SP' \textbf{ with } \sigma') \cup (SP_G \textbf{ with } \iota')$; see Section 5.2. Stability of $\sigma(F)$ follows as before. Let $A_G \in Mod[SP_G]$. Then by (i), $A_G \equiv_{OBS_G} B_G$ for some $B_G \in Mod[SP_G \cup (SP \textbf{ with } \sigma)]$. By Lemma 8.4.42, for some $B'_G \in Mod[SP' \textbf{ with } \sigma']$ with $B'_G|_{\iota'} = B_G \in Mod[SP_G]$, $\sigma(F)(B_G) \equiv_{OBS'_G} B'_G$. Hence $B'_G \in Abs_{OBS'_G}(Mod[SP'_G])$ by (ii). By stability of $\sigma(F)$, $\sigma(F)(A_G) \equiv_{OBS'_G} \sigma(F)(B_G) \equiv_{OBS'_G} B'_G$, and so $\sigma(F)(A_G) \in Abs_{OBS'_G}(Mod[SP'_G])$. □

Requirement (i) is perhaps stronger than expected. But note that it straightforwardly follows from the inclusion of model classes $Mod[SP_G] \subseteq Mod[SP \textbf{ with } \sigma]$ (or equivalently, $Mod[SP_G]|_\sigma \subseteq Mod[SP]$), which is often easy to verify.

Exercise 8.4.44. Show that (i) in Theorem 8.4.43 is strictly stronger than

(i′) $Mod[SP_G] \subseteq Abs_{OBS}(Mod[SP \textbf{ with } \sigma])$.

Check that the weaker condition (i′) is not sufficient for the above theorem, but if the model classes $Mod[SP_G]$ and $Mod[SP \textbf{ with } \sigma]$ are behaviourally closed — see Definition 8.3.30 — then (i′) is equivalent to (i) and therefore sufficient.

Check that even when $Mod[SP_G]$ and $Mod[SP \textbf{ with } \sigma]$ are not both behaviourally closed, the conclusion of the theorem follows from (i′) together with the following stronger version of (ii):

(ii′) $Abs_{OBS'_G}(Mod[SP' \textbf{ with } \sigma']) \cap Mod[SP_G \textbf{ with } \iota'] \subseteq Abs_{OBS'_G}(Mod[SP'_G])$.

An informal conclusion is that some way of passing information from SP_G to SP'_G that is independent of the behavioural interpretation of the generic module and its correctness is needed; this results in some inconvenience of verification on either the argument side or the result side. □

Corollary 8.4.45. *Under the notation of Theorem 8.4.43, if*

(i) $Mod[SP_G] \subseteq Mod[SP \textbf{ with } \sigma]$*; and*
(ii) $Mod[SP' +_{\iota,\sigma} SP_G] \subseteq Abs_{OBS'_G}(Mod[SP'_G])$

then $\sigma(F) \in Mod[SP_G\, {}_{OBS_G}\!\!\overset{\iota'}{\Rightarrow}_{OBS'_G} SP'_G]$. □

Exercise 8.4.46. Modify the proof of Theorem 8.4.43 to justify the stronger conclusion of Corollary 8.4.45. HINT: Under (i) of Corollary 8.4.45, we may take $B_G = A_G$. □

Corollary 8.4.45 offers a behavioural version of Fact 6.2.18. The following proposition for amalgamated union of constructors plays the same role for Fact 6.2.19:

Proposition 8.4.47. *Consider the following pushout diagram:*

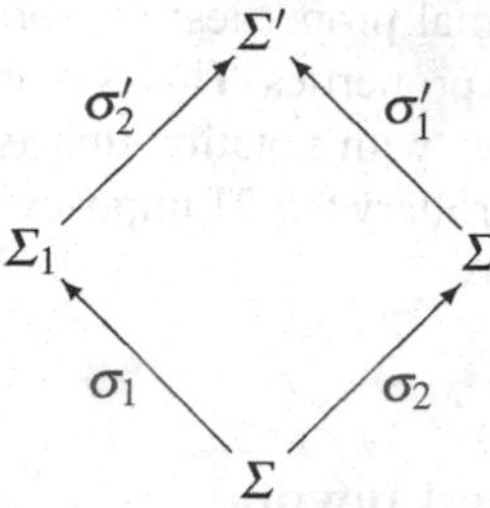

Let SP, SP_1 and SP_2 be Σ-, Σ_1- and Σ_2-specifications respectively, and let OBS, OBS_1 and OBS_2 be sets of observable sorts in Σ, Σ_1 and Σ_2 respectively. Consider behavioural models $F_1 \in Mod[SP_{\,OBS}\overset{\sigma_1}{\Longrightarrow}_{OBS_1} SP_1]$ and $F_2 \in Mod[SP_{\,OBS}\overset{\sigma_2}{\Longrightarrow}_{OBS_2} SP_2]$. Then the amalgamated union constructor of F_1 and F_2 as defined in Example 6.1.31 is a behavioural model $F_1 + F_2 \in Mod[SP_{\,OBS}\overset{\sigma}{\Longrightarrow}_{OBS'} SP']$ where $\sigma = \sigma_1;\sigma_2'$, $SP' = SP_1 +_{\sigma_1,\sigma_2} SP_2$ and $OBS' = \sigma_2'(OBS_1) \cup \sigma_1'(OBS_2)$.

Proof. $F_1 + F_2$ is a persistent and globally stable constructor by Exercise 8.4.25. To show behavioural correctness, consider any $A \in Mod[SP]$. By the assumptions, there are models $A_1 \in Mod[SP_1]$ and $A_2 \in Mod[SP_2]$ and correspondences $\rho_1 : A_1 \bowtie_{OBS_1} F_1(A)$ and $\rho_2 : A_2 \bowtie_{OBS_2} F_2(A)$ such that $\rho_1|_{\sigma_1} = id_{|A|}$ and $\rho_2|_{\sigma_2} = id_{|A|}$. Then amalgamating A_1 and A_2 yields a model $A' \in Mod[SP']$ and amalgamating ρ_1 and ρ_2 yields a correspondence $\rho' : A' \bowtie_{OBS'} (F_1 + F_2)(A)$ such that $\rho'|_{\sigma} = id_{|A|}$. □

8.4.5 Summary

The conclusion from this section is that to systematically develop a correct realisation of a specification of requirements, one should proceed via a sequence of behavioural implementation steps, using constructors that are behaviourally correct and stable. Behavioural correctness needs to be checked for each constructor on a case-by-case basis, relying in particular on Theorem 8.4.43 and Corollary 8.4.45 for the use of generic modules in a global context.

Stability is another matter: it should be viewed as a directive for language design, rather than as a condition to be checked for each constructor separately. Ultimately, constructors are generic modules defined in a programming language; recall the motivation for constructors as models of generic modules from Chapter 6. In a programming language with good modularisation facilities, all constructions that one can code should be (globally) stable. Technically, this is achieved by ensuring that basic generic modules extend correspondences and that this property is preserved by the various means we have for combining modules. The analysis and examples in Sections 8.4.3 and 8.4.4 do this for the examples of basic constructors and combinators from Chapter 6.

Stability corresponds to the programming language design principle that generic modules should respect encapsulation boundaries. That is, such modules may use the components provided by their imported parameters, but they must not take advantage of their particular internal properties: any branching in the code must be governed by directly observable properties. This is most directly visible in the richer context of **FPL** discussed below, with notation that is closer to the syntax of programming languages, where Corollary 8.5.21 imposes exactly this condition ("*OBS*-admissibility") to ensure stability.

8.5 To partial algebras and beyond

In the previous sections, we studied behavioural specifications in the simple framework of standard algebraic signatures and total algebras.

We will now consider other frameworks, beginning in Section 8.5.1 with the institution **FPL**; see Example 4.1.25. **FPL** provides a richer language of terms, and allows us to write more interesting examples than the standard algebraic framework considered so far in this chapter; see Section 8.5.2. However, before we can consider examples, we need to provide the infrastructure required for a behavioural view of **FPL**-specifications. The basic definitions of notions such as behavioural equivalence will require adjustment, and there are some subtle points that deserve special attention. Nevertheless, much of the material on behavioural specifications in **FPL** is repetitive with respect to the developments of the previous sections. We therefore often refrain from restating exactly those definitions and results for **FPL** that carry over directly from the previous sections. We then try to capture such commonalities by moving these developments to the context of an arbitrary institution, in Section 8.5.3.

8.5.1 Behavioural specifications in FPL

There are two crucial changes in **FPL** with respect to the standard algebraic framework that we have considered up to this point. First, we have a richer notion of model, where operations may be partial and some sorts are constrained to have carriers that are freely generated by sets of value constructors. Then we have a richer term language which includes recursive definitions and pattern matching on values of sorts constrained by sets of value constructors.

These changes need to be taken into account in giving a behavioural view of **FPL**-specifications. Some of the definitions will have to do this explicitly: for instance, the definition of behavioural equivalence will explicitly take undefinedness into account. Even where there is no change in wording, the notion captured may not be the same. For instance, behavioural equivalence is still defined via term evaluation, but now the language of terms is much richer. This gives more observational power, leading to a finer notion of behavioural equivalence; see Exercise 8.5.3 below.

Consider an **FPL**-signature $\mathsf{SIG} = \langle S, \Omega, D \rangle$ and a set $OBS \subseteq S$ of observable sorts.

Exercise 8.5.1 (Compatible models). As in Definition 8.2.1, define two SIG-models to be *OBS-compatible* if their carriers of sorts in *OBS* coincide.

We will deal below with behavioural equivalence of *OBS*-compatible models. Check that the "weak" versions of the concepts and results from Section 8.2.3 carry over to **FPL**. We will use them below whenever convenient without further explicit reference. □

Definition 8.5.2 (Behavioural equivalence). Two *OBS*-compatible SIG-models A, B are *behaviourally equivalent w.r.t. OBS*, written $A \equiv_{OBS} B$ (or simply, *behaviourally equivalent*, written $A \equiv B$, if *OBS* is clear), if for all sets of variables X_{OBS} (of observable sorts), terms $t \in |T_{\mathsf{SIG}}(X_{OBS})|$, and valuations $v : X_{OBS} \to |A|_{OBS}$ (so that $v : X_{OBS} \to |B|_{OBS}$ as well), $t_A(v)$ is defined iff $t_B(v)$ is defined, and if so then moreover $t_A(v) = t_B(v)$ when t is of a sort in *OBS*. □

This definition presumes that definedness of an *arbitrary* term is observable. While definedness of terms of observable sorts should clearly be observable, it is not so clear that definedness of other terms should be observable as well. But under the mild assumption that there is at least one observable sort $o \in OBS$ having a non-empty carrier, definedness of a term t of an arbitrary sort s is captured by the definedness of the following term of sort o: **let fun** $f(x{:}s){:}o = y$ **in** $f(t)$. **(Exercise:** Spell this out, taking explicit account of free variables.) This mild assumption may be discharged by requiring, for instance, that *Bool* be observable (recall that *Bool* with its constructors is in all **FPL**-signatures, and that all **FPL**-models are $\{Bool\}$-compatible); under such an assumption the free variable y in the above term may be replaced by one of the constructors of sort *Bool*.

Another point is that under certain conditions, testing definedness of terms subsumes testing whether the values of terms of observable sorts coincide. Namely, consider any two compatible SIG-models A and B, and suppose that SIG includes an equality test on an observable sort $o \in OBS$ in the form of an operation $eq : o \times o \to Bool$ that yields the value of *true* iff it is applied to identical arguments in both A and B. Then, for any term t of sort o with variables of observable sorts only, consider the term $t' =$ **case** $eq(t,x)$ **of** *true* => x | *false* => *undef* where *undef* is **let fun** $f(y{:}o){:}o = f(y)$ **in** $f(t)$ and x is a new variable of sort o. Now, if there exists a valuation v such that both $t_A(v)$ and $t_B(v)$ are defined but $t_A(v) \neq t_B(v)$, then $t'_A(v') = t_A(v)$ while $t'_B(v')$ is undefined, where v' extends v by $v'(x) = t_A(v)$. This reduction requires an equality test on observable sorts, which seems like a strong assumption to make. But there is no loss of generality in assuming this for observable sorts that are generated from other generated sorts; see Exercise 4.1.28.

Summarizing, under reasonable technical requirements we could define behavioural equivalence solely in terms of definedness of terms of observable sorts. We refrain from introducing such requirements, and instead use the above definition simply because it makes the following developments a little smoother.

Exercise 8.5.3. Any **FPL**-signature $\mathsf{SIG} = \langle S, \Omega, D\rangle$ contains an algebraic signature $\langle S, \Omega\rangle$. Moreover, some SIG-models, namely those with total operations, are $\langle S, \Omega\rangle$-algebras. Show that any such SIG-models that are behaviourally equivalent according to Definition 8.5.2 above are also behaviourally equivalent as $\langle S, \Omega\rangle$-algebras. Show that the opposite implication does not hold by giving an **FPL**-signature $\mathsf{SIG} = \langle S, \Omega, D\rangle$ and two SIG-models with total operations that are not behaviourally equivalent as SIG-models, but are behaviourally equivalent as $\langle S, \Omega\rangle$-algebras. HINT: Recall that **FPL**-terms may use pattern matching for values of sorts with value constructors.

A similar observation may be stated within **FPL**: given any **FPL**-signature $\mathsf{SIG} = \langle S, \Omega, D\rangle$ and $D' \subseteq D$, let $\mathsf{SIG}_{D'} = \langle S, \Omega, D'\rangle$ be the **FPL**-signature obtained from SIG by forgetting about the value constructor status of constructors not in D'. Then the identity on the underlying algebraic signature is an **FPL**-signature morphism $\iota_{D'}\colon \mathsf{SIG}_{D'} \to \mathsf{SIG}$ and the $\iota_{D'}$-reduct functor preserves but does not necessarily reflect behavioural equivalence of **FPL**-models w.r.t. any set $OBS \subseteq S$ of observable sorts.

View the above comparison with behavioural equivalence in the standard algebraic framework as a special case of what happens within **FPL** for $D' = \varnothing$. □

Definition 8.5.4 (Correspondence). Given two OBS-compatible SIG-models A, B, an *OBS-correspondence* (or simply, a *correspondence*, if OBS is clear) between A and B, written $\rho\colon A \bowtie_{OBS} B$, is a relation $\rho \subseteq |A| \times |B|$ such that:

- ρ is the identity on observable carriers, that is, for each observable sort $o \in OBS$, $\rho_o = id_{|A|_o} = id_{|B|_o}$;
- ρ is closed under the operations, that is, for each operation $f\colon s_1 \times \cdots \times s_n \to s$ in SIG and elements $a_1 \in |A|_{s_1}, \ldots, a_n \in |A|_{s_n}$ and $b_1 \in |B|_{s_1}, \ldots, b_n \in |B|_{s_n}$, if $a_1\ \rho_{s_1}\ b_1, \ldots, a_n\ \rho_{s_n}\ b_n$ then $f_A(a_1, \ldots, a_n)$ is defined iff $f_B(b_1, \ldots, b_n)$ is defined and if so then $f_A(a_1, \ldots, a_n)\ \rho_s\ f_B(b_1, \ldots, b_n)$; and
- ρ respects constructors, that is, for all sorts with value constructors $\langle d, \langle F_{w,d}\rangle_{w \in S^*}\rangle$ in SIG, for $c \in F_{s_1 \ldots s_n, d}$ and $c' \in F_{s'_1 \ldots s'_{n'}, d}$ and elements $a_1 \in |A|_{s_1}, \ldots, a_n \in |A|_{s_n}$ and $b_1 \in |B|_{s'_1}, \ldots, b_{n'} \in |B|_{s'_{n'}}$, if $c_A(a_1, \ldots, a_n)\ \rho_d\ c'_B(b_1, \ldots, b_{n'})$ then $c = c'$ with $s_1 \ldots s_n = s'_1 \ldots s'_{n'}$ and $a_1\ \rho_{s_1}\ b_1, \ldots, a_n\ \rho_{s_n}\ b_n$. □

The requirement that ρ respects constructors is needed because implicit (partial) "selector" and "discriminator" operations are introduced by **FPL**-terms with case analysis using pattern matching.

Theorem 8.5.5. *Two* SIG*-models* A, B *are behaviourally equivalent w.r.t.* OBS *if and only if there exists an* OBS*-correspondence between* A *and* B.

Proof.

($\Rightarrow$): As in the proof of Theorem 8.2.8, except for the minor complication of checking definedness of operations and the need to exhibit terms playing the role of selectors and discriminators to ensure that the relation defined respects constructors. (**Exercise:** Give such terms.)

($\Leftarrow$): Let $\rho{:}A \bowtie_{OBS} B$ be an OBS-correspondence. For any signature SIG' that extends SIG by new operations, any SIG'-models A' and B' that extend A and B, respectively, such that ρ is a correspondence between A' and B', and any set of variables X and term $t \in |T_{\mathsf{SIG}'}(X)|$, we show by induction on the structure of t that for all valuations $v_A{:}X \to |A|$ and $v_B{:}X \to |B|$ such that $v_A(x)\ \rho\ v_B(x)$ for all $x \in X$, $t_{A'}(v_A)$ is defined iff $t_{B'}(v_B)$ is defined, and if so then $t_{A'}(v_A)\ \rho\ t_{B'}(v_B)$. This implies the required definedness condition for terms having observable variables, and the required equality condition for such terms that are of observable sort. The stronger formulation of the induction hypothesis than in the proof of Theorem 8.2.8 is needed to handle terms of the form **let fun** $f(x_1{:}s_1,\ldots,x_n{:}s_n){:}s' = t'$ **in** t, where we need to prove that ρ is closed under the new operation corresponding to f and then use this fact when discussing the values of term t. □

Value constructors with pattern matching add observational power, as reflected in the requirement in the definition of correspondences above that value constructors be respected. For any signature SIG and set OBS of observable sorts in SIG, let $\widehat{OBS}$ be the least set of sorts in SIG containing OBS such that for each set of sorts with value constructors $\{\langle d_1,\mathscr{F}_1\rangle,\ldots,\langle d_n,\mathscr{F}_n\rangle\}$ in SIG, $d_1,\ldots,d_n \in \widehat{OBS}$ provided that for $1 \le i \le n$, each value constructor in $\mathscr{F}_i$ has all its argument sorts other than $d_1,\ldots,d_n$ in $\widehat{OBS}$.[5] In particular, notice that $\widehat{OBS}$ always includes $Bool$, so $Bool \in \widehat{\varnothing}$. The same holds for any other sort s with value constructors having argument sorts among $\{s\} \cup OBS$.

Proposition 8.5.6. *Given a signature* SIG *and a set OBS of observable sorts in* SIG*, any two* SIG*-models that are behaviourally equivalent w.r.t. OBS are also weakly behaviourally equivalent w.r.t.* $\widehat{OBS}$*.*

Proof. Since the carriers of sorts in $\widehat{OBS}$ are freely generated by the corresponding constructors from the carriers of sorts in OBS, every OBS-correspondence is a weak $\widehat{OBS}$-correspondence. □

If we assume that the values of sorts with value constructors are represented in some standard way, for instance as formal terms built using value constructor names (as in the semantics of Standard ML), then behavioural equivalence w.r.t. OBS also implies behavioural equivalence w.r.t. $\widehat{OBS}$. The opposite implication is immediate, so $\equiv_{OBS}$ then coincides with $\equiv_{\widehat{OBS}}$.

A related fact is that we can express equality between terms of sorts in $\widehat{OBS}$ in terms of equality tests on terms of sorts in OBS.

Exercise 8.5.7. Assume that for each observable sort $o \in OBS$ there is an operation $eq{:}o \times o \to Bool$. For any sort $s \in \widehat{OBS}$, by induction on the definition of $\widehat{OBS}$ define a term $t \in |T_{\mathsf{SIG}}(\{x,y:s\})|_{Bool}$ such that in any SIG-model A where $eq_A(c,d) = true_A$ iff $c = d$ for all $o \in OBS$ and $c,d \in |A|_o$, we have $t_A(x \mapsto a, y \mapsto b) = true_A$ iff $a = b$ for all $a,b \in |A|_s$. To handle mutually dependent sorts with value constructors, you will need to solve Exercise 4.1.27 first.

---

[5] This definition is a little complicated because of the possible presence of mutually dependent sorts with value constructors; cf. Example 4.1.25.

Do the same using a sort *unit* generated by a single constant in place of *Bool*. □

Exercise 8.5.8. If $o \in OBS$ is a sort with value constructors and one of the value constructors for o is unary with argument sort s, then every *OBS*-correspondence is bijective on s, even if $s \notin \widehat{OBS}$. Generalise this to n-ary value constructors (HINT: watch out for empty generated sorts such as **sort** *empty* **free with** $c(s, empty)$), expand the definition of $\widehat{OBS}$ correspondingly so that Proposition 8.5.6 still holds, and redo Exercise 8.5.7. □

Definition 8.5.9 (Behavioural submodel and morphism). Given an **FPL**-signature SIG, a set *OBS* of observable sorts in SIG and SIG-models A, B, a SIG-morphism $h: A \to B$ is *behavioural* (*w.r.t. OBS*) if it is strong and is the identity on carriers of observable sorts; it then follows that A and B are *OBS*-compatible. A is a *behavioural submodel* (*w.r.t. OBS*) of B if $|A| \subseteq |B|$ and the inclusion is a behavioural morphism. A is *OBS-generated* if it is the least behavioural submodel of B (or equivalently, of itself). The *OBS-generated submodel* of A will be denoted by $\langle A \rangle_{OBS}$. □

Exercise 8.5.10. Check that behavioural morphisms are correspondences and so are their kernels. State and prove analogues to Corollary 8.2.10, Theorem 8.2.14 and Exercise 8.2.15 for **FPL**. □

Exercise 8.5.11. Investigate how behavioural equivalence varies when the set of observable sorts changes, considering the cases where no sorts are observable and where all sorts are observable. Compare with the standard algebraic case; see Exercise 8.2.16. Check that your proof that $OBS \subseteq OBS'$ implies $\equiv_{OBS'} \subseteq \equiv_{OBS}$ from that exercise is still valid. □

Exercise 8.5.12. Confirm that Definition 8.2.18 of behavioural closure and behavioural abstraction, and their basic properties in Proposition 8.2.19, carry over to **FPL**. □

Definition 8.5.13 (Partial behavioural congruence). Let SIG be an **FPL**-signature and let *OBS* be a set of sorts in SIG. A *partial behavioural congruence w.r.t. OBS* on a SIG-model A is a relation $\simeq \subseteq |A| \times |A|$ that is an *OBS*-correspondence $\simeq: A \bowtie_{OBS} A$. □

Exercise 8.5.14. Spell out the above definition explicitly without making reference to the notion of correspondence. Check that for any SIG-model A there is a largest partial behavioural congruence on A (see Proposition 8.3.3) and that any partial behavioural congruence on A is reflexive on $\langle A \rangle_{OBS}$. □

Exercise 8.5.15. Generalise Definition 8.3.18 and Exercise 2.7.32, defining partial congruences on a partial algebra A (as transitive and symmetric relations on $|A|$ that are closed under defined operations in A) and the quotient of A by such a congruence. Notice that partial behavioural congruences must preserve and reflect definedness of operations and respect value constructors. Prove that a quotient of a SIG-model (viewed as a partial algebra) by a partial behavioural congruence is a SIG-model.

(A minor point is that you will need to define the quotient in such a way that the interpretation of the sort *Bool* is preserved.) We will call such quotients *behavioural*. □

Definition 8.5.16 (Contexts and behavioural indistinguishability). Let (observable) contexts and the value of a context on an element of a model under a valuation be as defined in Definitions 2.6.1 and 8.3.4 (but using **FPL**-terms, of course). Given an **FPL**-signature SIG with a set *OBS* of observable sorts and a SIG-model A, *behavioural indistinguishability on A w.r.t. OBS* is the relation $\sim_A^{OBS} \subseteq |A| \times |A|$ defined as follows: for $s \in S$ and $a, b \in |A|_s$, $a \sim_A^{OBS} b$ if and only if for all contexts $C \in |T_{\mathsf{SIG}}(X_{OBS} \uplus \{\Box{:}s\})|$ with observable variables and all valuations $v{:}X_{OBS} \to |A|_{OBS}$, $C_A^v[a]$ is defined iff $C_A^v[b]$ is defined, and if so then moreover $C_A^v[a] = C_A^v[b]$ when C is of a sort in *OBS*. □

Exercise 8.5.17. Prove that for any *OBS*-generated SIG-model A, behavioural indistinguishability is the largest partial behavioural congruence on A. (See Theorem 8.3.7.) □

Exercise 8.5.18. Define behavioural satisfaction and behavioural semantics of presentations along the lines of Definition 8.3.9, Exercise 8.3.10 and Definition 8.3.11. Check that Section 8.3.1 carries over, including the definitions of fully abstract model as well as behavioural expansion and behavioural closedness of a class of **FPL**-models. □

Exercise 8.5.19. Totality of an operation is not preserved under behavioural equivalence: if $A \equiv_{OBS} B$ then a total operation in A may be undefined in B on non-observable junk values. However, show then that the formula $\forall x \bullet \exists y \bullet f(x) = y$ behaviourally holds in A iff it behaviourally holds in B, since behavioural satisfaction quantifies over *OBS*-generated values only; see Corollary 8.3.29 via Exercise 8.5.18. □

Given the above definitions, the concept of behavioural implementation (Definition 8.4.2) can be used in **FPL** as stated there. Then Proposition 8.4.3 trivially holds in the present context. Furthermore, the ideas related to stability of constructors presented in Section 8.4.2 together with the specific results given there — Proposition 8.4.5, Theorem 8.4.6 (vertical composition) and Corollary 8.4.7 — carry over as well.

The concept of behaviourally trivial constructors (Definition 8.4.10) can also be adopted without essential change here. Typical examples of behaviourally trivial constructors are restriction to behavioural submodels and quotients via partial behavioural congruences as in Examples 8.4.14 and 8.4.16. The analysis of the role of behaviourally trivial constructors in behavioural implementations as captured by Propositions 8.4.12 and 8.4.13, and Exercises 8.4.15 and 8.4.19, could also be repeated here.

Exercise 8.5.20. Recall Example 7.2.9, which gives a constructor implementation

$$\text{NATSET} \underset{\kappa}{\rightsquigarrow} \text{NATLIST}$$

using the constructor $\kappa = ListSet;R_{Sig[\text{NATSET}]};(\_/SP)$ as defined there. Show that $\text{NATSET} \overset{\{Bool,Nat\}}{\underset{ListSet}{\rightsquigarrow}} \text{NATLIST}$ since $R_{Sig[\text{NATSET}]}$ and $\_/SP$ are behaviourally trivial. □

As in the standard algebraic framework — see Non-example 8.4.22 — translation of a stable constructor need not be stable here either. The situation in the framework of **FPL** is even more subtle since pushouts in the category of **FPL** need not exist. Consequently, translation of a persistent constructor may not even be well defined. However, there is an important class $\mathbf{I_{FPL}}$ of signature morphisms, namely injective renamings of sorts and operations that do not introduce value constructors for old sorts, which covers the cases of interest. As explained in Exercise 4.4.19, **FPL** is $\mathbf{I_{FPL}}$-semi-exact. Given a persistent constructor $F \in Mod(\mathsf{SIG} \overset{\iota}{\Rightarrow} \mathsf{SIG}')$, with $\iota \in \mathbf{I_{FPL}}$, for any **FPL**-signature morphism $\delta\colon \mathsf{SIG} \to \mathsf{SIG}_G$, the translation of F by δ, $\delta(F) \in Mod(\mathsf{SIG}_G \overset{\iota'}{\Rightarrow} \mathsf{SIG}'_G)$, is defined as in Example 6.1.28, where ι' and SIG'_G come from the following pushout diagram:

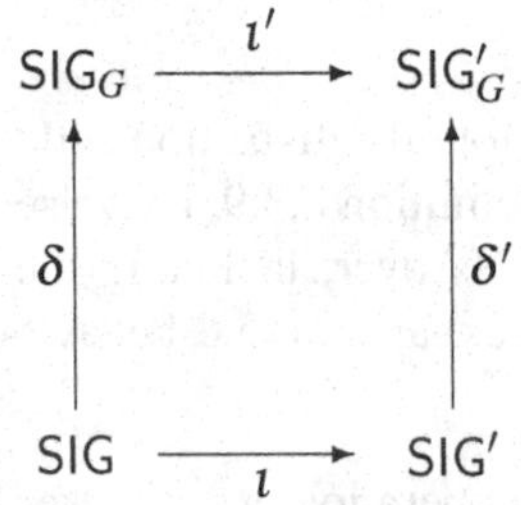

Under the assumption that $\iota \in \mathbf{I_{FPL}}$, the definition of globally stable constructors — Definition 8.4.24 — carries over. Due to Proposition 8.5.6, we may relax the condition $\sigma(OBS) \subseteq OBS_G$, replacing it with $\sigma(OBS) \subseteq \widehat{OBS_G}$. Then the characterisation of globally stable constructors as those that extend correspondences (Theorem 8.4.26) still holds; this uses the fact that correspondences amalgamate over the pushout considered, as in Exercise 8.4.27. Further developments carry over as well, including Exercise 8.4.28, Lemma 8.4.29, and Exercises 8.4.30, 8.4.32 (but you might want to somewhat simplify the construction given there, exploiting the possibility of using partial operations and sorts with value constructors) and 8.4.33. This leads, as in Section 8.4.4, to the conclusion that we wish to work with constructors that are persistent (along a morphism in $\mathbf{I_{FPL}}$) and extend correspondences.

Exercise 8.4.31 carries over as well. However, it is worth refining it further by considering **FPL**-signature morphisms mapping operation names to **FPL**-terms that may additionally contain conditionals as introduced in Exercise 4.1.28, generalised to permit equalities between terms of sorts in $\widehat{OBS}$ as conditions. We call such **FPL**-signature morphisms *OBS-admissible*. The following corollary may be proved along the lines of Exercise 8.4.31, starting with Proposition 8.5.6.

Corollary 8.5.21. *Let $\delta\colon \mathsf{SIG}' \to \mathsf{SIG}$ and $\iota\colon \mathsf{SIG} \to \mathsf{SIG}'$ be* **FPL**-*signature morphisms such that $\iota \in \mathbf{I_{FPL}}$ and $\iota;\delta = id_{\mathsf{SIG}}$. Let OBS and OBS' be sets of observable sorts in* SIG *and* SIG′ *respectively such that $\delta(OBS') \subseteq \widehat{OBS}$. If δ is OBS-admissible* *then the* δ*-reduct of* **FPL***-models* $\_|_\delta \in Mod(\mathsf{SIG} \overset{\iota}{\Rightarrow} \mathsf{SIG}')$ *extends correspondences and hence is globally stable w.r.t.* OBS *and* OBS'*.* □

Exercise 8.5.22. Give an example which shows that the *OBS*-admissibility requirement in Corollary 8.5.21 cannot be dropped. □

Corollary 8.5.23. *Let* $K \in Mod(\mathsf{SIG} \overset{j}{\Rightarrow} \mathsf{SIG}')$ *be defined using the notation of Example 6.1.9, where* SIG *is a subsignature of* SIG′ *and* j *is the signature inclusion. Recall that* K *is the composition of a free extension and a reduct,* $K = F_\iota;(\_|_\delta)$*,*

$$\mathsf{SIG} \xrightarrow{\iota} \mathsf{SIG}_{intermediate} \xleftarrow{\delta} \mathsf{SIG}'$$

where ι *and* δ *are introduced by the notation in use, with* $\iota = j;\delta$*. Then* K *is globally stable with respect to sets of observable sorts* OBS *and* OBS' *in* SIG *and* SIG′ *respectively, provided that* $\delta(OBS') \subseteq \iota(\widehat{OBS})$ *and that formulae in conditionals within the body are equalities between terms of sorts in* $\iota(\widehat{OBS})$*.*

Proof. The notation in use ensures that ι is an **FPL**-signature inclusion that only introduces new sorts with value constructors. Therefore, the free extension F_ι is total and naturally persistent by Exercise 6.1.27. The rest follows by Corollary 8.5.21 and a version of Exercise 8.4.30 for **FPL**. □

The definitions of behavioural correctness for (persistent) constructors (Definition 8.4.34) and of a behavioural model of a constructor specification (Definition 8.4.36) can be reused in the present context. It is easy to see that Exercise 8.4.39 and Proposition 8.4.40 carry over as well. So do the results providing conditions which ensure correctness of the use of behavioural models in a global context: Lemma 8.4.42, Theorem 8.4.43 and Corollary 8.4.45, where we assume as usual that ι as used there is in $\mathbf{I_{FPL}}$. Finally, Proposition 8.4.47 concerning amalgamated union of behavioural models holds here as well, although its scope is limited to **FPL**-signature morphisms σ_1 and σ_2 for which the pushout exists. In particular, this holds when σ_1 or σ_2 is in $\mathbf{I_{FPL}}$.

8.5.2 A larger example

We will now present the development of a realisation of a specification of sets of strings in terms of two-level hash tables, working in the institution **FPL**. The development is closely related to the one in Section 7.4 but it exploits the flexibility of behavioural interpretation of specifications, with some steps being behaviourally correct, i.e. constituting behavioural implementations, but not being "literally" correct, i.e. constituting constructor implementations in the sense of Chapter 7. We use many of the same names to facilitate comparison but the reader should be aware that some of the specifications are different. The diagram in Figure 8.1, which shows

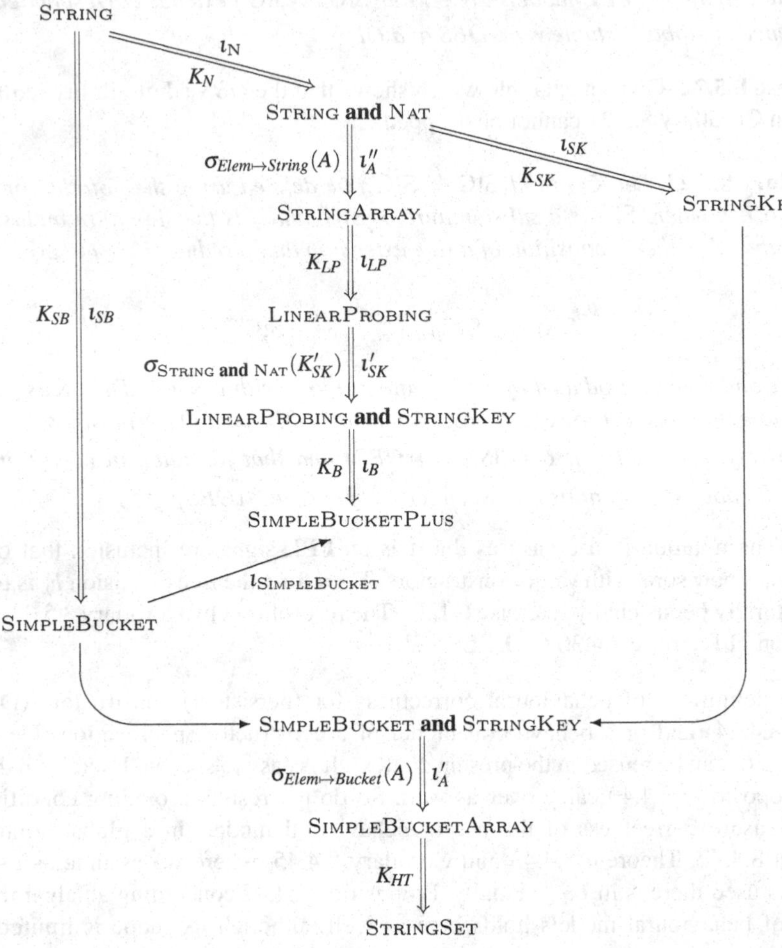

Fig. 8.1 Specifications, constructors and signature morphisms involved in the development

most of the constructors, specifications and signature morphisms that will arise, may be helpful in following the development.

In this example, all behavioural implementations will be with respect to the empty set of observable sorts. This is quite typical in institutions like **FPL**, but it departs from standard approaches to behavioural equivalence in the usual algebraic framework, where choosing a non-empty set of observable sorts is crucial for having any observations at all (cf. Exercises 8.2.16 and 8.5.11). In particular, recall from Proposition 8.5.6 and the discussion preceding it that generated sorts such as *Nat*, *String* and *Bool* can always be treated as observable. Moreover, even though it may seem necessary to vary observable sorts in the process of modular development, where some sorts must be locally considered as observable, we achieve the required effect with the empty set of observable sorts via the notions of behavioural correctness and behavioural model (Definitions 8.4.34 and 8.4.36).

Recall the **FPL**-specifications NAT and STRING from Section 7.4:

spec NAT =
  sorts *Nat* **free with** 0 | *succ*(*Nat*)
  ops $\_\_+\_\_ : Nat \times Nat \to Nat$
  $\_\_\leq\_\_ : Nat \times Nat \to Bool$
  $\forall m, n : Nat$
  • ...

spec STRING =
  sorts *String* **free with** ε | a^(*String*) | $\cdots$ | z^(*String*)

We start with a specification of sets of strings:

spec STRINGSET =
  STRING
  then
  sorts *NatSet*
  ops *empty*: *NatSet*
  add: $String \times NatSet \to NatSet$
  present: $String \times NatSet \to Bool$
  $\forall s, s' : String, S : NatSet$
  • $def(empty) \wedge def(add(s,S)) \wedge def(present(s,S))$
  • $add(s, add(s,S)) = add(s,S)$
  • $add(s, add(s',S)) = add(s', add(s,S))$
  • $present(s, empty) = false \wedge present(s, add(s,S)) = true$
  • $s \neq s' \Rightarrow present(s, add(s',S)) = present(s,S)$

Note that this specification imposes stronger requirements on *add* than those stated in the specification STRINGTABLE that we started with in Section 7.4. Also note that, even though the axioms require all the operations to be total, this requirement is not preserved under behavioural equivalence: in models that are behaviourally equivalent to models of STRINGSET, the operations need only be defined on reachable arguments; see Exercise 8.5.19. A similar comment applies to all of the specifications below.

The first design decision is to use direct chaining hash tables to represent these sets. This will lead to an implementation of STRINGSET by the specification SIMPLEBUCKETARRAY below.

spec STRINGKEY =
  NAT **and** STRING
  then
  ops *hash*: $String \to Nat$
  $\forall s : String \bullet def(hash(s)) \wedge hash(\varepsilon) = 0$

spec SIMPLEBUCKET = STRINGSET **with** $\sigma_{NatSet \mapsto Bucket}$

where $\sigma_{NatSet \mapsto Bucket}$ is a surjective signature morphism from Sig[STRINGSET] that renames $NatSet$ to $Bucket$ and is the identity otherwise.

spec SIMPLEBUCKETARRAY =
  SIMPLEBUCKET **and** STRINGKEY
  then
    sorts $Array[Bucket]$
    ops $empty{:}Array[Bucket]$
      $put{:}Nat \times Bucket \times Array[Bucket] \to Array[Bucket]$
      $get{:}Nat \times Array[Bucket] \to? Bucket$
      $used{:}Nat \times Array[Bucket] \to Bool$
    $\forall i, j{:}Nat, b, b'{:}Bucket, a{:}Array[Bucket]$
      • $def(empty) \wedge def(put(i,b,a)) \wedge def(used(i,b))$
      • $put(i,b,put(i,b',a)) = put(i,b,a)$
      • $i \neq j \Rightarrow put(i,b,put(j,b',a)) = put(j,b',put(i,b,a))$
      • $used(i, empty{:}Array[Bucket]) = false$
      • $used(i,put(i,b,a)) = true$
      • $i \neq j \Rightarrow used(i,put(j,b,a)) = used(i,a)$
      • $get(i,put(i,b,a)) = b$

Some of the operation names are overloaded in this specification, for instance *empty* (we have $empty{:}Bucket$ and $empty{:}Array[Bucket]$). The context of use helps to disambiguate, but in cases here and below where the overloading might be confusing we attach some explicit sort information. We have deliberately omitted a requirement of definedness for *get* because we do not want to require it to be defined, for instance, for the empty array. However, the last axiom, in combination with the third axiom, ensures that it is defined when a value has been put into the relevant position in the array.

Exercise 8.5.24. Prove that if *BA* is a *reachable* model of SIMPLEBUCKETARRAY then

$$BA \models \forall i{:}Nat, a{:}Array[Bucket] \bullet used(i,a) = true \Rightarrow \exists b{:}Bucket; a'{:}Array[Bucket] \bullet a = put(i,b,a')$$

and so also

$$BA \models \forall i{:}Nat, a{:}Array[Bucket] \bullet used(i,a) = true \Rightarrow def(get(i,a)).$$

(Actually, it is sufficient to require that the carrier $|BA|_{Array[Bucket]}$ is generated from the other carriers.) Give a model of SIMPLEBUCKETARRAY that does not satisfy these properties. □

We will now proceed to give a behavioural implementation of STRINGSET by SIMPLEBUCKETARRAY with respect to the empty set of observable sorts. We will establish

$$\text{STRINGSET} \overset{\varnothing}{\underset{K_{HT}}{\leadsto}} \text{SIMPLEBUCKETARRAY}$$

where the constructor K_{HT} is defined using the notation of Example 6.1.9 as follows:

constructor K_{HT} : $Sig[\textsc{SimpleBucketArray}] \Rightarrow Sig[\textsc{StringSet}] =$
  sorts $NatSet = Array[Bucket]$
  ops $empty{:}NatSet = empty{:}Array[Bucket]$
    fun $add(s{:}String, S{:}NatSet){:}NatSet =$
      let $i = hash(s)$
      in if $used(i,S) = true$ **then** $put(i, add(s, get(i,S){:}Bucket), S)$
        else $put(i, add(s, empty{:}Bucket), S)$
    fun $present(s{:}String, S{:}NatSet){:}Bool =$
      let $i = hash(s)$
      in if $used(i,S) = true$ **then** $present(s, get(i,S){:}Bucket)$
        else $false$

In fact, we nearly have a normal constructor implementation here. Namely, for any reachable model $BA \in Mod[\textsc{SimpleBucketArray}]$, $K_{HT}(BA)$ satisfies all the axioms in STRINGSET and so indeed, by the **FPL** version of Exercise 8.4.15,

$$\textsc{StringSet} \overset{\varnothing}{\underset{K_{HT}}{\leadsto}} \textsc{SimpleBucketArray}.$$

Exercise 8.5.25. Check the above claim for models *BA* of SIMPLEBUCKETARRAY that are reachable (notice that the reachability requirement cannot be omitted; see Exercise 8.5.24). To prove the idempotency and commutativity of *add* on sets we need both idempotency and commutativity of *add* on buckets as well as the corresponding properties of *put* on arrays. Conclude that correctness of the above behavioural implementation follows, since K_{HT} is globally stable (and therefore stable) by Corollary 8.5.21. □

Let specifications ELEM and ELEMARRAY be defined as follows:

spec ELEM =
  sorts *Elem*

spec ELEMARRAY =
  ELEM **and** NAT
  then
    sorts $Array[Elem]$
    ops $empty{:}Array[Elem]$
      $put{:}Nat \times Elem \times Array[Elem] \to Array[Elem]$
      $get{:}Nat \times Array[Elem] \to Elem$
      $used{:}Nat \times Array[Elem] \to Bool$
  $\forall i,j{:}Nat, e{:}Elem, a{:}Array[Elem]$
    • $def(empty) \wedge def(put(i,e,a)) \wedge def(used(i,a))$
    • $used(i, empty) = false$
    • $used(i, put(i,e,a)) = true$
    • $i \neq j \Rightarrow used(i, put(j,e,a)) = used(i,a)$
    • $get(i, put(i,e,a)) = e$
    • $i \neq j \Rightarrow get(i, put(j,e,a)) = get(i,a)$

Let us assume that we have (perhaps from a software library) a behavioural model of ELEM **and** $\text{NAT} \overset{\iota_A}{\Longrightarrow} \text{ELEMARRAY}$,

$$A \in \mathit{Mod}[\text{ELEM } \mathbf{and} \text{ NAT }_{\varnothing}\overset{\iota_A}{\Longrightarrow}_{\varnothing} \text{ ELEMARRAY}]$$

where ι_A is the signature inclusion. Recall that this implies that A is persistent, globally stable, and behaviourally correct. The next implementation step uses A, lifted to take any model of SIMPLEBUCKET **and** STRINGKEY as argument. This will be given by translation of A (see Example 6.1.28) by the signature morphism $\sigma_{Elem \mapsto Bucket}$ from $\mathit{Sig}[\text{ELEM } \mathbf{and} \text{ NAT}]$ to $\mathit{Sig}[\text{SIMPLEBUCKET } \mathbf{and} \text{ STRINGKEY}]$ which maps *Elem* to *Bucket* and is the inclusion otherwise. We can choose the target (pushout) signature for $\sigma_{Elem \mapsto Bucket}(A)$ to be $\mathit{Sig}[\text{SIMPLEBUCKETARRAY}]$:

$$\begin{array}{llcl}
\sigma_{Elem \mapsto Bucket}(A): & & & \\
& \text{SIMPLEBUCKET } \mathbf{and} \text{ STRINGKEY} & \overset{\iota'_A}{\Longrightarrow} & \text{SIMPLEBUCKETARRAY} \\
& \uparrow \sigma_{Elem \mapsto Bucket} & & \uparrow \sigma'_{Elem \mapsto Bucket} \\
A: & \text{ELEM } \mathbf{and} \text{ NAT} & \overset{\iota_A}{\Longrightarrow} & \text{ELEMARRAY}
\end{array}$$

Then $\sigma_{Elem \mapsto Bucket}(A)$ is a (globally) stable persistent constructor since A is globally stable. Moreover, we claim that $\sigma_{Elem \mapsto Bucket}(A)$ is a behavioural model of

$$\text{SIMPLEBUCKET } \mathbf{and} \text{ STRINGKEY} \overset{\iota'_A}{\Longrightarrow} \text{SIMPLEBUCKETARRAY}$$

and therefore

$$\text{SIMPLEBUCKETARRAY} \;{}^{\varnothing}\!\underset{\sigma_{Elem \mapsto Bucket}(A)}{\rightsquigarrow}\; \text{SIMPLEBUCKET } \mathbf{and} \text{ STRINGKEY}.$$

Exercise 8.5.26. Prove the above claim, using Corollary 8.4.45 adapted for **FPL**. HINT: The first requirement of the corollary follows trivially. The second one amounts to proving that, for any $\mathit{Sig}[\text{SIMPLEBUCKETARRAY}]$-model M formed by amalgamating models of ELEMARRAY and SIMPLEBUCKET **and** STRINGKEY, there is a behaviourally equivalent model $N \in \mathit{Mod}[\text{SIMPLEBUCKETARRAY}]$. The model N can be built by quotienting M by the congruence induced by

$$\begin{array}{l}
put(i,b,put(i,b',a)) = put(i,b,a) \\
i \neq j \Rightarrow put(i,b,put(j,b',a)) = put(j,b',put(i,b,a)).
\end{array}$$

The axioms of ELEMARRAY ensure that this congruence is the identity on sorts other than $\mathit{Array}[\mathit{Bucket}]$. □

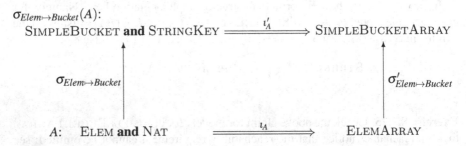

The specification of the constructor A is not sufficient to ensure that $\sigma_{Elem \mapsto Bucket}(A)$ is a model of SIMPLEBUCKET **and** STRINGKEY $\Rightarrow$ SIMPLEBUCKETARRAY because the axioms of ELEMARRAY are not sufficient to guarantee that the second and third axioms of SIMPLEBUCKETARRAY hold: there may be models $BA \in Mod[\text{SIMPLEBUCKET } \textbf{and} \text{ STRINGKEY}]$ such that $\sigma_{Elem \mapsto Bucket}(A)(BA)$ does not satisfy one or both of these two axioms, even if we additionally assume reachability. But never mind: any computation in $\sigma_{Elem \mapsto Bucket}(A)(BA)$ yielding a result that we can directly test (using, for instance, pattern matching on *Nat*, *String* or *Bool*) produces a result that is correct according to SIMPLEBUCKETARRAY. The stability of K_{HT} ensures, by Theorem 8.4.6 for **FPL**, that the two behavioural implementation steps can be composed, yielding

$$\text{STRINGSET} \overset{\varnothing}{\underset{\sigma_{Elem \mapsto Bucket}(A);K_{HT}}{\rightsquigarrow}} \text{SIMPLEBUCKET } \textbf{and} \text{ STRINGKEY}.$$

We will now implement SIMPLEBUCKET **and** STRINGKEY by implementing each of the component specifications separately and applying Proposition 8.4.47 for **FPL** to obtain an implementation of their sum. The shared part of SIMPLEBUCKET and STRINGKEY is STRING, so the implementations of these components will take models of STRING as arguments. A realisation of STRING can be given easily; as in Section 7.4, we will not pursue this here.

For both components, we begin by providing a model N for NAT; again, details are omitted. This is clearly necessary to build a model for STRINGKEY, and it will be useful in our construction of a model for SIMPLEBUCKET as in Section 7.4. Viewing the model N as a construction from the initial **FPL**-signature to NAT, a simple application of Proposition 8.4.47 for **FPL** yields a behavioural model K_N of $\text{STRING} \overset{\iota_N}{\Longrightarrow} \text{STRING } \textbf{and} \text{ NAT}$ where ι_N is the obvious signature inclusion.

For STRINGKEY, the development of a hash function for given realisations of STRING and NAT to give a realisation of STRINGKEY is left to the reader as in Section 7.4. This yields a behavioural model K_{SK} of $\text{STRING } \textbf{and} \text{ NAT} \overset{\iota_{SK}}{\Longrightarrow} \text{STRINGKEY}$ where ι_{SK} is the obvious signature inclusion.

For SIMPLEBUCKET, as in Section 7.4, we use hash tables with linear probing to resolve collisions.

spec STRINGARRAY =
  STRING **and** NAT
  then
    sorts *Array[String]*
    ops $empty{:}Array[String]$
      $put{:}Nat \times String \times Array[String] \to Array[String]$
      $get{:}Nat \times Array[String] \to String$
      $used{:}Nat \times Array[String] \to Bool$
    $\forall i,j{:}Nat, s,s'{:}String, a{:}Array[String]$
      • $def(empty) \wedge def(put(i,s,a)) \wedge def(used(i,s))$
      • $put(i,s,put(i,s',a)) = put(i,s,a)$
      • $i \neq j \Rightarrow put(i,s,put(j,s',a)) = put(j,s',put(i,s,a))$
      • $used(i,empty) = false$
      • $used(i,put(i,s,a)) = true$
      • $i \neq j \Rightarrow used(i,put(j,s,a)) = used(i,a)$
      • $get(i,put(i,s,a)) = s$

spec LINEARPROBING =
  STRINGARRAY
  then
    ops $putnear: Nat \times String \times Array[String] \rightarrow Array[String]$
    $isnear: Nat \times String \times Array[String] \rightarrow Bool$
    $\forall i: Nat, s: String, a, a': Array[String]$
    • $def(putnear(i,s,a)) \wedge def(isnear(i,s,a))$
    • $putnear(i,s,a) = a' \Leftrightarrow$
      $\exists j: Nat \bullet (used(i+j,a) = false \vee get(i+j,a) = s) \wedge$
      $(\forall k: Nat \bullet (j \leq k) = false \Rightarrow used(i+k,a) = true) \wedge$
      $a' = put(i+j,s,a)$
    • $isnear(i,s,a) = true \Leftrightarrow$
      $\exists j: Nat \bullet (\forall k: Nat \bullet (k \leq j) = true \Rightarrow used(i+k,a) = true) \wedge$
      $get(i+j,a) = s$

We will separate the behavioural implementation of SIMPLEBUCKET in terms of LINEARPROBING **and** STRINGKEY into two steps. The intermediate specification will be

spec SIMPLEBUCKETPLUS =
  SIMPLEBUCKET **and** LINEARPROBING **and** STRINGKEY

Let $\iota_{\text{SIMPLEBUCKET}}$: SIMPLEBUCKET $\rightarrow$ SIMPLEBUCKETPLUS be the obvious inclusion. We proceed now with the implementation of SIMPLEBUCKETPLUS.

constructor K_B : Sig[LINEARPROBING **and** STRINGKEY] $\Rightarrow$
  Sig[SIMPLEBUCKETPLUS] =
  sorts $Bucket = Array[String]$
  ops $empty{:}Bucket = empty{:}Array[String]$
    fun $add(s{:}String, b{:}Bucket){:}Bucket = putnear(hash(s), s, b)$
    fun $present(s{:}String, b{:}Bucket){:}Bool = isnear(hash(s), s, b)$

We claim that K_B is a behavioural model of LINEARPROBING **and** STRINGKEY $\overset{\iota_B}{\Longrightarrow}$ SIMPLEBUCKETPLUS (where ι_B is the obvious signature inclusion) and therefore

$$\text{SIMPLEBUCKETPLUS} \; \overset{\varnothing}{\underset{K_B}{\rightsquigarrow}} \; \text{LINEARPROBING } \textbf{and} \text{ STRINGKEY}.$$

Exercise 8.5.27. Prove the above claim. HINT: Consider an arbitrary model M of LINEARPROBING **and** STRINGKEY. First check that the following three axioms

$\forall s{:}String \bullet present(s, empty) = false$
$\forall s{:}String, b{:}Bucket \bullet present(s, add(s,b)) = true$
$\forall s, s'{:}String, b{:}Bucket \bullet s \neq s' \Rightarrow (present(s, add(s',b)) = present(s,b))$

hold in $K_B(M)$, using the definition of *empty*, *add* and *present* on *Bucket* via *empty*, *putnear* and *isnear* on *Array*[*String*], and using axioms for *putnear* and *isnear* as well as for STRINGARRAY, which M and therefore $K_B(M)$ satisfy. Then, consider the least congruence on $K_B(M)$ induced by

$\forall s{:}String, b{:}Bucket \bullet add(s, add(s,b)) = add(s,b)$
$\forall s, s'{:}String, b{:}Bucket \bullet add(s, add(s',b)) = add(s', add(s,b))$.

Prove that this congruence is the identity on all sorts other than *Bucket*. Quotienting $K_B(M)$ by this congruence yields a model of SIMPLEBUCKETPLUS with a behavioural correspondence between it and $K_B(M)$ which extends the identity on M. Persistency and global stability of K_B follow as usual. □

Now, similarly as before, we need an implementation (a behavioural model) for STRING **and** NAT $\overset{\iota_{SK}}{\Longrightarrow}$ STRINGKEY. If we use K_{SK} again then all of the strings in a bucket will have the same hash value, so we assume being given another behavioural model $K'_{SK} \in Mod[$STRING **and** NAT $_{\varnothing}\overset{\iota_{SK}}{\Longrightarrow}_{\varnothing}$ STRINGKEY$]$. We lift this to the context of LINEARPROBING via the specification inclusion $\sigma_{\text{STRING and NAT}}$: STRING **and** NAT $\to$ LINEARPROBING.

$$\begin{array}{lcl}
\sigma_{\text{STRING and NAT}}(K'_{SK}): \text{LINEARPROBING} & \overset{\iota'_{SK}}{\Longrightarrow} & \text{LINEARPROBING and STRINGKEY} \\
\uparrow \sigma_{\text{STRING and NAT}} & & \uparrow \\
K'_{SK}: \text{STRING and NAT} & \overset{\iota_{SK}}{\Longrightarrow} & \text{STRINGKEY}
\end{array}$$

Choosing the pushout signature appropriately, we thus obtain a behavioural model $\sigma_{\text{STRING and NAT}}(K'_{SK})$ of

$$\text{LINEARPROBING} \overset{\iota'_{SK}}{\Longrightarrow} \text{LINEARPROBING and STRINGKEY}$$

and therefore

$$\text{LINEARPROBING and STRINGKEY} \;{}^{\varnothing}\!\!\underset{\sigma_{\text{STRING and NAT}}(K'_{SK})}{\rightsquigarrow} \text{LINEARPROBING}.$$

Now we proceed with the implementation of LINEARPROBING as in Section 7.4:

spec STRINGFINITEARRAY =
  STRINGARRAY
  then
    $\forall a{:}Array[String] \bullet \exists n{:}Nat \bullet \forall j{:}Nat \bullet (n \le j) = true \Rightarrow used(j,a) = false$

constructor K_{LP} : Sig[STRINGFINITEARRAY] $\Rightarrow$ Sig[LINEARPROBING] =
  ops fun $putnear(n{:}Nat, s{:}String, a{:}Array[String]){:}Array[String] =$
      if $used(n,a) = false$ **then** $put(n,s,a)$
      else if $get(n,a) = s$ **then** a **else** $putnear(succ(n),s,a)$
    fun $isnear(n{:}Nat, s{:}String, a{:}Array[String]){:}Bool =$
      if $used(n,a) = false$ **then** $false$
      else if $get(n,a) = s$ **then** $true$ **else** $isnear(succ(n),s,a)$

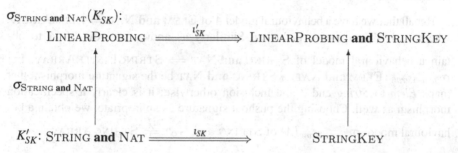

K_{LP} is a model of STRINGFINITEARRAY $\xrightarrow{\iota_{LP}}$ LINEARPROBING (where ι_{LP} is the obvious signature inclusion). Moreover, by Corollary 8.5.23, K_{LP} is globally stable and therefore stable. Consequently, K_{LP} is a behavioural model of STRINGFINITEARRAY $\xrightarrow{\iota_{LP}}$ LINEARPROBING. This yields

$$\text{LINEARPROBING} \underset{K_{LP}}{\overset{\varnothing}{\rightsquigarrow}} \text{STRINGFINITEARRAY}.$$

Exercise 8.5.28. Show thaat we have LINEARPROBING $\underset{K_{LP}}{\overset{\varnothing}{\rightsquigarrow}}$ STRINGARRAY as well, since up to behavioural equivalence, it doesn't matter that *putnear* may be undefined on some unreachable arrays. However, show that K_{LP} is *not* behaviourally correct and therefore not a behavioural model of STRINGARRAY $\xrightarrow{\iota_{LP}}$ LINEARPROBING. □

Recall that we have a behavioural model A of ELEM **and** NAT $\xrightarrow{\iota_A}$ ELEMARRAY, where ι_A is the signature inclusion. Similarly as above, we now lift A to obtain a behavioural model of STRING **and** NAT $\xrightarrow{\iota''_A}$ STRINGFINITEARRAY. Let $\sigma_{Elem\mapsto String}$: ELEM **and** NAT $\to$ STRING **and** NAT be the signature morphism that maps *Elem* to *String* and is the inclusion otherwise; it is clearly a specification morphism as well. Choosing the pushout signature as appropriate, we obtain a behavioural model $\sigma_{Elem\mapsto String}(A)$ of STRING **and** NAT $\xrightarrow{\iota''_A}$ STRINGARRAY.

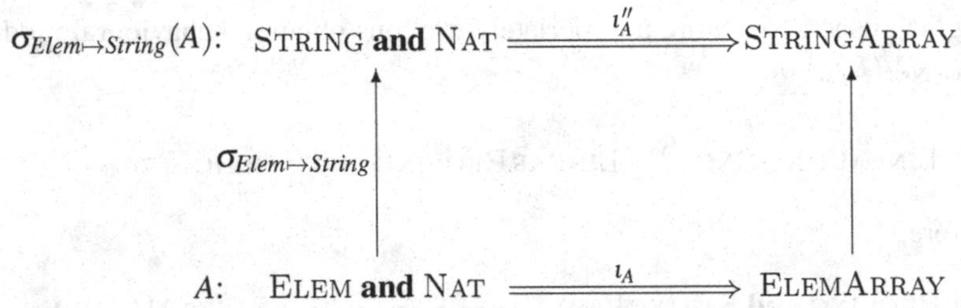

Moreover, $\sigma_{Elem\mapsto String}(A)$ is also a behavioural model of STRING **and** NAT $\xrightarrow{\iota''_A}$ STRINGFINITEARRAY since any model of STRINGARRAY is behaviourally equivalent to a model of STRINGFINITEARRAY (for instance to its submodel generated by *String*, *Nat* and *Bool*), with the behavioural equivalence witnessed by a correspondence which reduces to the identity on $Sig[\text{STRING } \textbf{and} \text{ NAT}]$.

We now have a sequence of behavioural models that can be composed to yield a behavioural model of STRING $\xrightarrow{\iota_N;\iota''_A;\iota_{LP};\iota'_{SK};\iota_B}$ SIMPLEBUCKETPLUS, namely

$$K_{SBP} = K_N;\sigma_{Elem\mapsto String}(A);K_{LP};\sigma_{\text{STRING and NAT}}(K'_{SK});K_B.$$

By Proposition 8.4.40 for **FPL**, the constructor $K_{SB} = K_{SBP};-|_{\iota_{\text{SIMPLEBUCKET}}}$ is a behavioural model of STRING $\xrightarrow{\iota_{SB}}$ SIMPLEBUCKET, where ι_{SB} is the signature inclusion. Next, by Proposition 8.4.47 for **FPL**, $K_{SB} + (K_N;K_{SK})$ is a behavioural model of STRING $\overset{\iota}{\Rightarrow}$ SIMPLEBUCKET **and** STRINGKEY (where ι is the signature inclusion), which gives

$$\text{SIMPLEBUCKET \textbf{and} STRINGKEY} \overset{\varnothing}{\underset{K_{SB}+(K_N;K_{SK})}{\rightsquigarrow}} \text{STRING \textbf{and} NAT}.$$

Composing this with earlier implementation steps gives

$$\text{STRINGSET} \overset{\varnothing}{\underset{(K_{SB}+(K_N;K_{SK}));\sigma_{Elem\mapsto Bucket}(A);K_{HT}}{\rightsquigarrow}} \text{STRING}.$$

To wrap up the example, we can choose any (behavioural) realisation S of STRING. This finally yields a model

$$((K_{SB}+(K_N;K_{SK}));\sigma_{Elem\mapsto Bucket}(A);K_{HT})(S)$$
$$\text{where } K_{SB} = K_N;\sigma_{Elem\mapsto String}(A);K_{LP};\sigma_{\text{STRING and NAT}}(K'_{SK});K_B;\_|_{\iota_{\text{SIMPLEBUCKET}}}.$$

Up to behavioural equivalence (with respect to the empty set of sorts) this is a model of STRINGSET.

Exercise 8.5.29. Consider the following version of the K_{LP} constructor, which implements *putnear* in such a way that the most recently added elements are kept closest to the hash position:

```
constructor K'_LP : Sig[STRINGFINITEARRAY] ⇒ Sig[LINEARPROBING] =
  ops fun putnear(n:Nat, s:String, a:Array[String]):Array[String] =
        let fun shift(m:Nat, v:String, a:Array[String]):Array[String] =
              if used(m,a) = false then put(m,v,a)
              else if get(m,a) = s then put(m,v,a)
                   else put(m,v,shift(succ(m),get(m,a),a))
        in shift(n,s,a)
      fun isnear(n:Nat, s:String, a:Array[String]):Bool = ...
```

Now, K'_{LP} is *not* behaviourally correct along ι_{LP} with respect to LINEARPROBING and STRINGFINITEARRAY. Show, however, that the composed constructor used above, where K_{LP} is replaced by K'_{LP}, namely

$$K'_{SB} = K_N;\sigma_{Elem\mapsto String}(A);K'_{LP};\sigma_{\text{STRING and NAT}}(K'_{SK});K_B;\_|_{\iota_{\text{SIMPLEBUCKET}}}$$

is a behavioural model of STRING $\overset{\iota_{SB}}{\Longrightarrow}$ SIMPLEBUCKET.

HINT: Notice that K'_{LP} is persistent and globally stable. Show that for any model SA of STRINGFINITEARRAY, there is a relation between $K_{LP}(SA)$ and $K'_{LP}(SA)$ that is the identity on *String*, *Nat* and *Bool* and is closed under "derived operations" $putnear(hash(s),s,a)$, $isnear(hash(s),s,a)$ and *empty*. Extend this to the corresponding property of the composed constructors with K_{LP} and K'_{LP} respectively.

Try to generalise the notions of correspondence, behavioural abstraction, behavioural implementation, stability, and so on, to a setting where we give a set of "observable contexts" rather than a set of observable sorts. This generalisation should capture the behavioural correctness of K'_{LP} with respect to an appropriate set of observable contexts. □

Exercise 8.5.30. The above example deliberately avoids the higher-order constructors of Section 6.4 used in the version of the example presented in Section 7.4. It is an interesting research topic to develop a theory of behavioural equivalence for higher-order constructors. Try this. Redo the example above in the style of Section 7.4 and use your framework to study it. □

8.5.3 Behavioural specifications in an arbitrary institution

In this subsection we will sketch a generalisation of the preceding material to the context of an arbitrary institution. We will proceed in three stages. First, we will discuss specifications and their implementations in the presence of an arbitrary equivalence relation on models. Second, we will introduce concepts that play the role of correspondences in witnessing such an equivalence relation on models, via a generalised notion of behavioural morphism. Finally, we will discuss general properties that may be used to characterise behavioural morphisms.

Throughout this section we will work in an arbitrary but fixed institution $\mathbf{INS} = \langle \mathbf{Sign}, \mathbf{Sen}, \mathbf{Mod}, \langle \models_\Sigma \rangle_{\Sigma \in |\mathbf{Sign}|} \rangle$. We will gradually impose requirements for further structure and properties as needed.

8.5.3.1 INS-behavioural abstraction and implementations

We have introduced two particular notions of behavioural equivalence on models above: Definition 8.2.2 for standard algebras, and Definition 8.5.2 for **FPL**-models. Both of these definitions were parameterised by a set of observable sorts and were expressed via term evaluation. In an arbitrary institution, we have neither sorts nor terms. Therefore for each signature $\Sigma \in |\mathbf{Sign}|$, instead of working with respect to a given set of observable sorts in Σ as hitherto, we will work directly with respect to a given equivalence relation on Σ-models. We will occasionally refer to this equivalence as an **INS**-*behavioural equivalence* in order to stress its intended role, and use a similar prefix to distinguish the abstract notions introduced here from their standard versions. Even in this very general framework, many of the definitions and results in Sections 8.2.2 and 8.4 carry over, so that their original versions are special cases of those below. The source of this equivalence is a matter that will be discussed below.

Definition 8.2.18 can be restated in a general form as follows:

Definition 8.5.31 (INS-behavioural closure and abstraction). For any class $\mathscr{M} \subseteq |\mathbf{Mod}(\Sigma)|$ of Σ-models, its **INS**-*behavioural closure w.r.t. an equivalence* $\equiv \subseteq |\mathbf{Mod}(\Sigma)| \times |\mathbf{Mod}(\Sigma)|$ is

$$Abs_{\equiv}(\mathscr{M}) = \{N \in |\mathbf{Mod}(\Sigma)| \mid N \equiv M \text{ for some } M \in \mathscr{M}\}.$$

If SP is a Σ-specification then **abstract** SP **wrt** $\equiv$ is a specification with the following semantics:

$$Sig[\textbf{abstract } SP \textbf{ wrt} \equiv] = \Sigma$$
$$Mod[\textbf{abstract } SP \textbf{ wrt} \equiv] = Abs_{\equiv}(Mod[SP]).$$ □

Exercise 8.5.32. Check that the simple properties of the behavioural closure operation given in Proposition 8.2.19 carry over, including analogous properties for the abstraction operation. □

Exercise 8.5.33. Following Exercise 8.2.23, for any class $\mathcal{M} \subseteq |\mathbf{Mod}(\Sigma)|$ of Σ-models, characterise its **INS**-behavioural closures with respect to two trivial equivalences: the identity, and the total relation on $|\mathbf{Mod}(\Sigma)|$. Prove that for any equivalences $\equiv$ and $\equiv'$ on $|\mathbf{Mod}(\Sigma)|$, if $\equiv \subseteq \equiv'$ then $Abs_{\equiv}(\mathcal{M}) \subseteq Abs_{\equiv'}(\mathcal{M})$. State consequences of these facts for abstraction. □

The following definition of **INS**-behavioural implementations captures in the present context the essence of behavioural implementations (Definition 8.4.2).

Definition 8.5.34 (INS-behavioural implementation). A specification SP *is* **INS**-*behaviourally implemented by a specification SP' via a constructor* $\kappa: Sig[SP'] \Rightarrow Sig[SP]$ *w.r.t. an equivalence* $\equiv$ *on* $|\mathbf{Mod}(Sig[SP])|$, written $SP \overset{\equiv}{\underset{\kappa}{\rightsquigarrow}} SP'$, if $dom(\kappa) \supseteq Mod[SP']$ and $Abs_{\equiv}(Mod[SP]) \supseteq \kappa(Mod[SP'])$. □

As before (see Proposition 8.4.3), constructor implementations in the sense of Definition 7.2.1 are **INS**-behavioural implementations with respect to an arbitrary equivalence on models.

As with behavioural implementations, **INS**-behavioural implementations via arbitrary constructors do not vertically compose (see Example 8.4.1 in Section 8.4.2). Again, stability of constructors comes to the rescue.

Definition 8.5.35 (INS-stable constructor). Let $\Sigma, \Sigma' \in |\mathbf{Sign}|$ be signatures and let $\equiv$ and $\equiv'$ be equivalences on $|\mathbf{Mod}(\Sigma)|$ and $|\mathbf{Mod}(\Sigma')|$, respectively. Let $\mathcal{M}' \subseteq |\mathbf{Mod}(\Sigma')|$ be a class of Σ'-models.

A constructor $\kappa: \Sigma' \Rightarrow \Sigma$ is **INS**-*stable on* $\mathcal{M}'$ *w.r.t.* $\equiv'$ *and* $\equiv$ if for all models $M' \in \mathcal{M}'$ and $N' \in |\mathbf{Mod}(\Sigma')|$, whenever $M' \equiv' N'$ then also $\kappa(M') \equiv \kappa(N')$.

We say that κ is **INS**-*stable* (*w.r.t.* $\equiv'$ *and* $\equiv$) if it is **INS**-stable on the class $dom(\kappa)$. □

Exercise 8.5.36. Notice that, as in Proposition 8.4.5, any **INS**-behavioural implementation via an **INS**-stable constructor amounts to a constructor implementation between the behavioural abstractions of the corresponding specifications. Then reformulate and prove the vertical composition property (Theorem 8.4.6 and Corollary 8.4.7). □

The concept of a behaviourally trivial constructor carries over as well.

Definition 8.5.37 (INS-behaviourally trivial constructor). Consider a signature $\Sigma \in |\mathbf{Sign}|$, an equivalence $\equiv$ on $|\mathbf{Mod}(\Sigma)|$, and a class $\mathscr{M} \subseteq |\mathbf{Mod}(\Sigma)|$ of Σ-models.

A constructor $\kappa: \Sigma \Rightarrow \Sigma$ is **INS**-*behaviourally trivial w.r.t.* $\equiv$ *on* $\mathscr{M}$ if for every model $M \in \mathscr{M}$, $M \equiv \kappa(M)$. $\kappa: \Sigma \Rightarrow \Sigma$ is **INS**-*behaviourally trivial* (*w.r.t.* $\equiv$) if it is so on the class $|\mathbf{Mod}(\Sigma)|$. □

INS-behaviourally trivial constructors are **INS**-stable (see Proposition 8.4.11). The role of behaviourally trivial constructors for behavioural implementations and their compositions carries over as well, and can be embodied by the appropriate adaptions of Propositions 8.4.12 and 8.4.13.

We will refrain from attempting to present general versions of restriction to a sort-generated subalgebra and of quotient that would correspond to the two standard examples of behaviourally trivial constructors in Section 8.4.3; see Examples 8.4.14 and 8.4.16 and Exercises 8.4.15 and 8.4.19. Even though their definitions can be given in institutions where model categories come equipped with the structure described in Section 4.5, their **INS**-behavioural triviality can only be justified for specific equivalences. But see Exercise 8.5.50 below.

Translation of an **INS**-stable constructor to a larger context need not be **INS**-stable here for reasons similar to those explained in Non-example 8.4.22. Again, a possible solution is to concentrate on constructors that are globally **INS**-stable in the following sense.

Let **I** be a class of morphisms in **Sign** such that **INS** is **I**-semi-exact; see Definition 4.4.18.

Notation. For any signature morphism $\sigma: \Sigma \to \Sigma'$ and equivalence $\equiv$ on $|\mathbf{Mod}(\Sigma)|$, $\equiv|_\sigma^{-1}$ is an equivalence on $|\mathbf{Mod}(\Sigma')|$ such that $M' \equiv|_\sigma^{-1} N'$ iff $M'|_\sigma \equiv N'|_\sigma$. □

Definition (Globally INS-stable constructor, first version). Let $\iota: \Sigma \to \Sigma'$ be a signature morphism in **I** and let and $\equiv$ and $\equiv'$ be equivalences on $|\mathbf{Mod}(\Sigma)|$ and $|\mathbf{Mod}(\Sigma')|$, respectively. Let $\mathscr{M} \subseteq |\mathbf{Mod}(\Sigma)|$ be a class of Σ-models.

A persistent constructor $F \in Mod(\Sigma \overset{\iota}{\Rightarrow} \Sigma')$ is *globally* **INS**-*stable on* $\mathscr{M}$ *w.r.t.* $\equiv$ *and* $\equiv'$ if for every pushout diagram in **Sign**

$$\begin{array}{ccc} \Sigma_G & \xrightarrow{\iota'} & \Sigma'_G \\ \uparrow{\scriptstyle\sigma} & & \uparrow{\scriptstyle\sigma'} \\ \Sigma & \xrightarrow[\iota]{} & \Sigma' \end{array}$$

that admits amalgamation, and equivalence $\equiv_G$ on $|\mathbf{Mod}(\Sigma_G)|$ such that $\equiv_G|_\sigma \subseteq \equiv$, the translation of F along σ, $\sigma(F) \in Mod(\Sigma_G \overset{\iota'}{\Rightarrow} \Sigma'_G)$, is **INS**-stable on $\mathscr{M}_G = \{M_G \in |\mathbf{Mod}(\Sigma_G)| \mid M_G|_\sigma \in \mathscr{M}\}$ w.r.t. $\equiv_G$ and $\equiv'_G$, where $\equiv'_G = \equiv_G|_{\iota'}^{-1} \cap \equiv'|_{\sigma'}^{-1}$. We say that F is *globally* **INS**-*stable* (*w.r.t.* $\equiv$ *and* $\equiv'$) if it is globally **INS**-stable on the class $dom(F)$. □

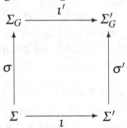

This definition is based on Definition 8.4.24, replacing conditions that relate sets of observable sorts over different signatures with conditions that directly relate the equivalences these sets would determine in the standard case. We will now compare what this definition yields in the standard algebraic framework with the original definition.

Let $\sigma\colon\Sigma \to \Sigma_G$ be an algebraic signature morphism and let OBS and OBS_G be sets of observable sorts in Σ and Σ_G respectively. By Example 8.4.9, if $\sigma(OBS) \subseteq OBS_G$ then $\equiv_{OBS_G}|_\sigma \subseteq \equiv_{OBS}$, that is, $\equiv_{OBS_G} \subseteq \equiv_{OBS}|_\sigma^{-1}$. The opposite implication holds as well: suppose that $\sigma(o) \notin OBS_G$ for some $o \in OBS$. Let A_G be the Σ_G-algebra with singleton carriers for all sorts. Let A'_G extend A_G by adding a new element to the carrier $|A_G|_{\sigma(o)}$, with all the operations continuing to yield results in the carrier of $|A_G|$ for all arguments. Then A_G is an OBS_G-behavioural subalgebra of A'_G but $A_G|_\sigma \not\equiv_{OBS} A'_G|_\sigma$. And so $\equiv_{OBS_G}|_\sigma \not\subseteq \equiv_{OBS}$.

Moreover, under the notation introduced in Definition 8.4.24, for any Σ'_G-models A'_G and B'_G, if $A'_G \equiv_{OBS'_G} B'_G$ then $A'_G|_{\iota'} \equiv_{OBS_G} B'_G|_{\iota'}$ and $A'_G|_{\sigma'} \equiv_{OBS'} B'_G|_{\sigma'}$, by a similar argument. Unfortunately, here the opposite implication fails.

Exercise 8.5.38. Recall from Exercise 8.2.16 that in general $\equiv_{OBS_1 \cup OBS_2} \neq \equiv_{OBS_1} \cap \equiv_{OBS_2}$. Study what properties of the morphisms and sets of sorts involved would be required to ensure $\equiv_{OBS'_G} = \equiv_{OBS_G}|_{\iota'}^{-1} \cap \equiv_{OBS'}|_{\sigma'}^{-1}$. □

A conclusion is that, in the standard algebraic framework, if a persistent constructor $F \in Mod(\Sigma \overset{\iota}{\Rightarrow} \Sigma')$ is globally stable w.r.t. OBS and OBS' then it is also globally **INS**-stable w.r.t. $\equiv_{OBS}$ and $\equiv_{OBS'}$. But the two notions do not coincide, as the following counterexample demonstrates.

Counterexample 8.5.39. Let Σ be the algebraic signature with sorts s, s' and constant $a\colon s$, with $OBS = \varnothing$. Let Σ' extend Σ by the constants $b, c\colon s$ with $\iota\colon\Sigma \to \Sigma'$ being the signature inclusion and $OBS' = \varnothing$. Consider the following two Σ-algebras:

$$A : |A|_s = \{0,1,*\}, |A|_{s'} = \{\clubsuit\}, a_A = *$$
$$B : |B|_s = \{0,1,*\}, |B|_{s'} = \{\diamondsuit\}, a_B = *$$

Define a (total) persistent constructor $F \in Mod(\Sigma \overset{\iota}{\Rightarrow} \Sigma')$ so that $b_{F(A)} = 0$ and $c_{F(A)} = 1$, while $b_{F(B)} = 1$ and $c_{F(B)} = 0$, with F extending arbitrarily other Σ-algebras.

Now, F does not extend correspondences and so it is not globally stable w.r.t. OBS and OBS'. For instance, let Σ_G extend Σ by the constant $z\colon s$, with $\sigma\colon\Sigma \to \Sigma_G$ being the inclusion and $OBS_G = \{s\}$, and let A_G and B_G extend A and B respectively by interpreting $z\colon s$ as 0. Then $A_G \equiv_{OBS_G} B_G$ but $\sigma(F)(A_G) \not\equiv_{OBS'_G} \sigma(F)(B_G)$, where $OBS'_G = \iota'(OBS_G) \cup \sigma'(OBS') = \{s\}$ (choosing the pushout signature $\Sigma'_G = \Sigma_G \cup \Sigma'$ with ι' and σ' being the obvious signature inclusions), since $z = b$ holds in $\sigma(F)(A_G)$ but does not hold in $\sigma(F)(B_G)$. (Alternatively, instead of making s directly observable we could have added an observable sort $Bool$ with an operation $p\colon s \to Bool$ that in A_G and B_G would distinguish between 0 and 1.)

On the other hand, F is globally **INS**-stable w.r.t. $\equiv_{OBS}$ and $\equiv_{OBS'}$. To see this, consider any algebraic signature morphism $\sigma : \Sigma \to \Sigma_G$ with set OBS_G of observable sorts in Σ_G. Then for any two Σ_G-algebras A_G and B_G, if $A_G \equiv_{OBS_G} B_G$ then $(\sigma(F)(A_G))|_{\iota'} \equiv_{OBS_G} (\sigma(F)(B_G))|_{\iota'}$ by persistency of $\sigma(F)$ (Example 6.1.28). Furthermore, $(\sigma(F)(A_G))|_{\sigma'} \equiv_{OBS'} (\sigma(F)(B_G))|_{\sigma'}$ since $OBS' = \varnothing$. Thus indeed $(\sigma(F)(A_G)) \equiv'_G (\sigma(F)(B_G))$ where $\equiv'_G = \equiv_{OBS_G}|_{\iota'}^{-1} \cap \equiv_{OBS'}|_{\sigma'}^{-1}$. □

Similar troubles arise in **FPL**. The above counterexample applies there as well, but we cannot even directly conclude that global stability of a persistent constructor w.r.t. sets of observable sorts implies its global **INS**-stability w.r.t. the behavioural equivalences they determine.

Exercise 8.5.40. Consider an **FPL**-signature morphism $\sigma : \mathsf{SIG} \to \mathsf{SIG}_G$ with observable sorts OBS and OBS_G in SIG and SIG_G respectively. If $\sigma(OBS) \subseteq OBS_G$ then $\equiv_{OBS_G}|_\sigma \subseteq \equiv_{OBS}$. But the opposite implication does not hold. Give a counterexample. Replace the requirement that $\sigma(OBS) \subseteq OBS_G$ by a weaker condition which ensures $\equiv_{OBS_G}|_\sigma \subseteq \equiv_{OBS}$, for instance $\sigma(OBS) \subseteq \widehat{OBS_G}$. Show that in general $\sigma(OBS) \subseteq \widehat{OBS_G}$ is still not equivalent to $\equiv_{OBS_G}|_\sigma \subseteq \equiv_{OBS}$. □

In principle, it is possible to generalise the definitions of behavioural correctness (Definition 8.4.34) and behavioural model (Definition 8.4.36) to the current framework, and then obtain results about correctness of behavioural models when used in a global context (Theorem 8.4.43 and Corollary 8.4.45). However, we view Counterexample 8.5.39 as rendering this line of development pointless since it shows that we would admit constructors that break encapsulation boundaries by varying their results depending on hidden properties of their arguments. Indeed, global **INS**-stability is a rather weak requirement which does not go beyond **INS**-stability.

Exercise 8.5.41. Check that if a persistent constructor $F \in Mod(\Sigma \overset{\iota}{\Rightarrow} \Sigma')$ is **INS**-stable on $\mathscr{M}$ w.r.t. $\equiv$ and $\equiv'$ (see Definition 8.5.35) then it is also globally **INS**-stable on $\mathscr{M}$ w.r.t. $\equiv$ and $\equiv'$. HINT: Under the notation of the definition of global **INS**-stability above, for $M_G, N_G \in \mathscr{M}_G$ such that $M_G \equiv_G N_G$, $\sigma(F)(M_G)|_{\iota'} = M_G$ and $\sigma(F)(N_G)|_{\iota'} = N_G$, so that $\sigma(F)(M_G) \equiv_G|_{\iota'}^{-1} \sigma(F)(N_G)$, and $\sigma(F)(M_G)|_{\sigma'} = F(M_G|_\sigma)$ and $\sigma(F)(N_G)|_{\sigma'} = F(N_G|_\sigma)$, so that $\sigma(F)(M_G) \equiv'|_{\sigma'}^{-1} \sigma(F)(N_G)$. □

Consequently, rather than trying to further refine this line of definitions, we will switch to a less abstract notion of **INS**-behavioural equivalence, adding enough extra structure to bring these results closer to the standard cases considered earlier.

8.5.3.2 INS-behavioural morphisms

In the above we have disregarded entirely the question of how an **INS**-behavioural equivalence over a given model class is determined. As a result, **INS**-behavioural equivalences as considered above came with no notion of witness to play a role comparable to that of correspondences for behavioural equivalence in the standard algebraic framework (Theorem 8.2.8) and in **FPL** (Theorem 8.5.5). To provide such witnesses, we will exploit the characterisation of correspondences in the standard algebraic framework as spans of certain model morphisms (Exercise 8.2.15). We will therefore base our definitions on a choice of morphisms in model categories of **INS** which we will view as **INS**-behavioural morphisms. We will use subclasses of these, corresponding to our choices of sets of observable sorts in the standard algebraic case, to define particular **INS**-behavioural equivalences.

Definition 8.5.42 (INS-behavioural morphism). A *family of* **INS**-*behavioural morphisms* is a $|\mathbf{Sign}|$-indexed family $\mathscr{H} = \langle \mathscr{H}_\Sigma \subseteq \mathbf{Mod}(\Sigma)\rangle_{\Sigma\in|\mathbf{Sign}|}$ where for each signature Σ, $\mathscr{H}_\Sigma$ is a class of morphisms between Σ-models that contains all the identities and is closed under composition (so that it forms a wide subcategory of $\mathbf{Mod}(\Sigma)$). Moreover, $\mathscr{H}$ is required to be closed under reducts and amalgamation:

- For any signature morphism $\sigma\colon \Sigma \to \Sigma'$, $\mathscr{H}_{\Sigma'}|_\sigma \subseteq \mathscr{H}_\Sigma$; and
- For any pushout in **Sign**

$$\begin{array}{ccc} \Sigma_1 & \xrightarrow{\sigma_2'} & \Sigma' \\ \sigma_1 \uparrow & & \uparrow \sigma_1' \\ \Sigma & \xrightarrow[\sigma_2]{} & \Sigma_2 \end{array}$$

that admits amalgamation and any Σ'-morphism $h \in \mathbf{Mod}(\Sigma')$, we have $h \in \mathscr{H}_{\Sigma'}$ provided that both $h|_{\sigma_1'} \in \mathscr{H}_{\Sigma_2}$ and $h|_{\sigma_2'} \in \mathscr{H}_{\Sigma_1}$. □

We will henceforth assume that **INS** is equipped with a family $\mathscr{H}$ of **INS**-behavioural morphisms.

Definition 8.5.43 (INS-behavioural equivalence). Given a signature $\Sigma \in |\mathbf{Sign}|$ and a class $\mathscr{B} \subseteq \mathscr{H}_\Sigma$ of **INS**-behavioural Σ-morphisms, the **INS**-*behavioural equivalence defined by* $\mathscr{B}$ is the least equivalence $\equiv_\mathscr{B} \subseteq |\mathbf{Mod}(\Sigma)| \times |\mathbf{Mod}(\Sigma)|$ on Σ-models such that for any two Σ-models $M,N \in |\mathbf{Mod}(\Sigma)|$, $M \equiv_\mathscr{B} N$ whenever we have a Σ-model $C \in |\mathbf{Mod}(\Sigma)|$ and a span of morphisms $h_M\colon C \to M$ and $h_N\colon C \to N$ with $h_M, h_N \in \mathscr{B}$. □

In the following we will assume that all classes $\mathscr{B}$ of **INS**-behavioural Σ-morphisms we consider include all identities on Σ-models.

Exercise 8.5.44. Simplify the above definition of **INS**-behavioural equivalence generated by a class of behavioural morphisms by noticing that $\equiv_\mathscr{B}$ is the least equivalence on $|\mathbf{Mod}(\Sigma)|$ such that $M \equiv_\mathscr{B} N$ whenever there is a behavioural morphism $h\colon M \to N$ in $\mathscr{B}$. Check if this helps to simplify further developments and proofs below in any essential way. □

The definition of **INS**-behavioural equivalence above could be given simply in terms of an arbitrary class $\mathscr{B}$ of Σ-morphisms, and so introducing the family $\mathscr{H}$ of

INS-behavioural morphisms may seem superfluous at this stage. However, restricting the choice of classes $\mathscr{B}$ to **INS**-behavioural morphisms only becomes important in later developments; see for instance Definition 8.5.51 below, where considering arbitrary classes of morphisms would yield a more restrictive notion, with some standard examples excluded. Typically the choice of $\mathscr{B}$ is determined by some additional parameter (like the set of observable sorts) and the class of **INS**-behavioural morphisms is the union of all such $\mathscr{B}$, over all choices of such a parameter. In frameworks like **EQ** and **FPL**, **INS**-behavioural morphisms are those that are behavioural with respect to the empty set of observable sorts:

Example 8.5.45. For the standard algebraic framework, for any algebraic signature Σ, take all Σ-homomorphisms to be **INS**-behavioural. Then, for any set *OBS* of observable sorts in Σ, let $\mathscr{B}_{OBS}$ be the class of all Σ-homomorphisms that are identities on the carriers of sorts in *OBS*. Then $\equiv_{\mathscr{B}_{OBS}}$ coincides with $\equiv_{OBS}$, the behavioural equivalence with respect to *OBS*, by Exercise 8.2.15.

For **FPL**, for any **FPL**-signature SIG, take all *strong* SIG-morphisms to be **INS**-behavioural. Then, for any set *OBS* of observable sorts in SIG, let $\mathscr{B}_{OBS}$ be the class of all strong SIG-morphisms that are identities on the carriers of sorts in *OBS*. Then $\equiv_{\mathscr{B}_{OBS}}$ coincides with $\equiv_{OBS}$, the behavioural equivalence on **FPL**-models with respect to *OBS*, by Exercise 8.5.10.

In both cases, allowing morphisms in $\mathscr{B}_{OBS}$ to be bijective on the carriers of sorts in *OBS* yields weak behavioural equivalence w.r.t. *OBS*; see Definitions 8.2.24 and 8.2.26 and Theorem 8.2.28. □

Exercise 8.5.46. Recall Exercise 8.2.17. For any algebraic signature Σ and sets *IN* and *OUT* of its sorts, let $\mathscr{B}_{IN,OUT}$ be the class of Σ-homomorphisms that are surjective on the carriers of sorts in *IN* and injective on the carriers of sorts in *OUT*. Check that the equivalence $\equiv_{\mathscr{B}_{IN,OUT}}$ coincides with the weak version of behavioural equivalence with respect to IN, OUT you defined in that exercise. □

Exercise 8.5.47. In the examples above, the existence of a span of behavioural morphisms defined an equivalence directly. Give an example of a class of morphisms where this is not the case, and the relation given by the existence of spans needs to be closed (under reflexivity and/or transitivity) to yield an equivalence. Check that if $\mathscr{B} \subseteq \mathscr{H}_\Sigma$ is closed under pullbacks in $\mathbf{Mod}(\Sigma)$ then $M \equiv_{\mathscr{B}} N$ iff there exists a span of morphisms $h_M : C \to M$ and $h_N : C \to N$ with $h_M, h_N \in \mathscr{B}$. Notice however that it would not be sufficient to introduce similar assumptions about the class $\mathscr{H}_\Sigma$ only. □

All of the definitions and results above concerning **INS**-behavioural abstraction and implementations can be parameterised by classes of **INS**-behavioural morphisms which determine **INS**-behavioural equivalences, rather than by **INS**-behavioural equivalences directly. For instance:

Definition 8.5.48 (INS-behavioural closure and abstraction). For any class $\mathscr{M} \subseteq |\mathbf{Mod}(\Sigma)|$ of Σ-models, its **INS**-*behavioural closure w.r.t.* $\mathscr{B} \subseteq \mathscr{H}_\Sigma$ is

$$Abs_{\mathscr{B}}(\mathscr{M}) = \{N \in |\mathbf{Mod}(\Sigma)| \mid N \equiv_{\mathscr{B}} M \text{ for some } M \in \mathscr{M}\}.$$

If SP is a Σ-specification then **abstract** SP **wrt** $\mathscr{B}$ is a specification with the following semantics:

$$\begin{aligned} Sig[\textbf{abstract}\ SP\ \textbf{wrt}\ \mathscr{B}] &= \Sigma \\ Mod[\textbf{abstract}\ SP\ \textbf{wrt}\ \mathscr{B}] &= Abs_{\equiv_{\mathscr{B}}}(Mod[SP]) \end{aligned}$$

□

Exercise 8.5.49. Apply a similar transformation to the other definitions and results. Check that choosing the classes of behavioural morphisms for sets of observable sorts as in Example 8.5.45 yields the corresponding specific notions and results in the standard algebraic framework and in **FPL**. □

Exercise 8.5.50. Suppose we impose the reachability structure defined for an arbitrary institution in Definition 4.5.4 in the current context. In the resulting framework, introduce notions of **INS**-behavioural submodel (Definition 8.2.9) and **INS**-behavioural quotient (the former as the source of a behavioural factorisation monomorphism, the latter as the target of a behavioural factorisation epimorphism), as well as partial **INS**-behavioural quotient (i.e. an **INS**-behavioural quotient of an **INS**-behavioural submodel). Define constructors for restriction to the behavioural generated submodel (Example 8.4.14) and behavioural quotient (Example 8.4.16) in this framework, and check that they are **INS**-behaviourally trivial.

Define a model to be **INS**-fully abstract (Definition 8.3.21) when it has no non-trivial partial **INS**-behavioural quotient, and require that every model have a partial **INS**-fully abstract behavioural quotient (Lemma 8.3.23). Strengthen the assumptions on **INS** and **INS**-behavioural morphisms in order to ensure that Theorem 8.3.24 carries over.

Then define **INS**-behavioural satisfaction as normal satisfaction in the partial **INS**-fully abstract behavioural quotient, thus using Theorem 8.3.26 as a definition. Reformulate Definition 8.3.30 in the resulting framework and check that the proofs of Corollaries 8.3.29 and 8.3.31 and Theorem 8.3.32 carry over, introducing further assumptions on the institution **INS** when necessary. □

We are now ready to retry generalising the definition of global stability, assuming again that **I** is a class of morphisms in **Sign** such that **INS** is **I**-semi-exact; see Definition 4.4.18.

Notation. For any signature morphism $\sigma: \Sigma \to \Sigma'$ and class $\mathscr{B} \subseteq \mathscr{H}_\Sigma$ of **INS**-behavioural Σ-morphisms, $\mathscr{B}|_\sigma^{-1} = \{h \in \mathscr{H}_{\Sigma'} \mid h|_\sigma \in \mathscr{B}\}$, that is, $\mathscr{B}|_\sigma^{-1}$ is the coimage of $\mathscr{B}$ under the reduct restricted to **INS**-behavioural morphisms, $\_|_\sigma : \mathscr{H}_{\Sigma'} \to \mathscr{H}_\Sigma$. □

Definition 8.5.51 (Globally INS-stable constructor). Let $\iota: \Sigma \to \Sigma'$ be a signature morphism in **I** and let and $\mathscr{B} \subseteq \mathscr{H}_\Sigma$ and $\mathscr{B}' \subseteq \mathscr{H}_{\Sigma'}$ be classes of **INS**-behavioural morphisms in $\mathbf{Mod}(\Sigma)$ and $\mathbf{Mod}(\Sigma')$, respectively. Let $\mathscr{M} \subseteq |\mathbf{Mod}(\Sigma)|$ be a class of Σ-models.

A persistent constructor $F \in Mod(\Sigma \overset{\iota}{\Rightarrow} \Sigma')$ is *globally* **INS**-*stable on* $\mathscr{M}$ *w.r.t.* $\mathscr{B}$ *and* $\mathscr{B}'$ if for every pushout diagram in **Sign**

$$\begin{array}{ccc} \Sigma_G & \xrightarrow{\iota'} & \Sigma'_G \\ \uparrow{\scriptstyle \sigma} & & \uparrow{\scriptstyle \sigma'} \\ \Sigma & \xrightarrow[\iota]{} & \Sigma' \end{array}$$

that admits amalgamation, and class $\mathscr{B}_G \subseteq \mathscr{H}_{\Sigma_G}$ of **INS**-behavioural Σ_G-morphisms, such that $\mathscr{B}_G|_\sigma \subseteq \mathscr{B}$, the translation of F along σ, $\sigma(F) \in Mod(\Sigma_G \stackrel{\iota'}{\Rightarrow} \Sigma'_G)$, is **INS**-stable on $\mathscr{M}_G = \{M_G \in |\mathbf{Mod}(\Sigma_G)| \mid M_G|_\sigma \in \mathscr{M}\}$ w.r.t. $\mathscr{B}_G$ and $\mathscr{B}'_G$, where $\mathscr{B}'_G = \mathscr{B}_G|_{\iota'}^{-1} \cap \mathscr{B}'|_{\sigma'}^{-1}$. We say that F is *globally* **INS**-*stable* (*w.r.t.* $\mathscr{B}$ *and* $\mathscr{B}'$) if it is globally **INS**-stable on the class $dom(F)$. □

Exercise 8.5.52. Instantiate this definition to the standard algebraic framework and to **FPL** via Example 8.5.45, defining a constructor $F \in Mod(\Sigma \stackrel{\iota}{\Rightarrow} \Sigma')$ to be globally stable w.r.t. sets OBS and OBS' of sorts in Σ and Σ', respectively, if it is globally **INS**-stable w.r.t. $\mathscr{B}_{OBS}$ and $\mathscr{B}_{OBS'}$. Show that this yields a notion that is at least as strong as that given by Definition 8.4.24 (carried over to **FPL** as well): a constructor that is globally stable in this sense is also globally stable in the sense of Definition 8.4.24. HINT: Under the notation of Definition 8.4.24 and Example 8.5.45, if $\sigma(OBS) \subseteq OBS_G$ then $\mathscr{B}_{OBS_G}|_\sigma \subseteq \mathscr{B}_{OBS}$ and $\mathscr{B}_{OBS'_G} = \mathscr{B}_{OBS_G}|_{\iota'}^{-1} \cap \mathscr{B}_{OBS'}|_{\sigma'}^{-1}$.

Prove that in fact these two notions coincide, at least on model classes that are closed under behavioural equivalence. HINT: Use Exercise 8.4.33 (and its analogue for **FPL**), and the fact that if $F \in Mod(\Sigma \stackrel{\iota}{\Rightarrow} \Sigma')$ extends correspondences then for any behavioural Σ-morphism $h: A \to B$ there is a span of behavioural Σ'-morphisms $h'': C' \to F(A)$, $h': C' \to F(B)$ such that $h''|_\iota = id_A$ and $h'|_\iota = h$. □

The key fact is now that Theorem 8.4.26 carries over as follows.

Theorem 8.5.53. *A constructor* $F \in Mod(\Sigma \stackrel{\iota}{\Rightarrow} \Sigma')$ *is globally stable on* $\mathscr{M} \subseteq |\mathbf{Mod}(\Sigma)|$ *w.r.t.* $\mathscr{B} \subseteq \mathscr{H}_\Sigma$ *and* $\mathscr{B}' \subseteq \mathscr{H}_{\Sigma'}$ *if it* extends **INS**-behavioural morphisms on $Abs_{\mathscr{B}}(\mathscr{M})$*, that is, for any* **INS**-*behavioural morphism* $h: M \to N$ *with* $M, N \in Abs_{\mathscr{B}}(\mathscr{M})$*, if* $h \in \mathscr{B}$ *then there exists an* **INS**-*behavioural morphism* $h': F(M) \to F(N)$ *such that* $h' \in \mathscr{B}'$ *and* $h = h'|_\iota$. □

Exercise 8.5.54. Check that the proof of Theorem 8.4.26 may be adapted to prove Theorem 8.5.53. HINT: First use the assumption that **INS**-behavioural morphisms are closed under amalgamation (see Definition 8.5.42) to show that for $\sigma: \Sigma \to \Sigma_G$, $\sigma(F)$ extends **INS**-behavioural Σ_G-morphisms $h_G: M_G \to N_G$ such that $h_G|_\sigma \in \mathscr{B}$ and $M_G|_\sigma, N_G|_\sigma \in Abs_{\mathscr{B}}(\mathscr{M})$. □

Exercise 8.5.55. Adapt Exercise 8.4.28, Lemma 8.4.29, Exercise 8.4.30 and Exercise 8.4.31 to the current framework. □

The following definitions generalise Definitions 8.4.34 and 8.4.36.

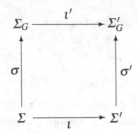

Definition 8.5.56 (INS-behaviourally correct constructor). Let $\iota{:}\Sigma \to \Sigma'$ be a signature morphism, let SP and SP' be Σ- and Σ'-specifications respectively, and let $\mathscr{B}' \subseteq \mathscr{H}_{\Sigma'}$ be a class of **INS**-behavioural Σ'-morphisms. A persistent constructor $F \in Mod(\Sigma \stackrel{\iota}{\Rightarrow} \Sigma')$ is *INS-behaviourally correct along ι w.r.t. SP and SP' for $\mathscr{B}'$* if for every $M \in Mod[SP]$ there exist $M'_0, C'_1, M'_1, C'_2, \ldots, C'_n, M'_n \in |\mathbf{Mod}(\Sigma')|$ and **INS**-behavioural Σ'-morphisms in $\mathscr{B}'$

$$F(M) = M'_0 \xleftarrow{h'_1} C'_1 \xrightarrow{h''_1} M'_1 \xleftarrow{h'_2} C'_2 \xrightarrow{h''_2} \cdots \xleftarrow{h'_n} C'_n \xrightarrow{h''_n} M'_n \in Mod[SP']$$

such that all of $h'_1|_\iota, h''_1|_\iota, \ldots, h'_n|_\iota, h''_n|_\iota$ are the identity (on M). □

Note that under additional assumptions about $\mathscr{B}$, the sequence of spans in the above definition may be replaced by a single span; see Exercise 8.5.47.

Definition 8.5.57 (INS-behavioural model of a constructor specification). Let $\iota{:}\Sigma \to \Sigma'$ be a signature morphism in **I**, let SP and SP' be Σ- and Σ'-specifications respectively, and let $\mathscr{B} \subseteq \mathscr{H}_\Sigma$ and $\mathscr{B}' \subseteq \mathscr{H}_{\Sigma'}$ be classes of **INS**-behavioural morphisms. We write $Mod[SP \,{}_{\mathscr{B}}\!\!\stackrel{\iota}{\Rightarrow}_{\mathscr{B}'} SP']$ for the class of all **INS**-*behavioural models of $SP \stackrel{\iota}{\Rightarrow} SP'$ w.r.t. $\mathscr{B}$ and $\mathscr{B}'$*, that is, constructors that are globally **INS**-stable w.r.t. $\mathscr{B}$ and $\mathscr{B}'$ and are **INS**-behaviourally correct along ι w.r.t. SP and SP' for $\mathscr{B}'$. □

The results about correctness of behavioural models when used in a global context (Theorem 8.4.43 and Corollary 8.4.45) may now be restated in the present context as follows.

Theorem 8.5.58. *Let $\iota{:}\Sigma \to \Sigma'$ be a signature morphism in* **I**, *let SP and SP' be Σ- and Σ'-specifications respectively, and let $\mathscr{B} \subseteq \mathscr{H}_\Sigma$ and $\mathscr{B}' \subseteq \mathscr{H}_{\Sigma'}$ be classes of* **INS**-*behavioural morphisms. Consider an* **INS**-*behavioural model $F \in Mod[SP \,{}_{\mathscr{B}}\!\!\stackrel{\iota}{\Rightarrow}_{\mathscr{B}'} SP']$ and a pushout diagram*

$$\begin{array}{ccc} \Sigma_G & \xrightarrow{\iota'} & \Sigma'_G \\ \uparrow{\scriptstyle \sigma} & & \uparrow{\scriptstyle \sigma'} \\ \Sigma & \xrightarrow[\iota]{} & \Sigma' \end{array}$$

that admits amalgamation. Let $\mathscr{B}_G \subseteq \mathscr{H}_{\Sigma_G}$ be a class of **INS**-*behavioural Σ_G-morphisms such that $\mathscr{B}_G|_\sigma \subseteq \mathscr{B}$ and let $\mathscr{B}'_G = \mathscr{B}_G|_{\iota'}^{-1} \cap \mathscr{B}'|_{\sigma'}^{-1}$. If:*

(i) $Mod[SP_G] \subseteq Abs_{\mathscr{B}_G}(Mod[SP_G \cup (SP \textbf{ with } \sigma)])$*; and*
(ii) $Mod[SP' +_{\iota,\sigma} SP_G] \subseteq Abs_{\mathscr{B}'_G}(Mod[SP'_G])$

then $\sigma(F)$ is stable w.r.t. $\mathscr{B}_G$ and $\mathscr{B}'_G$, and $SP'_G \stackrel{\mathscr{B}'_G}{\underset{\sigma(F)}{\rightsquigarrow}} SP_G$. □

Corollary 8.5.59. *Under the notation of Theorem 8.5.58, if:*

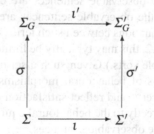

(i) $Mod[SP_G] \subseteq Mod[SP \textbf{ with } \sigma]$; *and*
(ii) $Mod[SP' +_{\iota,\sigma} SP_G] \subseteq Abs_{\mathscr{B}'_G}(Mod[SP'_G])$

then $\sigma(F) \in Mod[SP_G\, {}_{\mathscr{B}_G}\!\overset{\iota'}{\Longrightarrow}_{\mathscr{B}'_G} SP'_G]$. □

Exercise 8.5.60. Prove Theorem 8.5.58 and Corollary 8.5.59 adapting the proofs of Theorem 8.4.43 and of Corollary 8.4.45 as appropriate. In a similar way, adapt Proposition 8.4.47 and its proof to the current framework. □

8.5.3.3 Observing computations

The original motivation for behavioural specifications was phrased in terms of identifying models that displayed the same behaviour, understood as results obtained by observing computations in these models. The definitions of behavioural equivalence reflect this. In the standard algebraic case, we observe the values of computations given as algebraic terms of observable sorts. In **FPL**, we observe termination of computations given as **FPL**-terms, and the values of terminating computations given as **FPL**-terms of observable sorts. Even these two standard cases show that the notions of computations and observations we make on them may considerably differ from one framework to another. So far, the abstract definitions given above did not rely explicitly on any similar concept, even though it has clearly been our intention that the **INS**-behavioural equivalences morphisms considered be somehow linked to this informal idea.

Looking for the means to make such a concept precise, an obvious possibility is to consider observations as given by a set of sentences of the institution in question. Informally, we choose sentences that capture the elementary facts about computation in the given model that the user is able to directly observe. In the standard algebraic framework, these observable sentences are equations between terms of observable sorts. In **FPL**, the observable sentences are definedness assertions for arbitrary **FPL**-terms and equations between such terms of observable sorts. (As explained after Definition 8.5.2, this may typically be limited to definedness assertions for **FPL**-terms of observable sorts.) Given such a set of observable sentences, the natural choice for the class of behavioural morphisms would be the class of all model morphisms that preserve and reflect satisfaction of sentences in this set. Alternatively, we could go directly to the behavioural equivalence which links models that satisfy exactly the same observable sentences.

These ideas would lead to yet another version of the material above, this time parameterised by sets of sentences. The following exercise captures some of the possibilities in this direction and indicates some of the potential problems.

Exercise 8.5.61 (INS-behavioural morphisms from observable formulae). In the standard algebraic framework, for any algebraic signature Σ and set OBS of observable sorts in Σ, consider the set Φ_Σ^{OBS} of all equations between ground Σ-terms of sorts in OBS to be the set of observable sentences. Then a Σ-homomorphism preserves and reflects satisfaction of sentences in Φ_Σ^{OBS} iff it is injective on the reachable elements of the carriers of sorts in *OBS*. Taking this to be the class of behavioural morphisms determines a behavioural equivalence on Σ-algebras: two Σ-algebras are equivalent iff they satisfy exactly the same equations in Φ_Σ^{OBS}. (So in this case the behavioural equivalence determined by the set of behavioural morphisms and the one determined directly by the set of observations coincide.) Check that this behavioural equivalence is exactly behavioural equivalence w.r.t. IN, OUT as introduced in Exercise 8.2.17 for $IN = \varnothing$ and $OUT = OBS$.

For other choices of *IN*, e.g. $IN = OBS$, the resulting behavioural equivalence w.r.t. IN, OUT cannot in general be characterised by a set of Σ-sentences in a similar way. Check for instance that taking Σ-equations between terms of sorts in *OBS* with variables of sorts in *OBS* does not yield the standard behavioural equivalence. One solution to the problem this mismatch poses is to consider sets of observable *open formulae* in place of sets of observable sentences. For instance, here we can consider the set Ψ_Σ^{OBS} which consists of all $\Sigma(X_{OBS})$-equalities between terms of sorts in *OBS*, where X_{OBS} is a set of variables of sorts in *OBS*, and $\Sigma(X_{OBS})$ is the extension of Σ with these variables as constants. Then a Σ-homomorphism is Ψ_Σ^{OBS}-behavioural if all of its expansions to $\Sigma(X_{OBS})$-homomorphisms preserve and reflect satisfaction of $\Sigma(X_{OBS})$-sentences in Ψ_Σ^{OBS}.

Using the machinery of free variables in Section 4.4.2, such a definition may be rephrased in an arbitrary institution, subject to mild technical assumptions. Try to devise requirements on the family of sets of observable open formulae so that the family of classes of morphisms that preserve and reflect satisfaction of observable formulae satisfy the requirements in the definition of **INS**-behavioural morphisms (Definition 8.5.42).

Unfortunately, even in the standard framework, the definition above does not yield the standard notion of behavioural homomorphism w.r.t. *OBS*: the Ψ_Σ^{OBS}-behavioural morphisms are injective but not necessarily surjective on carriers of sorts in *OBS*. Try to strengthen the definition of Ψ_Σ^{OBS}-behavioural morphisms so that surjectivity on carriers of sorts in *OBS* is ensured as well. Make a similar adjustment to the institution-independent version of the definition.

Try to provide a similar treatment of behavioural morphisms and behavioural equivalence in **FPL**. □

Exercise 8.5.62 (INS-behavioural morphism as open morphism). An alternative to the logical view of behavioural equivalence sketched above is to take as primary the behaviour of a model, understood as a morphism from certain special models describing formal computations to the model in question. The first step is therefore to indicate, for each signature $\Sigma \in |\mathbf{Sign}|$, a subcategory $\mathbf{C}_\Sigma \subseteq \mathbf{Mod}(\Sigma)$ of Σ-models that capture formal computations. Given such a subcategory, the class of behavioural morphisms may then be determined by the requirement that they preserve (which is trivial) as well as reflect the actual computations in the models. This informal requirement is made precise by the notion of an "open morphism" as follows.

A Σ-morphism $h: M \to N$ is $\mathbf{C}_\Sigma$*-open* if for any commutative square in $\mathbf{Mod}(\Sigma)$

$$
\begin{array}{ccc}
C_2 & \xrightarrow{q} & N \\
{\scriptstyle m}\uparrow & & \uparrow{\scriptstyle h} \\
C_1 & \xrightarrow[p]{} & M
\end{array}
$$

where $m{:}C_1 \to C_2$ is in $\mathbf{C}_\Sigma$, there exists a morphism $q'{:}C_2 \to M$ in $\mathbf{Mod}(\Sigma)$ such that $q';h = q$ and $m;q' = p$. (This captures the intuition that h reflects the behaviour q of N to the behaviour q' of M.)

For instance, in the standard algebraic framework, for any algebraic signature Σ and set OBS of observable sorts in Σ, the formal computations we may want to observe are captured by Σ-term algebras with variables of sorts in OBS. Formally, we take $\mathbf{C}_\Sigma^{OBS}$ to have objects of the form $T_\Sigma(X_{OBS})$, with morphisms generated by renamings of variables. Check that $\mathbf{C}_\Sigma^{OBS}$-open morphisms are exactly Σ-homomorphisms that are bijective on the carriers of sorts in OBS. Conclude that **INS**-behavioural equivalence generated by taking this class as the **INS**-behavioural Σ-morphisms coincides with the standard behavioural equivalence w.r.t. OBS.

Come up with a similar notion of observing subcategory to generate the standard behavioural equivalences in **FPL**.

Try to devise requirements on $\langle \mathbf{C}_\Sigma \rangle_{\Sigma \in |\mathbf{Sign}|}$ so that the induced classes of open morphisms satisfy the requirements in the definition of **INS**-behavioural morphisms (Definition 8.5.42). □

8.6 Bibliographical remarks

The literature on behavioural aspects of specifications and related topics is vast, including for instance indistinguishability of states in a deterministic automaton [Moo56] as used in the minimisation algorithm, bisimilarity and testing equivalences in process algebra [Mil89], contextual equivalence in λ-calculus [Mit96], and universal coalgebra [Rut00]. Giving comprehensive references to all of this material would take us far beyond the scope of these notes, and we therefore restrict ourselves to comments on directly relevant work in algebraic specification.

There is no standard terminology for the concepts that we refer to here using the terms "behavioural equivalence" and "behavioural indistinguishability". This applies also to our own work; for instance, the terminology in [BST08] clashes with that used here. The reader should be aware that these terms (and their variants with "observational" in place of "behavioural") are sometimes used in the algebraic specification literature to refer to similarly motivated but technically different concepts.

Ideas concerning behavioural interpretation of specifications originated with [GGM76] and [Rei81]. The former was a reaction to work on the initial algebra approach to specification in [GTW76] — see [ST08] for a retrospective discussion

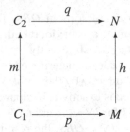

— while the latter was motivated by regarding algebras as a natural generalisation of finite state automata; see also [GM82]. Similar ideas were behind work on the final algebra approach to specification [Wan79]. Other early work includes [SW83] and the proof-oriented approach in [Gan83]. The presentation here is based on our own work [ST87], [ST88b], [ST89], [ST97], [BST08], which was heavily influenced by the work of Schoett [Sch87], and that of Bidoit and Hennicker [BHW95], [BH06a]. The latter paper describes the "constructor-based observational logic" COL which nicely marries reachability and observability aspects of algebraic specifications. Another perspective on some of this material is in the work on hidden algebra, which started with [Gog91b], [GD94b] and [GM00] and took a restricted view in which observers have at most one argument of non-observable sort. The hidden algebra approach was later generalised in [RG98] and [Roş00] to remove this restriction.

The presentation of behavioural equivalence in Section 8.2.1 follows [Sch87], including the notion of correspondence (Definition 8.2.5) and the characterisation of behavioural equivalence in terms of correspondences in Theorem 8.2.8 and via spans of behavioural homomorphisms in Exercise 8.2.15. Precursors to the notion of correspondence include weak homomorphisms in [Gin68] and simulations in [Mil71]. An extension of correspondences to multialgebras is presented in [Nip86]. For a higher-order version of correspondences developed later, making a connection with the notion of logical relations in λ-calculus [Mit96], see prelogical relations [HS02]. Correspondences are presented as a method for proving behavioural equivalence in [Sch90].

Exercises 8.2.17 and 8.3.8 hint at some of the variations on the definition of behavioural equivalence that have been considered. In particular, two extremes have been studied, both of which we now consider inappropriate. In the first, no inputs are allowed ($IN = \varnothing$, in the notation of Exercises 8.2.17) as in [ST87], which has to be overridden when the use of generic constructions in a larger context is taken into account, as in Theorem 8.4.43. In the second, arbitrary inputs are allowed ($IN = S$) as in [Rei85], which fails to identify algebras that differ only in their behaviour on "junk" elements of non-observable sorts. This problem is shared by the definition of behavioural equivalence in [Rei81] and [GM82], which requires a "sink" of surjective behavioural homomorphisms (compare this with Theorem 8.2.14). The same behavioural equivalence is used in some later papers for the sake of simplicity as a stepping stone to our definition of behavioural equivalence — see for instance [BHW94], which is a preliminary version of [BHW95] — since the notion of behavioural indistinguishability to which it corresponds, in the sense of Theorem 8.3.24, is an ordinary congruence rather than a partial congruence.

Behavioural abstraction first appeared in ASL [SW83]. The same essential idea is implicit in work on abstract model specifications; see Z [Spi92] and VDM [Jon80]. One advantage of using abstract model specifications is the availability of a possible model of the system under development as an aid to intuition and to validate the requirements imposed on the system's behaviour. Disadvantages are that the observable behaviour of the specified system needs to be completely defined at an early stage, and that the intuition provided by the model may be misleading.

The distinction between behavioural and weak behavioural equivalence has not been belaboured in the literature; one definition or the other has simply been used without discussion of the other alternative. And rightly so, according to our analysis in Section 8.2.3, provided that all of the classes of algebras considered are closed under isomorphism (Corollary 8.2.34 and Exercise 8.2.35).

The notions of indistinguishability and behavioural satisfaction originate with [Rei85]. Results such as Theorems 8.3.24, 8.3.26 and 8.3.32 that link these with behavioural equivalence for the case of first-order logic (Exercises 8.3.27 and 8.3.34) are due to [BHW95], with precursors for (conditional) equational logic in [Rei85] and [NO88]. The notion of full abstraction (Definition 8.3.21) is originally from the study of the relationship between denotational semantics and operational semantics [Mil77], [Plo77].

The fact that behavioural satisfaction does not lead in a natural way to an institution (Exercise 8.3.16) has been (re)discovered several times beginning with [Rei85]. Various devices have been used to circumvent this problem; one is weaker notions of institution, including specification logics [EPO89] and pre-institutions [SS93], [SS96]. The other approach is to restrict signature morphisms to the unproblematic cases but stopping short of Exercises 8.3.17 and 8.3.35 by limiting attention to signature morphisms that introduce no new observers for old sorts; see [Gog91b] and [BH06a].

The presentation of behavioural implementation in Section 8.4.1 follows [ST88b], where the term "abstractor implementation" was used. A related notion is used in [ONS93] to classify various earlier approaches to implementation of specifications. The idea of allowing implementation steps to realise specifications up to behavioural equivalence is from [SW83], where it was the main motivation behind the inclusion of behavioural abstraction as a specification-building operation. A number of earlier approaches implicitly incorporated this possibility to some extent via the use of quotient steps or surjective homomorphisms in the notion of implementation, starting with [Hoa72]. This is captured here by the concept of behavioural homomorphism; see Definition 8.2.9, Corollary 8.2.10 and the subsequent discussion. Example 8.2.13 shows that the use of behavioural implementation gives a more general approach.

The material on stability in Sections 8.4.2–8.4.5, originates in [Sch87], having been further elaborated on in the context of specific specification frameworks in [ST89] and [BST08]; the presentation here follows the latter. The notion of behavioural triviality (Definition 8.4.10) and the elimination of behaviourally trivial constructors from chains of behavioural implementation steps are from [ST88b]. The phenomenon in Exercise 8.4.37 was first pointed out in [Ber87].

Behavioural equivalence in the context of partial algebras was considered by [Rei85], and an appropriate notion of correspondence is given in [Sch87]. The presentation of behavioural equivalence and behavioural implementation in **FPL** in Section 8.5.1 largely follows the presentation for CASL in [BST08], where partiality was also an issue. The generalisation hinted at towards the end of Exercise 8.5.29 originates in [SW83], [ST87] and [ST88b], where equivalence was determined by an arbitrary set of terms as observers.

The material in Section 8.5.3 on behavioural specifications in an arbitrary institution is more speculative than that in the rest of the chapter and should be viewed largely as suggestions for future research. A first attempt in this direction was in [ST87], which was based on elementary equivalence of models with respect to a set of observable formulae, an idea which emerges here in Exercise 8.5.61. An abstract view of the material relating behavioural satisfaction and behavioural equivalence in Section 8.3 was given in [BT96]. A generalisation to the specific case of higher-order logic is presented in [HS96]. Exercise 8.5.62 originates in [JNW96] via [Las98].

Some further topics remain for future research. One is to give an account of behavioural equivalence for (higher-order) constructors as introduced in Section 6.4; see Exercise 8.5.30. It seems clear that the notion of logical relations [Mit96] is relevant, whereby behavioural equivalence would lift to arbitrary constructor signatures by defining $F_1 \equiv F_2$ for $F_1, F_2 \in Mod(\mathcal{S}_1 \Rightarrow \mathcal{S}_2)$ iff for all $A_1, A_2 \in Mod(\mathcal{S}_1)$, $A_1 \equiv A_2$ implies $F_1(A_1) \equiv F_2(A_2)$. From the form of the definition, there is already an obvious connection with stability, but further ramifications are unclear.

Another open issue concerns the interplay between non-determinism and underspecification, especially in the context of changes in the level of abstraction arising in the course of implementation steps involving data representation. A paradigmatic example is the operation $choose : set \to elem$ in the specification of sets with the requirement that $S \neq \varnothing \Rightarrow choose(S) \in S$. Now, consider an implementation step in which sets are represented by unordered lists, in which *choose* returns the first element in its argument. A corresponding constructor implementation would require a quotient step by an equivalence that is not compatible with *choose*. Essentially, *choose* is not a function when viewed on the abstract level of sets but is one only when viewed on the underlying representation in terms of lists. An attempt to view this as a behavioural implementation runs into the same problem, with the indistinguishability relation that is determined by the rest of the signature not being compatible with *choose*. We know no satisfactory answer to this conundrum, but one possible approach is via multialgebras [Nip86], in which a non-deterministic operation delivers a set of possible outcomes; another would involve relaxing the requirement that indistinguishability be a congruence with respect to operations like *choose*; see [QG93].

Chapter 9
Proofs for specifications

The approach throughout the preceding chapters has been overwhelmingly semantic and model-theoretic: specifications describe classes of models, refinement between specifications is captured by model class inclusion, programs are identified with models, and correctness of a program P with respect to a specification SP refers to membership of the model corresponding to P in the class of models of SP.

This is mathematically precise, pleasingly elegant and abstract, but it lacks a crucial dimension. What is missing are formal proofs, whereby syntax of specifications, sentences and programs are directly manipulated, without reference to models, to establish specification properties, specification refinements, and program correctness. Such proofs are of obvious central importance in actually using specifications in software engineering. For example, in order to be convinced that an implementation step $SP \rightsquigarrow SP'$ is correct, a proof is required; since a direct comparison of model classes is infeasible, this must be based on manipulation of SP and SP' as formal expressions and comparison of the properties required by SP with those guaranteed by SP'. Given the complexity of practical specifications, such activity requires machine support, for example using a theorem prover which applies proof rules, decision procedures and heuristic search methods to specifications and sentences.

We need proofs at four levels, at least. First, we need to know how to prove that a sentence follows from a set of axioms: $\Phi \vdash \varphi$. Second, we need to extend this to the level of (structured) specifications: $SP \vdash \varphi$. Third, we need entailment between specifications, to support proofs of correctness of simple implementations: $SP \vdash SP'$. Finally, we need a proof system for proving correctness of constructor implementations: $SP \vdash_{\kappa} SP'$. These notations will be formally defined below. In addition, in order to take the developments of Chapter 8 into account, at each of these levels we need to consider both a "literal" version, written with $\vdash$, and a behavioural version, written with $\vdash_{OBS}$. (In fact, it turns out to be a little more complicated than that; details below.) In line with our institution-based treatment, we show how a basic proof system for an institution (proving $\Phi \vdash \varphi$) "lifts" to proof systems at the other levels. For the behavioural version, this lifting is much more difficult; in general, the increased power of behavioural specifications must be paid for in significantly increased difficulty of proofs.

In software engineering practice, specifications and programs are written in different languages. It then seems that none of the above four levels deals with the issue of proving that a program satisfies a specification, which would require a separate program logic. However, in our framework this is covered by proofs of correctness of constructor implementations, since programs arise by composing constructors. A specific example which is reminiscent of the Standard ML programming language is given by the institution **FPL** (Example 4.1.25), with constructors corresponding to ML functors (Example 6.1.9). Proofs of correctness of such constructor implementations will emerge largely from the institution-independent lifting of a proof system $\Phi \vdash \varphi$ for **FPL** (which incorporates a program logic for functional programs encoded as **FPL**-terms) through $SP \vdash \varphi$ to $SP \vdash SP'$ and $SP \vdash_{\kappa} SP'$.

Some proof-theoretic aspects of our framework have been covered earlier, in Section 2.4 on the equational calculus, in Section 2.6 on term rewriting, and in Section 5.4 on a property-oriented semantics of structured specifications. Here we proceed further by studying the problem of giving proof calculi for the four levels mentioned above. In each case our starting point is a corresponding model-theoretic relation that has been introduced in earlier chapters, which serves as a standard that we aim to soundly approximate by proof-theoretic means.

The deliberate decision to build a theory of specification and formal development on a model-theoretic basis, with proofs taking a subordinate role, might appear to some readers to lead to an imbalance which becomes most apparent in this chapter. Although, as Section 9.6.5 below will demonstrate, the proof methods presented are adequate to deal with non-trivial examples, there are a few specification constructs for which we have no adequate proof-theoretic treatment, while for others there are proof methods that are useful in many cases but not all. Our excuse for this situation is our conviction that "reality" is on the level of models, with any mismatch between models and proof techniques being unfortunate but unavoidable. One reason it is unavoidable is that completeness results for formal logical systems are unobtainable under all circumstances of practical interest: by Gödel's incompleteness theorem, it is already impossible to provide a sound and complete proof system for ordinary arithmetic, and we show that similar impossibility results apply for different reasons to other specification constructs. Beyond soundness, which is non-negotiable, all that we can expect is to minimise the number of cases that cannot be handled, and (sometimes) to provide easily checkable conditions that allow the problematic cases to be recognized.

9.1 Entailment systems

The starting point of our proof-theoretic development is the abstract concept of entailment, expressing the relationship between a set of assumptions and a conclusion that can be drawn from them.

Definition 9.1.1 (Entailment relation). An *entailment relation* on a set $\mathbb{S}$ of sentences is a binary relation $\vdash \subseteq \mathcal{P}(\mathbb{S}) \times \mathbb{S}$ satisfying the following properties:

1. (reflexivity) $\{\varphi\} \vdash \varphi$;
2. (weakening) if $\Phi \vdash \varphi$ then $\Phi \cup \Psi \vdash \varphi$; and
3. (transitivity) if $\Phi \vdash \psi$ and $\Psi_\varphi \vdash \varphi$ for each $\varphi \in \Phi$ then $\bigcup_{\varphi\in\Phi} \Psi_\varphi \vdash \psi$

for all sentences $\varphi, \psi \in \mathbb{S}$ and sets of sentences $\Phi, \Psi \subseteq \mathbb{S}$. □

The properties required by an entailment system are exactly the basic properties of the semantic consequence relation in an arbitrary institution; see Proposition 4.2.6.

Entailment systems are signature-indexed families of entailment relations closed under translation along signature morphisms.

Definition 9.1.2 (Entailment system). Let $\mathbf{Sen}: \mathbf{Sign} \to \mathbf{Set}$ be a functor. By an *entailment system* for **Sen** we mean a $|\mathbf{Sign}|$-indexed family of entailment relations $\langle \vdash_\Sigma \subseteq \mathcal{P}(\mathbf{Sen}(\Sigma)) \times \mathbf{Sen}(\Sigma)\rangle_{\Sigma\in|\mathbf{Sign}|}$ such that for each morphism $\sigma: \Sigma \to \Sigma'$ in **Sign**, sentence $\varphi \in \mathbf{Sen}(\Sigma)$ and set $\Phi \subseteq \mathbf{Sen}(\Sigma)$, if $\Phi \vdash_\Sigma \varphi$ then $\mathbf{Sen}(\sigma)(\Phi) \vdash_{\Sigma'} \mathbf{Sen}(\sigma)(\varphi)$, where $\mathbf{Sen}(\sigma)(\Phi)$ denotes the image of Φ under $\mathbf{Sen}(\sigma)$.

Given an institution $\mathbf{INS} = \langle \mathbf{Sign}, \mathbf{Sen}, \mathbf{Mod}, \langle \models_\Sigma \rangle_{\Sigma\in|\mathbf{Sign}|}\rangle$, an *entailment system for* **INS** is an entailment system $\langle \vdash_\Sigma \rangle_{\Sigma\in|\mathbf{Sign}|}$ for **Sen** that is sound with respect to semantic consequence, that is, for each signature Σ, sentence $\varphi \in \mathbf{Sen}(\Sigma)$ and set $\Phi \subseteq \mathbf{Sen}(\Sigma)$, if $\Phi \vdash_\Sigma \varphi$ then $\Phi \models_\Sigma \varphi$. It is *complete for* **INS** if the opposite implications hold as well, that is, if $\Phi \models_\Sigma \varphi$ then $\Phi \vdash_\Sigma \varphi$. □

The following examples outline entailment systems for some of the institutions encountered in earlier chapters. The presentation is necessarily sketchy and incomplete, as the engineering of proof systems is a topic that is orthogonal to the main concerns of this book. Our aim is merely to provide examples that are sufficient to illustrate the definitions and to support reasoning in the concrete verification examples that come later.

Example 9.1.3. The equational calculus of Definition 2.4.1 defines a complete entailment system for the institution **EQ** of equational logic. Exercise 2.4.2 shows that for any $\Sigma \in |\mathbf{AlgSig}|$, $\vdash_\Sigma$ as defined there is an entailment relation, and Exercise 2.4.3 together with Theorem 2.4.6 shows that $\langle \vdash_\Sigma \rangle_{\Sigma\in|\mathbf{AlgSig}|}$ is an entailment system for **EQ**. Theorem 2.4.12 gives its completeness. □

Example 9.1.4. An entailment system for the institution **PEQ** of partial equational logic is obtained from the equational calculus by replacing the instantiation rule with

$$\frac{\Phi \vdash_\Sigma \forall X \bullet t = t' \qquad \Phi \vdash_\Sigma \forall Y \bullet def(\theta(x)) \text{ for all } x \in X}{\Phi \vdash_\Sigma \forall Y \bullet t[\theta] = t'[\theta]} \quad \theta: X \to |T_\Sigma(Y)|$$

and adding the following rules for definedness:

$$\frac{}{\Phi \vdash_\Sigma \forall X \bullet def(x)} \quad X_s \subseteq \mathscr{X} \text{ for all } s \in S, \text{ and } x \in X$$

$$\frac{\Phi \vdash_\Sigma \forall X \bullet def(f(t_1,\ldots,t_n))}{\Phi \vdash_\Sigma \forall X \bullet def(t_i)} \quad i = 1,\ldots,n$$

$$\frac{\Phi \vdash_\Sigma \forall X \bullet t = t' \qquad \Phi \vdash_\Sigma \forall X \bullet def(t)}{\Phi \vdash_\Sigma \forall X \bullet def(t')}$$

$$\frac{\Phi \vdash_\Sigma \forall X \bullet def(t) \qquad \Phi \vdash_\Sigma \forall Y \bullet def(\theta(x)) \text{ for all } x \in X}{\Phi \vdash_\Sigma \forall Y \bullet def(t[\theta])} \quad \theta{:}X \to |T_\Sigma(Y)|$$

It is easy to verify that this yields an entailment system for **PEQ** by modifying the proofs for **EQ** as appropriate. (**Exercise:** Do this!) □

Exercise 9.1.5. Consider a signature Σ with sort s, constant $a{:}s$ and unary operations $f,g{:}s \to s$. Show that whenever a partial Σ-algebra satisfies $\forall x{:}s\bullet\ f(x) = g(x)$, it also satisfies $f(a) = g(a)$. However, there is no way to derive $\{\forall x{:}s\bullet\ f(x) = g(x)\} \vdash_\Sigma f(a) = g(a)$ using the rules suggested in Example 9.1.4, which shows that the entailment system defined there is not complete for the institution **PEQ** of partial equational logic. It is possible to adjust the modified instantiation rule in Example 9.1.4 to take care of cases like this (essentially allowing substitution of all terms for variables that occur on both sides of an equation), but we will take a different route, postponed until Exercise 9.1.10.

For now, adapt the equational calculus to the institution $\mathbf{PEQ}^e$ of partial existence equational logic; see Exercise 4.1.8. Among the rules of the equational calculus in Definition 2.4.1, reflexivity has to be limited to equations between variables only, with the symmetry, transitivity and congruence rules remaining as they are there. The substitutivity rule requires an adjustment similar to that in Example 9.1.4, where definedness of a term t can be captured as $t \stackrel{e}{=} t$. Additional rules are required to assert that whenever an (existential) equation can be derived, both sides of the equation are defined, and all their subterms are defined as well. Check that these rules define a (sound) entailment system for $\mathbf{PEQ}^e$. Prove that it is complete. HINT: Proceed similarly as in the proof sketch for Theorem 2.4.12 (completeness of equational calculus) but limit consideration to *minimally defined* partial algebras of terms, i.e. partial algebras built on terms t such that $t \stackrel{e}{=} t$ can be proved. □

Example 9.1.6. An entailment system for the institution **PROP** of propositional logic may be defined in a number of possible ways. Perhaps the simplest is to refer almost directly to the definition of semantic consequence and define entailment by considering directly all possible valuations of propositional variables in $\{\bot,\top\}$. That is, for each set $P \in |\mathbf{Set}|$ of propositional variables, set $\Phi \subseteq \mathbf{Sen}_{\mathbf{PROP}}(P)$ of propositional sentences, and sentences $\varphi \in \mathbf{Sen}_{\mathbf{PROP}}(P)$, define $\Phi \vdash_P \varphi$ to hold exactly when for all valuations $M{:}P \to \{\bot,\top\}$, either $M \models_P \varphi$ or $M \not\models_P \psi$ for some $\psi \in \Phi$. Note that if P is finite then we have only a finite (although exponential in the size of P) number of valuations to check, and for finite Φ this yields a finitary decision procedure for entailment. Clearly, this defines an entailment system that is sound and complete for **PROP**.

However, it may be more convenient to present the entailment relation for propositional logic in a more standard way, for instance using the following rules:[1]

(axiom) $$\frac{}{\Phi \vdash_P \varphi} \quad \varphi \in \Phi$$

(true-I) $$\frac{}{\Phi \vdash_P \mathsf{true}}$$

(false-I) $$\frac{\Phi_1 \vdash \varphi \qquad \Phi_2 \vdash_P \neg\varphi}{\Phi_1 \cup \Phi_2 \vdash_P \mathsf{false}}$$

(false-E) $$\frac{\Phi \vdash_P \mathsf{false}}{\Phi \vdash_P \psi}$$

($\wedge$-I) $$\frac{\Phi_1 \vdash_P \varphi \qquad \Phi_2 \vdash_P \psi}{\Phi_1 \cup \Phi_2 \vdash_P \varphi \wedge \psi}$$

($\wedge$-E) $$\frac{\Phi \vdash_P \varphi \wedge \psi}{\Phi \vdash_P \varphi} \qquad \frac{\Phi \vdash_P \varphi \wedge \psi}{\Phi \vdash_P \psi}$$

($\vee$-I) $$\frac{\Phi \vdash_P \varphi}{\Phi \vdash_P \varphi \vee \psi} \qquad \frac{\Phi \vdash_P \psi}{\Phi \vdash_P \varphi \vee \psi}$$

($\vee$-E) $$\frac{\Phi \vdash_P \varphi_1 \vee \varphi_2 \qquad \Psi_1 \cup \{\varphi_1\} \vdash_P \psi \qquad \Psi_2 \cup \{\varphi_2\} \vdash_P \psi}{\Phi \cup \Psi_1 \cup \Psi_2 \vdash_P \psi}$$

($\Rightarrow$-I) $$\frac{\Phi \cup \{\varphi\} \vdash_P \psi}{\Phi \vdash_P \varphi \Rightarrow \psi}$$

($\Rightarrow$-E) $$\frac{\Phi \vdash_P \varphi \Rightarrow \psi \qquad \Phi \vdash_P \varphi}{\Phi \vdash_P \psi}$$

($\neg$-I) $$\frac{\Phi \cup \{\varphi\} \vdash_P \mathsf{false}}{\Phi \vdash_P \neg\varphi}$$

($\neg\neg$-E) $$\frac{\Phi \vdash_P \neg\neg\psi}{\Phi \vdash_P \psi}$$

Exercise. Show that the rules given above define an entailment relation $\vdash_P \subseteq \mathcal{P}(\mathbf{Sen}_{\mathbf{PROP}}(P)) \times \mathbf{Sen}_{\mathbf{PROP}}(P)$, that is, that $\vdash_P$ is closed under weakening and transitivity. Then show that these entailment relations are closed under renamings of propositional variables, and are sound w.r.t. semantic consequence $\models_{\mathbf{PROP},P}$, so that

[1] The names for the rules indicate the connective concerned and whether it is "introduced" or "eliminated" by the rule.

they form an entailment system $\langle \vdash_P \rangle_{P \in |\mathbf{Set}|}$ for **PROP**. HINT: All these claims follow by induction on the derivation of $\Phi \vdash_P \varphi$.

Exercise. Show that the entailment system $\langle \vdash_P \rangle_{P \in |\mathbf{Set}|}$ for **PROP** is complete. HINT: To prove that if $\Phi \not\vdash_P \varphi$ then $\Phi \not\models_P \varphi$, build a Boolean algebra with elements that are equivalence classes of propositional sentences with variables in P w.r.t. the relation $\Leftrightarrow_\Phi$ defined by $\psi_1 \Leftrightarrow_\Phi \psi_2$ iff $\Phi \vdash_P (\psi_1 \Rightarrow \psi_2) \wedge (\psi_2 \Rightarrow \psi_1)$. Check that this defines an equivalence relation on $\mathbf{Sen}_{\mathbf{PROP}}(P)$, and moreover, that all propositional connectives preserve $\Leftrightarrow_\Phi$. Check that the equivalence classes of $\Leftrightarrow_\Phi$ with the propositional connectives as operations form a Boolean algebra (that is, that the axioms listed in Example 2.2.4 hold in the resulting algebra). Show that $\psi \Leftrightarrow_\Phi \mathtt{true}$ iff $\Phi \vdash_P \psi$. Suppose $\Phi \not\vdash_P \varphi$; then $\varphi \not\Leftrightarrow_\Phi \mathtt{true}$. Now, use Exercise 4.1.10: consider the function that maps each propositional variable to its equivalence class to show that in $\mathbf{PROP}^{\mathbf{BA}}$, and hence in **PROP** as well, $\Phi \not\models_P \varphi$. □

Exercise 9.1.7. Exercise 4.1.11 defines the institution $\mathbf{PROP}^{\mathbf{I}}$ of intuitionistic propositional logic. Check that the rules in Example 9.1.6 with the important exception of double-negation elimination ($\neg\neg$-E) remain sound for intuitionistic propositional logic and yield a sound entailment system for $\mathbf{PROP}^{\mathbf{I}}$. In fact, this entailment system is also complete for $\mathbf{PROP}^{\mathbf{I}}$; see Chapter 2 in [SU06]. □

Example 9.1.8. Recall the institution **FOP** of first-order predicate logic, which uses propositional connectives and quantifiers to build sentences out of atomic predicate formulae. It is no surprise that an entailment system for this institution may be built by extending the rules for propositional connectives as given in Example 9.1.6 by rules to deal with quantifiers. We will refrain from repeating here the rules that are essentially in Example 9.1.6, but note that formally they have to be rephrased now for entailments indexed by first-order signatures and applied to sentences of first-order predicate logic.

Before we present the rules for quantifiers, recall that for any first-order signature $\Theta = \langle S, \Omega, \Pi \rangle$, we write T_Θ for the algebra of terms over the algebraic signature $\langle S, \Omega \rangle$. Then, for any S-sorted set of variables X (with sets of variables for different sorts that are mutually disjoint and that are disjoint from the operation and predicate names in Θ), $\Theta(X)$ is the signature Θ extended by the variables X as constants of the appropriate sorts. We tacitly identify terms and sentences over $\Theta(X)$ with terms and open formulae over Θ with variables from X. Finally, we rely on the usual notion of substitution of terms for free variables in first-order logic formulae, written $\varphi[\theta]$, where φ is a first-order formula and θ maps some variables to Θ-terms, preserving their sorts. This is defined so that no clashes of free variables in terms in the range of θ with bound variables in φ occur; for instance, require that the bound variables in φ are distinct from the variables used in the terms in the range of θ.

Equipped with these notations, we can add rules for quantified formulae to the rules for propositional connectives:

$$(\forall\text{-I}) \quad \frac{\Phi \vdash_{\Theta(\{x:s\})} \varphi}{\Phi \vdash_{\Theta} \forall x{:}s \bullet \varphi} \quad \Phi \subseteq \mathbf{Sen_{FOP}}(\Theta), \varphi \in \mathbf{Sen_{FOP}}(\Theta(\{x:s\}))$$

$$(\forall\text{-E}) \quad \frac{\Phi \vdash_{\Theta} \forall x{:}s \bullet \varphi}{\Phi \vdash_{\Theta} \varphi[x \mapsto t]} \quad t \in |T_{\Theta}|_s$$

$$(\exists\text{-I}) \quad \frac{\Phi \vdash_{\Theta} \varphi[x \mapsto t]}{\Phi \vdash_{\Theta} \exists x{:}s \bullet \varphi} \quad t \in |T_{\Theta}|_s$$

$$(\exists\text{-E}) \quad \frac{\Phi_1 \vdash_{\Theta} \exists x{:}s \bullet \varphi \quad \Phi_2 \cup \{\varphi\} \vdash_{\Theta(\{x:s\})} \psi}{\Phi_1 \cup \Phi_2 \vdash_{\Theta} \psi} \quad \begin{array}{l}\Phi_1, \Phi_2, \{\psi\} \subseteq \mathbf{Sen_{FOP}}(\Theta), \\ \varphi \in \mathbf{Sen_{FOP}}(\Theta(\{x:s\}))\end{array}$$

We vary the signature over which the entailment is defined to capture the standard assumptions on variables that can be quantified over when a universal quantification is introduced in (∀-I) or used as "free variables" when an existential quantification is eliminated in (∃-E). In particular, this means that rather than defining here an entailment relation for each signature separately, we define a family of such relations simultaneously for all signatures.

Exercise. Prove that the above rules generate an entailment system that is sound and complete for **FOP**. A warning: any completeness proof here is considerably more complex than in Example 9.1.6. HINT: Given a set Φ of formulae such that $\Phi \not\vdash \varphi$, provide a universe of terms that is rich enough to witness all existential sentences that follow from Φ, by simultaneously extending Φ to Φ' such that we still have $\Phi' \not\vdash \varphi$ and adding a new constant $c{:}s$ for each formula ψ such that whenever $\Phi' \vdash \exists x{:}s \bullet \psi$, we have $\Phi' \vdash \psi[x \mapsto c]$. Then build a model for Φ' on this universe in which φ does not hold. A considerably more detailed sketch may be found in [Sha08], and a complete proof in any monograph on logic, for instance [RS63]. □

Exercise 9.1.9. Expand further the system of rules from Examples 9.1.6 and 9.1.8 to introduce equational reasoning as required for consequence in the institution **FOPEQ** of first-order logic with equality. For this, adapt the rules from Definition 2.4.1 as appropriate. Notice though that this time we do not need to deal explicitly with quantification over (finite) sets of variables, as the rules to introduce and eliminate quantification are already available (as (∀-I) and (∀-E) respectively in Example 9.1.8). In particular, no new rule for instantiation is needed. However, an additional rule to ensure that equal terms may be substituted for each other in atomic predicate formulae is necessary. This extra rule together with the usual congruence rule from Definition 2.4.1 may be equivalently replaced by a single rule that allows for substitution of "equals for equals" in an arbitrary formula.

Check that the obtained system of rules does indeed yield a (sound) entailment system for **FOPEQ**. Its completeness is again more difficult to prove, although only a relatively straightforward modification of the completeness proof sketched in the hint in Example 9.1.8 above is needed. □

Exercise 9.1.10. The rules for propositional logic in Example 9.1.6 remain valid also for the versions of first-order logic with partial operations introduced in Exercises 4.1.17 and 4.1.18. This is not true though for the rules for quantifiers in

Example 9.1.8 — the rule (∀-E) that eliminates universal quantification by substituting a term for the quantified variable must in addition require that the term have a defined value, and a similar problem affects the rule (∃-I), where the term witnessing the existential formula must be proved to be defined as well. Modify these rules as in Example 9.1.4, and adapt the rules for definedness of terms given there to obtain a (sound) entailment system for the institution **PFOP** of partial first-order predicate logic. Prove its completeness. HINT: The proof essentially follows the completeness proof for first-order predicate logic outlined in Example 9.1.8, viewing definedness as an additional predicate and then limiting the model constructed to only those terms for which definedness can be derived.

Then extend the proof system further by the rules for equational reasoning to obtain a (sound) entailment system for the institution **PFOPEQ** of partial first-order logic with equality. Unfortunately, the counterexample in Exercise 9.1.5 applies here as well, and so such an entailment system is not complete for **PFOPEQ**. As mentioned in Exercise 9.1.5, adjustments to the substitutivity rule are possible, but it is perhaps simpler to resort to existential equations.

Consider a version of partial first-order predicate logic with existential equality — this can easily be defined as an institution **PFOPEQ**^{e}. (**Exercise:** Do this!) Now, adapt the rules for existential equations sketched in Exercise 9.1.5 to the context of **PFOPEQ**^{e} and combine them with the rules for **PFOP** to obtain a (sound) entailment system for **PFOPEQ**^{e}. Prove its completeness. HINT: The hints above and in Exercise 9.1.5 apply.

Strong equality may be added to **PFOPEQ**^{e} as an abbreviation, with $t = t'$ standing for $(t \stackrel{e}{=} t') \vee (\neg def(t) \wedge \neg def(t'))$, so that a sound and complete entailment system for **PFOPEQ** may be obtained from the sound and complete entailment system for **PFOPEQ**^{e}. □

Example 9.1.11. An entailment system for **FPL** can be built on top of the entailment system for **PFOEQ** by adding rules to deal with value constructors and the two additional term forms in **FPL**.

Sorts freely generated by value constructors require axioms which state that constructors are total, injective, and have disjoint ranges (**Exercise:** Spell this out!). Moreover, for each set of sorts with value constructors $\{\langle d_1, \mathscr{F}_1\rangle, \ldots, \langle d_n, \mathscr{F}_n\rangle\}$ in SIG and family of formulae P_s with free variable $x{:}s$, for each $s \in S$, where $P_s(x)$ is a true formula (for instance, $\forall y{:}s \bullet y = y$) for $s \notin \{d_1, \ldots, d_n\}$, we add the following structural induction rules for proving properties of elements of sort d_j, $j = 1, \ldots, n$:

$$\frac{\text{for each } i = 1, \ldots, n \text{ and } f{:}s_1 \times \cdots \times s_k \to d_i \in \mathscr{F}_i, \quad \Phi \vdash_{\mathsf{SIG}} \forall x_1{:}s_1 \bullet \cdots \forall x_n{:}s_n \bullet (P_{s_1}(x_1) \wedge \cdots \wedge P_{s_k}(x_k) \Rightarrow P_{d_i}(f(x_1, \ldots, x_k)))}{\Phi \vdash_{\mathsf{SIG}} \forall x{:}d_j \bullet P_{d_j}(x)}$$

Exercise. Check that the standard induction rule for natural numbers is an instance of this rule for any signature that contains the sort declaration

sort *Nat* **free with** 0 | *succ*(*Nat*).

Exercise. Show that the soundness of the above induction rule relies on the fact that value constructors are total.

Under the assumption that the patterns $pat_1,\ldots,pat_n$ do not overlap — meaning that in every SIG-model, no value matches both pat_i and pat_j for $i \neq j$ — the following rule captures the computational behaviour of pattern-matching case analysis:

$$\frac{\Phi \vdash_{\mathsf{SIG}} \forall X \bullet t = pat_i[\theta_i]}{\Phi \vdash_{\mathsf{SIG}} \forall X \bullet (\mathbf{case}\ t\ \mathbf{of}\ pat_1 \Rightarrow t_1 \mid \cdots \mid pat_n \Rightarrow t_n) = t_i[\theta_i]}$$

where X_i is the set of all variables occurring in pat_i, $\theta_i\colon X_i \to |T_{\mathsf{SIG}}(X)|$, and $i = 1,\ldots,n$. (**Exercise:** Generalise this rule to take account of overlapping patterns.)

The following rules deal with the use of locally defined functions. They allow, respectively, declarations of unused functions to be removed, function calls to be expanded to the function body, and local declarations to be moved to the place they are used (thus permitting evaluation of terms containing function calls).

$$\frac{}{\Phi \vdash_{\mathsf{SIG}} \forall X \bullet (\mathbf{let\ fun}\ f(x_1{:}s_1,\ldots,x_n{:}s_n){:}s' = t'\ \mathbf{in}\ t) = t}$$

where $t' \in |T_{\mathsf{SIG}(f)}(X \cup \{x_1{:}s_1,\ldots,x_n{:}s_n\})|_{s'}$, $t \in |T_{\mathsf{SIG}}(X)|$, and f is not in SIG.

$$\frac{\Phi \vdash_{\mathsf{SIG}(f)} \forall X \bullet t'[x_1 \mapsto t_1,\ldots,x_n \mapsto t_n] = \widehat{t}}{\begin{array}{r}\Phi \vdash_{\mathsf{SIG}} \forall X \bullet \mathbf{let\ fun}\ f(x_1{:}s_1,\ldots,x_n{:}s_n){:}s' = t'\ \mathbf{in}\ f(t_1,\ldots,t_n)\\ = \mathbf{let\ fun}\ f(x_1{:}s_1,\ldots,x_n{:}s_n){:}s' = t'\ \mathbf{in}\ \widehat{t}\end{array}}$$

where $\mathsf{SIG}(f)$ is SIG extended by $f\colon s_1 \times \cdots \times s_n \to s'$, $\widehat{t} \in |T_{\mathsf{SIG}(f)}(X)|_{s'}$, $t' \in |T_{\mathsf{SIG}(f)}(X \cup \{x_1{:}s_1,\ldots,x_n{:}s_n\})|_{s'}$, and $t_1 \in |T_{\mathsf{SIG}(f)}(X)|_{s_1},\ldots,t_n \in |T_{\mathsf{SIG}(f)}(X)|_{s_n}$.

$$\frac{\Phi \vdash_{\mathsf{SIG}(f)} \forall X \bullet t[z \mapsto \mathbf{let\ fun}\ f(x_1{:}s_1,\ldots,x_n{:}s_n){:}s' = t'\ \mathbf{in}\ f(t_1,\ldots,t_n)] = \widehat{t}}{\begin{array}{r}\Phi \vdash_{\mathsf{SIG}} \forall X \bullet \mathbf{let\ fun}\ f(x_1{:}s_1,\ldots,x_n{:}s_n){:}s' = t'\ \mathbf{in}\ t[z \mapsto f(t_1,\ldots,t_n)]\\ = \mathbf{let\ fun}\ f(x_1{:}s_1,\ldots,x_n{:}s_n){:}s' = t'\ \mathbf{in}\ \widehat{t}\end{array}}$$

where $\mathsf{SIG}(f)$ is SIG extended by $f\colon s_1 \times \cdots \times s_n \to s'$, $t \in |T_{\mathsf{SIG}(f)}(X \cup \{z{:}s'\})|$, $\widehat{t} \in |T_{\mathsf{SIG}(f)}(X)|_{s'}$, $t' \in |T_{\mathsf{SIG}(f)}(X \cup \{x_1{:}s_1,\ldots,x_n{:}s_n\})|_{s'}$, and $t_1 \in |T_{\mathsf{SIG}(f)}(X)|_{s_1},\ldots,t_n \in |T_{\mathsf{SIG}(f)}(X)|_{s_n}$.

Finally, the rules of the equational calculus need to be modified to take account of the fact that the two additional term forms in **FPL** add a number of new contexts. Apart from the usual congruence rule we need the following rule for terms with local function definitions (**Exercise:** Spell out the signature and variables for each of the terms occurring in these rules)

$$\frac{\Phi \vdash_{\mathsf{SIG}(f)} \forall X \cup \{x_1{:}s_1,\ldots,x_n{:}s_n\} \bullet t' = t'_1 \qquad \Phi \vdash_{\mathsf{SIG}(f)} \forall X \bullet t = t_1}{\begin{array}{r}\Phi \vdash_{\mathsf{SIG}} \forall X \bullet \mathbf{let\ fun}\ f(x_1{:}s_1,\ldots,x_n{:}s_n){:}s' = t'\ \mathbf{in}\ t\\ = \mathbf{let\ fun}\ f(x_1{:}s_1,\ldots,x_n{:}s_n){:}s' = t'_1\ \mathbf{in}\ t_1\end{array}}$$

and the following rule for terms with pattern-matching **case** expressions:

$$\frac{\Phi \vdash_{\mathsf{SIG}} \forall X \bullet t = t' \qquad \Phi \vdash_{\mathsf{SIG}} \forall X \cup X_1 \bullet t_1 = t_1' \quad \cdots \quad \Phi \vdash_{\mathsf{SIG}} \forall X \cup X_n \bullet t_n = t_n'}{\begin{array}{l}\Phi \vdash_{\mathsf{SIG}} \forall X \bullet \textbf{case}\ t\ \textbf{of}\ pat_1 \Rightarrow t_1 \mid \cdots \mid pat_n \Rightarrow t_n \\ \qquad\qquad = \textbf{case}\ t'\ \textbf{of}\ pat_1 \Rightarrow t_1' \mid \cdots \mid pat_n \Rightarrow t_n'\end{array}}$$

where X_i is the set of all variables occurring in pat_i, for $i = 1, \ldots, n$.

The above rules use the notation $|T_{\mathsf{SIG}}(X)|$ for the set of all SIG-terms with variables X. The instantiation rule of **PFOPEQ** has to be adapted accordingly. They also require an obvious generalisation of the notion of substitution of **FPL**-terms for variables, which takes proper account of the scope of variables and local function names in terms with local function definitions and pattern-matching **case** expressions. Moreover, the obvious rules of α-conversion (renaming bound identifiers, e.g. variables bound by patterns, variables naming local function parameters, names of locally defined functions) should be added. (**Exercise:** Spell this out.)

Exercise. The following equality (over a one-sorted signature in which f and g are constructors, h is a unary function and a is a constant) is not derivable using the rules above: **case** a **of** $f(g(x)) \Rightarrow h(x) =$ **case** a **of** $f(y) \Rightarrow$ **case** y **of** $g(x) \Rightarrow h(x)$. Try to devise additional (sound) rules for manipulating **case** expressions that would allow equations like this one to be derived.

In spite of our best efforts, the entailment system given by the proof rules outlined above is not complete. The presence of sorts freely generated by value constructors implies that there is no finitary rule-based entailment system that is complete for **FPL**, as in Theorem 2.5.26. □

Example 9.1.12. We illustrate the use of the above entailment system for **FPL** using a specification from Example 7.1.4.

Let $\mathsf{SIG} = Sig[\text{INS}]$ and let Φ be the set containing all the axioms concerning *le*, *append* and *is_in* listed in NATORD and NATLIST, together with the following definitional axiom for *insert* (as in constructor *K2* in Example 7.2.8):

$$\begin{array}{l}\textbf{fun}\ insert(n{:}Nat, l{:}NatList){:}NatList = \\ \quad \textbf{case}\ l\ \textbf{of}\ nil \Rightarrow cons(n, nil) \\ \qquad\qquad \mid cons(m, l') \Rightarrow \textbf{if}\ le(n,m) = true\ \textbf{then}\ cons(n,l) \\ \qquad\qquad\qquad\qquad\qquad\qquad \textbf{else}\ cons(m, insert(n, l'))\end{array}$$

Recall that this axiom is an abbreviation for the following:

$$\begin{array}{l}\forall n{:}Nat, l{:}NatList \\ \bullet\ insert(n,l) = \\ \qquad \textbf{let fun}\ insert'(n'{:}Nat, l'{:}NatList){:}NatList = \\ \qquad\qquad \textbf{case}\ l'\ \textbf{of}\ nil \Rightarrow cons(n', nil) \\ \qquad\qquad\qquad \mid cons(m, l'') \Rightarrow \\ \qquad\qquad\qquad\qquad \textbf{case}\ le(n', m)\ \textbf{of}\ true \Rightarrow cons(n', l') \\ \qquad\qquad\qquad\qquad\qquad \mid false \Rightarrow cons(m, insert'(n', l'')) \\ \qquad \textbf{in}\ insert'(n, l)\end{array}$$

We will now prove a number of consequences of Φ. Formal derivations for these proofs would be incomprehensible and would not fit on the page. We therefore give rigorous arguments from which the reader is encouraged to reconstruct the formal proofs employing solely the proof rules given above.

As a warming-up exercise, we prove the following lemma:

$$\Phi \vdash_{\mathsf{SIG}} \forall n{:}Nat \bullet insert(n,nil) = cons(n,nil)$$

The essence of the proof is the following chain of equalities:

$$\begin{array}{l}
insert(n,nil)\\
\quad = \textbf{let fun } insert'(n'{:}Nat, l'{:}NatList){:}NatList =\\
\qquad\qquad \textbf{case } l' \textbf{ of } nil \Rightarrow cons(n',nil)\\
\qquad\qquad\qquad \mid cons(m,l'') \Rightarrow\\
\qquad\qquad\qquad\qquad \textbf{case } le(n',m) \textbf{ of } true \Rightarrow cons(n',l')\\
\qquad\qquad\qquad\qquad\qquad \mid false \Rightarrow cons(m, insert'(n',l''))\\
\qquad \textbf{in } insert'(n,nil)\\
\quad = \textbf{let fun } insert'(n'{:}Nat, l'{:}NatList){:}NatList = \ldots\\
\qquad \textbf{in case } nil \textbf{ of } nil \Rightarrow cons(n',nil)\\
\qquad\qquad \mid cons(m,l'') \Rightarrow\\
\qquad\qquad\qquad \textbf{case } le(n,m) \textbf{ of } true \Rightarrow cons(n,nil)\\
\qquad\qquad\qquad\qquad \mid false \Rightarrow cons(m, insert'(n,l''))\\
\quad = \textbf{let fun } insert'(n'{:}Nat, l'{:}NatList){:}NatList = \ldots\\
\qquad \textbf{in } cons(n,nil)\\
\quad = cons(n,nil)
\end{array}$$

Now we prove the following lemma:

$$\begin{array}{l}
\Phi \vdash_{\mathsf{SIG}} \forall l{:}NatList, k, n{:}Nat \bullet\\
\qquad insert(n, cons(k,l)) = \textbf{case } le(n,k) \textbf{ of } true \Rightarrow cons(n, cons(k,l))\\
\qquad\qquad\qquad\qquad\qquad\qquad \mid false \Rightarrow cons(k, insert(n,l))
\end{array}$$

Again, this essentially follows from the following chain of equalities:

$$\begin{array}{l}
insert(n, cons(k,l))\\
\quad = \textbf{let fun } insert'(n'{:}Nat, l'{:}NatList){:}NatList =\\
\qquad\qquad \textbf{case } l' \textbf{ of } nil \Rightarrow cons(n',nil)\\
\qquad\qquad\qquad \mid cons(m,l'') \Rightarrow\\
\qquad\qquad\qquad\qquad \textbf{case } le(n',m) \textbf{ of } true \Rightarrow cons(n',l')\\
\qquad\qquad\qquad\qquad\qquad \mid false \Rightarrow cons(m, insert'(n',l''))\\
\qquad \textbf{in } insert'(n, cons(k,l))\\
\quad = \textbf{let fun } insert'(n'{:}Nat, l'{:}NatList){:}NatList = \ldots\\
\qquad \textbf{in case } cons(k,l) \textbf{ of } nil \Rightarrow cons(n,nil)\\
\qquad\qquad \mid cons(m,l'') \Rightarrow\\
\qquad\qquad\qquad \textbf{case } le(n,m) \textbf{ of } true \Rightarrow cons(n, cons(k,l))\\
\qquad\qquad\qquad\qquad \mid false \Rightarrow cons(m, insert'(n,l''))\\
\quad = \textbf{let fun } insert'(n'{:}Nat, l'{:}NatList){:}NatList = \ldots\\
\qquad \textbf{in case } le(n,k) \textbf{ of } true \Rightarrow cons(n, cons(k,l))\\
\qquad\qquad \mid false \Rightarrow cons(k, insert'(n,l))
\end{array}$$

$$
\begin{aligned}
&= \textbf{let fun}\ insert'(n'{:}Nat, l'{:}NatList){:}NatList = \ldots\\
&\quad \textbf{in case}\ le(n,k)\ \textbf{of}\ true \Rightarrow cons(n, cons(k,l))\\
&\qquad\qquad\qquad\qquad \mid false \Rightarrow\\
&\qquad\qquad\qquad\qquad\qquad cons(k, \textbf{let fun}\ insert'(n'{:}Nat, l'{:}NatList){:}NatList = \ldots\\
&\qquad\qquad\qquad\qquad\qquad\qquad \textbf{in}\ insert'(n,l))\\
&= \textbf{case}\ le(n,k)\ \textbf{of}\ true \Rightarrow cons(n, cons(k,l))\\
&\qquad\qquad\qquad \mid false \Rightarrow\\
&\qquad\qquad\qquad\qquad cons(k, \textbf{let fun}\ insert'(n'{:}Nat, l'{:}NatList){:}NatList = \ldots\\
&\qquad\qquad\qquad\qquad\qquad \textbf{in}\ insert'(n,l))\\
&= \textbf{case}\ le(n,k)\ \textbf{of}\ true \Rightarrow cons(n, cons(k,l))\\
&\qquad\qquad\qquad \mid false \Rightarrow cons(k, insert(n,l))
\end{aligned}
$$

Exercise. Check that all equalities in these chains can be justified by the rules in Example 9.1.11 (and the definitional axiom for *insert*).

We will now prove

$$\Phi \vdash_{\mathsf{SIG}} \forall l{:}NatList, n{:}Nat \bullet def(insert(n,l)).$$

The proof proceeds by induction on l. From the first lemma above, using one of the definedness rules, we immediately obtain the premise of the induction rule for $l = nil$:

$$\Phi \vdash_{\mathsf{SIG}} \forall n{:}Nat \bullet def(insert(n,nil)).$$

The premise of the induction rule for $l = cons(k,l')$ is

$$
\begin{aligned}
\Phi \vdash_{\mathsf{SIG}} \forall k{:}Nat, l'{:}NatList \bullet\ &(\forall n{:}Nat \bullet def(insert(n,l')) \Rightarrow\\
&\quad \forall n{:}Nat \bullet def(insert(n, cons(k,l')))).
\end{aligned}
$$

This follows, since by assuming $\forall n{:}Nat \bullet def(insert(n,l'))$ and applying the second lemma above, we get $\forall n{:}Nat \bullet def(insert(n,cons(k,l')))$. Notice that this relies on definedness of $le(n,k)$ and follows by case analysis on its value, justified in turn by the induction rule for sort *Bool*. The conclusion

$$\Phi \vdash_{\mathsf{SIG}} \forall l{:}NatList, n{:}Nat \bullet def(insert(n,l))$$

follows now by the induction rule for sort *NatList*.

We now prove

$$\Phi \vdash_{\mathsf{SIG}} \forall l{:}NatList \bullet P(l)$$

where $P(l)$ is the following formula:

$$
\begin{aligned}
&\forall n{:}Nat \bullet \exists l_1, l_2{:}NatList \bullet\\
&\quad insert(n,l) = append(l_1, cons(n,l_2)) \wedge l = append(l_1,l_2)\\
&\quad \wedge (\forall l_1'{:}NatList, m{:}Nat \bullet l_1 = append(l_1', cons(m,nil)) \Rightarrow le(m,n) = true)\\
&\quad \wedge (\forall l_2'{:}NatList, m{:}Nat \bullet l_2 = cons(m,l_2') \Rightarrow le(n,m) = true)
\end{aligned}
$$

Again, the proof proceeds by induction on l. The two premises of the induction rule are $P(nil)$ and $\forall k{:}Nat, l'{:}NatList \bullet (P(l') \Rightarrow P(cons(k,l')))$.

For $P(nil)$, notice that:

- $append(nil, cons(n, nil)) = cons(n, nil)$
- $nil = append(nil, nil)$
- $\forall l'_1{:}NatList, m{:}Nat \bullet \neg(nil = append(l'_1, cons(m, nil)))$
- $\forall l'_2{:}NatList, m{:}Nat \bullet \neg(nil = cons(m, l'_2))$

(Formal proofs of these rely on the definitional axiom for *append*.) Thus, putting $l_1 = nil$ and $l_2 = nil$, we obtain $P(nil)$ using the first lemma above.

For the second premise, $\forall k{:}Nat, l'{:}NatList \bullet (P(l') \Rightarrow P(cons(k, l')))$, given k and l' we assume $P(l')$, so that for any $n{:}Nat$ we have $l'_1{:}NatList$ and $l'_2{:}NatList$ such that:

(a) $insert(n, l') = append(l'_1, cons(n, l'_2))$;
(b) $l' = append(l'_1, l'_2)$;
(c) $\forall l''_1{:}NatList, m{:}Nat \bullet l'_1 = append(l''_1, cons(m, nil)) \Rightarrow le(m, n) = true$; and
(d) $\forall l''_2{:}NatList, m{:}Nat \bullet l'_2 = cons(m, l''_2) \Rightarrow le(n, m) = true$.

Now, for arbitrary $n{:}Nat$, we consider two cases.

Assume $le(n, k) = true$: Then, by the second lemma above, $insert(n, cons(k, l')) = cons(n, cons(k, l'))$. Moreover:

- $append(nil, cons(n, cons(k, l'))) = cons(n, cons(k, l'))$;
- $cons(k, l') = append(nil, cons(k, l'))$;
- $\forall l''_1{:}NatList, m{:}Nat \bullet \neg(nil = append(l''_1, cons(m, nil)))$; and
- $\forall l''_2{:}NatList, m{:}Nat \bullet cons(k, l') = cons(m, l''_2) \Rightarrow le(n, m) = true$

which shows $P(cons(k, l'))$ by putting $l_1 = nil$ and $l_2 = cons(k, l')$.

Assume $le(n, k) = false$: Then, by the second lemma above, $insert(n, cons(k, l')) = cons(k, insert(n, l'))$. Moreover:

(a′) $append(cons(k, l'_1), cons(n, l'_2)) = cons(k, insert(n, l'))$;
(b′) $cons(k, l') = append(cons(k, l'_1), l'_2)$;
(c′) $\forall l''_1{:}NatList, m{:}Nat \bullet cons(k, l'_1) = append(l''_1, cons(m, nil)) \Rightarrow le(m, n) = true$; and
(d′) $\forall l''_2{:}NatList, m{:}Nat \bullet l'_2 = cons(m, l''_2) \Rightarrow le(n, m) = true$.

(a′) and (b′) follow from (a) and (b), respectively, using the following fact about *append*:

$$\forall u, v{:}NatList, i{:}Nat \bullet append(cons(i, u), v) = cons(i, append(u, v))$$

Similarly, by the axioms for *append*, if $cons(k, l'_1) = append(l''_1, cons(m, nil))$ then either $l'_1 = nil = l''_1$ and $k = m$, and then $le(n, m) = false$, and hence $le(m, n) = true$; or $l''_1 = cons(k, u)$ and $l'_1 = append(u, cons(m, nil))$, and hence by (c), $le(m, n) = true$. In both cases, (c′) follows. Finally, (d′) is just (d). This shows $P(cons(k, l'))$ by putting $l_1 = cons(k, l'_1)$ and $l_2 = l'_2$.

From the induction rule, we now conclude

$$\Phi \vdash_{\mathrm{SIG}} \forall l{:}NatList \bullet P(l).$$ □

Exercise 9.1.13. The final paragraph of Example 9.1.11 attributes the incompleteness of the entailment system for **FPL** to the presence of sorts freely generated by value constructors. But there are also other sources of incompleteness. Convince yourself that it is not possible to prove that $\neg def(f(5))$ follows from **fun** $f(x{:}Nat){:}Nat = f(x)$ using the rules given in Example 9.1.11. (Note, however, that $\neg def(g(5))$ does follow from **fun** $g(x{:}Nat){:}Nat = succ(g(x))$.) Add rules to fix this. HINT: See [Pau87], but note that completeness for sentences of this form is unachievable because divergence of general recursive functions is not semi-decidable. □

The standard way of presenting an entailment relation is via proof rules, and that is what we have done in the examples above. This is not the only way to define an entailment system though, as the following exercise illustrates.

Exercise 9.1.14. Given any institution **INS**, the family of semantic consequence relations forms an entailment system which is sound and complete for **INS**, by Propositions 4.2.6 and 4.2.9. Show that any entailment system can be obtained in this way from some institution. HINT: See Example 4.1.36, but consider only sets of sentences that are closed under entailment. □

Throughout the rest of this section, let $\mathbf{INS} = \langle \mathbf{Sign}, \mathbf{Sen}, \mathbf{Mod}, \langle \models_\Sigma \rangle_{\Sigma \in |\mathbf{Sign}|} \rangle$ be an institution with entailment system $\langle \vdash_\Sigma \rangle_{\Sigma \in |\mathbf{Sign}|}$.

All of the concepts that have been defined in a model-theoretic style in previous chapters using the semantic consequence relation $\models$ have proof-theoretic variants, obtained by replacing semantic consequence by entailment $\vdash$. If the entailment system in question is complete, then the proof-theoretic and model-theoretic concepts coincide; otherwise discrepancies may arise. The notions of theory and (conservative) theory morphism provide an example.

Exercise 9.1.15. Show that a theory as defined in Definition 4.2.1 is closed under entailment. Choose an institution with an (incomplete) entailment system and give an example of a set of sentences closed under entailment (that is, a "theory" in a proof-theoretic sense) that is not a theory in the sense of Definition 4.2.1. □

Exercise 9.1.16. Consider a signature morphism $\sigma{:}\Sigma \to \Sigma'$ and sets $\Phi \subseteq \mathbf{Sen}(\Sigma)$ and $\Phi' \subseteq \mathbf{Sen}(\Sigma')$. Consider the following properties:

1. For all $\varphi \in \mathbf{Sen}(\Sigma)$, $\Phi \models_\Sigma \varphi$ implies $\Phi' \models_{\Sigma'} \sigma(\varphi)$.
2. For all $\varphi \in \mathbf{Sen}(\Sigma)$, $\Phi \vdash_\Sigma \varphi$ implies $\Phi' \vdash_{\Sigma'} \sigma(\varphi)$.
3. For all $\varphi \in \mathbf{Sen}(\Sigma)$, $\Phi' \models_{\Sigma'} \sigma(\varphi)$ implies $\Phi \models_\Sigma \varphi$.
4. For all $\varphi \in \mathbf{Sen}(\Sigma)$, $\Phi' \vdash_{\Sigma'} \sigma(\varphi)$ implies $\Phi \vdash_\Sigma \varphi$.

Property 1 is that $\sigma{:}Cl_\Sigma(\Phi) \to Cl_{\Sigma'}(\Phi')$ is a theory morphism; if property 3 also holds then σ is conservative. Properties 2 and 4 capture the corresponding proof-theoretic concepts. Show that these model-theoretic and proof-theoretic properties are independent of each other in general. □

Definition 9.1.17 (Compactness). An entailment system $\langle \vdash_\Sigma \rangle_{\Sigma \in |\mathbf{Sign}|}$ for a functor $\mathbf{Sen}{:}\,\mathbf{Sign} \to \mathbf{Set}$ is *compact* if for every signature Σ and all $\Phi \subseteq \mathbf{Sen}(\Sigma)$ and $\varphi \in \mathbf{Sen}(\Sigma)$, whenever $\Phi \vdash_\Sigma \varphi$ then $\Phi_{fin} \vdash_\Sigma \varphi$ for some finite $\Phi_{fin} \subseteq \Phi$. □

Clearly, if an entailment system is given by a system of finitary rules, it is necessarily compact.

Exercise 9.1.18. Show that the model-theoretic and proof-theoretic notions of compactness in Definitions 4.2.7 and 9.1.17, respectively, are independent of each other: give an example of an institution **INS** with an entailment system $\langle\vdash_\Sigma\rangle_{\Sigma\in|\mathbf{Sign}|}$ for **INS** such that **INS** is compact but $\langle\vdash_\Sigma\rangle_{\Sigma\in|\mathbf{Sign}|}$ is not; give another example showing that the opposite implication may fail. □

We have deliberately refrained from formalising the notion of proof rule and how an entailment system is generated from a set of proof rules. Clearly, there may be many sets of proof rules that generate the same entailment system, and there may be many entailment systems for a given institution. The following exercise indicates that interaction between various entailment systems for the same institution (and between the proof rules they may be generated by) is more delicate than it might appear to be at first glance.

Exercise 9.1.19. Show that if $\langle\vdash_\Sigma\rangle_{\Sigma\in|\mathbf{Sign}|}$ and $\langle\vdash'_\Sigma\rangle_{\Sigma\in|\mathbf{Sign}|}$ are entailment systems for **INS**, then $\langle\vdash_\Sigma\cap\vdash'_\Sigma\rangle_{\Sigma\in|\mathbf{Sign}|}$ is also an entailment system for **INS**. Generalise this to an arbitrary collection of entailment systems, and conclude that for any family $R=\langle R_\Sigma\subseteq\mathcal{P}(\mathbf{Sen}(\Sigma))\times\mathbf{Sen}(\Sigma)\rangle_{\Sigma\in|\mathbf{Sign}|}$ there exists a least entailment system that (componentwise) includes R. This allows us to define the union of entailment systems $\langle\vdash_\Sigma\rangle_{\Sigma\in|\mathbf{Sign}|}$ and $\langle\vdash'_\Sigma\rangle_{\Sigma\in|\mathbf{Sign}|}$ as the least entailment system that includes $\langle\vdash_\Sigma\cup\vdash'_\Sigma\rangle_{\Sigma\in|\mathbf{Sign}|}$. Prove that the union of any family of sound entailment systems for **INS** is sound for **INS** as well.

Notice that if entailment systems $\vdash$ and $\vdash'$ are generated by sets of proof rules $\mathscr{R}$ and $\mathscr{R}'$, respectively, in the style used in the examples above, then the entailment system generated by $\mathscr{R}\cup\mathscr{R}'$ may be larger than the union of $\vdash$ and $\vdash'$. Find $\mathscr{R}$ and $\mathscr{R}'$ which (separately) generate entailment systems that are sound for some institution **INS**, but $\mathscr{R}\cup\mathscr{R}'$ generates an unsound entailment system. □

There are properties other than soundness and completeness that are of interest in practical proof systems, such as efficiency (size of proofs) and clarity, that we do not touch on here.

9.2 Proof in structured specifications

The previous section introduced entailment systems as a proof-theoretic counterpart of the notion of semantic consequence for sets of sentences, or, equivalently, for flat specifications. Now we consider structured specifications as introduced in Chapter 5. We fix an arbitrary institution $\mathbf{INS}=\langle\mathbf{Sign},\mathbf{Sen},\mathbf{Mod},\langle\models_\Sigma\rangle_{\Sigma\in|\mathbf{Sign}|}\rangle$ equipped with an entailment system $\langle\vdash_\Sigma\rangle_{\Sigma\in|\mathbf{Sign}|}$.

We now provide rules for the proof-theoretic counterpart of the semantic consequence relation, as introduced in Definition 5.4.1. This defines a (signature-indexed) relation $\vdash$ between specifications and sentences, called *entailment for specifications* and written $SP \vdash \varphi$, which requires φ to be a $Sig[SP]$-sentence. Unlike in the previous section, many of these rules are institution-independent since they make no reference to any particular form of sentences, exploiting the institution-independent nature of our specification-building operations.

Definition 9.2.1 (Soundness and completeness of entailment for specifications). A relation $\vdash$ of entailment for specifications is *sound* if $SP \vdash \varphi$ implies $SP \models \varphi$ for all specifications SP and $Sig[SP]$-sentences φ. It is *complete* if the opposite implications hold. □

Soundness is typically proved by showing that each rule is sound in the following sense: a rule of the form[2]

$$\frac{SP_1 \vdash \varphi_1 \qquad \cdots \qquad SP_n \vdash \varphi_n}{SP \vdash \varphi}$$

is *sound* if, whenever $SP_1 \models \varphi_1$ and ... and $SP_n \models \varphi_n$, we also have $SP \models \varphi$. Then it is easy to see that a set of sound rules gives rise to a sound relation of entailment for specifications. Completeness is a more delicate matter and typically cannot be proved on a rule-by-rule basis.

First of all, any system of rules for entailment for specifications must be based on an entailment system for the underlying institution. The following rule provides the natural connection between the two systems:

$$\frac{SP \vdash \varphi_1 \quad \cdots \quad SP \vdash \varphi_n \qquad \{\varphi_1, \ldots, \varphi_n\} \vdash_{Sig[SP]} \varphi}{SP \vdash \varphi} \qquad (*)$$

It is obvious that this rule is sound since the underlying entailment system is sound by definition. Strictly speaking, completeness of the underlying entailment relation is not necessary, even to achieve completeness of entailment for specifications: rules for entailments for specifications that manipulate sentences without reference to the underlying entailment system are possible in principle. However, it is hard to imagine building a good proof system that way, so for the analysis of completeness of entailment for specifications we will focus on the case where the underlying entailment system is complete.

Moreover, most of the time we will also require that the underlying entailment system be compact. This is consistent with the standard decision to restrict attention to finitary proof rules and finitary proofs. Most facts below that rely on compactness could be generalised by dropping this assumption, provided that the rule $(*)$ above is replaced by the following infinitary version:

$$\frac{SP \vdash \varphi \text{ for each } \varphi \in \Phi \qquad \Phi \vdash_{Sig[SP]} \varphi}{SP \vdash \varphi}$$

[2] We will also use rules with side conditions and with premises involving other judgement forms, for which the notion of soundness extends in the obvious way.

Any relation $\vdash$ of entailment for specifications determines a property-oriented semantics $\mathscr{T}_{\vdash}$ of specifications in the sense of Section 5.4, as follows:

Definition 9.2.2 ($\mathscr{T}_{\vdash}$). Given an entailment relation $\vdash$ for specifications, we define a function $\mathscr{T}_{\vdash}$ taking Σ-specifications to Σ-theories for each $\Sigma \in |\mathbf{Sign}|$ as follows:

$$\mathscr{T}_{\vdash}[SP] = Cl_{Sig[SP]}(\{\varphi \mid SP \vdash \varphi\}).$$

Then, $\vdash$ is *monotone*, *compositional*, *closed complete* for a specification-building operation **sbo**, *non-absent-minded*, and *flat complete*, respectively, if $\mathscr{T}_{\vdash}$ is so in the sense of Definition 5.4.5. □

The closure in the definition of $\mathscr{T}_{\vdash}$ is only required in order to guarantee that the result is indeed a theory and so may be omitted in the presence of the rule $(*)$ above, provided that the underlying entailment system is complete and compact. Compositionality may then be ensured by defining $\vdash$ using proof rules in which consequences of any specification are determined only by consequences of its immediate constituents.

Exercise 9.2.3. According to Definition 9.2.2, $\vdash$ is closed complete for a unary specification-building operation **sbo** if for all specifications SP such that $\mathbf{sbo}(SP)$ is well formed and $Mod_{Sig[SP]}(\mathscr{T}_{\vdash}[SP]) = Mod[SP]$, $\mathscr{T}_{\vdash}[\mathbf{sbo}(SP)] = Th[\mathbf{sbo}(SP)]$. Spell out what closed completeness means in terms of $\vdash$ and $\models$, without using $\mathscr{T}_{\vdash}$ or Th. □

Any complete relation $\vdash$ of entailment for specifications is closed complete for all specification-building operations. This fact suggests that one way of structuring a proof of completeness for $\vdash$ might be to consider closed completeness for each specification-building operation separately, and reason about how the rules allow consequences of specifications built using that operation to be derived from consequences of its immediate constituent specifications. However, this may fail: closed completeness of $\vdash$ for all specification-building operations is not in general sufficient to guarantee completeness of $\vdash$. Nevertheless, given a complete and compact entailment system for the underlying institution and the rule $(*)$ above to incorporate it into entailment for specifications, a closed complete $\vdash$ is the best that is soundly achievable using a compositional and non-absent-minded (see Exercise 5.4.13) entailment relation for specifications, according to Theorem 5.4.10. If such an entailment relation is incomplete in general, it may still be complete if we restrict attention to a certain class of institutions — or even to a particular institution of interest — and/or to a certain class of specifications.

We use the following rules for proving consequences of specifications built from flat specifications using union, translation and hiding:

$$\frac{}{\langle \Sigma, \Phi \rangle \vdash \varphi} \quad \varphi \in \Phi$$

$$\frac{SP_1 \vdash \varphi}{SP_1 \cup SP_2 \vdash \varphi} \qquad \frac{SP_2 \vdash \varphi}{SP_1 \cup SP_2 \vdash \varphi}$$

$$\frac{SP \vdash \varphi}{SP \textbf{ with } \sigma \vdash \sigma(\varphi)}$$

$$\frac{SP \vdash \sigma(\varphi)}{SP \textbf{ hide via } \sigma \vdash \varphi}$$

Proposition 9.2.4 (Soundness). *The rules above define a sound relation of entailment for specifications.* □

Proposition 9.2.5. *Entailment for specifications, as defined by the rules above, is closed complete for flat specifications (viewed as nullary specification-building operations), union, translation and hiding, provided that the underlying entailment system is complete and compact.* □

The presence of the rule ($*$) above, and completeness and compactness of the underlying proof system, are essential for closed completeness for flat specifications and hiding.

If the underlying entailment system is complete and compact then the above system of rules determines the property-oriented semantics $\mathscr{T}_0$ of specifications built from flat specifications using union, translation and hiding given in Definition 5.4.3, in the sense that $\mathscr{T}_\vdash = \mathscr{T}_0$. Then Counterexample 5.4.8 shows that this system is incomplete for **EQ**. Therefore, by Corollary 5.4.12, there can be no sound and complete compositional and non-absent-minded (see Exercise 5.4.13) proof system for structured specifications over **EQ**. On the other hand, when the specification SP in that counterexample is taken to be a specification in **FOEQ**, the critical sentence $b = c$ is easily derivable, as explained there; see Example 9.2.14 below. In fact, the above system of rules yields a complete relation of entailment for specifications in institutions that satisfy some basic properties (although **FOEQ** does not satisfy these properties as stated here — see Exercise 4.4.23 — a similar and quite satisfactory completeness result may be derived for it; see Exercise 9.2.9):

Theorem 9.2.6 (Completeness). *Let* $\mathbf{INS} = \langle \mathbf{Sign}, \mathbf{Sen}, \mathbf{Mod}, \langle \models_\Sigma \rangle_{\Sigma \in |\mathbf{Sign}|} \rangle$ *be an institution equipped with an entailment system* $\langle \vdash_\Sigma \rangle_{\Sigma \in |\mathbf{Sign}|}$. *Suppose that* **INS** *is finitely exact (Definition 4.4.6) and has the Craig interpolation property (Definition 4.4.21), truth, implication and conjunction (Definition 4.1.42). If* $\langle \vdash_\Sigma \rangle_{\Sigma \in |\mathbf{Sign}|}$ *is complete for* **INS** *then entailment for specifications, as defined by the rules above, is complete for specifications built from finitary flat specifications using union, translation and hiding.*

Proof. Assume that $SP \models \varphi$ for a finitary specification SP built from flat specifications using union, translation and hiding, with $Sig[SP] = \Sigma$ and $\varphi \in \mathbf{Sen}(\Sigma)$. We show that $SP \vdash \varphi$ by induction on the structure of SP:

- Let SP be $\langle\Sigma,\Phi\rangle$ for some finite $\Phi \subseteq \mathbf{Sen}(\Sigma)$. Then $\Phi \models_\Sigma \varphi$ and so, since $\langle\vdash_\Sigma\rangle_{\Sigma\in|\mathbf{Sign}|}$ is complete for **INS**, we get $\Phi \vdash_\Sigma \varphi$. Using the rule for flat specifications and $(*)$, we obtain $SP \vdash \varphi$.
- Let SP be SP' **hide via** σ for some finitary specification SP' with $Sig[SP'] = \Sigma'$ and $\sigma\colon\Sigma \to \Sigma'$. Then $SP' \models \sigma(\varphi)$. By the inductive hypothesis, $SP' \vdash \sigma(\varphi)$, and so using the rule for hiding, we obtain $SP \vdash \varphi$.
- Let SP be SP' **with** σ for some finitary specification SP' with $Sig[SP'] = \Sigma'$ and $\sigma\colon\Sigma' \to \Sigma$. By Theorem 5.6.10, $SP' \equiv \langle\Sigma_1,\Phi_1\rangle$ **hide via** σ_1 for some signature $\Sigma_1 \in |\mathbf{Sign}|$, finite set $\Phi_1 \subseteq \mathbf{Sen}(\Sigma_1)$ of Σ_1-sentences, and $\sigma_1\colon\Sigma' \to \Sigma_1$. Let $\bigwedge\Phi_1$ be the conjunction of Φ_1 (or true if Φ_1 is empty). Then, as in the proof of Theorem 5.6.10, $SP \equiv \langle\widehat{\Sigma},\{\sigma'(\bigwedge\Phi_1)\}\rangle$ **hide via** σ_1', where the following is a pushout in **Sign**:

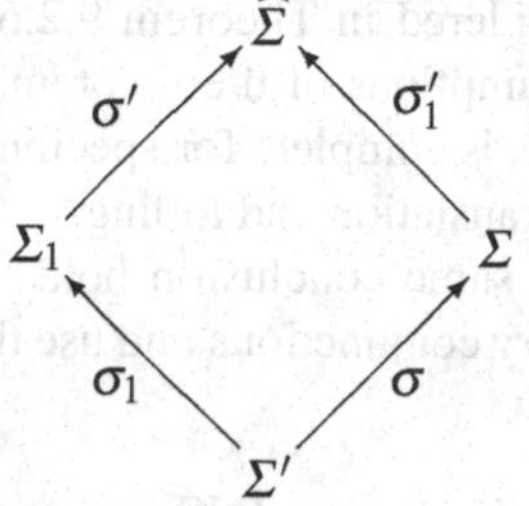

  $SP \models \varphi$ implies $\sigma'(\bigwedge\Phi_1) \models_{\widehat{\Sigma}} \sigma_1'(\varphi)$. Hence, by the Craig interpolation property, there is an interpolant $\theta \in \mathbf{Sen}(\Sigma')$ such that $\bigwedge\Phi_1 \models_{\Sigma_1} \sigma_1(\theta)$ and $\sigma(\theta) \models_\Sigma \varphi$. The former yields $SP' \models \theta$, so by the inductive hypothesis, $SP' \vdash \theta$, and using the rule for translation we get $SP \vdash \sigma(\theta)$. Since $\langle\vdash_\Sigma\rangle_{\Sigma\in|\mathbf{Sign}|}$ is complete for **INS**, the latter yields $\sigma(\theta) \vdash_\Sigma \varphi$, and so using the rule $(*)$, we get $SP \vdash \varphi$.
- Let SP be $SP_1 \cup SP_2$ for finitary specifications SP_1 and SP_2 with $Sig[SP_1] = Sig[SP_2] = \Sigma$. By Theorem 5.6.10, $SP_1 \equiv \langle\Sigma_1,\Phi_1\rangle$ **hide via** σ_1 for some signature $\Sigma_1 \in |\mathbf{Sign}|$, finite set $\Phi_1 \subseteq \mathbf{Sen}(\Sigma_1)$ of Σ_1-sentences, and $\sigma_1\colon\Sigma \to \Sigma_1$, and $SP_2 \equiv \langle\Sigma_2,\Phi_2\rangle$ **hide via** σ_2 for some signature $\Sigma_2 \in |\mathbf{Sign}|$, finite set $\Phi_2 \subseteq \mathbf{Sen}(\Sigma_2)$ of Σ_2-sentences, and $\sigma_2\colon\Sigma \to \Sigma_2$. Let $\bigwedge\Phi_1$ be the conjunction of Φ_1 (or true if Φ_1 is empty), and $\bigwedge\Phi_2$ be the conjunction of Φ_2 (or true if Φ_2 is empty). Then $SP \equiv \langle\widehat{\Sigma},\{\sigma_2'(\bigwedge\Phi_1),\sigma_1'(\bigwedge\Phi_2)\}\rangle$ **hide via** $\sigma_2;\sigma_1'$, as in the proof of Theorem 5.6.10, where the following is a pushout in **Sign**:

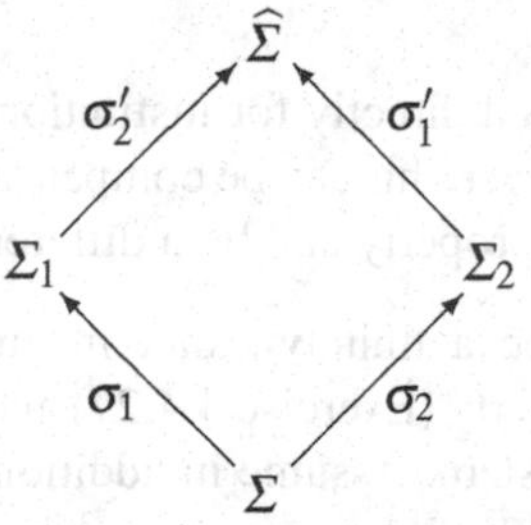

  $SP \models \varphi$ implies $\{\sigma_2'(\bigwedge\Phi_1),\sigma_1'(\bigwedge\Phi_2)\} \models_{\widehat{\Sigma}} \sigma_1'(\sigma_2(\varphi))$. Then, by Exercise 4.1.43, $\sigma_2'(\bigwedge\Phi_1) \models_{\widehat{\Sigma}} \sigma_1'(\bigwedge\Phi_2 \Rightarrow \sigma_2(\varphi))$. Hence, by the Craig interpolation property, there is an interpolant $\theta \in \mathbf{Sen}(\Sigma)$ such that $\bigwedge \Phi_1 \models_{\Sigma_1} \sigma_1(\theta)$ and $\sigma_2(\theta) \models_{\Sigma} \bigwedge \Phi_2 \Rightarrow \sigma_2(\varphi)$. The former yields $SP_1 \models \theta$, so by the inductive hypothesis, $SP_1 \vdash \theta$, and using the rule for union we get $SP \vdash \theta$. Using Exercise 4.1.43 again (and the properties of implication), the latter gives $\bigwedge \Phi_2 \models_{\Sigma} \sigma_2(\theta \Rightarrow \varphi)$, which yields $SP_2 \models \theta \Rightarrow \varphi$. Thus, by the inductive hypothesis, $SP_2 \vdash \theta \Rightarrow \varphi$, and using the rule for union we get $SP \vdash \theta \Rightarrow \varphi$. Since $\{\theta, \theta \Rightarrow \varphi\} \models_{\Sigma} \varphi$, and $\langle \vdash_{\Sigma} \rangle_{\Sigma \in |\mathbf{Sign}|}$ is complete for **INS**, we have $\{\theta, \theta \Rightarrow \varphi\} \vdash_{\Sigma} \varphi$, and so using the rule $(*)$ we obtain $SP \vdash \varphi$. □

Exercise 9.2.7. Give a counterexample to show that Theorem 9.2.6 does not hold for infinitary specifications.

Check that if the institution **INS** is compact (Definition 4.2.7) then the assumption that specifications considered in Theorem 9.2.6 are finitary may be dropped. That is, under the other assumptions of the theorem, entailment for specifications, as defined by the rules above, is complete for specifications built from *arbitrary* flat specifications using union, translation and hiding.

Similarly, check that the same conclusion holds if instead of compactness we assume that **INS** has *infinitary* conjunctions and use the infinitary version of the rule $(*)$. □

Exercise 9.2.8. Consider an institution **INS** and a collection **I** of signature morphisms in **INS** such that **INS** is **I**-semi-exact (see Definition 4.4.18) and has the Craig interpolation property for all pushouts along morphisms in **I**. Prove that if **INS** has truth, implication and conjunction and $\langle \vdash_{\Sigma} \rangle_{\Sigma \in |\mathbf{Sign}|}$ is complete for **INS** then entailment for specifications, as defined by the rules above, is complete for specifications built from finitary flat specifications using union, translation and hiding that involve hiding only with respect to morphisms in **I**. HINT: Use Exercise 5.6.12 and check that the proof above goes through. □

Exercise 9.2.9. Consider the institution **FOPEQ** equipped with the complete entailment system outlined in Exercise 9.1.9. Show that entailment for specifications, as defined by the rules above, is complete for specifications built from finitary flat specifications using union, translation and hiding that involve hiding only with respect to morphisms that are injective on sorts. HINT: Use the exercise above and Exercise 4.4.23. □

Theorem 9.2.6 cannot be used directly for institutions where propositional connectives are not available. However, this can be compensated for by the use of a different version of the interpolation property and by a different style of reasoning.

Exercise 9.2.10. Let **INS** be a finitely exact institution that satisfies the Craig-Robinson interpolation property (Exercise 4.4.27) and comes equipped with a sound and complete entailment system. Assume in addition that semantic consequence in **INS** is reflected by translation of sentences along signature morphisms (for instance because all of its reduct functors are surjective on models; cf. Proposition 4.2.17). Adapt the proof of Theorem 9.2.6 to show that entailment defined by the proof rules above with the infinitary version of $(*)$ is complete for specifications built from flat specifications using union, translation and hiding.

Construct a counterexample to show that the assumption that entailment is reflected by translation of sentences cannot be dropped in general. □

Exercise 9.2.11. Rather than proving consequences of specifications, one may want to carry out proofs of entailment between sentences *in the context of* specifications. For any institution **INS**, given a Σ-specification *SP*, set $\Phi \subseteq \mathbf{Sen}(\Sigma)$ of Σ-sentences and Σ-sentence $\varphi \in \mathbf{Sen}(\Sigma)$, write SP **then** $\Phi \vdash \varphi$ to mean that φ is a consequence of Φ in the context of *SP*, i.e. for all models $M \in Mod[SP]$, if $M \models \Phi$ then $M \models \varphi$ as well. In general this cannot be captured by entailment for specifications as discussed above, but for institutions that have appropriate conjunction and implication, it should be equivalent to $SP \vdash \bigwedge \Phi \Rightarrow \varphi$.

Adapt the above compositional rules defining entailment for specifications to prove judgements of the above form as follows:

$$\frac{}{\langle \Sigma, \Phi \rangle \ \mathbf{then}\ \varnothing \vdash \varphi} \quad \varphi \in \Phi$$

$$\frac{SP_1 \ \mathbf{then}\ \Phi \vdash \varphi}{(SP_1 \cup SP_2)\ \mathbf{then}\ \Phi \vdash \varphi} \qquad \frac{SP_2 \ \mathbf{then}\ \Phi \vdash \varphi}{(SP_1 \cup SP_2)\ \mathbf{then}\ \Phi \vdash \varphi}$$

$$\frac{SP\ \mathbf{then}\ \Phi \vdash \varphi}{(SP\ \mathbf{with}\ \sigma)\ \mathbf{then}\ \sigma(\Phi) \vdash \sigma(\varphi)}$$

$$\frac{SP\ \mathbf{then}\ \sigma(\Phi) \vdash \sigma(\varphi)}{(SP\ \mathbf{hide\ via}\ \sigma)\ \mathbf{then}\ \Phi \vdash \varphi}$$

Add a rule to assert SP **then** $\Phi \vdash \varphi$ for $\varphi \in \Phi$ and a rule corresponding to $(*)$, thus incorporating entailment as given for the underlying institution. Show that the proof system so built is sound.

Then adapt the proof of Theorem 9.2.6 to show that in any finitely exact institution equipped with a complete entailment system and satisfying the Craig-Robinson interpolation property, this proof system is complete for entailment between sentences in the context of specifications built from flat specifications using union, translation and hiding. □

The following exercise shows that the interaction of hiding with union and translation is responsible for all difficulties with completeness:

Exercise 9.2.12. Let $\mathbf{INS} = \langle \mathbf{Sign}, \mathbf{Sen}, \mathbf{Mod}, \langle \models_\Sigma \rangle_{\Sigma \in |\mathbf{Sign}|} \rangle$ be an institution with a complete entailment system $\langle \vdash_\Sigma \rangle_{\Sigma \in |\mathbf{Sign}|}$. Show that, under no further assumptions, entailment for specifications as defined by the rules above is complete for specifications built from finitary flat specifications using union and translation. Relate this to Proposition 5.6.9.

Generalise this further by allowing successive outermost applications of hiding. HINT: This follows from closed completeness of entailment, since any number of successive applications of hiding may be equivalently replaced by a single such application, preserving derivability of sentences. □

Exercise 9.2.13. Recall the specification-building operations (export, enrichment, sum) defined in Section 5.2 in terms of flat specifications, union, translation and hiding. Check that the definition of enrichment justifies the following proof rules by composition of the rules above for flat specifications, union and translation, under the assumption that sentences do not change under translation along signature inclusions:

$$\frac{SP \vdash \varphi}{SP\ \mathbf{then\ sorts}\ S'\ \mathbf{ops}\ \Omega' \bullet \Phi' \vdash \varphi}$$

$$\frac{}{SP\ \mathbf{then\ sorts}\ S'\ \mathbf{ops}\ \Omega' \bullet \Phi' \vdash \varphi'}\ \ \varphi' \in \Phi'$$

Provide rules for export and sum. □

Example 9.2.14. Consider the following specifications in **FOEQ**:

$$\begin{array}{lll} \mathbf{spec}\ SP_0 = & \mathbf{sorts} & s, s' \\ & \mathbf{ops} & a{:}\,s \\ & & b{:}\,s' \\ & & c{:}\,s' \end{array}$$

$\mathbf{spec}\ SP_1 = SP_0\ \mathbf{hide\ ops}\ a{:}\,s$

$\mathbf{spec}\ SP = SP_1\ \mathbf{then}\ \forall x{:}s \bullet b = c$

These specifications are exactly the same as in Counterexample 5.4.8, but considered in a different institution. We get

$$\dfrac{\dfrac{\dfrac{\dfrac{\vdash_{Sig[SP_0]} \exists x{:}s \bullet x = x}{SP_0 \vdash \exists x{:}s \bullet x = x}}{SP_1 \vdash \exists x{:}s \bullet x = x}}{SP \vdash \exists x{:}s \bullet x = x} \qquad \dfrac{}{SP \vdash \forall x{:}s \bullet b = c} \qquad \begin{array}{l}\{\exists x{:}s \bullet x = x, \\ \quad \forall x{:}s \bullet b = c\} \vdash_{Sig[SP]} b = c\end{array}}{SP \vdash b = c}$$

Exercise: Identify the rules used in the above derivation. □

Example 9.2.15. Recall the specifications in Example 7.1.4, written in the institution **FPL**. We will derive

$\textsc{InsDone} \vdash \forall l{:}NatList, n{:}Nat \bullet def(insert(n,l))$
$\textsc{InsDone} \vdash \forall l{:}NatList \bullet P(l)$

where $P(l)$ is the following formula:

$$\begin{array}{l} \forall n{:}Nat \bullet \exists l_1, l_2{:}NatList \bullet \\ \quad insert(n,l) = append(l_1, cons(n, l_2)) \wedge l = append(l_1, l_2) \\ \quad \wedge (\forall l'_1{:}NatList, m{:}Nat \bullet l_1 = append(l'_1, cons(m, nil)) \Rightarrow le(m,n) = true) \\ \quad \wedge (\forall l'_2{:}NatList, m{:}Nat \bullet l_2 = cons(m, l'_2) \Rightarrow le(n,m) = true) \end{array}$$

The essence of these proofs is in Example 9.1.12, but there we did not make reference to the structured specifications from which the axioms used were obtained.

The derivation proceeds as follows. We first extract the axioms listed in NATORD using the second rule for enrichment, obtaining

$$\text{NATORD} \vdash \forall m,n{:}Nat \bullet def(le(m,n))$$
$$\text{NATORD} \vdash \forall m{:}Nat \bullet le(m,m) = true$$

and similarly for the other three axioms. This is also the place to derive some of the consequences of these axioms that will be needed later, using the rule $(*)$ and referring to the underlying entailment system. For instance,

$$\text{NATORD} \vdash \forall m,n{:}Nat \bullet le(m,n) = false \Rightarrow le(n,m) = true.$$

Then the consequences of NATORD are promoted to consequences of NATLIST, using the first rule for enrichment, obtaining for instance

$$\text{NATLIST} \vdash \forall m,n{:}Nat \bullet def(le(m,n))$$
$$\text{NATLIST} \vdash \forall m{:}Nat \bullet le(m,m) = true$$
$$\text{NATLIST} \vdash \forall m,n{:}Nat \bullet le(m,n) = false \Rightarrow le(n,m) = true.$$

We can also derive properties of lists. For instance, since

$$\vdash_{Sig[\text{NATLIST}]} \forall l_2'{:}NatList, m{:}Nat \bullet \neg(nil = cons(m,l_2')),$$

by the rule $(*)$ we get

$$\text{NATLIST} \vdash \forall l_2'{:}NatList, m{:}Nat \bullet \neg(nil = cons(m,l_2')).$$

Moreover, the definitional axiom for *append* can be extracted:

$$\text{NATLIST} \vdash \textbf{fun } append(l{:}NatList, l'{:}NatList){:}NatList =$$
$$\textbf{case } l \textbf{ of } nil \Rightarrow l' \mid cons(n,l'') \Rightarrow cons(n, append(l'',l'))$$

Consequences of this axiom may now be derived in the underlying entailment system for **FPL** and promoted to consequences of NATLIST using the rule $(*)$. For instance,

$$\text{NATLIST} \vdash \forall l{:}NatList \bullet append(nil,l) = l$$
$$\text{NATLIST} \vdash \forall l_1'{:}NatList, m{:}Nat \bullet \neg(nil = append(l_1', cons(m,nil)))$$
$$\text{NATLIST} \vdash \forall u,v{:}NatList, i{:}Nat \bullet append(cons(i,u),v) = cons(i, append(u,v)).$$

Now that we have derived these consequences of the definitional axiom for *append*, we will not need to make further reference to it.

These consequences of NATLIST are promoted to consequences of INSDONE using the first rule for enrichment, for instance,

$$\text{INSDONE} \vdash \forall l{:}NatList \bullet append(nil,l) = l$$
$$\text{INSDONE} \vdash \forall l_1'{:}NatList, m{:}Nat \bullet \neg(nil = append(l_1', cons(m,nil)))$$
$$\text{INSDONE} \vdash \forall u,v{:}NatList, i{:}Nat \bullet append(cons(i,u),v) = cons(i, append(u,v))$$

and similarly for the properties of *le*:

$\text{InsDone} \vdash \forall m,n{:}Nat \bullet def(le(m,n))$
$\text{InsDone} \vdash \forall m{:}Nat \bullet le(m,m) = true$
$\text{InsDone} \vdash \forall m,n{:}Nat \bullet le(m,n) = false \Rightarrow le(n,m) = true$

The definitional axiom for *insert* may be extracted using the second rule for enrichment, giving

$\text{InsDone} \vdash$ **fun** $insert(n{:}Nat, l{:}NatList){:}NatList =$
case l **of** $nil \Rightarrow cons(n, nil)$
$\mid cons(m, l') \Rightarrow$ **if** $le(n,m) = true$ **then** $cons(n,l)$
else $cons(m, insert(n, l'))$

Finally, as sketched in Example 9.1.12, the entailment system for **FPL** allows us to derive $\forall l{:}NatList, n{:}Nat \bullet def(insert(n,l))$ and $\forall l{:}NatList \bullet P(l)$ from the above consequences of InsDone. The desired conclusions then follow by rule $(*)$. □

The lack of completeness of the set of rules above for consequences of structured specifications in institutions like **EQ** may be a cause for concern. One alternative is to give up compositionality and use the normal form theorem (Theorem 5.6.10) to extract a set of sentences from *SP*, and then use the entailment system for the underlying institution to prove that the sentence of interest is a consequence of that set. This strategy is captured by the following non-compositional rule:

$$\frac{\Phi \vdash_{\Sigma} \sigma(\varphi)}{SP \vdash \varphi} \qquad \mathsf{nf}(SP) = \langle \Sigma, \Phi \rangle \textbf{ hide via } \sigma$$

where nf denotes the construction in the proof of the normal form theorem.

Theorem 9.2.16. *Let* **INS** *be a finitely exact institution equipped with an entailment system* $\langle \vdash_{\Sigma} \rangle_{\Sigma \in |\mathbf{Sign}|}$. *The relation of entailment for specifications given by the single rule above is sound. Moreover, it is complete for specifications built from flat specifications using union, translation and hiding provided that the underlying entailment system is complete for* **INS**.

Proof. Recall from Theorem 5.6.10 that $SP \equiv \mathsf{nf}(SP)$. Soundness is then straightforward. For completeness, let φ be such that $SP \models \varphi$. Then $\mathsf{nf}(SP) \models \varphi$ and by the satisfaction condition, $\Phi \models_{\Sigma} \sigma(\varphi)$. Hence, by completeness of the underlying entailment system, $\Phi \vdash_{\Sigma} \sigma(\varphi)$, and so $SP \vdash \varphi$. □

Example 9.2.17. Consider once again the specification in Counterexample 5.4.8, in the institution **EQ**. We have

spec $\mathsf{nf}(SP) =$
  sorts s, s'
  ops $a{:}s$
    $b{:}s'$
    $c{:}s'$
  $\forall x{:}s \bullet b = c$
  hide ops $a{:}s$

Since we have $\forall x{:}s \bullet b = c \vdash_\Sigma b = c$, where Σ is the signature of the specification inside the application of hiding in $\mathsf{nf}(SP)$ above (with the constant $a{:}s$), the non-compositional rule above yields $SP \vdash b = c$. □

A disadvantage of this approach is that the set of axioms Φ that is collected in the normal form of a specification *SP* may be large since it contains sentences corresponding to all the axioms in *SP*. An advantage of compositional proof is that that a proof of $SP \vdash \varphi$ follows the structure of *SP*. This suggests the possibility of using the structure of *SP* to guide the search for a proof. As Example 9.2.15 above illustrates, proof in a structured specification can involve frequent changes of context, where proof fragments in the context of "small" specifications correspond to the proofs of lemmas which are then brought to bear on the main proof via translation to the context of an appropriate "larger" specification. In contrast, proving a lemma via the normal form construction involves gathering together all of the axioms in the specification, even if only a few of them are required for the proof. In presenting proofs like the one in Example 9.1.12, it is good style to draw attention only to relevant premises, but the formal system makes no distinction between what is relevant and what is irrelevant. Specification structure provides the scaffolding for making such distinctions, and the use of compositional rules focuses proof search on relevant contexts.

The rule above derived from the normal form theorem is justified by the obvious fact that equivalent specifications have the same consequences. Therefore, whenever we can show that $SP \equiv SP'$, it is sound to conclude from $SP \vdash \varphi$ that $SP' \vdash \varphi$. Here, $\equiv$ is equivalence of specifications as defined in Section 5.6, and so each of the algebraic laws there (see e.g. Proposition 5.6.2) gives rise to a corresponding rule.

Exercise 9.2.18. The following rule

$$\frac{\Phi \vdash_\Sigma \sigma(\varphi)}{SP \vdash \varphi} \quad SP \equiv \langle \Sigma, \Phi \rangle \textbf{ hide via } \sigma$$

is sound in any institution **INS**. Moreover, when the underlying entailment system is complete for **INS**, the relation of entailment for specifications given by this single rule is complete for all specifications *SP* for which a "normal form" of the shape $\langle \Sigma, \Phi \rangle$ **hide via** σ can be given.

In particular, consider a collection **I** of signature morphisms in **INS** such that **INS** is **I**-semi-exact; see Definition 4.4.18. Use Exercise 5.6.12 to show that the above rule is sound and complete for structured specifications built from flat specifications using union, translation and hiding which involve hiding only with respect to morphisms in **I**, provided the underlying entailment system is complete for **INS**. □

Sometimes we have only inclusion of model classes rather than specification equivalence; see Exercise 5.6.3. This is enough: if $Sig[SP] = Sig[SP']$ then in order to infer $SP' \vdash \varphi$ from $SP \vdash \varphi$, it is enough to require $Mod[SP] \supseteq Mod[SP']$. Another way of checking that two specifications satisfy this condition is by using a proof system for entailment between specifications, discussed in the next section.

Exercise 9.2.19. Show that

$$(SP\ \mathbf{hide\ via}\ \sigma) \cup SP' \equiv (SP \cup SP'\ \mathbf{with}\ \sigma)\ \mathbf{hide\ via}\ \sigma$$

where $\sigma: Sig[SP'] \to Sig[SP]$. (This equivalence can be seen as a special case of Proposition 5.6.7.) Use this to justify the soundness of the following:

$$\frac{SP \vdash \varphi \qquad SP' \vdash \varphi' \qquad \{\varphi, \sigma(\varphi')\} \vdash_{Sig[SP]} \sigma(\psi)}{(SP\ \mathbf{hide\ via}\ \sigma) \cup SP' \vdash \psi}$$

Use this rule to show that the specification in Counterexample 5.4.8 entails $b = c$, noting that the use of enrichment there does not change the signature. Conclude that this rule cannot be derived from the compositional rules above. □

Exercise 9.2.20. Use Proposition 5.6.5 to justify the soundness of the following rule:

$$\frac{SP \vdash \varphi_1 \quad \cdots \quad SP \vdash \varphi_n \qquad \{\sigma'(\varphi_1), \ldots, \sigma'(\varphi_n)\} \vdash_{\widehat{\Sigma}} \tau'(\psi')}{(SP\ \mathbf{hide\ via}\ \sigma)\ \mathbf{with}\ \tau \vdash \psi'}$$

where $Sig[SP] = \Sigma$ and the following pushout of signatures admits amalgamation:

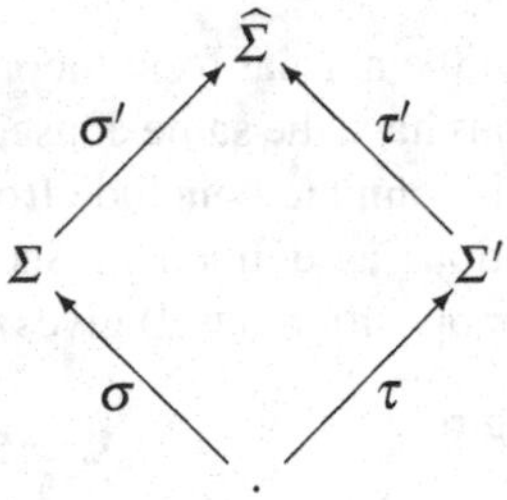

Check that given the signature pushout as above and $Sig[SP'] = \Sigma'$, by Proposition 5.6.7 we obtain the following rule:

$$\frac{\begin{array}{c} SP \vdash \varphi_1 \quad \cdots \quad SP \vdash \varphi_n \qquad SP' \vdash \varphi'_1 \quad \cdots \quad SP' \vdash \varphi'_m \\ \{\sigma'(\varphi_1), \ldots, \sigma'(\varphi_n), \tau'(\varphi'_1), \ldots, \tau'(\varphi'_m)\} \vdash_{\widehat{\Sigma}} \sigma'(\sigma(\psi)) \end{array}}{(SP\ \mathbf{hide\ via}\ \sigma) \cup (SP'\ \mathbf{hide\ via}\ \tau) \vdash \psi}$$

□

Exercise 9.2.21. Justify the following rule:

$$\frac{\begin{array}{c} SP \vdash \varphi_1 \quad \cdots \quad SP \vdash \varphi_n \qquad SP' \vdash \varphi'_1 \quad \cdots \quad SP' \vdash \varphi'_m \\ \{\sigma'(\varphi_1), \ldots, \sigma'(\varphi_n), \tau'(\varphi'_1), \ldots, \tau'(\varphi'_m)\} \vdash_{\widehat{\Sigma}} \tau'(\psi') \end{array}}{((SP\ \mathbf{hide\ via}\ \sigma)\ \mathbf{with}\ \tau) \cup SP' \vdash \psi'}$$

where the morphisms and $\widehat{\Sigma}$ are as in the pushout diagram (required to admit amalgamation) in Exercise 9.2.20. □

Example 9.2.22. Consider the following specification in the institution **FPL**:

spec SORTINSCOUNT =
  SORTINS
  then
    ops fun $count(n{:}Nat, l{:}NatList){:}Nat =$
        case l **of** nil => 0
            | $cons(m,l')$ => **if** $n = m$ **then** $succ(count(n,l'))$
                else $count(n,l')$

where SORTINS is as in Example 7.1.4; for future reference let φ_{count} be the definitional axiom for *count*, as it occurs above as well as in the specification SORTCOUNT given in that example.

We will derive

$$\text{SORTINSCOUNT} \vdash \forall n{:}Nat, l{:}NatList \bullet count(n,l) = count(n, sort(l)).$$

Unfortunately, this derivation cannot easily be done using only the compositional proof rules given earlier. Informally, to derive the conclusion we need both the axiom for *count* and the definitional axiom for *sort*. The latter also involves *insert*, so its axioms will be needed as well. No subspecification of SORTINSCOUNT has a signature that includes all of these operations. (**Exercise:** Can you think of a way to circumvent these difficulties using a local function definition?)

We appeal instead to a rule that allows information to be extracted from multiple subspecifications. Luckily, the rule in Exercise 9.2.21 is exactly what is needed here. The structure of SORTINSCOUNT (after expanding the enrichment and export buried in SORTINS) matches the form of specification in the conclusion of this rule. All the signature morphisms involved are inclusions in $\mathbf{I_{FPL}}$, with no new value constructors added for old sorts — this is entirely typical, if care is taken to avoid name clashes. By Exercise 4.4.19, the required pushout exists and admits amalgamation. The pushout signature $\widehat{\Sigma}$ expands the signature of NATLIST by the operations *insert*, *sort* and *count*. The instance of the rule that is required is the following (omitting translation of sentences by signature inclusions):

$$\frac{\begin{array}{c}\text{SORTBYINSERT} \vdash \varphi_1 \cdots \text{SORTBYINSERT} \vdash \varphi_n \\ \langle Sig[\text{SORTINSCOUNT}], \{\varphi_{count}\}\rangle \vdash \varphi'_1 \cdots \langle Sig[\text{SORTINSCOUNT}], \{\varphi_{count}\}\rangle \vdash \varphi'_m \\ \{\varphi_1, \ldots, \varphi_n, \varphi'_1, \ldots, \varphi'_m\} \vdash_{\widehat{\Sigma}} \psi'\end{array}}{\text{SORTINSCOUNT} \vdash \psi'}$$

For $\varphi'_1, \ldots, \varphi'_m$ we will take just φ_{count} with $m = 1$. For $\varphi_1, \cdots, \varphi_n$ we will take whatever is needed from SORTBYINSERT, derived as in Example 9.2.15: the definitional axioms for *sort* from SORTBYINSERT, the axioms for *insert* from INS, properties of *append* from NATLIST, and so on. Then it is enough to derive in the underlying entailment system for **FPL** that

$$\{\varphi_1, \ldots, \varphi_n, \varphi_{count}\} \vdash_{\widehat{\Sigma}} \forall n{:}Nat, l{:}NatList \bullet count(n,l) = count(n, sort(l)).$$

The (rather complicated) proof by induction uses techniques similar to those in Example 9.1.12, and is omitted here. □

For the specification-building operations that have been discussed so far, it was possible to give proof rules that are institution-independent in the sense that they apply in any institution, or in any institution that satifies only mild additional assumptions. This is not always possible; in some cases the only institution-independent rules available are too weak to be useful.

Definition 9.2.23 (Selection operation). A unary specification-building operation **sbo** is a *selection operation* if $Sig[\mathbf{sbo}(SP)] = Sig[SP]$ and $Mod[\mathbf{sbo}(SP)] \subseteq Mod[SP]$ for any SP such that $\mathbf{sbo}(SP)$ is well formed. □

Then we have the following proof rule for any selection operation **sbo**:

$$\frac{SP \vdash \varphi}{\mathbf{sbo}(SP) \vdash \varphi}$$

This rule is obviously sound, but typically it is too weak to capture the properties that are used to make the selection.

One such selection operation is **free**; see Exercise 5.1.9. The corresponding proof rule is the following:

$$\frac{SP \vdash \varphi}{\mathbf{free}\ SP\ \mathbf{wrt}\ \sigma \vdash \varphi}$$

This seems to be all that can be provided in an institution-independent way. Much more can be said for specific institutions and under assumptions concerning the form of the specification at hand. For instance, in **FOEQ**, if SP is a flat specification in which the axioms involve only outermost universal quantification, an induction rule would be sound.

Exercise 9.2.24. Prove that, under the above assumptions on SP, every model of **free** SP **wrt** σ is reachable with respect to σ (see Exercise 5.1.10). Follow the pattern of the induction rule for **FPL** in Example 9.1.12 to formulate an appropriate induction rule, and show that it is sound. HINT: The new sorts are the generated ones, with all operations taken as value constructors.

Note that the same rule is sound for the **reachable** specification-building operation, with no assumptions required on the form of SP. □

The induction rule cannot lead to a complete entailment system for specifications containing the **free** specification-building operation. A special case of **free** is **initial**, and Theorem 2.5.26 says that there is no complete proof system for equational specifications of the form **initial** $\langle\Sigma, \Phi\rangle$. The proof of this theorem applies to any specification language that is powerful enough to specify the standard model of the natural numbers up to isomorphism, or a class of models having the same theory, so it also applies to non-equational Φ, to specifications of the form **reachable** $\langle\Sigma, \Phi\rangle$ **on** S, and so to practically any specification language involving either **reachable** or **free**.

Exercise 9.2.25. For **quotient** in **EQ** (see Exercise 5.1.12), check that the following rule is sound and closed complete

$$\frac{SP \vdash e}{\textbf{quotient } SP \textbf{ preserving } S \vdash e}$$

even though **quotient** is not a selection operation. Notice that this rule is not sound in institutions with the same signatures and models as **EQ** but with more powerful sentences, for instance **FOEQ**.

For **unquotient** in **EQ**, check that the following rule is sound and closed complete:

$$\frac{SP \vdash e}{\textbf{unquotient } SP \textbf{ preserving } S \vdash e} \quad \begin{array}{l} e \text{ is an equation between} \\ \text{terms of a sort in } S \end{array}$$

□

Exercise 9.2.26. Recall (Exercise 5.1.8) the **close**$_{\simeq}$ specification-building operation that closes the class of models of a Σ-specification SP under an equivalence $\simeq_\Sigma \subseteq |\mathbf{Mod}(\Sigma)| \times |\mathbf{Mod}(\Sigma)|$. Assume that satisfaction in **INS** is preserved under $\simeq$, that is, for any $\Sigma \in |\mathbf{Sign}|$ and $M, M' \in |\mathbf{Mod}(\Sigma)|$ such that $M \simeq_\Sigma M'$, for all sentences $\varphi \in \mathbf{Sen}(\Sigma)$, we have $M \models_\Sigma \varphi$ iff $M' \models_\Sigma \varphi$. Show that the following rule is then sound and closed complete:

$$\frac{SP \vdash \varphi}{\textbf{close}_{\simeq}\, SP \vdash \varphi}$$

In particular, for institutions in which satisfaction is preserved under model isomorphism (this is the abstraction condition of Definition 4.5.13) we have the following sound and closed complete rule:

$$\frac{SP \vdash \varphi}{\textbf{iso-close } SP \vdash \varphi}$$

□

9.3 Entailment between specifications

As before, we work within an institution $\mathbf{INS} = \langle \mathbf{Sign}, \mathbf{Sen}, \mathbf{Mod}, \langle \models_\Sigma \rangle_{\Sigma \in |\mathbf{Sign}|} \rangle$ equipped with an entailment system $\langle \vdash_\Sigma \rangle_{\Sigma \in |\mathbf{Sign}|}$.

It is now time to switch to the next level of proofs, *entailment between specifications*, written $SP \vdash SP'$ where $Sig[SP] = Sig[SP']$. The semantic relationship between specifications that this is designed to approximate is given by inclusion of model classes, $Mod[SP] \subseteq Mod[SP']$. This corresponds to $SP' \rightsquigarrow SP$ (see Definition 7.1.1, and note the change of order!), so in fact this yields a system for proving correctness of simple implementations, and in the discussion below we will switch freely between these two terminologies. Inclusion of model classes generalises semantic consequence for specifications since $SP \models \varphi$ iff $Mod[SP] \subseteq Mod[\langle Sig[SP], \{\varphi\} \rangle]$ for any specification SP and $Sig[SP]$-sentence φ. By analogy with this case, given $SP \vdash SP'$ we refer to SP' as a *consequence* of SP; we also sometimes refer to SP as the *antecedent* of $SP \vdash SP'$. We will also say that SP' is a *semantic consequence* of SP whenever $Mod[SP] \subseteq Mod[SP']$.

Definition 9.3.1 (Soundness and completeness of entailment between specifications). A relation $\vdash$ of entailment between specifications is *sound* if for all specifications SP and SP' such that $Sig[SP] = Sig[SP']$, $SP \vdash SP'$ implies $Mod[SP] \subseteq Mod[SP']$. It is *complete* if the opposite implications hold. □

In other words, entailment between specifications is sound if every consequence of any specification is its semantic consequence, and vice versa for completeness. As before, soundness is typically proved by showing soundness of each rule separately, while completeness is relevant to sets of rules.

Proof rules for the relation of entailment between specifications build on entailment for specifications ($SP \vdash \varphi$) as discussed in the previous section, so we assume that this is given, and that it is sound. Completeness of entailment for specifications is difficult to achieve, so we will assume it only when required in the analysis of completeness of entailment between specifications. In fact it will emerge that very little can be said about completeness at the level of entailment between specifications, with Theorem 9.3.3 below as a notable exception.

The link between the two entailment relations is provided by the following rule:

$$\frac{SP \vdash \varphi_1 \quad \cdots \quad SP \vdash \varphi_n}{SP \vdash \langle Sig[SP], \{\varphi_1, \ldots, \varphi_n\}\rangle}$$

That this rule is sound follows directly from the definition. Clearly, it only allows the proof of entailments with finitary flat specifications as consequences. The following infinitary version is needed to derive entailments with arbitrary flat specifications as consequences:

$$\frac{SP \vdash \varphi \text{ for each } \varphi \in \Phi}{SP \vdash \langle Sig[SP], \Phi\rangle}$$

We use the following rules for proving entailment between specifications with consequences built using union, translation and hiding:

$$\frac{SP \vdash SP_1 \qquad SP \vdash SP_2}{SP \vdash SP_1 \cup SP_2}$$

$$\frac{SP' \textbf{ hide via } \sigma \vdash SP}{SP' \vdash SP \textbf{ with } \sigma}$$

$$\frac{\widehat{SP} \vdash SP'}{SP \vdash SP' \textbf{ hide via } \sigma} \qquad \begin{array}{c}\sigma : SP \to \widehat{SP} \text{ admits} \\ \text{model expansion}\end{array}$$

The side condition on the last rule that the specification morphism $\sigma : SP \to \widehat{SP}$ admits model expansion (see Definition 5.5.6) is discussed below.

Proposition 9.3.2 (Soundness). *The rules above define a sound relation of entailment between specifications.* □

Theorem 9.3.3 (Completeness). *Given a sound and complete relation $\vdash$ of entailment for specifications, the relation of entailment between specifications, as defined by the rules above, is complete for consequences built from finitary flat specifications using union, translation and hiding.*

Proof. Assume that $Mod[SP] \subseteq Mod[SP']$ for specifications SP and SP' with a common signature $Sig[SP] = Sig[SP'] = \Sigma$, where SP' is built from finitary flat specifications using union, translation and hiding. We show that $SP \vdash SP'$ by induction on the structure of SP':

- Let SP' be $\langle \Sigma, \Phi \rangle$ for some finite $\Phi \subseteq \mathbf{Sen}(\Sigma)$. Since $Mod[SP] \subseteq Mod[SP']$, for each $\varphi \in \Phi$, $SP \models \varphi$, and so, by completeness of entailment for specifications, we have $SP \vdash \varphi$. Using the (finitary) rule for consequences built as flat specifications, we obtain $SP \vdash SP'$.
- Let SP' be $SP_1 \cup SP_2$ for specifications SP_1 and SP_2 with $Sig[SP_1] = Sig[SP_2] = \Sigma$. Since $Mod[SP] \subseteq Mod[SP']$ and $Mod[SP'] = Mod[SP_1] \cap Mod[SP_2]$, we have $Mod[SP] \subseteq Mod[SP_1]$ and $Mod[SP] \subseteq Mod[SP_2]$. Hence, by the inductive hypothesis, $SP \vdash SP_1$ and $SP \vdash SP_2$, and so, by the rule for consequences built by union, $SP \vdash SP'$.
- Let SP' be SP_1 **with** σ for some specification SP_1 with $Sig[SP_1] = \Sigma_1$ and $\sigma : \Sigma_1 \to \Sigma$. Since $Mod[SP] \subseteq Mod[SP']$ and $Mod[SP'$ **hide via** $\sigma] \subseteq Mod[SP_1]$ (see Exercise 5.6.3(2)) it follows that $Mod[SP$ **hide via** $\sigma] \subseteq Mod[SP_1]$ by monotonicity of hiding (see Exercise 5.1.4). Hence, by the inductive hypothesis, SP **hide via** $\sigma \vdash SP_1$, and using the rule for consequences built by translation, we obtain $SP \vdash SP'$.
- Let SP' be SP_1 **hide via** σ for some specification SP_1 with $Sig[SP_1] = \Sigma_1$ and $\sigma : \Sigma \to \Sigma_1$. Consider the Σ_1-specification $\widehat{SP}$ defined as $(SP$ **with** $\sigma) \cup SP_1$. Then $\sigma : SP \to \widehat{SP}$ is a specification morphism (since if $M_1 \in Mod[\widehat{SP}]$ then $M_1 \in Mod[SP$ **with** $\sigma]$ and so $M_1|_\sigma \in Mod[SP]$) and admits model expansion (since if $M \in Mod[SP] \subseteq Mod[SP']$ then there is $M_1 \in Mod[SP_1]$ with $M_1|_\sigma = M$; moreover $M_1 \in Mod[SP$ **with** $\sigma]$ and so $M_1 \in Mod[\widehat{SP}]$). Trivially, $Mod[\widehat{SP}] \subseteq Mod[SP_1]$. Thus, by the inductive hypothesis, $\widehat{SP} \vdash SP_1$, and so using the rule for consequences built by hiding, we obtain $SP \vdash SP'$. □

Exercise 9.3.4. Check that with the infinitary version of the rule for consequences built as flat specifications, the assumption that the specifications considered as consequences in Theorem 9.3.3 are finitary may be dropped. That is, under the other assumptions of the theorem, the relation of entailment between specifications, as defined by the rules above, is complete for consequences built from arbitrary flat specifications using union, translation and hiding. □

Exercise 9.3.5. As in Exercise 9.2.9, consider the institution **FOPEQ** equipped with the complete entailment system outlined in Exercise 9.1.9 and entailment for specifications as defined by the rules in Section 9.2. Show that the relation of entailment between specifications, as defined by the rules above, is complete for entailment between specifications built from finitary flat specifications using union, translation and hiding which involve only morphisms that are injective on sorts.

Note that completeness holds under a restriction of the form of specifications on *both* sides of the entailment relation. While Theorem 9.3.3 restricts the specification-building operations used in the consequence only, Theorem 9.2.6 additionally restricts the specification-building operations used in the antecedent, and Exercise 9.2.9 requires injectivity on sorts of signature morphisms used in applications of hiding in the antecedent and (via the rule above for translation) in applications of translation in the consequence. □

Exercise 9.3.6. Recall the export, enrichment and sum operations defined in Section 5.2 in terms of flat specifications, union, translation and hiding. Check that the definition of enrichment justifies the following proof rule by composition of the rules above for flat specifications, union and translation, under the assumption that sentences do not change under translation along signature inclusions:

$$\frac{SP\ \mathbf{hide\ sorts}\ S'\ \mathbf{ops}\ \Omega' \vdash SP' \qquad SP \vdash \varphi'_1 \quad \cdots \quad SP \vdash \varphi'_n}{SP \vdash SP'\ \mathbf{then\ sorts}\ S'\ \mathbf{ops}\ \Omega' \bullet \{\varphi'_1, \ldots, \varphi'_n\}} \quad \text{names from } S' \text{ or } \Omega' \text{ are not in } Sig[SP']$$

Provide rules for export and sum. □

The rules above are compositional with respect to the structure of the consequence: when read backwards, each rule reduces the task of proving that $SP \vdash SP'$ to proving that immediate subspecifications of SP' are consequences of specifications built from SP. Thus, simplification of the consequence is achieved by means of possible additional complication of the antecedent. Eventually this reduces the proof task to entailment for specifications, for a specification that is typically more complex than the original antecedent.

Although at the surface no new axioms are added to the antecedent in the rules above, the essential additional complexity comes from the use of the rule for hiding via the side condition. This requires construction of a specification $\widehat{SP}$ so that $\sigma: Sig[SP] \to Sig[SP']$ becomes a specification morphism $\sigma: SP \to \widehat{SP}$ that admits model expansion. (An apparently weaker requirement, namely that every model of SP have a σ-expansion to a model of $\widehat{SP}$ — dropping the requirement that $\sigma: SP \to \widehat{SP}$ is a specification morphism — would be equally good here; however, given a specification that satisfies this weaker requirement, we can consider its union with the translation of SP, obtaining a specification that satisfies the side condition of the rule as stated.) Discharging this condition is outside the formal system we present. In general it is as hard as proving entailment between specifications, so where possible we attempt to ensure it by construction of $\widehat{SP}$. The construction used in the proof of Theorem 9.3.3, which takes $\widehat{SP} = SP' \cup (SP\ \mathbf{with}\ \sigma)$, is useless in practice since discharging the side condition amounts to proving the entailment at hand, $SP \vdash SP'\ \mathbf{hide\ via}\ \sigma$. But typically, in $SP'\ \mathbf{hide\ via}\ \sigma$, hiding is used to remove auxiliary components (sorts and operations, in algebraic signatures) that are introduced only to specify the components of actual interest. Then, $\widehat{SP}$ adds these auxiliary components to SP in such a way that the model expansion property is ensured — this is often guaranteed by the form of their defining axioms in SP', which can be reused in $\widehat{SP}$ — and the proof task is reduced to showing that, under these definitions, the other properties required by SP' hold.

Exercise 9.3.7. Consider a specification SP in **FPL**. Let $\widehat{SP}$ be its enrichment by a new operation $f{:}s_1 \times \cdots \times s_n \to s$ together with a *definitional axiom* of the form

$$\mathbf{fun}\ f(x_1{:}s_1,\ldots,x_n{:}s_n){:}s = t$$

where $t \in |T_{Sig[\widehat{SP}]}(\{x_1{:}s_1,\ldots,x_n{:}s_n\})|_s$. Show that the inclusion $SP \hookrightarrow \widehat{SP}$ is a specification morphism that admits model expansion. Generalise this to extensions by a sequence of operations (not a set, to allow for hierarchical dependencies between the new operations) and by new sorts. Link this with the notation introduced in Example 6.1.9.

State similar facts for institutions with standard signatures, like **EQ**, **FOPEQ**, etc., where a notion of *definitional axiom* can be easily given as well. □

Example 9.3.8. Recall the specifications in Example 7.1.4. We sketch a derivation for SORTINS ⊢ SORTPERM, which proves SORTPERM $\rightsquigarrow$ SORTINS.

Recall that

spec SORTPERM = SORTCOUNT **hide ops** *count*: *Nat* × *NatList* → *Nat*.

To use the corresponding rule for hiding we need a specification $\widehat{\text{SORTINS}}$ which expands SORTINS with the operation *count* so that model expansion is admitted. As suggested above, such a specification can be obtained by taking the definitional axiom for *count* from SORTCOUNT. So, we put $\widehat{\text{SORTINS}}$ = SORTINSCOUNT, reusing the specification SORTINSCOUNT introduced in Example 9.2.22. The specification inclusion from SORTINS to SORTINSCOUNT admits model expansion because the axiom is definitional (see Exercise 9.3.7).

Then we have to derive SORTINSCOUNT ⊢ SORTCOUNT. By the rule for enrichment, this requires:

- SORTINSCOUNT ⊢ φ_{count}, where φ_{count} is the definitional axiom for *count* — this follows trivially by the rule for enrichment (for entailment for specifications);
- SORTINSCOUNT ⊢ $\forall n{:}Nat, l{:}NatList \bullet count(n,l) = count(n,sort(l))$ — which was derived in Example 9.2.22; and
- SORTINS′ ⊢ SORT, where

  spec SORTINS′ = SORTINSCOUNT **hide ops** *count*: *Nat* × *NatList* → *Nat*

  We derive this now. (Note that SORTINS′ is equivalent to SORTINS, but we will not make use of the equivalence in order to stay within the realm of the proof rules above.)

  To show that SORTINS′ ⊢ SORT we need again to come up with an enrichment $\widehat{\text{SORTINS}'}$ of SORTINS′ by *is_sorted* that admits model expansion, such that

$\widehat{\text{SORTINS}'} \vdash$ SORT1. Unfortunately, the axioms for *is_sorted* in SORT1 are not definitional. We could try to prove that they ensure the model expansion property, but in effect this would involve the same work as giving a definitional axiom for *is_sorted* and then showing that the definition entails the axioms for *is_sorted* in SORT1. This can be done as follows:

```
spec SORTINS'^ =
    SORTINS'
then
    ops fun is_sorted(l:NatList):Bool =
            let fun check(n:Nat, l':NatList):Bool =
                case l' of nil => true
                        | cons(m, l'') =>
                            if le(n,m) = true then check(n, l'') else false
            in case l of nil => true
                    | cons(n, l') =>
                        if check(n, l') = true then is_sorted(l') else false
```

We now need to show that $\widehat{\text{SORTINS}'} \vdash$ SORT1. The following premises are sufficient to apply the rule for enrichment and complete the proof:

- $\widehat{\text{SORTINS}'}$ **hide ops** *sort*, *is_sorted* $\vdash$ NATLIST — although formally this requires traversing the structure of NATLIST, it is obvious that it will work since NATLIST is already a part of $\widehat{\text{SORTINS}'}$. Additional proof methods that allow such cases to be dealt with more directly are discussed below (see Exercise 9.3.9).
- $\widehat{\text{SORTINS}'} \vdash \varphi$ for each of the five axioms φ in the enrichment that defines SORT1 in Example 7.1.4 — here, techniques similar to those in Example 9.2.15 and Example 9.1.11 need to be used, referring to the definitional axioms for *is_sorted*, *sort*, *is_in* and *append* as well as the axioms for *insert* and *le*. (**Exercise:** Do the proof.) □

When proving entailment between specifications, it is always possible to replace the specifications at hand, both in the consequence and in the antecedent, by equivalent ones. In this way, a number of useful rules may be obtained by applying the algebraic laws in Section 5.6. Similarly, this allows specifications to be reduced to normal form. Such reduction might be useful on the side of the consequence, since it allows structure to be eliminated in one step rather than transferred, layer by layer, to the antecedent. Such a one-step elimination of the structure of the consequence specification is captured by the following rule:

$$\frac{\widehat{SP} \vdash \Phi'}{SP \vdash SP'} \quad \begin{array}{l}\mathsf{nf}(SP') = \langle \Sigma', \Phi' \rangle \textbf{ hide via } \sigma, \text{ where} \\ \sigma : SP \to \widehat{SP} \text{ admits model expansion}\end{array}$$

When the algebraic laws in Section 5.6 provide model class inclusions rather than equalities, they can be usefully exploited here as well. For example:

Exercise 9.3.9. Use Exercise 5.6.3(2) to justify the following rule:

$$\frac{}{(SP \textbf{ with } \sigma) \textbf{ hide via } \sigma \vdash SP} \quad \sigma : Sig[SP] \to \Sigma'$$

Combine this with other rules to obtain the following:

$$\frac{}{(SP \textbf{ then sorts } S' \textbf{ ops } \Omega' \bullet \Phi') \textbf{ hide sorts } S' \textbf{ ops } \Omega' \vdash SP}$$ □

Example 9.3.10. Referring to specifications in Example 7.1.4 again, the rule above allows one to show directly that SORTPERM ⊢ SORT; this proves the simple implementation step SORT ⤳ SORTPERM. □

Entailment between specifications may be used to show equivalence between specifications:

$$\frac{SP \vdash SP' \qquad SP' \vdash SP}{SP \equiv SP'}$$

Proving algebraic laws like those stated in Section 5.6 would sometimes also involve the use of induction on the structure of specifications. Such algebraic laws, and the proof rules that are derived from them, would be limited to specifications built using the specification-building operations considered.

The above discussion of entailment between specifications and the proof rules we gave largely concern specifications built from flat specifications using union, translation and hiding. For this class of specifications we have a proof system for the relation of entailment between specifications that is sound (Proposition 9.3.2) and complete under reasonable assumptions (Theorem 9.3.3).

The situation with other specification-building operations is much less elegant. We can state sound rules — see below — but these are far from what would be necessary to ensure completeness. The **reachable** and **free** operations are serious stumbling blocks: once either of these is present, completeness is not achievable in standard institutions as Theorem 2.5.26 made clear. In the presence of such operations, the strategy described above of reducing a proof task $SP \vdash SP'$ to entailment for specifications may fail. In such cases it is sometimes possible to succeed by adopting a more ad hoc strategy. For this purpose we state below a number of additional rules. Many of them are useful as shortcuts even in the context of the complete proof system above.

For instance, for any monotone specification-building operation **sbo** (see Exercises 5.1.4 and 5.1.5) the following rule is clearly sound:

$$\frac{SP_1 \vdash SP'_1 \qquad \cdots \qquad SP_n \vdash SP'_n}{\textbf{sbo}(SP_1, \ldots, SP_n) \vdash \textbf{sbo}(SP'_1, \ldots, SP'_n)}$$

The rule is nicely compositional, but its applicability is limited to cases where the structure of the antecedent and of the consequence match. The discussion on horizontal composability of simple implementations in Section 7.1 (see Proposition 7.1.6 and Examples 7.1.7 and 7.1.8) is relevant here and shows that the rule above, although useful in many cases, cannot be sufficient in general.

The following "identity rule" is clearly sound:

$$\frac{}{SP \vdash SP}$$

Another useful rule is that of transitivity of entailment between specifications:

$$\frac{SP \vdash SP' \qquad SP' \vdash SP''}{SP \vdash SP''}$$

Suppose that **sbo** is a selection operation (Definition 9.2.23). Then we have the following:

$$\frac{SP \vdash SP'}{\mathbf{sbo}(SP) \vdash SP'}$$

Selection operations used in the consequence specification cause more trouble: we need to take the "selection condition" explicitly into account. So, suppose that the selection operation **sbo** is defined by a property P of models, i.e. $Mod[\mathbf{sbo}(SP)] = \{M \in Mod[SP] \mid P(M)\}$. Then we have the following:

$$\frac{SP \vdash SP'}{SP \vdash \mathbf{sbo}(SP')} \quad P(M) \text{ for all } M \in Mod[SP]$$

Methods for discharging the side condition depend on the nature of P. Sometimes it may be internalized as a set of sentences Φ of **INS**; for instance, in an algebraic signature with a finite number of terms $\{t_1, \ldots, t_n\}$ of sort s, reachability on s as used in the **reachable** specification-building operation (Exercise 5.1.10) is captured by the single sentence $\forall x{:}s \bullet x = t_1 \vee \cdots \vee x = t_n$ in **FOEQ**. In such a case, discharging the side condition amounts to proving $SP \vdash \varphi$ for each $\varphi \in \Phi$. Sometimes it may be convenient to extend the underlying institution to make these side conditions expressible. For instance, to deal with reachability in the general case we may extend **FOEQ** to include infinitary disjunction. Even if P cannot be internalized, for some forms of SP the side condition can be discharged using model-theoretic reasoning, leading to rules like the following sound rule for the **reachable** operation:

$$\frac{\mathbf{reachable}\ SP\ \mathbf{on}\ S \vdash SP'}{\mathbf{reachable}\ SP\ \mathbf{on}\ S \vdash \mathbf{reachable}\ SP'\ \mathbf{on}\ S'} \quad S \supseteq S'$$

In some cases, the selection property depends on the argument specification; then the rule must take the following more general form:

$$\frac{SP \vdash SP'}{SP \vdash \mathbf{sbo}(SP')} \quad P_{SP'}(M) \text{ for all } M \in Mod[SP]$$

For instance, the rule for the **free** operation (Exercise 5.1.9) would be as follows:

$$\frac{SP \vdash SP'}{SP \vdash \textbf{free } SP' \textbf{ wrt } \sigma} \quad \begin{array}{l}\text{for all } M \in Mod[SP], \text{ for} \\ M' \in Mod[SP'], f\colon M|_\sigma \to M'|_\sigma, \\ f = f^\sharp|_\sigma \text{ for unique } f^\sharp\colon M \to M'\end{array}$$

An institution of higher-order logic would internalize the freeness side condition as a sentence that would need to be a consequence of *SP*, in the same sense as infinitary disjunction may be used to internalize reachability.

Exercise 9.3.11. **FPL** internalizes the "absolute freeness" of ML-style datatypes, but in its signatures rather than its sentences. Show how to use the rule above to reduce entailments of the form $SP \vdash \textbf{free } \langle \Sigma, \varnothing \rangle \textbf{ wrt } \iota$ in **FPL** for any signature inclusion $\iota\colon \Sigma_0 \to \Sigma$ to the requirement that the specification morphism $j\colon SP \to SP \textbf{ with } j$ admit model expansion, with j adding to the signature of *SP* the requirement that the sorts not in Σ_0 come with the appropriate value constructors. □

Exercise 9.3.12. Check that the following rules for closure under model equivalence (cf. Exercise 5.1.8) are sound:

$$\frac{}{SP \vdash \textbf{close}_\simeq SP} \qquad \frac{SP \vdash SP'}{\textbf{close}_\simeq SP \vdash \textbf{close}_\simeq SP'}$$

In particular, these rules apply to the **iso-close** operation; see Exercise 5.1.7. □

Example 9.3.13. We build on Example 9.2.15 to show the entailment SORTDONE ⊢ SORTINS between specifications from Example 7.1.4; this proves the simple implementation SORTINS ⇝ SORTDONE.

Both specifications, SORTDONE and SORTINS, arise by application of hiding of the *insert* operation. Since hiding is monotone, by the general rule for monotone specification-building operations it is enough to derive SORTBYINSERTDONE ⊢ SORTBYINSERT. Then, relying on the monotonicity of enrichment, this can be shown by deriving INSDONE ⊢ INS.

Using the rule for enrichment, this requires showing the two entailments for INSDONE proved in Example 9.2.15, and SORTBYINSERTDONE **hide ops** *insert* ⊢ NATLIST, which follows easily by Exercise 9.3.9.

The proof above and the proofs in Examples 9.3.8 and 9.3.10 justify the simple implementations claimed at the end of Example 7.1.4:

SORT ⇝ SORTPERM ⇝ SORTINS ⇝ SORTDONE

Notice how the proofs here formalise the semantic arguments given in Example 7.1.4. □

9.4 Correctness of constructor implementations

We continue to work within an institution $\mathbf{INS} = \langle \mathbf{Sign}, \mathbf{Sen}, \mathbf{Mod}, \langle \models_\Sigma \rangle_{\Sigma \in |\mathbf{Sign}|} \rangle$ equipped with an entailment system $\langle \vdash_\Sigma \rangle_{\Sigma \in |\mathbf{Sign}|}$.

We now turn to proving *constructor entailment between specifications*, written $SP \vdash_\kappa SP'$, for any specifications SP, SP' and constructor $\kappa \in Mod(Sig[SP] \Rightarrow Sig[SP'])$. This is designed to approximate the semantic correctness of constructor implementations, that is, $\kappa \in Mod[SP \Rightarrow SP']$, which corresponds to $SP' \underset{\kappa}{\rightsquigarrow} SP$ (see Definitions 6.2.1 and 7.2.1). Inclusion of model classes as discussed in the previous section is covered as a special case by taking the identity constructor.

Definition 9.4.1 (Soundness and completeness of constructor entailment). A (ternary) relation $\_ \vdash_{\_} \_$ of constructor entailment between specifications is *sound* if $SP \vdash_\kappa SP'$ implies $\kappa(Mod[SP]) \subseteq Mod[SP']$ and $Mod[SP] \subseteq dom(\kappa)$ for all specifications SP and SP' with $\kappa \in Mod(Sig[SP] \Rightarrow Sig[SP'])$. It is *complete* if the opposite implications hold. □

A proof system for proving constructor entailments between specifications may be obtained directly from a proof system for entailments between specifications via the following rule:

$$\frac{\kappa(SP) \vdash SP'}{SP \vdash_\kappa SP'} \quad Mod[SP] \subseteq dom(\kappa)$$

where, in $\kappa(SP)$, the constructor κ is regarded as a specification-building operation; see Exercise 6.1.15.

The side condition is necessary: without it, the rule is unsound. We regard it as outside the scope of the system we provide here, since treating it formally would require us to develop an institution-independent modular "programming language" for defining constructors.

However, two important special cases dominate in practice, and these are not difficult to handle. The first is when the constructor κ is total on the class of models over its argument signature. Then the side condition amounts to a "static" check that the signatures match, and the rule takes the following form:

$$\frac{\kappa(SP) \vdash SP'}{SP \vdash_\kappa SP'} \quad \kappa \in Mod(Sig[SP] \Rightarrow Sig[SP']) \text{ is total}$$

Some of the constructors presented in Section 6.1 are always total. One example is reducts, including reducts along derived signature morphisms; cf. Example 6.1.3. A particularly simple case is that of the identity constructor, with $SP \vdash_{id} SP'$ following from $SP \vdash SP'$. On the other hand, free extension constructors (cf. Example 6.1.7) need not be total, not even when the specification involved has no axioms. This includes the case of **FPL** where, however, it is possible to give a simple syntactic condition on the signatures involved which guarantees that the free extension for specifications without axioms is total; see Example 6.1.7. The notation in Example 6.1.9 and its extension in Exercise 7.3.5 always yields total constructors.

The second case is when κ comes with its specification as discussed in Section 6.2, either as a hypothesis or derived using the system in Section 9.5 below, where for closed specifications SP and SP', a judgement $\Gamma \blacktriangleright E : SP \Rightarrow SP'$ yields $[\![E]\!]_\rho \in Mod[SP \Rightarrow SP']$ for any Γ-environment ρ — see Exercise 9.5.4 — which amounts to $SP \vdash_{[\![E]\!]_\rho} SP'$.

The following weakening rule allows a constructor's interface specifications to be adjusted to fit the context of use:

$$\frac{SP \vdash_\kappa SP' \qquad SP_1 \vdash SP \qquad SP' \vdash SP_2}{SP_1 \vdash_\kappa SP_2}$$

This can be derived from the following transitivity rule by taking (twice) one of the constructors to be the identity:

$$\frac{SP \vdash_\kappa SP' \qquad SP' \vdash_{\kappa'} SP''}{SP \vdash_{\kappa;\kappa'} SP''}$$

Notice that this rule is another way of asserting that constructor implementations compose vertically; see Proposition 7.2.4.

Constructor entailment is closed under specification equivalence, so the algebraic laws of Section 5.6 for specification equivalence and model class inclusion give rise to rules for constructor entailment in a way similar to that in the previous sections.

The main idea in proofs of constructor entailment between specifications is to reduce the problem to proving entailment between specifications. But unfortunately, the rules in Sections 9.2 and 9.3 do not explicitly deal with antecedents of the form $\kappa(SP)$. We will rely on the fact that specifications of this form will typically be equivalent to specifications built using other specification-building operations. There are two ways to exploit this fact. The first way would be to extend the system for entailment between specifications in Section 9.3 to capture these equivalences as additional rules, and use the first rule in this section to derive corresponding constructor entailments. The second way, which we will follow below, is to use these equivalences to justify additional rules for constructor entailment between specifications for various constructors, without explicitly referring to specifications of the form $\kappa(SP)$.

We now consider how to do this for the two most important classes of constructors κ. Reducts are simple:

Exercise 9.4.2. Let κ be the reduct constructor determined by a signature morphism $\sigma : \Sigma \to \Sigma'$; see Example 6.1.3. Check that for any Σ'-specification SP', $\kappa(SP') \equiv SP'$ **hide via** σ. Since reduct is a total constructor, this justifies the following rule:

$$\frac{SP' \textbf{ hide via } \sigma \vdash SP}{SP' \vdash_{\_|_\sigma} SP}$$

□

Exercise 9.4.3. Consider the free extension constructor $F_{\sigma,SP'}$ determined by a signature morphism $\sigma : \Sigma \to \Sigma'$ and Σ'-specification SP'; see Example 6.1.7. Check that the following equivalence holds for any Σ-specification SP:

$$\textbf{iso-close}\,(F_{\sigma,SP'}(SP)) \equiv (\textbf{free}\ SP'\ \textbf{wrt}\ \sigma) \cup (SP\ \textbf{with}\ \sigma)$$

Convince yourself that there is no specification expression that is equivalent to $F_{\sigma,SP'}(SP)$ that does not involve $F_{\sigma,SP'}$ itself, in general. (This is a somewhat delicate point which is related to the fact that even though free extension is defined only up to isomorphism, a constructor is a function on models and so it chooses some particular element in this isomorphism class — and so the closure under isomorphism in the above equivalence is necessary.)

The equivalence above does not appear to be entirely satisfactory, since we give a specification that is equivalent not to $F_{\sigma,SP'}(SP)$, but only to **iso-close** $(F_{\sigma,SP'}(SP))$. But this is not a problem: we have $F_{\sigma,SP'}(SP) \vdash \textbf{iso-close}\,(F_{\sigma,SP'}(SP))$ by Exercise 9.3.12, and so if we can prove $\textbf{iso-close}\,(F_{\sigma,SP'}(SP)) \vdash SP''$ then we also get $F_{\sigma,SP'}(SP) \vdash SP''$. Since $dom(F_{\sigma,SP'})$ is specified by $(\textbf{free}\ SP'\ \textbf{wrt}\ \sigma)\ \textbf{hide via}\ \sigma$, this justifies the following rule:

$$\frac{SP \vdash (\textbf{free}\ SP'\ \textbf{wrt}\ \sigma)\ \textbf{hide via}\ \sigma \qquad (\textbf{free}\ SP'\ \textbf{wrt}\ \sigma) \cup (SP\ \textbf{with}\ \sigma) \vdash SP''}{SP \vdash_{F_{\sigma,SP'}} SP''}$$

□

The rule in the above exercise involves the **free** specification-building operation for which no closed complete rule is available; see Exercise 9.2.24 and the surrounding discussion. Fortunately, this problem can often be resolved satisfactorily in particular institutions, at least for the cases that arise in practice.

Exercise 9.4.4. The **FPL**-constructor notation introduced in Example 6.1.9 gives rise to constructors of the form $F_\iota;(\_|_\delta)$ where $\iota\colon \mathsf{SIG} \to \mathsf{SIG}''$ is an **FPL**-signature inclusion which adds new sorts, each with a set of value constructors, and $\delta\colon \mathsf{SIG}' \to \mathsf{SIG}''$ is an **FPL**-signature morphism. Here, F_ι is the absolutely free extension constructor $F_{\iota,\langle \mathsf{SIG}'',\varnothing\rangle}$, which is total; see Example 6.1.7. Similarly as in the discussion above, notice that for any SIG-specification SP, $F_\iota(SP)|_\delta \vdash (\textbf{iso-close}\ F_\iota(SP))|_\delta$ and so it will be sufficient to eliminate the constructors in the consequence of this entailment. To this end, show that

$$(\textbf{iso-close}\ F_\iota(SP))|_\delta \equiv ((\textbf{free}\ \langle \mathsf{SIG}'',\varnothing\rangle\ \textbf{wrt}\ \iota) \cup (SP\ \textbf{with}\ \iota))\ \textbf{hide via}\ \delta.$$

But since $\iota\colon \mathsf{SIG} \to \mathsf{SIG}''$ merely introduces new sorts, each with a set of value constructors, every SIG''-model is a free extension of its ι-reduct. Show that then

$$(\textbf{iso-close}\ F_\iota(SP))|_\delta \equiv (SP\ \textbf{with}\ \iota)\ \textbf{hide via}\ \delta$$

which justifies the following rule:

$$\frac{(SP\ \textbf{with}\ \iota)\ \textbf{hide via}\ \delta \vdash SP'}{SP \vdash_{F_\iota;(\_|_\delta)} SP'}$$

This achieves our goal since the rules in Section 9.3 suffice for dealing with the entailment in the premise, building on the rules in Section 9.2 for proving entailment for specifications and the rules in Example 9.1.11 for entailment in **FPL**.

Taking this further, it is sometimes convenient to take advantage of the particular way that δ arises in the **FPL**-constructor notation of Example 6.1.9. Namely, consider

constructor $K : \mathsf{SIG} \Rightarrow \mathsf{SIG}' = body$

and suppose that $\mathsf{SIG} \subseteq \mathsf{SIG}'$ and *body* lists new sorts, each with a set of value constructors, and defines new operations, all in SIG', so all the new sorts and operations in SIG' are defined and no auxiliary sorts or operations are introduced. Show that for any SIG-specification *SP*,

$$\textbf{iso-close}\ K(SP) \equiv SP\ \textbf{then}\ body.$$

Under those assumptions on the form of *body*, this yields the following rule:

$$\frac{SP\ \textbf{then}\ body \vdash SP'}{SP \vdash_K SP'}$$

In the general situation, when SIG contains, or *body* introduces, auxiliary sorts and/or operations which are not in SIG', check that we have

$$\textbf{iso-close}\ K(SP) \equiv (SP\ \textbf{then}\ body)\ \textbf{reveal}\ \mathsf{SIG}'$$

which yields the following rule:

$$\frac{(SP\ \textbf{then}\ body)\ \textbf{reveal}\ \mathsf{SIG}' \vdash SP'}{SP \vdash_K SP'}$$

Check that the previous rule is derivable from this version when no auxiliary sorts or operations are introduced. □

This situation is typical: often $\kappa(SP)$ (or its closure under isomorphism) amounts to an enrichment of *SP* by definitions that are more or less explicit in the form of κ.

Example 9.4.5. Recall the chain of constructor implementations in Example 7.2.8, relating specifications in **FPL** introduced in Example 7.1.4. We can now prove the correctness of the consecutive constructor implementations using the techniques and rules discussed above.

- The trivial constructor *K0* has the form

  constructor $K0 : Sig[\textsc{SortPerm}] \Rightarrow Sig[\textsc{Sort}] = body$

  where *body* is empty. Then from SORTPERM **then** *body* $\vdash$ SORTPERM (trivially) and SORTPERM $\vdash$ SORT (by Example 9.3.10), by transitivity of entailment between specifications and the penultimate rule in Exercise 9.4.4 above, we get SORTPERM $\vdash_{K0}$ SORT, i.e. SORT $\underset{K0}{\rightsquigarrow}$ SORTPERM.

- The constructor *K1* has the form

  constructor *K1* : *Sig*[INS] ⇒ *Sig*[SORTPERM] =
  ops fun *sort*(*l*:*NatList*):*NatList* = ...

  and INS **then ops fun** *sort*(*l*:*NatList*):*NatList* = ... is SORTBYINSERT while SORTBYINSERT **hide ops** *insert* is SORTINS. Since SORTINS ⊢ SORTPERM (by Example 9.3.8), by the final rule in Exercise 9.4.4 above we get INS $\vdash_{K1}$ SORTPERM, that is, SORTPERM $\underset{K1}{\leadsto}$ INS.
- The constructor *K2* has the form

  constructor *K2* : *Sig*[NATLIST] ⇒ *Sig*[INS] =
  ops fun *insert*(*n*:*Nat*, *l*:*NatList*):*NatList* = ...

  and NATLIST **then ops fun** *insert*(*n*:*Nat*, *l*:*NatList*):*NatList* = ... is INSDONE. Since INSDONE ⊢ INS (from the derivation within Example 9.3.13), by the penultimate rule in Exercise 9.4.4 above we get NATLIST $\vdash_{K2}$ INSDONE, that is, INSDONE $\underset{K2}{\leadsto}$ NATLIST.

This differs from the reasoning in Example 9.3.13 by allowing us to avoid proving that SORTDONE ⊢ SORTINS: we can stop with INSDONE ⊢ INS, in which the constructor *K1* has been "peeled off" as in the diagram at the beginning of Section 7.2.

In Example 7.2.8, we go one step further and choose an implementation for NATLIST:

- The constructor *K3* has the form

  constructor *K3* : *Sig*[NATLIST] ⇒ *Sig*[INS] =
  sorts *Nat* **free with** 0 | *succ*(*Nat*)
  NatList **free with** *nil* | *cons*(*Nat*, *NatList*)
  ops **fun** *le*(*m*:*Nat*, *n*:*Nat*):*Bool* = ...
  fun *append*(*l*:*NatList*, *l′*:*NatList*):*NatList* = ...
  fun *is_in*(*n*:*Nat*, *l*:*NatList*):*Bool* = ...

  Let us now put

  spec NATLISTLE =
  EMPTY
  then
  sorts *Nat* **free with** 0 | *succ*(*Nat*)
  NatList **free with** *nil* | *cons*(*Nat*, *NatList*)
  ops **fun** *le*(*m*:*Nat*, *n*:*Nat*):*Bool* = ...
  fun *append*(*l*:*NatList*, *l′*:*NatList*):*NatList* = ...
  fun *is_in*(*n*:*Nat*, *l*:*NatList*):*Bool* = ...

  Now, using the techniques from Sections 9.2 and 9.3, one can easily derive NATLISTLE ⊢ NATLIST. The non-trivial part of the derivation is an inductive proof that *le* as defined by the code in NATLISTLE satisfies the axioms within

NATLIST that require le to be a linear order. By the penultimate rule in Exercise 9.4.4 above, this gives $\text{EMPTY} \vdash_{K3} \text{NATLIST}$, i.e. $\text{NATLIST} \underset{K3}{\rightsquigarrow} \text{EMPTY}$. □

9.5 Proof and parameterisation

In this section we build on the proof systems in the previous sections to give a system of rules for proving correctness of higher-order constructors with respect to constructor specifications as presented in Section 6.4. We disregard the (higher-order) parameterised specifications of Definition 6.4.5, but see Exercise 9.5.5 below. This material is somewhat more tentative than that in the previous sections of this chapter.

We continue to work within an institution $\mathbf{INS} = \langle \mathbf{Sign}, \mathbf{Sen}, \mathbf{Mod}, \langle \models_\Sigma \rangle_{\Sigma \in |\mathbf{Sign}|} \rangle$ equipped with an entailment system $\langle \vdash_\Sigma \rangle_{\Sigma \in |\mathbf{Sign}|}$.

We will be deriving three main forms of judgement:

- $\Gamma \blacktriangleright SP \vdash \varphi$ for entailment for specifications, slightly generalising the system in Section 9.2.
- $\Gamma \blacktriangleright SP \vdash SP'$ for entailment between constructor specifications, generalising the system in Section 9.3; and
- $\Gamma \blacktriangleright E : SP$ for correctness of constructors with respect to specifications, generalising the system for constructor entailment between specifications in Section 9.4.

We will also use two other judgement forms, having an auxiliary role:

- Γ is well formed for well-formedness of contexts; and
- $\Gamma \blacktriangleright SP : Spec(\mathscr{S})$ for well-formedness of specifications (as constructor specifications for constructor signature $\mathscr{S}$).

The two well-formedness judgements above generalise the corresponding concepts in Section 6.4, but capture more complex properties, which may depend on the correctness and entailments expressed by the other judgements. As a result, the system of rules for each of the judgement forms refers to the others, and relies on auxiliary definitions which are introduced in Exercise 9.5.4 below, which in turn refer to the judgements. Therefore the rules, definitions and properties below constitute a single large definition/theorem. For ease of understanding, we present it in chunks that can be understood in relative isolation.

Contexts

A *context* Γ is a sequence of the form $X_1{:}SP_1, \ldots, X_n{:}SP_n$, $n \geq 0$, where $X_1, \ldots, X_n$ are distinct variables and each SP_i is a specification (which includes constructor signatures, via Exercise 6.4.15). We write $dom(\Gamma)$ for $\{X_1, \ldots, X_n\}$ and $\Gamma(X_i)$ for SP_i. Well-formedness of contexts is given by the judgement: Γ is well formed.

$$\frac{}{\text{is well formed}} \qquad \frac{\Gamma \text{ is well formed} \quad \Gamma \blacktriangleright SP : Spec(\mathscr{S}_{any})}{\Gamma, X{:}SP \text{ is well formed}} \; X \notin dom(\Gamma)$$

The first rule states that the empty context is well formed. In the second rule, we use the premise $\Gamma \blacktriangleright SP : Spec(\mathscr{S}_{any})$ to capture the requirement that SP be a well-formed constructor specification in the context Γ (for an arbitrary constructor signature $\mathscr{S}_{any}$).

Well-formedness of specifications

$$\frac{\Gamma \text{ is well formed}}{\Gamma \blacktriangleright \langle\Sigma, \Phi\rangle : Spec(\Sigma)} \quad \Sigma \in |\mathbf{Sign}|, \Phi \subseteq \mathbf{Sen}(\Sigma)$$

$$\frac{\Gamma \blacktriangleright SP : Spec(\Sigma)}{\Gamma \blacktriangleright SP \textbf{ with } \sigma : Spec(\Sigma')} \quad \sigma{:}\Sigma \to \Sigma'$$

$$\frac{\Gamma \blacktriangleright SP' : Spec(\Sigma')}{\Gamma \blacktriangleright SP' \textbf{ hide via } \sigma : Spec(\Sigma)} \quad \sigma{:}\Sigma \to \Sigma'$$

... and similarly for other specification-building operations ...

The system in Section 6.4 provides three new forms of constructor specifications for arbitrary constructor signatures: union, which subsumes the union of Σ-specifications, singleton, and Π-specifications. We provide rules for well-formedness of such specifications:

$$\frac{\Gamma \blacktriangleright SP_1 : Spec(\mathscr{S}) \qquad \Gamma \blacktriangleright SP_2 : Spec(\mathscr{S})}{\Gamma \blacktriangleright SP_1 \cup SP_2 : Spec(\mathscr{S})}$$

$$\frac{\Gamma \blacktriangleright E : SP \qquad \Gamma \blacktriangleright SP : Spec(\mathscr{S})}{\Gamma \blacktriangleright \{E\} : Spec(\mathscr{S})}$$

$$\frac{\Gamma \blacktriangleright SP : Spec(\mathscr{S}) \qquad \Gamma, X{:}SP \blacktriangleright SP' : Spec(\mathscr{S}')}{\Gamma \blacktriangleright \Pi X{:}SP \bullet SP' : Spec(\mathscr{S} \Rightarrow \mathscr{S}')}$$

It may be shown (see Exercise 9.5.1 below) that the second premise in the rule for well-formedness of singleton specification is superfluous in the sense that whenever $\Gamma \blacktriangleright E : SP$ can be derived then $\Gamma \blacktriangleright SP : Spec(\mathscr{S})$ can be derived for a unique constructor signature $\mathscr{S}$. We leave this premise here just to directly indicate the generalised signature to be used in the conclusion. A similar remark applies to some of the rules below as well.

Entailment for Σ-specifications

$$\frac{\Gamma \blacktriangleright \langle\Sigma, \Phi\rangle : Spec(\Sigma)}{\Gamma \blacktriangleright \langle\Sigma, \Phi\rangle \vdash \varphi} \quad \varphi \in \Phi$$

$$\frac{\Gamma \blacktriangleright SP : Spec(\Sigma) \qquad \Gamma \blacktriangleright SP \vdash \varphi}{\Gamma \blacktriangleright SP \textbf{ with } \sigma \vdash \sigma(\varphi)} \quad \sigma{:}\Sigma \to \Sigma', \varphi \in \mathbf{Sen}(\Sigma)$$

$$\frac{\Gamma \blacktriangleright SP' : Spec(\Sigma') \qquad \Gamma \blacktriangleright SP' \vdash \sigma(\varphi)}{\Gamma \blacktriangleright SP' \text{ **hide via** } \sigma \vdash \varphi} \quad \sigma{:}\Sigma \to \Sigma', \varphi \in \mathbf{Sen}(\Sigma)$$

... and similarly for other proof rules in Section 9.2 ...

The last remark covers, for instance, the rule $(*)$ of Section 9.2, which now takes the following form:

$$\frac{\Gamma \blacktriangleright SP \vdash Spec(\Sigma) \qquad \Gamma \blacktriangleright SP \vdash \varphi_1 \quad \cdots \quad \Gamma \blacktriangleright SP \vdash \varphi_n \qquad \{\varphi_1, \ldots, \varphi_n\} \vdash_\Sigma \varphi}{\Gamma \blacktriangleright SP \vdash \varphi}$$

With respect to Section 9.2, a new form of Σ-specifications is singleton specifications. For a singleton specification of the form $\{[M]^\Sigma\}$ where M is a Σ-model, we need to resort to semantic reasoning:

$$\frac{\Gamma \text{ is well formed}}{\Gamma \blacktriangleright \{[M]^\Sigma\} \vdash \varphi} \quad M \models_\Sigma \varphi$$

From the rules below for entailment between specifications with a singleton specification as antecedent and for transitivity of entailment, we can derive the following:

$$\frac{\Gamma \blacktriangleright E : SP \qquad \Gamma \blacktriangleright SP \vdash \varphi}{\Gamma \blacktriangleright \{E\} \vdash \varphi}$$

This justifies why the semantic rule for entailment for singleton specifications above was limited to Σ-models, whose syntax is external to the system. There should be no need to resort to semantic reasoning for other constructor expressions: we offer rules below for deriving specifications for such expressions, and then the last rule applies.

Entailment between specifications

$$\frac{\Gamma \blacktriangleright SP : Spec(\Sigma) \qquad \Gamma \blacktriangleright SP \vdash \varphi \text{ for each } \varphi \in \Phi}{\Gamma \blacktriangleright SP \vdash \langle \Sigma, \Phi \rangle}$$

$$\frac{\Gamma \blacktriangleright SP' \text{ **hide via** } \sigma \vdash SP}{\Gamma \blacktriangleright SP' \vdash SP \text{ **with** } \sigma}$$

$$\frac{\Gamma \blacktriangleright SP : Spec(\Sigma) \quad \Gamma \blacktriangleright SP' : Spec(\Sigma') \quad \Gamma \blacktriangleright \widehat{SP} \vdash SP'}{\Gamma \blacktriangleright SP \vdash SP' \text{ **hide via** } \sigma} \quad \begin{array}{l}\sigma{:}\Sigma \to \Sigma', \text{ for each} \\ \Gamma\text{-environment } \rho, \\ [\![SP]\!]_\rho = ([\![\widehat{SP}]\!]_\rho)|_\sigma\end{array}$$

... and similarly for other proof rules in Section 9.3 ...

The side condition for the last rule above captures the side condition of the corresponding rule in Section 9.3 that $\sigma: SP \to \widehat{SP}$ is a specification morphism which admits model expansion. We refer here to the semantics of specification expressions as defined in Definition 6.4.11, which is justified below by Exercise 9.5.4. The comments in Section 9.3 relating to methods for discharging this side condition still apply here.
By Exercise 9.5.1 below, the premise of the second rule above (for translation) ensures that SP and SP' are well-formed specifications of signatures in $|\mathbf{Sign}|$, and that σ is a signature morphism between their signatures.

$$\frac{\Gamma \blacktriangleright SP \vdash SP_1 \qquad \Gamma \blacktriangleright SP \vdash SP_2}{\Gamma \blacktriangleright SP \vdash SP_1 \cup SP_2}$$

$$\frac{\Gamma \blacktriangleright SP_1 \vdash SP \quad \Gamma \blacktriangleright SP_1 \cup SP_2 : Spec(\mathscr{S}_{any})}{\Gamma \blacktriangleright SP_1 \cup SP_2 \vdash SP}$$

$$\frac{\Gamma \blacktriangleright SP_2 \vdash SP \quad \Gamma \blacktriangleright SP_1 \cup SP_2 : Spec(\mathscr{S}_{any})}{\Gamma \blacktriangleright SP_1 \cup SP_2 \vdash SP}$$

$$\frac{\Gamma \blacktriangleright E : SP}{\Gamma \blacktriangleright \{E\} \vdash SP}$$

$$\frac{\Gamma \blacktriangleright SP \vdash SP_1 \qquad \Gamma, X{:}SP \blacktriangleright SP'_1 \vdash SP'}{\Gamma \blacktriangleright \Pi X{:}SP_1 \bullet SP'_1 \vdash \Pi X{:}SP \bullet SP'}$$

Constructor correctness

The judgement $\Gamma \blacktriangleright E : SP_1$ generalises constructor entailment between specifications as studied in Section 9.4: $SP \vdash_\kappa SP'$ corresponds to the judgement $\blacktriangleright E : \Pi X{:}SP \bullet SP'$ where X does not occur in SP' and $[\![E]\!]_\varnothing = \kappa$. We will not directly import the rules given in Section 9.4. One reason is that we refrain from extending the syntax for specifications given in Section 6.4 to explicitly cover specifications of the form $\kappa(SP)$, as used in the main rule in that section. Rules for specific constructors (see Exercises 9.4.2, 9.4.3 and 9.4.4) will take a different form here.

$$\frac{\Gamma \text{ is well formed}}{\Gamma \blacktriangleright X : \Gamma(X)} \quad X \in dom(\Gamma)$$

$$\frac{\Gamma \blacktriangleright E : SP_1 \qquad \Gamma \blacktriangleright E : SP_2}{\Gamma \blacktriangleright E : SP_1 \cup SP_2} \qquad \frac{\Gamma \blacktriangleright E : SP_{any}}{\Gamma \blacktriangleright E : \{E\}}$$

$$\frac{\Gamma, X{:}SP \blacktriangleright E' : SP'}{\Gamma \blacktriangleright \lambda X{:}SP \bullet E' : \Pi X{:}SP \bullet SP'} \qquad \frac{\Gamma \blacktriangleright E : \Pi X{:}SP_1 \bullet SP'_1 \qquad \Gamma \blacktriangleright E_1 : SP_1}{\Gamma \blacktriangleright E(E_1) : SP'_1[E_1/X]}$$

$$\frac{\Gamma \blacktriangleright SP : Spec(\mathscr{S})}{\Gamma \blacktriangleright [F]^{\mathscr{S}} : SP} \quad F \in Mod(\mathscr{S}), \text{ and for each } \Gamma\text{-environment } \rho, F \in [\![SP]\!]_\rho$$

The side condition in the last rule calls for external semantic reasoning to verify correctness of the constructor $F \in Mod(\mathscr{S})$ against the specification SP. This allows us to use the techniques developed in Section 9.4 for constructor entailment between specifications to justify correctness judgements for first-order constructors: given a Σ-specification SP and Σ'-specification SP', if $SP \vdash_\kappa SP'$ for $\kappa \in Mod(\Sigma \Rightarrow \Sigma')$, then $\kappa \in [\![SP \Rightarrow SP']\!]$, and so by the above rule we have $\blacktriangleright [\kappa]^{\Sigma \Rightarrow \Sigma'} : SP \Rightarrow SP'$. (As before, $SP \Rightarrow SP'$ stands for $\Pi X{:}SP\bullet SP'$, where X does not occur in SP'.)

Similarly as discussed in Section 6.4, an alternative is to limit this rule to Σ-models only

$$\frac{\Gamma \blacktriangleright SP : Spec(\Sigma)}{\Gamma \blacktriangleright [M]^{\Sigma} : SP} \quad \begin{array}{l} M \in Mod(\Sigma), \text{ and for each} \\ \Gamma\text{-environment } \rho,\ M \in [\![SP]\!]_\rho \end{array}$$

and introduce a particular set of constructors either as constants, with their specifications defined by rules like

$$\frac{\Gamma \text{ is well formed}}{\Gamma \blacktriangleright \_|_\sigma : \Pi X{:}\langle \Sigma', \varnothing\rangle \bullet \{X\} \textbf{ hide via } \sigma} \quad \sigma{:}\Sigma \to \Sigma'$$

or as additional syntax, with rules like this:

$$\frac{\Gamma \blacktriangleright E : SP' \qquad \Gamma \blacktriangleright SP' : Spec(\Sigma')}{\Gamma \blacktriangleright E|_\sigma : SP' \textbf{ hide via } \sigma} \quad \sigma{:}\Sigma \to \Sigma'$$

Note that each of these two rules subsumes the rule for constructor entailment for reduct in Exercise 9.4.2.

Exercise. Devise similar rules for the free extension constructor as defined in Example 6.1.7 and introduced into our syntax in Exercise 6.4.3. For instance, the second rule might be the following:

$$\frac{\begin{array}{c}\Gamma \blacktriangleright E : SP \qquad \Gamma \blacktriangleright SP' : Spec(\Sigma') \\ \Gamma \blacktriangleright SP \vdash (\textbf{free } SP' \textbf{ wrt } \sigma) \textbf{ hide via } \sigma\end{array}}{\Gamma \blacktriangleright F_{\sigma,SP'}(E) : (\textbf{free } SP' \textbf{ wrt } \sigma) \cup (SP \textbf{ with } \sigma)} \quad \sigma{:}\Sigma \to \Sigma'$$

Compare these rules with the one in Exercise 9.4.3.

Exercise. Suppose that the following pushout diagram in **Sign** admits amalgamation:

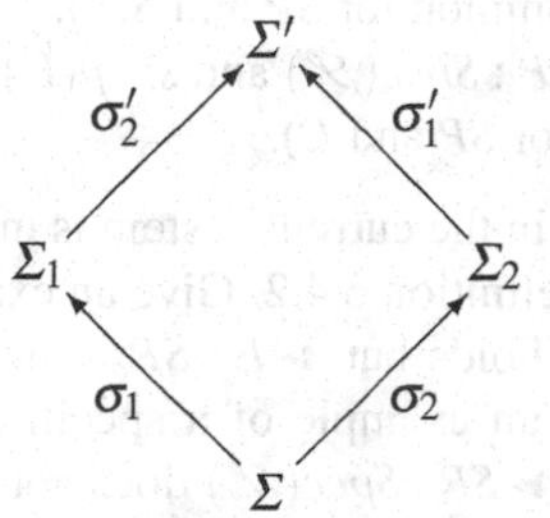

Devise similar rules for the amalgamated union constructor as defined in Example 6.1.32 and introduced into our syntax in Exercise 6.4.4. For instance, the first rule might be the following:

$$\frac{\Gamma \text{ is well formed}}{\Gamma \blacktriangleright \_+_{\sigma_1,\sigma_2}\_ : \Pi X_1{:}\langle\Sigma_1,\varnothing\rangle \bullet \Pi X_2{:}(\{X_1\} \textbf{ hide via } \sigma_1) \textbf{ with } \sigma_2 \bullet (\{X_1\} \textbf{ with } \sigma_2') \cup (\{X_2\} \textbf{ with } \sigma_1')} \quad \text{for above pushout}$$

Transitivity and reflexivity

Finally, for each of the main judgement forms, we add rules that capture (reflexivity and) transitivity of entailment between constructor specifications:

$$\frac{\Gamma \blacktriangleright SP \vdash SP' \qquad \Gamma \blacktriangleright SP' \vdash \varphi}{\Gamma \blacktriangleright SP \vdash \varphi}$$

$$\frac{\Gamma \blacktriangleright SP : Spec(\mathscr{S}_{any})}{\Gamma \blacktriangleright SP \vdash SP} \qquad \frac{\Gamma \blacktriangleright SP \vdash SP' \qquad \Gamma \blacktriangleright SP' \vdash SP''}{\Gamma \blacktriangleright SP \vdash SP''}$$

$$\frac{\Gamma \blacktriangleright E : SP \qquad \Gamma \blacktriangleright SP \vdash SP'}{\Gamma \blacktriangleright E : SP'}$$

Whenever convenient we will simply write a judgement rather than explicitly spell out that it is derivable by the above rules.

Exercise 9.5.1. We can now relate the verification system given above to the typing system of Definition 6.4.2 (with judgements of the form $\Gamma \rhd W : \mathscr{G}$, for a simpler notion of context). By induction on derivations in the entire system for all judgement forms in use, define a function that for each well-formed context Γ yields a (well-formed) context $strip(\Gamma)$ in the sense of Definition 6.4.1, and prove the following:

- If Γ is the empty context, then $strip(\Gamma)$ stands for the empty context as well. If $\Gamma,X{:}SP$ is well formed, then define $strip(\Gamma,X{:}SP)$ to be $strip(\Gamma),X{:}\mathscr{S}$ where $\Gamma \blacktriangleright SP : Spec(\mathscr{S})$; such a constructor signature $\mathscr{S}$ is unique by the next property and Exercise 6.4.7.
- If $\Gamma \blacktriangleright SP : Spec(\mathscr{S})$ then $strip(\Gamma) \rhd SP : Spec(\mathscr{S})$.
- If $\Gamma \blacktriangleright SP \vdash \varphi$ then $\Gamma \blacktriangleright SP : Spec(\Sigma)$ for some signature $\Sigma \in |\mathbf{Sign}|$ and $\varphi \in \mathbf{Sen}(\Sigma)$.
- If $\Gamma \blacktriangleright SP \vdash SP'$ then $\Gamma \blacktriangleright SP : Spec(\mathscr{S})$ and $\Gamma \blacktriangleright SP' : Spec(\mathscr{S})$ for some constructor signature $\mathscr{S}$ (common for SP and SP').
- If $\Gamma \blacktriangleright E : SP$ then $\Gamma \blacktriangleright SP : Spec(\mathscr{S})$ and $strip(\Gamma) \rhd E : \mathscr{S}$ for some constructor signature $\mathscr{S}$ (common for SP and E).

However, well-formedness in the current system is more demanding than typability defined by the system in Definition 6.4.2. Give an example of a constructor expression E such that $\rhd E : \mathscr{S}$ holds but $\blacktriangleright E : SP_{any}$ does not hold for any specification SP_{any}. Similarly, give an example of a specification expression SP such that $\rhd SP : Spec(\mathscr{S})$ holds but $\blacktriangleright SP : Spec(\mathscr{S})$ does not hold. □

Exercise 9.5.2. Mimicking Exercise 6.4.8, try to check that the following substitutivity properties hold in the system above. One problem is that since specifications used in contexts may now contain variables, contexts cannot be arbitrarily permuted (in contrast with the contexts of Definition 6.4.1). Only a limited form of permutation is available: show that if a judgement may be derived for a context $\Gamma, X{:}SP, X'{:}SP', \Gamma'$ and the variable X is not free in SP', then the same judgement may be derived for the context $\Gamma, X'{:}SP', X{:}SP, \Gamma'$. Even with this, stronger properties than in Exercise 6.4.8 are required here:

- If $\Gamma, X{:}SP_1, \Gamma'$ is well formed and $\Gamma \blacktriangleright E_1 : SP_1$ then $\Gamma, \Gamma'[E_1/X]$ is well formed, where $\Gamma'[E_1/X]$ is the context obtained from Γ' by substituting E_1 for X in all the specifications for variables in Γ', i.e. $dom(\Gamma'[E_1/X]) = dom(\Gamma')$ and for $X' \in dom(\Gamma')$, $\Gamma'[E_1/X](X') = \Gamma'(X')[E_1/X]$.
- $\Gamma, \Gamma'[E_1/X] \blacktriangleright SP[E_1/X] : Spec(\mathscr{S})$ whenever $\Gamma, X{:}SP_1, \Gamma' \blacktriangleright SP : Spec(\mathscr{S})$ and $\Gamma \blacktriangleright E_1 : SP_1$.
- $\Gamma, \Gamma'[E_1/X] \blacktriangleright SP[E_1/X] \vdash \varphi$ whenever $\Gamma, X{:}SP_1, \Gamma' \blacktriangleright SP \vdash \varphi$ and $\Gamma \blacktriangleright E_1 : SP_1$.
- $\Gamma, \Gamma'[E_1/X] \blacktriangleright SP[E_1/X] \vdash SP'[E_1/X]$ whenever $\Gamma, X{:}SP_1, \Gamma' \blacktriangleright SP \vdash SP'$ and $\Gamma \blacktriangleright E_1 : SP_1$.
- $\Gamma, \Gamma'[E_1/X] \blacktriangleright E[E_1/X] : SP[E_1/X]$ whenever $\Gamma, X{:}SP_1, \Gamma' \blacktriangleright E : SP$ and $\Gamma \blacktriangleright E_1 : SP_1$. □

Exercise 9.5.3. In the spirit of Exercise 6.4.9, check that the following subject reduction properties hold for the fragment of the system obtained by excluding singleton specifications:

- If $\Gamma \blacktriangleright E : SP$ and $E \rightarrow_\beta E'$ then $\Gamma \blacktriangleright E' : SP$.
- If $\Gamma \blacktriangleright E : SP$ and $E \rightarrow_\eta E'$ then $\Gamma \blacktriangleright E' : SP$.

Give an example of a constructor expression E in the full system such that $\Gamma \blacktriangleright E : \{E\}$ but for some E' with $E \rightarrow_\beta E'$, $\Gamma \blacktriangleright E' : \{E\}$ does not hold. The essence of the problem here is that the judgement $\Gamma \blacktriangleright E' : \{E\}$ captures equality between E and E', but the system does not contain rules that deal properly with equality, whether via β-reduction or otherwise. (But note that if E and E' are over a signature $\Sigma \in |\mathbf{Sign}|$, then semantic equality between them can be used to discharge the side condition needed to obtain $\Gamma \blacktriangleright \{E'\} \vdash \{E\}$ **hide via** id_Σ, which with $\Gamma \blacktriangleright E' : \{E'\}$ and $\Gamma \blacktriangleright \{E\}$ **hide via** $id_\Sigma \vdash \{E\}$ yields $\Gamma \blacktriangleright E' : \{E\}$.)

One remedy would be to extend the system with rules for equality. Investigate this, making reference to [Asp97].

Another possibility is to relegate reasoning about equality to an external system via the use of a semantic side condition as in the following rule, which properly generalises the corresponding one above:

$$\frac{\Gamma \blacktriangleright E : SP_{any} \qquad \Gamma \blacktriangleright E' : SP_{any}}{\Gamma \blacktriangleright E' : \{E\}} \quad \begin{array}{l}[\![E]\!]_\rho = [\![E']\!]_\rho \text{ for each}\\ \Gamma\text{-environment } \rho\end{array}$$

Check that adding this rule does not violate the properties in Exercises 9.5.1 and 9.5.4. Check that all judgements in the extended system are preserved under β-reduction as well as well-formed β-expansion.

Note that η-reduction may change the semantics of a constructor by enlarging its domain: for a closed specification SP with $Mod[SP] \subset dom(F)$, $\lambda X{:}SP\bullet F(X)$ restricts the domain of F to $Mod[SP]$. Thus it would be unsound to derive (we omit decoration on the uses of F as a constructor expression) $\blacktriangleright F : \{\lambda X{:}SP\bullet F(X)\}$ from $\blacktriangleright \lambda X{:}SP\bullet F(X) : \{\lambda X{:}SP\bullet F(X)\}$ or vice versa, and the rule above does not permit this. □

Exercise 9.5.4. Show that the system of rules above is sound with respect to the semantics in Definition 6.4.11. That is, by induction on derivations in the entire system for all judgement forms in use, and relying on Exercise 9.5.1 above, define a notion of an environment for a well-formed context and prove the following:

- If Γ is the empty context then define any (partial) assignment of constructors to variables to be a Γ-environment. If $\Gamma, X{:}SP$ is well formed then define $\Gamma, X{:}SP$-environments to be all environments of the form $\rho[X \mapsto F]$ for Γ-environment ρ and $F \in [\![SP]\!]_\rho$. $[\![SP]\!]_\rho$ is defined by the next property, since we have $\Gamma \blacktriangleright SP : Spec(\mathscr{S})$ for a (unique) constructor signature $\mathscr{S}$. Moreover, all Γ-environments are $strip(\Gamma)$-environments in the sense of Definition 6.4.10.
- If $\Gamma \blacktriangleright SP : Spec(\mathscr{S})$ then for any Γ-environment ρ, $[\![SP]\!]_\rho$ is defined and $[\![SP]\!]_\rho \subseteq Mod(\mathscr{S})$.
- If $\Gamma \blacktriangleright SP \vdash \varphi$ then for any Γ-environment ρ, $[\![SP]\!]_\rho$ is defined, $[\![SP]\!]_\rho \subseteq Mod(\Sigma)$, and $[\![SP]\!]_\rho \models_\Sigma \varphi$, where $\Gamma \blacktriangleright SP : Spec(\Sigma)$.
- If $\Gamma \blacktriangleright SP \vdash SP'$ then for any Γ-environment ρ, $[\![SP]\!]_\rho$ and $[\![SP']\!]_\rho$ are defined, and $[\![SP]\!]_\rho \subseteq [\![SP']\!]_\rho \subseteq Mod(\mathscr{S})$, where $\Gamma \blacktriangleright SP : Spec(\mathscr{S})$ and $\Gamma \blacktriangleright SP' : Spec(\mathscr{S})$.
- If $\Gamma \blacktriangleright E : SP$ then for any Γ-environment ρ, $[\![E]\!]_\rho$ and $[\![SP]\!]_\rho$ are defined, and $[\![E]\!]_\rho \in [\![SP]\!]_\rho \subseteq Mod(\mathscr{S})$, where $\Gamma \blacktriangleright SP : Spec(\mathscr{S})$ and $strip(\Gamma) \rhd E : \mathscr{S}$.

The use of abbreviated notation for the semantics of specification and constructor expressions, e.g. $[\![SP]\!]_\rho$ rather than $[\![strip(\Gamma) \rhd SP : Spec(\mathscr{S})]\!]_\rho$, is justified by Exercise 6.4.7. □

This concludes the large definition/theorem that started at the beginning of the section.

Exercise 9.5.5. Extend the system above by adding parameterised specifications, following Definition 6.4.5, with no attempt to provide specifications for (parameterised) specifications other than specification types. This requires extending contexts by allowing variables to be mapped to specification types, and adding a new judgement form $\Gamma \blacktriangleright P : \mathscr{T}$. Explicit rules for β- and η-reduction and well-formed β- and η-expansion of expressions involving parameterised specifications will be required; they do not arise from the rule discussed in Exercise 9.5.3, and for instance the judgements $\Gamma \blacktriangleright E : (\lambda X{:}\mathscr{T}\bullet P)(P')$ and $\Gamma \blacktriangleright E : P[P'/X]$ are independent of each other (assuming well-formedness of $(\lambda X{:}\mathscr{T}\bullet P)(P')$ in the context Γ). □

Exercise 9.5.6. Try extending the system obtained in the last exercise by dropping the stratification between constructor signatures and specification types, as suggested in Exercise 6.4.20. □

Example 9.5.7. Suppose that the following pushout diagram in **Sign** admits amalgamation:

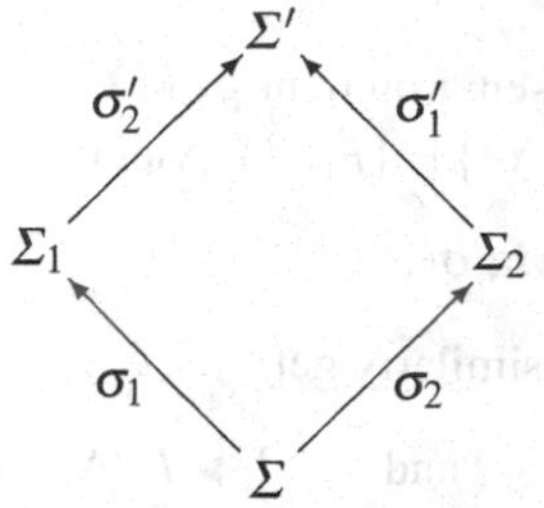

Note that the specification for the amalgamated union constructor

$$\blacktriangleright \_ +_{\sigma_1,\sigma_2} \_ : \Pi X_1{:}\langle \Sigma_1, \varnothing \rangle \bullet \Pi X_2{:}(\{X_1\}\ \textbf{hide via}\ \sigma_1)\ \textbf{with}\ \sigma_2 \bullet$$
$$(\{X_1\}\ \textbf{with}\ \sigma_2') \cup (\{X_2\}\ \textbf{with}\ \sigma_1')$$

captures the expected requirement of sharing between its arguments. To verify any application, we have to check that this requirement is satisfied. A typical way to ensure this is by using persistent constructors to build these arguments from the common shared part, thus ensuring that they share this common part and so can be put together. Here is a schematic example which illustrates how this works.

Consider the following context Γ:

$$F_1{:}SP \overset{\sigma_1}{\Longrightarrow} SP_1, F_2{:}SP \overset{\sigma_2}{\Longrightarrow} SP_2, X{:}SP$$

where $\blacktriangleright SP : Spec(\Sigma)$, $\blacktriangleright SP_1 : Spec(\Sigma_1)$ and $\blacktriangleright SP_2 : Spec(\Sigma_2)$. We will show that in this context, $F_1(X) +_{\sigma_1,\sigma_2} F_2(X)$ is well formed and that the constructor $\lambda X{:}SP \bullet F_1(X) +_{\sigma_1,\sigma_2} F_2(X)$ is correct with respect to the expected specification.

Recall from Example 6.2.9 that $SP \overset{\sigma}{\Longrightarrow} SP'$ abbreviates $\Pi X{:}SP \bullet (\{X\}\ \textbf{with}\ \sigma) \cup SP'$. Hence, from $\Gamma \blacktriangleright F_1(X) : SP \overset{\sigma_1}{\Longrightarrow} SP_1$ we get

$$\Gamma \blacktriangleright F_1(X) : (\{X\}\ \textbf{with}\ \sigma_1) \cup SP_1$$

from which it follows that

$$\Gamma \blacktriangleright F_1(X) : \{X\}\ \textbf{with}\ \sigma_1 \qquad \text{and} \qquad \Gamma \blacktriangleright F_1(X) : SP_1.$$

The latter yields

$$\Gamma \blacktriangleright F_1(X) : \langle \Sigma_1, \varnothing \rangle$$

and so

$$\Gamma \blacktriangleright F_1(X) +_{\sigma_1,\sigma_2} \_ : \Pi X_2{:}(\{F_1(X)\}\ \textbf{hide via}\ \sigma_1)\ \textbf{with}\ \sigma_2 \bullet$$
$$(\{F_1(X)\}\ \textbf{with}\ \sigma_2') \cup (\{X_2\}\ \textbf{with}\ \sigma_1').$$

From the former we get

$$\Gamma \blacktriangleright F_1(X)|_{\sigma_1} : (\{X\}\ \textbf{with}\ \sigma_1)\ \textbf{hide via}\ \sigma_1$$

and since $\Gamma \blacktriangleright (\{X\}\ \mathbf{with}\ \sigma_1)\ \mathbf{hide\ via}\ \sigma_1 \vdash \{X\}$, we derive

$$\Gamma \blacktriangleright F_1(X)|_{\sigma_1} : \{X\}.$$

This implies that for every Γ-environment ρ, $[\![\{X\}]\!]_\rho = \{\rho(X)\} = [\![\{F_1(X)\}]\!]_\rho|_{\sigma_1}$, which together with $\Gamma \blacktriangleright \{F_1(X)\} \vdash \{F_1(X)\}$ yields

$$\Gamma \blacktriangleright \{X\} \vdash \{F_1(X)\}\ \mathbf{hide\ via}\ \sigma_1.$$

For the application of F_2, we similarly get

$$\Gamma \blacktriangleright F_2(X) : \{X\}\ \mathbf{with}\ \sigma_2 \qquad \text{and} \qquad \Gamma \blacktriangleright F_2(X) : SP_2.$$

The former entails now

$$\Gamma \blacktriangleright F_2(X) : (\{F_1(X)\}\ \mathbf{hide\ via}\ \sigma_1)\ \mathbf{with}\ \sigma_2.$$

Consequently

$$\Gamma \blacktriangleright F_1(X) +_{\sigma_1,\sigma_2} F_2(X) : (\{F_1(X)\}\ \mathbf{with}\ \sigma_2') \cup (\{F_2(X)\}\ \mathbf{with}\ \sigma_1')$$

from which we get

$$\Gamma \blacktriangleright F_1(X) +_{\sigma_1,\sigma_2} F_2(X) : (SP_1\ \mathbf{with}\ \sigma_2') \cup (SP_2\ \mathbf{with}\ \sigma_1'),$$

that is, using sum with explicit sharing as defined in Section 5.2,

$$\Gamma \blacktriangleright F_1(X) +_{\sigma_1,\sigma_2} F_2(X) : SP_1 +_{\sigma_1,\sigma_2} SP_2.$$

We also get

$$\Gamma \blacktriangleright F_1(X) +_{\sigma_1,\sigma_2} F_2(X) : (\{X\}\ \mathbf{with}\ \sigma_1)\ \mathbf{with}\ \sigma_2' \cup (\{X\}\ \mathbf{with}\ \sigma_2)\ \mathbf{with}\ \sigma_1'$$

and thus

$$\Gamma \blacktriangleright F_1(X) +_{\sigma_1,\sigma_2} F_2(X) : \{X\}\ \mathbf{with}\ \sigma_1;\sigma_2',$$

which gives

$$\Gamma \blacktriangleright F_1(X) +_{\sigma_1,\sigma_2} F_2(X) : (\{X\}\ \mathbf{with}\ \sigma_1;\sigma_2') \cup (SP_1 +_{\sigma_1,\sigma_2} SP_2).$$

Putting all this together, we conclude

$$F_1{:}SP \stackrel{\sigma_1}{\Longrightarrow} SP_1, F_2{:}SP \stackrel{\sigma_2}{\Longrightarrow} SP_2 \blacktriangleright \\ \lambda X{:}SP\bullet F_1(X) +_{\sigma_1,\sigma_2} F_2(X) : SP \xrightarrow{\sigma_1;\sigma_2'} SP_1 +_{\sigma_1,\sigma_2} SP_2.$$

Exercise. Identify the rules used in the above derivation. Some of them have not been given explicitly in this section but are inherited from Section 9.3 (e.g. rules for monotonicity of specification-building operations and rules corresponding to the algebraic laws in Section 5.6). □

Exercise 9.5.8. Recall the specifications and constructors introduced in Section 6.5. Proceeding similarly as in Example 9.5.7 above, show that

- $\blacktriangleright F_{\text{BUCKET}}$: CONSBUCKET
- F_{BUCKET}:CONSBUCKET $\blacktriangleright F_{\text{TABLE}}$: CONSTABLE
  (this requires you to spell out the definition of CONSTABLE)
- $\blacktriangleright F_{\text{USERSTRINGHASHTABLE}} : SP$
  where SP is an appropriate Π-specification. □

Exercise 9.5.9. Recall the specifications and constructors given in Examples 7.2.8 and 7.3.3. Write out specifications for the constructors *K4–K7*, adding appropriate sharing requirements for the arguments of *K4*. Prove the correctness of the constructors with respect to your specifications and then verify that

$$\blacktriangleright K4(K1(K2(K7(\mathsf{empty}))))(K5(K6(K7(\mathsf{empty})))) : \text{SORTONCE}.$$

Then, assuming that the constructor $K4'$ of Example 7.3.4 satisfies the specification that is implicit in the constructor implementation step in that example, write out specifications for C, C' and C'' of Example 7.3.7 and Exercise 7.3.8 and prove the correctness of the constructors with respect to your specifications.

Finally, try to provide proof rules that are appropriate for the notation for higher-order constructors in **FPL** introduced in Exercise 7.3.5. Make sure that the rules are sufficient to prove the correctness of the constructor $K4'$. □

9.6 Proving behavioural properties

The preceding sections in this chapter dealt with proof systems corresponding to the standard satisfaction relation between models and sentences. Chapter 8 introduced a more permissive interpretation of sentences and specifications whereby additional models are admitted provided they exhibit the same externally observable behaviour as that of the usual models. Along with such a change of interpretation of sentences and specifications comes the need for a corresponding change in proof systems.

We will proceed similarly as in the preceding sections, starting with a discussion of (behavioural) entailment between sentences, and then proceeding to entailment for behavioural specifications, and finally to entailment between behavioural specifications and correctness of behavioural implementations.

As in Chapter 8, we will abandon working in an arbitrary institution and instead consider the standard case of algebraic signatures and many-sorted total algebras, looking in particular at the institution **FOEQ** of first-order equational logic, sometimes restricting attention just to equations. We then move to the institution **FPL** in Section 9.6.5.

In this section we will refrain from explicitly introducing notation for the various levels of proof-theoretic entailment. Instead, we will study properties of the corresponding notions of semantic consequence, and leave it as an exercise for the reader to turn these properties into proof rules.

9.6.1 Behavioural consequence

In Definition 8.3.9 we introduced behavioural satisfaction between standard algebras and equations. This was extended to first-order equational sentences (without predicates) in Exercise 8.3.10. Recall that for an arbitrary signature Σ and set *OBS* of observable sorts in Σ, we write $A \models_{OBS} \varphi$ when a Σ-algebra A behaviourally satisfies (w.r.t. *OBS*) a Σ-sentence φ. (We extend this to classes of Σ-algebras as usual: $\mathscr{A} \models_{OBS} \varphi$ if $A \models_{OBS} \varphi$ for all $A \in \mathscr{A}$.) As with any satisfaction relation, this induces a natural semantic consequence relation.

Definition 9.6.1 (Behavioural consequence $\Phi \models_{OBS} \varphi$). For any signature Σ with observable sorts *OBS*, set Φ of first-order Σ-sentences, and first-order Σ-sentence φ, we define $\Phi \models_{OBS} \varphi$ to mean that for every Σ-algebra A, $A \models_{OBS} \Phi$ implies $A \models_{OBS} \varphi$. Equivalently, $\Phi \models_{OBS} \varphi$ iff $Mod_{OBS}(\langle\Sigma,\Phi\rangle) \models_{OBS} \varphi$; see Definition 8.3.11. □

Before we launch into a study of this notion of consequence, let us remark that we will later need a different relation for proving correctness of behavioural implementations, where the premises are interpreted literally and only the consequent is interpreted behaviourally, written $\Phi \Vdash_{OBS} \varphi$. We will study this relation in the second part of this section. Similar mixed relations will also be of interest in subsequent sections. The reason for this is that the methods we use to reduce behavioural consequence and correctness of behavioural implementations to simpler relations all yield relations of this kind.

Exercise 9.6.2. Show that behavioural consequence is not compact, even for equational logic: give a signature Σ with observable sorts *OBS*, an infinite set $\mathscr{E}$ of Σ-equations, and a Σ-equation e, such that $\mathscr{E} \models_{OBS} e$ but there is no finite subset $\mathscr{E}' \subseteq \mathscr{E}$ such that $\mathscr{E}' \models_{OBS} e$. Conclude that there can be no finitary sound and complete proof system for behavioural consequence, either in equational logic or in first-order logic. □

Exercise 9.6.3. Use Exercise 8.3.16 to show that $\Phi \models_{OBS} \varphi$ is not always closed under translation along signature morphisms, even for equational logic. Using Exercise 8.3.17, try to characterise the signature morphisms $\sigma : \Sigma \to \Sigma'$ with sets of observable sorts *OBS* and *OBS'* in Σ and Σ', respectively, such that $\mathscr{E} \models_{OBS} e$ implies $\sigma(\mathscr{E}) \models_{OBS'} \sigma(e)$ for any set $\mathscr{E}$ of Σ-equations and Σ-equation e. Would considering arbitrary first-order sentences change anything here? □

By Theorem 8.3.26 (and Exercise 8.3.27), $\Phi \models_{OBS} \varphi$ is the same as requiring that for every Σ-algebra A, if $A/{\approx_A} \models \Phi$ then $A/{\approx_A} \models \varphi$. This gives us one way of reducing the problem of proving behavioural consequence to that of proving ordinary entailments between sentences. Other such reductions will follow, and this will be our strategy in the development of proof techniques for behavioural consequence. One straightforward observation is that if $\Phi \models \varphi$ then $\Phi \models_{OBS} \varphi$, which means that all the proof techniques of Section 9.1 that apply to institutions **EQ** and **FOEQ** are sound here as well. The opposite implication fails in general.

Exercise 9.6.4. Give an example of a signature Σ with observable sorts OBS and Σ-equation e such that $\varnothing \models_{OBS} e$ but $\varnothing \not\models e$. Try to think of an example for which $OBS \neq \varnothing$. $\square$

This gap is easy to fill in the case of **EQ**, albeit with an infinitary proof rule.

Theorem 9.6.5. *Consider any signature Σ with observable sorts OBS, set $\mathscr{E}$ of Σ-equations and Σ-equation $\forall X \bullet t = t'$. Then $\mathscr{E} \models_{OBS} \forall X \bullet t = t'$ iff for all substitutions $\theta: X \to |T_\Sigma(X_{OBS})|$ and all observable contexts $C \in |T_\Sigma(X_{OBS} \uplus \{\square{:}s\})|_{OBS}$, $\mathscr{E} \models \forall X_{OBS} \bullet C[t[\theta]] = C[t'[\theta]]$.*

Proof. For the "only if" part, consider $A \in Mod(\mathscr{E})$. By Exercise 8.3.13, $A \models_{OBS} \mathscr{E}$, and hence $A \models_{OBS} \forall X \bullet t = t'$, i.e. (by Definition 8.3.9) for all valuations $v: X \to |\langle A \rangle_{OBS}|$ into the OBS-generated subalgebra of A, $t_A(v) \sim_A^{OBS} t'_A(v)$. Now, consider $v_{OBS}: X_{OBS} \to |A|$ and a substitution $\theta: X \to |T_\Sigma(X_{OBS})|$. Define $v_\theta: X \to |\langle A \rangle_{OBS}|$ by $v_\theta(x) = \theta(x)_A(v_{OBS})$. Then $(t[\theta])_A(v_{OBS}) = t_A(v_\theta) \sim_A^{OBS} t'_A(v_\theta) = (t'[\theta])_A(v_{OBS})$. Hence, for any observable context $C \in |T_\Sigma(X_{OBS} \uplus \{\square{:}s\})|_{OBS}$, $C_A^{v_{OBS}}[t_A(v_\theta)] = C_A^{v_{OBS}}[t'_A(v_\theta)]$. This yields $(C[t[\theta]])_A(v_{OBS}) = (C[t'[\theta]])_A(v_{OBS})$, which proves $A \models \forall X_{OBS} \bullet C[t[\theta]] = C[t'[\theta]]$.

For the "if" part, let $A \models_{OBS} \mathscr{E}$. Then $A/{\approx_A} \models \mathscr{E}$ by Theorem 8.3.26, and so for all substitutions $\theta: X \to |T_\Sigma(X_{OBS})|$ and all observable contexts $C \in |T_\Sigma(X_{OBS} \uplus \{\square{:}s\})|_{OBS}$, $A/{\approx_A} \models \forall X_{OBS} \bullet C[t[\theta]] = C[t'[\theta]]$. Since $A/{\approx_A}$ is OBS-generated, this implies that for all valuations $v: X \to |A/{\approx_A}|$ and for all observable contexts $C \in |T_\Sigma(X_{OBS} \uplus \{\square{:}s\})|_{OBS}$, $C^v_{A/\approx_A}[t_{A/\approx_A}(v)] = C^v_{A/\approx_A}[t'_{A/\approx_A}(v)]$. Hence $t_{A/\approx_A}(v) \sim^{OBS}_{A/\approx_A} t'_{A/\approx_A}(v)$. Now, since $A/{\approx_A}$ is fully abstract by Lemma 8.3.23, this yields $t_{A/\approx_A}(v) = t'_{A/\approx_A}(v)$ and so $A/{\approx_A} \models \forall X \bullet t = t'$; thus $A \models_{OBS} \forall X \bullet t = t'$ by Theorem 8.3.26. $\square$

The above theorem can be split into two equivalences, one dealing with restricting valuations to the OBS-generated part of algebras, and the other dealing with the use of contexts to determine indistinguishability.

Corollary 9.6.6. *Consider any signature Σ with observable sorts OBS, set Φ of first-order Σ-sentences and Σ-sentence of the form $\forall X_{OBS} \bullet \forall x{:}s \bullet \varphi$ for a sort $s \notin OBS$, where φ is a first-order Σ-formula with free variables in $X_{OBS} \cup \{x{:}s\}$. Then $\Phi \models_{OBS} \forall X_{OBS} \bullet \forall x{:}s \bullet \varphi$ iff for all terms $t \in |T_\Sigma(X_{OBS})|_s$, $\Phi \models_{OBS} \forall X_{OBS} \bullet \varphi[x \mapsto t]$. Similarly, $\Phi \models_{OBS} \forall X_{OBS} \bullet \exists x{:}s \bullet \varphi$ if (not iff in general!) for some term $t \in |T_\Sigma(X_{OBS})|_s$, $\Phi \models_{OBS} \forall X_{OBS} \bullet \varphi[x \mapsto t]$.* $\square$

Corollary 9.6.7. *Consider any signature Σ with observable sorts OBS, set $\mathscr{E}$ of Σ-equations and Σ-equation $\forall X_{OBS} \bullet t = t'$. Then $\mathscr{E} \models_{OBS} \forall X_{OBS} \bullet t = t'$ iff for all observable contexts $C \in |T_\Sigma(X_{OBS} \uplus \{\square{:}s\})|_{OBS}$, $\mathscr{E} \models \forall X_{OBS} \bullet C[t] = C[t']$.* $\square$

Exercise 9.6.8. Prove the above corollaries by extracting appropriate arguments from the proof of Theorem 9.6.5, and then prove the theorem as a direct consequence of these two corollaries, at least for the case where the set X is finite. $\square$

Exercise 9.6.9. Show that the extension of Theorem 9.6.5 as well as Corollary 9.6.7 to **FOEQ** fails: in general we cannot allow the set $\mathscr{E}$ to contain arbitrary first-order sentences; see Exercise 8.3.28. □

Corollary 9.6.6 justifies the use of induction when proving behavioural consequences.

Exercise 9.6.10. Formulate a proof rule to capture such inductive arguments. For inspiration, see Example 9.1.11 for an induction rule formulated in the context of sorts with value constructors in **FPL**. □

Corollary 9.6.7 justifies the use of *context induction* for proving behavioural consequences in equational logic.

Lemma 9.6.11 (Context induction). *Given a signature Σ with observable sorts OBS, let ξ be a property of observable contexts $C \in |T_\Sigma(X_{OBS} \uplus \{\square{:}s\})|_{OBS}$. If:*

- *ξ holds for the contexts $\square{:}s$ with $s \in OBS$; and*
- *for all operations $f{:}s_1 \times \cdots \times s_i \times \cdots \times s_n \to o$ with $o \in OBS$, contexts $C \in |T_\Sigma(X_{OBS} \uplus \{\square{:}s\})|_{s_i}$, and terms $t_1 \in |T_\Sigma(X_{OBS})|_{s_1}$, ..., $t_{i-1} \in |T_\Sigma(X_{OBS})|_{s_{1-1}}$, $t_{i+1} \in |T_\Sigma(X_{OBS})|_{s_{1+1}}$, ..., $t_n \in |T_\Sigma(X_{OBS})|_{s_n}$, whenever ξ holds for all observable contexts C' that are subterms of C, then ξ also holds for the context $f(t_1,\ldots,t_{i-1},C,t_{i+1},\ldots,t_n)$,*

then ξ holds for all observable contexts. □

Exercise 9.6.12. Justify the soundness of the above context induction principle by noticing that the subterm relation is a well-founded (Noetherian) ordering. □

Exercise 9.6.13. Show how context induction can be used to prove behavioural consequences in equational logic. HINT: Use Corollary 9.6.7, taking the property ξ of an observable context C to be $\mathscr{E} \models \forall X_{OBS} \bullet C[t] = C[t']$. □

Exercise 9.6.14. Generalise Corollary 9.6.7 to conditional equations with premises of observable sorts and show that the context induction principle can be used for proving behavioural consequences for such sentences as well. □

In spite of the last exercise, the proof techniques for behavioural consequence presented above do not apply much beyond equational logic. In particular (see Exercise 9.6.9) some of them are already unsound for **FOEQ**.

An alternative to techniques based on context induction is to explicitly exploit the definition of behavioural satisfaction (Definition 8.3.9) by encoding the indistinguishability predicate.

Definition 9.6.15 (Behavioural relativisation). For any algebraic signature Σ, let $\ulcorner\Sigma\urcorner$ be its extension to a first-order signature by the addition of a predicate symbol $\asymp_s{:}s \times s$ for each sort s in Σ. Then for each first-order formula φ, let $\ulcorner\varphi\urcorner$ denote the *behavioural relativisation* of φ, obtained by replacing equality by $\asymp$ and relativising quantifiers to the domain of $\asymp$ as follows:

- $\ulcorner t = t' \urcorner$ is $t \asymp_s t'$ for terms t, t' of sort s;
- $\ulcorner \varphi_1 \wedge \varphi_2 \urcorner$ is $\ulcorner \varphi_1 \urcorner \wedge \ulcorner \varphi_2 \urcorner$, and analogously for other connectives; and
- $\ulcorner \forall x{:}s \bullet \psi \urcorner$ is $\forall x{:}s \bullet x \asymp_s x \Rightarrow \ulcorner \psi \urcorner$ and $\ulcorner \exists x{:}s \bullet \psi \urcorner$ is $\exists x{:}s \bullet x \asymp_s x \wedge \ulcorner \psi \urcorner$.

For a set Φ of first-order Σ-sentences, $\ulcorner \Phi \urcorner$ is the set obtained by taking the behavioural relativisation of all of the sentences in Φ. □

For any signature $\Sigma = \langle S, \Omega \rangle$ with observable sorts OBS, each $\ulcorner \Sigma \urcorner$-model consists of a Σ-algebra A and a family of binary relations $\asymp_A = \langle \asymp_{A,s} \subseteq |A|_s \times |A|_s \rangle_{s \in S}$. We write such a $\ulcorner \Sigma \urcorner$-model as a pair $\langle A, \asymp_A \rangle$.

The following lemma follows directly from the definition of behavioural satisfaction; see Exercise 8.3.10.

Lemma 9.6.16. *For any signature Σ with observable sorts OBS, Σ-algebra A and first-order Σ-sentence φ, $A \models_{OBS} \varphi$ iff $\langle A, \approx_A^{OBS} \rangle \models \ulcorner \varphi \urcorner$, where $\approx_A^{OBS}$ is the partial behavioural indistinguishability congruence on A; see Definition 8.3.19.* □

The idea is now to axiomatise $\asymp$ so as to capture the partial behavioural indistinguishability congruence. The axiomatisation consists of five parts:

- $EQUIV(\asymp)$ states that $\asymp$ is a partial equivalence:

$$\bigwedge_{s \text{ in } \Sigma} (\forall x, y{:}s \bullet x \asymp_s y \Rightarrow y \asymp_s x) \wedge (\forall x, y, z{:}s \bullet x \asymp_s y \wedge y \asymp_s z \Rightarrow x \asymp_s z)$$

  This is a first-order $\ulcorner \Sigma \urcorner$-sentence provided that Σ has finitely many sorts.
- $CONG(\asymp)$ asserts the Σ-congruence property for $\asymp$:

$$\bigwedge_{f{:}s_1 \times \cdots \times s_n \to s \text{ in } \Sigma} \forall x_1, y_1{:}s_1, \cdots, x_n, y_n{:}s_n \bullet x_1 \asymp_{s_1} y_1 \wedge \cdots \wedge x_n \asymp_{s_n} y_n \Rightarrow f(x_1, \ldots, x_n) \asymp_s f(y_1, \ldots, y_n)$$

  This is a first-order $\ulcorner \Sigma \urcorner$-sentence provided that Σ has finitely many operations.
- $IDOBS(\asymp)$ states that $\asymp$ is identity on observable sorts:

$$\bigwedge_{s \in OBS} \forall x, y{:}s \bullet x \asymp_s y \Leftrightarrow x = y$$

  Again, this is a first-order $\ulcorner \Sigma \urcorner$-sentence provided that OBS is finite.
- $OBSREACH(\asymp)$ states that the domain of $\asymp$ is OBS-generated:

$$\bigwedge_{s \text{ in } \Sigma} \forall x{:}s \bullet x \asymp_s x \Leftrightarrow \bigvee_{\substack{X_{OBS} \text{ finite} \\ t \in |T_\Sigma(X_{OBS})|_s}} \exists X_{OBS} \bullet x = t$$

  This is a $\ulcorner \Sigma \urcorner$-sentence in the infinitary logic $L_{\omega_1 \omega}$ which extends first-order logic by allowing countably infinite conjunction and disjunction; see Exercise 4.1.14. It is not a first-order sentence because even for finite signatures Σ the disjunction involved is in general (countably) infinite: there may be infinitely many Σ-terms of some sort s, even when terms are considered up to renaming of variables. In some cases the set of terms involved may be finite (up to renaming of variables), and then we can rewrite the formula to limit the choice of variables X_{OBS} to variables from a pre-defined finite set, yielding a first-order $\ulcorner\Sigma\urcorner$-sentence.

- *INDIST*$(\asymp)$ links $\asymp$ with indistinguishability by observable contexts:

$$\bigwedge_{s \text{ in } \Sigma} \forall x,y{:}s \bullet x \asymp_s y \Leftrightarrow \left(x \asymp_s x \wedge y \asymp_s y \wedge \bigwedge_{\substack{X_{OBS} \text{ finite}, x,y \notin X_{OBS} \\ C \in |T_\Sigma(X_{OBS} \uplus \{\square{:}s\})|_{OBS}}} \forall X_{OBS} \bullet C[x] = C[y] \right)$$

  The same comments apply as for *OBSREACH*$(\asymp)$; here the source of infiniteness is the potentially infinite number of observable contexts for a given non-observable sort. Again, in some cases this set of contexts is finite (up to renaming of variables), and then this amounts to a first-order $\ulcorner\Sigma\urcorner$-sentence.

Lemma 9.6.17. *For any signature Σ with observable sorts OBS and $\ulcorner\Sigma\urcorner$-model $\langle A, \asymp_A \rangle$, $\langle A, \asymp_A \rangle$ satisfies the $\ulcorner\Sigma\urcorner$-sentence $EQUIV(\asymp) \wedge CONG(\asymp) \wedge IDOBS(\asymp) \wedge OBSREACH(\asymp) \wedge INDIST(\asymp)$ iff $\asymp_A$ is the partial indistinguishability congruence on A.* □

Exercise 9.6.18. Convince yourself that the sentences above directly capture the definition of $\approx_A^{OBS}$; this proves the lemma. □

Corollary 9.6.19. *For any signature Σ with observable sorts OBS and $\ulcorner\Sigma\urcorner$-model $\langle A, \asymp_A \rangle$, $\langle A, \asymp_A \rangle \models OBSREACH(\asymp) \wedge INDIST(\asymp)$ iff $\asymp_A$ is the partial indistinguishability congruence on A.*

Proof. Follows from Lemma 9.6.17, since

$$OBSREACH(\asymp), INDIST(\asymp) \models EQUIV(\asymp) \wedge CONG(\asymp) \wedge IDOBS(\asymp). \quad \square$$

Putting Lemma 9.6.16 and Corollary 9.6.19 together, we obtain:

Theorem 9.6.20. *Consider any signature Σ with observable sorts OBS, set Φ of first-order Σ-sentences and first-order Σ-sentence φ. Then $\Phi \models_{OBS} \varphi$ iff*

$$\ulcorner\Phi\urcorner, OBSREACH(\asymp), INDIST(\asymp) \models \ulcorner\varphi\urcorner. \quad \square$$

Example 9.6.21. In Example 2.7.7 we gave a specification of sets of natural numbers. Here is a variant of that specification, assuming a specification BOOL of booleans such that all models of BOOL are isomorphic to the standard model, and a specification NAT of natural numbers:

spec SETNAT =
  BOOL **and** NAT
  then
    ops $\varnothing$: *NatSet*
      $add: Nat \times NatSet \to NatSet$
      $\_\_ \in \_\_: Nat \times NatSet \to Bool$
    $\forall n, m: Nat, S: NatSet$
    • $n \in \varnothing = false$
    • $n \in add(n, S) = true$
    • $m \neq n \Rightarrow m \in add(n, S) = m \in S$

As discussed in Example 2.7.7, neither commutativity nor idempotency of *add* is a consequence of SETNAT. However, both of these properties are behavioural consequences of the axioms in SETNAT if we take *Bool* and *Nat* as the only observable sorts. We will prove commutativity using Theorem 9.6.20. Let $\Sigma = Sig[\text{SETNAT}]$ and $OBS = \{Bool, Nat\}$, and let Φ be the axioms of SETNAT, including those inherited from NAT and BOOL.

We have to show

$$\ulcorner\Phi\urcorner, OBSREACH(\asymp), INDIST(\asymp) \models \ulcorner\forall m, n{:}Nat, S{:}NatSet \bullet add(m, add(n, S)) = add(n, add(m, S))\urcorner$$

where $\ulcorner\forall m, n{:}Nat, S{:}NatSet \bullet add(m, add(n, S)) = add(n, add(m, S))\urcorner$ is

$$\forall m, n{:}Nat, S{:}NatSet \bullet (m \asymp_{Nat} m \wedge n \asymp_{Nat} n \wedge S \asymp_{NatSet} S) \Rightarrow add(m, add(n, S)) \asymp_{NatSet} add(n, add(m, S)).$$

So, assume $\ulcorner\Phi\urcorner$, $OBSREACH(\asymp)$, $INDIST(\asymp)$, $m \asymp_{Nat} m$, $n \asymp_{Nat} n$ and $S \asymp_{NatSet} S$. First note that by $OBSREACH(\asymp)$ and $INDIST(\asymp)$, $\asymp_{Nat}$ and $\asymp_{Bool}$ are the identities. It follows that $m \asymp_{Nat} m$ and $n \asymp_{Nat} n$ are superfluous. All observable Σ-contexts for *NatSet* are boolean combinations of terms of the form $n_0 \in add(n_1, \ldots add(n_k, \Box{:}NatSet)\ldots)$ where $k \geq 0$ and $n_0, n_1, \ldots, n_k$ are terms of sort *Nat*. Given that BOOL unambiguously determines the interpretation of the boolean connectives, by $INDIST(\asymp)$, $add(m, add(n, S)) \asymp_{NatSet} add(n, add(m, S))$ holds iff for all $k \geq 0$ and terms $n_0, n_1, \ldots, n_k$ of sort *Nat*,

$$n_0 \in add(n_1, \ldots add(n_k, add(m, add(n, S)))\ldots) = n_0 \in add(n_1, \ldots add(n_k, add(n, add(m, S)))\ldots).$$

By repeated use of the axioms in $\ulcorner\Phi\urcorner$ that link $\in$ and *add*, and relying on the fact that $\asymp_{Nat}$ and $\asymp_{Bool}$ are the identities as well as on the assumption $S \asymp_{NatSet} S$, one can easily show that

$$n_0 \in add(n_1, \ldots add(n_k, add(m, add(n, S)))\ldots) = true$$

iff

$$n_0 = n_1 \vee \cdots \vee n_0 = n_k \vee n_0 = m \vee n_0 = n \vee n_0 \in S = true,$$

and likewise (because disjunction is commutative) for

$n_0 \in add(n_1, \ldots add(n_k, add(n, add(m, S))) \ldots) = true.$

This completes the proof.

Exercise. Prove idempotency of *add* in a similar way. □

Exercise 9.6.22. Using higher-order logic, in which one can quantify over relations, give an alternative definition of $OBSREACH(\asymp)$ which states that $\asymp$ is defined on the least set that contains all values of observable sorts and is closed under the application of the operations. Similarly, give an alternative definition of $INDIST(\asymp)$ which states that $\asymp$ is the largest partial behavioural congruence satisfying $OBSREACH(\asymp)$. Redo Example 9.6.21 using these versions of $OBSREACH(\asymp)$ and $INDIST(\asymp)$. □

Exercise 9.6.23. For any Σ-algebra A, show that A is fully abstract iff $\langle A, =_{|A|}\rangle \models OBSREACH(\asymp) \wedge INDIST(\asymp)$. Combine this with Lemma 8.3.23, Theorem 8.3.26 and Exercise 8.3.27 to show that $\Phi \models_{OBS} \varphi$ iff $\Phi, OBSREACH(=), INDIST(=) \models \varphi$. Here, $OBSREACH(=)$ and $INDIST(=)$ are the infinitary equational formulae obtained by syntactically replacing $\asymp_s$ by $=$, for each sort s, in $OBSREACH(\asymp)$ and $INDIST(\asymp)$ respectively. □

As stated at the beginning of the section, we will also need a different notion of consequence, whereby the premises are interpreted literally instead of behaviourally.

Definition 9.6.24 (Behavioural consequence $\Phi \Vvdash_{OBS} \varphi$). For any signature Σ with observable sorts OBS, set Φ of first-order Σ-sentences, and first-order Σ-sentence φ, we define $\Phi \Vvdash_{OBS} \varphi$ to mean that for every Σ-algebra A, $A \models \Phi$ implies $A \models_{OBS} \varphi$. Equivalently, $\Phi \Vvdash_{OBS} \varphi$ iff $Mod[\langle\Sigma, \Phi\rangle] \models_{OBS} \varphi$. □

Proposition 9.6.25. *For any signature Σ with observable sorts OBS, set Φ of first-order Σ-sentences, and first-order Σ-sentence φ, if $\Phi \Vvdash_{OBS} \varphi$ then $\Phi \models_{OBS} \varphi$.*

Proof. For any Σ-algebra A, if $A \models_{OBS} \Phi$ then $A/{\approx_A^{OBS}} \models \Phi$ and so $A/{\approx_A^{OBS}} \models_{OBS} \varphi$. Since $A/{\approx_A^{OBS}}$ is fully abstract, we have $A/{\approx_A^{OBS}} \models \varphi$ and so $A \models_{OBS} \varphi$. □

Exercise 9.6.26. Show that a sufficient condition for the opposite implication is that $Mod(\Phi)$ is behaviourally closed (Definition 8.3.30): if $Mod(\Phi)$ is behaviourally closed then $\Phi \models_{OBS} \varphi$ implies $\Phi \Vvdash_{OBS} \varphi$ for any sentence φ. Conclude that the two notions of behavioural consequence introduced in Definitions 9.6.1 and 9.6.24 coincide for equational logic, which allows for the use of e.g. context induction (Lemma 9.6.11) for proving $\Phi \Vvdash_{OBS} \varphi$ in **EQ**. □

Exercise 9.6.27. Show that behavioural closure of $Mod(\Phi)$ is also a necessary condition for $\Phi \models_{OBS} \varphi$ and $\Phi \Vvdash_{OBS} \varphi$ to coincide for all φ. □

Exercise 9.6.28. Show that $\Vvdash_{OBS}$ is also distinct from the usual consequence relation, $\models$. Give an example of a set Φ of sentences and a sentence φ such that $\Phi \Vvdash_{OBS} \varphi$ but not $\Phi \models \varphi$, and another example of Φ and φ such that $\Phi \models \varphi$ but not $\Phi \Vvdash_{OBS} \varphi$. The former example can easily be given in equational logic; the latter requires a set Φ such that $Mod(\Phi)$ is not behaviourally closed. □

Although $\Vdash_{OBS}$ and $\models$ do not coincide in general, there is an important class of consequences for which they are the same.

Definition 9.6.29 (Observable sentence). Consider a signature Σ with set OBS of observable sorts. A Σ-sentence φ is *observable* if for some Σ-sentence ψ, φ is equivalent to $\ulcorner\psi\urcorner$ on the class of $\ulcorner\Sigma\urcorner$-models that satisfy $OBSREACH(\asymp) \wedge INDIST(\asymp)$, that is, $A \models \varphi$ iff $\langle A, \asymp_A\rangle \models \ulcorner\psi\urcorner$ for all $\ulcorner\Sigma\urcorner$-models $\langle A, \asymp_A\rangle$ such that $\langle A, \asymp_A\rangle \models OBSREACH(\asymp) \wedge INDIST(\asymp)$. □

Exercise 9.6.30. Show that all equations between terms of observable sorts that involve only variables of observable sorts are observable. Then show that observable first-order sentences are closed under propositional connectives. What about quantifiers? □

Proposition 9.6.31. *Let Σ be a signature with a set OBS of observable sorts. For any set Φ of observable Σ-sentences, $Mod_\Sigma(\Phi)$ is behaviourally closed and closed under behavioural expansion.*

Proof. It is enough to prove that for any observable Σ-sentence φ and Σ-algebra A, $A \models \varphi$ iff $A/{\approx_A^{OBS}} \models \varphi$. So, assume that φ is equivalent to $\ulcorner\psi\urcorner$ on $\ulcorner\Sigma\urcorner$-models that satisfy $OBSREACH(\asymp) \wedge INDIST(\asymp)$. Since by Corollary 9.6.19, $\langle A, \approx_A^{OBS}\rangle \models OBSREACH(\asymp) \wedge INDIST(\asymp)$, we have $A \models \varphi$ iff $\langle A, \approx_A^{OBS}\rangle \models \ulcorner\psi\urcorner$ iff $A \models_{OBS} \psi$ iff $A/{\approx_A^{OBS}} \models \psi$ iff $\langle A/{\approx_A^{OBS}}, =\rangle \models \ulcorner\psi\urcorner$ iff $A/{\approx_A^{OBS}} \models \varphi$. □

Consequently, observable sentences are preserved under behavioural equivalence, which justifies the use of the same terminology ("observable sentence") here and in Section 8.5.3.3.

Proposition 9.6.32. *A Σ-sentence φ is observable iff for all Σ-algebras A, $A \models \varphi$ is equivalent to $A \models_{OBS} \varphi$.*

Proof. The "only if" part follows by Proposition 9.6.31, since $A/{\approx_A^{OBS}} \models \varphi$ iff $A \models_{OBS} \varphi$. For the "if" part, suppose that for all Σ-algebras A, $A \models \varphi$ is equivalent to $A \models_{OBS} \varphi$. We show that φ is equivalent to $\ulcorner\varphi\urcorner$ on all $\ulcorner\Sigma\urcorner$-models that satisfy $OBSREACH(\asymp) \wedge INDIST(\asymp)$. Let $\langle A, \asymp_A\rangle$ be such a model; by Corollary 9.6.19, $\asymp_A$ coincides with $\approx_A^{OBS}$ and therefore we have $A \models \varphi$ iff $A \models_{OBS} \varphi$ iff $\langle A, \approx^{OBS}\rangle \models \ulcorner\varphi\urcorner$ iff $\langle A, \asymp_A\rangle \models \ulcorner\varphi\urcorner$. □

Corollary 9.6.33. *If all of the sentences in Φ are observable then $\Phi \models_{OBS} \phi$ is equivalent to $\Phi \Vdash_{OBS} \phi$.* □

There can be no finitary sound and complete proof system for $\Vdash_{OBS}$ because it is not compact, see Exercise 9.6.2, where a counterexample can be given using an infinite set Φ of observable sentences.

The analogue of Theorem 9.6.20 now takes the following form:

Theorem 9.6.34. *Consider any signature Σ with observable sorts OBS, set Φ of first-order Σ-sentences and first-order Σ-sentence φ. Then $\Phi \Vdash_{OBS} \varphi$ iff*

$$\Phi, OBSREACH(\asymp), INDIST(\asymp) \models \ulcorner\varphi\urcorner. \qquad \square$$

In many examples, the above axiomatisation of indistinguishability may be considerably simplified by eliminating redundancy in the infinite sets (of terms and contexts, respectively) involved in $OBSREACH(\asymp)$ and $INDIST(\asymp)$.

Exercise 9.6.35. Consider a signature Σ with observable sorts OBS and a set Φ of first-order Σ-sentences. Let $\mathbb{T}$ be a family of sets of terms with $\mathbb{T}_{X_{OBS},s} \subseteq |T_\Sigma(X_{OBS})|_s$ for each finite set X_{OBS} of observable variables and sort $s \notin OBS$. Define $OBSREACH_{\mathbb{T}}(\asymp)$ by limiting the infinite disjunction $OBSREACH(\asymp)$ to terms in $\mathbb{T}$ for sorts $s \notin OBS$, and explicitly requiring $\asymp$ to be total on sorts in OBS:

$$\left(\bigwedge_{s \in OBS} \forall x{:}s \bullet x \asymp_s x\right) \wedge \bigwedge_{\substack{s \text{ in } \Sigma,\\ s \notin OBS}} \forall x{:}s \bullet x \asymp_s x \Leftrightarrow \bigvee_{\substack{X_{OBS} \text{ finite}\\ t \in \mathbb{T}_{X_{OBS},s}}} \exists X_{OBS} \bullet x = t$$

Suppose that

$$\Phi, OBSREACH_{\mathbb{T}}(\asymp) \models \bigwedge_{f:s_1 \times \cdots \times s_n \to s \text{ in } \Sigma} \forall x_1{:}s_1, \cdots, x_n{:}s_n \bullet x_1 \asymp_{s_1} x_1 \wedge \cdots \wedge x_n \asymp_{s_n} x_n \Rightarrow f(x_1, \ldots, x_n) \asymp_s f(x_1, \ldots, x_n)$$

Under this assumption, prove that Lemma 9.6.17 still holds for $\ulcorner\Sigma\urcorner$-models $\langle A, \asymp_A\rangle$ such that $A \models \Phi$ if we replace $OBSREACH(\asymp)$ by $OBSREACH_{\mathbb{T}}(\asymp)$. □

Exercise 9.6.36. Consider a signature Σ with observable sorts OBS and a set Φ of first-order Σ-sentences. Let $\mathbb{C}$ be a family of sets of contexts with $\mathbb{C}_{X_{OBS},s} \subseteq |T_\Sigma(X_{OBS} \uplus \{\Box{:}s\})|_{OBS}$ for each finite set X_{OBS} of observable variables and sort $s \notin OBS$.

Define $INDIST_{\mathbb{C}}(\asymp)$ by limiting the infinite conjunction in $INDIST(\asymp)$ to contexts in $\mathbb{C}$ for sorts $s \notin OBS$, and explicitly requiring $\asymp$ to be the identity for sorts in OBS:

$$IDOBS(\asymp) \wedge \bigwedge_{\substack{s \text{ in } \Sigma,\\ s \notin OBS}} \forall x, y{:}s \bullet x \asymp_s y \Leftrightarrow \left(x \asymp_s x \wedge y \asymp_s y \wedge \bigwedge_{\substack{X_{OBS} \text{ finite}\\ C \in \mathbb{C}_{X_{OBS},s}}} \forall X_{OBS} \bullet C[x] = C[y] \right)$$

Suppose that

$$\Phi, OBSREACH(\asymp), INDIST_{\mathbb{C}}(\asymp) \models CONG(\asymp).$$

Under this assumption, prove that Lemma 9.6.17 still holds for $\ulcorner\Sigma\urcorner$-models $\langle A, \asymp_A\rangle$ such that $A \models \Phi$ if we replace $INDIST(\asymp)$ by $INDIST_{\mathbb{C}}(\asymp)$. □

Exercise 9.6.37. Use Exercises 9.6.35 and 9.6.36 to show that, under the assumptions in the exercises, we can replace $OBSREACH(\asymp)$ by $OBSREACH_{\mathbb{T}}(\asymp)$ and $INDIST(\asymp)$ by $INDIST_{\mathbb{C}}(\asymp)$ in Theorem 9.6.34. □

Example 9.6.38. We revisit Example 9.6.21, adding an operation for union of sets:

spec SETNAT$'$ =
  BOOL **and** NAT
  then
    ops $\varnothing$:*NatSet*
      add: $Nat \times NatSet \to NatSet$
      $\_\_\cup\_\_$: $NatSet \times NatSet \to NatSet$
      $\_\_\in\_\_$: $Nat \times NatSet \to Bool$
    $\forall n,m{:}Nat, S,S'{:}NatSet$
      • $n \in \varnothing = false$
      • $n \in add(n,S) = true$
      • $m \neq n \Rightarrow m \in add(n,S) = m \in S$
      • $\varnothing \cup S = S$
      • $add(n,S) \cup S' = add(n, S \cup S')$

Commutativity of $\cup$ is not a consequence of SETNAT$'$. However, it is a behavioural consequence of the axioms in SETNAT$'$ if we take *Bool* and *Nat* as the only observable sorts. One way of proving this is to use Theorem 9.6.34 directly, but Exercise 9.6.37 allows the proof to be simplified considerably. Let $\Sigma' = Sig[\text{SETNAT}']$ and $OBS = \{Bool, Nat\}$, and let Φ' be the axioms of SETNAT$'$, including those inherited from NAT and BOOL.

OBSREACH($\asymp$) can be simplified using Exercise 9.6.35. First we replace the conjuncts for the observable sorts *Bool* and *Nat* by $\varphi_{Bool} = \forall x{:}Bool \bullet x \asymp_{Bool} x$ and $\varphi_{Nat} = \forall x{:}Nat \bullet x \asymp_{Nat} x$, respectively. For *Bool*, this replaces

$$\begin{array}{l}\forall x{:}Bool \bullet x \asymp_{Bool} x \Leftrightarrow \\ \qquad (x = true \vee \ldots \vee \exists b{:}Bool \bullet x = b \vee \exists b{:}Bool \bullet x = \neg b \vee \ldots \\ \qquad\quad \vee \exists n{:}Nat \bullet x = (n \in \varnothing) \vee \exists n,m{:}Nat \bullet x = (n \in add(m,\varnothing)) \vee \ldots)\end{array}$$

and similarly for *Nat*. Next, and more significantly, we can take

$$\mathbb{T}_{NatSet} = \{\varnothing, add(n_1,\varnothing), \ldots, add(n_k, \ldots add(n_1,\varnothing)\ldots), \ldots\},$$

with variables $n_1, \ldots, n_k{:}Nat$. This excludes terms involving $\cup$ as well as the additional clutter of operations from NAT. We can now replace the conjunct for *NatSet* by

$$\begin{array}{l}\varphi_{NatSet} = \forall x{:}NatSet \bullet x \asymp_{NatSet} x \Leftrightarrow \\ \qquad (x = \varnothing \vee \exists n_1{:}Nat \bullet x = add(n_1,\varnothing) \vee \ldots \vee \\ \qquad\ \exists n_1,\ldots,n_k{:}Nat \bullet x = add(n_k, \ldots add(n_1,\varnothing)\ldots) \vee \ldots)\end{array}$$

where the disjunction refers to the terms in $\mathbb{T}_{NatSet}$. This follows from the fact that in the context of the axioms of SETNAT$'$, φ_{NatSet} implies that

$$\forall x_1,x_2{:}NatSet \bullet x_1 \asymp_{NatSet} x_1 \wedge x_2 \asymp_{NatSet} x_2 \Rightarrow (x_1 \cup x_2) \asymp_{NatSet} (x_1 \cup x_2).$$

To see this, note that we get $x_1 = add(n_k, \ldots add(n_1,\varnothing)\ldots)$ for some $n_1, \ldots, n_k$ from φ_{NatSet} and $x_1 \asymp_{NatSet} x_1$. Then, repeated use of the axioms for $\cup$ in SETNAT$'$ yields $x_1 \cup x_2 = add(n_1, \ldots add(n_k, x_2)\ldots)$. Using φ_{NatSet} again (in the opposite direction) yields $(x_1 \cup x_2) \asymp_{NatSet} (x_1 \cup x_2)$. This proves the condition in Exercise 9.6.35. Then

$OBSREACH_{\mathbb{T}}(\asymp) = \varphi_{Bool} \wedge \varphi_{Nat} \wedge \varphi_{NatSet}$.

$INDIST(\asymp)$ can be likewise simplified using Exercise 9.6.36. First we replace the conjuncts for the observable sorts *Bool* and *Nat* by $\psi_{Bool} = \forall x,y{:}Bool\bullet x \asymp_{Bool} y \Leftrightarrow x = y$ and $\psi_{Nat} = \forall x,y{:}Nat\bullet x \asymp_{Nat} y \Leftrightarrow x = y$. For *Bool*, this replaces a formula which involves all observable contexts for *Bool*, and similarly for *Nat*. Next, and more significantly, we can take

$$\mathbb{C}_{NatSet} = \{m \in \square{:}NatSet\}$$

for some variable *m*:*Nat*. This excludes infinitely many contexts which start by manipulating the set to which the context is applied, and only then test for membership, as well as the additional clutter of operations from NAT and BOOL. The conjunct of $INDIST(\asymp)$ for *NatSet* can now be replaced by

$$\psi_{NatSet} = \forall x,y{:}NatSet\bullet x \asymp_{NatSet} y \Leftrightarrow (x \asymp_{NatSet} x \wedge y \asymp_{NatSet} y \wedge \forall n{:}Nat\bullet (n \in x) = (n \in y)).$$

This follows from the fact that in the context of the axioms of SETNAT$'$ and $OBSREACH_{\mathbb{T}}(\asymp)$,

$$INDIST_{\mathbb{C}}(\asymp) = \psi_{Bool} \wedge \psi_{Nat} \wedge \psi_{NatSet}$$

implies the congruence property for $\asymp$, which we now show for each of the operations in Σ'.

For the operations in NAT and BOOL, the congruence property is trivial since $\asymp_{Bool}$ and $\asymp_{Nat}$ are then the identities. The remaining operations are $\varnothing$ (for which the congruence property follows from φ_{NatSet}), $\in$ (for which the congruence property follows directly from ψ_{NatSet} and the fact that $\asymp_{Nat}$ is the identity), *add* and $\cup$.

For *add*, we need to show that (for any x and y such that $x \asymp_{NatSet} x$ and $y \asymp_{NatSet} y$) if $\forall n{:}Nat\bullet (n \in x) = (n \in y)$ then $\forall n,m{:}Nat\bullet (n \in add(m,x)) = (n \in add(m,y))$, which follows from the axioms that link $\in$ and *add* in SETNAT$'$.

For $\cup$, we first show an auxiliary fact, namely that for any *x*:*NatSet* such that $x \asymp_{NatSet} x$, any *y*:*NatSet*, and any *n*:*Nat*, the axioms of SETNAT$'$ together with $OBSREACH_{\mathbb{T}}(\asymp)$ entail that

$$n \in (x \cup y) = true \quad \text{iff} \quad n \in x = true \text{ or } n \in y = true.$$

This follows since we have $x = add(n_k, \ldots add(n_1, \varnothing)\ldots)$ for some $n_1,\ldots,n_k$ from $x \asymp_{NatSet} x$ and $OBSREACH_{\mathbb{T}}(\asymp)$, and repeated use of the axioms for $\cup$ in SETNAT$'$ yields $x \cup y = add(n_1, \ldots add(n_k, y)\ldots)$. Then, by the axioms that link $\in$ and *add* in SETNAT$'$, $n \in (x \cup y) = true$ iff $n = n_1$ or $\cdots$ or $n = n_k$ or $n \in y = true$, which is equivalent to $n \in x = true$ or $n \in y = true$.

The congruence property for $\cup$ then follows easily. Suppose that $x \asymp_{NatSet} x'$ and $y \asymp_{NatSet} y'$, that is, for all *n*:*Nat*, $n \in x = true$ iff $n \in x' = true$, and likewise for y and y'. By the above fact, we get $n \in (x \cup y) = true$ iff $n \in (x' \cup y') = true$, and so by ψ_{NatSet}, $(x \cup y) \asymp_{NatSet} (x' \cup y')$. This concludes the proof of the condition in Exercise 9.6.36.

Now, to show $\Phi' \models_{OBS} \forall S,S'{:}NatSet\bullet S \cup S' = S' \cup S$, it is enough to show that

$$\Phi', OBSREACH_{\mathbb{T}}(\asymp), INDIST_{\mathbb{C}}(\asymp) \models \ulcorner \forall S, S'{:}NatSet \bullet S \cup S' = S' \cup S \urcorner$$

where $\ulcorner \forall S, S'{:}NatSet \bullet S \cup S' = S' \cup S \urcorner$ is

$$\forall S, S'{:}NatSet \bullet S \asymp_{NatSet} S \wedge S' \asymp_{NatSet} S' \Rightarrow (S \cup S') \asymp_{NatSet} (S' \cup S).$$

Assuming $S \asymp_{NatSet} S$ and $S' \asymp_{NatSet} S'$, by $INDIST_{\mathbb{C}}(\asymp)$, $(S \cup S') \asymp_{NatSet} (S' \cup S)$ is the same as $m \in (S \cup S') = true$ iff $m \in (S' \cup S) = true$ for all $m{:}Nat$. By the auxiliary fact above, both $m \in (S \cup S') = true$ and $m \in (S' \cup S) = true$ hold iff $m \in S = true$ or $m \in S' = true$, which completes the proof.

Exercise. Prove associativity and idempotency of $\cup$ in a similar way. □

Exercise 9.6.39. Following the pattern of the simplifications spelled out in Exercises 9.6.35, 9.6.36 and 9.6.37, modify Theorem 9.6.20 to state that $\Phi \models_{OBS} \varphi$ if and only if

$$\ulcorner \Phi \urcorner, OBSREACH_{\mathbb{T}}(\asymp), INDIST_{\mathbb{C}}(\asymp) \models \ulcorner \varphi \urcorner$$

under conditions analogous to those in Exercises 9.6.35 and 9.6.36. Check if this can be used to show that $\Phi \models_{OBS} \forall S, S'{:}NatSet \bullet S \cup S' = S' \cup S$. Independently of this, note that $\Phi \models_{OBS} \forall S, S'{:}NatSet \bullet S \cup S' = S' \cup S$ follows directly from the conclusion of Example 9.6.38 by Proposition 9.6.25. □

Exercise 9.6.40. Note that for any signature Σ with observable sorts OBS and $\ulcorner \Sigma \urcorner$-model $\langle A, \asymp_A \rangle$, $\langle A, \asymp_A \rangle \models EQUIV(\asymp) \wedge CONG(\asymp) \wedge IDOBS(\asymp) \wedge OBSREACH(\asymp)$ iff $\asymp_A$ is a behavioural congruence on $\langle A \rangle_{OBS}$. Use Corollary 8.3.20 to show that if

$$\Phi, EQUIV(\asymp), IDOBS(\asymp), CONG(\asymp), OBSREACH(\asymp) \models \ulcorner e \urcorner$$

then $\Phi \Vdash_{OBS} e$ for any set Φ of first-order sentences and equation e. Informally, if we can prove that an equation holds up to a congruence that is finer than the behavioural indistinguishability relation, then it also holds up to behavioural indistinguishability. This argument breaks down if equality is used in a negative context, e.g. is negated or is the left-hand side of an implication.

Formalise the notion of negative and positive positions in a first-order formula, and use this to generalise the above result to first-order sentences ϕ where all equations used in a negative position are of observable sorts.

Check that the simplification spelled out in Exercise 9.6.35 can be applied here as well. □

9.6.2 Behavioural consequence for specifications

We now turn to reasoning about behavioural consequences for structured specifications.

Definition 9.6.41 (Behavioural consequence for specifications). We say that a Σ-sentence φ is a *behavioural consequence* of a Σ-specification SP *w.r.t.* OBS, written $SP \Vdash_{OBS} \varphi$, if $Mod[SP] \models_{OBS} \varphi$. □

As above, there are other possible definitions of behavioural consequence that differ from this one, and from each other.

Exercise 9.6.42. Define $SP \models_{OBS} \varphi$ by $Abs_{OBS}(Mod[SP]) \subseteq Abs_{OBS}(Mod[\langle\Sigma,\varphi\rangle])$, and $SP \Vvdash_{OBS} \varphi$ by $Abs_{OBS}(Mod[SP]) \models_{OBS} \varphi$. Show that these coincide with $SP \Vdash_{OBS} \varphi$ provided that $Mod[SP]$ and $Mod[\langle\Sigma,\varphi\rangle]$ are behaviourally closed, but may differ otherwise. □

Exercise 9.6.43. Note that, in general, $SP \models \varphi$ does not imply $SP \Vdash_{OBS} \varphi$. (Find a counterexample.) Show that the implication holds if the class of models of φ is behaviourally closed, for instance, when φ is an equation or conditional equation with premises of observable sorts. □

Compositional proofs of consequence for structured specifications make use of the rules given in Section 9.2 for deriving consequences of complex specifications from consequences of their constituents. We now examine the applicability of these rules for deriving behavioural consequences.

Flat specifications

By definition, $\langle\Sigma,\Phi\rangle \Vdash_{OBS} \varphi$ iff $\Phi \Vdash_{OBS} \varphi$, and then methods from Section 9.6.1 apply.

Union

It is easy to see that the rules for union from Section 9.2 carry over, that is, if $SP_1 \Vdash_{OBS} \varphi$ then $SP_1 \cup SP_2 \Vdash_{OBS} \varphi$ for all Σ-specifications SP_1, SP_2 and Σ-sentences φ, and likewise when $SP_2 \Vdash_{OBS} \varphi$.

Translation

The rule for translation from Section 9.2 does not carry over.

Exercise 9.6.44. Show that SP **with** $\sigma \Vdash_{\sigma(OBS)} \sigma(\varphi)$ does not necessarily hold even when $SP \Vdash_{OBS} \varphi$. HINT: Use Exercise 8.3.16. □

The essence of the problem is that for some signature morphisms, reducts of models and translation of sentences are not compatible with behavioural satisfaction. Given a signature morphism $\sigma:\Sigma \to \Sigma'$, Σ-specification SP, a set OBS of observable sorts in Σ and a Σ-equation e, a sufficient condition to ensure that $SP \Vdash_{OBS} e$ implies

SP **with** $\sigma \models_{\sigma(OBS)} \sigma(e)$ is to require that all observable contexts in Σ' for sorts in $\sigma(\Sigma)$ are observable contexts from Σ translated by σ. To generalise this to arbitrary first-order sentences, we would also have to assume that every Σ'-term of a sort in $\sigma(\Sigma)$ with variables in $\sigma(OBS)$ is a σ-translation of a Σ-term. Both requirements rarely hold in typical examples. Fortunately, they can be relaxed somewhat when the translation is immediately joined with another specification that links the new operations with the old ones, as is often the case when the translation arises from an application of enrichment.

Exercise 9.6.45. Consider $\sigma:\Sigma \to \Sigma'$, *SP* and *OBS* as above. Assume that we have a Σ'-specification SP' such that for every observable context $C' \in |T_{\Sigma'}(\sigma(X_{OBS}) \uplus \{\square{:}\sigma(s)\})|_{\sigma(o)}$ there exists $C \in |T_\Sigma(X_{OBS} \uplus \{\square{:}s\})|_o$ such that $SP' \cup (SP\ \mathbf{with}\ \sigma) \models \forall \square{:}\sigma(s) \bullet \forall \sigma(X_{OBS}) \bullet C' = \sigma(C)$. Here, s is a sort in Σ, X_{OBS} is a set of variables of observable sorts in Σ as usual, and $\sigma(X_{OBS})$ is the same set of variables of observable sorts with the sorts renamed by σ; for simplicity, we assume that X_{o_1} and X_{o_2} are disjoint for all distinct $o_1, o_2 \in OBS$. Show that this is a sufficient condition to ensure that for any Σ-equation e, $SP \models_{OBS} e$ implies $SP' \cup (SP\ \mathbf{with}\ \sigma) \models_{\sigma(OBS)} \sigma(e)$.

Show that the sufficient condition can be slightly weakened by limiting quantification over $\square{:}\sigma(s)$ to values generated from sorts in *OBS* by operations in Σ, replacing the universally quantified formula above with the infinitary conjunction

$$\bigwedge_{t \in |T_\Sigma(Y_{OBS})|} \forall \sigma(X_{OBS} \cup Y_{OBS}) \bullet C'[\sigma(t)] = \sigma(C[t]).$$

Try to refine the condition further to allow the choice of the context C to depend on the term t. Formulate a similar condition to guarantee that the implication holds for first-order Σ-sentences. Check that these conditions can be further simplified using the techniques of Exercises 9.6.35 and 9.6.36. HINT: Start by looking at Exercise 8.3.35. □

Exercise 9.6.46. Use Exercise 9.6.45 to formulate a rule for proving behavioural consequences for specifications of the form *SP* **then** *body'*. □

Hiding

In view of the difficulties with translation above, it is perhaps surprising that the rule for hiding from Section 9.2 carries over for equational logic. Given a signature morphism $\sigma:\Sigma \to \Sigma'$, a Σ'-specification SP', a set *OBS* of observable sorts in Σ and a Σ-equation e, $SP' \models_{\sigma(OBS)} \sigma(e)$ implies SP' **hide via** $\sigma \models_{OBS} e$ by Proposition 8.3.15. The same holds by Exercise 8.3.16 if e is a conditional equation with premises of observable sorts.

Exercise 9.6.47. Following Exercise 9.6.45, devise a sufficient condition which ensures that $SP' \models_{\sigma(OBS)} \sigma(\varphi)$ implies SP' **hide via** $\sigma \models_{OBS} \varphi$ for any first-order sentence φ. □

We can use normalisation (Theorem 5.6.10) to reduce any specification built from flat specifications using union, translation and hiding to an equivalent one of the form $\langle \Sigma, \Phi \rangle$ **hide via** σ. This gives a (non-compositional) proof system for structured specifications as in the discussion preceding Theorem 9.2.16. But, in contrast to the situation with ordinary consequence for specifications, the resulting proof system is not complete, since the hiding rule is not closed complete for behavioural consequence (Definition 9.2.2): in general, SP' **hide via** $\sigma \models\!\!\!\models_{OBS} \varphi$ does not imply that $SP' \models\!\!\!\models_{\sigma(OBS)} \sigma(\varphi)$, even when φ is a Σ-equation.

Exercise 9.6.48. Study conditions which ensure closed completeness of the rule for hiding for behavioural consequence, in the style of Exercise 9.6.45. □

Recall from Section 9.2 that the rules for particular specification-building operations had to be augmented by a rule called $(*)$ that made a connection between entailment for specifications and entailment in the underlying institution. The following implication plays the same role for behavioural consequence:

$$\frac{SP \models\!\!\!\models_{OBS} \varphi_1 \quad \cdots \quad SP \models\!\!\!\models_{OBS} \varphi_n \qquad \{\varphi_1, \ldots, \varphi_n\} \models_{OBS} \varphi}{SP \models\!\!\!\models_{OBS} \varphi} \qquad (**)$$

This refers to behavioural consequence of the form $\{\varphi_1, \ldots, \varphi_n\} \models_{OBS} \varphi$ rather than $\{\varphi_1, \ldots, \varphi_n\} \models\!\!\!\models_{OBS} \varphi$. The latter can also be used since it implies $\{\varphi_1, \ldots, \varphi_n\} \models_{OBS} \varphi$ by Proposition 9.6.25. In general the resulting implication would be weaker, but sometimes the two relations coincide, for instance when $Mod(\{\varphi_1, \ldots, \varphi_n\})$ is behaviourally closed (Exercise 9.6.26), or in particular when all of the sentences $\varphi_1, \ldots, \varphi_n$ are observable (Proposition 9.6.31). Note also that when φ_i is observable we can take $SP \models \varphi_i$ instead of $SP \models\!\!\!\models_{OBS} \varphi_i$, for any $i = 1, \ldots, n$. Similar remarks apply to the use of $\{\varphi_1, \ldots, \varphi_n\} \models \varphi$ instead of $\{\varphi_1, \ldots, \varphi_n\} \models_{OBS} \varphi$.

Exercise 9.6.49. Check that Corollary 9.6.6 can be generalised to entailments of the form $SP \models\!\!\!\models_{OBS} \forall X_{OBS} \bullet \forall x{:}s \bullet \varphi$ and $SP \models\!\!\!\models_{OBS} \forall X_{OBS} \bullet \exists x{:}s \bullet \varphi$. As in Exercise 9.6.10, formulate a proof rule to capture the corresponding inductive arguments.

Using Example 8.4.14, prove the following somewhat more general reformulation of the above: for any signature Σ with set OBS of observable sorts, Σ-specification SP, Σ-sentence φ and set S of sorts in Σ that is disjoint from OBS, $SP \models\!\!\!\models_{OBS} \varphi$ if $R_S(SP) \models\!\!\!\models_{OBS} \varphi$. (Here R_S is the specification-building operation corresponding to the constructor that restricts any Σ-algebra to its subalgebra generated by its carriers of sorts not in S; see Example 6.1.13 and Exercise 6.1.15.)

Generalise this further to any behaviourally trivial constructor (Definition 8.4.10). □

Exercise 9.6.50. Show that the relativisation techniques for proving behavioural consequence, as captured by Theorem 9.6.34, apply here as well: $SP \models\!\!\!\models_{OBS} \varphi$ iff

$$(SP \textbf{ with } \iota) \cup \langle \ulcorner\Sigma\urcorner, \{OBSREACH(\asymp), INDIST(\asymp)\} \rangle \models \ulcorner\varphi\urcorner$$

where $\Sigma = Sig[SP]$ and $\iota{:}\, \Sigma \to \ulcorner\Sigma\urcorner$ is the signature inclusion.

Furthermore, the simplifications spelled out in Exercises 9.6.35, 9.6.36 and 9.6.37 can be adapted for use here as well to replace $OBSREACH(\asymp)$ and/or $INDIST(\asymp)$ by $OBSREACH_{\mathbb{T}}(\asymp)$ and $INDIST_{\mathbb{C}}(\asymp)$ respectively, under the conditions in Exercises 9.6.35 and 9.6.36, adjusted in the obvious way.

Repeat Example 9.6.38 to show that $\textsc{SetNat}' \Vdash_{\{Bool,Nat\}} \forall S,S'{:}NatSet \bullet S \cup S' = S' \cup S$, viewing $\textsc{SetNat}'$ as a structured specification. □

9.6.3 Behavioural consequence between specifications

The next level of reasoning concerns behavioural consequence between structured specifications.

Definition 9.6.51 (Behavioural consequence between specifications). Given a signature Σ with a set OBS of observable sorts, we say that a Σ-specification SP' is a *behavioural consequence* of a Σ-specification SP *w.r.t.* OBS, written $SP \Vdash_{OBS} SP'$, if $Mod[SP] \subseteq Abs_{OBS}(Mod[SP'])$. □

The condition $Mod[SP] \subseteq Abs_{OBS}(Mod[SP'])$ is equivalent to $Abs_{OBS}(Mod[SP]) \subseteq Abs_{OBS}(Mod[SP'])$; see Proposition 8.2.19.

Although trivial, the following fact is probably the most useful in practice, allowing the use of methods developed in Section 9.3 for proving behavioural consequence between specifications.

Proposition 9.6.52. *Given any signature Σ with set OBS of observable sorts and Σ-specifications SP and SP', if $SP \vdash SP'$ then $SP \Vdash_{OBS} SP'$.* □

In general, reasoning about behavioural consequence between specifications is difficult. The rules in Section 9.3 carry over only under restrictive conditions.

Flat specifications

The rule for flat specifications from Section 9.3 carries over.

Proposition 9.6.53. *Given any signature Σ with set OBS of observable sorts, Σ-specification SP and Σ-sentences $\varphi_1,\ldots,\varphi_n$, if $SP \Vdash_{OBS} \varphi_1$ and ... and $SP \Vdash_{OBS} \varphi_n$ then $SP \Vdash_{OBS} \langle\Sigma,\{\varphi_1,\ldots,\varphi_n\}\rangle$.*

Proof. Let $A \in Mod[SP]$. Then, by the assumptions and Theorem 8.3.26 with Exercise 8.3.27, we get $A/{\approx_A^{OBS}} \in Mod[\langle\Sigma,\{\varphi_1,\ldots,\varphi_n\}\rangle]$. Hence, since $A \stackrel{w}{\equiv}_{OBS} A/{\approx_A^{OBS}}$ and $Mod[\langle\Sigma,\{\varphi_1,\ldots,\varphi_n\}\rangle]$ is closed under isomorphism, we conclude that $A \in Abs_{OBS}(Mod[\langle\Sigma,\{\varphi_1,\ldots,\varphi_n\}\rangle])$. □

Note that the argument in the proof above would not work if, instead of $SP \Vdash_{OBS} \varphi_i$, we had assumed $SP \models_{OBS} \varphi_i$; see Exercise 9.6.42.

Union

The rule for union from Section 9.3 carries over only under additional assumptions about the specifications involved.

Proposition 9.6.54. *For any signature Σ with set OBS of observable sorts and Σ-specifications SP, SP_1, SP_2, assume that $SP \models\!\!\models_{OBS} SP_1$, $SP \models\!\!\models_{OBS} SP_2$ and both $Mod[SP_1]$ and $Mod[SP_2]$ are behaviourally closed and closed under isomorphism. Then $SP \models\!\!\models_{OBS} SP_1 \cup SP_2$.*

Proof. Let $A \in Mod[SP]$. By the assumptions, there is $A_1 \in Mod[SP_1]$ such that $A \equiv_{OBS} A_1$. Since $Mod[SP_1]$ is behaviourally closed, $A_1/{\approx_{A_1}^{OBS}} \in Mod[SP_1]$. By Theorem 8.3.24, $A/{\approx_A^{OBS}}$ and $A_1/{\approx_{A_1}^{OBS}}$ are isomorphic, and so since $Mod[SP_1]$ is closed under isomorphism $A/{\approx_A^{OBS}} \in Mod[SP_1]$. Similarly, $A/{\approx_A^{OBS}} \in Mod[SP_2]$. Therefore $A/{\approx_A^{OBS}} \in Mod[SP_1 \cup SP_2]$, which yields $A \in Abs_{OBS}(Mod[SP_1 \cup SP_2])$. □

Exercise 9.6.55. Give a counterexample showing that the requirement that $Mod[SP_1]$ and $Mod[SP_2]$ be behaviourally closed in Proposition 9.6.54 cannot be dropped in general. □

Translation

The rule for translation from Section 9.3 carries over only under additional assumptions about the specifications involved.

Proposition 9.6.56. *For any signature Σ with set OBS of observable sorts, signature morphism $\sigma: \Sigma \to \Sigma'$, Σ-specification SP and Σ'-specification SP', whenever SP' **hide via** $\sigma \models\!\!\models_{OBS} SP$ then $SP' \models\!\!\models_{\sigma(OBS)} SP$ **with** σ provided that*

1. *$Mod[SP]$ is behaviourally closed and closed under isomorphism; and*
2. *for each $A' \in Mod[SP']$, $(A'|_\sigma)/{\approx_{A'|_\sigma}^{OBS}}$ and $(A'/{\approx_{A'}^{\sigma(OBS)}})|_\sigma$ are isomorphic.*

Proof. Let $A' \in Mod[SP']$. Then $A'|_\sigma \in Mod[SP'$ **hide via** $\sigma]$ and so by the assumptions, for some $A \in Mod[SP]$, $A'|_\sigma \equiv_{OBS} A$. It follows that $A'|_\sigma/{\approx_{A'|_\sigma}^{OBS}} \in Mod[SP]$. By (2), $(A'/{\approx_{A'}^{\sigma(OBS)}})|_\sigma \in Mod[SP]$ since $Mod[SP]$ is closed under isomorphism. Consequently, $A'/{\approx_{A'}^{\sigma(OBS)}} \in Mod[SP$ **with** $\sigma]$ and so $A' \in Abs_{\sigma(OBS)}(Mod[SP$ **with** $\sigma])$, since $Mod[SP$ **with** $\sigma]$ is closed under isomorphism whenever $Mod[SP]$ is. □

Exercise 9.6.57. Use the technique of Exercise 9.6.45 to replace requirement (2) in Proposition 9.6.56 above by a more proof-theoretic condition. □

Exercise 9.6.58. Give counterexamples showing that the implication in Proposition 9.6.56 may fail when either of the requirements does not hold. □

Hiding

The rule for hiding from Section 9.3 carries over only under additional assumptions about the specifications involved.

Proposition 9.6.59. *For any signature Σ with set OBS of observable sorts, signature morphism $\sigma: \Sigma \to \Sigma'$ which is injective on sorts, Σ-specifications SP and Σ'-specifications SP′ and $\widehat{SP}$, such that $\sigma: SP \to \widehat{SP}$ is a specification morphism admitting model expansion, if $\widehat{SP} \models_{\sigma(OBS)} SP'$ then $SP \models_{OBS} SP'$* **hide via** *σ provided that*

1. $Mod[SP']$ is behaviourally closed and closed under isomorphism; and
2. for each $\widehat{A} \in Mod[\widehat{SP}]$, $(\widehat{A}|_\sigma)/\approx^{OBS}_{\widehat{A}|_\sigma}$ and $(\widehat{A}/\approx^{\sigma(OBS)}_{\widehat{A}})|_\sigma$ are isomorphic.

Proof. Let $A \in Mod[SP]$. Let $\widehat{A} \in Mod[\widehat{SP}]$ be such that $\widehat{A}|_\sigma = A$. It follows that $\widehat{A}/\approx^{\sigma(OBS)}_{\widehat{A}} \in Mod[SP']$, and so $(\widehat{A}/\approx^{\sigma(OBS)}_{\widehat{A}})|_\sigma \in Mod[SP'$ **hide via** $\sigma]$. Since $\widehat{A}|_\sigma = A$ and the assumptions imply that $Mod[SP'$ **hide via** $\sigma]$ is closed under isomorphism, $A/\approx^{OBS}_A \in Mod[SP'$ **hide via** $\sigma]$ by (2), yielding $A \in Abs_{OBS}(Mod[SP'$ **hide via** $\sigma])$. □

The injectivity of σ is a technical assumption to ensure that $Mod[SP'$ **hide via** $\sigma]$ is closed under isomorphism. We could get rid of this assumption by explicitly closing the class of models of this specification under isomorphism; see Exercise 5.1.7. The requirement that $\sigma: SP \to \widehat{SP}$ admit model expansion is the same as in Section 9.3 and all comments there apply in the present context as well.

Exercise 9.6.60. Use the technique of Exercise 9.6.45 to replace requirement (2) in Proposition 9.6.59 above by a more proof-theoretic condition. □

Exercise 9.6.61. Give counterexamples showing that the implication in Proposition 9.6.59 may fail when either of the requirements does not hold. □

This completes the study of the rules for consequence between structured specification using our core specification-building operations.

Exercise 9.6.62. Section 9.3 lists a number of auxiliary rules for entailment between structured specifications. Check that at least the identity and transitivity rules and the rule for selection operations carry over to the current context as well. □

Exercise 9.6.63. Given a signature Σ with set *OBS* of observable sorts and behaviourally trivial constructor $F: \Sigma \Rightarrow \Sigma$ (Definition 8.4.10), show that for any Σ-specifications *SP* and *SP′*, $SP \models_{OBS} SP'$ if $F(SP) \models_{OBS} SP'$, where F is viewed as the specification-building operation defined by the constructor F; see Exercise 6.1.15.

As an example, take F to be $R_S: \Sigma \Rightarrow \Sigma$, the constructor that restricts any Σ-algebra to its subalgebra generated by its carriers of sorts not in S — see Example 6.1.13 — with a set S of sorts in Σ that is disjoint from *OBS*. □

The conditions on the implications in Propositions 9.6.54, 9.6.56 and 9.6.59 are quite restrictive in practice. In particular, the assumption that the class of models of a specification is behaviourally closed essentially requires the specification to be built in equational logic. Attempts at a proof-theoretic treatment of behavioural consequence between structured specifications have not been fruitful beyond this rather simple case. In the general case, we often need to resort to an essentially model-theoretic technique. Assuming as usual that *SP* and *SP'* are specifications over a signature Σ with a set *OBS* of observable sorts, to show that $SP \Vvdash_{OBS} SP'$, for each model $A \in Mod[SP]$ one has to exhibit a model $A' \in Mod[SP']$ together with a correspondence $\rho : A \bowtie_{OBS} A'$. This technique can be turned into a formal proof method by internalizing the semantic requirement above as a specification.

Theorem 9.6.64. *Let Σ be a finite signature with a set OBS of observable sorts. For any Σ-specification SP such that $Mod[SP]$ is closed under isomorphism, there is a Σ-specification $\ulcorner SP \urcorner$ such that $Mod[\ulcorner SP \urcorner] = Abs_{OBS}(Mod[SP])$.*

Proof. Let $i : \Sigma \to \Sigma'$ be an isomorphism such that all the symbols in Σ' are distinct from those in $\Sigma = \langle S, \Omega \rangle$. Let Θ be the first-order signature that extends $\Sigma \cup \Sigma'$ by a predicate symbol $p_s : s \times i(s)$ for each sort s in Σ. It is now easy to write out axioms which state that for each sort $s \in OBS$, p_s is a bijection and that the family of predicates $\langle p_s \rangle_{s \in S}$ is closed under the operations in Σ and Σ' respectively:

- *BIJOBS*(p) is

$$\bigwedge_{s \in OBS} \forall x{:}s \bullet \exists y{:}i(s) \bullet (p_s(x,y) \wedge \forall y'{:}i(s) \bullet p_s(x,y') \Rightarrow y = y') \wedge \\ \forall y{:}i(s) \bullet \exists x{:}s \bullet (p_s(x,y) \wedge \forall x'{:}s \bullet p_s(x',y) \Rightarrow x = x')$$

- *CORR*(p) is

$$\bigwedge_{f:s_1 \times \cdots \times s_n \to s \text{ in } \Sigma} \forall x_1{:}s_1, y_1{:}i(s_1), \cdots, x_n{:}s_n, y_n{:}i(s_n) \bullet p_{s_1}(x_1,y_1) \wedge \cdots \wedge p_{s_n}(x_n,y_n) \Rightarrow \\ p_s(f(x_1,\ldots,x_n), i(f)(y_1,\ldots,y_n))$$

Let $\iota : \Sigma \to \Theta$ and $\iota' : \Sigma' \to \Theta$ be the signature inclusions. Let $\ulcorner SP \urcorner$ be the following specification:

$$((SP \textbf{ with } \iota) \cup \langle \Theta, \{BIJOBS(p), CORR(p)\} \rangle) \textbf{ hide via } i;\iota'$$

Every model in $Mod[(SP \textbf{ with } \iota) \cup \langle \Theta, \{BIJOBS(p), CORR(p)\} \rangle]$ essentially consists of a model $A \in Mod[SP]$, an algebra $A' \in |\mathbf{Alg}(\Sigma')|$, and an interpretation of the predicates p_s which forms a (weak) *OBS*-correspondence between A and A'. Consequently, models of $\ulcorner SP \urcorner$ are Σ-algebras A' such that for some model $A \in Mod[SP]$ there exists a weak *OBS*-correspondence between A and A'. □

Exercise 9.6.65. $\ulcorner SP \urcorner$ as constructed above is a structured specification in **FOPEQ**, even if *SP* is an equational specification. Using Exercise 8.2.15, give an alternative construction of $\ulcorner SP \urcorner$ which is algebraic in the sense that it does not use signatures containing predicates, and involves only conditional equations as additional axioms. □

Corollary 9.6.66. *Let Σ be a signature with a set OBS of observable sorts. For any Σ-specifications SP and SP′ such that Mod[SP] is closed under isomorphism, $SP' \Vdash_{OBS} SP$ iff $SP' \models \ulcorner SP \urcorner$ where $\ulcorner SP \urcorner$ is given by Theorem 9.6.64.* □

Exercise 9.6.67. Convince yourself that the above corollary really captures the model-theoretic proof method discussed above by checking the following. To prove that $SP' \models \ulcorner SP \urcorner$, one exhibits a Θ-specification $\widehat{SP'}$ such that $i;\iota' : SP' \to \widehat{SP'}$ is a specification morphism that admits model expansion, and $\widehat{SP'} \vdash (SP \textbf{ with } \iota) \cup \langle \Theta, \{BIJOBS(p), CORR(p)\} \rangle$, using the notation of the proof of Theorem 9.6.64. To show that $i;\iota'$ admits model expansion, in practice $\widehat{SP'}$ has to be given as a definitional extension of SP' where for each model $A' \in Mod[SP']$ the additional explicit definitions define a model $A \in Mod[SP]$ and a weak *OBS*-correspondence between A and A'. □

9.6.4 Correctness of behavioural implementations

The motivation for the behavioural interpretation of specifications was the desire to permit more flexibility in developing programs from specifications. For that purpose, behavioural implementations were introduced in Section 8.4. The ultimate goal of behavioural reasoning is thus to prove the correctness of such implementations. The practical importance of the developments in Sections 9.6.1–9.6.3 is in the use of these methods to support such proofs.

Let SP and SP' be specifications, OBS' be a set of observable sorts in $Sig[SP']$, and $\kappa : Sig[SP] \Rightarrow Sig[SP']$ be a constructor. Using a similar notation as in the preceding sections, we write $SP \Vdash_{OBS',\kappa} SP'$ whenever SP' is behaviourally implemented by SP via κ w.r.t. OBS', that is if $Mod[SP] \subseteq dom(\kappa)$ and $\kappa(Mod[SP]) \subseteq Abs_{OBS'}(Mod[SP'])$; see Definition 8.4.2.

Constructor implementations are behavioural implementations, which justifies the following easy fact.

Proposition 9.6.68. *Let SP and SP′ be specifications, OBS′ be a set of observable sorts in Sig[SP′], and $\kappa : Sig[SP] \Rightarrow Sig[SP']$ be a constructor. If $SP \vdash_\kappa SP'$ then $SP \Vdash_{OBS',\kappa} SP'$.* □

Although trivial, this deals with the most frequent case encountered in practice.

The basic method of proving correctness of constructor implementations in Section 9.4 was to reduce the problem to entailment between specifications. This carries over to the present context as well.

Proposition 9.6.69. *Let SP and SP′ be specifications, OBS′ be a set of observable sorts in Sig[SP′], and $\kappa : Sig[SP] \Rightarrow Sig[SP']$ be a constructor. If $Mod[SP] \subseteq dom(\kappa)$ and $\kappa(SP) \Vdash_{OBS'} SP'$ then $SP \Vdash_{OBS',\kappa} SP'$. Here, in $\kappa(SP)$, the constructor κ is regarded as a specification-building operation; see Exercise 6.1.15.* □

Exercise 9.6.70. Show that the rule from Exercise 9.4.2 concerning the reduct constructor adapts directly to the context of behavioural implementations. □

The weakening and transitivity rules from Section 9.4 carry over in the expected way, assuming stability of constructors (see Definition 8.4.4) where appropriate.

Exercise 9.6.71. Check that the stability assumption cannot be dropped in any of the propositions below. □

Proposition 9.6.72. *Let SP, SP', SP_1 and SP_2 be specifications such that $Sig[SP] = Sig[SP_1]$ and $Sig[SP'] = Sig[SP_2]$, let OBS be a set of observable sorts in $Sig[SP]$ and OBS' be a set of observable sorts in $Sig[SP']$, and let $\kappa\colon Sig[SP] \Rightarrow Sig[SP']$ be a constructor that is stable on $Mod[SP]$ w.r.t. OBS and OBS'. If $SP \models_{OBS',\kappa} SP'$, $SP_1 \models_{OBS} SP$ and $SP' \models_{OBS'} SP_2$ then $SP_1 \models_{OBS',\kappa} SP_2$.* □

Proposition 9.6.73. *Let SP, SP' and SP'' be specifications, let OBS' be a set of observable sorts in $Sig[SP']$ and OBS'' be a set of observable sorts in $Sig[SP'']$, and let $\kappa\colon Sig[SP] \Rightarrow Sig[SP']$ be a constructor and $\kappa'\colon Sig[SP'] \Rightarrow Sig[SP'']$ be a constructor that is stable on $Mod[SP']$ w.r.t. OBS' and OBS''. If $SP \models_{OBS',\kappa} SP'$ and $SP' \models_{OBS'',\kappa'} SP''$ then $SP \models_{OBS'',\kappa;\kappa'} SP''$.* □

Note that Proposition 9.6.73 is just vertical composability of behavioural implementations; see Theorem 8.4.6.

In Section 7.3, we have discussed the possibility of constructor implementation proceeding along the structure of a specification when the top-level specification-building operation in use arose from a constructor. For correctness of constructor implementations, the rule

$$\frac{SP \vdash SP'}{SP \vdash_{\kappa} \kappa(SP')} \quad Mod[SP] \subseteq dom(\kappa)$$

can be easily derived by monotonicity of κ regarded as a specification-building operation. For behavioural implementations, we have to additionally impose a stability requirement.

Proposition 9.6.74. *Let SP_1 and SP_2 be specifications such that $Sig[SP_1] = Sig[SP_2]$, let OBS be a set of observable sorts in $Sig[SP_1]$ and OBS' be a set of observable sorts in Σ', and let $\kappa\colon Sig[SP_1] \Rightarrow \Sigma'$ be a constructor that is stable on $Mod[SP_2]$ w.r.t. OBS and OBS'. If $SP_1 \models_{OBS} SP_2$ then $SP_1 \models_{OBS',\kappa} \kappa(SP_2)$.* □

Exercise 9.6.75. Let SP, SP' be specifications, let OBS be a set of observable sorts in $Sig[SP]$ and OBS' be a set of observable sorts in $Sig[SP']$, and let $\kappa\colon Sig[SP] \Rightarrow Sig[SP']$ be a constructor that is stable on $Mod[SP]$ w.r.t. OBS and OBS'. As in Exercise 9.6.63, consider a behaviourally trivial constructor $F\colon \Sigma \Rightarrow \Sigma$ (Definition 8.4.10) and show that $SP \models_{OBS',\kappa} SP'$ if $F(SP) \models_{OBS',\kappa} SP'$, where F is viewed as the specification-building operation defined by the constructor F; see Exercise 6.1.15.

As an example, take F to be R_S, the constructor that restricts any $Sig[SP]$-algebra to its subalgebra generated by its carriers of sorts not in S — see Example 6.1.13 — with a set S of sorts in $Sig[SP]$ that is disjoint from OBS. Put this together with

Exercise 9.6.63 to show that $SP \models_{OBS',\kappa} SP'$ if $R_{S'}(\kappa(R_S(SP))) \models_{OBS'} SP'$ where S' is a set of sorts in $Sig[SP']$ that is disjoint from OBS'. Link this with Exercise 8.4.15. □

Correctness of behavioural implementations as discussed here covers behavioural correctness of constructors — see Definition 8.4.34, via Exercise 8.4.35 — since $SP \models_{\iota(sorts(\Sigma))\cup OBS',F} SP'$ exactly captures $F(Mod[SP]) \subseteq Abs_{\iota(sorts(\Sigma))\cup OBS'}(SP')$ and $Mod[SP] \subseteq dom(F)$. It does not cover persistency or global stability, and so it does not yield a proof that a constructor is a behavioural model of a constructor specification; see Definition 8.4.36. However, as discussed in Section 8.4.5, these properties are often introduced by the programming language notation used to define constructors, and so are satisfied by all constructors rather than needing to be checked on a case-by-case basis. In particular, for **FPL**, these properties are satisfied by constructors defined using the notation of Example 6.1.9, under the (syntactic) conditions spelled out in Corollary 8.5.23.

Exercise 9.6.76. Given the comments above, adapt Exercises 8.4.38 and 8.4.39, Proposition 8.4.40, Lemma 8.4.42, Theorem 8.4.43, Corollary 8.4.45 and Proposition 8.4.47 to the current context. □

9.6.5 A larger example, revisited

We will now revisit the example of Section 8.5.2 and argue about correctness of the developments there somewhat more formally using the methods presented in the preceding sections. An immediate issue is that the example is in **FPL** while so far in Section 9.6 our attention has been restricted to institutions **EQ** and **FOEQ**. Given the machinery and terminology developed in Section 8.5.1, all of the preceding developments carry over straightforwardly with no special adaptation necessary, at least for the purposes of the example at hand. Some facts specific to the notation used for defining **FPL**-constructors are the only exception.

The arguments in Section 8.5.2 concerning translation, composition and amalgamation of constructors and their correctness were already rather formal, so there is no need to repeat them here, taking advantage of Exercise 9.6.76. What we will do, however, is to prove that constructors used in this example defined using the notation of Example 6.1.9 are indeed behaviourally correct, as argued in Section 8.5.2 using semantic reasoning. That is, we will outline the proofs of the following properties, where all of the specifications and constructors are as given in Section 8.5.2.

1. STRINGSET $\overset{\varnothing}{\underset{K_{HT}}{\rightsquigarrow}}$ SIMPLEBUCKETARRAY.
2. $\sigma_{Elem\mapsto Bucket}(A)$ is a behavioural model of

$$\text{SIMPLEBUCKET } \mathbf{and} \text{ STRINGKEY} \overset{\iota'_A}{\Longrightarrow} \text{SIMPLEBUCKETARRAY}$$

and so

$$\textsc{SimpleBucketArray} \overset{\varnothing}{\underset{\sigma_{Elem \mapsto Bucket}(A)}{\rightsquigarrow}} \textsc{SimpleBucket}\ \textbf{and}\ \textsc{StringKey}.$$

3. K_B is a behavioural model of

$$\textsc{LinearProbing}\ \textbf{and}\ \textsc{StringKey} \overset{\iota_B}{\Longrightarrow} \textsc{SimpleBucketPlus}$$

and so $\textsc{SimpleBucketPlus} \overset{\varnothing}{\underset{K_B}{\rightsquigarrow}} \textsc{LinearProbing}\ \textbf{and}\ \textsc{StringKey}$.

4. K_{LP} is a behavioural model of

$$\textsc{StringFiniteArray} \overset{\iota_{LP}}{\Longrightarrow} \textsc{LinearProbing}$$

and so $\textsc{LinearProbing} \overset{\varnothing}{\underset{K_{LP}}{\rightsquigarrow}} \textsc{StringFiniteArray}$.

5. $\sigma_{Elem \mapsto String}(A)$ is a behavioural model of

$$\textsc{String}\ \textbf{and}\ \textsc{Nat} \overset{\iota''_A}{\Longrightarrow} \textsc{StringFiniteArray}$$

and so $\textsc{StringFiniteArray} \overset{\varnothing}{\underset{\sigma_{Elem \mapsto String}(A)}{\rightsquigarrow}} \textsc{String}\ \textbf{and}\ \textsc{Nat}$.

Even though we take here the empty set of observable sorts, by Proposition 8.5.6 the sorts *String*, *Nat*, and *Bool*, which are in $\widehat{\varnothing}$, can be regarded as observable; we will do so throughout this section.

Some of these proofs will be based on the following facts which adapt the rules for such constructors from Section 9.4:

Proposition 9.6.77. *Let* $K{:}\mathsf{SIG} \Rightarrow \mathsf{SIG}'$ *be defined using the notation of Example 6.1.9. Recall that* K *is the composition of a free extension and a reduct,* $K = F_\iota;(\_|_\delta)$

$$\mathsf{SIG} \xrightarrow{\ \iota\ } \mathsf{SIG}_{intermediate} \xleftarrow{\ \delta\ } \mathsf{SIG}'$$

where ι *and* δ *are introduced by the notation in use. Then for any* SIG*-specification* SP *and* SIG'*-specification* SP'*, and set* OBS' *of observable sorts in* SIG', $SP \models_{OBS',K} SP'$ *provided that*

$$(SP\ \textbf{with}\ \iota)\ \textbf{hide via}\ \delta \models_{OBS'} SP'.$$

Proof. Given a model $M \in Mod[SP]$, $K(M) \in Mod[(SP\ \textbf{with}\ \iota)\ \textbf{hide via}\ \delta]$ by the argument in Exercise 9.4.4. Hence, by the assumption, $K(M) \in Abs_{OBS'}(SP')$. □

Exercise 9.6.78. Under the notation of Proposition 9.6.77, assume that there is a signature inclusion $\iota'{:}\mathsf{SIG} \to \mathsf{SIG}'$ such that $\iota';\delta = \iota$. Show that K is then a behavioural model of $SP \overset{\iota'}{\Longrightarrow} SP'$ w.r.t. OBS' provided

$$(SP\ \textbf{with}\ \iota)\ \textbf{hide via}\ \delta \models_{\iota'(sorts(\mathsf{SIG})) \cup OBS'} SP'.$$ □

Exercise 9.6.79. Modify Proposition 9.6.77 to take advantage of the particular way that δ arises in the **FPL**-constructor notation of Example 6.1.9, as in Exercise 9.4.4. Namely, consider

constructor $K : \mathsf{SIG} \Rightarrow \mathsf{SIG}' = body$

and suppose that $\mathsf{SIG} \subseteq \mathsf{SIG}'$ (with the signature inclusion ι') and that *body* lists new sorts, each with a set of value constructors, and defines new operations. Show that, for any SIG-specification SP and SIG'-specification SP', and set OBS' of observable sorts in SIG', $SP \models_{OBS',K} SP'$ provided that

$$(SP\ \mathbf{then}\ body)\ \mathbf{reveal}\ \mathsf{SIG}' \models_{OBS'} SP'.$$

Following Exercise 9.6.78, strengthen the requirements so that they guarantee that K is a behavioural model of $SP \overset{\iota'}{\Longrightarrow} SP'$ w.r.t. OBS'. □

We begin by proving $\textsc{StringSet} \overset{\varnothing}{\underset{K_{HT}}{\rightsquigarrow}} \textsc{SimpleBucketArray}$. By Exercise 9.6.75, it is enough to prove

$$R_{Array[Bucket]}(\textsc{SimpleBucketArray}) \models_{\varnothing,K_{HT}} \textsc{StringSet}.$$

By Proposition 9.6.68, this follows from

$$R_{Array[Bucket]}(\textsc{SimpleBucketArray}) \vdash_{K_{HT}} \textsc{StringSet}$$

which can be derived using the techniques of Sections 9.4, following the semantic argument in Section 8.5.2 (in and around Exercise 8.5.25). Note that the application of the restriction constructor $R_{Array[Bucket]}$ ensures that Exercise 8.5.24 applies.

We now turn to the proof that $\sigma_{Elem \mapsto Bucket}(A)$ is a behavioural model of

$$\textsc{SimpleBucket}\ \mathbf{and}\ \textsc{StringKey} \overset{\iota'_A}{\Longrightarrow} \textsc{SimpleBucketArray}$$

where A is a behavioural model of $\textsc{Elem}$ **and** $\textsc{Nat} \overset{\iota_A}{\Longrightarrow} \textsc{ElemArray}$, and ι_A is the signature inclusion.

We first refine the specification for A by showing that any behavioural model of $\textsc{Elem}$ **and** $\textsc{Nat} \overset{\iota_A}{\Longrightarrow} \textsc{ElemArray}$ is a behavioural model of $\textsc{Elem}$ **and** $\textsc{Nat} \overset{\iota_A}{\Longrightarrow} \textsc{ElemArray}'$, where

spec $\textsc{ElemArray}' =$
  $\textsc{ElemArray}$
  then
    $\forall i,j{:}Nat, b,b'{:}Elem, a{:}Array[Elem]$
    • $put(i,b,put(i,b',a)) = put(i,b,a)$
    • $i \neq j \Rightarrow put(i,b,put(j,b',a)) = put(j,b',put(i,b,a))$

By Exercise 8.4.38, it is enough to show that $\textsc{ElemArray} \models_{\{Elem\}} \textsc{ElemArray}'$. $\textsc{ElemArray}'$ is the union of $\textsc{ElemArray}$ and a flat specification containing the two axioms in $\textsc{ElemArray}'$ above. Since the axioms involved in both of these specifications are equations or conditional equations with premises of observable sorts, their model classes are behaviourally closed, and therefore by Propositions 9.6.54 and 9.6.53 it is enough to prove:

- $\mathrm{ELEMARRAY} \models_{\{Elem\}} \mathrm{ELEMARRAY}$, which is trivial;
- $\mathrm{ELEMARRAY} \models_{\{Elem\}} \forall i{:}Nat, e, e'{:}Elem, a{:}Array[Elem] \bullet$
$put(i,e,put(i,e',a)) = put(i,e,a)$; and
- $\mathrm{ELEMARRAY} \models_{\{Elem\}} \forall i,j{:}Nat, e, e'{:}Elem, a{:}Array[Elem] \bullet$
$i \neq j \Rightarrow put(i,e,put(j,e',a)) = put(j,e',put(i,e,a))$.

The last two facts can be proved in the same style as illustrated in Example 9.6.38, with the small twist discussed in Exercise 9.6.50.

Exercise. Carry out the proofs. Take the simplified set of contexts

$$\mathbb{C} = \{used(i, \Box{:}Array[Elem]), get(i, \Box{:}Array[Elem])\}. \qquad \Box$$

This shows that $A \in Mod[\mathrm{ELEM}\ \mathbf{and}\ \mathrm{NAT}\ {}_{\varnothing}\!\overset{\iota_A}{\Longrightarrow}_{\varnothing} \mathrm{ELEMARRAY}']$. Now, the proof that $\sigma_{Elem\mapsto Bucket}(A)$ is a behavioural model of

$$\mathrm{SIMPLEBUCKET}\ \mathbf{and}\ \mathrm{STRINGKEY} \overset{\iota'_A}{\Longrightarrow} \mathrm{SIMPLEBUCKETARRAY}$$

relies on Corollary 8.4.45. The first requirement of the corollary holds here trivially. The second amounts to

$$\mathrm{ELEMARRAY}' +_{\iota_A, \sigma_{Elem\mapsto Bucket}} (\mathrm{SIMPLEBUCKET}\ \mathbf{and}\ \mathrm{STRINGKEY}) \models_{\varnothing} \mathrm{SIMPLEBUCKETARRAY}.$$

SIMPLEBUCKETARRAY has the form

$$\mathrm{SIMPLEBUCKET}\ \mathbf{and}\ \mathrm{STRINGKEY}\ \mathbf{then\ sorts}\ Array[Bucket]\ \mathbf{ops}\ \Omega_{SBA} \bullet \Phi_{SBA}$$

where Ω_{SBA} introduces the operations *empty*, *put*, *get* and *used*, and Φ_{SBA} lists their axioms. Luckily, all the axioms involved in SIMPLEBUCKETARRAY are equations or conditional equations with premises of observable sorts. Consequently, the class of models of SIMPLEBUCKET **and** STRINGKEY, translated to the signature $\mathsf{SIG}_{SBA} = Sig[\mathrm{SIMPLEBUCKETARRAY}]$, as well as the class of models of $\langle \mathsf{SIG}_{SBA}, \Phi_{SBA}\rangle$, are behaviourally closed, which allows for the use of Proposition 9.6.54. Therefore, it is enough to show that

$$\mathrm{ELEMARRAY}' +_{\iota_A, \sigma_{Elem\mapsto Bucket}} (\mathrm{SIMPLEBUCKET}\ \mathbf{and}\ \mathrm{STRINGKEY}) \models_{\varnothing} (\mathrm{SIMPLEBUCKET}\ \mathbf{and}\ \mathrm{STRINGKEY})\ \mathbf{with}\ \iota'_A$$

and

$$\mathrm{ELEMARRAY}' +_{\iota_A, \sigma_{Elem\mapsto Bucket}} (\mathrm{SIMPLEBUCKET}\ \mathbf{and}\ \mathrm{STRINGKEY}) \models_{\varnothing} \langle \mathsf{SIG}_{SBA}, \Phi_{SBA}\rangle.$$

The former requirement follows easily by Proposition 9.6.52 since we trivially have

$$\mathrm{ELEMARRAY}' +_{\iota_A, \sigma_{Elem\mapsto Bucket}} (\mathrm{SIMPLEBUCKET}\ \mathbf{and}\ \mathrm{STRINGKEY}) \vdash (\mathrm{SIMPLEBUCKET}\ \mathbf{and}\ \mathrm{STRINGKEY})\ \mathbf{with}\ \iota'_A$$

because $\mathrm{ELEMARRAY}' +_{\iota_A, \sigma_{Elem\mapsto Bucket}} (\mathrm{SIMPLEBUCKET}\ \mathbf{and}\ \mathrm{STRINGKEY})$ stands for

$$(\textsc{ElemArray}' \textbf{ with } \sigma'_{Elem\mapsto Bucket}) \cup ((\textsc{SimpleBucket} \textbf{ and } \textsc{StringKey}) \textbf{ with } \iota'_A)$$

(where $\sigma'_{Elem\mapsto Bucket}$ and ι'_A were defined in Section 8.5.2 as the pushout morphisms for ι_A and $\sigma_{Elem\mapsto Bucket}$).

For the latter requirement, let Φ be the set of all axioms in ELEMARRAY′, SIMPLEBUCKET and STRINGKEY (translated to SIG_{SBA} via the appropriate signature morphisms, which rename only sort names). We trivially have

$$\textsc{ElemArray}' +_{\iota_A,\sigma_{Elem\mapsto Bucket}} (\textsc{SimpleBucket} \textbf{ and } \textsc{StringKey}) \vdash \Phi.$$

Moreover, since all of the sentences in Φ are equations or conditional equations with premises of observable sorts, it follows by Exercise 9.6.43 that

$$\textsc{ElemArray}' +_{\iota_A,\sigma_{Elem\mapsto Bucket}} (\textsc{SimpleBucket} \textbf{ and } \textsc{StringKey}) \models_{\varnothing} \Phi.$$

By Proposition 9.6.53 and the implication $(**)$ of Section 9.6.2, to conclude the proof we have to show that $\Phi \models_{\varnothing} \varphi$ for all $\varphi \in \Phi_{SBA}$, but this is trivial since all of the axioms in Φ_{SBA} are included in Φ.

We now show that K_B is a behavioural model of

$$\textsc{LinearProbing} \textbf{ and } \textsc{StringKey} \overset{\iota_B}{\Longrightarrow} \textsc{SimpleBucketPlus}.$$

Recall that SIMPLEBUCKETPLUS is

$$\textsc{SimpleBucket} \textbf{ and } \textsc{LinearProbing} \textbf{ and } \textsc{StringKey}.$$

Given the form of the definition of K_B, by Exercise 9.6.78 it is enough to show that for $OBS = sorts(\textsc{LinearProbing} \textbf{ and } \textsc{StringKey})$

$$(\textsc{LinearProbing} \textbf{ and } \textsc{StringKey}) \textbf{ hide via } \delta \models_{OBS} \textsc{SimpleBucketPlus}$$

where δ is an **FPL**-signature morphism that maps *Bucket* to *Array*[*Elem*], *empty*, *add* and *present* to their defining terms as given in the body of K_B and is identity on the signature of LINEARPROBING **and** STRINGKEY. Even though not all of the axioms in the constituents of SIMPLEBUCKETPLUS are equations or conditional equations with premises of observable sorts — see, for instance, LINEARPROBING — the classes of models of these constituents are behaviourally closed with respect to *OBS*. Therefore, by Proposition 9.6.54 it is now enough to prove

$$(\textsc{LinearProbing} \textbf{ and } \textsc{StringKey}) \textbf{ hide via } \delta \models_{OBS} (\textsc{LinearProbing} \textbf{ and } \textsc{StringKey}) \textbf{ with } \iota_2$$

and

$$(\textsc{LinearProbing} \textbf{ and } \textsc{StringKey}) \textbf{ hide via } \delta \models_{OBS} \textsc{SimpleBucket} \textbf{ with } \iota_1$$

where ι_1 and ι_2 are the obvious **FPL**-signature inclusions.

The former is easy: since $\iota_2;\delta = id_{Sig[\text{LINEARPROBING and STRINGKEY}]}$, by the compositional rule for translation in Section 9.3, we get

$$(\textsc{LinearProbing}\ \textbf{and}\ \textsc{StringKey})\ \textbf{hide via}\ \delta \models (\textsc{LinearProbing}\ \textbf{and}\ \textsc{StringKey})\ \textbf{with}\ \iota_2.$$

By Proposition 9.6.52, this yields

$$(\textsc{LinearProbing}\ \textbf{and}\ \textsc{StringKey})\ \textbf{hide via}\ \delta \Vdash_{OBS} \textsc{LinearProbing}\ \textbf{and}\ \textsc{StringKey}\ \textbf{with}\ \iota_2.$$

As for the latter, first recall that SIMPLEBUCKET, and hence its translation by ι_1, is essentially a flat specification with the following axioms:

$$\Psi : \begin{cases} \forall s{:}String, b{:}Bucket \bullet def(empty) \wedge def(add(s,b)) \wedge def(present(s,b)) \\ \forall s{:}String \bullet present(s, empty) = false \\ \forall s{:}String, b{:}Bucket \bullet present(s, add(s,b)) = true \\ \forall s,s'{:}String, b{:}Bucket \bullet s \neq s' \Rightarrow (present(s, add(s',b)) = present(s,b)) \end{cases}$$
$$\alpha_1 : \forall s{:}String, b{:}Bucket \bullet add(s, add(s,b)) = add(s,b)$$
$$\alpha_2 : \forall s,s'{:}String, b{:}Bucket \bullet add(s, add(s',b)) = add(s', add(s,b))$$

By Proposition 9.6.53, it is enough to show that each of these axioms is a behavioural consequence of (LINEARPROBING **and** STRINGKEY) **hide via** δ w.r.t. *OBS*. Since for each ψ in Ψ we have LINEARPROBING **and** STRINGKEY $\models \delta(\psi)$ (**Exercise:** Prove this) we obtain

$$(\textsc{LinearProbing}\ \textbf{and}\ \textsc{StringKey})\ \textbf{hide via}\ \delta \models \psi.$$

By Exercise 9.6.43, this implies

$$(\textsc{LinearProbing}\ \textbf{and}\ \textsc{StringKey})\ \textbf{hide via}\ \delta \Vdash_{OBS} \Psi$$

since each $\psi \in \Psi$ is an equation or conditional equation with premises of observable sorts.

The remaining two axioms are translated by δ to equations between terms of a sort in *OBS* which do not follow from LINEARPROBING **and** STRINGKEY, and so a different approach is required here.

We will use Exercise 9.6.50 to obtain

$$(\textsc{LinearProbing}\ \textbf{and}\ \textsc{StringKey})\ \textbf{hide via}\ \delta \Vdash_{OBS} \alpha_1.$$

Since

$$(\textsc{LinearProbing}\ \textbf{and}\ \textsc{StringKey})\ \textbf{hide via}\ \delta \models \Psi,$$

it is enough to show

$$\Psi, OBSREACH(\asymp), INDIST_{\mathbb{C}}(\asymp) \models CONG(\asymp)$$

and

$$\Psi, OBSREACH(\asymp), INDIST_{\mathbb{C}}(\asymp) \models \ulcorner \alpha_1 \urcorner$$

where $\mathbb{C} = \{present(s, \Box{:}Bucket)\}$. The former follows easily. For the latter, first note that $\Psi \models \varphi_1$ where φ_1 is the sentence

$$\forall s, s'{:}String, b{:}Bucket \bullet present(s', add(s, add(s, b))) = present(s', add(s, b))$$

and then $\varphi_1, INDIST_{\mathbb{C}}(\asymp) \models \ulcorner \alpha_1 \urcorner$ (**Exercise:** Prove this).

The proof of

$$(\text{LINEARPROBING } \mathbf{and} \text{ STRINGKEY}) \ \mathbf{hide\ via}\ \delta \Vdash_{OBS} \alpha_2$$

follows the same pattern. This completes the proof of correctness of K_B.

We now show that K_{LP} is a behavioural model of

$$\text{STRINGFINITEARRAY} \overset{\iota_{LP}}{\Longrightarrow} \text{LINEARPROBING}.$$

For this we need to show that

$$\text{STRINGFINITEARRAY} \Vdash_{sorts(\text{STRINGFINITEARRAY}), K_{LP}} \text{LINEARPROBING}.$$

Indeed, K_{LP} is a model of STRINGFINITEARRAY $\Rightarrow$ LINEARPROBING, as indicated in Section 8.5.2 and earlier in Section 7.4.

Exercise. Use the techniques of Section 9.4, as illustrated in Example 9.4.5, to prove STRINGFINITEARRAY $\vdash_{K_{LP}}$ LINEARPROBING. □

An application of Proposition 9.6.68 now completes the proof of correctness of K_{LP}.

Finally, we show that $\sigma_{Elem \mapsto String}(A)$ is a behavioural model of

$$\text{STRING } \mathbf{and} \text{ NAT} \overset{\iota''_A}{\Longrightarrow} \text{STRINGFINITEARRAY}.$$

To begin with, using Corollary 8.4.45, we get that $\sigma_{Elem \mapsto String}(A)$ is a behavioural model of

$$\text{STRING } \mathbf{and} \text{ NAT} \overset{\iota''_A}{\Longrightarrow} \text{STRINGARRAY}.$$

(**Exercise:** Check that no difficulties arise in the proof.) By Exercise 8.4.38, it is now enough to show that STRINGARRAY $\Vdash_{\varnothing}$ STRINGFINITEARRAY. By Corollary 9.6.66, this follows if STRINGARRAY $\models \ulcorner$STRINGFINITEARRAY$\urcorner$, which can be proved using the techniques of Section 9.3, formalising the corresponding argument in Section 8.5.2. (**Exercise:** Carry out the proof.)

Exercise 9.6.80. Recall Exercise 8.5.30, concerning behavioural equivalence for higher-order constructors. Enrich the theory you developed there by adding an account of proofs, and use this to formalise the behavioural correctness arguments in the corresponding version of the example. □

9.7 Bibliographical remarks

The organization of this chapter is different from that of traditional work on proof systems because our emphasis is on dealing with the issues raised by structured specifications, and on proofs of relationships between such specifications. In logic and in nearly all work on automated reasoning, reasoning takes place in the context of a fixed and unstructured set of axioms, ignoring the problems covered in Sections 9.2–9.5. Our emphasis on the complications introduced by structured specifications means that we devote relatively little space to Section 9.1, which is the area where there is by far the largest quantity of related work. See [BCH99] for a presentation along similar lines which touches on many of the topics that we cover here.

The notion of entailment relation is rather standard; see [Avr91] (where the term is "consequence relation") for a survey. The way in which entailment relations are indexed by signatures to form an entailment system (Definition 9.1.2) is taken from [Mes89]; see also [FS88] and [HST94]. Specific entailment relations, and the engineering of proof systems that give rise to them, are a central focus of study in logic, with their efficient mechanisation a major topic in Artificial Intelligence; see for instance the *Journal of Automated Reasoning*. See also [Pad99]. We avoid any in-depth study of this topic here, omitting discussion of (semi-)decision procedures that support or automate reasoning within particular logical systems (for instance those based on the DPLL algorithm [DLL62] for propositional logic, or on resolution [BG01] or tableaux [Häh01] methods for propositional and first-order logics) and just provide rules and entailment systems for some specific institutions that will feature in later examples.

An exhaustive exposition of partial algebra, including rules for equational logic with existential equality ($\mathbf{PEQ}^e$) as hinted at in Exercise 9.1.5, is in [Bur86]; the rules for partial equational logic with strong equality (**PEQ**) in Example 9.1.4 may be derived from these. The problem of dealing with partial functions arises in most logics for reasoning about programs, and an alternative approach is via the use of three-valued logic; see for instance [KTB91] and [Fit08]. The rules for **PROP** in Example 9.1.6 are standard — see for instance [Sha08] — and similarly for the rules for **FOP** in Example 9.1.8. Example 9.1.11 relates to work on reasoning about functional programs; see for instance [BD77] and [Tho89]. For an entry point to the literature on formalisations of the notion of proof rule, see [HHP93].

The idea of generalising entailment from the form $\Phi \vdash \varphi$ to the form $SP \vdash \varphi$, replacing the set of assumptions on the left-hand side of the turnstile with a structured specification, is from [SB83], where the main topic was how proof search can take advantage of specification structure; see also [HST94]. Closely related to this are problems arising in theorem proving systems that have facilities for building a library of theories with dependencies between them; see [FGT92] and [MAH06]. The institution-independent compositional proof rules given near the beginning of Section 9.2 for translation, hiding and union are from [ST88a]. The completeness proof (Theorem 9.2.6) is from [Bor02], which is an institution-independent version of a proof for a **PFOPEQ**-like institution in [Cen94]. The notion of consequence in the context of specifications sketched in Exercise 9.2.11 comes from [Dia08], where the completeness result based on Craig-Robinson interpolation is given. Use of this form of interpolation in connection with Theorem 9.2.6 (Exercise 9.2.10) appears to be new. The suggestion to use the normal form theorem to give a proof system that is complete even for institutions in which interpolation fails (Theorem 9.2.16) is from [Wir93]. The disadvantage of this approach is that it completely flattens the structure of the specification in question. An intermediate possibility is to perform local transformations on the specification as in Exercises 9.2.18–9.2.21. Interpolation still seems to be required to achieve completeness in this approach, but this obstacle can be overcome via rules that "reach arbitrarily deeply" into the structure of specifications. A systematic approach along these lines is the development graphs of [MAH06]. Another non-compositional approach, which uses additional axioms and rules that are derived from the form of the specification in question, is [HWB97].

Entailment between specifications $SP \vdash SP'$ has received comparatively little attention in the literature, except for the simple case in which SP' is a flat specification, in which case the problem reduces immediately to repeated use of entailment for specifications $SP \vdash \varphi$. The problem of proving correctness of specification morphisms [Smi93] may be reduced to that of entailment between specifications, and the problem of proving correctness of simple implementations is equivalent since $SP \vdash SP'$ iff $SP' \rightsquigarrow SP$. The first works to deal in a systematic way with the case where SP' is a structured specification are [Far89], [Far90] and [Far92], which focused on specifications SP' of the form SP'' **hide via** σ, and, somewhat independently, [Wir93] followed by [Cen94]. The same basic ideas may be described using development graphs [MAH06]. Theorem 9.3.3 is from [Bor02], which is based on a proof in [Cen94]. Algebraic laws for transforming specifications to specifications with the same or a smaller class of models are given in [Wir82] and [SW83], with a systematic study in [BHK90]; see Exercise 9.3.9 and the preceding discussion for their relevance to proof of entailment between specifications.

Methods for proving correctness of implementation steps are an essential component of any comprehensive approach to formal development of software from specifications. Many implementation concepts (see Section 7.5 for an assortment of references) include some work on this topic, with [Gan83] giving a proof-oriented formulation of (behavioural) implementation. As discussed earlier, each of these approaches can be seen as a special case of our notion of constructor implementation, and so these methods address constructor entailment between specifications $SP \vdash_{\kappa} SP'$ for specific kinds of constructors κ. VDM [Jon80] provides an explicit recipe for verifying the correctness of implementation steps in the style of [Hoa72], where the constructor involves a composition of a restrict step and a quotient step. The general problem of proving correctness of constructor implementations is addressed in [Far89] and [Far92]. By definition, a constructor implementation $SP \underset{\kappa}{\rightsquigarrow} SP'$ is correct iff $\kappa \in Mod[SP' \Rightarrow SP]$, which makes a link to work on proving correctness of generic modules in Goguen's approach to parameterised programming [Tra93] and Extended ML [KST97].

The material in Section 9.5 on proving correctness of higher-order constructors with respect to constructor specifications is from [SST92], but we build on a stratified version (Section 6.4) of the system given there. The original system was investigated in [Asp97], and there is some related material in the COLD-K specification language [Jon89] — see [Fei89] — and in [Cen94], but neither of these contains Π-abstraction. Clearly, the overall approach and presentation here are strongly influenced by work on dependent type theories; see, e.g., [HHP93], [AH05].

Proof of behavioural properties (Section 9.6) is a topic of active research. Our main influences have been [ST87], [HS96], and especially [BH96] and [BH98]; see also proof techniques developed for Bidoit and Hennicker's constructor-based observational logic COL [BH06a], [BH06b]. Context induction (Lemma 9.6.11) is from [Hen91]. The circular coinduction proof method [GLR00] corresponds to context induction with an appropriately chosen context induction scheme; see [BH06a]. The idea hinted at in Exercise 9.6.22 is from [HS96]. Exercise 9.6.40 hints at coinductive techniques for proving indistinguishability; see [Gor95] and [JR97]. The idea of using correspondences as a method for proving behavioural equivalence in [Sch90] underlies the proof method expressed by Corollary 9.6.66. The combination of higher-order parameterisation and behavioural equivalence hinted at in Exercises 8.5.30 and 9.6.80 has not yet received attention in the literature. An interesting issue there concerns the consequences of global stability (which, via Exercise 8.4.33, is equivalent to what has been called "local stability" in the literature; see [BST08]), where additional properties may emerge when constructors are required to be globally stable; see [Pet10].

Chapter 10
Working with multiple logical systems

The preceding chapters dealt with various issues in the theory and methodology of software specification and development in the context of an arbitrary logical system formalised as an institution. As argued in the introduction to Chapter 4, the possibility of choosing the logical system most appropriate for the task at hand, without the need to rebuild the entire framework from scratch, is of considerable importance. Formulating definitions and proving theorems at this level of abstraction also lends crucial insight by tending to expose the issues of fundamental importance, uncluttered by extraneous details pertaining to the logical system at hand.

One advantage of this way of proceeding has not yet been exploited. Namely, all developments have been presented in the context of an arbitrary *but fixed* institution. However, it should also be possible to move smoothly between different logical systems and to work with a number of logical systems simultaneously. The purpose of this chapter is to examine some issues related to this idea.

The possibility of using more than one institution in a single specification or development task is potentially of great practical importance. Different institutions are often relevant to the description of different aspects of a system, or to the construction or description of its different parts. For example, we might want to combine the full expressive power of first-order logic with the "constructiveness" of equational logic under initial semantics. To take another example, we might want to use the framework of continuous algebras as a technical tool to describe infinitary objects like infinite lists or trees, but then forget the underlying ordering on carriers when dealing with non-continuous or even non-monotone operations. Perhaps most importantly, we want to implement systems in executable programming languages but specify them beforehand by means of some non-executable logic. We discuss this issue in Section 10.1 and propose mechanisms that enable the use of different institutions in the same specification and development process.

Using multiple institutions in the same task in a non-trivial way requires the institutions in question to be somehow related to each other. In Section 10.2 we present the notion of an *institution morphism* as an attempt to capture one way of comparing and relating institutions. This notion proves too rich and hence too restrictive for the purpose of using multiple institutions in a single specification or development task. Therefore, Section 10.1 is based on *institution semi-morphisms*, a simpler notion of maps between (parts of) institutions.

Institutions and their morphisms form a category, the basic properties of which are studied in Section 10.3. As an interesting side remark and a convenient technical tool, a more compact (albeit less transparent) definition of an institution is given. The category of institutions provides a rudimentary framework for putting institutions together: the limit construction may be used to combine a number of simpler logical systems related by institution morphisms, much as colimits in the category of specifications over any particular institution may be used for putting together simpler specifications to obtain more complex ones (cf. Sections 5.5 and 6.3). Although this is one possibility, it seems that it is more useful to combine logical systems at the level of their presentations; see Section 10.5 for hints on this.

There are many possible notions of a map between institutions, other than institution morphisms. We suggest a possible classification in Section 10.4, and focus there on one such notion, *institution comorphisms*, comparing institution comorphisms with institution morphisms and using them to move specifications and their consequences from one institution to another.

This chapter is not much more than a teaser: many of the topics discussed here are still areas of active ongoing research, and are far from being fully explored and from achieving much stability. Pointers to some recent related work are given in Section 10.5.

10.1 Moving specifications between institutions

The purpose of this section is to introduce some tools for using specifications (and in particular, sentences) of one institution to describe models of another institution. The key observation is that, of course, the two institutions involved must be somehow related to each other, but this relationship does not have to cover all of the institution components. Consistently with our overall view that the ultimate meaning of a specification is given by the class of its models (over its signature), in order to reinterpret specifications from one institution in another institution it is necessary and sufficient to relate the signatures and models of the two institutions. Intuitively, we will require that models of one institution **INS**, which is viewed as a richer, more detailed framework, be mapped to models of the other institution **INS**$'$, which is viewed as a more primitive, less detailed, and hence more abstract framework. It may help to think of this mapping as extracting simpler models belonging to **INS**$'$ from the more complex models belonging to **INS**. We will need a family of such mappings indexed by signatures, and again, signatures of **INS** will be mapped to signatures of **INS**$'$. Note, however, that this "gradation of detail" applies only to the semantic components required by signatures and provided by models, and not necessarily to the logical notation available to describe properties of models: we will not require any relationship between the sentences of the two institutions. This may be considered unsatisfactory for the purposes of relating institutions viewed as complete logical systems. In Section 10.2 we will introduce the notion of *institution morphism*, relating not just the signatures and models, but also the sentences of two institutions. For now, a simpler notion will be sufficient.

10.1.1 Institution semi-morphisms

Definition 10.1.1 (Institution semi-morphism). Consider two institutions $\mathbf{INS} = \langle \mathbf{Sign}, \mathbf{Sen}, \mathbf{Mod}, \langle \models_\Sigma \rangle_{\Sigma \in |\mathbf{Sign}|} \rangle$ and $\mathbf{INS}' = \langle \mathbf{Sign}', \mathbf{Sen}', \mathbf{Mod}', \langle \models'_{\Sigma'} \rangle_{\Sigma' \in |\mathbf{Sign}'|} \rangle$.

An *institution semi-morphism* $\mu \colon \mathbf{INS} \to_{semi} \mathbf{INS}'$ consists of:

- a functor $\mu^{Sign} \colon \mathbf{Sign} \to \mathbf{Sign}'$; and
- a natural transformation $\mu^{Mod} \colon \mathbf{Mod} \to (\mu^{Sign})^{op};\mathbf{Mod}'$,[1] that is, for each $\Sigma \in |\mathbf{Sign}|$, a functor $\mu^{Mod}_\Sigma \colon \mathbf{Mod}(\Sigma) \to \mathbf{Mod}'(\mu^{Sign}(\Sigma))$ such that the following diagram commutes for every $\sigma \colon \Sigma_1 \to \Sigma_2$ in $\mathbf{Sign}$:

$$
\begin{array}{ccccc}
\Sigma_2 & \quad & \mathbf{Mod}(\Sigma_2) & \xrightarrow{\mu^{Mod}_{\Sigma_2}} & \mathbf{Mod}'(\mu^{Sign}(\Sigma_2)) \\
\uparrow{\scriptstyle \sigma} & & \downarrow{\scriptstyle \mathbf{Mod}(\sigma)} & & \downarrow{\scriptstyle \mathbf{Mod}'(\mu^{Sign}(\sigma))} \\
\Sigma_1 & & \mathbf{Mod}(\Sigma_1) & \xrightarrow[\mu^{Mod}_{\Sigma_1}]{} & \mathbf{Mod}'(\mu^{Sign}(\Sigma_1))
\end{array}
$$

□

Let us start with a few simple examples of institution semi-morphisms, each of which leads from an obviously richer institution to a more primitive one. We will present their extensions to full institution morphisms, covering also translations of sentences, in Section 10.2.

Example 10.1.2. There is an obvious institution semi-morphism $\mu \colon \mathbf{FOPEQ} \to_{semi} \mathbf{EQ}$ mapping structures of first-order logic to algebras of equational logic (Examples 4.1.12 and 4.1.4), where:

- $\mu^{Sign} \colon \mathbf{FOSig} \to \mathbf{AlgSig}$ is the functor that maps any first-order signature $\Theta = \langle S, \Omega, \Pi \rangle \in |\mathbf{FOSig}|$ to the algebraic signature $\mu^{Sign}(\Theta) = \langle S, \Omega \rangle \in |\mathbf{AlgSig}|$, and any first-order signature morphism $\theta \colon \langle S, \Omega, \Pi \rangle \to \langle S', \Omega', \Pi' \rangle$ to the algebraic signature morphism $\mu^{Sign}(\theta) = \langle \theta_{sorts}, \theta_{ops} \rangle \colon \langle S, \Omega \rangle \to \langle S', \Omega' \rangle$, where $\theta = \langle \theta_{sorts}, \theta_{ops}, \theta_{preds} \rangle$.
- For each first-order signature $\Theta \in |\mathbf{FOSig}|$, $\mu^{Mod}_\Theta \colon \mathbf{FOStr}(\Theta) \to \mathbf{Alg}(\mu^{Sign}(\Theta))$ is the functor that maps any first-order Θ-structure $A \in |\mathbf{FOStr}(\Theta)|$ to the $\mu^{Sign}(\Theta)$-algebra $\mu^{Mod}_\Theta(A) \in |\mathbf{Alg}(\mu^{Sign}(\Theta))|$ having the same carrier sets and

---

[1] Recall that the functor $(\mu^{Sign})^{op} \colon \mathbf{Sign}^{op} \to (\mathbf{Sign}')^{op}$ is the "same" as $\mu^{Sign} \colon \mathbf{Sign} \to \mathbf{Sign}'$ but considered between the opposite categories (Example 3.4.5).

the same interpretation of operation names as A, ignoring the interpretation of predicate names. Then, any first-order Θ-morphism $h{:}A \to B$ is mapped to the $\mu^{Sign}(\Theta)$-homomorphism $\mu^{Mod}_{\Theta}(h){:}\mu^{Mod}_{\Theta}(A) \to \mu^{Mod}_{\Theta}(B)$ which is identical to $h{:}A \to B$ viewed as a family of maps between carrier sets.

It is easy to check that the family of functors $\mu^{Mod}_{\Theta}{:}\mathbf{FOStr}(\Theta) \to \mathbf{Alg}(\mu^{Sign}(\Theta))$, for $\Theta \in |\mathbf{FOSig}|$, forms a natural transformation $\mu^{Mod}{:}\mathbf{FOStr} \to (\mu^{Sign})^{op};\mathbf{Alg}$. □

Example 10.1.3. Another obvious institution semi-morphism is $\mu{:}\mathbf{CEQ} \to_{semi} \mathbf{EQ}$, mapping continuous algebras as models in the institution **CEQ** of equational logic for continuous algebras (Example 4.1.22) to algebras of equational logic:

- $\mu^{Sign}{:}\mathbf{Sign}_{\mathbf{CEQ}} \to \mathbf{Sign}_{\mathbf{EQ}}$ is the identity functor ($\mathbf{Sign}_{\mathbf{CEQ}} = \mathbf{Sign}_{\mathbf{EQ}} = \mathbf{AlgSig}$).
- For each algebraic signature $\Sigma \in |\mathbf{AlgSig}|$, $\mu^{Mod}_{\Sigma}{:}\mathbf{CAlg}(\Sigma) \to \mathbf{Alg}(\Sigma)$ is the functor that forgets the ordering on carriers: it maps any continuous Σ-algebra $A \in |\mathbf{CAlg}(\Sigma)|$ to the Σ-algebra $\mu^{Mod}_{\Sigma}(A) \in |\mathbf{Alg}(\Sigma)|$ whose carriers are the underlying sets of the cpo carriers of A and whose operations are the same functions as the operations of A. This mapping extends trivially to continuous homomorphisms.

The family of functors $\mu^{Mod}_{\Sigma}{:}\mathbf{CAlg}(\Sigma) \to \mathbf{Alg}(\Sigma)$, for $\Sigma \in |\mathbf{AlgSig}|$, forms a natural transformation $\mu^{Mod}{:}\mathbf{CAlg} \to \mathbf{Alg}$. □

Exercise 10.1.4. Define an institution semi-morphism from the institution **PFOPEQ** of partial first-order predicate logic with equality (Exercise 4.1.17) to the institution **PEQ** of partial equational logic (Example 4.1.6). Also, define an institution semi-morphism from the institution **3FOPEQ** of three-valued first-order predicate logic with equality (Example 4.1.24) to **PEQ**. In both cases, the translations of signatures would be the same as in Example 10.1.2, and the translations of models would be similar. □

Example 10.1.5. A more interesting (certainly less obvious) example of an institution semi-morphism arises from the idea of "totalisation" of partial algebras (Example 3.4.18).

Consider the institution $\mathbf{PEQ}_{\mathbf{str}}$ of partial equational logic with strong homomorphisms (Example 4.1.6 and Exercise 4.1.20) and the institution **EQ** of equational logic (Example 4.1.4). An institution semi-morphism $\mu{:}\mathbf{PEQ}_{\mathbf{str}} \to_{semi} \mathbf{EQ}$ may be defined as follows:

- $\mu^{Sign}{:}\mathbf{AlgSig} \to \mathbf{AlgSig}$ is the identity functor.
- For each algebraic signature $\Sigma \in |\mathbf{AlgSig}|$, $\mu^{Mod}_{\Sigma}{:}\mathbf{PAlg}_{\mathbf{str}}(\Sigma) \to \mathbf{Alg}(\Sigma)$ is the totalisation functor $\mathbf{Tot}_{\Sigma}$ defined in Example 3.4.18. □

Example 10.1.6. There is also a trivial institution semi-morphism $\mu{:}\mathbf{EQ} \to_{semi} \mathbf{PEQ}$ mapping total algebras to partial algebras:

- $\mu^{Sign}{:}\mathbf{AlgSig} \to \mathbf{AlgSig}$ is the identity functor.
- For each algebraic signature $\Sigma \in |\mathbf{AlgSig}|$, $\mu^{Mod}_{\Sigma}{:}\mathbf{Alg}(\Sigma) \to \mathbf{PAlg}(\Sigma)$ is the inclusion of the category of (total) Σ-algebras into the category of partial Σ-algebras.

Note that this also yields μ: **EQ** $\rightarrow_{semi}$ $\mathbf{PEQ_{str}}$. □

As we have already mentioned, each of the above examples of institution semi-morphisms will be extended in Section 10.2 to sentences. There are, however, situations where we can relate models of two institutions without any hope of relating their sentences. First, a trivial observation:

Example 10.1.7. Given two institutions **INS** and **INS**$'$ such that $\mathbf{Sign_{INS}} = \mathbf{Sign_{INS'}}$ and $\mathbf{Mod_{INS}} = \mathbf{Mod_{INS'}}$, there is a trivial institution semi-morphism from **INS** to **INS**$'$ (and vice versa) consisting of the identity functor on the common category of signatures and the identity natural transformation on the common model functor. The sentences of **INS** and **INS**$'$ need not be related in any way. Thus, in particular, such trivial institution semi-morphisms exist, in both directions, between **GEQ** and **EQ** (Examples 4.1.3 and 4.1.4), **PGEQ** and **PEQ** (Examples 4.1.7 and 4.1.6), and so on. For any institution **INS**, there are such trivial institution semi-morphisms, again in both directions, between **INS** and its closure under conjunction $\mathbf{INS}^{\wedge}$ (Example 4.1.38) and under other logical connectives $\mathbf{INS}^{prop}$ (Example 4.1.41), as well as its extensions by initiality constraints $\mathbf{INS}^{init}$ (Definition 4.3.5) and by data constraints $\mathbf{INS}^{data}$ (Exercise 4.3.10).

Moreover, if we then have an institution semi-morphism relating some institution **INS**$''$ with **INS**, the same institution semi-morphism relates **INS**$''$ with **INS**$'$. For example, the institution semi-morphism μ: **FOPEQ** $\rightarrow_{semi}$ **EQ** defined in Example 10.1.2 is also an institution semi-morphism from **FOPEQ** to **GEQ**, $\mathbf{EQ}^{\wedge}$, $\mathbf{EQ}^{data}$, and so on. □

Somewhat more surprisingly, there also exist institution semi-morphisms taking the signatures and models of an intuitively simpler institution to those of a richer one. Here again we cannot expect that an extension to a translation of sentences would be possible.

Example 10.1.8. There exists an institution semi-morphism μ: **EQ** $\rightarrow_{semi}$ **FOPEQ**, where μ^{Sign}: **AlgSig** $\rightarrow$ **FOSig** is the inclusion of the category of algebraic signatures into the category of first-order signatures (formally, adding to any algebraic signature the empty set of predicate names), and then μ^{Mod}: **Alg** $\rightarrow (\mu^{Sign})^{op}$;**FOStr** is the identity natural transformation (the category of first-order Σ-structures coincides with the category of Σ-algebras for any algebraic signature $\Sigma \in |\mathbf{AlgSig}| \subseteq |\mathbf{FOSig}|$).

Similarly, there is an institution semi-morphism μ: **PEQ** $\rightarrow_{semi}$ **PFOPEQ**, which again consists of the inclusion of signature categories and of the identity natural transformation on model functors. □

Perhaps the most important examples of institution semi-morphisms that cannot be extended to sentences arise when we relate institutions describing programming languages with the more usual logical systems of specification frameworks:

Example 10.1.9. Recall the institution $\mathbf{IMP}_{DT}$ of a simple imperative programming language with types over an algebra DT of built-in data types (Example 4.1.32). There is an institution semi-morphism μ: $\mathbf{IMP}_{DT} \rightarrow_{semi}$ **PEQ** from $\mathbf{IMP}_{DT}$ to the institution **PEQ** of partial equational logic:

- μ^{Sign}: $\mathbf{Sign_{IMP_{DT}}} \to \mathbf{AlgSig}$ maps any signature $\Pi \in |\mathbf{Sign_{IMP_{DT}}}|$ to the algebraic signature $\mu^{Sign}(\Pi) = \Sigma_{DT} \cup \Pi$, where Σ_{DT} is the algebraic signature of the algebra DT of built-in data types; $\mu^{Sign}(\Pi)$ was written as Π_{DT} in Example 4.1.32. This extends to signature morphisms in the obvious way: for any signature morphism $\sigma: \Pi \to \Pi'$ in $\mathbf{Sign_{IMP_{DT}}}$, $\mu^{Sign}(\sigma): \Sigma_{DT} \cup \Pi \to \Sigma_{DT} \cup \Pi'$ is the algebraic signature morphism that extends σ by the identity on Σ_{DT}.
- Consider a signature $\Pi = \langle T, P\rangle \in |\mathbf{Sign_{IMP_{DT}}}|$. Any model $M \in |\mathbf{Mod_{IMP_{DT}}}(\Pi)|$ determines a carrier $|M|_s$ for each sort s in $\mu^{Sign}(\Pi)$ ($|M|_s = |DT|_s$ if s is a sort in Σ_{DT}) and maps any procedure name $p: s_1, \ldots, s_n \to s$ in P to the set of possible computations of p on admissible arguments, which determines a partial function $p_M: |M|_{s_1} \times \cdots \times |M|_{s_n} \to |M|_s$. This allows us to define $\mu^{Mod}_{\Pi}(M) \in |\mathbf{PAlg}(\Sigma_{DT} \cup \Pi)|$ as the unique partial $(\Sigma_{DT} \cup \Pi)$-algebra such that $\mu^{Mod}_{\Pi}(M)|_{\Sigma_{DT}} = DT$, $|\mu^{Mod}_{\Pi}(M)|_s = |M|_s$ for each $s \in T$, and for each $p \in P$, $p_{\mu^{Mod}_{\Pi}(M)} = p_M$.

There is no natural translation of an arbitrary equation over $\Sigma_{DT} \cup \Pi$, where $\Pi \in |\mathbf{Sign_{IMP_{DT}}}|$, to type and procedure definitions (Π-sentences in $\mathbf{IMP}_{DT}$) which would determine exactly the functions that satisfy the equation; nor is it possible to replace an arbitrary set of type and procedure definitions by a set of algebraic equations. This shows that we cannot hope here for an extension of $\mu: \mathbf{IMP}_{DT} \to_{semi} \mathbf{PEQ}$ to sentences. □

Exercise 10.1.10. Recall the institution **FPL** of a simple functional language (Example 4.1.25) and the institution **FProg** which captures its "programming fragment" (Exercise 4.1.30).

Define an institution semi-morphism $\mu: \mathbf{FProg} \to_{semi} \mathbf{PEQ}$ from **FProg** to the institution **PEQ** of partial equational logic that extracts algebraic signatures out of **FPL**-signatures (omitting the sets of value constructors) and is just an inclusion on model categories. Again, we cannot expect any extension to sentences.

The above institution semi-morphism does not extend to an institution semi-morphism from **FPL**, where quite complex derived signature morphisms are used in the category of signatures. Define another institution $\mathbf{FPL_{stnd}}$, a version of **FPL** with signature morphisms limited to "standard" morphisms that map operation names essentially to operation names (as in **FProg**), and check that the above definition of $\mu: \mathbf{FProg} \to_{semi} \mathbf{PEQ}$ also yields an institution semi-morphism $\mu: \mathbf{FPL_{stnd}} \to_{semi} \mathbf{PEQ}$. □

As expected, institution semi-morphisms may be composed, and in fact, institutions with institution semi-morphisms form a complete category. We will not, however, treat this topic in detail here. All the relevant definitions and constructions are quite straightforward, and may be extracted from the corresponding parts of the study of institution morphisms in Sections 10.2 and 10.3.

Exercise 10.1.11. Define composition of institution semi-morphisms. Check that your definition coincides with the appropriate fragments of Definition 10.3.1 below. Prove that composition is associative and has identities, giving a category $\mathcal{INS}_{semi}$ of institutions with their semi-morphisms. Try to characterise isomorphisms in $\mathcal{INS}_{semi}$

(HINT: Institution semi-morphisms disregard sentences). Give a construction of products and equalisers in $\mathcal{INS}_{semi}$ and conclude that $\mathcal{INS}_{semi}$ is complete. HINT: See Exercise 3.4.67. □

10.1.2 Duplex institutions

An institution semi-morphism $\mu\colon \mathbf{INS} \to_{semi} \mathbf{INS}'$ gives a way of extracting $\mathbf{INS}'$-models from $\mathbf{INS}$-models. This is sufficient to allow $\mathbf{INS}'$-sentences to be used to describe the properties of $\mathbf{INS}$-models: an $\mathbf{INS}'$-sentence φ may be defined to hold in an $\mathbf{INS}$-model if φ holds in its translation under μ. Since the sentences of $\mathbf{INS}$ and $\mathbf{INS}'$ need not be related at all, reinterpreting the sentences of $\mathbf{INS}'$ in this way may increase the specification power of $\mathbf{INS}$.

Definition 10.1.12 (Duplex institution). Consider an institution semi-morphism $\mu\colon \mathbf{INS}_1 \to_{semi} \mathbf{INS}_2$. We define the *duplex institution* $\mathbf{INS}_1$ **plus** $\mathbf{INS}_2$ **via** $\mu = \langle \mathbf{Sign}, \mathbf{Sen}, \mathbf{Mod}, \langle \models_\Sigma \rangle_{\Sigma \in |\mathbf{Sign}|} \rangle$, which enriches $\mathbf{INS}_1$ by $\mathbf{INS}_2$-sentences reinterpreted via μ, as follows:

- $\mathbf{INS}_1$ **plus** $\mathbf{INS}_2$ **via** μ has the same signatures as $\mathbf{INS}_1$: $\mathbf{Sign} = \mathbf{Sign}_{\mathbf{INS}_1}$.
- For each $\Sigma \in |\mathbf{Sign}|$, the set $\mathbf{Sen}(\Sigma)$ of Σ-sentences of $\mathbf{INS}_1$ **plus** $\mathbf{INS}_2$ **via** μ includes Σ-sentences of $\mathbf{INS}_1$ as well as $\mu^{Sign}(\Sigma)$-sentences of $\mathbf{INS}_2$, where the latter are written in the form φ_2 **via** μ, for $\varphi_2 \in \mathbf{Sen}_{\mathbf{INS}_2}(\mu^{Sign}(\Sigma))$ (we assume that sentences of this form do not occur in $\mathbf{Sen}_{\mathbf{INS}_1}(\Sigma)$).
  For each $\sigma\colon \Sigma \to \Sigma'$ in $\mathbf{Sign}$, $\mathbf{Sen}(\sigma)\colon \mathbf{Sen}(\Sigma) \to \mathbf{Sen}(\Sigma')$ is defined as $\mathbf{Sen}_{\mathbf{INS}_1}(\sigma)$ on Σ-sentences in $\mathbf{Sen}_{\mathbf{INS}_1}(\Sigma) \subseteq \mathbf{Sen}(\Sigma)$, and then for any φ_2 **via** $\mu \in \mathbf{Sen}_{\mathbf{INS}}(\Sigma)$, where $\varphi_2 \in \mathbf{Sen}_{\mathbf{INS}_2}(\mu^{Sign}(\Sigma))$, $\mathbf{Sen}(\sigma)(\varphi_2 \textbf{ via } \mu) = \mu^{Sign}(\sigma)(\varphi_2)$ **via** μ.
- $\mathbf{INS}_1$ **plus** $\mathbf{INS}_2$ **via** μ has the same models as $\mathbf{INS}_1$: $\mathbf{Mod} = \mathbf{Mod}_{\mathbf{INS}_1}$.
- For each signature $\Sigma \in |\mathbf{Sign}|$, the satisfaction relation $\models_\Sigma$ is defined to coincide with $\models_{\mathbf{INS}_1,\Sigma}$ for Σ-sentences in $\mathbf{Sen}_{\mathbf{INS}_1}(\Sigma)$, while for $\varphi_2 \in \mathbf{Sen}_{\mathbf{INS}_2}(\mu^{Sign}(\Sigma))$ and $M \in |\mathbf{Mod}(\Sigma)|$, $M \models_\Sigma \varphi_2$ **via** μ iff $\mu^{Mod}_\Sigma(M) \models_{\mathbf{INS}_2,\mu^{Sign}(\Sigma)} \varphi_2$.

Exercise. Check that the satisfaction condition for $\mathbf{INS}_1$ **plus** $\mathbf{INS}_2$ **via** μ follows from the satisfaction conditions for $\mathbf{INS}_1$ and $\mathbf{INS}_2$. □

Example 10.1.13. An institution semi-morphism $\mu\colon \mathbf{FOPEQ} \to_{semi} \mathbf{EQ}$ mapping first-order structures of first-order logic to algebras of equational logic is given in Example 10.1.2. Thus, we can construct a new institution **FOPEQ plus EQ via** μ, which allows equations to be used to specify properties of first-order structures. Of course, there is little that is interesting about this, since equations are in fact already present in **FOPEQ**, as will be formalised by the extension of μ to an institution morphism in Example 10.2.2 below.

However, as remarked in Example 10.1.7, μ is also an institution semi-morphism from **FOPEQ** to the institution $\mathbf{EQ}^{data}$ of data constraints in **EQ**. Thus, we can form another duplex institution **FOPEQ plus** $\mathbf{EQ}^{data}$ **via** μ, which in addition to the usual first-order sentences allows equational data constraints to be used to specify first-order structures. This is quite interesting: unlike **FOPEQ**, the institution **EQ** of equational logic is liberal, and so it provides a more natural framework for dealing with initiality and data constraints (even though they can in fact be expressed directly as data constraints over **FOPEQ** in the institution $\mathbf{FOPEQ}^{data}$). In particular, **EQ** might be equipped with powerful term rewriting and inductive reasoning tools which are not available for theories of first-order logic.

Exercise. Simulate the equational data constraints of **FOPEQ plus** $\mathbf{EQ}^{data}$ **via** μ by data constraints of $\mathbf{FOPEQ}^{data}$ using the obvious embedding of $\mu^{Sign}(\Theta)$-equations into the set of first-order Θ-sentences, for any first-order signature $\Theta \in |\mathbf{FOSig}|$. HINT: This is not quite trivial, as the naive translation may yield data constraints in $\mathbf{FOPEQ}^{data}$ that restrict the interpretation of predicates in Θ more severely than the original equational data constraints in **FOPEQ plus** $\mathbf{EQ}^{data}$ **via** μ. □

Example 10.1.14. Example 10.1.9 introduces an institution semi-morphism from the institution $\mathbf{IMP}_{DT}$ of a simple imperative programming language to the institution **PEQ** of partial equational logic. We can compose this with the institution semi-morphism from **PEQ** to the institution **PFOPEQ** of partial first-order logic given in Example 10.1.8, obtaining an institution semi-morphism $\mu: \mathbf{IMP}_{DT} \to_{semi} \mathbf{PFOPEQ}$. The duplex institution $\mathbf{IMP}_{DT}$ **plus PFOPEQ via** μ allows for an arbitrary mixture of explicit definitions of functional procedures using the sentences of $\mathbf{IMP}_{DT}$ and their loose specifications using (partial) first-order sentences.

Similarly, by Exercise 10.1.10 we have an institution semi-morphism from the institution **FProg** of simple functional programs to **PEQ**, which can be composed with a semi-morphism to **PFOPEQ** as above, yielding an institution semi-morphism $\mu': \mathbf{FProg} \to_{semi} \mathbf{PFOPEQ}$. This can be used to build the duplex institution **FProg plus PFOPEQ via** μ' which allows for an arbitrary mixture of function definitions from **FProg** and their loose specifications using (partial) first-order sentences. This is already present in the institution **FPL**, which in fact offers considerably richer sentences that may involve in their terms quite complex functional "code", rather than just use functions with "code" provided separately as in the duplex institution **FProg plus PFOPEQ via** μ'. Thus, the duplex institution is a weaker (but in some sense also cleaner) mixture of functional programs and logical formulae than **FPL**. □

Exercise 10.1.15. For any institution semi-morphism $\mu_1: \mathbf{INS} \to_{semi} \mathbf{INS}_1$, adding to **INS** the sentences of $\mathbf{INS}_1$ reinterpreted via μ_1 does not change its category of signatures or its model functor. Thus, as remarked in Example 10.1.7, any other institution semi-morphism $\mu_2: \mathbf{INS} \to_{semi} \mathbf{INS}_2$ is also an institution semi-morphism $\mu_2: \mathbf{INS}$ **plus** $\mathbf{INS}_1$ **via** $\mu_1 \to_{semi} \mathbf{INS}_2$, and so we can form the institution (**INS plus** $\mathbf{INS}_1$ **via** μ_1) **plus** $\mathbf{INS}_2$ **via** μ_2, which extends **INS** by adding sentences of both $\mathbf{INS}_1$ and $\mathbf{INS}_2$ reinterpreted via μ_1 and μ_2 respectively. By iterating this construction, for any finite set of institution semi-morphisms $\mu_i: \mathbf{INS} \to_{semi} \mathbf{INS}_i$, $i = 1, \ldots, n$, define the "multiplex" institution **INS plus** $\mathbf{INS}_1 \ldots \mathbf{INS}_n$ **via** $\mu_1 \ldots \mu_n$.

This can be generalised using the constructions hinted at in Exercise 10.1.11. Given an arbitrary diagram $\mathscr{D}$ in $\mathcal{INS}_{semi}$ with nodes $\mathbf{INS}_i$, $i = 1, \ldots, n$ (we restrict attention to finite diagrams only for notational convenience), construct in $\mathcal{INS}_{semi}$ its limit **INS** with projections μ_i: **INS** $\to_{semi}$ **INS**$_i$. Intuitively, each signature of **INS** is a combination of the signatures of **INS**$_i$, $i = 1, \ldots, n$, that is compatible with the institution semi-morphisms in $\mathscr{D}$, and similarly for models. A combination of the institutions in $\mathscr{D}$ may be defined as **INS plus INS**$_1 \ldots$**INS**$_n$ **via** $\mu_1 \ldots \mu_n$. Compare this construction with the construction of limits in the category $\mathcal{INS}$ of institutions with full institution morphisms to be given in Section 10.3 (Theorem 10.3.9). □

10.1.3 Migrating specifications

As shown in the definition of a duplex institution, an institution semi-morphism μ: **INS** $\to_{semi}$ **INS**$'$ can be used to reinterpret in **INS** the sentences of **INS**$'$. This cannot be lifted immediately to an arbitrary structured specification over **INS**$'$. Any Σ'-sentence $\varphi' \in \mathbf{Sen}_{\mathbf{INS}'}(\Sigma')$ may give rise to many sentences φ' **via** μ in the duplex institution **INS plus INS**$'$ **via** μ, one for each signature $\Sigma \in |\mathbf{Sign}_{\mathbf{INS}}|$ such that $\mu^{Sign}(\Sigma) = \Sigma'$. The choice of the signature Σ is implicit in the classification of the sentences in **INS plus INS**$'$ **via** μ by their signatures. When dealing with specifications, this choice must be made explicit. This leads to the following inter-institutional specification-building operation of *translation by an institution semi-morphism*.

Definition 10.1.16 (Translating specifications by an institution semi-morphism). Let μ: **INS** $\to_{semi}$ **INS**$'$ be an institution semi-morphism, SP' an **INS**$'$-specification with signature $Sig[SP'] \in |\mathbf{Sign}_{\mathbf{INS}'}|$ and models $Mod[SP'] \subseteq |\mathbf{Mod}_{\mathbf{INS}'}(Sig[SP'])|$, and $\Sigma \in |\mathbf{Sign}_{\mathbf{INS}}|$ an **INS**-signature with $\mu^{Sign}(\Sigma) = Sig[SP']$. Then SP' **with** μ **to** Σ is a specification over **INS** with the following semantics:

$$Sig[SP' \textbf{ with } \mu \textbf{ to } \Sigma] = \Sigma$$
$$Mod[SP' \textbf{ with } \mu \textbf{ to } \Sigma] = \{M \in |\mathbf{Mod}_{\mathbf{INS}}(\Sigma)| \mid \mu_\Sigma^{Mod}(M) \in Mod[SP']\}.$$

Thus, extending the notation introduced in Section 5.1 to identify explicitly the institution over which specifications are interpreted,

$$\_\textbf{ with } \mu \textbf{ to } \Sigma\colon Spec_{\mathbf{INS}'}(\mu^{Sign}(\Sigma)) \to Spec_{\mathbf{INS}}(\Sigma).$$ □

This operation may be used together with other specification-building operations to construct specifications with structure that spans a number of institutions related by institution semi-morphisms. The construction of such a specification may "migrate" through a number of institutions, join specifications coming from different institutions, and so on. The resulting specifications are *heterogeneous* in the sense that they use a number of institutions linked by institution semi-morphisms. However, each such specification eventually focuses on a single institution in which it "ends up". This is the institution where the overall semantics, given as before in terms of the specification's signature and model class, resides. In a way, the other institutions involved may be viewed as auxiliary, used only to conveniently describe some parts (or rather, the signature and model class) of the final specification. As in Definition 10.1.16 above, we say that a (heterogeneous) specification *SP* is *over* an institution **INS** to indicate the institution in which the semantics of *SP* is given; the notation $Spec_{\mathbf{INS}}$ (perhaps with the obvious extra qualification by a signature) stands for the class of all specifications over **INS** considered. Since the semantics of such specifications is given via signatures and model classes in **INS**, all the concepts and results presented for specifications (so far presumably built entirely in **INS**) in terms of their semantics still apply to heterogeneous specifications over **INS** as well.

For example, recall the institution semi-morphism $\mu{:}\mathbf{IMP}_{DT} \to_{semi} \mathbf{PFOPEQ}$ used in Example 10.1.14 to build the institution $\mathbf{IMP}_{DT}$ **plus PFOPEQ via** μ. We may now move specifications over **PFOPEQ** to $\mathbf{IMP}_{DT}$, thus obtaining specifications of functional procedures in terms of the logical notation of **PFOPEQ**, which is rather more convenient for writing requirement specifications than that provided by the institution $\mathbf{IMP}_{DT}$. This is similar to building specifications over the duplex institution $\mathbf{IMP}_{DT}$ **plus PFOPEQ via** μ, but is still more flexible.

Example 10.1.17. Consider the following simple specification (over the institution **PFOPEQ** of partial first-order logic) of finite bags of natural numbers, built assuming some standard specifications NAT and BOOL of natural numbers and booleans respectively and using the **reachable** specification-building operation introduced in Exercise 5.1.10 to restrict to reachable models.

spec BAGNAT =
  reachable
    NAT **and** BOOL
    then
      sorts *Bag*
      ops $\varnothing{:}Bag$
        $add{:}Nat \times Bag \to Bag$
        $isempty{:}Bag \to Bool$
        $count{:}Nat \times Bag \to Nat$
      $\forall a,b{:}Nat, B{:}Bag$
        • $def(\varnothing) \wedge def(add(a,B))$
        • $add(a,add(b,B)) = add(b,add(a,B))$
        • $isempty(\varnothing) = true$
        • $isempty(add(a,B)) = false$
        • $count(a,\varnothing) = 0$
        • $count(a,add(a,B)) = 1 + count(a,B)$
        • $a \neq b \Rightarrow count(b,add(a,B)) = count(b,B)$
  on $\{Bag\}$

Let $\mu{:}\mathbf{IMP}_{DT} \to_{semi} \mathbf{PFOPEQ}$ be the institution semi-morphism sketched in Example 10.1.14. We would like to translate the specification BAGNAT to the institution $\mathbf{IMP}_{DT}$, obtaining a specification of a set of functional procedures in the programming language underlying $\mathbf{IMP}_{DT}$. The necessary prerequisite for this is that we have a signature $\Pi \in \mathbf{Sign}_{\mathbf{IMP}_{DT}}$ such that $\mu^{Sign}(\Pi) = Sig[\text{BAGNAT}]$. Since $\mu^{Sign}{:}\mathbf{Sign}_{\mathbf{IMP}_{DT}} \to \mathbf{FOSig}$ always yields signatures containing the signature Σ_{DT}

of the underlying data type of the programming language, $Sig[\mathrm{BAGNAT}]$ may not be in the range of μ^{Sign}. We can, however, translate the specification BAGNAT to the "closest" signature in **PFOPEQ** of the required form, simply by adding (unconstrained) the types and operations of DT. To avoid this extra step, for simplicity of presentation here, let us assume that the built-in data types of the programming language are just the natural numbers and booleans with operations as in NAT and BOOL and that the algebra DT is a (standard) model of NAT **and** BOOL. Then the types and operations of BAGNAT which have to be implemented are given in the following signature in $\mathbf{IMP}_{DT}$:

$$\begin{array}{lll} \Pi_{\mathrm{BAGNAT}} = & \textbf{types} & Bag \\ & \textbf{procedures} & \varnothing\colon \to Bag \\ & & add\colon Nat, Bag \to Bag \\ & & isempty\colon Bag \to Bool \\ & & count\colon Nat, Bag \to Nat \end{array}$$

We now have $\mu^{Sign}(\Pi_{\mathrm{BAGNAT}}) = Sig[\mathrm{BAGNAT}]$ and so can form the following specification over $\mathbf{IMP}_{DT}$:

spec $\mathrm{BAGNAT}_{\mathbf{IMP}}$ = BAGNAT **with** μ **to** Π_{BAGNAT}

$\mathrm{BAGNAT}_{\mathbf{IMP}}$ is an $\mathbf{IMP}_{DT}$-specification with signature Π_{BAGNAT} and with models which implement the type *Bag* and operations $\varnothing$, *add*, *isempty* and *count* so that the (partial) functions they determine satisfy the axioms in BAGNAT. More precisely, the models of $\mathrm{BAGNAT}_{\mathbf{IMP}}$ must determine partial algebras that are models of BAGNAT, which in particular requires them to be reachable on the sort *Bag*. Notice that the (discrete) categorical structure of models in $\mathbf{IMP}_{DT}$, and hence in the duplex institution $\mathbf{IMP}_{DT}$ **plus PFOPEQ via** μ, is not rich enough to capture this property (in the style of Section 4.5).

$\mathrm{BAGNAT}_{\mathbf{IMP}}$ may be used in the same way as any other specification in $\mathbf{IMP}_{DT}$. In particular, it may be combined with other specifications using all the available specification-building operations. Perhaps most importantly, it may also be further refined in $\mathbf{IMP}_{DT}$. Consider:

spec LISTNAT-FOR-$\mathrm{BAGNAT}_{\mathbf{IMP}}$ =
  types *List*
  procedures $nil\colon \to List$
    $null\colon List \to Bool$
    $nth\colon Nat, List \to Nat$
    $incr\colon Nat, List \to List$
  • **type** $List = \mathtt{unit} + Nat \times List$
  • **proc** nil = **result** $\langle\rangle{:}List$
  • **proc** $null(L{:}List)$ = **result** $\mathtt{is\text{-}in}_1(L){:}Bool$
  • **proc** $nth(n{:}Nat, L{:}List)$ =
    `while` $n > 0$ `and` $\mathtt{is\text{-}in}_2(L)$ `do`
      $n := n-1$; $L := \mathtt{proj}_2(L)$ `od`;
    `if` $\mathtt{is\text{-}in}_1(L)$ `then` $r := 0$ `else` $r := \mathtt{proj}_1(L)$ `fi`;
    result $r{:}Nat$

- **proc** $incr(n{:}Nat, L{:}List) =$
  $L1 := \langle\rangle$;
  `while` $n > 0$ **and** `is-in`$_2(L)$ **do**
  $L1 := \langle$`proj`$_1(L), L1\rangle$; $n := n-1$; $L := $`proj`$_2(L)$ **od**;
  `if is-in`$_1(L)$ `then`
  $v := 0$; `while` $n > 0$ **do** $n := n-1$; $L1 := \langle 0, L1\rangle$ **od**
  `else` $v := $`proj`$_1(L)$; $L := $`proj`$_2(L)$ `fi`;
  $L := \langle v+1, L\rangle$;
  `while is-in`$_2(L1)$ **do**
  $L := \langle$`proj`$_1(L1), L\rangle$; $L1 := $`proj`$_2(L1)$ **od**;
  result $L{:}List$

For help reading the above "code", note that in many conventional languages, *list* would be a built-in data type of lists of natural numbers, with $\langle\rangle$ typically written as `nil`, `is-in`$_1(L)$ as `null`(L), `proj`$_1(L)$ as `head`(L), `proj`$_2(L)$ as `tail`(L), and $\langle v, L\rangle$ as `cons`(v, L).

The overall idea here is to consider lists of natural numbers, with $nth(n, L)$ extracting the $(n+1)$th element from the list L (or 0 if no such element exists) and $incr(n, L)$ increasing the $(n+1)$th element of L by 1 (padding the list with 0's if the length of L is smaller than n). Such lists can be used directly to represent bags of natural numbers by storing the number of occurrences of n in a bag B at the nth position of the list that represents B. To state this intention formally, let $\Pi_{LN\text{-}for\text{-}BN} = Sig[\text{LISTNAT-FOR-BAGNAT}_{\mathbf{IMP}}]$ and let $\sigma{:}\,\Pi_{\text{BAGNAT}} \to \Pi_{LN\text{-}for\text{-}BN}$ be the signature morphism given by the map $\{Bag \mapsto List, \varnothing \mapsto nil, add \mapsto incr, isempty \mapsto null, count \mapsto nth\}$. We then have the following simple implementation (see Section 7.1):

$$\text{BAGNAT}_{\mathbf{IMP}} \rightsquigarrow \text{LISTNAT-FOR-BAGNAT}_{\mathbf{IMP}} \textbf{ hide via } \sigma$$

This formally captures the claim that LISTNAT-FOR-BAGNAT$_{\mathbf{IMP}}$ provides an implementation for BAGNAT built in $\mathbf{IMP}_{DT}$. □

Example 10.1.18. Example 10.1.17 can be redone in a functional programming style using the institution **FProg**. Recall the institution semi-morphism $\mu'{:}\,\mathbf{FProg} \to \mathbf{PFOPEQ}$ introduced in Example 10.1.14. Consider the **FPL**-signature $\mathsf{SIG}_{\text{BAGNAT}}$, which is essentially the signature $Sig[\text{BAGNAT}]$ with no value constructors (other than those for sort *Bool*). Since $(\mu')^{Sign}(\mathsf{SIG}_{\text{BAGNAT}}) = Sig[\text{BAGNAT}]$, we can form the following specification over **FProg**:

spec BAGNAT$_{\mathbf{FProg}}$ = BAGNAT **with** μ **to** $\mathsf{SIG}_{\text{BAGNAT}}$

BAGNAT$_{\mathbf{FProg}}$ may be used in the same way as any other specification in **FProg**; most importantly, it may be further refined to build an implementation of BAGNAT in **FProg**. Consider:

spec LISTNAT-FOR-BAGNAT$_{\mathbf{FProg}}$ =
sorts *Nat* **free with** 0 | *succ*(*Nat*)
List **free with** *nil* | *cons*(*Nat*, *List*)
... operations from NAT **and** BOOL and their definitions omitted ...
ops **fun** *null*(*L*:*List*):*NatList* =
case *L* **of** *nil* => *true* | *cons*(*n*, *L'*) => *false*
fun *nth*(*n*:*Nat*, *L*:*List*):*Nat* =
case *L* **of** *nil* => 0
| *cons*(*k*, *L'*) => **case** *n* **of** 0 => *k*
| *succ*(*m*) => *nth*(*m*, *L'*)
fun *incr*(*n*:*Nat*, *L*:*List*):*List* =
case *n* **of** 0 => **case** *L* **of** *nil* => *cons*(*succ*(0), *nil*)
| *cons*(*k*, *L'*) => *cons*(*succ*(*k*), *L'*)
| *succ*(*m*) => **case** *L* **of** *nil* => *cons*(0, *incr*(*m*, *nil*))
| *cons*(*k*, *L'*) => *cons*(*k*, *incr*(*m*, *L'*))

This is a version of the "imperative code" from Example 10.1.17 in a functional programming style.

Let $\mathsf{SIG}_{\textit{LN-for-BN}} = \textit{Sig}[\text{LISTNAT-FOR-BAGNAT}_{\mathbf{FProg}}]$ and let $\sigma'\colon \mathsf{SIG}_{\text{BAGNAT}} \to \mathsf{SIG}_{\textit{LN-for-BN}}$ be the signature morphism determined by the map $\{\textit{Bag} \mapsto \textit{List}, \varnothing \mapsto \textit{nil}, \textit{add} \mapsto \textit{incr}, \textit{isempty} \mapsto \textit{null}, \textit{count} \mapsto \textit{nth}\}$. We then have

$$\text{BAGNAT}_{\mathbf{FProg}} \rightsquigarrow \text{LISTNAT-FOR-BAGNAT}_{\mathbf{FProg}} \ \mathbf{hide\ via}\ \sigma'.$$

This formally captures the claim that LISTNAT-FOR-BAGNAT$_{\mathbf{FProg}}$ provides an implementation for BAGNAT built in **FProg**. □

Translation by an institution semi-morphism $\mu\colon \mathbf{INS} \to_{semi} \mathbf{INS}'$ allows us to change the institution in which specifications are built by reinterpreting any specification dealing with simpler **INS**$'$-models in the richer framework of **INS**-models. There is a dual operation which translates specifications over **INS** to the simpler context of **INS**$'$.

Definition 10.1.19 (Hiding specifications via an institution semi-morphism). Let $\mu\colon \mathbf{INS} \to_{semi} \mathbf{INS}'$ be an institution semi-morphism, *SP* an **INS**-specification with signature $\textit{Sig}[SP] \in |\mathbf{Sign}_{\mathbf{INS}}|$ and models $\textit{Mod}[SP] \subseteq |\mathbf{Mod}_{\mathbf{INS}}(\textit{Sig}[SP])|$. Then *SP* **hide via** μ is a specification over **INS**$'$ with the following semantics:[2]

$$\begin{aligned} \textit{Sig}[SP\ \mathbf{hide\ via}\ \mu] &= \mu^{Sign}(\textit{Sig}[SP]) \\ \textit{Mod}[SP\ \mathbf{hide\ via}\ \mu] &= \{\mu^{Mod}_{\textit{Sig}[SP]}(M) \mid M \in \textit{Mod}[SP]\}. \end{aligned}$$

Thus, for each signature $\Sigma \in |\mathbf{Sign}_{\mathbf{INS}}|$,

$$\_\ \mathbf{hide\ via}\ \mu\colon \textit{Spec}_{\mathbf{INS}}(\Sigma) \to \textit{Spec}_{\mathbf{INS}'}(\mu^{Sign}(\Sigma)).$$ □

[2] CASL notation: if μ is an institution morphism, see Definition 10.2.1, then this is written *SP* **hide** μ in HETCASL, the language of HETS [MML07].

This yields another specification-building operation which may be arbitrarily mixed with other operations to stretch the structure of specifications over a number of institutions. As indicated in Examples 10.1.17 and 10.1.18, this allows us to consider developments via simple implementations which, although formally taking place in a single institution, may in fact involve complex specifications from other institutions as well. It is perhaps more desirable to explicitly change institution in the course of the development process. This may be captured analogously to change of signature during the development process as enabled by constructor implementations (Definition 7.2.1), by generalising in a rather obvious way the notion of a constructor from Definition 6.1.1.

Definition 10.1.20 (Generalised constructor). Consider two institutions $\mathbf{INS} = \langle \mathbf{Sign}, \mathbf{Sen}, \mathbf{Mod}, \langle \models_\Sigma \rangle_{\Sigma \in |\mathbf{Sign}|} \rangle$ and $\mathbf{INS}' = \langle \mathbf{Sign}', \mathbf{Sen}', \mathbf{Mod}', \langle \models'_{\Sigma'} \rangle_{\Sigma' \in |\mathbf{Sign}'|} \rangle$.

- A *constructor* $\kappa \colon \langle \mathbf{INS}', \Sigma' \rangle \Rightarrow \langle \mathbf{INS}, \Sigma \rangle$, where $\Sigma' \in |\mathbf{Sign}'|$ and $\Sigma \in |\mathbf{Sign}|$, is a partial function mapping Σ'-models in $\mathbf{INS}'$ to Σ-models in $\mathbf{INS}$, $\kappa \colon |\mathbf{Mod}'(\Sigma')| \to |\mathbf{Mod}(\Sigma)|$.
- A specification SP over $\mathbf{INS}$ *is implemented by* a specification SP' over $\mathbf{INS}'$ *via* a constructor $\kappa \colon \langle \mathbf{INS}', Sig[SP'] \rangle \Rightarrow \langle \mathbf{INS}, Sig[SP] \rangle$, written $SP \underset{\kappa}{\rightsquigarrow} SP'$, if $Mod[SP'] \subseteq dom(\kappa)$ and $Mod[SP] \supseteq \kappa(Mod[SP'])$.
- A *constructor operation* determined by a constructor $\kappa \colon \langle \mathbf{INS}', \Sigma' \rangle \Rightarrow \langle \mathbf{INS}, \Sigma \rangle$, where $\Sigma' \in |\mathbf{Sign}'|$ and $\Sigma \in |\mathbf{Sign}|$, is a specification-building operation which for any Σ'-specification SP' over $\mathbf{INS}'$ yields a specification $\kappa(SP')$ over $\mathbf{INS}$ with the following semantics:

$$\begin{aligned} Sig[\kappa(SP')] &= \Sigma \\ Mod[\kappa(SP')] &= \kappa(Mod[SP']). \end{aligned}$$ □

It should be obvious that this is indeed a generalisation of the corresponding definitions in Sections 6.1 and 7.2: if $\mathbf{INS} = \mathbf{INS}'$ then we obtain the concepts discussed there in the framework of an arbitrary but fixed institution. It should also be clear that the discussion about the role of constructor implementations in the development process as well as the results presented there carry over to the present more general framework, where the development process may gradually migrate from one institution to another. Typically, constructors leading from one institution to another are components of an institution semi-morphism and extract simpler models from more complex ones. Consequently, development will proceed from institutions with a simpler notion of a model to those in which models carry gradually more and more information.

Example 10.1.21. For any institution semi-morphism $\mu \colon \mathbf{INS}' \to_{semi} \mathbf{INS}$ and any signature $\Sigma' \in |\mathbf{Sign}_{\mathbf{INS}'}|$, $\mu^{Mod}_{\Sigma'} \colon \mathbf{Mod}_{\mathbf{INS}'}(\Sigma') \to \mathbf{Mod}_{\mathbf{INS}}(\mu^{Sign}(\Sigma'))$ is a constructor $\mu^{Mod}_{\Sigma'} \colon \langle \mathbf{INS}', \Sigma' \rangle \Rightarrow \langle \mathbf{INS}, \mu^{Sign}(\Sigma') \rangle$. These constructors are *total* functions, so we do not need to check the requirement concerning their definedness when using them in implementation steps. The constructor operation determined by $\mu^{Mod}_{\Sigma'}$ coincides with $\_\_$ **hide via** μ.

For example, recall the institution semi-morphism $\mu\colon \mathbf{IMP}_{DT} \to_{semi} \mathbf{PFOPEQ}$ and the specifications and notation introduced in Example 10.1.17. Then

$$\text{BAGNAT} \underset{\kappa}{\rightsquigarrow} \text{LISTNAT-FOR-BAGNAT}_{\mathbf{IMP}}$$

where the constructor $\kappa\colon \langle \mathbf{IMP}_{DT}, \Pi_{LN\text{-}for\text{-}BN} \rangle \Rightarrow \langle \mathbf{PFOPEQ}, Sig[\text{BAGNAT}] \rangle$ is given by composition of the following two constructors:

$$\begin{aligned} -|_\sigma &: \langle \mathbf{IMP}_{DT}, \Pi_{LN\text{-}for\text{-}BN} \rangle \Rightarrow \langle \mathbf{IMP}_{DT}, \Pi_{\text{BAGNAT}} \rangle \\ \mu^{Mod}_{\Pi_{\text{BAGNAT}}} &: \langle \mathbf{IMP}_{DT}, \Pi_{\text{BAGNAT}} \rangle \Rightarrow \langle \mathbf{PFOPEQ}, Sig[\text{BAGNAT}] \rangle. \end{aligned}$$

Similarly, given the institution semi-morphism $\mu'\colon \mathbf{FProg} \to_{semi} \mathbf{PFOPEQ}$ and the specifications and notation introduced in Example 10.1.18, we have

$$\text{BAGNAT} \underset{\kappa'}{\rightsquigarrow} \text{LISTNAT-FOR-BAGNAT}_{\mathbf{FProg}}$$

with the constructor $\kappa'\colon \langle \mathbf{FProg}, \mathsf{SIG}_{LN\text{-}for\text{-}BN} \rangle \Rightarrow \langle \mathbf{PFOPEQ}, Sig[\text{BAGNAT}] \rangle$ given by composition of the following two constructors:

$$\begin{aligned} -|_{\sigma'} &: \langle \mathbf{FProg}, \mathsf{SIG}_{LN\text{-}for\text{-}BN} \rangle \Rightarrow \langle \mathbf{FProg}, \mathsf{SIG}_{\text{BAGNAT}} \rangle \\ (\mu')^{Mod}_{\mathsf{SIG}_{\text{BAGNAT}}} &: \langle \mathbf{FProg}, \mathsf{SIG}_{\text{BAGNAT}} \rangle \Rightarrow \langle \mathbf{PFOPEQ}, Sig[\text{BAGNAT}] \rangle. \end{aligned}$$ □

Exercise 10.1.22. In the spirit of Section 5.6, check if the two operations of changing institution via institution semi-morphisms commute with other specification-building operations. For example, for any institution semi-morphism $\mu\colon \mathbf{INS} \to_{semi} \mathbf{INS}'$, prove that (under appropriate assumptions about the signatures of specifications and the signature morphisms involved)

$$\begin{aligned} &(SP_1' \textbf{ with } \mu^{Sign}(\sigma)) \textbf{ with } \mu \textbf{ to } \Sigma_2 \equiv (SP_1' \textbf{ with } \mu \textbf{ to } \Sigma_1) \textbf{ with } \sigma \\ &(SP_2 \textbf{ hide via } \sigma) \textbf{ hide via } \mu \equiv (SP_2 \textbf{ hide via } \mu) \textbf{ hide via } \mu^{Sign}(\sigma). \end{aligned}$$

Show that we also have

$$\begin{aligned} &Mod[(SP_1 \textbf{ with } \sigma) \textbf{ hide via } \mu] \subseteq Mod[(SP_1 \textbf{ hide via } \mu) \textbf{ with } \mu^{Sign}(\sigma)] \\ &Mod[(SP_2' \textbf{ hide via } \mu^{Sign}(\sigma)) \textbf{ with } \mu \textbf{ to } \Sigma_1] \supseteq \\ &\qquad\qquad Mod[(SP_2' \textbf{ with } \mu \textbf{ to } \Sigma_2) \textbf{ hide via } \sigma]. \end{aligned}$$

Check that the opposite inclusions hold as well if the "naturality squares" for μ^{Mod} (see Definition 10.1.1) are pullbacks in **Cat**.

Since sentences are not related at all by institution semi-morphisms, we cannot expect any result relating the operation of changing institution via an institution semi-morphism to flat specifications. Show, however, that if the institution semi-morphism extends to a full institution morphism as described in Definition 10.2.1 below then the situation is analogous to that with the translation and hiding specification-building operations within a single institution (cf. Proposition 5.6.2(2) and Exercise 5.6.3).

We also have, as in Exercise 5.6.3, for any specification SP over **INS** with $Sig[SP] = \Sigma$ and specification SP' over $\mathbf{INS}'$ such that $\mu^{Sign}(\Sigma) = Sig[SP']$,

$$Mod[(SP' \textbf{ with } \mu \textbf{ to } \Sigma) \textbf{ hide via } \mu] \subseteq Mod[SP'],$$
$$Mod[(SP \textbf{ hide via } \mu) \textbf{ with } \mu \textbf{ to } \Sigma] \supseteq Mod[SP].$$

Show that the first inclusion (but in general not the second) may be reversed if μ^{Mod}_{Σ} is surjective. □

Exercise 10.1.23. Show — for instance using the last two inclusions indicated in Exercise 10.1.22 above — that given an institution semi-morphism μ: **INS** → **INS**′, the specification-building operations _ **with** μ **to** Σ and _ **hide via** μ form a Galois connection between classes of Σ-specifications over **INS** and $\mu^{Sign}(\Sigma)$-specifications over **INS**′, considered up to equivalence and ordered by the inclusion of model classes.

In particular this implies that for any specifications *SP* over **INS** and *SP*′ over **INS**′ such that $\mu^{Sign}(Sig[SP]) = Sig[SP']$, SP' **with** μ **to** $Sig[SP] \rightsquigarrow SP$ if and only if $SP' \rightsquigarrow SP$ **hide via** μ. Conclude then (informally) that translation of specifications by an institution semi-morphism when used on its own does not add new possibilities to the development process beyond those given by the constructor _ **hide via** _.

Show that when translation by an institution semi-morphism is used in combination with other specification-building operations, it may not be possible to similarly replace the use of _ **with** μ **to** _ by appropriate applications of _ **hide via** μ. □

The concepts and results on behavioural implementations in Section 8.4 carry over to this more general framework as well.

Example 10.1.24. Recall the specification of bags of natural numbers built in the institution **PFOPEQ** in Example 10.1.17 and its constructor implementation built essentially in Example 10.1.18 using the institution semi-morphism μ': **FProg** → **PFOPEQ**, as presented in Example 10.1.21.

Consider another attempt at implementing BAGNAT in **FProg**:

```
spec LISTNAT-FOR-BAGNAT'_FProg =
  sorts Nat free with 0 | succ(Nat)
        List free with nil | cons(Nat, List)
      ... operations from NAT and BOOL and their definitions omitted ...
  ops  fun eq(n:Nat, m:Nat):Bool =
         case n of 0 => case m of 0 => true
                                  | succ(m') => false
                 | succ(n') => case m of 0 => false
                                        | succ(m') => eq(n', m')
       fun null(L:List):NatList =
         case L of nil => true | cons(n, L') => false
       fun count(n:Nat, L:List):Nat =
         case L of nil => 0
                 | cons(k, L') => case eq(k, n) of true => succ(count(n, L'))
                                                 | false => count(n, L')
       fun add(n:Nat, L:List):List = cons(n, L)
```

Let $\mathsf{SIG}'_{\textit{LN-for-BN}} = \textit{Sig}[\textsc{ListNat-for-BagNat}'_{\mathbf{FProg}}]$, with the signature inclusion $\iota: \mathsf{SIG}_{\textsc{BagNat}} \to \mathsf{SIG}'_{\textit{LN-for-BN}}$. Consider the constructor κ'' given by composition of the following two constructors:

$$-|_\iota : \langle \mathbf{FProg}, \mathsf{SIG}'_{\textit{LN-for-BN}} \rangle \Rightarrow \langle \mathbf{FProg}, \mathsf{SIG}_{\textsc{BagNat}} \rangle$$
$$(\mu')^{\textit{Mod}}_{\mathsf{SIG}'_{\textsc{BagNat}}} : \langle \mathbf{FProg}, \mathsf{SIG}_{\textsc{BagNat}} \rangle \Rightarrow \langle \mathbf{PFOPEQ}, \textit{Sig}[\textsc{BagNat}] \rangle.$$

It is easy to check now that for any model $M \in \textit{Mod}[\textsc{ListNat-for-BagNat}'_{\mathbf{FProg}}]$, $\kappa''(M)$ does *not* satisfy $\textit{add}(0, \textit{add}(\textit{succ}(0), \varnothing)) = \textit{add}(\textit{succ}(0), \textit{add}(0, \varnothing))$ and so is *not* a model of $\textsc{BagNat}$. Consequently, we do *not* have a constructor implementation $\textsc{BagNat} \underset{\kappa''}{\leadsto} \textsc{ListNat-for-BagNat}'_{\mathbf{FProg}}$.

Recall, however, the behavioural interpretation of specifications of partial algebras as presented in Section 8.5.1.[3] It is easy to see that $\kappa''(M)$ is a behavioural model of $\textsc{BagNat}$, $\kappa''(M) \in \textit{Abs}_{\{\textit{Bool},\textit{Nat}\}}(\textit{Mod}[\textsc{BagNat}])$, so that, as intuitively expected, we do have a behavioural implementation:

$$\textsc{BagNat} \; {}^{\{\textit{Bool},\textit{Nat}\}}\!\underset{\kappa''}{\leadsto} \textsc{ListNat-for-BagNat}'_{\mathbf{FProg}}.$$

Exercise. Prove the correctness of this behavioural implementation, relying as much as possible on the proof techniques presented in Chapter 9. Good luck! □

10.2 Institution morphisms

Institution semi-morphisms entirely disregard the sentences and satisfaction relations of the institutions involved. Therefore they are not really satisfactory as a tool to relate and compare different logical systems presented as institutions. For this purpose, we will introduce here the concept of an institution morphism, which extends an institution semi-morphism by adding translations of sentences subject to the requirement that satisfaction be preserved.

Let us first have a closer look at the archetypical relationship between the institution **FOPEQ** of first-order logic with equality (cf. Example 4.1.12) and the institution **EQ** of equational logic (cf. Example 4.1.4). **FOPEQ** is a "richer" logical system, which may be intuitively viewed as some kind of an enrichment obtained by building on top of the "more primitive" logical system **EQ**. Very informally, we build **FOPEQ** from **EQ** by adding some extra components to signatures, then expanding models to interpret these new components, and finally enlarging the sets of sentences. This indicates three parts of such a relationship, which we will now consider in more detail.

First, any first-order signature is "built on top of" an algebraic signature: from any first-order signature we can extract the algebraic signature it is "built on top of"

[3] The presentation in Section 8.5.1 was given in the institution **FPL**. This is directly applicable to models in **PFOPEQ**, since the **PFOPEQ** signatures involved have no predicates and can be considered as **FPL**-signatures with no value constructors.

by simply forgetting all the predicate symbols, and this mapping extends to signature morphisms in the obvious way. Similarly, any first-order structure is "built on top of" an algebra: from any first-order structure we can extract the algebra it is "built on top of" (which is over the algebraic signature extracted from the first-order signature of this structure) by simply forgetting all the predicates. Again, this extends to structure morphisms in the obvious way; no change at all is needed in this case. This is captured by the institution semi-morphism from **FOPEQ** to **EQ** presented in Example 10.1.2. In addition, any equation may be viewed as a first-order sentence by simply regarding the universal quantifier used in the notation for equations as the universal quantifier of first-order logic (iterated in an arbitrary order for all the variables involved). Thus, the correspondence between **FOPEQ** and **EQ** consists of three essential components: a mapping from **FOPEQ**-signatures to **EQ**-signatures; a family of mappings (one for each **FOPEQ**-signature) from **FOPEQ**-models to **EQ**-models; and a family of mappings (one for each **FOPEQ**-signature) from **EQ**-sentences to **FOPEQ**-sentences — notice the change of direction. A crucial property of these mappings is that they are compatible with the translations of sentences and models induced by signature morphisms in each of the institutions, and that they preserve the satisfaction relations.

These three mappings constitute an institution morphism from **FOPEQ** to **EQ**.

Definition 10.2.1 (Institution morphism). An *institution morphism* $\mu\colon \mathbf{INS} \to \mathbf{INS}'$ from an institution $\mathbf{INS} = \langle \mathbf{Sign}, \mathbf{Sen}, \mathbf{Mod}, \langle \models_\Sigma \rangle_{\Sigma \in |\mathbf{Sign}|} \rangle$ to an institution $\mathbf{INS}' = \langle \mathbf{Sign}', \mathbf{Sen}', \mathbf{Mod}', \langle \models'_{\Sigma'} \rangle_{\Sigma' \in |\mathbf{Sign}'|} \rangle$ consists of:

- a functor $\mu^{Sign}\colon \mathbf{Sign} \to \mathbf{Sign}'$,
- a natural transformation $\mu^{Sen}\colon \mu^{Sign};\mathbf{Sen}' \to \mathbf{Sen}$, that is, for each $\Sigma \in |\mathbf{Sign}|$, a function $\mu^{Sen}_\Sigma\colon \mathbf{Sen}'(\mu^{Sign}(\Sigma)) \to \mathbf{Sen}(\Sigma)$ such that the following diagram commutes for every $\sigma\colon \Sigma_1 \to \Sigma_2$ in **Sign**

$$\begin{array}{ccccc}
\Sigma_2 & & \mathbf{Sen}'(\mu^{Sign}(\Sigma_2)) & \xrightarrow{\mu^{Sen}_{\Sigma_2}} & \mathbf{Sen}(\Sigma_2) \\
\uparrow \sigma & & \uparrow \mathbf{Sen}'(\mu^{Sign}(\sigma)) & & \uparrow \mathbf{Sen}(\sigma) \\
\Sigma_1 & & \mathbf{Sen}'(\mu^{Sign}(\Sigma_1)) & \xrightarrow[\mu^{Sen}_{\Sigma_1}]{} & \mathbf{Sen}(\Sigma_1)
\end{array}$$

- a natural transformation $\mu^{Mod}\colon \mathbf{Mod} \to (\mu^{Sign})^{op};\mathbf{Mod}'$, that is, for each $\Sigma \in |\mathbf{Sign}|$, a functor $\mu^{Mod}_\Sigma\colon \mathbf{Mod}(\Sigma) \to \mathbf{Mod}'(\mu^{Sign}(\Sigma))$ such that the following diagram commutes for every $\sigma\colon \Sigma_1 \to \Sigma_2$ in **Sign**

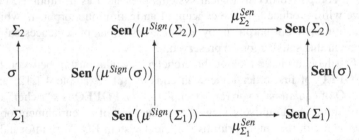

$$
\begin{array}{ccccc}
\Sigma_2 & & \mathbf{Mod}(\Sigma_2) & \xrightarrow{\mu^{Mod}_{\Sigma_2}} & \mathbf{Mod}'(\mu^{Sign}(\Sigma_2)) \\
\uparrow \sigma & & \downarrow \mathbf{Mod}(\sigma) & & \downarrow \mathbf{Mod}'(\mu^{Sign}(\sigma)) \\
\Sigma_1 & & \mathbf{Mod}(\Sigma_1) & \xrightarrow[\mu^{Mod}_{\Sigma_1}]{} & \mathbf{Mod}'(\mu^{Sign}(\Sigma_1))
\end{array}
$$

such that for any signature $\Sigma \in |\mathbf{Sign}|$, the translations $\mu^{Sen}_{\Sigma}: \mathbf{Sen}'(\mu^{Sign}(\Sigma)) \to \mathbf{Sen}(\Sigma)$ of sentences and $\mu^{Mod}_{\Sigma}: \mathbf{Mod}(\Sigma) \to \mathbf{Mod}'(\mu^{Sign}(\Sigma))$ of models preserve the satisfaction relation, that is, for any $\varphi' \in \mathbf{Sen}'(\mu^{Sign}(\Sigma))$ and $M \in |\mathbf{Mod}(\Sigma)|$

$$M \models_{\Sigma} \mu^{Sen}_{\Sigma}(\varphi') \iff \mu^{Mod}_{\Sigma}(M) \models'_{\mu^{Sign}(\Sigma)} \varphi' \qquad \text{[Satisfaction condition].}$$

□

We say that the institution morphism $\mu: \mathbf{INS} \to \mathbf{INS}'$ as above *extends* the institution semi-morphism $\langle \mu^{Sign}, \mu^{Mod} \rangle: \mathbf{INS} \to_{semi} \mathbf{INS}'$. Whenever convenient, we will use an institution morphism μ in contexts where the institution semi-morphism that μ extends is required, as for instance in inter-institutional specification-building operations as introduced by Definitions 10.1.16 and 10.1.19.

Example 10.2.2. As indicated in the remarks preceding the above definition, there is a natural institution morphism relating first-order logic to equational logic, $\mu: \mathbf{FOPEQ} \to \mathbf{EQ}$. It extends the semi-morphism $\langle \mu^{Sign}, \mu^{Mod} \rangle: \mathbf{FOPEQ} \to_{semi} \mathbf{EQ}$ defined in Example 10.1.2 as follows:

- For $\Theta \in |\mathbf{FOSig}|$, $\mu^{Sen}_{\Theta}: \mathbf{Sen_{EQ}}(\mu^{Sign}(\Theta)) \to \mathbf{Sen_{FOPEQ}}(\Theta)$ maps any $\mu^{Sign}(\Theta)$-equation of the form $\forall X \bullet t = t'$ to the corresponding first-order Θ-sentence $\forall x_1{:}s_1, \cdots, x_n{:}s_n \bullet t = t'$, where X is the family of variables $\{x_1{:}s_1, \ldots, x_n{:}s_n\}$. (This relies on the assumption that only finite sets of variables are used in equations — cf. Definition 2.1.1. Moreover, the order of variables used in this translation, albeit arbitrary, must be chosen consistently for all equations.)

It is easy to check that the family $\mu^{Sen}_{\Theta}: \mathbf{Sen_{EQ}}(\mu^{Sign}(\Theta)) \to \mathbf{Sen_{FOPEQ}}(\Theta)$, for $\Theta \in |\mathbf{FOSig}|$, forms a natural transformation $\mu^{Sen}: \mu^{Sign};\mathbf{Sen_{EQ}} \to \mathbf{Sen_{FOPEQ}}$.

The satisfaction condition follows from the trivial observation that the value of a term in a first-order structure does not depend on the interpretation of predicate symbols. □

Example 10.2.3. A similarly obvious institution morphism $\mu: \mathbf{CEQ} \to \mathbf{EQ}$ extends as follows the institution semi-morphism $\langle \mu^{Sign}, \mu^{Mod} \rangle: \mathbf{CEQ} \to_{semi} \mathbf{EQ}$ defined in Example 10.1.3:

- For each algebraic signature $\Sigma \in |\mathbf{AlgSig}|$, $\mu^{Sen}_{\Sigma}: \mathbf{Sen_{EQ}}(\Sigma) \to \mathbf{Sen_{CEQ}}(\Sigma)$ is the inclusion of the (finitary) Σ-equations into the set of infinitary Σ-equations.

The family $\mu^{Sen}_{\Sigma}: \mathbf{Sen_{EQ}}(\Sigma) \to \mathbf{Sen_{CEQ}}(\Sigma)$, $\Sigma \in |\mathbf{AlgSig}|$, forms a natural transformation $\mu^{Sen}: \mathbf{Sen_{EQ}} \to \mathbf{Sen_{CEQ}}$. The satisfaction condition is easy to check using the observation that the value of any finitary term in a continuous algebra, and hence the meaning of any finitary equation, does not depend on the ordering. □

Exercise 10.2.4. Following Example 10.2.2, extend the institution semi-morphism $\langle \mu^{Sign}, \mu^{Mod}\rangle : \mathbf{PFOPEQ} \rightarrow_{semi} \mathbf{PEQ}$ of Exercise 10.1.4 to define an institution morphism $\mu : \mathbf{PFOPEQ} \rightarrow \mathbf{PEQ}$. □

Example 10.2.5. Here are a number of rather straightforward examples of institution morphisms:

1. There is a trivial institution morphism from **EQ** to **GEQ** extending the trivial institution semi-morphism mentioned in Example 10.1.7: the translations of sentences are inclusions of the set of ground equations into the set of all equations over a given algebraic signature.
2. For any institution **INS**, there is an institution morphism from $\mathbf{INS}^{\wedge}$ (Example 4.1.38) to **INS** extending the trivial institution semi-morphism mentioned in Example 10.1.7: the translations of sentences map any sentence φ of **INS** to the one-element set $\{\varphi\}$, which is a sentence of $\mathbf{INS}^{\wedge}$. Similarly, there is an institution morphism from $\mathbf{INS}^{prop}$ (Example 4.1.41) to **INS** which extends the trivial institution semi-morphism mentioned in Example 10.1.7 by translations of sentences which are just inclusions.
3. For any institution **INS**, there is an institution morphism from **INS** to the institution $\mathbf{INS}^{\mathbf{Sign}^+}$ with signatures enriched by sentences (cf. Example 4.1.39): the translation of signatures "enriches" any signature of **INS** by the empty set of sentences, thus yielding a signature of $\mathbf{INS}^{\mathbf{Sign}^+}$; the translations of sentences and of models are just the identities.
   Notice that the signature translation functor here is not surjective.
   There is also an institution morphism going in the opposite direction, from $\mathbf{INS}^{\mathbf{Sign}^+}$ to **INS**: the signature translation extracts the original signature component of the enriched signatures of $\mathbf{INS}^{\mathbf{Sign}^+}$, the sentence translations are the identities, and the model translations are inclusions.
4. For any institution **INS**, there is an institution morphism from **INS** to the institution $\mathbf{INS}^{\mathbf{Mod}^+}$ with enlarged model categories (cf. Example 4.1.40): the translations of signatures and sentences are just the identities, and for any signature Σ, the translation of Σ-models of **INS** to Σ-models of $\mathbf{INS}^{\mathbf{Mod}^+}$ maps any model $M \in |\mathbf{Mod}_{\mathbf{INS}}(\Sigma)|$ to its trivial expansion $\langle id_\Sigma, M\rangle \in |\mathbf{Mod}_{\mathbf{INS}^{\mathbf{Mod}^+}}(\Sigma)|$.
   Notice that the translations of models here are not surjective.
   Perhaps more interestingly, there is an institution morphism going in the opposite direction, from $\mathbf{INS}^{\mathbf{Mod}^+}$ to **INS**: the translations of signatures and sentences are the identities, and for any signature Σ, the translation of Σ-models of $\mathbf{INS}^{\mathbf{Mod}^+}$ to Σ-models of **INS** maps any "model" $\langle \sigma : \Sigma \rightarrow \Sigma', M' \in |\mathbf{Mod}_{\mathbf{INS}}(\Sigma')|\rangle$ in $|\mathbf{Mod}_{\mathbf{INS}^{\mathbf{Mod}^+}}(\Sigma)|$ to $M'|_\sigma$ in $|\mathbf{Mod}_{\mathbf{INS}}(\Sigma)|$ (this obviously extends to a functor).

The proof of the satisfaction condition for the institution morphisms listed above amounts essentially to the remark that the satisfaction relations in the "richer" institution (the source of the institution morphism) extend the corresponding satisfaction relations in the "more primitive" institution (the target of the institution morphism). □

Exercise 10.2.6. For any institution **INS** with model functor **Mod**, Example 4.1.35 introduced the institution $\mathbf{INS}^{\mathbf{Sen}(\mathbf{Mod})}$, where "sentences" are classes of **INS**-models. Define an institution morphism from $\mathbf{INS}^{\mathbf{Sen}(\mathbf{Mod})}$ to **INS**.

In a sense dually, for any institution **INS** with a sentence functor **Sen**, Example 4.1.36 introduced the institution $\mathbf{INS}^{\mathbf{Mod}(\mathbf{Sen})}$, where "models" are sets of **INS**-sentences. Try to define an institution morphism from **INS** to $\mathbf{INS}^{\mathbf{Mod}(\mathbf{Sen})}$ which would map any model to its theory. Does this always work? HINT: Show that the mapping of models to their theories extends to a functor for the institution **GEQ**, but does not naturally extend to a functor for the institution **EQ**.

Give a sufficient condition under which a homomorphism between algebras induces an inclusion of their equational theories. Formulate an analogous condition for morphisms between first-order structures and their first-order theories. Are the conditions you have given necessary? □

Example 10.2.7. In Example 10.1.5 we have defined an institution semi-morphism $\langle \mu^{Sign}, \mu^{Mod} \rangle : \mathbf{PEQ_{str}} \rightarrow_{semi} \mathbf{EQ}$, using the idea of "totalisation" of partial functions. We will not extend this to an institution morphism from $\mathbf{PEQ_{str}}$ to **EQ** — see the exercise below. However, by Example 10.1.7, it is also an institution semi-morphism $\langle \mu^{Sign}, \mu^{Mod} \rangle : \mathbf{PEQ_{str}} \rightarrow_{semi} \mathbf{GEQ}$, which extends to an institution morphism $\mu : \mathbf{PEQ_{str}} \rightarrow \mathbf{GEQ}$ in the obvious way:

- For each algebraic signature $\Sigma \in |\mathbf{AlgSig}|$, $\mu^{Sen}_{\Sigma} : \mathbf{Sen_{GEQ}}(\Sigma) \rightarrow \mathbf{Sen_{PEQ_{str}}}(\Sigma)$ is the inclusion of the set of ground Σ-equations into the set of all Σ-equations and Σ-definedness formulae.

Proving the satisfaction condition relies on the fact that for any algebraic signature $\Sigma \in |\mathbf{AlgSig}|$, the value of a Σ-term in a partial Σ-algebra $A \in |\mathbf{PAlg}(\Sigma)|$ is defined if and only if its value in the Σ-algebra $\mu^{Mod}_{\Sigma}(A) \in |\mathbf{Alg}(\Sigma)|$ is not $\perp$, and moreover, if this is the case, then the value of this term is the same in A and in $\mu^{Mod}_{\Sigma}(A)$. Given this, the proof of the satisfaction condition proceeds by simple case analysis: an equation $t = t'$ holds in A if and only if either the values of t and t' in A are defined and equal — and then they are also equal in $\mu^{Mod}_{\Sigma}(A)$ — or the values of t and t' are undefined in A — and then the values of t and t' in $\mu^{Mod}_{\Sigma}(A)$ are equal to $\perp$ (and hence equal as well).

Exercise. Complete the proof of the satisfaction condition. Try to generalise the above institution morphism to an institution morphism from $\mathbf{PEQ_{str}}$ to **EQ**, and point out where the proof of the satisfaction condition would break down. HINT: See Example 10.2.8. □

Example 10.2.8. Consider a new institution $\mathbf{PEQ_{pval}}$ which is just like $\mathbf{PEQ_{str}}$ except that the satisfaction relation takes into account *partial* valuations of variables as well. That is, for any algebraic signature $\Sigma \in |\mathbf{AlgSig}|$, partial Σ-algebra $A \in |\mathbf{PAlg}(\Sigma)|$ and Σ-equation $\forall X \bullet t = t'$, $A \models_{\mathbf{PEQ_{pval}},\Sigma} \forall X \bullet t = t'$ if for all *partial* functions $v : X \rightarrow |A|$, either the values of both t and t' under v in A are undefined or they are defined and equal. The satisfaction of definedness formulae is defined analogously. Notice that this yields a different satisfaction relation than $\models_{\mathbf{PEQ_{str}},\Sigma}$.

The institution semi-morphism $\langle \mu^{Sign}, \mu^{Mod} \rangle : \mathbf{PEQ_{str}} \to_{semi} \mathbf{EQ}$ presented in Example 10.1.5 is also an institution semi-morphism $\langle \mu^{Sign}, \mu^{Mod} \rangle : \mathbf{PEQ_{pval}} \to_{semi} \mathbf{EQ}$ (by Example 10.1.7). It extends to an institution morphism $\mu : \mathbf{PEQ_{pval}} \to \mathbf{EQ}$ in the obvious way, so that the satisfaction condition is ensured:

- For each algebraic signature $\Sigma \in |\mathbf{AlgSig}|$, $\mu^{Sen}_{\Sigma} : \mathbf{Sen_{EQ}}(\Sigma) \to \mathbf{Sen_{PEQ_{pval}}}(\Sigma)$ is the inclusion of Σ-equations into the set of Σ-equations and Σ-definedness formulae.

Exercise. Following the same pattern, define a new institution $\mathbf{PFOPEQ_{pval}}$ of partial first-order logic with equality based on partial valuations of variables, and then construct an institution morphism $\mu : \mathbf{PFOPEQ_{pval}} \to \mathbf{FOPEQ}$. □

Example 10.2.9. There is also an institution morphism $\mu : \mathbf{EQ} \to \mathbf{PEQ}$ which extends the institution semi-morphism $\langle \mu^{Sign}, \mu^{Mod} \rangle : \mathbf{EQ} \to_{semi} \mathbf{PEQ}$ defined in Example 10.1.6 as follows:

- For each algebraic signature $\Sigma \in |\mathbf{AlgSig}|$, $\mu^{Sen} : \mathbf{Sen_{PEQ}}(\Sigma) \to \mathbf{Sen_{EQ}}(\Sigma)$ is the identity on Σ-equations and maps each Σ-definedness formula of the form $\forall X \bullet def(t)$ to the Σ-equation $\forall X \bullet t = t$ (in fact, any trivially true Σ-equation would do).

The satisfaction condition obviously holds, since in any algebra in the range of the translation of models — that is, in any partial algebra in which all operations are totally defined — equations are interpreted in the same way in **EQ** and in **PEQ**, and definedness formulae are always satisfied. □

Example 10.2.10. There is an institution morphism $\mu : \mathbf{3FOPEQ} \to \mathbf{PEQ}$ which extends the institution semi-morphism $\langle \mu^{Sign}, \mu^{Mod} \rangle : \mathbf{3FOPEQ} \to_{semi} \mathbf{PEQ}$ of Exercise 10.1.4 as follows:

- For each $\Theta \in |\mathbf{FOSig}|$, the translation of sentences $\mu^{Sen}_{\Theta} : \mathbf{Sen_{PEQ}}(\mu^{Sign}(\Theta)) \to \mathbf{Sen_{3FOPEQ}}(\Theta)$ is defined as follows:
  - Each definedness statement $\forall X \bullet def(t) \in \mathbf{Sen_{PEQ}}(\mu^{Sign}(\Theta))$ is translated to the sentence $(\forall x_1{:}s_1, \cdots, x_n{:}s_n \bullet def(t))$ **is** $tt \in \mathbf{Sen_{3FOPEQ}}(\Theta)$, where X is $\{x_1{:}s_1, \ldots, x_n{:}s_n\}$.[4]
  - Each equation $\forall X \bullet t = t' \in \mathbf{Sen_{PEQ}}(\mu^{Sign}(\Theta)$ is translated to the sentence

$$(\forall x_1{:}s_1, \cdots, x_n{:}s_n \bullet t = t' \vee (\neg def(t) \wedge \neg def(t'))) \text{ **is** } tt \in \mathbf{Sen_{3FOPEQ}}(\Theta)$$

    where again X is $\{x_1{:}s_1, \ldots, x_n{:}s_n\}$.

The proof of the satisfaction condition relies on the property that for every first-order signature $\Theta \in |\mathbf{FOSig}|$, the value (whether defined or not) of a term t in a

[4] As in Example 10.2.2, the order of variables may be chosen in an arbitrary but uniform way.

partial first-order Θ-structure $A \in |\mathbf{Mod_{3FOPEQ}}(\Theta)|$ does not depend on the interpretation of predicates in A. Notice that because of the different interpretation of equations in **PEQ** and in **3FOPEQ**, we had to use an appropriate "encoding" of (strong) equations in **PEQ** as disjunctions of formulae in **3FOPEQ**. □

Exercise 10.2.11. Recall the institution semi-morphism $\langle \mu^{Sign}, \mu^{Mod} \rangle : \mathbf{IMP}_{DT} \to_{semi} \mathbf{PEQ}$ defined in Example 10.1.9. Show that it cannot be extended to an institution morphism.

Similarly, show that the institution semi-morphisms from **EQ** to **FOPEQ** and from **PEQ** to **PFOPEQ** defined in Example 10.1.8 cannot be extended to institution morphisms. □

Exercise 10.2.12. Recall the institution semi-morphism $\langle \mu^{Sign}, \mu^{Mod} \rangle : \mathbf{FProg} \to_{semi} \mathbf{PEQ}$ defined in Exercise 10.1.10. Show that it cannot be extended to an institution morphism. However, as mentioned in Exercise 10.1.10 as well, this is also an institution semi-morphism $\langle \mu^{Sign}, \mu^{Mod} \rangle : \mathbf{FPL_{stnd}} \to_{semi} \mathbf{PEQ}$, and so by Example 10.1.7, also an institution semi-morphism $\langle \mu^{Sign}, \mu^{Mod} \rangle : \mathbf{FPL_{stnd}} \to_{semi} \mathbf{PFOPEQ}$. Show that this extends to an institution morphism $\mu : \mathbf{FPL_{stnd}} \to \mathbf{PFOPEQ}$, with first-order sentences (with algebraic terms only) viewed as a subset of **FPL**-sentences.

Show also that the trivial semi-morphism from $\mathbf{FPL_{stnd}}$ to **FProg** (given by Example 10.1.7) also extends to an institution morphism in an obvious way. □

For the rest of this section, let $\mathbf{INS} = \langle \mathbf{Sign}, \mathbf{Sen}, \mathbf{Mod}, \langle \models_\Sigma \rangle_{\Sigma \in |\mathbf{Sign}|} \rangle$ and $\mathbf{INS}' = \langle \mathbf{Sign}', \mathbf{Sen}', \mathbf{Mod}', \langle \models'_{\Sigma'} \rangle_{\Sigma' \in |\mathbf{Sign}'|} \rangle$ be institutions linked by an institution morphism $\mu : \mathbf{INS} \to \mathbf{INS}'$.

Using the consequences of the satisfaction condition for institution morphisms in much the same way as we did in Section 4.2 for the satisfaction condition in a single institution, one can show that the semantic consequence relation is preserved when sentences are translated by an institution morphism. A number of related facts then follow in a similar way.

Proposition 10.2.13. *Consider an* **INS**-*signature* $\Sigma \in |\mathbf{Sign}|$ *and let* $\Sigma' = \mu^{Sign}(\Sigma) \in |\mathbf{Sign}'|$. *Then, for any sets* $\Phi' \subseteq \mathbf{Sen}'(\Sigma')$ *of* Σ'-*sentences and* $\Phi \subseteq \mathbf{Sen}(\Sigma)$ *of* Σ-*sentences, and* Σ'-*sentence* $\varphi' \in \mathbf{Sen}'(\Sigma')$, *we have:*

1. $\mu^{Sen}_\Sigma(Cl_{\Sigma'}(\Phi')) \subseteq Cl_\Sigma(\mu^{Sen}_\Sigma(\Phi'))$. *In other words, if* $\Phi' \models'_{\Sigma'} \varphi'$ *then* $\mu^{Sen}_\Sigma(\Phi') \models_\Sigma \mu^{Sen}_\Sigma(\varphi')$.
2. $Cl_\Sigma(\mu^{Sen}_\Sigma(Cl_{\Sigma'}(\Phi'))) = Cl_\Sigma(\mu^{Sen}_\Sigma(\Phi'))$.
3. $Cl_{\Sigma'}((\mu^{Sen}_\Sigma)^{-1}(\Phi)) \subseteq (\mu^{Sen}_\Sigma)^{-1}(Cl_\Sigma(\Phi))$.
4. $Cl_{\Sigma'}((\mu^{Sen}_\Sigma)^{-1}(\Phi)) \subseteq Th_{\Sigma'}(\mu^{Mod}_\Sigma(Mod_\Sigma(\Phi)))$.

Proof. The straightforward proof in each case proceeds exactly as the proof of a corresponding proposition or corollary in Section 4.2, using the satisfaction condition for institution morphisms in place of the satisfaction condition for institutions.

Exercise. Prove each of the above statements directly. Then identify the corresponding results in Section 4.2 and compare the proofs. □

The preservation of theories (closed sets of sentences) of **INS** under institution morphisms deserves special attention:

Proposition 10.2.14. *Consider any* **INS**-*signature* $\Sigma \in |\mathbf{Sign}|$, $\mu^{Sign}(\Sigma) = \Sigma'$, *and set* $\Phi \subseteq \mathbf{Sen}(\Sigma)$ *of* Σ-*sentences. Then* $(\mu^{Sen}_{\Sigma})^{-1}(Cl_{\Sigma}(\Phi)) = Th_{\Sigma'}(\mu^{Mod}_{\Sigma}(Mod_{\Sigma}(\Phi)))$. *Consequently, if* Φ *is closed then* $(\mu^{Sen}_{\Sigma})^{-1}(\Phi) \subseteq \mathbf{Sen}'(\Sigma')$ *is a closed set of* Σ'-*sentences in* **INS**$'$.

Proof. Again, we refer to the proof of Proposition 4.2.15. □

The above proposition determines a mapping of theories in **INS** to theories in **INS**$'$. Moreover, the naturality of the translations of sentences ensures that this mapping extends to a functor.

Definition 10.2.15 (Theory functor induced by an institution morphism). Every institution morphism $\mu\colon \mathbf{INS} \to \mathbf{INS}'$ determines a functor $\mu^{Th}\colon \mathbf{Th}_{\mathbf{INS}} \to \mathbf{Th}_{\mathbf{INS}'}$ between the categories of theories of **INS** and of **INS**$'$, respectively:

- For any **INS**-theory $\langle \Sigma, \Phi \rangle \in |\mathbf{Th}_{\mathbf{INS}}|$, $\mu^{Th}(\langle \Sigma, \Phi \rangle) = \langle \mu^{Sign}(\Sigma), (\mu^{Sen}_{\Sigma})^{-1}(\Phi)\rangle \in |\mathbf{Th}_{\mathbf{INS}'}|$.
- For any **INS**-theory morphism $\sigma\colon T_1 \to T_2$, $\mu^{Th}(\sigma) = \mu^{Sign}(\sigma)\colon \mu^{Th}(T_1) \to \mu^{Th}(T_2)$ is an **INS**$'$-theory morphism. □

As in Section 4.2, none of the implications or inclusions in Proposition 10.2.13 may be reversed in general. However, the situation is again different if model translation is surjective (by this we refer to the surjectivity of the translation functors on models; surjectivity on model morphisms is not required).

Proposition 10.2.16. *Consider an* **INS**-*signature* $\Sigma \in |\mathbf{Sign}|$ *and let* $\Sigma' = \mu^{Sign}(\Sigma) \in |\mathbf{Sign}'|$. *Suppose that the translation of models* $\mu^{Mod}_{\Sigma}\colon \mathbf{Mod}(\Sigma) \to \mathbf{Mod}'(\Sigma')$ *is surjective on models. Then, for any set* $\Phi' \subseteq \mathbf{Sen}'(\Sigma')$ *of* Σ'-*sentences and* Σ'-*sentence* $\varphi' \in \mathbf{Sen}'(\Sigma')$, *we have:*

1. $\Phi' \models'_{\Sigma'} \varphi'$ *if and only if* $\mu^{Sen}_{\Sigma}(\Phi') \models_{\Sigma} \mu^{Sen}_{\Sigma}(\varphi')$.
2. $Cl_{\Sigma'}(\Phi') = (\mu^{Sen}_{\Sigma})^{-1}(Cl_{\Sigma}(\mu^{Sen}_{\Sigma}(\Phi')))$.

Proof. See the proof of Proposition 4.2.17 for part (1). Part (2) corresponds to Corollary 4.2.18. □

The last proposition suggests that institution morphisms in which all translations of models are surjective may be especially useful. In particular, not only do semantic consequences in the more primitive institution **INS**$'$ remain valid when translated to the richer institution **INS** (this is Proposition 10.2.13(1)), but also whatever we can deduce in the richer logic **INS** about the sentences of the more primitive logic **INS**$'$ is valid in the context of the more primitive logic **INS**$'$ as well. This means that any proof system or theorem prover developed for **INS** may be used for **INS**$'$ as well; for instance, a resolution-based theorem prover for first-order logic with equality may be used for deriving consequences in equational logic. The following example illustrates that this is not possible for an arbitrary institution morphism.

Example 10.2.17. Example 10.2.9 defines an institution morphism $\mu:\mathbf{EQ}\to\mathbf{PEQ}$. The translation of **PEQ**-sentences to **EQ** preserves semantic consequence, but does not reflect it. For example, let Σ be an algebraic signature with a sort s and a constant $a{:}s$. Consider the Σ-definedness formula $def(a)$. Then $\mu^{Sen}_\Sigma(def(a))$ is the equation $a=a$, and we trivially have $\varnothing\models_{\mathbf{EQ},\Sigma} a=a$. Consequently, we can deduce that $def(a)$ holds in any partial Σ-algebra *in the range of the model translation functor* $\mu^{Mod}_\Sigma:\mathbf{Alg}(\Sigma)\to\mathbf{PAlg}(\Sigma)$ (which is trivial to check directly as well). It does not follow that the same is true for all partial Σ-algebras; indeed, $\varnothing\not\models_{\mathbf{PEQ},\Sigma} def(a)$. □

Exercise 10.2.18. Consider the institution morphism $\mu:\mathbf{PEQ_{pval}}\to\mathbf{EQ}$ sketched in Example 10.2.8, where $\mathbf{PEQ_{pval}}$ is the institution of partial equational logic with partial valuations of variables. Show that the translation of sentences does not reflect semantic consequence. HINT: The sentence $\forall x{:}s\bullet x=a$, where a is a constant of sort s, defines in $\mathbf{PEQ_{pval}}$ those models having an empty carrier of sort s. □

The following generalisation of Proposition 10.2.16(1) corresponds to Proposition 4.2.21.

Proposition 10.2.19. *Consider an* **INS**-*signature* $\Sigma\in|\mathbf{Sign}|$ *and let* $\Sigma'=\mu^{Sign}(\Sigma)\in|\mathbf{Sign}'|$.

Suppose that the translations of Σ*-models in* **INS** *that satisfy a set* $\Gamma_\Sigma\subseteq\mathbf{Sen}(\Sigma)$ *of* Σ*-sentences are exactly characterised in* **INS**$'$ *by a set* $\Gamma'_\Sigma\subseteq\mathbf{Sen}'(\Sigma')$ *of* Σ'*-sentences: that is,* $Mod_{\Sigma'}(\Gamma'_\Sigma)=\mu^{Mod}_\Sigma(Mod_\Sigma(\Gamma_\Sigma))$*. Then for any set* $\Phi'\subseteq\mathbf{Sen}'(\Sigma')$ *of* Σ'*-sentences and* Σ'*-sentence* $\varphi'\in\mathbf{Sen}'(\Sigma')$, $\Phi'\cup\Gamma'_\Sigma\models'_{\Sigma'}\varphi'$ *if and only if* $\mu^{Sen}_\Sigma(\Phi')\cup\Gamma_\Sigma\models_\Sigma\mu^{Sen}_\Sigma(\varphi')$.

Proof. See the proof of Proposition 4.2.21. □

Corollary 10.2.20. *Consider an* **INS**-*signature* $\Sigma\in|\mathbf{Sign}|$ *and let* $\Sigma'=\mu^{Sign}(\Sigma)\in|\mathbf{Sign}'|$.

Suppose that the translations of Σ*-models are exactly characterised in* **INS**$'$ *by a set* $\Gamma'_\Sigma\subseteq\mathbf{Sen}'(\Sigma')$ *of* Σ'*-sentences: that is,* $Mod_{\Sigma'}(\Gamma'_\Sigma)=\mu^{Mod}_\Sigma(|\mathbf{Mod}(\Sigma)|)$*. Then for any set* $\Phi'\subseteq\mathbf{Sen}'(\Sigma')$ *of* Σ'*-sentences and* Σ'*-sentence* $\varphi'\in\mathbf{Sen}'(\Sigma')$, $\Phi'\cup\Gamma'_\Sigma\models'_{\Sigma'}\varphi'$ *if and only if* $\mu^{Sen}_\Sigma(\Phi')\models_\Sigma\mu^{Sen}_\Sigma(\varphi')$. □

In the same way as the surjectivity of the reduct functor was not sufficient in Section 4.2 to ensure that the coimage of sets of sentences w.r.t. a signature morphism commutes with the closure operator (Exercise 4.2.20), the coimage of sets of sentences w.r.t. an institution morphism does not commute with the closure operator either, even if the translations of models are surjective. It is also somewhat disappointing to notice that the functor on theories determined by an institution morphism (Definition 10.2.15) does not preserve colimits of theories in general.

Example 10.2.21. Recall the institution morphism $\mu:\mathbf{FOPEQ}\to\mathbf{EQ}$ defined in Example 10.2.2. Notice that for any first-order signature $\Theta\in|\mathbf{FOSig}|$, the corresponding translation of models $\mu^{Mod}_\Theta:\mathbf{FOStr}(\Theta)\to\mathbf{Alg}(\mu^{Sign}(\Theta))$ is surjective (every $\mu^{Sign}(\Theta)$-algebra may be expanded to a Θ-structure, for example by interpreting the predicates as always false). Notice also that **FOPEQ** is an exact institution and moreover the translation of signatures μ^{Sign}: **FOSig** $\rightarrow$ **AlgSig** is cocontinuous.

Inconsistent sets of first-order sentences $\Phi \subseteq \mathbf{Sen}_{\mathbf{FOPEQ}}(\Theta)$ (for some first-order signature Θ with $\Sigma = \mu^{Sign}(\Theta)$) provide counterexamples which show that in general $Cl_\Sigma((\mu^{Sen}_\Theta)^{-1}(\Phi)) \neq (\mu^{Sen}_\Theta)^{-1}(Cl_\Theta(\Phi))$ (and hence $Cl_\Sigma((\mu^{Sen}_\Theta)^{-1}(\Phi)) \neq Th_\Sigma(\mu^{Mod}_\Theta(Mod_\Theta(\Phi)))$), and that the translation of theories μ^{Th}: $\mathbf{Th}_{\mathbf{FOPEQ}} \rightarrow \mathbf{Th}_{\mathbf{EQ}}$ does not preserve colimits. For example:

Let Θ be a first-order signature with a sort s, two constants $a, b{:}\,s$, and a unary predicate $p{:}\,s$, and let $\Sigma = \mu^{Sign}(\Theta)$. Let $\Phi = \{\neg p(a), p(a)\} \subseteq \mathbf{Sen}_{\mathbf{FOPEQ}}(\Theta)$. Then $Cl_\Theta(\Phi) = \mathbf{Sen}_{\mathbf{FOPEQ}}(\Theta)$, while $(\mu^{Sen})^{-1}(\Phi) = \varnothing$, and so $Cl_\Sigma((\mu^{Sen}_\Theta)^{-1}(\Phi)) \neq (\mu^{Sen}_\Theta)^{-1}(Cl_\Theta(\Phi))$. Similarly, $Cl_\Sigma((\mu^{Sen}_\Theta)^{-1}(\Phi)) \neq Th_\Sigma(\mu^{Mod}_\Theta(Mod_\Theta(\Phi)))$ (since $Mod_\Theta(\Phi) = \varnothing$).

Then, consider $\Phi_0 = \varnothing$, $\Phi_1 = \{p(a)\}$, and $\Phi_2 = \{\neg p(a)\}$. The identity on Θ is a theory morphism from $Cl_\Theta(\Phi_0)$ to $Cl_\Theta(\Phi_1)$ as well as from $Cl_\Theta(\Phi_0)$ to $Cl_\Theta(\Phi_2)$. The pushout of these two morphisms in $\mathbf{Th}_{\mathbf{FOPEQ}}$ yields the theory $Cl_\Theta(\Phi)$ (again over Θ). Then, $\mu^{Th}(\langle\Theta, Cl_\Theta(\Phi)\rangle) = \langle\Sigma, \mathbf{Sen}_{\mathbf{EQ}}(\Sigma)\rangle$. However, since $\mu^{Th}(\langle\Theta, Cl_\Theta(\Phi_1)\rangle) = \mu^{Th}(\langle\Theta, Cl_\Theta(\Phi_2)\rangle) = \langle\Sigma, Cl_\Sigma(\varnothing)\rangle$ (which is easy to check directly), the pushout of the corresponding diagram in $\mathbf{Th}_{\mathbf{EQ}}$ yields $\langle\Sigma, Cl_\Sigma(\varnothing)\rangle$, which is distinct from the image of the pushout in $\mathbf{Th}_{\mathbf{FOPEQ}}$. □

The essence of the problem indicated by the above example is that given an institution morphism μ: **INS** $\rightarrow$ **INS**$'$ we have tried to "simulate" a construction performed in the richer institution **INS** by the same construction in the more primitive institution **INS**$'$ — and this cannot really be expected to work. On the other hand, whenever a construction in the more primitive framework of **INS**$'$ can be embedded into the richer framework of **INS**, the results should be preserved. The following exercise illustrates how this might work; see also Section 10.4, Proposition 10.4.25.

Exercise 10.2.22. Consider an institution morphism μ: **INS** $\rightarrow$ **INS**$'$ together with a functor **inv**: **Sign**$' \rightarrow$ **Sign** such that **inv**;$\mu^{Sign} = \mathbf{Id}_{\mathbf{Sign}'}$ (think of $\mathbf{inv}(\Sigma') \in |\mathbf{Sign}|$ as a "canonical" choice of expansion of each **INS**$'$-signature $\Sigma' \in |\mathbf{Sign}'|$ to an **INS**-signature).

Define a functor on theories $\mu^{Th}_{\mathbf{inv}}$: $\mathbf{Th}_{\mathbf{INS}'} \rightarrow \mathbf{Th}_{\mathbf{INS}}$ which maps any **INS**$'$-theory $\langle\Sigma', \Phi'\rangle \in |\mathbf{Th}_{\mathbf{INS}'}|$ to the **INS**-theory $\langle\mathbf{inv}(\Sigma'), Cl_{\mathbf{inv}(\Sigma')}(\mu^{Sen}_{\mathbf{inv}(\Sigma')}(\Phi'))\rangle \in |\mathbf{Th}_{\mathbf{INS}}|$. This map extends to theory morphisms by the naturality of μ^{Sen}: μ^{Sign};**Sen**$' \rightarrow$ **Sen** and by Proposition 4.2.26.

Prove that if **inv**: **Sign**$' \rightarrow$ **Sign** is (finitely) cocontinuous then so is $\mu^{Th}_{\mathbf{inv}}$: $\mathbf{Th}_{\mathbf{INS}'} \rightarrow \mathbf{Th}_{\mathbf{INS}}$. HINT: Use the explicit construction of colimits in the category of theories given in the proof of Theorem 4.4.1, and then rely on the naturality of the translations of sentences and on Proposition 10.2.13(1, 2) as well as Proposition 4.2.9.

Look through the examples of institution morphisms μ: **INS** $\rightarrow$ **INS**$'$ given earlier in this section and check for which of them such a "canonical" choice of signature expansions **inv**: **Sign**$' \rightarrow$ **Sign** may be given. □

10.3 The category of institutions

It is easy to see that institution morphisms as introduced in Section 10.2 compose in a natural, componentwise way, and in fact institutions with institution morphisms form a category:

Definition 10.3.1 (Category of institutions). Consider three institutions $\mathbf{INS} = \langle \mathbf{Sign}, \mathbf{Sen}, \mathbf{Mod}, \langle \models_\Sigma \rangle_{\Sigma \in |\mathbf{Sign}|} \rangle$, $\mathbf{INS}' = \langle \mathbf{Sign}', \mathbf{Sen}', \mathbf{Mod}', \langle \models'_{\Sigma'} \rangle_{\Sigma' \in |\mathbf{Sign}'|} \rangle$, $\mathbf{INS}'' = \langle \mathbf{Sign}'', \mathbf{Sen}'', \mathbf{Mod}'', \langle \models''_{\Sigma''} \rangle_{\Sigma'' \in |\mathbf{Sign}''|} \rangle$. Given institution morphisms $\mu_1 : \mathbf{INS} \to \mathbf{INS}'$ and $\mu_2 : \mathbf{INS}' \to \mathbf{INS}''$, their *composition* $\mu_1;\mu_2 : \mathbf{INS} \to \mathbf{INS}''$ is defined as follows:

- The translation of signatures is defined as the composition of the translation of signatures of the two institution morphisms:

  $$(\mu_1;\mu_2)^{Sign} = \mu_1^{Sign};\mu_2^{Sign} \quad : \mathbf{Sign} \to \mathbf{Sign}''$$

- The translations of sentences are defined by the pointwise composition of the translations of sentences of the two institution morphisms, where the translation of signatures is used to identify the appropriate components:[5]

  $$(\mu_1;\mu_2)^{Sen} = (\mu_1^{Sign} \cdot \mu_2^{Sen});\mu_1^{Sen} \quad : \mu_1^{Sign};\mu_2^{Sign};\mathbf{Sen}'' \to \mathbf{Sen}$$

  That is, for each $\Sigma \in |\mathbf{Sign}|$,

  $$(\mu_1;\mu_2)^{Sen}_\Sigma = (\mu_2^{Sen})_{\mu_1^{Sign}(\Sigma)};(\mu_1^{Sen})_\Sigma : \mathbf{Sen}''(\mu_2^{Sign}(\mu_1^{Sign}(\Sigma))) \to \mathbf{Sen}(\Sigma).$$

- The translations of models are defined by the pointwise composition of the translations of models of the two institution morphisms, where the translation of signatures is used to identify the appropriate components:[5]

  $$(\mu_1;\mu_2)^{Mod} = \mu_1^{Mod};((\mu_1^{Sign})^{op} \cdot \mu_2^{Mod}) \quad : \mathbf{Mod} \to (\mu_1^{Sign};\mu_2^{Sign})^{op};\mathbf{Mod}''$$

  That is, for each $\Sigma \in |\mathbf{Sign}|$,

  $$(\mu_1;\mu_2)^{Mod}_\Sigma = (\mu_1^{Mod})_\Sigma;(\mu_2^{Mod})_{\mu_1^{Sign}(\Sigma)} : \mathbf{Mod}(\Sigma) \to \mathbf{Mod}''(\mu_2^{Sign}(\mu_1^{Sign}(\Sigma))).$$

This yields a category $\mathcal{INS}$ of institutions and their morphisms. □

Exercise 10.3.2. Prove that the composition of institution morphisms does indeed yield an institution morphism (in particular, prove the satisfaction condition). Check that composition is associative. What are the identities in $\mathcal{INS}$? □

The category $\mathcal{INS}$ of institutions and their morphisms may be viewed as a framework for putting together institutions, in much the same way as we viewed the category of theories as a framework for putting together theories (Theorem 4.4.1), and

[5] See Definitions 3.4.42 and 3.4.46 for vertical composition of natural transformations and for multiplication of a natural transformation by a functor.

similarly for structured specifications (Theorem 5.5.11). In particular, we show below that the category $\mathcal{INS}$ of institutions is complete, and hence limits are available as a tool for combining institutions. Since we have chosen — somewhat arbitrarily — to define institution morphisms so that they go from a richer, more powerful institution to a simpler, more primitive one, it is the limit construction which is of interest here; in the case of specifications and theories, morphisms go from simpler to richer objects, and so objects are put together via the colimit construction.

The proof of completeness of $\mathcal{INS}$ is not very difficult in principle. However, institutions and their morphisms as presently defined are rather complex, and using these definitions directly in the proof would entail quite a heavy notational burden. We therefore use some simple categorical concepts to reformulate the definition of the category of institution in a much more compact (but also less transparent) form.

Consider an institution $\mathbf{INS} = \langle \mathbf{Sign}, \mathbf{Sen}, \mathbf{Mod}, \langle \models_\Sigma \rangle_{\Sigma \in |\mathbf{Sign}|} \rangle$. For any signature $\Sigma \in |\mathbf{Sign}|$, the satisfaction relation $\models_\Sigma \subseteq |\mathbf{Mod}(\Sigma)| \times \mathbf{Sen}(\Sigma)$ may be identified with a function $\mathbf{sat}_\Sigma : |\mathbf{Mod}(\Sigma)| \to [\mathbf{Sen}(\Sigma) \to \mathbf{B}]$, where $\mathbf{B} = \{ff, tt\}$ is a two-element set and $[\mathbf{Sen}(\Sigma) \to \mathbf{B}]$ stands for the set of all functions from $\mathbf{Sen}(\Sigma)$ to $\mathbf{B}$, as in Exercise 3.4.16. For any Σ-model $M \in |\mathbf{Mod}(\Sigma)|$, $\mathbf{sat}_\Sigma(M) \in [\mathbf{Sen}(\Sigma) \to \mathbf{B}]$ is a function which maps any Σ-sentence $\varphi \in \mathbf{Sen}(\Sigma)$ to its boolean meaning in M:

$$\mathbf{sat}_\Sigma(M)(\varphi) = \begin{cases} tt & \text{if } M \models_\Sigma \varphi \\ ff & \text{otherwise.} \end{cases}$$

Then, for any signature morphism $\sigma : \Sigma \to \Sigma'$, we have a functor $\_|_\sigma : \mathbf{Mod}(\Sigma') \to \mathbf{Mod}(\Sigma)$ and a function $\sigma : \mathbf{Sen}(\Sigma) \to \mathbf{Sen}(\Sigma')$ such that for any Σ'-model $M' \in |\mathbf{Mod}(\Sigma')|$ and Σ-sentence $\varphi \in \mathbf{Sen}(\Sigma)$ we have $\mathbf{sat}_\Sigma(M'|_\sigma)(\varphi) = \mathbf{sat}_{\Sigma'}(M')(\sigma(\varphi))$ (the satisfaction condition). The translation of sentences $\sigma : \mathbf{Sen}(\Sigma) \to \mathbf{Sen}(\Sigma')$ determines by "pre-composition" a function $\sigma;(\_) : [\mathbf{Sen}(\Sigma') \to \mathbf{B}] \to [\mathbf{Sen}(\Sigma) \to \mathbf{B}]$ (written as $[\sigma \to \mathbf{B}]$ in Exercise 3.4.16), and the satisfaction condition may then be expressed as the requirement that the following diagram commute:

$$\begin{array}{ccc}
|\mathbf{Mod}(\Sigma)| & \xrightarrow{\ \mathbf{sat}_\Sigma\ } & [\mathbf{Sen}(\Sigma) \to \mathbf{B}] \\
{\scriptstyle |_|_\sigma|} \uparrow & & \uparrow {\scriptstyle \sigma;(_)} \\
|\mathbf{Mod}(\Sigma')| & \xrightarrow[\ \mathbf{sat}_{\Sigma'}\ ]{} & [\mathbf{Sen}(\Sigma') \to \mathbf{B}]
\end{array}$$

In other words, for any signature $\Sigma \in |\mathbf{Sign}|$, the triple $\langle \mathbf{Mod}(\Sigma), \mathbf{sat}_\Sigma, \mathbf{Sen}(\Sigma) \rangle$ is an object of the comma category $(|\_|, [\_ \to \mathbf{B}])$ (see Definition 3.4.50), where $|\_| : \mathbf{Cat} \to \mathbf{Set}$ is the functor which for any category yields the collection of its objects (see Example 3.4.28) and $[\_ \to \mathbf{B}] : \mathbf{Set}^{op} \to \mathbf{Set}$ is the functor which for any set X yields the set of all functions from X to $\mathbf{B}$; see Exercise 3.4.16. Moreover, for any signature morphism $\sigma : \Sigma \to \Sigma'$, the pair $\langle \_|_\sigma : \mathbf{Mod}(\Sigma') \to \mathbf{Mod}(\Sigma), \sigma : \mathbf{Sen}(\Sigma) \to \mathbf{Sen}(\Sigma') \rangle$ is a morphism from $\langle \mathbf{Mod}(\Sigma'), \mathbf{sat}_{\Sigma'}, \mathbf{Sen}(\Sigma') \rangle$ to $\langle \mathbf{Mod}(\Sigma), \mathbf{sat}_\Sigma, \mathbf{Sen}(\Sigma) \rangle$ in this comma category. We will refer to the triples $\langle \mathbf{Mod}(\Sigma), \mathbf{sat}_\Sigma, \mathbf{Sen}(\Sigma)\rangle$ as *boolean rooms*, and to their morphisms as *room morphisms*.

Definition 10.3.3 (Category of boolean rooms). The category of (boolean) *rooms* $\mathbf{Room}(\mathbf{B})$ is the comma category $(|\_|, [\_ \to \mathbf{B}])$. □

Summing up the above discussion, any institution **INS** is (i.e. may be viewed as) a functor

$$\mathbf{INS}\colon \mathbf{Sign}^{op} \to \mathbf{Room}(\mathbf{B}).$$

Exercise 10.3.4. Show the converse: any functor $\mathbf{I}\colon \mathbf{Sign}_\mathbf{I}^{op} \to \mathbf{Room}(\mathbf{B})$ is (i.e. may be viewed as) an institution.

HINT: For any comma category $(\mathbf{F},\mathbf{G})$, where $\mathbf{F}\colon \mathbf{K1} \to \mathbf{K}$ and $\mathbf{G}\colon \mathbf{K2} \to \mathbf{K}$, define two functors $\mathbf{Left}\colon (\mathbf{F},\mathbf{G}) \to \mathbf{K1}$ and $\mathbf{Right}\colon (\mathbf{F},\mathbf{G}) \to \mathbf{K2}$, and a mapping $\mathbf{Middle}\colon |(\mathbf{F},\mathbf{G})| \to \mathbf{K}$ in the obvious way. Then put $\mathbf{Mod_I} = \mathbf{I};\mathbf{Left}\colon \mathbf{Sign}_\mathbf{I}^{op} \to \mathbf{Cat}$, $\mathbf{Sen_I} = (\mathbf{I};\mathbf{Right})^{op}\colon \mathbf{Sign_I} \to \mathbf{Set}$, and $\mathbf{sat}_{\mathbf{I},\Sigma} = \mathbf{Middle}(\mathbf{I}(\Sigma))\colon |\mathbf{Mod_I}(\Sigma)| \to [\mathbf{Sen_I}(\Sigma) \to \mathbf{B}]$ for $\Sigma \in |\mathbf{Sign_I}|$. Check the satisfaction condition! □

Consider institutions $\mathbf{INS}\colon \mathbf{Sign}^{op} \to \mathbf{Room}(\mathbf{B})$ and $\mathbf{INS}'\colon (\mathbf{Sign}')^{op} \to \mathbf{Room}(\mathbf{B})$ explicitly given as, respectively, $\mathbf{INS} = \langle \mathbf{Sign}, \mathbf{Sen}, \mathbf{Mod}, \langle \models_\Sigma\rangle_{\Sigma\in|\mathbf{Sign}|}\rangle$ and $\mathbf{INS}' = \langle \mathbf{Sign}', \mathbf{Sen}', \mathbf{Mod}', \langle \models'_{\Sigma'}\rangle_{\Sigma'\in|\mathbf{Sign}'|}\rangle$ and an institution morphism $\mu\colon \mathbf{INS} \to \mathbf{INS}'$ between them. Then μ consists of a functor $\mu^{Sign}\colon \mathbf{Sign} \to \mathbf{Sign}'$ and a family of pairs $\langle \mu_\Sigma^{Mod}\colon \mathbf{Mod}(\Sigma) \to \mathbf{Mod}'(\mu^{Sign}(\Sigma)), \mu_\Sigma^{Sen}\colon \mathbf{Sen}'(\mu^{Sign}(\Sigma)) \to \mathbf{Sen}(\Sigma)\rangle$, for $\Sigma \in |\mathbf{Sign}|$, such that the satisfaction condition holds. As in the case of the satisfaction condition in an institution, the satisfaction condition for μ may be expressed by requiring the following diagram to commute, where we put $\Sigma' = \mu^{Sign}(\Sigma)$, and $\mathbf{sat}_\Sigma$ and $\mathbf{sat}'_{\Sigma'}$ represent $\models_\Sigma$ and $\models'_{\Sigma'}$, respectively:

$$\begin{array}{ccc}
|\mathbf{Mod}'(\Sigma')| & \xrightarrow{\ \mathbf{sat}'_{\Sigma'}\ } & [\mathbf{Sen}'(\Sigma') \to \mathbf{B}] \\
{\scriptstyle |\mu_\Sigma^{Mod}|}\Big\uparrow & & \Big\uparrow{\scriptstyle \mu_\Sigma^{Sen};(_)} \\
|\mathbf{Mod}(\Sigma)| & \xrightarrow[\ \mathbf{sat}_\Sigma\ ]{} & [\mathbf{Sen}(\Sigma) \to \mathbf{B}]
\end{array}$$

This means that the pair $\langle \mu_\Sigma^{Mod}, \mu_\Sigma^{Sen}\rangle$ is a morphism in the comma category $\mathbf{Room}(\mathbf{B})$ from $\langle \mathbf{Mod}(\Sigma), \mathbf{sat}_\Sigma, \mathbf{Sen}(\Sigma)\rangle$ to $\langle \mathbf{Mod}'(\Sigma'), \mathbf{sat}'_{\Sigma'}, \mathbf{Sen}'(\Sigma')\rangle$. Moreover, since μ^{Mod} and μ^{Sen} are natural transformations, these pairs form a natural transformation from $\mathbf{INS}\colon \mathbf{Sign}^{op} \to \mathbf{Room}(\mathbf{B})$ to $(\mu^{Sign})^{op};\mathbf{INS}'\colon \mathbf{Sign}^{op} \to \mathbf{Room}(\mathbf{B})$. Thus, the institution morphism $\mu\colon \mathbf{INS} \to \mathbf{INS}'$ is (i.e. may be viewed as) a pair: a functor $\mu^{Sign}\colon \mathbf{Sign} \to \mathbf{Sign}'$ and a natural transformation $\mu^{nt}\colon \mathbf{INS} \to (\mu^{Sign})^{op};\mathbf{INS}'$.

Exercise 10.3.5. For any functors $\mathbf{I}\colon \mathbf{Sign}_\mathbf{I}^{op} \to \mathbf{Room}(\mathbf{B})$ and $\mathbf{I}'\colon \mathbf{Sign}_{\mathbf{I}'}^{op} \to \mathbf{Room}(\mathbf{B})$, show that any functor $\Phi\colon \mathbf{Sign_I} \to \mathbf{Sign_{I'}}$ together with a natural transformation $\tau\colon \mathbf{I} \to \Phi^{op};\mathbf{I}'$ is (i.e. may be viewed as) an institution morphism from $\mathbf{I}$ to $\mathbf{I}'$ viewed as institutions (Exercise 10.3.4). □

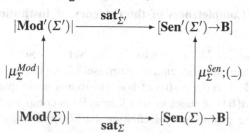

Recall that in Exercise 3.4.66, for any category **K**, we have defined the category $\mathbf{Funct}(\mathbf{K})$ of functors into **K**. Objects of $\mathbf{Funct}(\mathbf{K})$ are functors from an arbitrary category to **K**, and morphisms in $\mathbf{Funct}(\mathbf{K})$ from a functor $\mathbf{F}\colon\mathbf{K1} \to \mathbf{K}$ to a functor $\mathbf{G}\colon\mathbf{K2} \to \mathbf{K}$ consist of a functor $\boldsymbol{\Phi}\colon\mathbf{K1} \to \mathbf{K2}$ and a natural transformation $\tau\colon\mathbf{F} \to \boldsymbol{\Phi};\mathbf{G}$.

The view of institutions as functors into the category of rooms and of institution morphisms as morphisms between such functors may thus be summed up by the following definition:

Definition 10.3.6 (Category of institutions (revisited)). The category $\mathcal{INS}$ of institutions and their morphisms is the category $\mathbf{Funct}(\mathbf{Room}(\mathbf{B}))$ of functors into the category $\mathbf{Room}(\mathbf{B})$ of boolean rooms. □

Exercise 10.3.7. Using Exercises 10.3.4 and 10.3.5, show that the above definition is equivalent to Definition 10.3.1. □

Exercise 10.3.8. It is quite natural to generalise the above construction by replacing boolean rooms by **V**-rooms for an arbitrary set **V** (or even, for an arbitrary category **V**) where the category $\mathbf{Room}(\mathbf{V})$ of **V**-rooms is the comma category $(|\_|, [\_{\to}\mathbf{V}])$. Work out this generalisation in detail. Show how this yields a notion of institution with multi-valued satisfaction relations. Present the institution **3FOPEQ** of three-valued logic (Example 4.1.24) as an institution based on $\mathbf{V} = \{tt, \mathit{ff}, \mathit{undef}\}$. Try to think of situations in which it would be natural to use such multi-valued logical systems. □

The reformulated definition of the category $\mathcal{INS}$ of institutions given above offers a convenient way to prove completeness:

Theorem 10.3.9 (Completeness of the category of institutions). *The category $\mathcal{INS}$ is complete.*

Proof. By Exercise 3.4.24 the functor $[\_{\to}\mathbf{B}]\colon\mathbf{Set}^{op} \to \mathbf{Set}$ is continuous. Since **Set** is both complete and cocomplete (Exercise 3.2.53(1)), by Exercise 3.4.54 the category $\mathbf{Room}(\mathbf{B}) = (|\_|, [\_{\to}\mathbf{B}])$ of boolean rooms is complete. Thus, the category $\mathbf{Funct}(\mathbf{Room}(\mathbf{B}))$ of functors into $\mathbf{Room}(\mathbf{B})$ is complete by Exercise 3.4.67. This completes the proof, as by Definition 10.3.6 (and Exercise 10.3.7) $\mathcal{INS} = \mathbf{Funct}(\mathbf{Room}(\mathbf{B}))$.

Exercise. Complete the proof outlined above. In particular, you may want to go through the exercises involved and use the hints given there to complete proofs of the results we rely on here. □

The above result ensures that any diagram of institutions has a limit in $\mathcal{INS}$. The construction of this limit may be described informally as follows:

- First, construct the limit in **Cat** of the categories of signatures of the institutions involved.
- Then, for each signature in the resulting category, construct a boolean room for it as a limit of the corresponding diagram in $\mathbf{Room}(\mathbf{B})$, that is:

- The category of models is given as a limit in **Cat** of the corresponding diagram of the categories of models of the institutions involved.
- The set of sentences is given as a colimit in **Set** of the corresponding diagram of the sets of sentences of the institutions involved.
- The satisfaction relation is determined by the satisfaction relations of the institutions involved and by the satisfaction condition for institution morphisms.

- Finally, for each signature morphism in the resulting category, the induced translation of boolean rooms is given by the limit property of the room for the source signature:
  - The reduct functor on models is given by the limit property of the category of models over the source signature.
  - The translation of sentences is given by the colimit property of the set of sentences over the source signature.

Exercise 10.3.10. Check in detail that the above informal description coincides with the construction implied by the proof of Theorem 10.3.9. □

Example 10.3.11. The product institution $\mathbf{INS}_1 \times \mathbf{INS}_2$ defined in Example 4.1.45 for any institutions $\mathbf{INS}_1$ and $\mathbf{INS}_2$ is their categorical product in $\mathcal{INS}$. □

Example 10.3.12. Consider two simple extensions of the institution **EQ**:

- $\mathbf{EQ^{inh}}$: an extension of **EQ** by statements that a particular carrier set is inhabited. More formally, $\mathbf{EQ^{inh}}$ has the same signatures and models as **EQ**, and then for each algebraic signature $\Sigma \in |\mathbf{AlgSig}|$, a Σ-sentences of $\mathbf{EQ^{inh}}$ is either an ordinary equation (with translations along signature morphisms and satisfaction inherited from **EQ**) or an inhabitation assertion of the form $\mathbf{inh}(s)$ for some sort name s in Σ. For any Σ-algebra $A \in |\mathbf{Alg}(\Sigma)|$, $A \models_{\mathbf{EQ^{inh}}} \mathbf{inh}(s)$ if and only if $|A|_s \neq \varnothing$. The translation of inhabitation assertions along signature morphisms is defined in the obvious way, and it is trivial to check the satisfaction condition. There is an obvious institution morphism, $\mu_{\mathbf{inh}} : \mathbf{EQ^{inh}} \to \mathbf{EQ}$.
- $\mathbf{EQ^{onto}}$: an extension of **EQ** by statements that a given operation is surjective (this is expressible in **FOEQ**, but not in **EQ**). More formally, $\mathbf{EQ^{onto}}$ has the same signatures and models as **EQ**, and then for each algebraic signature $\Sigma \in |\mathbf{AlgSig}|$, a Σ-sentence of $\mathbf{EQ^{onto}}$ is either an ordinary equation (with translations along signature morphisms and satisfaction inherited from **EQ**) or a surjectivity assertion of the form f **is onto** for some operation name f in Σ. For any Σ-algebra $A \in |\mathbf{Alg}(\Sigma)|$, $A \models_{\mathbf{EQ^{inh}}} f$ **is onto** if and only if f_A is a surjective function. The translation of surjectivity assertions along signature morphisms is defined in the obvious way, and it is trivial to check the satisfaction condition. There is an obvious institution morphism $\mu_{\mathbf{onto}} : \mathbf{EQ^{onto}} \to \mathbf{EQ}$.

It is easy to construct the pullback of $\mu_{\mathbf{inh}} : \mathbf{EQ^{inh}} \to \mathbf{EQ}$ and $\mu_{\mathbf{onto}} : \mathbf{EQ^{onto}} \to \mathbf{EQ}$ in $\mathcal{INS}$:

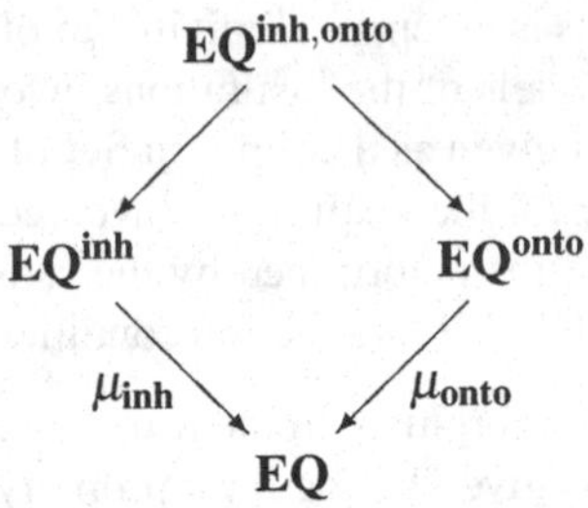

The pullback institution $\mathbf{EQ}^{\mathbf{inh},\mathbf{onto}}$ has the same signatures and models as $\mathbf{EQ}$ (and as $\mathbf{EQ}^{\mathbf{inh}}$ and $\mathbf{EQ}^{\mathbf{onto}}$). A sentence of $\mathbf{EQ}^{\mathbf{inh},\mathbf{onto}}$ is either an ordinary equation (with translation along signature morphisms and satisfaction inherited from $\mathbf{EQ}$), an inhabitation assertion (with translation along signature morphisms and satisfaction inherited from $\mathbf{EQ}^{\mathbf{inh}}$), or a surjectivity assertion (with translation along signature morphisms and satisfaction inherited from $\mathbf{EQ}^{\mathbf{onto}}$). □

Example 10.3.13. For any institution $\mathbf{INS}$, its closure under (infinite) conjunction $\mathbf{INS}^{\wedge}$ was defined in Example 4.1.38 and its closure under (finitary) propositional connectives $\mathbf{INS}^{prop}$ in Example 4.1.41.

These extensions of $\mathbf{INS}$ come with institution morphisms $\mu_{\wedge}\colon \mathbf{INS}^{\wedge} \to \mathbf{INS}$ and $\mu_{prop}\colon \mathbf{INS}^{prop} \to \mathbf{INS}$; see Example 10.2.5(2). Consider the pullback of $\mu_{\wedge}$ and μ_{prop} in $\mathcal{INS}$:

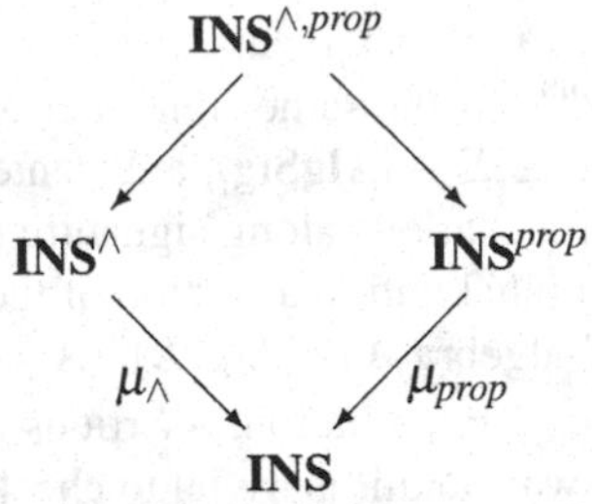

The result, which may be described similarly as in Example 10.3.12, is not very satisfactory. The pullback institution $\mathbf{INS}^{\wedge,prop}$ contains the original sentences of $\mathbf{INS}$, their infinite conjunctions (or sets) and their negations, disjunctions, negations of their disjunctions, and other propositional combinations, but not, say, negations of the infinite conjunctions or infinite conjunctions involving the negations. □

Exercise 10.3.14. Notice that for any institution $\mathbf{INS}$ the construction in Example 10.3.13 resulted in a new institution $\mathbf{INS}^{\wedge,prop}$ together with an institution morphism $\mu^{\mathbf{INS}}_{\wedge,prop}\colon \mathbf{INS}^{\wedge,prop} \to \mathbf{INS}$. Iterate this construction to obtain the following chain of institution morphisms in $\mathcal{INS}$:

$$\mathbf{INS} \xleftarrow{\mu^{\mathbf{INS}}_{\wedge,prop}} \mathbf{INS}^{\wedge,prop} \xleftarrow{\mu^{\mathbf{INS}^{\wedge,prop}}_{\wedge,prop}} (\mathbf{INS}^{\wedge,prop})^{\wedge,prop} \xleftarrow{\mu^{(\mathbf{INS}^{\wedge,prop})^{\wedge,prop}}_{\wedge,prop}} \cdots$$

Calculate the limit of this diagram in $\mathcal{INS}$. HINT: The resulting sets of sentences need *not* be closed under all infinite conjunctions. □

Example 10.3.15. Recall the institution morphisms $\mu_{\mathbf{FOEQ}}\colon \mathbf{FOPEQ} \to \mathbf{EQ}$ (Example 10.2.2) and $\mu_{\mathbf{CEQ}}\colon \mathbf{CEQ} \to \mathbf{EQ}$ (Example 10.2.3).

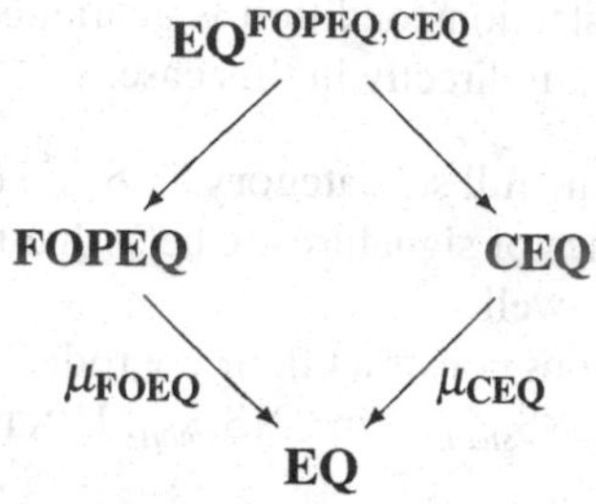

The pullback in $\mathcal{INS}$ of these two morphisms is the institution $\mathbf{EQ}^{\mathbf{FOPEQ},\mathbf{CEQ}}$, to be explicitly defined below; this can be calculated using the informal description of limits in $\mathcal{INS}$ following Theorem 10.3.9.

The category of signatures of $\mathbf{EQ}^{\mathbf{FOPEQ},\mathbf{CEQ}}$ is $\mathbf{Sign}_{\mathbf{FOPEQ}}$, the category of first-order signatures (since the signatures of **CEQ** coincide with those of **EQ** and $\mu^{Sign}_{\mathbf{CEQ}}$ is the identity functor).

Let $\Theta = \langle S, \Omega, \Pi \rangle \in |\mathbf{FOSig}|$ be a first-order signature, and let $\Sigma = \langle S, \Omega \rangle \in |\mathbf{AlgSig}|$ be the algebraic signature it contains. Recall that $\mu^{Sign}_{\mathbf{FOEQ}}(\Theta) = \Sigma$. Then:

- The set of Θ-sentences in $\mathbf{EQ}^{\mathbf{FOPEQ},\mathbf{CEQ}}$ is the union of the set of Θ-sentences in **FOPEQ** with the set of Σ-sentences in **CEQ** modulo the identification of Σ-sentences in **EQ**. That is, Θ-sentences in $\mathbf{EQ}^{\mathbf{FOPEQ},\mathbf{CEQ}}$ are either closed first-order formulae over Θ or infinitary Σ-equations, where universally quantified equalities of first-order logic and infinitary equations which happen to be finitary are identified (and are identified with the corresponding Σ-equations). Notice that, much as Example 10.3.13 above, infinitary equations cannot occur as atomic formulae in the first-order sentences of $\mathbf{EQ}^{\mathbf{FOPEQ},\mathbf{CEQ}}$.
- Θ-models in $\mathbf{EQ}^{\mathbf{FOPEQ},\mathbf{CEQ}}$ are pairs $\langle M, A \rangle$, where $M \in |\mathbf{FOStr}(\Theta)|$ is a first-order Θ-structure and $A \in |\mathbf{CAlg}(\Sigma)|$ is a continuous Σ-algebra such that forgetting the orderings in A yields the same algebra which results from M by forgetting the interpretation of predicates. So Θ-models in $\mathbf{EQ}^{\mathbf{FOPEQ},\mathbf{CEQ}}$ can be viewed as first-order structures with carriers equipped with a complete partial ordering where the operations are required to be continuous, but with no continuity restriction on the interpretation of predicates. Model morphisms are defined analogously: they are those first-order Θ-morphisms which are continuous.
- Θ-satisfaction in $\mathbf{EQ}^{\mathbf{FOPEQ},\mathbf{CEQ}}$ is induced in the expected way: the interpretation of first-order formulae is determined by the first-order part of $\mathbf{EQ}^{\mathbf{FOPEQ},\mathbf{CEQ}}$-models, and the interpretation of infinitary equations is determined by their continuous-algebra part. This definition is unambiguous for universally quantified equalities of first-order logic and finitary Σ-equations of the equational logic for continuous algebras, which are identified in $\mathbf{EQ}^{\mathbf{FOPEQ},\mathbf{CEQ}}$, since in both cases their interpretation coincides with the interpretation of ordinary Σ-equations in the algebra part of first-order structures and continuous algebras.

Finally, for any first-order signature morphism $\theta : \Theta \to \Theta'$, the θ-translation of $\mathbf{EQ}^{\mathbf{FOPEQ},\mathbf{CEQ}}$-sentences and the θ-reduct functor on $\mathbf{EQ}^{\mathbf{FOPEQ},\mathbf{CEQ}}$-models are induced by the corresponding translations and reduct functors of **FOPEQ** and **CEQ** in the obvious way. The satisfaction condition is guaranteed by the construction although it is also easy to check it directly in this case. □

Exercise 10.3.16. Consider the full subcategory $\mathcal{INS}_{small}$ of $\mathcal{INS}$ that consists of institutions with small categories of signatures. Check that the proof of completeness of $\mathcal{INS}$ applies to $\mathcal{INS}_{small}$ as well.

For any family of institutions construct their coproduct in $\mathcal{INS}$ and show that the coproduct of institutions in $\mathcal{INS}_{small}$ is in $\mathcal{INS}_{small}$. HINT: This is easy; see Example 4.1.44.

Give a construction for coequalisers in $\mathcal{INS}_{small}$. HINT: This is rather difficult. You will need coequalisers in **Cat** (cf. Exercise 3.4.36) to determine the category of signatures of the coequaliser institution, limits of certain diagrams of the sets of sentences, and colimits of certain diagrams of the categories of models. Fortunately, the satisfaction condition will then determine the satisfaction relation unambiguously.

Check where this construction may break down if the signature categories of the institutions involved are not small.

Conclude though that the category $\mathcal{INS}_{small}$ of institutions with small signature categories is (complete and) cocomplete. □

10.4 Institution comorphisms

Institution morphisms are by far not the only possible kinds of maps between institutions. Very informally, considering the obvious idea that such maps need to link the three layers of institutions (signatures, models and sentences), there are four basic concepts of such a map to consider, using different combinations of directions of the maps at each layer:[6]

morphism:

$$\begin{array}{ccc} \mathbf{INS} & \longrightarrow & \mathbf{INS}' \\ \hline \mathbf{Sen} & \longleftarrow & \mathbf{Sen}' \\ \mathbf{Sign} & \longrightarrow & \mathbf{Sign}' \\ \mathbf{Mod} & \longrightarrow & \mathbf{Mod}' \end{array}$$

comorphism:

$$\begin{array}{ccc} \mathbf{INS} & \longrightarrow & \mathbf{INS}' \\ \hline \mathbf{Sen} & \longrightarrow & \mathbf{Sen}' \\ \mathbf{Sign} & \longrightarrow & \mathbf{Sign}' \\ \mathbf{Mod} & \longleftarrow & \mathbf{Mod}' \end{array}$$

forward morphism:

$$\begin{array}{ccc} \mathbf{INS} & \longrightarrow & \mathbf{INS}' \\ \hline \mathbf{Sen} & \longrightarrow & \mathbf{Sen}' \\ \mathbf{Sign} & \longrightarrow & \mathbf{Sign}' \\ \mathbf{Mod} & \longrightarrow & \mathbf{Mod}' \end{array}$$

forward comorphism:

$$\begin{array}{ccc} \mathbf{INS} & \longrightarrow & \mathbf{INS}' \\ \hline \mathbf{Sen} & \longleftarrow & \mathbf{Sen}' \\ \mathbf{Sign} & \longrightarrow & \mathbf{Sign}' \\ \mathbf{Mod} & \longleftarrow & \mathbf{Mod}' \end{array}$$

[6] Terminology from [GR02].

We will not discuss here in any detail "forward" (co)morphisms, in which translations of sentences and models are covariant — they are somewhat esoteric, even though they have received some attention in the literature. However, institution comorphisms are important and useful. Informally, they capture how the "simpler" source institution is represented or "encoded" in the "more complex" target institution. The reader is encouraged to compare this with intuitions concerning institution morphisms, which indicate how the "richer" source institution is built over the "more primitive" target institution.

Definition 10.4.1 (Institution comorphism and semi-comorphism). Let $\mathbf{INS} = \langle \mathbf{Sign}, \mathbf{Sen}, \mathbf{Mod}, \langle \models_\Sigma \rangle_{\Sigma \in |\mathbf{Sign}|} \rangle$ and $\mathbf{INS}' = \langle \mathbf{Sign}', \mathbf{Sen}', \mathbf{Mod}', \langle \models'_{\Sigma'} \rangle_{\Sigma' \in |\mathbf{Sign}'|} \rangle$ be institutions.

An *institution comorphism* $\rho : \mathbf{INS} \to \mathbf{INS}'$ consists of:

- a functor $\rho^{Sign} : \mathbf{Sign} \to \mathbf{Sign}'$,
- a natural transformation $\rho^{Sen} : \mathbf{Sen} \to \rho^{Sign} ; \mathbf{Sen}'$, that is, for each $\Sigma \in |\mathbf{Sign}|$, a function $\rho^{Sen}_\Sigma : \mathbf{Sen}(\Sigma) \to \mathbf{Sen}'(\rho^{Sign}(\Sigma))$ such that the following diagram commutes for every $\sigma : \Sigma_1 \to \Sigma_2$ in **Sign**

$$\begin{array}{ccccc}
\Sigma_2 & \quad & \mathbf{Sen}(\Sigma_2) & \xrightarrow{\rho^{Sen}_{\Sigma_2}} & \mathbf{Sen}'(\rho^{Sign}(\Sigma_2)) \\
\uparrow \sigma & & \uparrow \mathbf{Sen}(\sigma) & & \uparrow \mathbf{Sen}'(\rho^{Sign}(\sigma)) \\
\Sigma_1 & & \mathbf{Sen}(\Sigma_1) & \xrightarrow[\rho^{Sen}_{\Sigma_1}]{} & \mathbf{Sen}'(\rho^{Sign}(\Sigma_1))
\end{array}$$

- a natural transformation $\rho^{Mod} : (\rho^{Sign})^{op} ; \mathbf{Mod}' \to \mathbf{Mod}$, that is, for each $\Sigma \in |\mathbf{Sign}|$, a functor $\rho^{Mod}_\Sigma : \mathbf{Mod}'(\rho^{Sign}(\Sigma)) \to \mathbf{Mod}(\Sigma)$ such that the following diagram commutes for every $\sigma : \Sigma_1 \to \Sigma_2$ in **Sign**

$$\begin{array}{ccccc}
\Sigma_2 & \quad & \mathbf{Mod}'(\rho^{Sign}(\Sigma_2)) & \xrightarrow{\rho^{Mod}_{\Sigma_2}} & \mathbf{Mod}(\Sigma_2) \\
\uparrow \sigma & & \downarrow \mathbf{Mod}'(\rho^{Sign}(\sigma)) & & \downarrow \mathbf{Mod}(\sigma) \\
\Sigma_1 & & \mathbf{Mod}'(\rho^{Sign}(\Sigma_1)) & \xrightarrow[\rho^{Mod}_{\Sigma_1}]{} & \mathbf{Mod}(\Sigma_1)
\end{array}$$

such that for any $\Sigma \in |\mathbf{Sign}|$, the translations $\rho^{Sen}_\Sigma : \mathbf{Sen}(\Sigma) \to \mathbf{Sen}'(\rho^{Sign}(\Sigma))$ of sentences and $\rho^{Mod}_\Sigma : \mathbf{Mod}'(\rho^{Sign}(\Sigma)) \to \mathbf{Mod}(\Sigma)$ of models preserve the satisfaction relation, that is, for any $\varphi \in \mathbf{Sen}(\Sigma)$ and $M' \in |\mathbf{Mod}'(\rho^{Sign}(\Sigma))|$

$$\rho^{Mod}_\Sigma(M') \models_\Sigma \varphi \iff M' \models'_{\rho^{Sign}(\Sigma)} \rho^{Sen}_\Sigma(\varphi) \qquad \text{[Satisfaction condition]}.$$

An *institution semi-comorphism* ρ: **INS** $\to_{semi}$ **INS**$'$ consists of two components: a functor ρ^{Sign}: **Sign** $\to$ **Sign**$'$ and a natural transformation ρ^{Mod}: $(\rho^{Sign})^{op}$;**Mod**$'$ $\to$ **Mod** as above. □

As with institution morphisms and comorphisms, we say that an institution comorphism ρ: **INS** $\to$ **INS**$'$ as above *extends* $\langle\rho^{Sign},\rho^{Mod}\rangle$: **INS** $\to_{semi}$ **INS**$'$, and whenever convenient use an institution comorphism ρ in contexts where the institution semi-comorphism that ρ extends is required.

Example 10.4.2. There is an obvious institution comorphism ρ: **EQ** $\to$ **FOPEQ** that embeds the institution **EQ** of equational logic into the institution **FOPEQ** of first-order logic with equality (Examples 4.1.12 and 4.1.4) as follows:

- ρ^{Sign}: **AlgSig** $\to$ **FOSig** includes algebraic signatures and their morphisms into the category of first-order signatures by equipping them with the empty set of predicate symbols.
- For each signature $\Sigma \in |$**AlgSig**$|$, ρ^{Sen}_{Σ}: $\mathbf{Sen}_{\mathbf{EQ}}(\Sigma) \to \mathbf{Sen}_{\mathbf{FOPEQ}}(\rho^{Sign}(\Sigma))$ maps any Σ-equation to the corresponding universally quantified equality as a first-order sentence (remarks in Example 10.2.2 still apply).
- For each signature $\Sigma \in |$**AlgSig**$|$, ρ^{Mod}_{Σ}: $\mathbf{FOStr}(\rho^{Sign}(\Sigma)) \to \mathbf{Alg}(\Sigma)$ is the identity functor.

Clearly, the satisfaction condition holds. □

Exercise 10.4.3. Define an institution comorphism ρ: **PEQ** $\to$ **PFOPEQ** which embeds the institution **PEQ** of partial equational logic into the institution **PFOPEQ** of partial first-order logic with equality (Example 4.1.6 and Exercise 4.1.17) following the pattern in Example 10.4.2. □

Example 10.4.4. An institution comorphism ρ: **EQ** $\to$ **CEQ** from the institution **EQ** of equational logic to the institution **CEQ** of equational logic for continuous algebras (cf. Example 4.1.22) is the identity on their common category of signatures, embeds (finitary) equations of **EQ** into the set of (potentially infinitary) equations of **CEQ**, and extracts standard algebras out of continuous algebras by forgetting about the ordering of their carriers. □

Exercise 10.4.5. Define an institution comorphism from the institution **PEQ** of partial equational logic to the institution **EQ** of equational logic which is the identity on signatures, is the inclusion on model categories, maps equalities of **PEQ** to equalities of **EQ** and maps definedness formulae of **PEQ** to any equation that holds in all algebras. Do not forget to check the satisfaction condition.

Then define an institution comorphism from **EQ** to the institution $\mathbf{PEQ}_{\mathbf{pval}}$ of partial equational logic with partial valuations of variables (Example 10.2.8) which again is the identity on the category of signatures, is the inclusion on sets of sentences, and maps any partial algebra to its "totalisation" as defined in Example 3.4.18. □

Exercise 10.4.6. Building on Exercise 4.1.16, capture the representation of predicates in first-order structures as boolean functions by defining an institution comorphism from the institution **FOPEQ** of first-order predicate logic with equality to the institution **FOEQBool** of first-order equational logic with boolean values.

Similarly, building on Exercise 4.1.19, define an institution comorphism from the institution **PFOPEQ** of partial first-order predicate logic with equality to the institution **PFOEQTruth** of partial first-order equational logic with truth, encoding predicates as partial functions into a one-element sort.

Try to define institution morphisms from **FOPEQ** to **FOEQBool** and from **PFOPEQ** to **PFOEQTruth**, respectively, if necessary using the versions of the institutions with strong morphisms between their models; see Exercise 4.1.20. HINT: Translation of sentences may be tricky, but it is possible: for instance, equality between terms of sort *Bool* in **FOEQBool** may be captured as equivalence of the corresponding predicate applications; quantification over *Bool* may be eliminated since *Bool* is generated by two constants. □

Exercise 10.4.7. View the obvious embedding into the institution **FPL** (Example 4.1.25) of its "programming fragment" **FProg** (Exercise 4.1.30) as an institution comorphism.

Somewhat more interestingly, define a comorphism that embeds into **FPL** the institution **PFOEQ** of partial first-order logic with equalities as its only atomic formulae (so, no predicates; see Exercise 4.1.18) by viewing algebraic signatures with no predicate names as **FPL**-signatures with no value constructors.

Finally, mimicking Exercise 10.4.6, define a comorphism from the institution **PFOPEQ** of partial first-order logic with predicates and equality to **FPL** by encoding predicates as partial functions into a special, one-element sort, as in Exercise 4.1.19. The extra advantage now is that we can use the power of **FPL** to capture directly in **FPL**-signatures the requirement on the models in **PFOEQTruth** given in Exercise 4.1.19.

Spelling this out: a first-order signature Θ would map to the **FPL**-signature which extends the algebraic part of Θ by a new sort *Truth* with a single value constructor *true*: *Truth*, and adds predicate names from Θ as operations into *Truth*. Consequently, all **FPL**-models over the so-obtained **FPL**-signature have a carrier of sort *Truth* with the value of *true* as its only element, and a partial function into this carrier for each predicate in Θ. Such models can then be mapped back to first-order Θ-structures by forgetting the carrier of sort *Truth* and interpreting each predicate as holding exactly when the corresponding partial operation is defined. Sentences are then transformed by the comorphism so that each predicate application is replaced by definedness (equality to *true*) of the corresponding (partial) operation application. □

Exercise 10.4.8. In Section 4.1.2 we defined a number of constructions which given an institution leave its category of signatures and model functor intact, but enlarge its sets of sentences. In all these cases, there is an obvious comorphism from the original institution to the new one. In particular, for any institution **INS** define a comorphism from **INS** to the institution $\mathbf{INS}^{\wedge}$, which closes the sets of sentences under infinite conjunctions (Example 4.1.38), and into the institution $\mathbf{INS}^{prop}$, which closes the sets of sentences under propositional connectives (Example 4.1.41). Similarly, given an institution semi-morphism $\mu:\mathbf{INS} \to \mathbf{INS}'$, define a comorphism from **INS** to the duplex institution **INS plus INS′ via** μ (Definition 10.1.12). □

Exercise 10.4.9. In Exercise 4.1.13 we have introduced institutions **FOP** of first-order predicate logic and **FOEQ** of first-order logic with equality (without predicates) as "sublogics" of the full institution **FOPEQ** of first-order logic with predicates and equality. Similarly, various "sublogics" of the institution **PFOPEQ** of partial first-order logic with equality were introduced in Exercise 4.1.18. We have also referred to the institutions **GEQ** of ground equational logic (Example 4.1.3) and **FProg** of simple functional program definitions (Example 4.1.30) as "subinstitutions" of **EQ** and **FPL**, respectively. Capture the concept of subinstitution using institution comorphisms with inclusions as translations of signature categories and of sets of sentences and with isomorphisms as translations of model categories. Make sure that it covers at least all the cases mentioned above and check which of the institution comorphisms introduced in the current section identify subinstitutions in this sense.

Note that it may be difficult to capture this concept adequately using institution morphisms: if the signature categories coincide and the translation of model categories are isomorphisms then such an institution comorphism to an institution **INS** from its subinstitution $\mathbf{INS}_{sub}$ yields also an institution morphism from **INS** to $\mathbf{INS}_{sub}$ — but when the category of $\mathbf{INS}_{sub}$-signatures is a proper subcategory of **INS**-signatures, there may be no institution morphism from **INS** to $\mathbf{INS}_{sub}$. □

Exercise 10.4.10. Following the pattern of Definition 10.3.1, define composition of institution comorphisms. Show that this yields a category $\mathcal{INS}^{co}$ of institutions and their comorphisms, and show that this category is complete.

Then similarly as in Exercise 10.3.16, define the full subcategory $\mathcal{INS}^{co}_{small}$ of $\mathcal{INS}^{co}$ that consists of institutions with small categories of signatures. Show that $\mathcal{INS}^{co}_{small}$ is also complete (this is easy) and cocomplete (this is difficult, as hinted at in the case of $\mathcal{INS}_{small}$). □

Exercise 10.4.11. As discussed in Section 10.1, for many purposes it is sufficient to link institutions by relating their signatures and models only, disregarding sentences. Institution semi-comorphisms (Definition 10.4.1) capture such a relationship in a way that is similar to but different from the way institution semi-morphisms do, as introduced in Definition 10.1.1.

Give examples of institution semi-comorphisms other than those extended to institution comorphisms above. Look for examples of institution semi-comorphisms that capture relationships between institutions that would not be naturally captured by institution semi-morphisms. In particular, define an institution semi-comorphism from the institution **PFOEQ** of partial first-order logic with equality to the institution $\mathbf{IMP}_{DT}$ of a simple imperative programming language (Example 4.1.32). □

Like institution semi-morphisms, institution semi-comorphisms may be used to move specifications between institutions. This may be achieved using two new inter-institutional specification-building operations:

Definition 10.4.12 (Translating and hiding specifications via an institution semi-comorphism). Consider two institutions $\mathbf{INS} = \langle \mathbf{Sign}, \mathbf{Sen}, \mathbf{Mod}, \langle \models_\Sigma \rangle_{\Sigma \in |\mathbf{Sign}|} \rangle$ and $\mathbf{INS}' = \langle \mathbf{Sign}', \mathbf{Sen}', \mathbf{Mod}', \langle \models'_{\Sigma'} \rangle_{\Sigma' \in |\mathbf{Sign}'|} \rangle$, and an institution semi-comorphism $\rho \colon \mathbf{INS} \rightarrow_{semi} \mathbf{INS}'$.

If SP is a specification over **INS** with signature $Sig[SP] = \Sigma \in |\mathbf{Sign}|$ and models $Mod[SP] \subseteq |\mathbf{Mod}(\Sigma)|$, then SP **with** ρ is a specification over $\mathbf{INS}'$ with the following semantics:[7]

$$\begin{aligned} Sig[SP \ \mathbf{with}\ \rho] &= \rho^{Sign}(\Sigma) \\ Mod[SP \ \mathbf{with}\ \rho] &= \{M' \in |\mathbf{Mod}'(\rho^{Sign}(\Sigma))| \mid \rho^{Mod}_\Sigma(M') \in Mod[SP]\}. \end{aligned}$$

Thus, for each signature $\Sigma \in |\mathbf{Sign}|$,

$$\_\_ \ \mathbf{with}\ \rho \colon Spec_{\mathbf{INS}}(\Sigma) \rightarrow Spec_{\mathbf{INS}'}(\rho^{Sign}(\Sigma)).$$

If SP' is a specification over $\mathbf{INS}'$ with signature $Sig[SP'] \in |\mathbf{Sign}'|$ and models $Mod[SP'] \subseteq |\mathbf{Mod}'(Sig[SP'])|$, and $\Sigma \in |\mathbf{Sign}|$ is an **INS**-signature such that $\rho^{Sign}(\Sigma) = Sig[SP']$, then SP' **hide via** ρ **to** Σ is a specification over **INS** with the following semantics:

$$\begin{aligned} Sig[SP \ \mathbf{hide\ via}\ \rho \ \mathbf{to}\ \Sigma] &= \Sigma \\ Mod[SP \ \mathbf{hide\ via}\ \rho \ \mathbf{to}\ \Sigma] &= \{\rho^{Mod}_\Sigma(M') \mid M' \in Mod[SP']\}. \end{aligned}$$

Thus:

$$\_\_ \ \mathbf{hide\ via}\ \rho \ \mathbf{to}\ \Sigma \colon Spec_{\mathbf{INS}'}(\rho^{Sign}(\Sigma)) \rightarrow Spec_{\mathbf{INS}}(\Sigma).$$ □

It is worth observing that while in the case of the two specification-building operations determined by institution semi-morphisms, hiding via an institution semi-morphism is in a sense "more natural", since it does not require the target signature to be indicated explicitly, the situation with the operations determined by institution semi-comorphisms is dual: now the translation seems "more natural", while hiding via an institution semi-comorphism requires the target signature to be chosen explicitly.

We have thus extended our repertoire of inter-institutional specification-building operations that can be combined with the intra-institutional specification-building operations introduced in Chapters 5 and 8. This results in an even larger class of heterogeneous specifications that may be built given an "environment" of logical systems formalised as institutions and linked by institution semi-morphisms and semi-comorphisms.

Exercise 10.4.13. Use the inter-institutional specification-building operations determined by the institution semi-comorphism from **PFOEQ** to $\mathbf{IMP}_{DT}$ suggested in Exercise 10.4.11 to redo Example 10.1.17. Then redo Example 10.1.18 using the institution semi-comorphism from **PFOPEQ** to **FProg** which essentially can be extracted from the institution comorphism sketched in Exercise 10.4.7. Finally, notice

[7] CASL notation: this is written using the same syntax in HETCASL, the language of HETS [MML07], when ρ is an institution comorphism; see Definition 10.4.1.

that institution semi-comorphisms determine generalised constructors in the sense of Definition 10.1.20, and check that Examples 10.1.21 and 10.1.24 also carry over when using such generalised constructors. □

The above exercise shows how institution (semi-)comorphisms may sometimes be used in place of institution (semi-)morphisms. In the rest of this section we study the relationships between the various concepts these notions induce, and check when and how they can replace each other.

Exercise 10.4.14. As in Exercise 10.1.22, check whether the two operations for moving specifications between institutions linked by institution semi-comorphisms commute with other specification-building operations. For example, for any institution semi-comorphism $\rho\colon \mathbf{INS} \to_{semi} \mathbf{INS}'$, prove that (under appropriate assumptions about the signatures of specifications and the signature morphisms involved)

$$(SP_1 \textbf{ with } \sigma) \textbf{ with } \rho \equiv (SP_1 \textbf{ with } \rho) \textbf{ with } \rho^{Sign}(\sigma)$$
$$(SP'_2 \textbf{ hide via } \rho^{Sign}(\sigma)) \textbf{ hide via } \rho \textbf{ to } \Sigma_1 \equiv (SP'_2 \textbf{ hide via } \rho \textbf{ to } \Sigma_2) \textbf{ hide via } \sigma.$$

Show that we also have

$$Mod[(SP'_1 \textbf{ with } \rho^{Sign}(\sigma)) \textbf{ hide via } \rho \textbf{ to } \Sigma_2] \subseteq Mod[(SP'_1 \textbf{ hide via } \rho \textbf{ to } \Sigma_1) \textbf{ with } \sigma]$$
$$Mod[(SP_2 \textbf{ hide via } \sigma) \textbf{ with } \rho] \supseteq Mod[(SP_2 \textbf{ with } \rho) \textbf{ hide via } \rho^{Sign}(\sigma)].$$

Check that the opposite inclusions hold as well if the "naturality squares" for ρ^{Mod} (see Definition 10.4.1) are pullbacks in **Cat**.

We also have, as in Exercises 5.6.3 and 10.1.22, for any specification SP over **INS** with $Sig[SP] = \Sigma$ and specification SP' over $\mathbf{INS}'$ such that $\rho^{Sign}(\Sigma) = Sig[SP']$,

$$Mod[(SP \textbf{ with } \rho) \textbf{ hide via } \rho \textbf{ to } \Sigma] \subseteq Mod[SP],$$
$$Mod[(SP' \textbf{ hide via } \rho \textbf{ to } \Sigma) \textbf{ with } \rho] \supseteq Mod[SP'].$$

Show that the first inclusion (but in general not the second) may be reversed if ρ^{Mod}_{Σ} is surjective.

Finally, as in Exercise 10.1.23, check that the specification-building operations __ **with** ρ and __ **hide via** μ **to** Σ form a Galois connection between classes of Σ-specifications over **INS** and $\rho^{Sign}(\Sigma)$-specifications over $\mathbf{INS}'$, considered up to equivalence and ordered by the inclusion of model classes. □

A number of the above examples of institution comorphisms are closely related to the examples of institution morphisms given in Section 10.2. In fact, it is quite often the case that an institution morphism determines an institution comorphism (in the opposite direction) and vice versa.

Proposition 10.4.15. *Consider an institution morphism $\mu\colon \mathbf{INS}' \to \mathbf{INS}$, functor $\rho^{Sign}\colon \mathbf{Sign} \to \mathbf{Sign}'$ and natural transformation $\eta\colon \mathbf{Id_{Sign}} \to \rho^{Sign};\mu^{Sign}$. Then there is an institution comorphism $\rho = \langle \rho^{Sign}, \rho^{Sen}, \rho^{Mod}\rangle\colon \mathbf{INS} \to \mathbf{INS}'$, where for $\Sigma \in |\mathbf{Sign}|$, $\rho^{Sen}_{\Sigma} = \mathbf{Sen}(\eta_\Sigma);\mu^{Sen}_{\rho^{Sign}(\Sigma)}$ and $\rho^{Mod}_{\Sigma} = \mu^{Mod}_{\rho^{Sign}(\Sigma)};\mathbf{Mod}(\eta_\Sigma)$.*

Consider an institution comorphism $\rho\colon \mathbf{INS} \to \mathbf{INS}'$*, functor* $\mu^{Sign}\colon \mathbf{Sign}' \to \mathbf{Sign}$ *and natural transformation* $\varepsilon\colon \mu^{Sign};\rho^{Sign} \to \mathbf{Id}_{\mathbf{Sign}'}$*. Then there is an institution morphism* $\mu = \langle \mu^{Sign}, \mu^{Sen}, \mu^{Mod} \rangle\colon \mathbf{INS}' \to \mathbf{INS}$*, where for* $\Sigma' \in |\mathbf{Sign}'|$*,* $\mu^{Sen}_{\Sigma'} = \rho^{Sen}_{\mu^{Sign}(\Sigma')};\mathbf{Sen}'(\varepsilon_{\Sigma'})$ *and* $\mu^{Mod}_{\Sigma'} = \mathbf{Mod}'(\varepsilon_{\Sigma'});\rho^{Mod}_{\mu^{Sign}(\Sigma')}$*.*

Proof sketch. We outline a proof for the first part only; the second part follows in essentially the same way. To show naturality of ρ^{Sen} and ρ^{Mod} simply present them in terms of composition of natural transformations and multiplication by functors (Definitions 3.4.42, 3.4.44 and 3.4.46): $\rho^{Sen} = (\eta \cdot \mathbf{Sen});(\rho^{Sign} \cdot \mu^{Sen})$ and $\rho^{Mod} = (\rho^{Sign} \cdot \mu^{Mod});(\eta \cdot \mathbf{Mod})$. The satisfaction condition can be checked easily, using the satisfaction condition for μ and for **INS** (w.r.t. η_Σ). □

The general situation of Proposition 10.4.15 arises in many examples when the translation of signatures involved in an institution morphism has a left adjoint or, dually, when the translation of signatures involved in an institution comorphism has a right adjoint. Then we have a bijection between institution morphisms and comorphisms with adjoint signature translations. The following corollary covers such cases directly:

Corollary 10.4.16. *Consider an institution morphism* $\mu\colon \mathbf{INS}' \to \mathbf{INS}$ *such that* $\mu^{Sign}\colon \mathbf{Sign}' \to \mathbf{Sign}$ *has a left adjoint* $\rho^{Sign}\colon \mathbf{Sign} \to \mathbf{Sign}'$ *with unit* $\eta\colon \mathbf{Id}_{\mathbf{Sign}} \to \rho^{Sign};\mu^{Sign}$*. Then* $L(\mu) = \langle \rho^{Sign}, \rho^{Sen}, \rho^{Mod} \rangle$*, where* $\rho^{Sen}_\Sigma = \mathbf{Sen}(\eta_\Sigma);\mu^{Sen}_{\rho^{Sign}(\Sigma)}$ *and* $\rho^{Mod}_\Sigma = \mu^{Mod}_{\rho^{Sign}(\Sigma)};\mathbf{Mod}(\eta_\Sigma)$ *for* $\Sigma \in |\mathbf{Sign}|$*, is an institution comorphism* $L(\mu)\colon \mathbf{INS} \to \mathbf{INS}'$*.*

Consider an institution comorphism $\rho\colon \mathbf{INS} \to \mathbf{INS}'$ *such that* $\rho^{Sign}\colon \mathbf{Sign} \to \mathbf{Sign}'$ *has a right adjoint* $\mu^{Sign}\colon \mathbf{Sign}' \to \mathbf{Sign}$ *with counit* $\varepsilon\colon \mu^{Sign};\rho^{Sign} \to \mathbf{Id}_{\mathbf{Sign}'}$*. Then* $R(\rho) = \langle \mu^{Sign}, \mu^{Sen}, \mu^{Mod} \rangle$*, where for* $\Sigma' \in |\mathbf{Sign}'|$*,* $\mu^{Sen}_{\Sigma'} = \rho^{Sen}_{\mu^{Sign}(\Sigma')};\mathbf{Sen}'(\varepsilon_{\Sigma'})$ *and* $\mu^{Mod}_{\Sigma'} = \mathbf{Mod}'(\varepsilon_{\Sigma'});\rho^{Mod}_{\mu^{Sign}(\Sigma')}$*, is an institution morphism* $R(\rho)\colon \mathbf{INS}' \to \mathbf{INS}$*.*

Moreover, R and L can be chosen so that $R(L(\mu)) = \mu$ *and* $L(R(\rho)) = \rho$*.* □

Exercise 10.4.17. Given an institution morphism $\mu\colon \mathbf{INS}' \to \mathbf{INS}$ with a functor $\rho^{Sign}\colon \mathbf{Sign} \to \mathbf{Sign}'$ that is left adjoint to $\mu^{Sign}\colon \mathbf{Sign}' \to \mathbf{Sign}$ with unit $\eta\colon \mathbf{Id}_{\mathbf{Sign}} \to \rho^{Sign};\mu^{Sign}$, consider the institution comorphism $\rho = L(\mu)\colon \mathbf{INS} \to \mathbf{INS}'$ determined as in the first part of Corollary 10.4.16. Show that for any signature $\Sigma' \in |\mathbf{Sign}'|$, $\mu^{Mod}_{\Sigma'} = \mathbf{Mod}(\varepsilon_{\Sigma'});\rho^{Mod}_{\mu^{Sign}(\Sigma')}$. HINT: We have $id_{\mu^{Sign}(\Sigma')} = \eta_{\mu^{Sign}(\Sigma')};\mu^{Sign}(\varepsilon_{\Sigma'})$ by the properties of adjunctions, and $\mu^{Mod}_{\Sigma'};\mathbf{Mod}'(\mu^{Sign}(\varepsilon_{\Sigma'})) = \mathbf{Mod}(\varepsilon_{\Sigma'});\mu^{Mod}_{\rho^{Sign}(\mu^{Sign}(\Sigma'))}$ by naturality of μ^{Mod}.

Similarly, show that under the circumstances captured by the second part of Corollary 10.4.16, for any signature $\Sigma \in |\mathbf{Sign}|$, $\rho^{Mod}_\Sigma = \mu^{Mod}_{\rho^{Sign}(\Sigma)};\mathbf{Mod}(\eta_\Sigma)$. □

The number of possible kinds of map between institutions may seem scary. Proposition 10.4.15 and its Corollary 10.4.16 show one way of multiplying any existing maps between institutions. This means that once we have a morphism between institutions, we may generate a comorphism between them, and vice versa — a key point being that for many purposes we may then forget about the original morphism. For instance:

Exercise 10.4.18. Consider an institution morphism $\mu{:}\,\mathbf{INS}' \to \mathbf{INS}$ and an institution comorphism $\rho{:}\,\mathbf{INS} \to \mathbf{INS}'$ related as in Corollary 10.4.16, so that the functors μ^{Sign} and ρ^{Sign} form an adjunction with unit $\eta{:}\,\mathbf{Id_{Sign}} \to \rho^{Sign};\mu^{Sign}$ and counit $\varepsilon{:}\,\mu^{Sign};\rho^{Sign} \to \mathbf{Id_{Sign'}}$, and $L(\mu) = \rho$, $R(\rho) = \mu$. Show that the inter-institutional specification-building operations determined by μ may then be expressed using those determined by $L(\mu)$ and the usual intra-institutional specification-building operations, and similarly, the inter-institutional specification-building operations determined by ρ may be expressed using those determined by $R(\rho)$ and the usual intra-institutional specification-building operations.

HINT: Under appropriate assumptions about the signatures of the specifications and the signature morphisms involved, use Exercise 10.4.17 to prove that:

- SP' **hide via** $\mu \equiv (SP'$ **hide via** $\varepsilon_{\Sigma'})$ **hide via** ρ **to** Σ
- SP **with** μ **to** $\Sigma' \equiv (SP$ **with** $\rho)$ **hide via** $\varepsilon_{\Sigma'}$
- SP' **hide via** ρ **to** $\Sigma \equiv (SP'$ **hide via** $\mu)$ **hide via** η_Σ
- SP **with** $\rho \equiv (SP$ **with** $\eta_\Sigma)$ **with** μ **to** Σ'

Note though that the weaker assumptions of Proposition 10.4.15 are not in general sufficient to obtain similar equivalences. □

Another way to master the diversity of maps between institutions is to try to reduce them in some way to a common concept. Unfortunately, it seems that such a natural and directly useful concept is difficult to devise, in spite of the following exercise.

Exercise 10.4.19. Consider an institution morphism $\mu{:}\,\mathbf{INS} \to \mathbf{INS}'$. Build an "intermediate institution" by re-indexing $\mathbf{INS}'$ using the signature translation of μ: $\mathbf{INS}'_0 = \langle \mathbf{Sign}, \mu^{Sign};\mathbf{Sen}', \mu^{Sign};\mathbf{Mod}', \langle \models'_{\mu^{Sign}(\Sigma)} \rangle_{\Sigma \in |\mathbf{Sign}|} \rangle$. Check that two obvious institution comorphisms then arise, $\rho_1 = \langle \mathbf{Id}, \mu^{Sen}, \mu^{Mod} \rangle{:}\,\mathbf{INS}'_0 \to \mathbf{INS}$ and $\rho_2 = \langle \mu^{Sign}, \mathbf{Id}, \mathbf{Id} \rangle{:}\,\mathbf{INS}'_0 \to \mathbf{INS}'$, which yields a span of institution comorphisms:

$$\mathbf{INS} \xleftarrow{\ \rho_1\ } \mathbf{INS}'_0 \xrightarrow{\ \rho_2\ } \mathbf{INS}'$$

Argue that the span captures the same relationship between the two institutions as the original morphism. Check also that any institution semi-morphism and semi-comorphism may be captured in the same way, as a span of comorphisms.

To show that there is no undue preference here, proceed in the opposite direction: present an arbitrary comorphism between institutions as a span of morphisms. Do the same for institution semi-morphisms and semi-comorphisms (that is, present them as spans of morphisms).

Show also that by using such spans we do not reduce the class of expressible specifications; check for instance that the inter-institutional specification-building operations determined by an institution morphism $\mu{:}\,\mathbf{INS} \to \mathbf{INS}'$ may be expressed using the operations determined by the comorphisms in the span for μ as defined above, namely:

- SP' **with** μ **to** $\Sigma \equiv (SP'$ **hide via** ρ_2 **to** $\Sigma)$ **with** ρ_1, for any $\mathbf{INS}'$-specification SP' and signature $\Sigma \in |\mathbf{Sign}|$ such that $\mu^{Sign}(\Sigma) = Sig[SP']$, and

- SP **hide via** $\mu \equiv$ (SP **hide via** ρ_1 **to** $Sig[SP]$) **with** ρ_2, for any **INS**-specification SP. □

Exercise 10.4.20. Recall the informal hints at the beginning of this section concerning institution forward morphisms and forward comorphisms. Spell out their definitions. Show that any forward morphism between institutions may be presented as a span of (co)morphisms, and similarly for forward comorphisms. □

The considerations above seem to suggest that we do not get much more out of institution comorphisms than we have already squeezed out of institution morphisms. This is not quite the case.

First, it is sometimes difficult to map *all* models of a rich institution to models of an institution that we want to represent in it. It is sometimes the case that we can recover models of the simpler institution **INS**′ only from the models of a richer institution **INS** that satisfy certain properties. This additional "selection of models" can often be captured by an institution comorphism from **INS** to its enrichment $\mathbf{INS}^{\mathbf{Sign}^+}$ that adds axioms to its signatures; see Example 4.1.39. We will refer to comorphisms from **INS**′ to $\mathbf{INS}^{\mathbf{Sign}^+}$ as *theoroidal* comorphisms from **INS**′ to **INS**.[8] Notice that while a theoroidal comorphism from **INS**′ to **INS** may be given even if there is no (ordinary) comorphism from **INS**′ to **INS**, the analogous "trick" for institution morphisms does not work: any institution morphism from $\mathbf{INS}^{\mathbf{Sign}^+}$ to **INS**′ determines an institution morphism from **INS** to **INS**′.

Exercise 10.4.21. Recall institutions **FOPEQ** of first-order logic with equality (Example 4.1.12) and **FOP** of first-order predicate logic (no built-in equality; Exercise 4.1.13).

Try to give an institution comorphism from **FOPEQ** to **FOP** that would view equality as a family of new, two-argument predicates (one for each sort). To ensure the satisfaction condition, the model functors should map first-order structures to their quotients by the relations determined by the equality predicates. But this will not work unless these relations are guaranteed to be congruences — and so such a "naive" construction fails.

Give a theoroidal comorphism from **FOPEQ** to **FOP**, mapping each first-order signature Θ to a theory with signature that extends Θ by new binary equality predicates for each sort and with axioms stating that these predicates form a congruence. □

Another reason for interest in institution comorphisms is that results concerning the reuse of proofs in a richer logic for a logic that can be represented in it via an institution comorphism turn out to be more useful than those given for institutions related by an institution morphism in Section 10.2.

For the rest of this section, let $\mathbf{INS} = \langle \mathbf{Sign}, \mathbf{Sen}, \mathbf{Mod}, \langle \models_\Sigma \rangle_{\Sigma \in |\mathbf{Sign}|} \rangle$ and $\mathbf{INS}' = \langle \mathbf{Sign}', \mathbf{Sen}', \mathbf{Mod}', \langle \models'_{\Sigma'} \rangle_{\Sigma' \in |\mathbf{Sign}'|} \rangle$ be institutions linked by an institution comorphism $\rho \colon \mathbf{INS} \to \mathbf{INS}'$.

[8] Terminology from [GR02].

Using the consequences of the satisfaction condition for institution morphisms in much the same way as we did in Section 4.2 for the satisfaction condition in a single institution, one can show that the semantic consequence relation is preserved when sentences are translated by an institution comorphism. A number of related facts then follow in a similar way; see also Section 10.2. We refrain from essentially repeating the proofs here.

Proposition 10.4.22. *Consider an* **INS**-*signature* $\Sigma \in |\mathbf{Sign}|$ *and let* $\Sigma' = \rho^{Sign}(\Sigma) \in |\mathbf{Sign}'|$. *For any sets* $\Phi \subseteq \mathbf{Sen}(\Sigma)$ *of* Σ*-sentences and* $\Phi' \subseteq \mathbf{Sen}'(\Sigma')$ *of* Σ'*-sentences, and* Σ*-sentence* $\varphi \in \mathbf{Sen}(\Sigma)$*, we have:*

1. $\rho^{Sen}_\Sigma(Cl_\Sigma(\Phi)) \subseteq Cl_{\Sigma'}(\rho^{Sen}_\Sigma(\Phi))$. *In other words, if* $\Phi \models_\Sigma \varphi$ *then* $\rho^{Sen}_\Sigma(\Phi) \models'_{\Sigma'} \rho^{Sen}_\Sigma(\varphi)$.
2. $Cl_{\Sigma'}(\rho^{Sen}_\Sigma(Cl_\Sigma(\Phi))) = Cl_{\Sigma'}(\rho^{Sen}_\Sigma(\Phi))$.
3. $Cl_\Sigma((\rho^{Sen}_\Sigma)^{-1}(\Phi')) \subseteq (\rho^{Sen}_\Sigma)^{-1}(Cl_{\Sigma'}(\Phi'))$.
4. $Cl_\Sigma((\rho^{Sen}_\Sigma)^{-1}(\Phi')) \subseteq Th_\Sigma(\rho^{Mod}_\Sigma(Mod_{\Sigma'}(\Phi')))$. □

Proposition 10.4.23. *For any* **INS**-*signature* $\Sigma \in |\mathbf{Sign}|$ *and set* $\Phi' \subseteq \mathbf{Sen}'(\rho^{Sign}(\Sigma))$ *of* $\rho^{Sign}(\Sigma)$*-sentences,* $(\rho^{Sen}_\Sigma)^{-1}(Cl_{\rho^{Sign}(\Sigma)}(\Phi')) = Th_\Sigma(\rho^{Mod}_\Sigma(Mod_{\rho^{Sign}(\Sigma)}(\Phi')))$.

Consequently, if Φ' *is closed then* $(\rho^{Sen}_\Sigma)^{-1}(\Phi') \subseteq \mathbf{Sen}(\Sigma)$ *is a closed set of* Σ*-sentences in* **INS**. □

Institution comorphisms directly determine a functor between the categories of theories of the institutions they link, perhaps even more naturally than in the case of institution morphisms (Definition 10.2.15):

Definition 10.4.24 (Theory functor induced by an institution comorphism). For every institution comorphism $\rho\colon \mathbf{INS} \to \mathbf{INS}'$ we define a functor $\rho^{Th}\colon \mathbf{Th_{INS}} \to \mathbf{Th_{INS'}}$ between the categories of theories of **INS** and of **INS**$'$, respectively:

- $\rho^{Th}(\langle \Sigma, \Phi \rangle) = \langle \rho^{Sign}(\Sigma), Cl_{\rho^{Sign}(\Sigma)}(\rho^{Sen}_\Sigma(\Phi)) \rangle \in |\mathbf{Th_{INS'}}|$, for any **INS**-theory $\langle \Sigma, \Phi \rangle \in |\mathbf{Th_{INS}}|$, and
- $\rho^{Th}(\sigma) = \rho^{Sign}(\sigma)\colon \rho^{Th}(T_1) \to \rho^{Th}(T_2)$, for any **INS**-theory morphism $\sigma\colon T_1 \to T_2$. □

In contrast to theory functors given by institution morphisms (see Example 10.2.21), the theory functors determined by institution comorphisms typically preserve colimits:

Proposition 10.4.25. *If the category* **Sign** *of signatures in* **INS** *is cocomplete and* $\rho^{Sign}\colon \mathbf{Sign} \to \mathbf{Sign}'$ *is cocontinuous then the theory functor* $\rho^{Th}\colon \mathbf{Th_{INS}} \to \mathbf{Th_{INS'}}$ *is cocontinuous as well.*

Proof. The proof essentially follows the hint in Exercise 10.2.22.

Consider a diagram D in $\mathbf{Th_{INS}}$ with nodes $D_n = \langle \Sigma_n, \Phi_n \rangle$, $n \in N$, and its colimit $\langle \alpha_n\colon \langle \Sigma_n, \Phi_n \rangle \to \langle \Sigma, Cl_\Sigma(\bigcup_{n\in N} \alpha_n(\Phi_n)) \rangle \rangle_{n\in N}$ as constructed in the proof of Theorem 4.4.1. Since $\rho^{Sign}\colon \mathbf{Sign} \to \mathbf{Sign}'$ is cocontinuous, the colimit in **Sign**$'$ of the underlying signature diagram for $\rho^{Th}(D)$ is $\langle \rho^{Sign}(\alpha_n)\colon \rho^{Sign}(\Sigma_n) \to \rho^{Sign}(\Sigma) \rangle_{n\in N}$.

All we have to show now is that $Cl_{\rho^{Sign}(\Sigma)}(\rho^{Sen}_{\Sigma}(Cl_{\Sigma}(\bigcup_{n\in N}\alpha_n(\Phi_n))))$ coincides with $Cl_{\rho^{Sign}(\Sigma)}(\bigcup_{n\in N}\rho^{Sign}(\alpha_n)(Cl_{\rho^{Sign}(\Sigma_n)}(\rho^{Sen}_{\Sigma_n}(\Phi_n))))$, which follows easily by the naturality of ρ^{Sen}, Corollary 4.2.11 and Proposition 10.4.22(2). □

The key assumption in the above proposition, the cocontinuity of signature translation, is ensured for instance when the institution comorphism is linked with an institution morphism as in Corollary 10.4.16, when the translation of signatures is a left adjoint, by Exercise 3.5.17.

As in Sections 4.2 and 10.2, none of the inclusions in Proposition 10.4.22 may be reversed in general. However, the situation is again different if model translation is surjective.

Proposition 10.4.26. *Let $\Sigma \in |\mathbf{Sign}|$ be an* **INS**-*signature. Suppose that the functor $\rho^{Mod}_{\Sigma}: \mathbf{Mod}'(\rho^{Sign}(\Sigma)) \to \mathbf{Mod}(\Sigma)$ is surjective on models. Then for any set $\Phi \subseteq \mathbf{Sen}(\Sigma)$ of Σ-sentences and Σ-sentence $\varphi \in \mathbf{Sen}(\Sigma)$, we have:*

1. *$\Phi \models_{\Sigma} \varphi$ if and only if $\rho^{Sen}_{\Sigma}(\Phi) \models'_{\rho^{Sign}(\Sigma)} \rho^{Sen}_{\Sigma}(\varphi)$.*
2. *$Cl_{\Sigma}(\Phi) = (\rho^{Sen}_{\Sigma})^{-1}(Cl_{\rho^{Sign}(\Sigma)}(\rho^{Sen}_{\Sigma}(\Phi)))$.* □

Exercise 10.4.27. Following Example 10.2.17, show that the requirement that ρ^{Mod}_{Σ} be surjective in Proposition 10.4.26 cannot be dropped. □

Exercise 10.4.28. Carry over Proposition 10.2.19 and its Corollary 10.2.20 to institution comorphisms, thus generalising Proposition 10.2.16(1) to the case where the translation of models ρ^{Mod}_{Σ} is not surjective, but the image of the class of all $\rho^{Sign}(\Sigma)$-models under ρ^{Mod}_{Σ} is definable in **INS**. Notice that the more general form, corresponding to Proposition 10.2.19 (originating in Proposition 4.2.21), becomes crucial when theoroidal institution comorphisms are considered. □

Proposition 10.4.26 suggests that institution comorphisms in which all translations of models are surjective may be especially useful. In particular, not only do semantic consequences in the more primitive institution **INS** remain valid when translated to the richer institution **INS**′ (this is Proposition 10.4.22(1)), but also whatever we can deduce in the richer logic **INS**′ about the sentences of the more primitive logic **INS** is valid there as well. This means that any proof system or theorem prover available for **INS**′ may be used for **INS** as well.

A crucial novelty with respect to Section 10.2 is that now we can extend this to arbitrary structured specifications, since institution comorphisms offer a natural way to translate (typical) structured specifications:

Exercise 10.4.29. Consider the class $Spec^{UTH}_{\mathbf{INS}}$ of specifications in an institution **INS** built from flat specifications using union, translation and hiding. Extend any institution comorphism $\rho: \mathbf{INS} \to \mathbf{INS}'$ to translate specifications in $Spec^{UTH}_{\mathbf{INS}}$ to **INS**′ by inductively defining a function $\rho^{Spec}: Spec^{UTH}_{\mathbf{INS}} \to Spec^{UTH}_{\mathbf{INS}'}$ as follows:

- $\rho^{Spec}(\langle\Sigma,\Phi\rangle) = \langle\rho^{Sign}(\Sigma), \rho^{Sen}_{\Sigma}(\Phi)\rangle$
- $\rho^{Spec}(SP_1 \cup SP_2) = \rho^{Spec}(SP_1) \cup \rho^{Spec}(SP_2)$

- $\rho^{Spec}(SP \textbf{ with } \sigma) = \rho^{Spec}(SP) \textbf{ with } \rho^{Sign}(\sigma)$
- $\rho^{Spec}(SP \textbf{ hide via } \sigma) = \rho^{Spec}(SP) \textbf{ hide via } \rho^{Sign}(\sigma)$

Prove that for any specification $SP \in Spec^{UTH}_{\mathbf{INS}}$, $\rho^{Mod}_{Sig[SP]}(Mod[\rho^{Spec}(SP)]) \subseteq Mod[SP]$. Give counterexamples to show that the opposite inclusion fails in general.

Then assume that the naturality squares for ρ^{Mod} (see Definition 10.4.1) are pullbacks in **Cat**. Prove by induction on the structure of $SP \in Spec^{UTH}_{\mathbf{INS}}$ that for any model $M' \in |\mathbf{Mod}'(\rho^{Sign}(\Sigma))|$, if $\rho^{Mod}_{\Sigma}(M') \in Mod[SP]$ then $M' \in Mod[\rho^{Spec}(SP)]$, where $\Sigma = Sig[SP]$. You may want to weaken the assumption concerning the naturality squares by dropping requirements concerning model morphisms and then dropping the requirement that the "amalgamated" models are determined uniquely.

Note that this shows that, in this case, the inter-institutional operation to translate specifications by an institution comorphism can be reduced to syntactic translation of specifications:

$$SP \textbf{ with } \rho \equiv \rho^{Spec}(SP).$$

Conclude also that if in addition the translation ρ^{Mod}_{Σ} is surjective on models then

$$\rho^{Mod}_{Sig[SP]}(Mod[\rho^{Spec}(SP)]) = Mod[SP].$$

Finally, note that this implies that for any sentence $\varphi \in \mathbf{Sen}(\Sigma)$, $SP \models_{\Sigma} \varphi$ if and only if $\rho^{Spec}(SP) \models_{\rho^{Sign}(\Sigma)} \rho^{Sen}_{\Sigma}(\varphi)$. □

10.5 Bibliographical remarks

The need for linking institutions by maps of some sort was already emphasized in the original institutions paper [GB92] and its early version [GB84a]; many later developments elaborate on the basic notion of an institution morphism introduced there and on its many variants.

The idea of moving specifications between institutions was first introduced in [ST88b], which used institution semi-morphisms (called semi-institution morphisms there) for this purpose, defined the inter-institutional specification-building operation _ **hide via** μ (written as **change institution of** _ **via** μ there), and applied it to capture developments migrating between institutions much as presented in Section 10.1.3. Translation of specifications by an institution semi-morphism was introduced in [Tar96], and similar operations determined by institution comorphisms were given in [Tar00]. Similar ideas were approached from a different angle in [Dia02]. Given a diagram of institutions linked by institution morphisms, a version of the Grothendieck construction (Definition 3.4.59) may be used to construct a single "heterogeneous" institution that incorporates all the institutions involved and enriches them with "inter-institutional" signature morphisms that use institution morphisms in the diagram. Inter-institutional specification-building operations determined by institution morphisms, as defined in Section 10.1.3, can then be captured using the usual intra-institutional hiding and translation operations in this

Grothendieck institution. This approach underlies the development of CafeOBJ, which supports the use of a cube of eight logics formalised as institutions and linked by institution morphisms [DF98], [DF02]. A similar trick may be used for institutions linked by institution comorphisms and the operations they determine [Mos02].

Definition 10.1.12 is inspired by duplex institutions in [GB92], which are essentially institutions of the form $\mathbf{INS}_1$ **plus** $\mathbf{INS}_2^{data}$ **via** μ, where $\mu : \mathbf{INS}_1 \rightarrow_{semi} \mathbf{INS}_2$ was unnecessarily required to extend to an institution morphism. Exercise 10.1.15 corresponds to multiplex institutions in [GB92].

Section 10.2 is based on [GB92], with Definition 10.2.1 and some of the basic facts concerning preservation of semantic consequence taken from there. In particular, Exercise 10.2.22 is Theorem 39 of [GB92].

Section 10.3 follows [Tar86a], where the completeness of the category of institutions was proved (see also [TBG91]), although some of the terminology used here comes from [May85] and [GB86]. Examples 10.3.12, 10.3.13 and 10.3.15 hint at the idea of using limits in the category of institutions to systematically combine logical systems; both this idea and its inherent limitations were pointed out in [Tar86a]. The problem the examples identify is that sentences in institutions are viewed as unstructured entities and so universal constructions in the category of institutions combine sets of sentences rather than combining the ways that sentences are built (via closure properties of these sets). One way to deal with this is to introduce more detailed presentations of institutions, for instance as *parchments* [GB86], where in essence an abstract context-free grammar (in the form of an algebraic signature) is used to define sentences. Combining such grammars typically yields a presentation of the expected sets of sentences, closed under the logical constructs used in each of the combined logics. But then the meaning of the essentially new sentences so obtained may not be defined unambiguously, and universal constructions on parchments may yield structures that do not define institutions without further "manual" adjustments; see [MTP97], [MTP98], [CMRS01], [CGR03] for further details. Another option is to combine logics together with their (compatible) representations in a "universal logic", where manual adjustments to define the meaning of new combinations of logical constructs are replaced by the standard combination of their representations in the universal logic; see [Tar96], [MTP97], [MTP98], [Tar00]. These ideas provide a potential alternative to a similar programme concerning "fibring" of logics [Gab98], [CSS05].

The taxonomy of possible mappings between institutions classified by the direction of translations of signatures, sentences and models in Section 10.4 was first spelled out in [Tar96] and then systematically studied in [GR02], which also introduced the now widely adopted terminology we follow here. Some variants of maps between institutions escape this classification: we have mentioned theoroidal comorphisms ("simple maps of institutions" in [Mes89]) but did not discuss for instance model translations that map individual models to model classes (or model translations given as partial functions on model classes), as in [SS93], or various ways of weakening the satisfaction condition; see e.g. [Tar87], [Tar96]. It is worth noting that theoroidal institution comorphisms give one example of a wide class of maps between institutions that can be derived from the basic notions as Kleisli morphisms for a monad over the category of institutions [Mos96a].

Institution comorphisms were introduced in [Mes89] as plain maps of institutions and in [Tar87], [Tar96] as institution representations, the latter term perhaps best reflecting the intuition that institution comorphisms capture one way of representing the source institution in the target.

The notion of subinstitution sketched in Exercise 10.4.9 follows [Mes89]; this is just a starting point for using institution comorphisms to compare relative "power" of logical systems captured as institutions; see [MTD09].

The importance of institution comorphisms was stressed in [CM97], where they were used to "borrow" consequence relations (and so proof systems, as discussed in Chapter 9, and related support tools) from a richer institution to any institution mapped into it by an institution comorphism. This was based on facts like Proposition 10.4.26 and the more general properties indicated in Exercise 10.4.28. In [Tar00] it is pointed out that other properties (like specification equivalence or refinements between specifications) and their proofs may be transported in such a way as well, using the inter-institutional operation _ **with** ρ to translate specifications by an institution comorphism. In this context Exercise 10.4.29, following the essence (though not all the technical details and practically motivated nuances) of developments in [Bor02], is of particular importance.

The overall picture emerging in this chapter is that in many practical situations one will be working in a *heterogeneous logical environment* which consists of a number of institutions linked by institution (semi-)morphisms, (semi-)comorphisms and other maps. This should allow the user to migrate between logical systems while building (heterogeneous) specifications and in the course of the development process. This activity can be supported by tools, like the HETS system [Mos05], [MML07], that offer support for defining and extending such logical environments, for building heterogeneous specifications, and for proving their consequences and entailments between them. More on this vision may be found in [MT09], which also admits truly distributed heterogeneous specifications that do not focus on any particular institution, capturing specifications in formalisms based on multiple views of systems such as UML [BRJ98] (see [CKTW08]). The diversity of the concepts of a map between institutions becomes a serious technical issue in this context. Results like Corollary 10.4.16, first given in [AF96] (although its form in Proposition 10.4.15 appears to be new), complemented by results like those in Exercise 10.4.18 (from [MT09]), point in this direction. The alternative and more universally applicable idea of using spans of maps of a certain kind to represent other maps between institutions (Exercises 10.4.19 and 10.4.20) has been noticed in [Mos03] and spelled out in [Mos05], but seems to originate in [MW98], which used institution forward morphisms (called "transformations" there) as a primary notion.

Institution forward morphisms were also introduced in [Tar00] (named "institution encodings" there) with results analogous to those pointed at in Exercises 10.4.28 and 10.4.29; these were used in [BH06a] to develop proof techniques for specifications built in the constructor-based observational logic COL.

Bibliography

[AC89] Egidio Astesiano and Maura Cerioli. On the existence of initial models for partial (higher-order) conditional specifications. In: Josep Díaz and Fernando Orejas, editors, *Proceedings of the International Joint Conference on Theory and Practice of Software Development, TAPSOFT'89*, Barcelona, *Lecture Notes in Computer Science*, Vol. 351, pages 74–88. Springer, 1989.

[AC01] David Aspinall and Adriana B. Compagnoni. Subtyping dependent types. *Theoretical Computer Science*, 266(1–2):273–309, 2001.

[ACEGG91] Jaume Agustí-Cullell, Francesc Esteva, Pere Garcia, and Lluis Godo. Formalizing multiple-valued logics as institutions. In: Bernadette Bouchon-Meunier, Ronald R. Yager, and Lotfi A. Zadeh, editors, *Proceedings of the 3rd International Conference on Information Processing and Management of Uncertainty in Knowledge-Based Systems, IPMU'90*, Paris, *Lecture Notes in Computer Science*, Vol. 521, pages 269–278. Springer, 1991.

[AF96] Mário Arrais and José Luiz Fiadeiro. Unifying theories in different institutions. In: Magne Haveraaen, Olaf Owe, and Ole-Johan Dahl, editors, *Recent Trends in Data Type Specification. Selected Papers from the 11th Workshop on Specification of Abstract Data Types*, Oslo, *Lecture Notes in Computer Science*, Vol. 1130, pages 81–101. Springer, 1996.

[AG97] Robert Allen and David Garlan. A formal basis for architectural connection. *ACM Transactions on Software Engineering and Methodology*, 6(3):213–249, 1997.

[AH05] David Aspinall and Martin Hofmann. Dependent types. In: Benjamin Pierce, editor, *Advanced Topics in Types and Programming Languages*, chapter 2, pages 45–86. MIT Press, 2005.

[AHS90] Jiří Adámek, Horst Herrlich, and George E. Strecker. *Abstract and Concrete Categories: The Joy of Cats*. Wiley, 1990.

[Ala02] Suad Alagic. Institutions: Integrating objects, XML and databases. *Information and Software Technology*, 44(4):207–216, 2002.

[AM75] Michael A. Arbib and Ernest G. Manes. *Arrows, Structures and Functors: The Categorical Imperative*. Academic Press, 1975.

[Asp95] David Aspinall. Subtyping with singleton types. In: Leszek Pacholski and Jerzy Tiuryn, editors, *Proceedings of the 8th International Workshop on Computer Science Logic, CSL'94*, Kazimierz, *Lecture Notes in Computer Science*, Vol. 933, pages 1–15. Springer, 1995.

[Asp97] David Aspinall. *Type Systems for Modular Programming and Specification*. Ph.D. thesis, University of Edinburgh, Department of Computer Science, 1997.

[Asp00] David Aspinall. Subtyping with power types. In: Peter Clote and Helmut Schwichtenberg, editors, *Proceedings of the 14th International Workshop on Computer Sci-

ence Logic, Fischbachau, *Lecture Notes in Computer Science*, Vol. 1862, pages 156–171. Springer, 2000.

[Avr91] Arnon Avron. Simple consequence relations. *Information and Computation*, 92:105–139, 1991.

[Awo06] Steve Awodey. *Category Theory*. Oxford University Press, 2006.

[Bar74] Jon Barwise. Axioms for abstract model theory. *Annals of Mathematical Logic*, 7:221–265, 1974.

[BBB+85] Friedrich L. Bauer, Rudolf Berghammer, Manfred Broy, Walter Dosch, Franz Geiselbrechtinger, Rupert Gnatz, Erich Hangel, Wolfgang Hesse, Bernd Krieg-Brückner, Alfred Laut, Thomas Matzner, Bernhard Möller, Friederike Nickl, Helmuth Partsch, Peter Pepper, Klaus Samelson, Martin Wirsing, and Hans Wössner. *The Munich Project CIP: Vol. 1: The Wide Spectrum Language CIP-L*, *Lecture Notes in Computer Science*, Vol. 183. Springer, 1985.

[BBC86] Gilles Bernot, Michel Bidoit, and Christine Choppy. Abstract data types with exception handling: An initial approach based on a distinction between exceptions and errors. *Theoretical Computer Science*, 46(1):13–45, 1986.

[BC88] Val Breazu-Tannen and Thierry Coquand. Extensional models for polymorphism. *Theoretical Computer Science*, 59(1–2):85–114, 1988.

[BCH99] Michel Bidoit, María Victoria Cengarle, and Rolf Hennicker. Proof systems for structured specifications and their refinements. In: Egidio Astesiano, Hans-Jörg Kreowski, and Bernd Krieg-Brückner, editors, *Algebraic Foundations of Systems Specification*, chapter 11, pages 385–433. Springer, 1999.

[BCH+04] Hubert Baumeister, Maura Cerioli, Anne Haxthausen, Till Mossakowski, Peter D. Mosses, Donald Sannella, and Andrzej Tarlecki. CASL semantics. In: [Mos04]. 2004.

[BD77] Rodney M. Burstall and John Darlington. A transformational system for developing recursive programs. *Journal of the Association for Computing Machinery*, 24(1):44–67, 1977.

[BDP+79] Manfred Broy, Walter Dosch, Helmuth Partsch, Peter Pepper, and Martin Wirsing. Existential quantifiers in abstract data types. In: Hermann A. Maurer, editor, *Proceedings of the 6th International Colloquium on Automata, Languages and Programming*, Graz, *Lecture Notes in Computer Science*, Vol. 71, pages 73–87. Springer, 1979.

[Bén85] Jean Bénabou. Fibred categories and the foundations of naïve category theory. *Journal of Symbolic Logic*, 50:10–37, 1985.

[Ber87] Gilles Bernot. Good functors ... are those preserving philosophy! In: David H. Pitt, Axel Poigné, and David E. Rydeheard, editors, *Proceedings of the 2nd Summer Conference on Category Theory and Computer Science*, Edinburgh, *Lecture Notes in Computer Science*, Vol. 283, pages 182–195. Springer, 1987.

[BF85] Jon Barwise and Solomon Feferman, editors. *Model-Theoretic Logics*. Springer, 1985.

[BG77] Rodney M. Burstall and Joseph A. Goguen. Putting theories together to make specifications. In: *Fifth International Joint Conference on Artificial Intelligence*, pages 1045–1058, Boston, 1977.

[BG80] Rodney M. Burstall and Joseph A. Goguen. The semantics of Clear, a specification language. In: Dines Bjørner, editor, *Proceedings of the 1979 Copenhagen Winter School on Abstract Software Specification*, *Lecture Notes in Computer Science*, Vol. 86, pages 292–332. Springer, 1980.

[BG81] Rodney M. Burstall and Joseph A. Goguen. An informal introduction to specifications using Clear. In: Robert S. Boyer and J Strother Moore, editors, *The Correctness Problem in Computer Science*, pages 185–213. Academic Press, 1981. Also in: Narain Gehani and Andrew D. McGettrick, editors, *Software Specification Techniques*. Addison-Wesley, 1986.

[BG01] Leo Bachmair and Harald Ganzinger. Resolution theorem proving. In: John Alan Robinson and Andrei Voronkov, editors, *Handbook of Automated Reasoning*, pages 19–99. Elsevier and MIT Press, 2001.

[BH96] Michel Bidoit and Rolf Hennicker. Behavioural theories and the proof of behavioural properties. *Theoretical Computer Science*, 165(1):3–55, 1996.

[BH98] Michel Bidoit and Rolf Hennicker. Modular correctness proofs of behavioural implementations. *Acta Informatica*, 35(11):951–1005, 1998.

[BH06a] Michel Bidoit and Rolf Hennicker. Constructor-based observational logic. *Journal of Logic and Algebraic Programming*, 67(1–2):3–51, 2006.

[BH06b] Michel Bidoit and Rolf Hennicker. Proving behavioral refinements of COL-specifications. In: Kokichi Futatsugi, Jean-Pierre Jouannaud, and José Meseguer, editors, *Algebra, Meaning and Computation: Essays Dedicated to Joseph A. Goguen on the Occasion of His 65th Birthday, Lecture Notes in Computer Science*, Vol. 4060, pages 333–354. Springer, 2006.

[BHK90] Jan A. Bergstra, Jan Heering, and Paul Klint. Module algebra. *Journal of the Association for Computing Machinery*, 37(2):335–372, 1990.

[BHW94] Michel Bidoit, Rolf Hennicker, and Martin Wirsing. Characterizing behavioural semantics and abstractor semantics. In: Donald Sannella, editor, *Proceedings of the 5th European Symposium on Programming*, Edinburgh, *Lecture Notes in Computer Science*, Vol. 788, pages 105–119. Springer, 1994.

[BHW95] Michel Bidoit, Rolf Hennicker, and Martin Wirsing. Behavioural and abstractor specifications. *Science of Computer Programming*, 25(2–3):149–186, 1995.

[Bir35] Garrett Birkhoff. On the structure of abstract algebras. *Proceedings of the Cambridge Philosophical Society*, 31:433–454, 1935.

[BL69] Rodney M. Burstall and Peter J. Landin. Programs and their proofs: An algebraic approach. In: Bernard Meltzer and Donald Michie, editors, *Machine Intelligence 4*, pages 17–43. Edinburgh University Press, 1969.

[BL70] Garrett Birkhoff and John Lipson. Heterogeneous algebras. *Journal of Combinatorial Theory*, 8(1):115–133, 1970.

[BM04] Michel Bidoit and Peter D. Mosses, editors. *CASL User Manual, Lecture Notes in Computer Science*, Vol. 2900. Springer, 2004. See also http://www.informatik.uni-bremen.de/cofi/wiki/index.php/CASL.

[BN98] Franz Baader and Tobias Nipkow. *Term Rewriting and All That*. Cambridge University Press, 1998.

[Bor94] Francis Borceaux. *Handbook of Categorical Algebra*. Cambridge University Press, 1994.

[Bor00] Tomasz Borzyszkowski. Higher-order logic and theorem proving for structured specifications. In: Didier Bert, Christine Choppy, and Peter D. Mosses, editors, *Recent Trends in Algebraic Development Techniques. Selected Papers from the 14th International Workshop on Algebraic Development Techniques*, Château de Bonas, *Lecture Notes in Computer Science*, Vol. 1827, pages 401–418. Springer, 2000.

[Bor02] Tomasz Borzyszkowski. Logical systems for structured specifications. *Theoretical Computer Science*, 286(2):197–245, 2002.

[Bor05] Tomasz Borzyszkowski. Generalized interpolation in first order logic. *Fundamenta Informaticae*, 66(3):199–219, 2005.

[BPP85] Edward K. Blum and Francesco Parisi-Presicce. The semantics of shared submodules specifications. In: Hartmut Ehrig, Christiane Floyd, Maurice Nivat, and James W. Thatcher, editors, *Mathematical Foundations of Software Development. Proceedings of the International Joint Conference on Theory and Practice of Software Development. Vol. 1: Colloquium on Trees in Algebra and Programming*, Berlin, *Lecture Notes in Computer Science*, Vol. 185, pages 359–373. Springer, 1985.

[BRJ98] Grady Booch, James Rumbaugh, and Ivar Jacobson. *The Unified Modeling Language User Guide*. Addison-Wesley, 1998.

[BS93] Rudolf Berghammer and Gunther Schmidt. Relational specifications. In: Cecylia Rauszer, editor, *Proc. XXXVIII Banach Center Semester on Algebraic Methods in Logic and their Computer Science Applications, Banach Center Publications*, Vol. 28, pages 167–190, Institute of Mathematics, Polish Academy of Sciences, 1993.

[BST02] Michel Bidoit, Donald Sannella, and Andrzej Tarlecki. Architectural specifications in CASL. *Formal Aspects of Computing*, 13:252–273, 2002.

[BST08] Michel Bidoit, Donald Sannella, and Andrzej Tarlecki. Observational interpretation of CASL specifications. *Mathematical Structures in Computer Science*, 18:325–371, 2008.

[BT87] Jan A. Bergstra and John V. Tucker. Algebraic specifications of computable and semicomputable data types. *Theoretical Computer Science*, 50(2):137–181, 1987.

[BT96] Michel Bidoit and Andrzej Tarlecki. Behavioural satisfaction and equivalence in concrete model categories. In: Hélène Kirchner, editor, *Proceedings of the 21st International Colloquium on Trees in Algebra and Programming*, Linköping, *Lecture Notes in Computer Science*, Vol. 1059, pages 241–256. Springer, 1996.

[Bur86] Peter Burmeister. *A Model Theoretic Oriented Approach to Partial Algebras*. Akademie-Verlag, 1986.

[BW82a] Friedrich L. Bauer and Hans Wössner. *Algorithmic Language and Program Development*. Springer, 1982.

[BW82b] Manfred Broy and Martin Wirsing. Partial abstract data types. *Acta Informatica*, 18(1):47–64, 1982.

[BW85] Michael Barr and Charles Wells. *Toposes, Triples and Theories, Grundlehren der mathematischen Wissenschaften*, Vol. 278. Springer, 1985.

[BW95] Michael Barr and Charles Wells. *Category Theory for Computing Science*. Prentice Hall, second edition, 1995.

[BWP84] Manfred Broy, Martin Wirsing, and Claude Pair. A systematic study of models of abstract data types. *Theoretical Computer Science*, 33(2–3):139–174, 1984.

[C++09] Working draft, Standard for programming language C++, June 2009. ISO/IEC report N2914. Available from http://www.open-std.org/jtc1/sc22/wg21/docs/papers/2009/n2914.pdf. Concepts and axioms were removed from later versions of this document.

[Car88] Luca Cardelli. Structural subtyping and the notion of power type. In: *Proceedings of the 15th ACM Symposium on Principles of Programming Languages*, San Diego, pages 70–79, 1988.

[CDE$^+$02] Manuel Clavela, Francisco Durán, Steven Eker, Patrick Lincoln, Narciso Martí-Oliet, José Meseguer, and José F. Quesada. Maude: Specification and programming in rewriting logic. *Theoretical Computer Science*, 285(2):187–243, 2002. See also http://maude.cs.uiuc.edu/.

[Cen94] María Victoria Cengarle. *Formal Specifications with Higher-Order Parameterization*. Ph.D. thesis, Ludwig-Maximilians-Universität München, Institut für Informatik, 1994.

[CF92] Robin Cockett and Tom Fukushima. About Charity. Technical Report No. 92/480/18, Department of Computer Science, University of Calgary, 1992.

[CGR03] Carlos Caleiro, Paula Gouveia, and Jaime Ramos. Completeness results for fibred parchments: Beyond the propositional base. In: Martin Wirsing, Dirk Pattinson, and Rolf Hennicker, editors, *Recent Trends in Algebraic Development Techniques. Selected Papers from the 16th International Workshop on Algebraic Development Techniques*, Frauenchiemsee, *Lecture Notes in Computer Science*, Vol. 2755, pages 185–200. Springer, 2003.

[Chu56] Alonzo Church. *Introduction to Mathematical Logic, Vol. 1*. Princeton University Press, 1956.

[Cîr02] Corina Cîrstea. On specification logics for algebra-coalgebra structures: Reconciling reachability and observability. In: *Proceedings of the 5th International Conference on Foundations of Software Science and Computation Structures. European Joint*

Conferences on Theory and Practice of Software (ETAPS 2002), Grenoble, *Lecture Notes in Computer Science*, Vol. 2303, pages 82–97. Springer, 2002.

[CJ95] Aurelio Carboni and Peter T. Johnstone. Connected limits, familial representability and Artin glueing. *Mathematical Structures in Computer Science*, 5(4):441–459, 1995.

[CK90] Chen-Chung Chang and H. Jerome Keisler. *Model Theory*. North-Holland, third edition, 1990.

[CK08a] María Victoria Cengarle and Alexander Knapp. An institution for OCL 2.0. Technical Report I0801, Institut für Informatik, Ludwig-Maximilians-Universität München, 2008.

[CK08b] María Victoria Cengarle and Alexander Knapp. An institution for UML 2.0 interactions. Technical Report I0808, Institut für Informatik, Ludwig-Maximilians-Universität München, 2008.

[CK08c] María Victoria Cengarle and Alexander Knapp. An institution for UML 2.0 static structures. Technical Report I0807, Institut für Informatik, Ludwig-Maximilians-Universität München, 2008.

[CKTW08] María Victoria Cengarle, Alexander Knapp, Andrzej Tarlecki, and Martin Wirsing. A heterogeneous approach to UML semantics. In: Pierpaolo Degano, Rocco de Nicola, and José Meseguer, editors, *Concurrency, Graphs and Models, Essays Dedicated to Ugo Montanari on the Occasion of His 65th Birthday*, *Lecture Notes in Computer Science*, Vol. 5065, pages 383–402. Springer, 2008.

[CM97] Maura Cerioli and José Meseguer. May I borrow your logic? (Transporting logical structures along maps). *Theoretical Computer Science*, 173(2):311–347, 1997.

[CMRM10] Mihai Codescu, Till Mossakowski, Adrían Riesco, and Christian Maeder. Integrating Maude into HETS. In: Michael Johnson and Dusko Pavlovic, editors, *Proceedings of the 13th International Conference on Algebraic Methodology and Software Technology*, Lac-Beauport, *Lecture Notes in Computer Science*, Vol. 6486. Springer, 2010.

[CMRS01] Carlos Caleiro, Paulo Mateus, Jaime Ramos, and Amílcar Sernadas. Combining logics: Parchments revisited. In: Maura Cerioli and Gianna Reggio, editors, *Recent Trends in Algebraic Development Techniques. Selected Papers from the 15th Workshop on Algebraic Development Techniques Joint with the CoFI WG Meeting*, Genova, *Lecture Notes in Computer Science*, Vol. 2267, pages 48–70. Springer, 2001.

[Coh65] Paul M. Cohn. *Universal Algebra*. Harper and Row, 1965.

[CS92] Robin Cockett and Dwight Spencer. Strong categorical datatypes I. In: Robert A.G. Seely, editor, *International Meeting on Category Theory 1991*, *Canadian Mathematical Society Proceedings*. American Mathematical Society, 1992.

[CSS05] Carlos Caleiro, Amílcar Sernadas, and Cristina Sernadas. Fibring logics: Past, present and future. In: Sergei N. Artemov, Howard Barringer, Artur S. d'Avila Garcez, Luís C. Lamb, and John Woods, editors, *We Will Show Them! Essays in Honour of Dov Gabbay, Vol. 1*, pages 363–388. College Publications, 2005.

[DF98] Răzvan Diaconescu and Kokichi Futatsugi. *CafeOBJ Report: The Language, Proof Techniques, and Methodologies for Object-Oriented Algebraic Specification*, *AMAST Series in Computing*, Vol. 6. World Scientific, 1998. See also http://www.ldl.jaist.ac.jp/cafeobj/.

[DF02] Răzvan Diaconescu and Kokichi Futatsugi. Logical foundations of CafeOBJ. *Theoretical Computer Science*, 285:289–318, 2002.

[DGS93] Răzvan Diaconescu, Joseph A. Goguen, and Petros Stefaneas. Logical support for modularisation. In: Gérard Huet and Gordon Plotkin, editors, *Logical Environments*, pages 83–130. Cambridge University Press, 1993.

[Dia00] Răzvan Diaconescu. Category-based constraint logic. *Mathematical Structures in Computer Science*, 10(3):373–407, 2000.

[Dia02] Răzvan Diaconescu. Grothendieck institutions. *Applied Categorical Structures*, 10(4):383–402, 2002.

[Dia08] Răzvan Diaconescu. *Institution-Independent Model Theory*. Birkhäuser, 2008.

[DJ90] Nachum Dershowitz and Jean-Pierre Jouannaud. Rewrite systems. In: Jan van Leeuwen, editor, *Handbook of Theoretical Computer Science. Vol. B (Formal Models and Semantics)*, pages 244–320. North-Holland and MIT Press, 1990.

[DLL62] Martin Davis, George Logemann, and Donald Loveland. A machine program for theorem-proving. *Communications of the ACM*, 5(7):394–397, 1962.

[DM00] Theodosis Dimitrakos and Thomas S.E. Maibaum. On a generalised modularization theorem. *Information Processing Letters*, 74(1–2):65–71, 2000.

[DMR76] Martin Davis, Yuri Matiyasevich, and Julia Robinson. Hilbert's tenth problem. Diophantine equations: Positive aspects of a negative solution. In: *Mathematical Developments Arising from Hilbert Problems, Proceedings of Symposia in Pure Mathematics*, Vol. 28, pages 323–378, American Mathematical Society, 1976.

[DP90] Brian A. Davey and Hilary A. Priestley. *Introduction to Lattices and Order*. Cambridge University Press, 1990.

[Ehr78] Hans-Dieter Ehrich. Extensions and implementations of abstract data type specifications. In: Józef Winkowski, editor, *Proceedings of the 7th Symposium on Mathematical Foundations of Computer Science*, Zakopane, *Lecture Notes in Computer Science*, Vol. 64, pages 155–164. Springer, 1978.

[Ehr81] Hans-Dieter Ehrich. On realization and implementation. In: Jozef Gruska and Michal Chytil, editors, *Proceedings of the 10th Symposium on Mathematical Foundations of Computer Science*, Štrbské Pleso, *Lecture Notes in Computer Science*, Vol. 118, pages 271–280. Springer, 1981.

[Ehr82] Hans-Dieter Ehrich. On the theory of specification, implementation and parametrization of abstract data types. *Journal of the Association for Computing Machinery*, 29(1):206–227, 1982.

[EKMP82] Hartmut Ehrig, Hans-Jörg Kreowski, Bernd Mahr, and Peter Padawitz. Algebraic implementation of abstract data types. *Theoretical Computer Science*, 20:209–263, 1982.

[EKT+83] Hartmut Ehrig, Hans-Jörg Kreowski, James W. Thatcher, Eric G. Wagner, and Jesse B. Wright. Parameter passing in algebraic specification languages. *Theoretical Computer Science*, 28(1–2):45–81, 1983.

[EM85] Hartmut Ehrig and Bernd Mahr. *Fundamentals of Algebraic Specification 1, EATCS Monographs on Theoretical Computer Science*, Vol. 6. Springer, 1985.

[Eme90] E. Allen Emerson. Temporal and modal logic. In: Jan van Leeuwen, editor, *Handbook of Theoretical Computer Science. Vol. B (Formal Models and Semantics)*, pages 995–1072. North-Holland and MIT Press, 1990.

[End72] Herbert B. Enderton. *A Mathematical Introduction to Logic*. Academic Press, 1972.

[EPO89] Hartmut Ehrig, Peter Pepper, and Fernando Orejas. On recent trends in algebraic specification. In: Giorgio Ausiello, Mariangiola Dezani-Ciancaglini, and Simona Ronchi Della Rocca, editors, *Proceedings of the 16th International Colloquium on Automata, Languages and Programming*, Stresa, *Lecture Notes in Computer Science*, Vol. 372, pages 263–288. Springer, 1989.

[EWT83] Hartmut Ehrig, Eric G. Wagner, and James W. Thatcher. Algebraic specifications with generating constraints. In: *Proceedings of the 10th International Colloquium on Automata, Languages and Programming*, Barcelona, *Lecture Notes in Computer Science*, Vol. 154, pages 188–202. Springer, 1983.

[Far89] Jordi Farrés-Casals. Proving correctness of constructor implementations. In: Antoni Kreczmar and Grazyna Mirkowska, editors, *Proceedings of the 14th Symposium on Mathematical Foundations of Computer Science*, Porabka-Kozubnik, *Lecture Notes in Computer Science*, Vol. 379, pages 225–235. Springer, 1989.

[Far90] Jordi Farrés-Casals. Proving correctness w.r.t. specifications with hidden parts. In: Hélène Kirchner and Wolfgang Wechler, editors, *Proceedings of the 2nd International Conference on Algebraic and Logic Programming*, Nancy, *Lecture Notes in Computer Science*, Vol. 463, pages 25–39. Springer, 1990.

[Far92] Jordi Farrés-Casals. *Verification in ASL and Related Specification Languages*. Ph.D. thesis, University of Edinburgh, Department of Computer Science, 1992.

[FC96] José Luiz Fiadeiro and José Félix Costa. Mirror, mirror in my hand: A duality between specifications and models of process behaviour. *Mathematical Structures in Computer Science*, 6(4):353–373, 1996.

[Fei89] Loe M.G. Feijs. The calculus $\lambda\pi$. In: Martin Wirsing and Jan A. Bergstra, editors, *Proceedings of the Workshop on Algebraic Methods: Theory, Tools and Applications*, Passau, *Lecture Notes in Computer Science*, Vol. 394, pages 307–328. Springer, 1989.

[FGT92] William M. Farmer, Joshua D. Guttman, and F. Javier Thayer. Little theories. In: Deepak Kapur, editor, *Proceedings of the 11th International Conference on Automated Deduction*, Saratoga Springs, *Lecture Notes in Artificial Intelligence*, Vol. 607, pages 567–581. Springer, 1992.

[Fia05] José Luiz Fiadeiro. *Categories for Software Engineering*. Springer, 2005.

[Fit08] John S. Fitzgerald. The typed logic of partial functions and the Vienna Development Method. In: Dines Bjørner and Martin Henson, editors, *Logics of Specification Languages*, pages 453–487. Springer, 2008.

[FJ90] John S. Fitzgerald and Cliff B. Jones. Modularizing the formal description of a database system. In: *Proceedings of the 3rd International Symposium of VDM Europe: VDM and Z, Formal Methods in Software Development*, Kiel, *Lecture Notes in Computer Science*, Vol. 428, pages 189–210. Springer, 1990.

[FS88] José Luiz Fiadeiro and Amílcar Sernadas. Structuring theories on consequence. In: Donald Sannella and Andrzej Tarlecki, editors, *Recent Trends in Data Type Specification. Selected Papers from the 5th Workshop on Specification of Abstract Data Types*, Gullane, *Lecture Notes in Computer Science*, Vol. 332, pages 44–72. Springer, 1988.

[Gab98] Dov M. Gabbay. *Fibring Logics, Oxford Logic Guides*, Vol. 38. Oxford University Press, 1998.

[Gan83] Harald Ganzinger. Parameterized specifications: Parameter passing and implementation with respect to observability. *ACM Transactions on Programming Languages and Systems*, 5(3):318–354, 1983.

[GB78] Joseph A. Goguen and Rodney M. Burstall. Some fundamental properties of algebraic theories: A tool for semantics of computation. Technical Report 53, Department of Artificial Intelligence, University of Edinburgh, 1978. Revised version appeared as [GB84b] and [GB84c].

[GB80] Joseph A. Goguen and Rodney M. Burstall. CAT, a system for the structured elaboration of correct programs from structured specifications. Technical Report CSL-118, Computer Science Laboratory, SRI International, 1980.

[GB84a] Joseph A. Goguen and Rodney M. Burstall. Introducing institutions. In: Edmund Clarke and Dexter Kozen, editors, *Proceedings of the Workshop on Logics of Programs*, Pittsburgh, *Lecture Notes in Computer Science*, Vol. 164, pages 221–256. Springer, 1984. Many revised versions were widely circulated, with [GB92] as the endpoint.

[GB84b] Joseph A. Goguen and Rodney M. Burstall. Some fundamental algebraic tools for the semantics of computation. Part 1: Comma categories, colimits, signatures and theories. *Theoretical Computer Science*, 31:175–209, 1984.

[GB84c] Joseph A. Goguen and Rodney M. Burstall. Some fundamental algebraic tools for the semantics of computation. Part 2: Signed and abstract theories. *Theoretical Computer Science*, 31:263–295, 1984.

[GB86] Joseph A. Goguen and Rodney M. Burstall. A study in the functions of programming methodology: Specifications, institutions, charters and parchments. In: David H. Pitt, Samson Abramsky, Axel Poigné, and David E. Rydeheard, editors, *Proceedings of the Tutorial and Workshop on Category Theory and Computer Programming*, Guildford, *Lecture Notes in Computer Science*, Vol. 240, pages 313–333. Springer, 1986.

[GB92] Joseph A. Goguen and Rodney M. Burstall. Institutions: Abstract model theory for specification and programming. *Journal of the Association for Computing Machinery*, 39(1):95–146, 1992.

[GD94a] Joseph A. Goguen and Răzvan Diaconescu. An Oxford survey of order sorted algebra. *Mathematical Structures in Computer Science*, 4(3):363–392, 1994.

[GD94b] Joseph A. Goguen and Răzvan Diaconescu. Towards an algebraic semantics for the object paradigm. In: Hartmut Ehrig and Fernando Orejas, editors, *Recent Trends in Data Type Specification. Selected Papers from the 9th Workshop on Specification of Abstract Data Types Joint with the 4th COMPASS Workshop*, Caldes de Malavella, *Lecture Notes in Computer Science*, Vol. 785, pages 1–29. Springer, 1994.

[GDLE84] Martin Gogolla, Klaus Drosten, Udo Lipeck, and Hans-Dieter Ehrich. Algebraic and operational semantics of specifications allowing exceptions and errors. *Theoretical Computer Science*, 34(3):289–313, 1984.

[GG89] Stephen J. Garland and John V. Guttag. An overview of LP, the Larch Prover. In: *Third International Conference on Rewriting Techniques and Applications*, Chapel Hill, *Lecture Notes in Computer Science*, Vol. 355, pages 137–151. Springer, 1989. See also http://nms.lcs.mit.edu/larch/LP/all.html.

[GGM76] Vincenzo Giarratana, Fabrizio Gimona, and Ugo Montanari. Observability concepts in abstract data type specifications. In: Antoni Mazurkiewicz, editor, *Proceedings of the 5th Symposium on Mathematical Foundations of Computer Science*, Gdańsk, *Lecture Notes in Computer Science*, Vol. 45, pages 567–578. Springer, 1976.

[GH78] John V. Guttag and James J. Horning. The algebraic specification of abstract data types. *Acta Informatica*, 10:27–52, 1978.

[GH93] John V. Guttag and James J. Horning. *Larch: Languages and Tools for Formal Specification*. Springer, 1993.

[Gin68] Abraham Ginzburg. *Algebraic Theory of Automata*. Academic Press, 1968.

[Gir87] Jean-Yves Girard. Linear logic. *Theoretical Computer Science*, 50:1–102, 1987.

[Gir89] Jean-Yves Girard. *Proofs and Types, Cambridge Tracts in Theoretical Computer Science*, Vol. 7. Cambridge University Press, 1989. Translated and with appendices by Paul Taylor and Yves Lafont.

[GJL$^+$03] Ronald Garcia, Jaakko Järvi, Andrew Lumsdaine, Jeremy Siek, and Jeremiah Willcock. A comparative study of language support for generic programming. In: *Proceedings of the 18th Annual ACM SIGPLAN Conference on Object-Oriented Programming, Systems, Languages, and Applications*, Anaheim, pages 115–134, 2003.

[GLR00] Joseph A. Goguen, Kai Lin, and Grigore Roşu. Circular coinductive rewriting. In: *Proceedings of the 15th International Conference on Automated Software Engineering*, Grenoble. IEEE Computer Society, 2000.

[GM82] Joseph A. Goguen and José Meseguer. Universal realization, persistent interconnection and implementation of abstract modules. In: Mogens Nielsen and Erik Meineche Schmidt, editors, *Proceedings of the 9th International Colloquium on Automata, Languages and Programming*, Aarhus, *Lecture Notes in Computer Science*, Vol. 140, pages 265–281. Springer, 1982.

[GM85] Joseph A. Goguen and José Meseguer. Completeness of many sorted equational deduction. *Houston Journal of Mathematics*, 11(3):307–334, 1985.

[GM92] Joseph A. Goguen and José Meseguer. Order-sorted algebra I: Equational deduction for multiple inheritance, overloading, exceptions and partial operations. *Theoretical Computer Science*, 105(2):217–273, 1992.

[GM00] Joseph A. Goguen and Grant Malcolm. A hidden agenda. *Theoretical Computer Science*, 245(1):55–101, 2000.

[Gog73] Joseph A. Goguen. Categorical foundations for general systems theory. In: Franz Pichler and Robert Trappl, editors, *Advances in Cybernetics and Systems Research*, London, pages 121–130. Transcripta Books, 1973.

[Gog74] Joseph A. Goguen. Semantics of computation. In: Ernest G. Manes, editor, *Proceedings of the 1st International Symposium on Category Theory Applied to Computation and Control*, San Francisco, *Lecture Notes in Computer Science*, Vol. 25, pages 151–163. Springer, 1974.

[Gog78] Joseph A. Goguen. Abstract errors for abstract data types. In: Erich Neuhold, editor, *Formal Description of Programming Concepts*, pages 491–526. North-Holland, 1978.

[Gog84] Martin Gogolla. Partially ordered sorts in algebraic specifications. In: *Proceedings of the 9th Colloquium on Trees in Algebra and Programming*, pages 139–153. Cambridge University Press, 1984.

[Gog85] Martin Gogolla. A final algebra semantics for errors and exceptions. In: Hans-Jörg Kreowski, editor, *Recent Trends in Data Type Specification. Selected Papers from the 3rd Workshop on Theory and Applications of Abstract Data Types*, Bremen, *Informatik-Fachberichte*, Vol. 116, pages 89–103. Springer, 1985.

[Gog91a] Joseph A. Goguen. A categorical manifesto. *Mathematical Structures in Computer Science*, 1(1):49–67, 1991.

[Gog91b] Joseph A. Goguen. Types as theories. In: George M. Reed, A. William Roscoe, and Ralph F. Wachter, editors, *Topology and Category Theory in Computer Science*, Oxford, pages 357–390. Oxford University Press, 1991.

[Gog96] Joseph A. Goguen. Parameterized programming and software architecture. In: Murali Sitaraman, editor, *Proceedings of the Fourth International Conference on Software Reuse*, pages 2–11. IEEE Computer Society Press, 1996.

[Gog11] Joseph A. Goguen. Information integration in institutions. Available from http://cseweb.ucsd.edu/users/goguen/projs/data.html; to appear in: Larry Moss, editor, *Thinking Logically: A Volume in Memory of Jon Barwise*, 2011.

[Gol06] Robert Goldblatt. *Topoi: The Categorial Analysis of Logic*. Dover, revised edition, 2006.

[Gor95] Andrew D. Gordon. Bisimilarity as a theory of functional programming. In: *Proceedings of the 11th Annual Conference on Mathematical Foundations of Programming Semantics. Electronic Notes in Theoretical Computer Science*, 1:232–252, 1995.

[GR02] Joseph A. Goguen and Grigore Roşu. Institution morphisms. *Formal Aspects of Computing*, 13(3–5):274–307, 2002.

[GR04] Joseph A. Goguen and Grigore Roşu. Composing hidden information modules over inclusive institutions. In: *From Object-Orientation to Formal Methods. Essays in Memory of Ole-Johan Dahl, Lecture Notes in Computer Science*, Vol. 2635, pages 96–123. Springer, 2004.

[Grä79] George A. Grätzer. *Universal Algebra*. Springer, second edition, 1979.

[GS90] Carl Gunter and Dana Scott. Semantic domains. In: Jan van Leeuwen, editor, *Handbook of Theoretical Computer Science. Vol. B (Formal Models and Semantics)*, pages 633–674. North-Holland and MIT Press, 1990.

[GTW76] Joseph A. Goguen, James W. Thatcher, and Eric G. Wagner. An initial algebra approach to the specification, correctness and implementation of abstract data types. Technical Report RC 6487, IBM Watson Research Center, Yorktown Heights NY, 1976. Also in: Raymond T. Yeh, editor, *Current Trends in Programming Methodology. Vol. IV (Data Structuring)*, pages 80–149. Prentice-Hall, 1978.

[GTWW73] Joseph A. Goguen, James W. Thatcher, Eric G. Wagner, and Jesse B. Wright. A junction between computer science and category theory, I: Basic concepts and examples (part 1). Technical Report RC 4526, IBM Watson Research Center, Yorktown Heights NY, 1973.

[GTWW75] Joseph A. Goguen, James W. Thatcher, Eric G. Wagner, and Jesse B. Wright. An introduction to categories, algebraic theories and algebras. Technical Report RC 5369, IBM Watson Research Center, Yorktown Heights NY, 1975.

[GTWW77] Joseph A. Goguen, James W. Thatcher, Eric G. Wagner, and Jesse B. Wright. Initial algebra semantics and continuous algebras. *Journal of the Association for Computing Machinery*, 24(1):68–95, 1977.

[Gut75] John V. Guttag. *The Specification and Application to Programming of Abstract Data Types*. Ph.D. thesis, University of Toronto, Department of Computer Science, 1975.

[Hag87] Tatsuya Hagino. *A Categorical Programming Language*. Ph.D. thesis, University of Edinburgh, Department of Computer Science, 1987.

[Häh01] Reiner Hähnle. Tableaux and related methods. In: John Alan Robinson and Andrei Voronkov, editors, *Handbook of Automated Reasoning*, pages 100–178. Elsevier and MIT Press, 2001.

[Hal70] Paul R. Halmos. *Naive Set Theory. Undergraduate Texts in Mathematics*. Springer, 1970.

[Hat82] William Hatcher. *The Logical Foundations of Mathematics. Foundations and Philosophy of Science and Technology*. Pergamon Press, 1982.

[Hay94] Susumu Hayashi. Singleton, union and intersection types for program extraction. *Information and Computation*, 109(1–2):174–210, 1994.

[Hee86] Jan Heering. Partial evaluation and ω-completeness of algebraic specifications. *Theoretical Computer Science*, 43:149–167, 1986.

[Hen91] Rolf Hennicker. Context induction: A proof principle for behavioural abstractions and algebraic implementations. *Formal Aspects of Computing*, 3(4):326–345, 1991.

[HHP93] Robert Harper, Furio Honsell, and Gordon Plotkin. A framework for defining logics. *Journal of the Association for Computing Machinery*, 40(1):143–184, January 1993.

[HHWT97] Thomas A. Henzinger, Pei-Hsin Ho, and Howard Wong-Toi. HYTECH: A model checker for hybrid systems. *Software Tools for Technology Transfer*, 1(1–2):110–122, 1997.

[Hig63] Phillip J. Higgins. Algebras with a scheme of operators. *Mathematische Nachrichten*, 27:115–132, 1963.

[HLST00] Furio Honsell, John Longley, Donald Sannella, and Andrzej Tarlecki. Constructive data refinement in typed lambda calculus. In: *Proceedings of the 3rd International Conference on Foundations of Software Science and Computation Structures. European Joint Conferences on Theory and Practice of Software (ETAPS 2000)*, Berlin, *Lecture Notes in Computer Science*, Vol. 1784, pages 161–176. Springer, 2000.

[Hoa72] C.A.R. Hoare. Proof of correctness of data representations. *Acta Informatica*, 1:271–281, 1972.

[HS73] Horst Herrlich and George E. Strecker. *Category Theory: An Introduction*. Allyn and Bacon, 1973.

[HS96] Martin Hofmann and Donald Sannella. On behavioural abstraction and behavioural satisfaction in higher-order logic. *Theoretical Computer Science*, 167:3–45, 1996.

[HS02] Furio Honsell and Donald Sannella. Prelogical relations. *Information and Computation*, 178:23–43, 2002.

[HST94] Robert Harper, Donald Sannella, and Andrzej Tarlecki. Structured presentations and logic representations. *Annals of Pure and Applied Logic*, 67:113–160, 1994.

[Hus92] Heinrich Hussmann. Nondeterministic algebraic specifications and nonconfluent term rewriting. *Journal of Logic Programming*, 12(1–4):237–255, 1992.

[HWB97] Rolf Hennicker, Martin Wirsing, and Michel Bidoit. Proof systems for structured specifications with observability operators. *Theoretical Computer Science*, 173(2):393–443, 1997.

[Jac99] Bart Jacobs. *Categorical Logic and Type Theory, Studies in Logic and the Foundations of Mathematics*, Vol. 141. Elsevier Science, 1999.

[JL87] Joxan Jaffar and Jean-Louis Lassez. Constraint logic programming. In: *Proceedings of the 14th ACM Symposium on Principles of Programming Languages*, Munich, pages 111–119, 1987.

[JNW96] André Joyal, Mogens Nielsen, and Glynn Winskel. Bisimulation from open maps. *Information and Computation*, 127(2):164–185, 1996.

[JOE95] Rosa M. Jiménez, Fernando Orejas, and Hartmut Ehrig. Compositionality and compatibility of parameterization and parameter passing in specification languages. *Mathematical Structures in Computer Science*, 5(2):283–314, 1995.

[Joh02] Peter T. Johnstone. *Sketches of an Elephant: A Topos Theory Compendium, Oxford Logic Guides*, Vol. 43 and 44. Clarendon Press, 2002.

[Jon80] Cliff B. Jones. *Software Development: A Rigorous Approach*. Prentice-Hall, 1980.

[Jon89] Hans B.M. Jonkers. An introduction to COLD-K. In: Martin Wirsing and Jan A. Bergstra, editors, *Proceedings of the Workshop on Algebraic Methods: Theory, Tools and Applications*, Passau, *Lecture Notes in Computer Science*, Vol. 394, pages 139–205. Springer, 1989.

[JR97] Bart Jacobs and Jan J.M.M. Rutten. A tutorial on (co)algebras and (co)induction. *Bulletin of the European Association for Theoretical Computer Science*, 62:222–259, 1997.

[KB70] Donald E. Knuth and Peter B. Bendix. Simple word problems in universal algebras. In: John Leech, editor, *Computational Problems in Abstract Algebra*, pages 263–297. Pergamon Press, 1970.

[Kir99] Hélène Kirchner. Term rewriting. In: Egidio Astesiano, Hans-Jörg Kreowski, and Bernd Krieg-Brückner, editors, *Algebraic Foundations of Systems Specification*, chapter 9, pages 273–320. Springer, 1999.

[KKM88] Claude Kirchner, Hélène Kirchner, and José Meseguer. Operational semantics of OBJ-3. In: Timo Lepistö and Arto Salomaa, editors, *Proceedings of the 15th International Colloquium on Automata, Languages and Programming*, Tampere, *Lecture Notes in Computer Science*, Vol. 317, pages 287–301. Springer, 1988.

[Klo92] Jan Willem Klop. Term rewriting systems. In: Samson Abramsky, Dov Gabbay, and Thomas S.E. Maibaum, editors, *Handbook of Logic in Computer Science. Vol. 2 (Background: Computational Structures)*, pages 1–116. Oxford University Press, 1992.

[KM87] Deepak Kapur and David R. Musser. Proof by consistency. *Artificial Intelligence*, 31(2):125–157, 1987.

[KR71] Heinz Kaphengst and Horst Reichel. Algebraische Algorithmentheorie. Technical Report WIB 1, VEB Robotron, Zentrum für Forschung und Technik, Dresden, 1971.

[Kre87] Hans-Jörg Kreowski. Partial algebras flow from algebraic specifications. In: Thomas Ottmann, editor, *Proceedings of the 14th International Colloquium on Automata, Languages and Programming*, Karlsruhe, *Lecture Notes in Computer Science*, Vol. 267, pages 521–530. Springer, 1987.

[KST97] Stefan Kahrs, Donald Sannella, and Andrzej Tarlecki. The definition of Extended ML: A gentle introduction. *Theoretical Computer Science*, 173:445–484, 1997.

[KTB91] Beata Konikowska, Andrzej Tarlecki, and Andrzej Blikle. A three-valued logic for software specification and validation. *Fundamenta Informaticae*, 14(4):411–453, 1991.

[Las98] Sławomir Lasota. Open maps as a bridge between algebraic observational equivalence and bisimilarity. In: Francesco Parisi-Presicce, editor, *Recent Trends in Data Type Specification. Selected Papers from the 12th International Workshop on Specification of Abstract Data Types*, Tarquinia, *Lecture Notes in Computer Science*, Vol. 1376, pages 285–299. Springer, 1998.

[Law63] F. William Lawvere. *Functorial Semantics of Algebraic Theories*. Ph.D. thesis, Columbia University, 1963.

[LB88] Butler Lampson and Rodney M. Burstall. Pebble, a kernel language for modules and abstract data types. *Information and Computation*, 76(2–3):278–346, 1988.

[LDF+10] Xavier Leroy, Damien Doligez, Alain Frisch, Jacques Garrigue, Didier Rémy, and Jérôme Vouillon. The Objective Caml system, release 3.12. Documentation and user's manual, 2010. Available from http://caml.inria.fr/pub/docs/manual-ocaml/.

[LEW96] Jacques Loeckx, Hans-Dieter Ehrich, and Markus Wolf. *Specification of Abstract Data Types*. John Wiley and Sons, 1996.

[Lin03] Kai Lin. *Machine Support for Behavioral Algebraic Specification and Verification*. Ph.D. thesis, University of California, San Diego, 2003.

[Lip83] Udo Lipeck. *Ein algebraischer Kalkül für einen strukturierten Entwurf von Datenabstraktionen*. Ph.D. thesis, Universität Dortmund, 1983.

[LLD06] Dorel Lucanu, Yuan-Fang Li, and Jin Song Dong. Semantic Web languages — towards an institutional perspective. In: Kokichi Futatsugi, Jean-Pierre Jouannaud, and José Meseguer, editors, *Algebra, Meaning and Computation: Essays Dedicated to Joseph A. Goguen on the Occasion of His 65th Birthday, Lecture Notes in Computer Science*, Vol. 4060, pages 99–123. Springer, 2006.

[LS86] Joachim Lambek and Philip J. Scott. *Introduction to Higher-Order Categorical Logic, Cambridge Studies in Advanced Mathematics*, Vol. 7. Cambridge University Press, 1986.

[LS00] Hugo Lourenço and Amílcar Sernadas. An institution of hybrid systems. In: Didier Bert, Christine Choppy, and Peter D. Mosses, editors, *Recent Trends in Algebraic Development Techniques. Selected Papers from the 14th International Workshop on Algebraic Development Techniques*, Château de Bonas, *Lecture Notes in Computer Science*, Vol. 1827, pages 219–236. Springer, 2000.

[Luo93] Zhaohui Luo. Program specification and data refinement in type theory. *Mathematical Structures in Computer Science*, 3(3):333–363, 1993.

[Mac71] Saunders Mac Lane. *Categories for the Working Mathematician*. Springer, 1971.

[Mac84] David MacQueen. Modules for Standard ML. In: *Proceedings of the 1984 ACM Conference on LISP and Functional Programming*, pages 198–207, 1984.

[MAH06] Till Mossakowski, Serge Autexier, and Dieter Hutter. Development graphs — proof management for structured specifications. *Journal of Logic and Algebraic Programming*, 67(1–2):114–145, 2006.

[Mai72] Thomas S.E. Maibaum. The characterization of the derivation trees of context free sets of terms as regular sets. In: *Proceedings of the 13th Annual IEEE Symposium on Switching and Automata Theory*, pages 224–230, 1972.

[Maj77] Mila E. Majster. Limits of the "algebraic" specification of abstract data types. *ACM SIGPLAN Notices*, 12(10):37–42, 1977.

[Mal71] Anatoly Malcev. Quasiprimitive classes of abstract algebras in the metamathematics of algebraic systems. In: *Mathematics of Algebraic Systems: Collected Papers, 1936–67, Studies in Logic and Mathematics*, Vol. 66, pages 27–31. North-Holland, 1971.

[Man76] Ernest G. Manes. *Algebraic Theories*. Springer, 1976.

[May85] Brian Mayoh. Galleries and institutions. Technical Report DAIMI PB-191, Aarhus University, 1985.

[Mei92] Karl Meinke. Universal algebra in higher types. *Theoretical Computer Science*, 100:385–417, 1992.

[Mes89] José Meseguer. General logics. In: Heinz-Dieter Ebbinghaus, editor, *Logic Colloquium '87*, Granada, pages 275–329. North-Holland, 1989.

[Mes92] José Meseguer. Conditional rewriting logic as a unified model of concurrency. *Theoretical Computer Science*, 96(1):73–155, 1992.

[Mes98] José Meseguer. Membership algebra as a logical framework for equational specification. In: Francesco Parisi-Presicce, editor, *Recent Trends in Data Type Specification. Selected Papers from the 12th International Workshop on Specification of Abstract Data Types*, Tarquinia, *Lecture Notes in Computer Science*, Vol. 1376, pages 18–61. Springer, 1998.

[Mes09] José Meseguer. Order-sorted parameterization and induction. In: Jens Palsberg, editor, *Semantics and Algebraic Specification: Essays Dedicated to Peter D. Mosses on the Occasion of His 60th Birthday, Lecture Notes in Computer Science*, Vol. 5700, pages 43–80. Springer, 2009.

[MG85] José Meseguer and Joseph A. Goguen. Initiality, induction and computability. In: Maurice Nivat and John C. Reynolds, editors, *Algebraic Methods in Semantics*, pages 459–541. Cambridge, 1985.

[MGDT07] Till Mossakowski, Joseph A. Goguen, Răzvan Diaconescu, and Andrzej Tarlecki. What is a logic? In: Jean-Yves Beziau, editor, *Logica Universalis: Towards a General Theory of Logic*, pages 111–135. Birkhäuser, 2007.

[MHST08] Till Mossakowski, Anne Haxthausen, Donald Sannella, and Andrzej Tarlecki. CASL — the common algebraic specification language. In: Dines Bjørner and Martin Henson, editors, *Logics of Specification Languages*, pages 241–298. Springer, 2008.

[Mid93] Aart Middeldorp. Modular properties of conditional term rewriting systems. *Information and Computation*, 104(1):110–158, 1993.

[Mil71] Robin Milner. An algebraic definition of simulation between programs. In: *Proceedings of the 2nd International Joint Conference on Artificial Intelligence*, pages 481–489, 1971.

[Mil77] Robin Milner. Fully abstract models of typed λ-calculi. *Theoretical Computer Science*, 4(1):1–22, 1977.

[Mil89] Robin Milner. *Communication and Concurrency*. Prentice-Hall, 1989.

[Mit96] John C. Mitchell. *Foundations of Programming Languages*. MIT Press, 1996.

[MM84] Bernd Mahr and Johann Makowsky. Characterizing specification languages which admit initial semantics. *Theoretical Computer Science*, 31:49–60, 1984.

[MML07] Till Mossakowski, Christian Maeder, and Klaus Lüttich. The heterogeneous tool set, HETS. In: Orna Grumberg and Michael Huth, editors, *Proceedings of the 13th International Conference on Tools and Algorithms for the Construction and Analysis of Systems. European Joint Conferences on Theory and Practice of Software (ETAPS 2007)*, Braga, *Lecture Notes in Computer Science*, Vol. 4424, pages 519–522. Springer, 2007. See also http://www.informatik.uni-bremen.de/cofi/hets/.

[Mog91] Eugenio Moggi. Notions of computation and monads. *Information and Computation*, 93:55–92, 1991.

[Moo56] Edward F. Moore. Gedanken-experiments on sequential machines. In: Claude E. Shannon and John McCarthy, editors, *Annals of Mathematics Studies 34, Automata Studies*, pages 129–153. Princeton University Press, 1956.

[Mos89] Peter D. Mosses. Unified algebras and modules. In: *Proceedings of the 16th ACM Symposium on Principles of Programming Languages*, Austin, pages 329–343, 1989.

[Mos93] Peter D. Mosses. The use of sorts in algebraic specifications. In: Michel Bidoit and Christine Choppy, editors, *Recent Trends in Data Type Specification. Selected Papers from the 8th Workshop on Specification of Abstract Data Types Joint with the 3rd COMPASS Workshop*, Dourdan, *Lecture Notes in Computer Science*, Vol. 655, pages 66–91. Springer, 1993.

[Mos96a] Till Mossakowski. Different types of arrow between logical frameworks. In: Friedhelm Meyer auf der Heide and Burkhard Monien, editors, *Proceedings of the 23rd International Colloquium Automata, Languages and Programming*, Paderborn, *Lecture Notes in Computer Science*, Vol. 1099, pages 158–169. Springer, 1996.

[Mos96b] Till Mossakowski. *Representations, Hierarchies and Graphs of Institutions*. Ph.D. thesis, Universität Bremen, 1996.

[Mos00] Till Mossakowski. Specification in an arbitrary institution with symbols. In: Didier Bert, Christine Choppy, and Peter D. Mosses, editors, *Recent Trends in Algebraic Development Techniques. Selected Papers from the 14th International Workshop on Algebraic Development Techniques*, Château de Bonas, *Lecture Notes in Computer Science*, Vol. 1827, pages 252–270. Springer, 2000.

[Mos02] Till Mossakowski. Comorphism-based Grothendieck logics. In: Krzysztof Diks and Wojciech Rytter, editors, *Proceedings of the 27th Symposium on Mathematical Foundations of Computer Science*, Warsaw, *Lecture Notes in Computer Science*, Vol. 2420, pages 593–604. Springer, 2002.

[Mos03] Till Mossakowski. Foundations of heterogeneous specification. In: Martin Wirsing, Dirk Pattinson, and Rolf Hennicker, editors, *Recent Trends in Algebraic Development Techniques. Selected Papers from the 16th International Workshop on Algebraic Development Techniques*, Frauenchiemsee, *Lecture Notes in Computer Science*, Vol. 2755, pages 359–375. Springer, 2003.

[Mos04] Peter D. Mosses, editor. *CASL Reference Manual*, *Lecture Notes in Computer Science*, Vol. 2960. Springer, 2004.

[Mos05] Till Mossakowski. *Heterogeneous Specification and the Heterogeneous Tool Set.* Habilitation thesis, Universität Bremen, 2005.

[MS85] David MacQueen and Donald Sannella. Completeness of proof systems for equational specifications. *IEEE Transactions on Software Engineering*, SE-11(5):454–461, 1985.

[MSRR06] Till Mossakowski, Lutz Schröder, Markus Roggenbach, and Horst Reichel. Algebraic-coalgebraic specification in CoCASL. *Journal of Logic and Algebraic Programming*, 67(1–2):146–197, 2006.

[MSS90] Vincenzo Manca, Antonino Salibra, and Giuseppe Scollo. Equational type logic. *Theoretical Computer Science*, 77(1–2):131–159, 1990.

[MST04] Till Mossakowski, Donald Sannella, and Andrzej Tarlecki. A simple refinement language for CASL. In: José Fiadeiro, editor, *Recent Trends in Algebraic Development Techniques. Selected Papers from the 17th International Workshop on Algebraic Development Techniques*, Barcelona, *Lecture Notes in Computer Science*, Vol. 3423, pages 162–185. Springer, 2004.

[MT92] Karl Meinke and John V. Tucker. Universal algebra. In: Samson Abramsky, Dov Gabbay, and Thomas S.E. Maibaum, editors, *Handbook of Logic in Computer Science. Vol. 1 (Background: Mathematical Structures)*, pages 189–409. Oxford University Press, 1992.

[MT93] V. Wiktor Marek and Mirosław Truszczyński. *Nonmonotonic Logics: Context-Dependent Reasoning*. Springer, 1993.

[MT94] David MacQueen and Mads Tofte. A semantics for higher-order functors. In: Donald Sannella, editor, *Proceedings of the 5th European Symposium on Programming*, Edinburgh, *Lecture Notes in Computer Science*, Vol. 788, pages 409–423. Springer, 1994.

[MT09] Till Mossakowski and Andrzej Tarlecki. Heterogeneous logical environments for distributed specifications. In: Andrea Corradini and Ugo Montanari, editors, *Recent Trends in Algebraic Development Techniques. Selected Papers from the 19th International Workshop on Algebraic Development Techniques*, Pisa, *Lecture Notes in Computer Science*, Vol. 5486, pages 266–289. Springer, 2009.

[MTD09] Till Mossakowski, Andrzej Tarlecki, and Răzvan Diaconescu. What is a logic translation? *Logica Universalis*, 3(1):95–124, 2009.

[MTHM97] Robin Milner, Mads Tofte, Robert Harper, and David MacQueen. *The Definition of Standard ML (Revised)*. MIT Press, 1997.

[MTP97] Till Mossakowski, Andrzej Tarlecki, and Wiesław Pawłowski. Combining and representing logical systems. In: Eugenio Moggi and Giuseppe Rosolini, editors, *Proceedings of the 7th International Conference on Category Theory and Computer Science*, Santa Margherita Ligure, *Lecture Notes in Computer Science*, Vol. 1290, pages 177–196. Springer, 1997.

[MTP98] Till Mossakowski, Andrzej Tarlecki, and Wiesław Pawłowski. Combining and representing logical systems using model-theoretic parchments. In: Francesco Parisi-Presicce, editor, *Recent Trends in Data Type Specification. Selected Papers from the 12th International Workshop on Specification of Abstract Data Types*, Tarquinia, *Lecture Notes in Computer Science*, Vol. 1376, pages 349–364. Springer, 1998.

[MTW88] Bernhard Möller, Andrzej Tarlecki, and Martin Wirsing. Algebraic specifications of reachable higher-order algebras. In: Donald Sannella and Andrzej Tarlecki, editors, *Recent Trends in Data Type Specification. Selected Papers from the 5th Workshop on Specification of Abstract Data Types*, Gullane, *Lecture Notes in Computer Science*, Vol. 332, pages 154–169. Springer, 1988.

[Mus80] David R. Musser. On proving inductive properties of abstract data types. In: *Proceedings of the 7th ACM Symposium on Principles of Programming Languages*, Las Vegas, pages 154–162, 1980.

[MW98] Alfio Martini and Uwe Wolter. A single perspective on arrows between institutions. In: Armando Haeberer, editor, *Proceedings of the 7th International Conference on*

Algebraic Methodology and Software Technology, Manaus, *Lecture Notes in Computer Science*, Vol. 1548, pages 486–501. Springer, 1998.

[Nel91] Greg Nelson, editor. *Systems Programming with Modula-3*. Prentice-Hall, 1991.

[Nip86] Tobias Nipkow. Non-deterministic data types: Models and implementations. *Acta Informatica*, 22(6):629–661, 1986.

[NO88] Pilar Nivela and Fernando Orejas. Initial behaviour semantics for algebraic specifications. In: Donald Sannella and Andrzej Tarlecki, editors, *Recent Trends in Data Type Specification. Selected Papers from the 5th Workshop on Specification of Abstract Data Types*, Gullane, *Lecture Notes in Computer Science*, Vol. 332, pages 184–207. Springer, 1988.

[Nou81] Farshid Nourani. On induction for programming logic: Syntax, semantics, and inductive closure. *Bulletin of the European Association for Theoretical Computer Science*, 13:51–64, 1981.

[Oka98] Chris Okasaki. *Purely Functional Data Structures*. Cambridge University Press, 1998.

[ONS93] Fernando Orejas, Marisa Navarro, and Ana Sánchez. Implementation and behavioural equivalence: A survey. In: Michel Bidoit and Christine Choppy, editors, *Recent Trends in Data Type Specification. Selected Papers from the 8th Workshop on Specification of Abstract Data Types Joint with the 3rd COMPASS Workshop*, Dourdan, *Lecture Notes in Computer Science*, Vol. 655, pages 93–125. Springer, 1993.

[Ore83] Fernando Orejas. Characterizing composability of abstract implementations. In: Marek Karpinski, editor, *Proceedings of the 1983 International Conference on Foundations of Computation Theory*, Borgholm, *Lecture Notes in Computer Science*, Vol. 158, pages 335–346. Springer, 1983.

[Pad85] Peter Padawitz. Parameter preserving data type specifications. In: Hartmut Ehrig, Christiane Floyd, Maurice Nivat, and James W. Thatcher, editors, *TAPSOFT'85: Proceedings of the International Joint Conference on Theory and Practice of Software Development. Vol. 2: Colloquium on Software Engineering*, Berlin, *Lecture Notes in Computer Science*, Vol. 186, pages 323–341. Springer, 1985.

[Pad99] Peter Padawitz. Proof in flat specifications. In: Egidio Astesiano, Hans-Jörg Kreowski, and Bernd Krieg-Brückner, editors, *Algebraic Foundations of Systems Specification*, chapter 10, pages 321–384. Springer, 1999.

[Pau87] Lawrence C. Paulson. *Logic and Computation: Interactive Proof with Cambridge LCF*. Cambridge University Press, 1987.

[Pau96] Lawrence C. Paulson. *ML for the Working Programmer*. Cambridge University Press, second edition, 1996.

[Paw96] Wiesław Pawłowski. Context institutions. In: Magne Haveraaen, Olaf Owe, and Ole-Johan Dahl, editors, *Recent Trends in Data Type Specification. Selected Papers from the 11th Workshop on Specification of Abstract Data Types*, Oslo, *Lecture Notes in Computer Science*, Vol. 1130, pages 436–457. Springer, 1996.

[Pet10] Marius Petria. *Generic Refinements for Behavioral Specifications*. Ph.D. thesis, University of Edinburgh, School of Informatics, 2010.

[Pey03] Simon Peyton Jones, editor. *Haskell 98 Language and Libraries: The Revised Report*. Cambridge University Press, 2003.

[Pho92] Wesley Phoa. An introduction to fibrations, topos theory, the effective topos and modest sets. Technical Report ECS-LFCS-92-208, LFCS, Department of Computer Science, University of Edinburgh, 1992.

[Pie91] Benjamin C. Pierce. *Basic Category Theory for Computer Scientists*. MIT Press, 1991.

[Plo77] Gordon Plotkin. LCF considered as a programming language. *Theoretical Computer Science*, 5(3):223–255, 1977.

[Poi86] Axel Poigné. On specifications, theories, and models with higher types. *Information and Control*, 68(1–3):1–46, 1986.

[Poi88] Axel Poigné. Foundations are rich institutions, but institutions are poor foundations. In: Hartmut Ehrig, Horst Herrlich, Hans-Jörg Kreowski, and Gerhard Preuß, editors, *Proceedings of the International Workshop on Categorical Methods in Computer Science with Aspects from Topology*, Berlin, *Lecture Notes in Computer Science*, Vol. 393, pages 82–101. Springer, 1988.

[Poi90] Axel Poigné. Parametrization for order-sorted algebraic specification. *Journal of Computer and System Sciences*, 40:229–268, 1990.

[Poi92] Axel Poigné. Basic category theory. In: Samson Abramsky, Dov Gabbay, and Thomas S.E. Maibaum, editors, *Handbook of Logic in Computer Science. Vol. 1 (Background: Mathematical Structures)*, pages 413–640. Oxford University Press, 1992.

[Pos47] Emil Post. Recursive unsolvability of a problem of Thue. *Journal of Symbolic Logic*, 12:1–11, 1947.

[PS83] Helmuth Partsch and Ralf Steinbrüggen. Program transformation systems. *ACM Computing Surveys*, 15(3):199–236, 1983.

[PŞR09] Andrei Popescu, Traian Florin Şerbănuţă, and Grigore Roşu. A semantic approach to interpolation. *Theoretical Computer Science*, 410(12–13):1109–1128, 2009.

[QG93] Xiaolei Qian and Allen Goldberg. Referential opacity in nondeterministic data refinement. *ACM Letters on Programming Languages and Systems*, 2(1–4):233–241, 1993.

[Qia93] Zhenyu Qian. An algebraic semantics of higher-order types with subtypes. *Acta Informatica*, 30(6):569–607, 1993.

[RAC99] Gianna Reggio, Egidio Astesiano, and Christine Choppy. CASL-LTL: A CASL extension for dynamic systems — summary. Technical Report DISI-TR-99-34, DISI, Università di Genova, 1999.

[RB88] David E. Rydeheard and Rodney M. Burstall. *Computational Category Theory*. Prentice Hall, 1988.

[Rei80] Horst Reichel. Initially-restricting algebraic theories. In: Piotr Dembiński, editor, *Proceedings of the 9th Symposium on Mathematical Foundations of Computer Science*, Rydzyna, *Lecture Notes in Computer Science*, Vol. 88, pages 504–514. Springer, 1980.

[Rei81] Horst Reichel. Behavioural equivalence — a unifying concept for initial and final specification methods. In: *Proceedings of the 3rd Hungarian Computer Science Conference*, pages 27–39, 1981.

[Rei85] Horst Reichel. Behavioural validity of equations in abstract data types. In: *Proceedings of the Vienna Conference on Contributions to General Algebra*, pages 301–324. Teubner-Verlag, 1985.

[Rei87] Horst Reichel. *Initial Computability, Algebraic Specifications, and Partial Algebras*. Oxford University Press, 1987.

[RG98] Grigore Roşu and Joseph A. Goguen. Hidden congruent deduction. In: Ricardo Caferra and Gernot Salzer, editors, *Proceedings of the 1998 Workshop on First-Order Theorem Proving*, Vienna, *Lecture Notes in Artificial Intelligence*, Vol. 1761, pages 251–266. Springer, 1998.

[RG00] Grigore Roşu and Joseph A. Goguen. On equational Craig interpolation. *Journal of Universal Computer Science*, 6(1):194–200, 2000.

[Rod91] Pieter Hendrik Rodenburg. A simple algebraic proof of the equational interpolation theorem. *Algebra Universalis*, 28:48–51, 1991.

[Rog06] Markus Roggenbach. CSP-CASL — a new integration of process algebra and algebraic specification. *Theoretical Computer Science*, 354(1):42–71, 2006.

[Roş94] Grigore Roşu. The institution of order-sorted equational logic. *Bulletin of the European Association for Theoretical Computer Science*, 53:250–255, 1994.

[Roş00] Grigore Roşu. *Hidden Logic*. Ph.D. thesis, University of California at San Diego, 2000.

[RRS00] Sergei Romanenko, Claudio Russo, and Peter Sestoft. Moscow ML owner's manual. Technical report, Royal Veterinary and Agricultural University, Copenhagen, 2000. Available from http://www.itu.dk/people/sestoft/mosml/manual.pdf.

[RS63] Helena Rasiowa and Roman Sikorski. *The Mathematics of Metamathematics, Monografie Matematyczne*, Vol. 41. Polish Scientific Publishers, 1963.

[Rus98] Claudio Russo. *Types for Modules*. Ph.D. thesis, University of Edinburgh, Department of Computer Science, 1998. Also in: *Electronic Notes in Theoretical Computer Science*, 60, 2003.

[Rut00] Jan J.M.M. Rutten. Universal coalgebra: A theory of systems. *Theoretical Computer Science*, 249(1):3–80, 2000.

[San82] Donald Sannella. *Semantics, Implementation and Pragmatics of Clear, a Program Specification Language*. Ph.D. thesis, University of Edinburgh, Department of Computer Science, 1982.

[SB83] Donald Sannella and Rodney M. Burstall. Structured theories in LCF. In: Giorgio Ausiello and Marco Protasi, editors, *Proceedings of the 8th Colloquium on Trees in Algebra and Programming*, L'Aquila, *Lecture Notes in Computer Science*, Vol. 159, pages 377–391. Springer, 1983.

[Sch86] David Schmidt. *Denotational Semantics: A Methodology for Language Development*. Allyn and Bacon, 1986.

[Sch87] Oliver Schoett. *Data Abstraction and the Correctness of Modular Programs*. Ph.D. thesis, University of Edinburgh, Department of Computer Science, 1987.

[Sch90] Oliver Schoett. Behavioural correctness of data representations. *Science of Computer Programming*, 14(1):43–57, 1990.

[Sch92] Oliver Schoett. Two impossibility theorems on behaviour specification of abstract data types. *Acta Informatica*, 29(6–7):595–621, 1992.

[Sco76] Dana Scott. Data types as lattices. *SIAM Journal of Computing*, 5(3):522–587, 1976.

[Sco04] Giuseppe Scollo. An institution isomorphism for planar graph colouring. In: Rudolf Berghammer, Bernhard Möller, and Georg Struth, editors, *Relational and Kleene-Algebraic Methods in Computer Science. Selected Papers from the 7th International Seminar on Relational Methods in Computer Science and 2nd International Workshop on Applications of Kleene Algebra*, Bad Malente, *Lecture Notes in Computer Science*, Vol. 3051, pages 252–264. Springer, 2004.

[SCS94] Amílcar Sernadas, José Félix Costa, and Cristina Sernadas. An institution of object behaviour. In: Hartmut Ehrig and Fernando Orejas, editors, *Recent Trends in Data Type Specification. Selected Papers from the 9th Workshop on Specification of Abstract Data Types Joint with the 4th COMPASS Workshop*, Caldes de Malavella, *Lecture Notes in Computer Science*, Vol. 785, pages 337–350. Springer, 1994.

[Sel72] Alan Selman. Completeness of calculi for axiomatically defined classes of algebras. *Algebra Universalis*, 2:20–32, 1972.

[SH00] Christopher A. Stone and Robert Harper. Deciding type equivalence in a language with singleton kinds. In: *Proceedings of the 27th ACM Symposium on Principles of Programming Languages*, Boston, pages 214–227, 2000.

[Sha08] Stewart Shapiro. Classical logic. In: Edward N. Zalta, editor, *The Stanford Encyclopedia of Philosophy*. CSLI, Stanford University, Fall 2008 edition, 2008. Available from http://plato.stanford.edu/archives/fall2008/entries/logic-classical/.

[SM09] Lutz Schröder and Till Mossakowski. HASCASL: Integrated higher-order specification and program development. *Theoretical Computer Science*, 410(12–13):1217–1260, 2009.

[Smi93] Douglas R. Smith. Constructing specification morphisms. *Journal of Symbolic Computation*, 15(5–6):571–606, 1993.

[Smi06] Douglas R. Smith. Composition by colimit and formal software development. In: Kokichi Futatsugi, Jean-Pierre Jouannaud, and José Meseguer, editors, *Algebra, Meaning, and Computation, Essays Dedicated to Joseph A. Goguen on the Occasion of His 65th Birthday*, *Lecture Notes in Computer Science*, Vol. 4060, pages 317–332. Springer, 2006.

[SML05] Lutz Schröder, Till Mossakowski, and Christoph Lüth. Type class polymorphism in an institutional framework. In: José Fiadeiro, editor, *Recent Trends in Algebraic Development Techniques. Selected Papers from the 17th International Workshop on Algebraic Development Techniques*, Barcelona, *Lecture Notes in Computer Science*, Vol. 3423, pages 234–248. Springer, 2005.

[Smo86] Gert Smolka. Order-sorted Horn logic: Semantics and deduction. Technical Report SR-86-17, Universität Kaiserslautern, Fachbereich Informatik, 1986.

[SMT+05] Lutz Schröder, Till Mossakowski, Andrzej Tarlecki, Bartek Klin, and Piotr Hoffman. Amalgamation in the semantics of CASL. *Theoretical Computer Science*, 331(1):215–247, 2005.

[Spi92] J. Michael Spivey. *The Z Notation: A Reference Manual*. Prentice Hall, second edition, 1992.

[SS93] Antonino Salibra and Guiseppe Scollo. A soft stairway to institutions. In: Michel Bidoit and Christine Choppy, editors, *Recent Trends in Data Type Specification. Selected Papers from the 8th Workshop on Specification of Abstract Data Types Joint with the 3rd COMPASS Workshop*, Dourdan, *Lecture Notes in Computer Science*, Vol. 655, pages 310–329. Springer, 1993.

[SS96] Antonino Salibra and Giuseppe Scollo. Interpolation and compactness in categories of pre-institutions. *Mathematical Structures in Computer Science*, 6(3):261–286, 1996.

[SST92] Donald Sannella, Stefan Sokołowski, and Andrzej Tarlecki. Toward formal development of programs from algebraic specifications: Parameterisation revisited. *Acta Informatica*, 29(8):689–736, 1992.

[ST85] Donald Sannella and Andrzej Tarlecki. Program specification and development in Standard ML. In: *Proceedings of the 12th ACM Symposium on Principles of Programming Languages*, New Orleans, pages 67–77, 1985.

[ST86] Donald Sannella and Andrzej Tarlecki. Extended ML: An institution-independent framework for formal program development. In: David H. Pitt, Samson Abramsky, Axel Poigné, and David E. Rydeheard, editors, *Proceedings of the Tutorial and Workshop on Category Theory and Computer Programming*, Guildford, *Lecture Notes in Computer Science*, Vol. 240, pages 364–389. Springer, 1986.

[ST87] Donald Sannella and Andrzej Tarlecki. On observational equivalence and algebraic specification. *Journal of Computer and System Sciences*, 34:150–178, 1987.

[ST88a] Donald Sannella and Andrzej Tarlecki. Specifications in an arbitrary institution. *Information and Computation*, 76(2–3):165–210, 1988.

[ST88b] Donald Sannella and Andrzej Tarlecki. Toward formal development of programs from algebraic specifications: Implementations revisited. *Acta Informatica*, 25:233–281, 1988.

[ST89] Donald Sannella and Andrzej Tarlecki. Toward formal development of ML programs: Foundations and methodology. In: Josep Díaz and Fernando Orejas, editors, *TAPSOFT'89: Proceedings of the International Joint Conference on Theory and Practice of Software Development. Vol. 2: Advanced Seminar on Foundations of Innovative Software Development II and Colloquium on Current Issues in Programming Languages*, Barcelona, *Lecture Notes in Computer Science*, Vol. 352, pages 375–389. Springer, 1989.

[ST97] Donald Sannella and Andrzej Tarlecki. Essential concepts of algebraic specification and program development. *Formal Aspects of Computing*, 9:229–269, 1997.

[ST06] Donald Sannella and Andrzej Tarlecki. Horizontal composability revisited. In: Kokichi Futatsugi, Jean-Pierre Jouannaud, and José Meseguer, editors, *Algebra, Meaning and Computation: Essays Dedicated to Joseph A. Goguen on the Occasion of His 65th Birthday*, *Lecture Notes in Computer Science*, Vol. 4060, pages 296–316. Springer, 2006.

[ST08] Donald Sannella and Andrzej Tarlecki. Observability concepts in abstract data type specification, 30 years later. In: Pierpaolo Degano, Rocco de Nicola, and José

Meseguer, editors, *Concurrency, Graphs and Models: Essays Dedicated to Ugo Montanari on the Occasion of his 65th Birthday*, *Lecture Notes in Computer Science*, Vol. 5065. Springer, 2008.

[Str67] Christopher Strachey. Fundamental concepts in programming languages. In: *NATO Summer School in Programming, Copenhagen*. 1967. Also in: *Higher-Order and Symbolic Computation* 13(1–2):11–49, 2000.

[SU06] Morten H. Sørensen and Paweł Urzyczyn. *Lectures on the Curry-Howard Isomorphism*, *Studies in Logic and the Foundations of Mathematics*, Vol. 149. Elsevier Science, 2006.

[SW82] Donald Sannella and Martin Wirsing. Implementation of parameterised specifications. In: Mogens Nielsen and Erik Meineche Schmidt, editors, *Proceedings of the 9th International Colloquium on Automata, Languages and Programming*, Aarhus, *Lecture Notes in Computer Science*, Vol. 140, pages 473–488. Springer, 1982.

[SW83] Donald Sannella and Martin Wirsing. A kernel language for algebraic specification and implementation. In: Marek Karpinski, editor, *Proceedings of the 1983 International Conference on Foundations of Computation Theory*, Borgholm, *Lecture Notes in Computer Science*, Vol. 158, pages 413–427. Springer, 1983.

[SW99] Donald Sannella and Martin Wirsing. Specification languages. In: Egidio Astesiano, Hans-Jörg Kreowski, and Bernd Krieg-Brückner, editors, *Algebraic Foundations of Systems Specification*, chapter 8, pages 243–272. Springer, 1999.

[Tar85] Andrzej Tarlecki. On the existence of free models in abstract algebraic institutions. *Theoretical Computer Science*, 37(3):269–304, 1985.

[Tar86a] Andrzej Tarlecki. Bits and pieces of the theory of institutions. In: David H. Pitt, Samson Abramsky, Axel Poigné, and David E. Rydeheard, editors, *Proceedings of the Tutorial and Workshop on Category Theory and Computer Programming*, Guildford, *Lecture Notes in Computer Science*, Vol. 240, pages 334–360. Springer, 1986.

[Tar86b] Andrzej Tarlecki. Quasi-varieties in abstract algebraic institutions. *Journal of Computer and System Sciences*, 33(3):333–360, 1986.

[Tar87] Andrzej Tarlecki. Institution representation. Unpublished note, Dept. of Computer Science, University of Edinburgh, 1987.

[Tar96] Andrzej Tarlecki. Moving between logical systems. In: Magne Haveraaen, Olaf Owe, and Ole-Johan Dahl, editors, *Recent Trends in Data Type Specification. Selected Papers from the 11th Workshop on Specification of Abstract Data Types*, Oslo, *Lecture Notes in Computer Science*, Vol. 1130, pages 478–502. Springer, 1996.

[Tar99] Andrzej Tarlecki. Institutions: An abstract framework for formal specification. In: Egidio Astesiano, Hans-Jörg Kreowski, and Bernd Krieg-Brückner, editors, *Algebraic Foundations of Systems Specification*, chapter 4, pages 105–130. Springer, 1999.

[Tar00] Andrzej Tarlecki. Towards heterogeneous specifications. In: Dov Gabbay and Maarten de Rijke, editors, *Frontiers of Combining Systems 2*, *Studies in Logic and Computation*, Vol. 7, pages 337–360. Research Studies Press, 2000.

[TBG91] Andrzej Tarlecki, Rodney M. Burstall, and Joseph A. Goguen. Some fundamental algebraic tools for the semantics of computation. Part 3: Indexed categories. *Theoretical Computer Science*, 91(2):239–264, 1991.

[Ter03] Terese. *Term Rewriting Systems*, *Cambridge Tracts in Theoretical Computer Science*, Vol. 55. Cambridge University Press, 2003.

[Tho89] Simon Thompson. A logic for Miranda. *Formal Aspects of Computing*, 1(4):339–365, 1989.

[TM87] Władysław M. Turski and Thomas S.E. Maibaum. *Specification of Computer Programs*. Addison-Wesley, 1987.

[Tra93] Will Tracz. Parametrized programming in LILEANNA. In: *Proceedings of the 1993 ACM/SIGAPP Symposium on Applied Computing*, Indianapolis, pages 77–86, 1993.

[TWW82] James W. Thatcher, Eric G. Wagner, and Jesse B. Wright. Data type specification: Parameterization and the power of specification techniques. *ACM Transactions on Programming Languages and Systems*, 4(4):711–732, 1982.

[Vra88] Jos L.M. Vrancken. The algebraic specification of semi-computable data types. In: Donald Sannella and Andrzej Tarlecki, editors, *Recent Trends in Data Type Specification. Selected Papers from the 5th Workshop on Specification of Abstract Data Types*, Gullane, *Lecture Notes in Computer Science*, Vol. 332, pages 249–259. Springer, 1988.

[Wad89] Philip Wadler. Theorems for free! In: *Proceedings of the 4th International ACM Conference on Functional Programming Languages and Computer Architecture*, London, pages 347–359, 1989.

[Wan79] Mitchell Wand. Final algebra semantics and data type extensions. *Journal of Computer and System Sciences*, 19:27–44, 1979.

[Wan82] Mitchell Wand. Specifications, models, and implementations of data abstractions. *Theoretical Computer Science*, 20(1):3–32, 1982.

[WB82] Martin Wirsing and Manfred Broy. An analysis of semantic models for algebraic specifications. In: Manfred Broy and Gunther Schmidt, editors, *Theoretical Foundations of Programming Methodology: Lecture Notes of an International Summer School, Marktoberdorf 1981*, pages 351–416. Reidel, 1982.

[WB89] Martin Wirsing and Manfred Broy. A modular framework for specification and implementation. In: Josep Díaz and Fernando Orejas, editors, *TAPSOFT'89: Proceedings of the International Joint Conference on Theory and Practice of Software Development. Vol. 1: Advanced Seminar on Foundations of Innovative Software Development I and Colloquium on Trees in Algebra and Programming*, Barcelona, *Lecture Notes in Computer Science*, Vol. 351, pages 42–73. Springer, 1989.

[WE87] Eric G. Wagner and Hartmut Ehrig. Canonical constraints for parameterized data types. *Theoretical Computer Science*, 50:323–349, 1987.

[Wec92] Wolfgang Wechler. *Universal Algebra for Computer Scientists, EATCS Monographs on Theoretical Computer Science*, Vol. 25. Springer, 1992.

[Wik] Wikipedia. Hash table. Available from http://en.wikipedia.org/wiki/Hash_table.

[Wir82] Martin Wirsing. Structured algebraic specifications. In: *Proceedings of the AFCET Symposium on Mathematics for Computer Science*, Paris, pages 93–107, 1982.

[Wir86] Martin Wirsing. Structured algebraic specifications: A kernel language. *Theoretical Computer Science*, 42(2):123–249, 1986.

[Wir90] Martin Wirsing. Algebraic specification. In: Jan van Leeuwen, editor, *Handbook of Theoretical Computer Science. Vol. B (Formal Models and Semantics)*, pages 675–788. North-Holland and MIT Press, 1990.

[Wir93] Martin Wirsing. Structured specifications: Syntax, semantics and proof calculus. In: Friedrich L. Bauer, Wilfried Brauer, and Helmut Schwichtenberg, editors, *Logic and Algebra of Specification: Proceedings of the NATO Advanced Study Institute, Marktoberdorf 1991*, pages 411–442. Springer, 1993.

[WM97] Michał Walicki and Sigurd Meldal. Algebraic approaches to nondeterminism: An overview. *ACM Computing Surveys*, 29(1):30–81, 1997.

[Zil74] Steven Zilles. Abstract specification of data types. Technical Report 119, Computation Structures Group, Massachusetts Institute of Technology, 1974.

[Zuc99] Elena Zucca. From static to dynamic abstract data-types: An institution transformation. *Theoretical Computer Science*, 216(1–2):109–157, 1999.

Index of categories and functors

Index of institutions

Index of notation

Index of concepts

D

S